Lecture Notes in Artificial Intelligence   2005

Subseries of Lecture Notes in Computer Science
Edited by J. G. Carbonell and J. Siekmann

Lecture Notes in Computer Science
Edited by G. Goos, J. Hartmanis and J. van Leeuwen

T0180426

Springer

Berlin
Heidelberg
New York
Barcelona
Hong Kong
London
Milan
Paris
Tokyo

Wojciech Ziarko   Yiyu Yao (Eds.)

Rough Sets and Current Trends in Computing

Second International Conference, RSCTC 2000
Banff, Canada, October 16-19, 2000
Revised Papers

 Springer

Series Editors

Jaime G. Carbonell,Carnegie Mellon University, Pittsburgh, PA, USA
Jörg Siekmann, University of Saarland, Saarbrücken, Germany

Volume Editors

Wojciech Ziarko
Yiyu Yao
Department of Computer Science
University of Regina
Regina, Saskatchewan, Canada S4S 0A2
E-mail:{ziarko/yyao}@cs.uregina.ca

Cataloging-in-Publication Data applied for

Die Deutsche Bibliothek - CIP-Einheitsaufnahme

Rough sets and current trends in computing : second international conference ;
revised papers / RSCTC 2000, Banff, Canada, October 16 - 19, 2000.
Wojciech Ziarko ; Yiyu Yao (ed.). - Berlin ; Heidelberg ; New York ;
Barcelona ; Hong Kong ; London ; Milan ; Paris ; Tokyo : Springer, 2001
  (Lecture notes in computer science ; Vol. 2005 : Lecture notes in
  artificial intelligence)
  ISBN 3-540-43074-1

CR Subject Classification (1998): I.2, F.1, F.4.1, I.5.1, I.4, H.2.8

ISSN 0302-9743
ISBN 3-540-43074-1 Springer-Verlag Berlin Heidelberg New York

Springer-Verlag Berlin Heidelberg New York
a member of BertelsmannSpringer Science+Business Media GmbH

http://www.springer.de

© Springer-Verlag Berlin Heidelberg 2001
Printed in Germany

Typesetting: Camera-ready by author, data conversion by PTP-Berlin, Stefan Sossna
Printed on acid-free paper   SPIN 10781967   06/3142   5 4 3 2 1 0

Preface

This volume contains the papers selected for presentation at the Second International Conference on Rough Sets and Current Trends in Computing RSCTC 2000 held in the beautiful Rocky Mountains resort town of Banff in Canada, October 16–19, 2000.

The main theme of the conference is centered around the theory of rough sets, its applications, and the theoretical development. The program also includes numerous research papers representing areas which are related to rough sets such as fuzzy sets, data mining, machine learning, pattern recognition, uncertain reasoning, neural nets, and genetic algorithms and some selected papers from other areas of computer science. This composition of various research areas is a reflection of the general philosophy of trying to bring together researchers representing different, but often closely related, research paradigms to enhance mutual understanding and to stimulate the exchange of ideas and sometimes of very diverse points of view on similar problems.

The conference, the second in the series, stems from annual workshops devoted to the topic of rough sets initiated in 1992 in Kiekrz, Poland (the second workshop was also held in Banff in 1993). The first conference was held in 1998 in Warszawa, Poland, followed by the most recent workshop organized in Yamaguchi, Japan in 1999.

It has been over twenty years now since the first introduction of basic ideas and definitions of rough set theory by Dr. Zdzislaw Pawlak. As with many other of Dr. Pawlak's ideas, the theory of rough sets now belongs to the standard vocabulary of Computer Science research, in particular research related to uncertain reasoning, data mining, machine learning, pattern recognition, just to mention a few.

In this context, one could ask the question as to what makes this theory so attractive in all these other areas which have already developed methodologies of their own. It seems that the universality of the theory is the keyword. It touches the very essence of the set definition, one of the fundamental notions of modern mathematics. The standard set theory is closely related to Boolean logic which in turn is at the heart of the operation of digital computers.

It is well known that many practical problems cannot be solved satisfactorily by programming existing computers, in particular problems related to learning, pattern recognition, some forms of control etc. The difficulty stems from the fact that it is often impossible to create black-and-white algorithmic descriptions of the objects of interest occurring in different application areas, for example wave form patterns occurring in sound analysis. The theory of rough sets and its extensions provide rigorous mathematical techniques for creating approximate descriptions of such objects, for analyzing, optimizing, and recognizing the limits of what can be effectively distinguished (i.e. classified) by means of the available object representation.

This is not to say that all these difficult complex object classification-related problems would be automatically solved with the adoption of the rough set approach. Instead, the rough set approach provides a common philosophical framework supported by precise mathematical language for dealing with these problems.

However, the details of specific solutions must be supplied by experts working in particular subject areas. Past experience indicates that the rough set approach is a team-oriented methodology. Usually a single individual does not have the expertise required for the effective application of the rough set approach to a practical problem. This means that developing practical applications of this methodology is difficult and costly. From the perspective of conference organizers it has also led to the relative rarity of application-oriented publications, and we would like to see more of these.

This imbalance was visible in previous workshops and conferences on the subject and is repeated here. We have a fine collection of good theoretical papers but application papers are few. Consequently, the proceedings are organized along the subject lines without further separation into theoretical versus practical papers. We sincerely hope that the current strong theoretical growth of rough set approaches, as demonstrated in this volume, will eventually lead to the parallel growth in the application side resulting in stronger participation of industrial users of the methodology.

The RSCTC 2000 program was further enhanced by invited keynote speakers: Setsuo Ohsuga, Zdzislaw Pawlak, and Lotfi A. Zadeh, and invited plenary speakers: Jerzy Grzymala-Busse, Roman Swiniarski, and Jan Zytkow.

The success of RSCTC 2000 was a result of the joint efforts of authors, Advisory Board, Program Committee, and referees. We want to thank the authors for deciding to publish their research at this conference and for their patience during the delays which occured when processing the submissions. The preparation of this volume would not have been possible without the help of referees and the members of the Advisory Board to whom we would like to express our thanks and appreciation for the time and effort they put into the refereeing and paper selection process. In particular, we would like to thank Program Committee Chairs: Andrzej Skowron (Europe) and Shusaku Tsumoto (Asia) for their kind support and advice on conference-related issues. We are grateful to our sponsors: the Faculty of Science, the President's office, and the Computer Science Department, University of Regina for the financial and organizational support. We would also like to express our thanks to Ms. Rita Racette for the secretarial and organizational help before and during the conference. Much of the research on rough sets and other topics presented in this volume was supported by research grants from Natural Sciences and the Engineering Research Council of Canada. This support is gratefully acknowledged.

March 2001   Wojciech Ziarko
                                                                    Yiyu Yao

RSCTC 2000 Conference Committee

Honorary Chairs: Zdzislaw Pawlak, Lotfi A. Zadeh

Conference Chair: Wojciech Ziarko

Program Chairs: Yiyu Yao, Andrzej Skowron (Europe), Shusaku Tsumoto (Asia)

Advisory Board:

Nick Cercone	Jerzy Grzymala-Busse	T.Y. Lin
Toshinori Munakata	Akira Nakamura	Setsuo Ohsuga
Lech Polkowski	Zbigniew Ras	Roman Slowinski
Paul P. Wang	Ning Zhong	

Publicity Chair: Ning Zhong

Local Organization Committee: Mengchi Liu (Chair), David Brindle, Cory Butz

Program Committee

Anna Buczak	Gianpiero Cattaneo	Juan-Carlos Cubero
Andrzej Czyzewski	Didier Dubois	Ivo Duentsch
Maria C. Fernandez	Salvatore Greco	Howard Hamilton
Howard Ho	Xiaohua Hu	Ryszard Janicki
Julia Johnson	Janusz Kacprzyk	Willi Kloesgen
Jan Komorowski	Jacek Koronacki	Bozena Kostek
Wojciech Kowalczyk	Marzena Kryszkiewicz	Pawan Lingras
Chunnian Liu	Jiming Liu	Mengchi Liu
Qing Liu	Tadeusz Luba	Solomon Marcus
Ernestina Menasalvas	Zbigniew Michalewicz	Ryszard Michalski
John Mordeson	Mikhail Moshkov	Hung Son Nguyen
Kanji Ohsima	Ewa Orlowska	Piero Pagliani
Witold Pedrycz	James F. Peters	Zdzislaw Piasta
Henri Prade	Mohamed Quafafou	Vijay Raghavan
Sheela Ramanna	Qiang Shen	Krzysztof Slowinski
Jerzy Stefanowski	Jaroslaw Stepaniuk	Zbigniew Suraj
Roman Swiniarski	Hans-Michael Voigt	Anita Wasilewska
S.K.M. Wong	Xindong Wu	Satoshi Yamane
Lizhu Zhou	Jan Zytkow	

Table of Contents

Introduction

Keynote Papers

Special Session: Granular Computing

Rough Sets and Systems

Fuzzy Sets and Systems

Rough Sets and Data Mining

Machine Learning and Data Mining

Non-classical Logics and Reasoning

Pattern Recognition and Image Processing

Neural Networks and Genetic Algorithms

Current Trends in Computing

Rough Sets: Trends, Challenges, and Prospects

Wojciech Ziarko

Computer Science Department
University of Regina
Regina, Saskatchewan, S4S 0A2
Canada

Abstract. The article presents a brief review of the past and the current state of the rough set-related research and provides some ideas about the perspectives of rough set methodology in the context of its likely impact on the future computing devices. The opinions presented are solely of the author and do not necessarily reflect the point of view of the majority of the rough set community.

The fundamentals of rough set theory have been laid out by Zdzislaw Pawlak about twenty years ago [1-3]. His original definition of the rough, or approximately described set in terms of some, already known and well defined disjoint classes of indistinguishable objects, seemed to capture in mathematical terms the essence of limits of machine perception and empirical learning processes. The mathematical theory of rough sets which was developed around this definition, with contributions from many mathematicians and logicians (see, for example [5,6,12-18,33-35]), resembles the classical set theory in its clarity and algebraic completeness. The introduction of this theory in 1980's coincided with the surge of interest in artificial intelligence (AI), machine learning, pattern recognition and expert systems. However, much of the research in those areas at that time was lacking comprehensive theoretical fundamentals, typically involving designing either classical logic-based or intuitive algorithms to deal with practical problems related to machine reasoning, perception or learning. The logic-based approaches turned out to be too strict to be practical whereas intuitive algorithms for machine learning, pattern recognition etc. lacked a unifying theoretical basis to understand their limitations and generally demonstrated inadequate performance.

To many researchers the theory of rough sets appeared as the missing common framework to conduct theoretical research on many seemingly diverse AI-related problems and to develop application-oriented algorithms incorporating the basic ideas of the rough set theory. Numerous algorithms and systems based on the rough set theory were developed during that early period, most of them for machine learning and data analysis tasks [4-7,13-16]. As the software for developing rough set-based applications was becoming more accessible, the experience with applying rough set methods to various applications was being gradually accumulated. The experience was also revealing the limitations of the rough set approach and inspiring new extensions of the rough set theory aimed

W. Ziarko and Y. Yao (Eds.): RSCTC 2000, LNAI 2005, pp. 1–7, 2001.

at overcoming these limitations (see, for instance [8,10-12,16,17,23,24,26]). In particular, the definitions of rough set approximation regions [2-3], originally based on the relation of set inclusion, turned out to be too restrictive in most applications involving using real life data for the purpose of deriving empirical models of complex systems. Consequently, typical extensions of the rough set model were aimed at incorporating probabilistic or fuzzy set-related aspects into rough set formalism to soften the definitions of the rough approximation regions to allow to handle broader class of practical problems [8,10,12,16,17,25,27]. This was particularly important in the context of data mining applications of rough set methodology where probabilistic data pattern discovery is common.

Another trend which emerged in recent years was focused on generalizing the notion of the equivalence relation defining the approximation space which forms the basis of the rough approximations [20,22].

When constructing models of data relationships using the methodology of rough sets it is often necessary to transform the original data into a derived form in which the original data items are replaced with newly defined secondary attributes. The secondary attributes are functions of the original data items. In typical applications they are derived via a discretization process in which the domain of a numeric variable is divided into a number of value ranges. The general question how to define the secondary attributes, or what is the best discretization of continuous variables has not been answered satisfactorily so far despite of valuable research contributions dealing with this problem [21,25]. It appears that the best discretization or secondary attribute definitions are provided by domain experts based on some prior experience.

Another tough problem waiting for the definite solution is the treatment of data records with missing or unknown data item values. Although a number of strategies have been developed in the past for handling missing values, some of them involving replacement of the missing value by a calculated value, the general problem is still open as none of the proposed solutions seems to address the problem of missing values satisfactorily.

The main challenge facing rough set community today is the development of real-life industrial applications of the rough set methodology. As it can be seen in this volume, the great majority of papers are theoretical ones with a few papers reporting practical implementations. Experimental applications of rough set techniques have been attempted since the first days of existence of rough set theory. They covered the whole spectrum of application domains ever attempted by AI-related methods. Before the advent of the discipline of data mining, the methods of rough sets were already in use, for example, for the analysis of medical or drug data, chemical process operation log data with the objective of better understanding the inter-data item interactions and to develop decision support models for medical diagnosis, drug design or process control (see, for example [5,9,18,27-32,36]). They were also used for experiments with speech and image classification, character recognition, musical sound classification, stock market prediction and many other similar applications. However, the applications which ended up in industrial use and brought concrete financial benefits are very rare.

This is probably due to the fact that the development of rough set applications requires team work and generally substantial funding, access to real-life data and commitment from all involved parties, including domain experts and rough set analysis specialists. It also requires quality software specifically tailored for the rough set-based processing of large amounts of data. Satisfying all this requirements is not easy and next to impossible for a single researcher with very limited budget resulting in the current lack of industrial applications of rough set approach. It appears that until rough set methodology acquires major industrial sponsors committed to implementing it in their products or processes we are going to observe the continuation of the current rough set research situation which is dominated by theoretical works with relatively few experimental application projects.

There are other challenges for the rough set methodology on the road to main stream acceptance and popularity. One of the important issues here appears to be the unavailability of popular books written in a way making them accessible to average computer scientists, programmers or even advanced computer users. Currently, the rough set literature is solely aimed at computer science researchers with advanced background and degrees. Clearly, it is seriously limiting the scope of potential users of this methodology. The related aspect is the lack of modern, easy to use popular software tools which would help novice and sophisticated users in familiarizing with rough sets and in developing simple applications. The tools of this kind should be extremely simple to use, with graphical user interface and should not require the user to do any programming in order to create an application.

To have users and developers employing rough set tools, the subject of rough sets should be taught at universities. Although it happens occasionally, it is far from common practice and computer science graduates normally know nothing about rough sets. The proper education is also important to ensure sustained growth of the discipline. This requires creation of educational programs incorporating rough set methodology, development of textbooks and convincing educational examples of rough set applications.

The establishment of permanent publicity mechanism for rough set research is the next major issue waiting for the solution. At the moment the publicity tools to propagate rough set methodology, results and applications are very limited. There is no specialized journal related to rough sets and there is lack of centralized well funded unit coordinating activities in this area. All publicity and organizational work in the rough set community is done by volunteers who devote their time and energy on the expense of other activities. It seems that some of the organizational tasks should be done by paid professionals, to ensure continuity and quality. This however requires financial resources which are currently not available.

As far as the future of rough set methodology is concerned, I personally believe that rough sets are going to shine in the following application areas.

- **Trainable control systems**
  Trainable control systems have their control algorithms derived from training session operation data when the system, or system simulator, is being controlled by skilled human operator or operators. It is well known that many complex control problems cannot be programmed due to lack of system behavioral models, complex nonlinearities etc. Problems of this kind appear in robotics, process control, animation etc. It appears, based on prior research and experience with laboratory experiments [9,27,28,31,36] that rough set theory provides practical solution to the problem of deriving compact working control algorithms from the complex system operation log data.
- **Development of predictive models from data to increase the likelihood of correct prediction (data-based decision support systems)**
  With the current flood of data accumulated in databases and representing various aspects of human activity, future decisions support systems will be increasingly relying on factual data in deriving their decision algorithms. This trend has been recognized by data mining researchers many of whom are adopting rough set approach. The role of rough sets in this area is expected to grow with the introduction of easy-to-use industrial strength software tools.
- **Pattern classification (recognition)**
  The technology of complex pattern recognition, including sound and image classification is not mature yet despite some products already on the market. It appears that rough set theory will have an important role to play here by significantly contributing to algorithms aimed at deriving pattern classification methods from training data.

In general, the rough set theory provides the mathematical fundamentals of a new kind of computing, data-based rather than human supplied algorithm-based. This data-based form of computing will become increasingly more important in the twenty first century as the limits of what can be programmed with standard algorithmic methods will become more apparent. A some sort of saturation will be observed in the class of problems solvable with the traditional techniques with the parallel unsaturated growth in the power of processors and ever growing human expectations. This will force the necessary shift to non-traditional system development methods in which rough set theory will play prominent role, similar to the role of Boolean algebra in contemporary computing devices.

Acknowledgment

The research reported in this article was partially supported by a research grant awarded by Natural Sciences and Engineering Research Council of Canada.

References

1. Pawlak, Z. Grzymała-Busse, J. Słowiński, R. and Ziarko, W. (1995). Rough sets. *Communications of the ACM*, 38, 88–95.

2. Pawlak, Z. (1991). *Rough Sets - Theoretical Aspects of Reasoning about Data.* Kluwer Academic Publishers, Boston, London, Dordrecht.
3. Pawlak, Z. (1982). Rough sets. *International Journal of Computer and Information Sciences*, 11, 341–356.
4. Son, N. (1997). Rule induction from continuous data. in: P.P. Wang (ed.), *Joint Conference of Information Sciences*,March 1-5, Duke University, Vol. 3, 81–84.
5. Słowiński, R. (ed.) (1992). *Intelligent Decision Support. Handbook of Applications and Advances of the Rough Set Theory*, Kluwer Academic Publishers, Boston, London, Dordrecht.
6. Ziarko, W. (ed.) (1994) Rough Sets, Fuzzy Sets and Knowledge Discovery, Springer Verlag, 326-334.
7. Yang, A., and Grzymała-Busse J. (1997). Modified algorithms LEM1 and LEM2 for rule induction form data with missing attribute values., In: P.P. Wang (ed.), *Joint Conference of Information Sciences*, March 1-5, Duke University, Vol. 3, 69–72.
8. Ziarko, W. (1993). Variable precision rough sets model.*Journal of Computer and Systems Sciences*, vol. 46, no. 1, 39-59.
9. Ziarko, W. Katzberg, J.(1993). Rough sets approach to system modelling and control algorithm acquisition. *Proceedings of IEEE WASCANEX 93 Conference*, Saskatoon, 154-163.
10. Ziarko, W. (1999) Decision making with probabilistic decision tables.*Proceedings of the 7th Intl Workshop on Rough Sets, Fuzzy Sets, Data Mining and Granular Computing*, RSFDGrC'99, Yamaguchi, Japan 1999, Lecture Notes in AI 1711, Springer Verlag, 463-471.
11. Pawlak, Z. (2000) Rough sets and decision algorithms.*Proceedings of the 2nd Intl Conference on Rough Sets and Current Trends in Computing*, RSCTC'2000, Banff, Canada, 1-16.
12. S. K. Pal and A. Skowron (eds.) (1999) *Rough Fuzzy Hybridization: A New Trend in Decision–Making*, Springer–Verlag, Singapore.
13. L. Polkowski and A. Skowron (eds.) (1998) *Rough Sets in Knowledge Discovery 1. Methodology and Applications*, this Series vol. 18, Physica–Verlag, Heidelberg.
14. L. Polkowski and A. Skowron (eds.) (1998) *Rough Sets in Knowledge Discovery 2. Applications, Case Studies and Software Systems*, this Series, vol. 19, Physica–Verlag, Heidelberg.
15. T. Y. Lin and N. Cercone (eds.)(1997) *Rough Sets and Data Mining. Analysis of Imprecise Data*, Kluwer Academic Publishers, Dordrecht.
16. N. Zhong, A. Skowron, and S. Ohsuga (eds.)(1999) *New Directions in Rough Sets, Data Mining, and Granular–Soft Computing*, Proceedings: the 7th International Workshop (RSFDGrC'99), Ube–Yamaguchi, Japan, November 1999, LNAI 1711, Springer–Verlag, Berlin.
17. M. Banerjee and S.K. Pal (1996) Roughness of a fuzzy set, *Information Science* 93(3/4)pp. 235–246.
18. S. Demri, E. Orłowska, and D. Vakarelov (1999) Indiscernibility and complementarity relations in Pawlak's information systems, in: *Liber Amicorum for Johan van Benthem's 50th Birthday.*
19. A. Czyżewski (1997) Learning algorithms for audio signal enhancement. Part 2: Implementation of the rough set method for the removal of hiss, *J. Audio Eng. Soc.* 45(11), pp. 931-943.
20. S. Greco, B. Matarazzo, and R. Słowiński (1999) Rough approximation of a preference relation by dominance relations, *European Journal of Operational Research* 117, 1999, pp. 63–83.

21. J. W. Grzymala–Busse and J. Stefanowski (1997) Discretization of numerical attributes by direct use of the rule induction algorithm LEM2 with interval extension, in: *Proceedings: the Sixth Symposium on Intelligent Information Systems* (IIS'97), Zakopane, Poland, pp. 149–158.
22. K. Krawiec, R. Słowiński, and D. Vanderpooten (1998) Learning of decision rules from similarity based rough approximations, in: L. Polkowski, A. Skowron (eds.), *Rough Sets in Knowledge Discovery 2. Applications, Case Studies and Software Systems*, Physica–Verlag, Heidelberg, pp. 37–54.
23. T. Y. Lin, Ning Zhong, J. J. Dong, and S. Ohsuga (1998) An incremental, probabilistic rough set approach to rule discovery, in: *Proceedings: the FUZZ-IEEE International Conference, 1998 IEEE World Congress on Computational Intelligence* (WCCI'98), Anchorage, Alaska.
24. E. Martienne and M. Quafafou (1998) Learning fuzzy relational descriptions using the logical framework and rough set theory, in: *Proceedings: the 7th IEEE International Conference on Fuzzy Systems* (FUZZ-IEEE'98), IEEE Neural Networks Council.
25. Nguyen Hung Son and Nguyen Sinh Hoa (1997) Discretization methods with back-tracking, in: *Proceedings: the 5th European Congress on Intelligent Techniques and Soft Computing* (EUFIT'97), Aachen, Germany, Verlag Mainz, Aachen, pp. 201–205.
26. W. Pedrycz (1999), Shadowed sets : bridging fuzzy and rough sets, in: S. K. Pal and A. Skowron (eds.), *Rough Fuzzy Hybridization: A New Trend in Decision–Making*, Springer–Verlag, Singapore, pp. 179–199.
27. J. E. Peters, A. Skowron, and Z. Suraj (1999) An application of rough set methods in control design, in : *Proceedings: the Workshop on Concurrency, Specification and Programming* (CS&P'99), Warsaw, Poland, pp.214–235.
28. Munakata, T. (1997) Rough control: a perspective, In: T. Y. Lin and N. Cercone (eds.), Rough Sets and Data Mining. Analysis for Imprecise Data. Kluwer Academic Publishers, Dordrecht, pp. 77-88.
29. Słowiński, K. (1992) Rough classification of HSV patients. In: R. Słowiński (ed.), Intelligent Decision Support. Handbook of Applications and Advances of the Rough Set Theory, Kluwer Academic Publishers, Dordrecht, pp. 77-93.
30. Słowiński, K., Sharif, E. S. (1993) Rough sets approach to analysis of data of diatnostic peritoneal lavage applied for multiple injuries patients. In: W. Ziarko (ed.), Rough Sets, Fuzzy Sets and Knowledge Discovery. Proceedings of the International Workshop on Rough Sets and Knowledge Discovery (RSKD'93), Banff, Alberta, Canada, October 12–15, Springer-Verlag, pp. 420-425.
31. Szladow, A., and Ziarko W. (1993) Adaptive process control using rough sets. Proceedings of the International Conference of Instrument Society of America, ISA/93, Chicago, pp. 1421-1430.
32. Tsumoto, S. Ziarko, W. Shan. N. Tanaka, H.(1995) Knowledge discovery in clinical databases based on variable precision rough sets model. Proc. of the Nineteenth Annual Symposium on Computer Applications in Medical Care, New Orleans, 1995, Journal of American Medical Informatics Association Supplement, pp. 270-274.
33. Wasilewska, A., Banerjee, M. (1995) Rough sets and topological quasi-Boolean algebras. Proceedings of CSC'95 Workshop on Rough Sets and Database Mining, Nashville, pp.54-59.
34. Wong, S. K. M., Wang, L. S., Yao, Y. Y.(1995) On modeling uncertainty with interval structures. Computational Intelligence: an International Journal $11/2$, pp. 406-426.

35. Vakarelov, D. (1991) A modal logic for similarity relations in Pawlak knowledge representation systems. Fundamenta Informaticae **15**, pp. 61-79.
36. Mrózek, A.(1992) Rough sets in computer implementation of rule-based control of industrial processes. In: R. Słowiński (ed.), Intelligent Decision Support. Handbook of Applications and Advances of the Rough Set Theory. Kluwer Academic Publishers, Dordrecht,pp. 19-31.

To What Extent Can Computers Aid Human Activity?
Toward Second Phase Information Technology

Setsuo Ohsuga

Faculty of Software & Information Science, Iwate Prefectural University
and
Department of Information and Computer Science, Waseda University
ohsuga@fd.catv.ne.jp

Abstract. A possibility of extending the scope of computers to aid human activity is discussed. Weak points of humans are discussed and a new information technology that can back up the human activity is proposed. It must be an intelligent system to enable a computer-led interactive system. A conceptual architecture of the system and then various technologies to make up the architecture are discussed. The main issues are; a modeling scheme to accept and represent wide area of problems, a method for externalizing human idea and of representing it as a model, a large knowledge base and generation of specific problem-solving system, autonomous problem decomposition and solving, program generation, integration of different information processing methods, and knowledge acquisition and discovery

1. Introduction

Today, various human activities have to do with computers. The computer's power is still growing and a computer is sharing a larger part of human activity forcing each activity to grow larger and more complex. As the result the whole social activity is expanding considerably. However, do they aid all aspects of the activity that human being need? The answer is "no". In some job fields such as the recent applications around Internet the computers can cope with many tasks very strongly and they are accelerating the emergence of many new activities. But in some other fields, e.g. in engineering design, the computer technology stays rather in the low level comparing to what are needed. This inequality of the computer's aids among human activities is growing. As the social activity expands the needs for such activities as design and development tasks also grow large without enough support method. Nevertheless expanding social activity requires the people the design and the development of many new systems. It is worried that it might bring many troubles in human society because the requirement might go beyond the human capability. For example, it was pointed out in [1] that a new type of problem was arising in the software development because of the increased size of software systems. In the other fields also the unexpected troubles are increasing such as the accidents in a nuclear plant, the failure of a rocket vehicle, etc.

W. Ziarko and Y. Yao (Eds.): RSCTC 2000, LNAI 2005, pp. 8-29, 2001.

There were many discussions on the causes of these accidents. Apart from those considered the primary causes, the lack of the person's capability to follow the increase of the scale and complexity of task lies as the basis of troubles in the conventional human-centered developing style. In other word human being create the troubles and nevertheless they can no more solve these troubles by themselves. Then is there some one that can do it in place of human? Only possibility for answering this question is to make computers more intelligent and change the development style by introducing them into system development. Such computers must be able to back up the weak points of human being and reduce the human load, for the human capability has already come to near the limit but computer's capability has the room to be enhanced by the development of new information technology. It does not mean that computer can be more intelligent than human being but that, as will be discussed below, the computer help human to make the best decision by undertaking the management tasks of environment of decision making. It is necessary for the purpose to analyze the human way of problem solving, to know the weak points of humans, to find a possible alternate method and then to develop new information technology that can back up and aid human activity. It needs AI technology but with the wider applicability than whatever has been developed so far [2].

In this paper the human activities are classified into two classes; the ones to which conventional information technology works effectively and the others. These are called the first type and the second type activities respectively. Corresponding to the types of activities the required information technologies are different. Roughly, it is said that the above classes correspond to whether an activity can be represented in an algorithmic way or not.

The effectiveness of using computers has been exhibited in many applications to which the computer's processing mechanism matches well with what is required by problem solving. These were of the first type. The typical examples are business applications, computations of the large mathematical functions in scientific and engineering applications, large-scale simulations and so on. Recently the new type of applications are increasing in which the computers deal with directly the signals generated in the surrounding systems without intervention of human being. The typical examples are information processing in network applications and the computer control systems embedded in various engineering systems like vehicle engine. These are however all the first type applications because each of them can be represented by an algorithmic method and, if the activity is specified, then the method of developing an information system to achieve the goal is decided without difficulty. These activities have been computerized. Along with the increase of the computing power, the number and the scale of these applications are rapidly increasing. This tendency urges the growth of the human activities in all aspects.

But from the viewpoint of the information processing technology, this class of applications covers only a part of whole areas of human activities and there are various problems that are not included in this class. Actually to develop a software-system for an application as above is an activity that is not classified into the first class. Their computerization has been difficult and these are mostly left to human being. The objective of this paper is to discuss the characteristics of the human activities and the information technology that can cover both types of activities, especially the second type.

2. Limitation of Human Capability and Expected Difficulties

What is the reason why human being could deal with successfully most activities so far but have troubles today? Why the current computers cannot support these activities well? It is necessary to answer these questions.

2. 1 Limitation of Human Capability

That human capability is limited is one of the reasons why new troubles occur. Human capability is limited in many aspects as follows.

(1) The limitation to manage very large scale objects (scale problem); A person cannot include too many objects in the scope of consideration at a time. If an object contains many components that exceed this limit, then some components may be ignored and the object cannot be well managed.

(2) The limitation to manage complex object (complexity problem); As the mutual relations between objects get large and dense, a change at an object propagates widely. A person cannot foresee its effect if the scope of the relation expands over the certain limit.

(3) The limitation to follow up the rapidly changing situations (time problem); There are two kinds of speed problems; the limit of the human physical activities and the limit of the human mental activity. The first problem has been a research object for long time in the science and technology and is abbreviated here. The latter is the problem of achieving a mental work in a short time. To develop a large system in a short time is a typical example.

(4) The limitation of self-management of making errors (error problem); Human being make errors in doing whatever the thing they may be. Sometimes it induces the big accidents. So far a number of methods to reduce the effects of human error have been developed. This situation may not change and the effort will continue in the same way as before.

(5) The limitation to understand many domains (multi-disciplinary problem); When an object grows large and complex, it has various aspects concerning different disciplines. A person who deals with such an object is required to have the multi-disciplinary knowledge. To acquire multi-disciplinary knowledge however is difficult for many individuals. It is not only because of the limitation of each individual's ability but also because the social education system is not suited for students the multi-disciplinary studies. It means that some aspects might be remained unconsidered well at problem solving and very often a large and complex object has been and will be designed and manipulated lacking the well-balanced view to the object. This multi-disciplinary problem is considered to be one of the most serious problems in near future.

(6) The limitation to understand to each other through communication (communication problem); A large problem, for example a large object design, is decomposed to a set of small problems and is distributed to many persons. In order to keep the consistency of the solution for the whole problem the relations between the separated sub-problems must be carefully dealt with. It requires the close communication between persons responsible to the sub-problems. Actually

it is sometimes incomplete. In particular when there are time lags between each sub-problem solving, to keep the well communication is the more difficult. Very often the serious problems occur when an existing system is to be modified by a person who is different from the ones who have designed the systems originally. It is because the transmission of information is incorrect between them. In turn it is because the recording of the decision-making in problem solving is insufficient.

(7) The limitation to see objectively the one's own activity (objectivity problem); It is difficult for many people to see objectively their own activities and record them. Therefore their activities cannot be traced faithfully afterward. It results in the loss of responsibility.

2. 2 Characteristics of New Problems

On the other hand, problems arising recently have the following characteristics.

(1) The scale of an object is growing and the number of components gets larger. Many artifacts recently developed, for example, airplane, buildings, airports, space vehicles and so on, are becoming bigger. Not only the artifacts but also many social systems such as government organizations, enterprises, hospitals, education systems, security systems and so on are growing larger. The subjects who manage these objects whatever they may be are persons.

(2) Not only the number of components is increasing but also the mutual relations among components in an object are becoming closer. Accordingly the object is becoming complex. For example, electronic systems are growing more and more complex. Also in the social systems, e.g. in a hospital, the different sections have been responsible to the different aspects of disease and relatively separated before. But today these are becoming closely related to each other.

(3) The time requirement is becoming more serious. Many social systems are changing dynamically. For example restructuring of enterprises including M and A (Merger and Acquisition) is progressing in many areas. Since every social system needs an information system, the development of a new information system in a short time becomes necessary. It is the new and truly difficult problems for human.

(4) The extent of expertise that are needed in an activity is expanding rapidly over the boarder of the different domains. In order to achieve the given goal of the activity, the subject of the activity is required to have every related expertise in a view. It requires the subject to have the multi-disciplinary view.

2. 3 Queries to Answer in Order for Avoiding Expected Unfavorable Results

Human cannot resolve the problems expected to arise. Consequently various unfavorable results are expected to occur. In order to avoid them a new method has to be developed. The method must be such as to be able to answer the following queries.

(1) A large problem has to be decomposed to smaller sub-problems and these sub-problems have to be distributed to the separated subjects of problem solving

(persons or computers) so that the role of each subject is confined to an appropriate size (scale problem, complexity problem).

Question: How to decompose a problem?

(2) The decomposed sub-problems are not always independent but usually related to each other. Then the sub-problem solvers must keep communication to each other. Sometimes the communication is incomplete (communication problem).

Question: How to assure the proper communications between the different problem solvers?

(3) The required speed of the system development is increasing. The human capability may not become able to follow this speed (time problem). Automatic system development, for example an automatic programming, becomes necessary.

Question: How to automate system development?

(4) A problem may concern various domains. The multi-disciplinary knowledge is required to the subjects of problem solving (multi-disciplinary problem).

Question: How to assure the use of knowledge in the wide areas?

(5) Each person may make decision in his/her problem solving. This decision may not be proper and it affects the final solution. The more persons concern a problem solving, the larger the possibility of the final solution becomes improper. The chance of the improper solution gets large therefore when a large problem is decomposed, distributed to the different people and the solutions for these sub-problems are integrated to the total solution. Roughly this probability is n times larger than that of the single problem solving provided a problem is decomposed into n sub-problems. Therefore the decision making in the history of each sub-problem solving must be recorded for being checked afterward. But very often it is not achieved properly (objectivity problem).

Question: How to record the decisions made by persons correctly?

3. An Approach toward Enlarging the Capability of Information System

Since many troubles occur because of the limitation of the human capability, the new method must be such as to reduce the human tasks as far as possible. If persons cannot do it themselves, what can do it in place of human? Only possibility for its resolution is to make computers more intelligent to back up the weak points of human. A new information technology is necessary based on a new idea for the purpose. It does not mean that computer can be more intelligent than human being. Human being has the high level intelligence with their capability of making decision of the best quality. They can make it if they are in an environment suited for them with respect to the scale of the target object, the time requirement, the scope of required knowledge, and so on. But very often they are forced to make decision in an unsuited environment. The possibility of resolving this problem is to make computers more intelligent to back up the human activity first in order to let them know their failure and more positively to reduce the rate of wrong decisions by providing the best environment for human for making decision.

3.1 Computer-Led Interactive Systems

The minimum requirement for such computers is to record the history of decisions made by human being. One of the problems involved in many large and complex problem solving is that a problem is decomposed and shared to many people. Then decisions made by them become invisible afterward. If some decision is improper the total system does not work properly but no one can check it. The system that satisfies the above requirement improves this situation. It leads us to changing the style of human-computer interaction from the conventional human-led interactive systems to computer-led interactive systems where X-led interactive system means that X has initiative in interaction and become able to manage problem-solving process. Then it can make records of individual decisions made by human being in this process. This system resolves the objectivity problem by making the record of the problem solving history instead of the human problem solver.

Generally speaking, a problem solving is composed of a number of stages such as problem generation, problem representation, problem understanding, solution planning, searching and deciding solution method, execution of the method, and displaying the solution. Currently computers join only partly in this process. Persons have to do most parts, i.e. from the beginning until deciding solution method and, in addition, making programs thereafter in order for using computers. Many decisions are made in this process. In order to record these decisions with the background information based on which these decisions are made, the process has to be managed by computers. Automating problem solving to a large extent is necessary for the purpose. It does not mean to let computers do everything but to let them manage the problem solving process.

Hence autonomy is primarily important for achieving the goal of computer-led interactive systems on one hand. If autonomy could be achieved to a large extent, on the other hand, the system can do not only making the record of decision making but also providing positively the proper environment for human decision making. For example, the system can decompose a large-scale problem into smaller-scale problems suited for many people and also prepare and use multi-disciplinary knowledge in various domains.

It is possible to develop an information technology to aid human activity in this manner. The basic requirements for the system design are derived by an effort to answer the queries given in 3.3.

3.2 Autonomous System

Autonomy is defined in [3] in relation with agency as an operation without direct intervention of human. It also says that autonomous system to have some kind of control over its actions and internal state. But this is a too broad definition for those who are going to realize autonomy actually. As every problem solving method is different by the problems to be solved, what is the strategy for controlling the operation to be determined for each problem?

Therefore, it is necessary to make it more concrete. Instead of this broad definition, the autonomy is defined in this paper as the capability of a computer to represent and process the problem solving structure required by the problem. The problem-solving structure means the structure of the operations for arriving at the solution of the given problem. It is different by each specific problem but the same type problems have the same skeletal problem solving structure. For example design as a non-deterministic problem solving is represented as the repetition of model analysis and model modification (Figure 1). Its structure is not algorithm-based as in the ordinary computerized methods. The detail of the operations for design is different by the domain of the problem and the extent to which the problem is matured. But the difference in the detail can be absorbed by the use of the domain specific knowledge base provided the method of using the domain knowledge base for problem solving is made the same even for the different problems. Then a unified problem solving structure is made to this type of problems. Conversely, the problem type is defined as a set of problems with the same problem solving structure. The necessary condition for a computer to be autonomous for wide classes of problems therefore is that the computer is provided with such a mechanism as to generate the proper problem-solving structures dynamically for the different types of problems.

When a problem is large, the problem decomposition is necessary before going into the detailed problem solving procedure for finding solution. The problem decomposition is problem dependent to a large extent but the decomposition method can be made common to many types of problems. System autonomy implies to decompose problems autonomously. Then the system resolves the size / complexity Problem.

4. System Architecture for Assuring System Autonomy

The autonomous system must be able first of all to represent problem solving structures for different problem types responding to the users request and deal with them autonomously. A special organization is required for the system. A conceptual architecture of the system and various component subsystems are discussed.

4.1 Conceptual Architecture of Autonomous System

From the viewpoint of information processing technology, there are two classes of problems; those that can be solved by the deterministic problem solving methods and the others. For the deterministic class problems the stage of providing a problem-solving method and that of executing it for obtaining a final solutions can be separated as the independent operations. For non-deterministic problem solving, to the contrary, these two operations cannot be separated but are closely related to each other. This requires a trial-and-error approach. In current computer's method to make a program and to execute it are separated as the different operations. It is suited only for the deterministic problem solving as an intrinsic nature of computers based on procedural language. It cannot be an autonomous system for the non-deterministic problems. A different architecture from this that enables the trial-and-error operations is necessary for realizing the autonomous system.

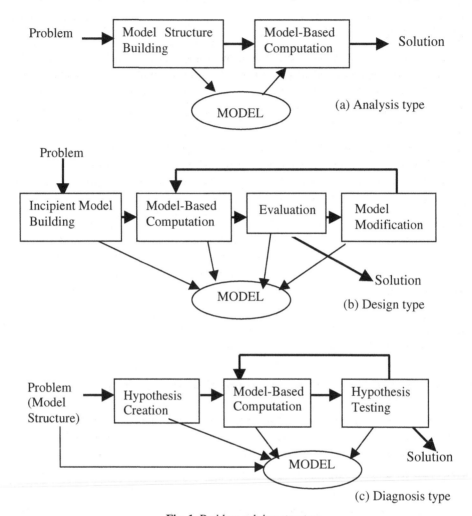

Fig. 1. Problem solving structure

Trial-and-error operations can be realized provided modularized representations of operations are available in computers. It is because the different combinations of operations can be generated to correspond to the different trials. For realizing this idea a completely declarative language and its processor must be provided on the conventional CPU with the procedural language. Figure 2 illustrates a conceptual architecture for achieving this idea. A CPU is the processor of procedural language. An inference engine as the processor for declarative language is implemented as a special procedural program.

Thus the activities in a computer are classified into two classes in relation with the required function in the computer; the activities described in the procedural form and those described only in declarative form. The classification principle is as follows. The activities that can be fixed to the detail in advance are represented in the procedural form such as operations around CPU including OS while the activities to adapt to the changing environment, mainly for applications, must be represented in the declarative form. There are some difficult cases to decide with this principle. Every activity requires a proper control. To represent the problem solving structure and the control of operations in the declarative form requires meta-rule. In many cases a meta-rule can cover wide scope of operations. That is, a single high-level control expression can control a scope of operations and a high-level operation can be fixed. In this paper however the activities that are defined in relation with applications are represented in the declarative form. It assures the flexibility of changing the control rule. In order to achieve this goal, a declarative language is designed to link easily with procedural operation.

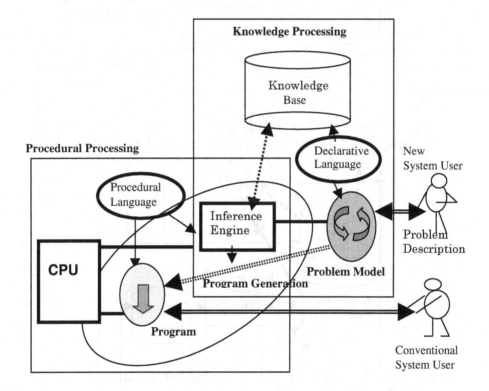

Fig. 2. Conceptual architecture of new information systems

4.2 Components of Autonomous System

The key issue is to clarify the requirement for the knowledge processing subsystem in Figure 2 and to develop a new technology to satisfy the requirement. Basic issues to constitute the technology are as follows (Figure 3).
1. New modeling method
2. Aiding human externalization
3. Large knowledge base and problem solving system generation

4. Autonomous problem decomposition and solving
5. Generation of program
6. Integration of different IPUs of different styles
7. Knowledge gathering, acquisition and discovery

Though these form the kernels of the system, only the outlines of the approaches the author's group are taking are introduced because of the lack of the paper. For the further details see [4].

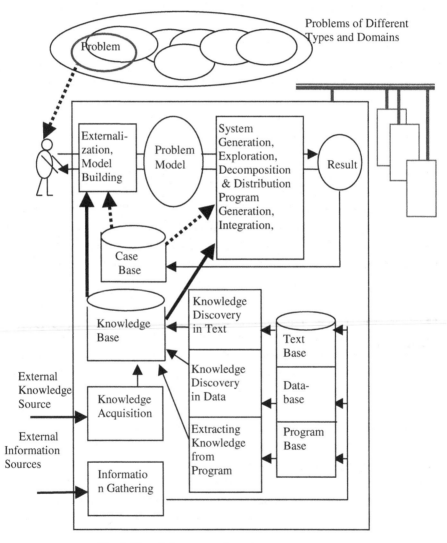

Fig. 3. Knowledge processing subsystem

4.2.1 New Modeling Method

As has been discussed, a problem solving is composed of a number of stages. Currently computers join only partly in this process. The objective of making computer-led interactive systems is to move most of the stages in this process to computer. Since problem is made in person's side, he/she has to represent it explicitly. This problem representation has to be accepted by the computer immediately. It is desirable that the computer helps persons to represent his/her problem and does everything afterward until to obtain solution.

Thus the difference between the conventional style and the new style in problem solving is in the location of a problem being transferred from human to computer and accordingly its representation. In the old style the problem is represented in the form of program but in the new style the problems must be represented in the form close to those generated in human brains, be transferred to computers as soon as they are created and be processed there. The formal representation of the problem is called here a problem model.

A problem is a concept created in a person's brain and a model is its explicit representation. The principle of problem modeling is to represent everything that relates problem solving explicitly so that no information remains in human brain in the form invisible to the others.

A method of problem model representation must be decided such that variety of problems can be represented in the same framework. Here is a big issue of ontology [5, 6]. As will be discussed in the next section, the problem model is represented as a compound of predicates and the structures of the conceptual entities that are also related with the language. In order to assure the common understanding of the model by many people, the structuring rules must be standardized first of all so that people come to the same understanding of the meaning of a structure. It is also necessary that people have the common understanding for the same language expressions [7]. In this paper it is assumed that these conditions are met. This paper does not go into the more detailed discussions on ontology.

Problem model is a formal representation of user's problem. It must be comprehensive for persons because it is created and represented by person. It must also be comprehensive for computers because it is to be manipulated by computers. It plays a key role in this new information technology. It is assumed in this paper that every problem is created in relation with some object in which a user has interest. If a representation of an object is not complete but some part lacked, then problems arise there. Therefore to represent a problem is to represent the object with some part lacking, and to solve problem is to fill the lacked part. This is named an object model. The basis of the object model is to represent the relation between the structure of the object being constructed from a set of components and functionality of every conceptual object (the object itself and its components). This is the definition of an object model in the problem representation in this paper.

Actually only the limited aspects of object within the scope of user's interest can be represented. Then different problems can arise from the same object depending on the different views from the users. It means that representations not only of an object but also of person's view to the object must be included in the problem model in order to represent the problem correctly.

A person may have interests in everything in the world, even in the other person's activity. This latter person being interested by the person may have an interest in still the other person. It implies that, if the persons are represented explicitly as subjects, a problem model forms a nest structure of the subject's interests. It is illustrated in Figure 4. For example, a problem of making programs needs this scheme. Program is a special type of automatic problem solver and three subjects at the different levels concern defining automatic programming. Subject S1 is responsible to execute a task in, for example, a business. Subject S2 makes program for the task of S1. For the purpose the subject S2 observes the S1's activity and makes its model as an object of interest before programming. Subject S3 observes the subject S2's activity as an object and automates S2's programming task. The activities of these subjects can be represented by the predicates such as processTrans(S1, Task), makeProgram(S2, processTrans(S1, Task), Program) and automateActivity (S3, makeProgram(S2, processTrans (S1, Task), Program), System), respectively. The upper activities need high order predicates. This kind of stratified objects/ activity is called the Multi-Strata Structure of Objects and Activities. A model of Multi-Strata Activities is called a Multi-Strata Model [8, 9].

A multi-strata model is composed from three different sub-models; Pure Object Model, User Subject Model and Non-User Subject Model. In the above example, S3 is the user who intends to solve a problem of making an automatic programming system. Looking from the user, the subjects S2 and S1 are in the objects being considered by the subjects S3 and S2 respectively and, therefore, the non-user subjects. The subject S1 does a task as a work on a pure object.

These are in the following relations (Figure 4).

Problem Model = Subject Model + Pure Object Model

Subject Model = User Subject Model + Non-User Subject Model

A problem model changes its state during problem solving. The model is classified into three classes according to the progress of problem solving; an incipient model, a sufficient model and a satisfactory model. An incipient model is the starting model with the least information given by user. The satisfactory model is a model that is provided with all the required information, i.e. solution.

The sufficient model is an intermediate-state model. Starting from the incipient model, information is added to the model until it reaches the state with sufficient information for deriving the satisfactory model.

It is desirable that the system collects as large as possible amount of information from inside and outside information sources in order to reduce the burden of the user to build a sufficient model with large amount of information. It is required to the system to gather the information using the information in the incipient model as the clue so that the incipient model, therefore the user's effort, can be made the smallest. After the model is given the enough information, the user can modify it incrementally to make it a sufficient model. The sufficient model made in this way however is not a correct model but is the model for starting problem solving in the narrow sense to make it the satisfactory model.

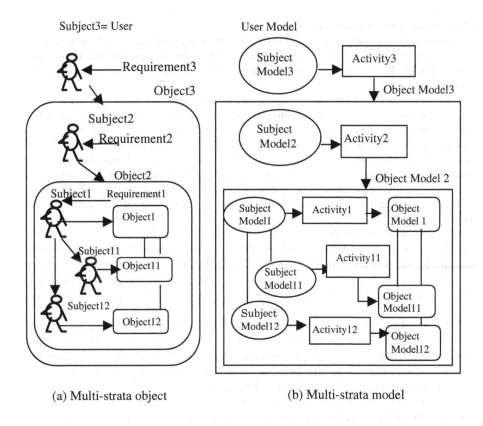

(a) Multi-strata object   (b) Multi-strata model

Fig. 4. Multi-strata object and model

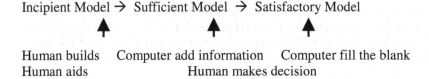

The major inside information source is a case base. It must be well structured for enabling the system to build a sufficient model as close as to what the user wishes. The outside information sources are in the web. The system must be provided with the search engine. But presently it is not easy to use outside information for autonomous problem solving. A new technology must be developed including ontology.

Various types of problems are defined depending on the lacked part in a model. For example, if some functionality of an entity in a pure object model is lacked, an analytic problem is generated. If some functionality is given as the requirement but the structure of entities in a pure object model is lacked, then a design type problem arises. If the structure of activities is lacked, then scheduling type problem arises.

By representing problems explicitly in this form information is made visible and therefore resolves objectivity and communication problems.

4.2.2 Aiding Human Externalization

Externalization is a computer's supports for (1) human cognitive process to help user for clarifying his/her idea and (2) model building in order to represent the idea in the form of model. It is a human-computer interface in the much broader sense than the ordinary ones.

Very often novice users are not accustomed to represent their ideas formally. Sometimes, their ideas are quite nebulous and they cannot represent the ideas in the correct sentences. How can the system help these users? This issue belongs to cognitive science. What the system can do is to stimulate the users to notice what they are intending. Some researches are made on this issue [10, 11] but these are not discussed in this paper any more. In the following it is assumed that the users have clear ideas on their problems. The system aids them to build models to represent the ideas. Easy problem specification, computer-led modeling and shared modeling are discussed.

(1) Easy problem specification

The model concerns the type of problem the user wants to do. The problem solving structure for the problem type must be made ready before going into problem solving. In general it is not easy for the novice users. The system allows the user to take such a method that the user says what he/she likes to do and the computer provides problem-solving structure.

Example; Let the user wants to do a design task in a specific domain, say #domain-j. The user selects Design as the type of problem and #domain-j as the domain. It is formalized to designObject(userName, Object, #domain-j). Then the system uses the following rule to generate the problem solving structure for design.

designObject(HumanSubject, Object, Domain):-
 designManagement(Computer, designObject(HumanSubject, Object, Domain))

designManagement represents the design-type problem solving structure. It is further expanded to the set of rules to represents the structure.

In reality, it is possible to define the more specific design-type problem like designMaterial if the latter has the specific problem solving structure for designing material. This structure is specialized to the design in a specific domain, say material in this case. In this way problem-type is classified as fine as possible and forms a hierarchy. This hierarchy of problem-types is presented to the user and the latter selects one or the more types. The problem domain is also divided to the sets of narrower domains and these domains form a hierarchy.

(2) Computer-led modeling and system building

In building a sufficient model the computer adds information to the model as much as possible. The main source of information in this case is the case base. In some special cases, special knowledge is used in order to make the model sufficient [4].

To allow persons to share model building

The model of any large-scale problem is also large. It is difficult to build up it alone but model building has to be shared by many people. In Figure 5 the user represents his/her intention in the form of the user model. At the same time he/she can makes the other part of the problem model that he/she wants to decide oneself. Figure 5 represents an example of modeling an enterprise. The user specifies a part of the model enclosed by the thick line other than the user model including the subjects with their activities. An externalization subsystem is provided to every subject to which

22   S. Ohsuga

human is assigned. After then the control moves downward to the lower nodes and human subjects behave as the users successively. That a human subject behaves as a user means that the externalization subsystem begins to work for the subject for aiding the human to build the still lower part of the model. But the activity of every lower human subject has already been decided by an upper human subject and is given to the subject. The subject is obliged to work to achieve the activity.

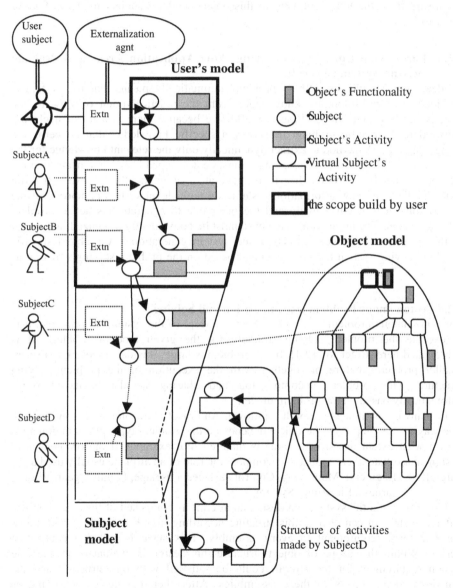

Fig. 5. An illustration of problem model

In this way the subjects SubjectA, SubjectB, SubjectC and so on in Figure 5 behave as the users one by one and extend the model downward. As will be described in 4.2.4, model building concerns closely with the object decomposition, and this model decomposition concerns the activity of the related subject. For example, let the subject be given a task of design as the required activity. The design-type problem-solving structure is prepared. According to this structure the object model is built downward from the top. Following to this object model a subject model is formed (refer to 5.2.4).

4.2.3 Large Knowledge Base Covering Wide Application Areas and Problem Solving System Generation

To deal with a multi-disciplinary problem automatically means that the system is provided with multi-domains knowledge. From the practical point of view to use directly the large knowledge base is ineffective because there is a lot of irrelevant knowledge. A specific problem uses only a specific knowledge that concerns the related domains. A method to extract dynamically only the relevant knowledge for the given problem in a large knowledge base must be developed [12]. This means that the real problem solving system has to be generated automatically for every specific problem. Thus the new information system is a multi-level system composed of the one to solve a problem actually and the other one to generate this actual problem solving system. The large knowledge base must be conveniently organized in order to facilitate this knowledge extraction and system generation. A problem specific problem-solving system is to be generated based on the problem solving structure as is shown in Figure 6

4.2.4 Autonomous Problem Decomposition and Solving

The autonomous problem solving is the key concept of the system. In principle, it is to execute the problem solving structure for the given type of problem. It is abbreviated here. Refer to [12,4]. If the problem is large, it is decomposed to a set of smaller problems before or together with the execution of the problem solving structure. This problem decomposition and sharing is also achieved semi-automatically based on the problem model.

In many cases models are built top-down. An object is decomposed from the top. The way for decomposing the object depends on the problem type, problem domains, and to what extent the problem area is matured. For example, if the problem is to design an airplane, there is such a decomposition rule of an airplane usually used as

Aircraft → {Engine(s), Main-wing, Control-surfaces, Fuselage, Landing-gear, Wire-harness, Electronic-System, ---}.

Using this rule the system can design an aircraft as composed of these assemblies. After then human can change this structure according his idea. He/she decides also the functional requirement to every assembly. This stage is called a conceptual design. Within this scope of object structure this design is evaluated whether the design requirement for the aircraft could be satisfied with this structure and the functional requirements to these assemblies. After then the design work moves downward to each assembly. The functional requirement decided as above to each assembly becomes the design requirement for the assembly and the assembly is decomposed further.

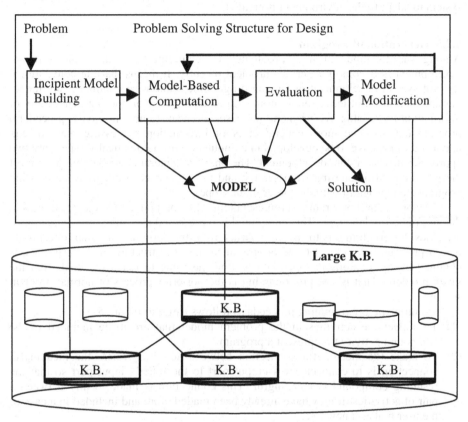

Fig. 6. Large knowledge base

Corresponding to the object decomposition a new subject is assigned to each new assembly resulting in a new subject structure. The same activity as that of the parent subject, i.e. 'designObject' in this case, is given to every new subject. Thus created pair of new object and new subject forms an agent in the system. That is this problem decomposition creates a new multi-agent system. Each agent shares a part of design.

Object Model Decomposition;   Object → {Object1, Object2, --- , ObjectN}
    Object-Subject Correspondence; | | |
Subject Model Formation;   Subject ← {Subject1, Subject2, --- , SubjectN}

The way for decomposing the object model is different by the problem. In the design type problem the object model is built in such a way that the required functionality of the object can be satisfied and a new person is assigned to a new object. In the other case the user at the top specifies the other activity. For example the user specifies an evolutionary rule to decompose an object automatically as an activity (instead of 'designObject') to the top subject. This rule is to decompose an object toward adapting the environment given by the user. Then an evolutionary system to adapt to the environment is created.

4.2.5 Generation of Program

Among various kinds of time problems the requirement for the rapid software development is becoming a serious problem recently. It means that the conventional style of system development by human programmer cannot be continued any longer but automation or, at least, semi-automation is necessary. Considering that variety of information-processing requirement increases in near future, not only procedural program but also the more general class of information processing method, e.g. neural-network, have to be developed in computers semi-automatically. There are two approaches to the system development. The one is to generate a new program to meet the given condition starting from scratch and the second is to integrate the existing programs to a program with the larger input scope.

Every procedural program is a representation of a compound of subject activities in the programming language. This compound is represented as a structure of the related activities. Programming is to specify a structure of the activities and then to translate it to a program. An object to be programmed, i.e. the structure of the activities, is obtained by the exploratory operation to find the route to reach the goal from the given problem. That is, the programming is the posterior process of normal problem solving.

The automatic programming proceeds as follows (refer to Figure 5).
(1) To collect the activities in the problem model that are to be included in the compound activity to represent a program .
(2) To make the meaningful connection between activities where the meaningful connection is to connect one's output port to the other's input port so that the resulting compound is meaningful. This connection is made locally between a pair of activities. It may have already been made before and included in a case or the user makes a new one.
(3) To define inputs of some activities and outputs of some activities as the global inputs from the outside and the global output ports to the outside respectively.
(4) To explore a route from the global inputs to the global outputs for the specific instance values substituted to the global input variables by inference resulting in a deduction tree.
(5) To generalize the deduction tree restoring the variables from the instance values resulting in the generalized deduction tree including necessary program structures for the target program.
(6) To translate the deduction tree into another structure of functions according to the program-structuring rule such as sequence, branch, loop, etc.
(7) To translate the structure of functions into a specific program code.
    Auxiliary stage; As an auxiliary procedure for the special case such as being required to meet the real time operation, the following stage is provided.
(8) To develop a simulator for estimating the execution time and to make a problem solving structure for real-time programming type problem to include this simulator.

The operation (8) is essential in case of embedded software. The embedded software, for example, the software embedded in vehicle engine, is becoming large and its role is increasing. Special attention must be paid for these embedded systems because of their peculiar developing condition [13].

4.2.6 Integration of Different Information Processing Methods

Integration is to merge two or more information processing units (IPUs hereafter) of the different styles such as a procedural program, a knowledge-based processing unit, a neural network and so on into an IPU with a larger scope. It needs a special method of transformation between different representation schemes. To find such a transformation as a general method is not easy but in many cases it has been performed manually in an ad hoc way for each specific pairs of IPUs to be merged. In this paper its automation is intended. For this purpose the integration problem is considered in two steps.

In the first step integration is defined as an operation of an IPU to take in the other IPU(s) and to use its function in order to realize a new function as a whole. The former is called the master IPU and the latter is the slave IPU. In order to combines the slave IPU in its operation, the master IPU must be able to represent the feature of the slave IPU in its own representation scheme. The master is also required to transform the format of some data in the master to the input of the slave and the output of the slave to the master in such a way that the operation of the slave can be activated properly in the master IPU. These representations are added to the master IPU from outside. Thus the master IPU has to be able to accept this additional information and expand its scope without disturbing its own processing. The expanded master IPU can decide the time to evoke the slave and translate the data to send to the slave. Receiving the output from the slave and translating it into the master scheme, the master IPU can continue the operation.

There can be a number of information-processing styles and for some of them it is difficult to meet this condition to be the master IPU. As the typical example, neural network, procedural processing and declarative processing are considered. A neural network has a fixed mechanism of processing and it is difficult to accept the information on the slave IPU from outside without changing the whole structure. A procedural processing should be more flexible than the neural network but in reality it is still difficult to add such information as above from outside without changing the main body of the program. That is, the time and way of evoking the slave and transformation routines must be programmed and embedded in the main program in advance. It is to rebuild the program and automation of integration adapting to the changing environment is difficult. Finally a declarative processing, for example a rule-based system using 'If-Then' rule, adding information from outside does not require the modification of the main part of the system but the system can expand the scope of processing by combining the added rules with the existing ones. This is an important characteristic of the declarative processing system. It makes the declarative processing potentially the best candidate of the master IPU. It can integrate various IPUs of the different styles. The descriptive information on the slave to be added to the master IPU must exist at integration and is added to the knowledge base of declarative processing system. This information on the slave is assumed created when the slave IPU was first developed.

In the second step this idea is generalized. Let two IPUs, say (Ia, Ma) and (Ib, Mb), of any information processing styles, Sa and Sb respectively, be to be merged. If at least one of them is a declarative IPU, then the integration as discussed above is possible by making this IPU the master. If both IPUa and IPUb are not the declarative ones, their direct integration is difficult. But their integration becomes still possible by introducing a new declarative IPU. In this case this third IPU is the master to both IPUa and IPUb, and the representations of both IPUa and IPUb are added thereto. Thus non-declarative IPUs can be integrated indirectly via the third declarative IPU. It is possible to prepare such a declarative IPU for the purpose of integrating arbitrary IPUs.

4.2.7 Acquiring Knowledge

The issues 4.2.1 through 4.2.6 are on the most fundamental parts of the system for achieving the autonomy. Therefore these parts must be developed in a relatively short time. But this system assumes the existing of the large amount of knowledge to be used for solving various problems. A method for providing the system with the knowledge is necessary. The amount and quality of knowledge contained in the systems will decide finally the value of the intelligent system. Generally speaking however this part of technology is not yet well developed. Though research interests to the methods for creating knowledge from various information sources are increasing recently and the new research groups are being formed, it is not yet established and is the long-term goal.

The major sources of knowledge are as follows.

(1) Human knowledge; Human is the major source of knowledge. Knowledge acquisition by interviewing to the experts is still the main method of human knowledge acquisition. In addition it is desirable that the system acquires knowledge made by user while using the system for problem solving.

(2) Text ; Published or recorded texts are also important knowledge sources. There are two kinds of texts; the edited text and the non-edited text.

    (i) Edited text ; Some texts have been edited and compiled as the knowledge for the purpose of making hard copied knowledge base such as handbooks in many technical fields, dictionary, etc.

    (ii) Unedited text; there are lot of texts that are generated and recorded in the daily activities such as the claims from the clients for the industrial products, the pilot's report after every flight including near-miss experience, etc. These texts are accumulated but hardly used because it takes a large man-hour to read and to use afterward. But the information included therein is quite important for improving the wrong situation unnoted before. A method to use the information autonomously is desirable. Today text mining and knowledge discovery from text are being made research for this class of texts.

(3) Data ; Huge amount of data are being accumulated in many areas of human activity and are still increasing. Some general information is hidden behind these data. Data mining and knowledge discovery in data (KDD) is the efforts to find hidden useful information.

(4) Program ; So far lot of programs have been developed. Every program was developed based on knowledge on some object at the time of development. After years the programmer leaves and the program remains being used but no one can understand the knowledge behind the program. To extract knowledge from the

program is becoming very important tasks. The knowledge extracted from programs can be used in many ways.

The issues (2) through (4) are the inverse problems and need the new technologies. Researches are being made in many places and recently there are many international conferences. But the technologies for discovering knowledge are not yet established. It will take more time. In this sense, these are the long-term goal.

5. Conclusion

It is worried that human capability cannot catch up the progress of increasing scale and complexity of problems to arise in near future and accordingly many troubles can happen in human society. Is there some one that can resolve these troubles in place of human being? Only possibility for answering this question is to make computers more intelligent so that human can be free from such works that are weak in achieving for them. The author has discussed in this paper a way to make computers more intelligent and expand the scope of information processing. It does not mean that computer can be more intelligent than human being but that the computer help human to make the best decision by undertaking the management tasks of environment of decision making. This paper analyzed the weak points of humans, found a possible alternate method and then to proposed a way to develop new information technology that can back up and aid human activity.

The major topics discussed in this paper were; (0) an overall software architecture for future information systems, (1) modeling scheme to accept and represent wide area of problems, (2) method for externalizing human idea and of representing it as a model, (3) large knowledge base and problem solving system generation, (4) autonomous problem decomposition and solving, (5) generation of program, (6) integration of different IPUs of different styles, and (7) knowledge acquisition and discovery.

A number of difficult problems were involved in this scope. Only an outline of a part of the research project on the way to resolve these difficulties has been presented. The research project started in 1998 under the sponsorship of The Japanese Science and Technology Agency and the issues (0) through (6) are being developed as the short period target (5 years). There remain many problems yet unsolved. Among them the issue of ontology will become the more serious problem in the future. It becomes more important when the system gather information widely spreading in the web. The issue (7) is included in the project but is not included directly in the short period target. It is included as the basic research to be continued in the following research.

A new representation scheme was necessary for representing these new concepts as shown above and a language/system MLL/KAUS (Multi-Layer Logic/ Knowledge Acquisition and Utilization System) suited for the purpose has been developed. It was not included in this paper. Refer to [14,15].

Acknowledgement.   The author expresses his sincere thanks to The Science and Technology Agency of Japanese Government for their support to the research. Without their support this research could not be continued.

References

[1]   W. Wayt Gibbs; Software Chronic Crisis, Scientific American, Volume 18, No.2, 1994
[2]   F. H. Ross; Artificial Intelligence, What Works and What Doesn't ? AI Magazine, Volume 18,
[3]   C. Castelfranchi; Intelligence Agents: Thories, Architectures, and Languages, in Guarantees for autonomy in cognitive agent architecture, (edited by M.Wooldridge and N.R.Jennings), Springer, 1995
[4]   S. Ohsuga ; How Can AI Systems Deal with Large and Complex Problems? First Pacific-Asia International Conference on Intelligent Agent Technology (invited paper), 1999 [10] K.Hori, A system for aiding creative concept formation, IEEE Transactions on Systems, Man and Cybernetics, Vol.24, No.6, 1994
[5]   T. Gruber; What-is-an-ontology?, http://www-ksl.stanford.edu/kst/what-is-ontology.html,
[6]   R. Mizoguchi, et.al.; Ontology for Modeling the World from Problem Solving Persoectives, Proc. of IJCAI-95 Workshop on Basic Ontological Issues in Knowledge Sharing, pp.1-12, 1995
[7]   R. Mizoguchi.;Knowledge Acquisition and Ontology, Proc. of the KB & KS, Tokyo, pp.121-128, 1993
[8]   S. Ohsuga; Multi-Strata Modeling to Automate Problem Solving Including Human Activity, Proc.Sixth European-Japanese Seminar on Information Modelling and Knowledge Bases,1996
[9]   S. Ohsuga ; A Modeling Scheme for New Information Systems -An Application to Enterprise Modeling and Program Specification, IEEE International Conference on Systems, Man and Cybernetics, 1999
[10] K.Hori; A System for Aiding Creative Concept Formation, IEEE Transactions on System, Man and Cybernetics, Vol.24, No.6, 1994
[11] Y.Sumi, et. al., Computer Aided Communications by Visualizing Thought Space Structure, Electronics and Communications in Japan, Part 3, Vol. 79. No.10, 11- 22, 1996.
[12] S. Ohsuga·Toward Truly Intelligent Information Systems - From Expert Systems To Automatic Programming, Knowledge Based Systems, Vol. 10, 1998
[13] E. Lee; What's Ahead for Embedded Software?, Computer, IEEE Computer Society, Sept, 2000
[14] S. Ohsuga and H.Yamauchi,Multi-Layer Logic - A Predicate Logic Including Data Structure As Knowledge Representation Language, New Generation Computing, 1985.
[15] H. Yamauchi, KAUS6 User's Manual, RCAST, Univ. of Tokyo, 1995.

Rough Sets and Decision Algorithms

Zdzisław Pawlak

Institute of Theoretical and Applied Informatics, Polish Academy of Sciences,
ul. Bałtycka 5, 44 000 Gliwice, Poland

Abstract. Rough set based data analysis starts from a data table, called an *information system*. The information system contains data about objects of interest characterized in terms of some attributes. Often we distinguish in the information system condition and decision attributes. Such information system is called a *decision table*. The decision table describes decisions in terms of conditions that must be satisfied in order to carry out the decision specified in the decision table. With every decision table a set of decision rules, called a *decision algorithm* can be associated. It is shown that every decision algorithm reveals some well known probabilistic properties, in particular it satisfies the Total Probability Theorem and the Bayes' Theorem. These properties give a new method of drawing conclusions from data, without referring to prior and posterior probabilities, inherently associated with Bayesian reasoning.

1 Introduction

Rough set based data analysis starts from a data table, called an *information system*. The information system contains data about objects of interest characterized in terms of some attributes. Often we distinguish in the information system condition and decision attributes. Such an information system is called a *decision table*. The decision table describes decisions in terms of conditions that must be satisfied in order to carry out the decision specified in the decision table. With every decision table we can associate a decision algorithm which is a set of *if... then...* decision rules. The decision rules can be also seen as a logical description of approximation of decisions, and consequently a decision algorithm can be viewed as a logical description of basic properties of the data. The decision algorithm can be simplified, what results in optimal description of the data, but this issue will not be discussed in this paper.

In the paper first basic notions of rough set theory will be introduced. Next the notion of the decision algorithm will be defined and some its basic properties will be shown. It is revealed that every decision algorithm has some well known probabilistic features, in particular it satisfies the Total Probability Theorem and the Bayes' Theorem [5]. These properties give a new method of drawing conclusions from data, without referring to prior and posterior probabilities, inherently associated with Bayesian reasoning. Three simple tutorial examples will be given to illustrate the above discussed ideas. The real-life examples are much more sophisticated and will not be presented here.

W. Ziarko and Y. Yao (Eds.): RSCTC 2000, LNAI 2005, pp. 30–45, 2001.

2   Approximation of Sets

Starting point of rough set based data analysis is a data set, called an information system.

An information system is a data table, whose columns are labeled by attributes, rows are labeled by objects of interest and entries of the table are attribute values.

Formally, by an *information system* we will understand a pair $S = (U, A)$, where U and A, are finite, nonempty sets called the *universe*, and the set of *attributes*, respectively. With every attribute $a \in A$ we associate a set V_a, of its *values*, called the *domain* of a. Any subset B of A determines a binary relation $I(B)$ on U, which will be called an *indiscernibility relation*, and defined as follows: $(x, y) \in I(B)$ if and only if $a(x) = a(y)$ for every $a \in A$, where $a(x)$ denotes the value of attribute a for element x. Obviously $I(B)$ is an equivalence relation. The family of all equivalence classes of $I(B)$, i.e., a partition determined by B, will be denoted by $U/I(B)$, or simply by U/B; an equivalence class of $I(B)$, i.e., block of the partition U/B, containing x will be denoted by $B(x)$.

If (x, y) belongs to $I(B)$ we will say that x and y are *B-indiscernible* (*indiscernible with respect to B*). Equivalence classes of the relation $I(B)$ (or blocks of the partition U/B) are referred to as *B-elementary sets* or *B-granules*.

If we distinguish in an information system two disjoint classes of attributes, called *condition* and *decision attributes*, respectively, then the system will be called a *decision table* and will be denoted by $S = (U, C, D)$, where C and D are disjoint sets of condition and decision attributes, respectively.

Suppose we are given an information system $S = (U, A)$, $X \subseteq U$, and $B \subseteq A$. Our task is to describe the set X in terms of attribute values from B. To this end we define two operations assigning to every $X \subseteq U$ two sets $B_*(X)$ and $B^*(X)$ called the *B-lower* and the *B-upper approximation* of X, respectively, and defined as follows:

$$B_*(X) = \bigcup_{x \in U} \{B(x) : B(x) \subseteq X\},$$

$$B^*(X) = \bigcup_{x \in U} \{B(x) : B(x) \cap X \neq \emptyset\}.$$

Hence, the *B*-lower approximation of a set is the union of all *B*-granules that are included in the set, whereas the *B-upper* approximation of a set is the union of all *B*-granules that have a nonempty intersection with the set. The set

$$BN_B(X) = B^*(X) - B_*(X)$$

will be referred to as the *B-boundary region* of X.

If the boundary region of X is the empty set, i.e., $BN_B(X) = \emptyset$, then X is *crisp* (*exact*) with respect to B; in the opposite case, i.e., if $BN_B(X) \neq \emptyset$, X is referred to as *rough* (*inexact*) with respect to B.

3   Decision Rules

In this section we will introduce a formal language to describe approximations in logical terms.

Let $S = (U, A)$ be an information system. With every $B \subseteq A$ we associate a formal language, i.e., a set of formulas $For(B)$. Formulas of $For(B)$ are built up from attribute-value pairs (a, v) where $a \in B$ and $v \in V_a$ by means of logical connectives $\wedge$ (*and*), $\vee$ (*or*), $\sim$ (*not*) in the standard way.

For any $\Phi \in For(B)$ by $\|\Phi\|_S$ we denote the set of all objects $x \in U$ satisfying Φ in S and refer to as the *meaning* of Φ in S.

The meaning $\|\Phi\|_S$ of Φ in S is defined inductively as follows:
$\|(a, v)\|_S = \{x \in U : a(v) = x\}$ for all $a \in B$ and $v \in V_a$, $\|\Phi \vee \Psi\|_S = \|\Phi\|_S \cup \|\Psi\|_S$, $\|\Phi \wedge \Psi\|_S = \|\Phi\|_S \cap \|\Psi\|_S$, $\| \sim \Phi\|_S = U - \|\Phi\|_S$.

A formula Φ is *true* in S if $\|\Phi\|_S = U$.

A *decision rule* in S is an expression $\Phi \to \Psi$, read *if Φ then Ψ*, where $\Phi \in For(C)$, $\Psi \in For(D)$ and C, D are condition and decision attributes, respectively; Φ and Ψ are referred to as *conditions* and *decisions* of the rule, respectively.

A decision rule $\Phi \to \Psi$ is *true* in S if $\|\Phi\|_S \subseteq \|\Psi\|_S$.

The number $supp_S(\Phi, \Psi) = card(\|\Phi \wedge \Psi\|_S)$ will be called the *support* of the rule $\Phi \to \Psi$ in S. We consider a probability distribution $p_U(x) = 1/card(U)$ for $x \in U$ where U is the (non-empty) universe of objects of S; we have $p_U(X) = card(X)/card(U)$ for $X \subseteq U$. For any formula Φ we associate its probability in S defined by

$$\pi_S(\Phi) = p_U(\|\Phi\|_S).$$

With every decision rule $\Phi \to \Psi$ we associate a conditional probability

$$\pi_S(\Psi|\Phi) = p_U(\|\Psi\|_S| \ \|\Phi\|_S)$$

that Ψ is true in S given Φ is true in S called the *certainty factor*, used first by Łukasiewicz [3] to estimate the probability of implications. We have

$$\pi_S(\Psi|\Phi) = \frac{card(\|\Phi \wedge \Psi\|_S)}{card(\|\Phi\|_S)}$$

where $\|\Phi\|_S \neq \emptyset$.

This coefficient is now widely used in data mining and is called *confidence coefficient*.

Obviously, $\pi_S(\Psi|\Phi) = 1$ if and only if $\Phi \to \Psi$ is true in S.

If $\pi_S(\Psi|\Phi) = 1$, then $\Phi \rightarrow \Psi$ will be called a *certain decision* rule; if $0 < \pi_S(\Psi|\Phi) < 1$ the decision rule will be referred to as a *uncertain decision* rule.

Besides, we will also use a *coverage factor* (used e.g. by Tsumoto [14] for estimation of the quality of decision rules) defined by

$$\pi_S(\Phi|\Psi) = p_U(\|\Phi\|_S| \,\|\Psi\|_S).$$

which is the conditional probability that Φ is true in S, given Ψ is true in S with the probability $\pi_S(\Psi)$. Obviously we have

$$\pi_S(\Phi|\Psi) = \frac{card(\|\Phi \wedge \Psi\|_S)}{card(\|\Psi\|_S)}.$$

The certainty factors in S can be also interpreted as the frequency of objects having the property Ψ in the set of objects having the property Φ and the coverage factor – as the frequency of objects having the property Φ in the set of objects having the property Ψ.

The number

$$\sigma_S(\Phi, \Psi) = \frac{supp_S(\Phi, \Psi)}{card(U)} = \pi_S(\Psi|\Phi) \cdot \pi_S(\Phi)$$

will be called the *strength* of the decision rule $\Phi \rightarrow \Psi$ in S.

4   Decision Algorithms

In this section we define the notion of a decision algorithm, which is a logical counterpart of a decision table.

Let $Dec(S) = \{\Phi_i \rightarrow \Psi_i\}_{i=1}^m$, $m \geq 2$, be a set of decision rules in a decision table $S = (U, C, D)$.

1) If for every $\Phi \rightarrow \Psi$, $\Phi' \rightarrow \Psi' \in Dec(S)$ we have $\Phi = \Phi'$ or $\|\Phi \wedge \Phi'\|_S = \emptyset$, and $\Psi = \Psi'$ or $\|\Psi \wedge \Psi'\|_S = \emptyset$, then we will say that $Dec(S)$ is the set of pairwise *mutually exclusive (independent)* decision rules in S.

2) If $\|\bigvee_{i=1}^m \Phi_i\|_S = U$ and $\|\bigvee_{i=1}^m \Psi_i\|_S = U$ we will say that the set of decision rules $Dec(S)$ *covers* U.

3) If $\Phi \rightarrow \Psi \in Dec(S)$ and $supp_S(\Phi, \Psi) \neq 0$ we will say that the decision rule $\Phi \rightarrow \Psi$ is *admissible* in S.

4) If $\bigcup_{X \in U/D} C_*(X) = \|\bigvee_{\Phi \rightarrow \Psi \in Dec^+(S)} \Phi\|_S$ where $Dec^+(S)$ is the set of all certain decision rules from $Dec(S)$, we will say that the set of decision rules $Dec(S)$ preserves the *consistency* of the decision table $S = (U, C, D)$.

The set of decision rules $Dec(S)$ that satisfies 1), 2) 3) and 4), i.e., is independent, covers U, preserves the consistency of S and all decision rules

$\varPhi \to \varPsi \in Dec(S)$ are admissible in S – will be called a *decision algorithm* in S.

Hence, if $Dec(S)$ is a decision algorithm in S then the conditions of rules from $Dec(S)$ define in S a partition of U. Moreover, the *positive region of D with respect to C*, i.e., the set

$$\bigcup_{X \in U/D} C_*(X)$$

is partitioned by the conditions of some of these rules, which are certain in S.

If $\varPhi \to \varPsi$ is a decision rule then the decision rule $\varPsi \to \varPhi$ will be called an *inverse* decision rule of $\varPhi \to \varPsi$.

Let $Dec^*(S)$ denote the set of all inverse decision rules of $Dec(S)$.

It can be shown that $Dec^*(S)$ satisfies 1), 2), 3) and 4), i.e., it is an decision algorithm in S.

If $Dec(S)$ is a decision algorithm then $Dec^*(S)$ will be called an *inverse* decision algorithm of $Dec(S)$.

The number

$$\eta(Dec(S)) = \sum_{\varPhi \to \varPsi \in Dec(S)} max\{\sigma_S(\varPhi, \varPsi)\}_{\varPsi \in D(\varPhi)}$$

where $D(\varPhi) = \{\varPsi : \varPhi \to \varPsi \in Dec(S)\}$ will be referred to as the *efficiency* of the decision algorithm $Dec(S)$ in S, and the sum is stretching over all decision rules in the algorithm.

The efficiency of a decision algorithm is the probability (ratio) of all objects of the universe, that are classified to decision classes, by means of decision rules $\varPhi \to \varPsi$ with maximal strength $\sigma_S(\varPhi, \varPsi)$ among rules $\varPhi \to \varPsi \in Dec(S)$ with satisfied $\varPhi$ on these objects. In other words, the efficiency says how well the decision algorithm classifies objects when the decision rules with maximal strength are used only.

5   Decision algorithms and approximations

Decision algorithms can be used as a formal language for describing approximations (see [5]).

Let $Dec(S)$ be a decision algorithm in S and let $\varPhi \to \varPsi \in Dec(S)$. By $C(\varPsi)$ we denote the set of all conditions of $\varPsi$ in $Dec(S)$ and by $D(\varPhi)$ - the set of all decisions of $\varPhi$ in $Dec(S)$.

Then we have the following relationships:

a) $C_*(\|\varPsi\|_S) = \| \bigvee_{\varPhi' \in C(\varPsi),\ \pi(\varPsi|\varPhi')=1} \varPhi'\|_S,$

b) $C^*(\|\varPsi\|_S) = \| \bigvee_{\varPhi' \in C(\varPsi),\ 0<\pi(\varPsi|\varPhi')\leq 1} \varPhi'\|_S,$

c) $BN_C(||\Psi||_S) = || \bigvee_{\Phi' \in C(\Psi),\ 0<\pi(\Psi|\Phi')<1} \Phi'||_S.$

From the above properties we can get the following definitions:

i) If $||\Phi||_S = C_*(||\Psi||_S)$, then formula Φ will be called the *C-lower approximation* of the formula Ψ and will be denoted by $C_*(\Psi)$;

ii) If $||\Phi||_S = C^*(||\Psi||_S)$, then the formula Φ will be called the *C-upper approximation* of the formula Φ and will be denoted by $C^*(\Psi)$;

iii) If $||\Phi||_S = BN_C(||\Psi||_S)$, then Φ will be called the *C-boundary* of the formula Ψ and will be denoted by $BN_C(\Psi)$.

The above properties say that any decision $\Psi \in Dec(S)$ can be uniquely described by the following certain and uncertain decision rules respectively:

$$C_*(\Psi) \to \Psi,$$

$$BN_C(\Psi) \to \Psi.$$

This property is an extension of some ideas given by Ziarko [16]. The approximations can also be defined more generally, as proposed in [15] by Ziarko, and consequently we obtain more general probabilistic decision rules.

6 Some properties of decision algorithms

Decision algorithms have interesting probabilistic properties which are discussed in this section.

Let $Dec(S)$ be a decision algorithm and let $\Phi \to \Psi \in Dec(S)$. Then the following properties are valid:

$$\sum_{\Phi' \in C(\Psi)} \pi_S(\Phi'|\Psi) = 1 \tag{1}$$

$$\sum_{\Psi' \in D(\Phi)} \pi_S(\Psi'|\Phi) = 1 \tag{2}$$

$$\pi_S(\Psi) = \sum_{\Phi' \in C(\Psi)} \pi_S(\Psi|\Phi') \cdot \pi_S(\Phi') \tag{3}$$

$$\pi_S(\Phi|\Psi) = \frac{\pi_S(\Psi|\Phi) \cdot \pi_S(\Phi)}{\sum_{\Phi' \in C(\Psi)} \pi_S(\Psi|\Phi') \cdot \pi_S(\Phi')} \tag{4}$$

That is, any decision algorithm, and consequently any decision table, satisfies (1), (2), (3) and (4). Observe that (3) is the well known *Total Probability Theorem* and (4) is the *Bayes' Theorem*. Note that we are not referring to prior and posterior probabilities – fundamental in Bayesian data analysis philosophy. The Bayes' Theorem in our case says that: if an implication $\Phi \to \Psi$

is true in the degree $\pi_S(\Psi|\Phi)$ then the inverse implication $\Psi \to \Phi$ is true in the degree $\pi_S(\Phi|\Psi)$.

Let us observe that the Total Probability Theorem can be presented in the form

$$\pi_S(\Psi) = \sum_{\Phi' \in C(\Psi)} \sigma_S(\Phi', \Psi) \tag{5}$$

and the Bayes' Theorem will assume the form

$$\pi_S(\Phi|\Psi) = \frac{\sigma_S(\Phi, \Psi)}{\displaystyle\sum_{\Phi' \in C(\Psi)} \sigma_S(\Phi', \Psi)} = \frac{\sigma_S(\Phi, \Psi)}{\pi_S(\Psi)} \tag{6}$$

Thus in order to compute the certainty and coverage factors of decision rules according to formula (6) it is enough to know the strength (support) of all decision rules in the decision algorithm only. The strength of decision rules can be computed from the data or can be a subjective assessment.

In other words, if we know the ratio of Φ_S in Ψ, thanks to the Bayes' Theorem, we can compute the ratio of Ψ_S in Φ.

7 Illustrative examples

In this section we will illustrate the concepts introduced previously by means of simple tutorial examples.

Example 1

Let us consider Table 1 in which data on the relationships between color of eyes and color of hair is given.

Table 1. Simple data table

Eyes	Hair	
	blond	dark
blue	16	0
brown	8	56

The above data can be presented as a decision table shown in Table 2.

Table 2. Decision table

Rule number	Eyes	Hair	Support
1	blue	blond	16
2	blue	dark	0
3	hazel	blond	8
4	hazel	dark	56

Assume that *Hair* is a decision attribute and *Eyes* is a condition attribute. The corresponding decision algorithm is given below:

1) *if (Eyes, blue) then (Hair, blond)*,
2) *if (Eyes, blue) then (Hair, dark)*,
3) *if (Eyes, hazel) then (Hair, blond)*,
4) *if (Eyes, hazel) then (Hair, dark)*.

The certainty and coverage factors for the decision rules are given in Table 3.

Table 3. Certainty and coverage factors

Rule number	Cert.	Cov.	Support	Strength
1	1.000	0.67	16	0.2
2	0.000	0.00	0	0.0
3	0.125	0.33	8	0.1
4	0.875	1.00	56	0.7

From the certainty factors of the decision rules we can conclude that:

- every person in the data table having *blue eyes* is for certain a *blond*,
- for certain there are no people in the data table having *blue eyes* who are *dark-haired*,
- the probability that a person having *hazel eyes* is a *blond* is 0.125,
- the probability that the person having *hazel eyes* is *dark-haired* equals to 0.875.

In other words the decision algorithm says that:

- *12,5%* persons with *hazel eyes* are *blond,*
- *87,5%* persons with *hazel eyes* are *dark-haired,*
- *100%* persons with *blue eyes* are *blond.*

From the above we can conclude that:

- people with *hazel eyes* are most probably *dark-haired,*
- people with *blue eyes* are for certain *blond.*

The efficiency of the decision algorithm is 0.9.
    The inverse decision algorithm is given below:

1') *if (Hair, blond) then (Eyes, blue),*
2') *if (Hair, dark) then (Eyes, blue),*
3') *if (Hair, blond) then (Eyes, hazel),*
4') *if (Hair, dark) then (Eyes, hazel).*

The coverage factors says that:

- the probability that a *blond* has *blue eyes* is 0.67,
- for certain there are no *dark-haired* people in the data table having *blue eyes,*
- the probability that a *blond* has *brown eyes* is 0.33,
- for certain every *dark-haired* person in the data table has *hazel eyes.*

In other words:

- *33% blond* have *hazel eyes,*
- *67% blond* have *blue eyes,*
- *100% dark-haired* persons have *hazel eyes.*

Thus we can conclude that:

- *blond* have most probably *blue eyes,*
- *dark-haired* people have for ceratin *hazel eyes.*

The efficiency of the inverse decision algorithm is 0.9.

Example 2

In Table 4 information about nine hundred people is represented. The population is characterized by the following attributes: *Height, Hair, Eyes* and *Nationality.*

Table 4. Characterization of nationalities

U	Height	Hair	Eyes	Nationality	Support
1	tall	blond	blue	Swede	270
2	medium	dark	hazel	German	90
3	medium	blond	blue	Swede	90
4	tall	blond	blue	German	360
5	short	red	blue	German	45
6	medium	dark	hazel	Swede	45

Suppose that *Height, Hair* and *Eyes* are condition attributes and *Nationality* is the decision attribute, i.e., we want to find description of each nationality in terms of condition attributes.

Below a decision algorithm associated with Table 4 is given:

1) *if (Height, tall) then (Nationality, Swede)*,
2) *if (Height, medium) and (Hair, dark) then (Nationality, German)*,
3) *if (Height, medium) and (Hair, blond) then (Nationality, Swede)*,
4) *if (Height, tall) then (Nationality, German)*,
5) *if (Height, short) then (Nationality, German)*,
6) *if (Height, medium) and (Hair, dark) then (Nationality, Swede)*.

The certainty and coverage factors for the decision rules are shown in Table 5.

Table 5. Certainty and coverage factors

Rule number	Cert.	Cov.	Support	Strength
1	0.43	0.67	270	0.3
2	0.67	0.18	90	0.1
3	1.00	0.22	90	0.1
4	0.57	0.73	360	0.4
5	1.00	0.09	45	0.05
6	0.33	0.11	45	0.05

From the certainty factors of the decision rules we can conclude that:

- 43% *tall* people are *Swede,*
- 57% *tall* people are *German,*
- 33% *medium and dark-haired* people are *Swede,*
- 67% *medium and dark-haired* people are *German,*
- 100% *medium and blond* people are *Swede,*
- 100% *short* people are *German.*

Summing up:

- *tall* people are most probably *German,*
- *medium and dark-haired* people are most probably *German,*
- *medium and blond* people are for certain *Swede,*
- *short* people are for certain *German.*

The efficiency of the above decision algorithm is 0.65.

The inverse algorithm is as follows:

1') *if (Nationality, Swede) then (Height, tall),*
2') *if (Nationality, German) then (Height, medium) and (Hair, dark),*
3') *if (Nationality, Swede) then (Height, medium) and (Hair, blond),*
4') *if (Nationality, German) then (Height, tall),*
5') *if (Nationality, German) then (Height, short),*
6') *if (Nationality, Swede) then (Height, medium) and (Hair, dark).*

From the coverage factors we get the following characterization of nationalities:

- *11% Swede are medium and dark-haired,*
- *22% Swede are medium and blond,*
- *67% Swede are tall,*
- *9% German are short,*
- *18% German are medium and dark-haired,*
- *73% German are tall.*

Hence we conclude that:

- *Swede* are most probably *tall,*
- *German* are most probably *tall.*

The efficiency of the inverse decision algorithm is 0.7.

Observe that there are no certain decision rules in the inverse decision algorithm nevertheless it can properly classify 70% objects.

Of course it is possible to find another decision algorithm from Table 4.

Observe that there are three methods of computation of the certainty and coverage factors: either directly from definition employing the data, or using formula (4) or (6).

Similarly, $\pi_S(\Psi)$ can be computed in three ways: using the definition and the data, or formula (3) or (5).

The obtained results are valid for the data only. In the case of another bigger data set the results may not be valid anymore .

Whether they are valid or not it dépends if Table 4 is a representative sample of a bigger population or not.

Example 3

Now we will consider an example taken from [12], which will show clearly the difference between the Bayesian and rough set approach to data analysis.

We will start from the data table presented below:

Table 6. Voting Intentions

Y_2	Y_3	Y_1			
		1	2	3	4
1	1	28	8	7	0
	2	153	114	53	14
	3	20	31	17	1
2	1	1	1	0	1
	2	165	86	54	6
	3	30	57	18	4

where Y_1 represents Voting Intentions (1 = Conservatives, 2 = Labour, 3 = Liberal Democrat, 4 = Others), Y_2 represents Sex (1 = male, 2 = female) and Y_3 represents Social Class (1 = high, 2 = middle, 3 = low).

Remark. In the paper [12] wrongly 1 = low and 3 = high instead of 1 = high and 3 = low.

We have to classify voters according to their Voting Intentions on the basis of Sex and Social Class.

First we create from Table 6 a decision table shown in Table 7:

Table 7. Voting Intentions

U	Y_2	Y_3	Y_1	Support	Strength
1	1	1	1	28	0.03
2	1	1	2	8	0.01
3	1	1	3	7	0.01
4	1	2	1	153	0.18
5	1	2	2	114	0.13
6	1	2	3	53	0.06
7	1	2	4	14	0.02
8	1	3	1	20	0.02
9	1	3	2	31	0.04
10	1	3	3	17	0.02
11	1	3	4	1	0.00
12	2	1	1	1	0.00
13	2	1	2	1	0.00
14	2	1	4	1	0.00
15	2	2	1	165	0.19
16	2	2	2	86	0.10
17	2	2	3	54	0.06
18	2	2	4	6	0.01
19	2	3	1	30	0.03
20	2	3	2	57	0.07
21	2	3	3	18	0.02
22	2	3	4	4	0.00

Next we simplify the decision table by employing only the decision rules with maximal strength, and we get the decision table presented in Table 8.

Table 8. Simplified Decision Table

U	Y_2	Y_3	Y_1	Support	Strength
1	1	1	1	28	0.07
2	1	2	1	153	0.35
3	1	3	2	31	0.07
4	2	2	1	165	0.38
5	2	3	2	57	0.13

It can be easly seen that the set of condition attributes can be reduct (see[4]) and the only reduced is the attribute Y_3 (Social Class).

Thus Table 8 can be repleaced by Table 9

Table 9. Reduced Decision Table

U	Y_3	Y_1	Strength	Certainty	Coverage
1	1	1	0.07(0.03)	1.00(0.60)	0.10(0.07)
2	2	1	0.73(0.37)	1.00(0.49)	0.90(0.82)
3	3	2	0.20(0.11)	1.00(0.55)	1.00(0.31)

The numbers in parenthesis refer to Table 7.

From this decision table we get the following decision algorithm:

	cer.
1. high class → Conservative party	0.60
2. middle class → Conservative party	0.49
3. lower class → Labour party	0.55

The efficiency of the decision algorithm is 0.51.

The inverse decision algorithm is given below:

	cer.
1'. Conservative party → high class	0.07
2'. Conservative party → middle class	0.82
3'. Labour party → lower class	0.31

The efficiency of the inverse decision algorithm is 0.48.

From the decision algorithm and the inverse decision algorithm we can conclude the following:

- *60% high class and 49% middle class intend to vote for the Conservative party*
- *55% lower class intend to vote for the Labour party*
- *7% intend to vote for the Conservative party belong to the high class*
- *82% intend to vote for the Conservative party belong to the middle class*
- *31% intend to vote for the Labour party belong to the lower class*

We advise the reader to examine the approach and results presented in [12] and compare them with that shown here.

Clearly, the rough set approach is much simpler and given better results then that discussed in [12].

8   Conclusions

The notion of a decision algorithm has been defined and its connection with decision table and other basic concepts of rough set theory discussed. Some probabilistic properties of decision algorithms have been revealed, in particular the relationship with the Total Probability Theorem and the Bayes' Theorem. These relationships give a new efficient method to draw conclusions from data, without referring to prior and posterior probabilities intrinsically associated with Bayesian reasoning.

Acknowledgments

Thanks are due to Prof. Andrzej Skowron and Prof. Wojciech Ziarko for their critical remarks.

References

1. Adams, E. W. (1975) The logic of conditionals, an application of probability to deductive logic. D. Reidel Publishing Company, Dordrecht, Boston
2. Grzymała-Busse J.(1991) Managing Uncertainty in Expert Systems. Kluwer Academic Publishers, Boston, Dordrecht
3. Łukasiewicz, J. Die logishen Grundlagen der Wahrscheinilchkeitsrechnung. Krakow, 1913. In: L. Borkowski (ed.), Jan Łukasiewicz – Selected Works, North Holland Publishing Company, Amsterdam, London, Polish Scientific Publishers, Warsaw, 1970
4. Pawlak, Z. (1991) Rough Sets – Theoretical Aspects of Reasoning about Data; Kluwer Academic Publishers: Boston, Dordrecht
5. Pawlak, Z.(1999) Decision rules, Bayes' rule and rough sets. In: N. Zhong, A. Skowron, S. Ohsuga (eds.), Proceedings of 7th International Workshop: New Directions in Rough Sets, Data Mining, and Granular –Soft Computing (RSFDGSC'99), Yamaguchi, Japan, November 1999, Lecture Notes in Artificial Intelligence 1711 Springer–Verlag, Berlin, 1–9
6. Pawlak, Z. Drawing conclusions from data – the rough set way (to appear)

7. Pawlak, Z. Combining rough sets and Bayes' rule (to appear)
8. Pawlak, Z.; Skowron, A. (1994) Rough membership functions. In: R.R. Yaeger, M. Fedrizzi, and J. Kacprzyk (eds.), Advances in the Dempster Shafer Theory of Evidence, John Wiley & Sons, Inc., New York, 251–271
9. Polkowski, L., Skowron, A. (1998) Proceedings of the First International Conference Rough Sets and Current Trends in Computing (RSCTC'98), Warsaw, Poland, June, Lecture Notes in Artificial Intelligence 1424, Springer–Verlag, Berlin
10. Polkowski, L., Skowron, A. (1998) Rough Sets in Knowledge Discovery Vol. 1-2, Physica-Verlag, Heidelberg
11. Polkowski, L., Tsumoto, S., Lin, T.Y. Rough Set Methods and Applications. New Development in Knowledge Discovery in Information Systems, Springer–Verlag (to appear)
12. Ramoni, M., Sebastiani, P. (1999) Bayasian Methods. In: M. Berthold, D. Hand (eds.), Intelligent Data Analysis, An Introduction, Springer, Berlin
13. Tsumoto, S.; Kobayashi, S. et al. (1996) (eds.), Proceedings of the Fourth International Workshop on Rough Sets, Fuzzy Sets, and Machine Discovery (RSFD'96). The University of Tokyo, November 6–8
14. Tsumoto, S.(1998) Modelling medical diagnostic rules based on rough sets. In: L. Polkowski, A. Skowron (eds.), Proceedings of the First International Conference Rough Sets and Current Trends in Computing (RSCTC'98), Warsaw, Poland, June, Lecture Notes in Artificial Intelligence 1424, Springer–Verlag, Berlin, 475–482
15. Ziarko, W. (1998) Variable precision rough set model. Journal of Computer and Systems Sciences 46, 39–59
16. Ziarko, W. (1998) Approximation region-based decision tables. In: L. Polkowski, A. Skowron (eds.), Rough Sets in Knowledge Discovery Vol.1-2, Physica-Verlag, Heidelberg, 178–185

Toward a Perception-Based Theory of Probabilistic Reasoning

L.A. Zadeh

Computer Science Division and the Electronic Research Laboratory
Department of EECS, University of California
Berkeley, CA 94720-1776

The past two decades have witnessed a dramatic growth in the use of probability-based methods in a wide variety of applications centering on automation of decision-making in an environment of uncertainty and incompleteness of information.

Successes of probability theory have high visibility. But what is not widely recognized is that successes of probability theory mask a fundamental limitation – the inability to operate on what may be called perception-based information. Such information is exemplified by the following. Assume that I look at a box containing balls of various sizes and form the perceptions: (a) there are about twenty balls; (b) most are large; and (c) a few are small. The question is: What is the probability that a ball drawn at random is neither large not small? Probability theory cannot answer this question because there is no mechanism within the theory to represent the meaning of perceptions in a form that lends itself to computation. The same problem arises in the examples:

Usually Robert returns from work at about 6 pm. What is the probability that Robert is home at 6:30 pm? I do not know Michelle's age but my perceptions are: (a) it is very unlikely that Michelle is old; and (b) it is likely that Michelle is not young. What is the probability that Michelle is neither young nor old? X is a normally distributed random variable with small mean and small variance. What is the probability that X is large? Given the data in insurance company database, what is the probability that my car may be stolen? In this case, the answer depends on perception-based information which is not in insurance company database.

In these simple examples – examples drawn from everyday experiences – the general problem is that of estimation of probabilities of imprecisely defined events, given a mixture of measurement-based and perception-based information. The crux of the difficulty is that perception-based information is usually described in a natural language – a language which probability theory cannot understand and hence is not equipped to handle.

To endow probability theory with a capability to operate on perception-based information, it is necessary to generalize it in three ways. To this end, let PT denote standard probability theory of the kind taught in university-level courses. The three modes of generalization are labeled: (a) f-generalization; (b) f.g-generalization: and (c) nl-generalization. More specifically: (a) f-generalization involves fuzzification, that is, progression from crisp sets to fuzzy sets, leading to a generalization of PT which is denoted as PT+. In PT+, probabilities,

W. Ziarko and Y. Yao (Eds.): RSCTC 2000, LNAI 2005, pp. 46–48, 2001.

functions, relations, measures and everything else are allowed to have fuzzy denotations, that is, be a matter of degree. In particular, probabilities described as low, high, not very high, etc. are interpreted as labels of fuzzy subsets of the unit interval or, equivalently, as possibility distributions of their numerical values. (b) f.g-generalization involves fuzzy granulation of variables, functions, relations, etc., leading to a generalization of PT which is denoted as PT++. By fuzzy granulation of a variable, X, what is meant is a partition of the range of X into fuzzy granules, with a granule being a clump of values of X which are drawn together by indistinguishability, similarity, proximity, or functionality. For example, fuzzy granulation of the variable Age partitions its vales into fuzzy granules labeled very young, young, middle-aged, old, very old, etc. Membership functions of such granules are usually assumed to be triangular or trapezoidal. Basically, granulation reflects the bounded ability of the human mind to resolve detail and store information. (c) Nl-generalization involves an addition to PT++ of a capability to represent the meaning of propositions expressed in a natural language, with the understanding that such propositions serve as descriptors of perceptions. Nl-generalization of PT leads to perception-based probability theory denoted as PTp.

An assumption which plays a key role in PTp is that the meaning of a proposition, p, drawn from a natural language may be represented as what is called a generalized constraint on a variable. More specifically, a generalized constraint is represented as X isr R, where X is the constrained variable; R is the constraining relation; and isr, pronounced ezar, is a copula in which r is an indexing variable whose value defines the way in which R constrains X. The principal types of constraints are: equality constraint, in which case isr is abbreviated to =; possibilistic constraint, with r abbreviated to blank; veristic constraint, with r=v; probabilistic constraint, in which case r=p, X is a random variable and R is its probability distribution; random-set constraint, r=rs, in which case X is set-valued random variable and R is its probability distribution; fuzzy-graph constraint, r=fg, in which case X is a function or a relation and R is its fuzzy graph; and usuality constraint, r=u, in which case X is a random variable and R is its usual – rather than expected – value.

The principal constraints are allowed to be modified, qualified, and combined, leading to composite generalized constraints. An example is: usually (X is small) and (X is large) is unlikely. Another example is: if (X is very small) then (Y is not very large) or if (X is large) then (Y is small).

The collection of composite generalized constraints forms what is referred to as the Generalized Constraint Language (GCL). Thus, in PTp, the Generalized Constraint Language serves to represent the meaning of perception-based information. Translation of descriptors of perceptions into GCL is accomplished through the use of what is called the constraint-centered semantics of natural languages (CSNL). Translating descriptors of perceptions into GCL is the first stage of perception-based probabilistic reasoning.

The second stage involves goal-directed propagation of generalized constraints from premises to conclusions. The rules governing generalized constraint propagation coincide with the rules of inference in fuzzy logic. The principal rule of inference is the generalized extension principle. In general, use of this principle reduces computation of desired probabilities to the solution of constrained problems in variational calculus or mathematical programming.

It should be noted that constraint-centered semantics of natural languages serves to translate propositions expressed in a natural language into GCL. What may be called the constraint-centered semantics of GCL, written as CSGCL, serves to represent the meaning of a composite constraint in GCL as a singular constraint X isr R. The reduction of a composite constraint to a singular constraint is accomplished through the use of rules which govern generalized constraint propagation.

Another point of importance is that the Generalized Constraint Language is maximally expressive, since it incorporates all conceivable constraints. A proposition in a natural language, NL, which is translatable into GCL is said to be admissible. The richness of GCL justifies the default assumption that any given proposition in NL is admissible. The subset of admissible propositions in NL constitutes what is referred to as a precisiated natural language, PNL. The concept of PNL opens the door to a significant enlargement of the role of natural languages in information processing, decision and control.

Perception-based theory of probabilistic reasoning suggests new problems and new directions in the development of probability theory. It is inevitable that in coming years there will be a progression from PT to PTp, since PTp enhances the ability of probability theory to deal with realistic problems in which decision-relevant information is a mixture of measurements and perceptions.

Situation Identification by Unmanned Aerial Vehicle

H.S. Nguyen, A.Skowron, M. Szczuka

Institute of Mathematics, Warsaw University
Banacha 2, 02–097, Warsaw, Poland
emails: {son,skowron,szczuka}@mimuw.edu.pl

Abstract. An approach to a multi-facet task of situation identification by Unmanned Aerial Vehicle (UAV) is presented. The concept of multi-layered identification system based on soft computing approach to reasoning with incomplete, imprecise or vague information is discussed.

1 Introduction

The task of controlling Unmanned Aerial Vehicle (UAV) in the traffic control applications arises a plethora of different problems (refer to [11]). Among the others the task of identifying the current road situation and deciding whether it should be considered as normal or potentially dangerous. The main source of information is the video system mounted on board of the UAV. It provides us with images of the situation underneath. Those images are gathered by digital video cameras working in visual and infrared band. With use of advanced techniques coming from the area of Image Processing and Computer Vision it is possible to identify and describe symbolically the objects existing within such as cars, road borders, cross-roads etc. Once the basic features are extracted from the image the essential process of identifying situation starts. The measurements taken from the processed image are matched against the set of decision rules. The rules that apply are contributing to the taking of final decision. In some cases it is necessary to go back to the image since more information should be drawn. The set of rules that match the image being processed at the time can identify a situation instantly or give us the choice of interpretations of what we see. In the latter case we may apply a higher level reasoning scheme in order to finally reach the decision. The entire cognition scheme that we construct is based on the concept of learning the basic notions and reasoning method form the set of pre-classified data (image sequences) that were given to us prior to putting the system into live.

Different tasks in the process of object identification require different learning techniques. In many applications, it is necessary to create some complex features from simple ones. This observation forces a hierarchical system of identification with multi–layered structure. In case of autonomous systems, some parameters of this layered structure are determined from experiment data in some learning processes called *layered learning*(see [9]).

W. Ziarko and Y. Yao (Eds.): RSCTC 2000, LNAI 2005, pp. 49–56, 2001.

Soft computing is one of many modern methods resolving some problems related to complex objects (eg. classification, identification and description) in real life systems. We emphasize two major directions: *computing with words* and *granular computing* aiming to build foundations for approximate layered reasoning (see [12], [13]).

In this paper we present an approach for layered learning based on soft computing techniques. We illustrate this idea by the problem of road situation identification. We also describe a method for automatic reasoning and decision making under uncertainty of information about objects (situation, measurement, etc.) using our layered structure.

2   The Problem of Complex Object Identification

In order to realise what are the problems behind the task of complex object classification/identification let us bring a simple example. Lets consider an image sequence showing two cars on the road turn, one taking over the other with high speed (see Figure 1). Now the question is how to mimick our perception of this situation with automatic system. What attributes in the image sequence should be taken into account and what is their range?

We may utilise background knowledge we have gained from human experts as well as general principles such as traffic regulations. However, the experts usually formulate their opinions in vague terms such us: "IF there is a turn nearby AND the car runs at high speed THEN this situation is rather dangerous".

The problem of finding out which attributes are being taken into account within our background knowledge is the first step. After that we have to identify the ranges for linguistic variables such as: "turn nearby", "high speed". The only reasonable way to do that is learning by example. The choice of proper learning scheme for basic attribute semantics is another problem that have to be overcame.

Yet another complication is the traditional trade-off between exactness and effectiveness. In case of real-time systems like UAV we need sometimes to omit some attributes since it takes too much time and effort to measure them. The proper choice of attributes for a particular situation is crucial, and we cannot expect to find a universal one. We should therefore possess a mechanism for dynamic update of feature vector as situation below evolves.

From the very beginning we allow our model to use notions formulated vaguely, imprecisely or fuzzy. That triggers the need for constructing the inference mechanism that is capable of using such a constructs as well as to maintain and propagate the levels of uncertainty.

The mechanism of propagating uncertainty should work both top-down and bottom-up. It is important to be able to get a conclusion with a good level of certainty provided we have measurements certain enough. But, it is equally important to have possibility of solving this equality other way - having the requirements for final answer determine allowable uncertainty level in the lower layers of reasoning scheme. And once again learning by example seem to be the best way to do that. Of course, the character of data (spatio-temporal) must be utilised.

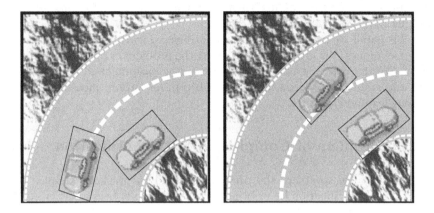

Fig. 1. The example of image sequence.

3 Construction of Identification System

Construction of Identification Structure includes:

- Learning of basic concepts (on different layers) from sensor measurements and expert knowledge (see [7], [8]).
- Synthesis of interfaces between different learning layers (in particular, for uncertainty coefficients propagation, decomposition of specifications and uncertainty coefficients as in [4]).

The domain knowledge can be presented by levels of concepts. In case of UAV, we propose three-layer knowledge structure.

The first layer is built from the information gained using image processing techniques. We can get information about color blobs in the image, contours, edges and distances between the objects identified. It is possible with advanced computer vision techniques and additional information about placement and movements of UAV (see [5], [2]) to get the readings that are highly resistant to scaling, rotation and unwanted effects caused by movement of both target and the UAV. Information at this stage is exact, but may be incomplete. Some of this information, however expressed in exact units may be by definition imprecise e.g. car speed estimated from the image analysis.

The next layer incorporates terms that are expressed vaguely, inexactly. At this level we use linguistic variables describing granules of information like "Object A moves fast". We also describe interactions between objects identified.

Those interactions such as "Object A is too close to Object B" are vaguely expressed as well.

Third layer is devoted to inference of final response of the system. In the basic concept, the inference is based on the set of decision rules extracted from the knowledge we have with regard to the set of training examples. In Figure 2 we present an example of knowledge structure for "danger overtaking maneuver".

Very crucial for operation of the entire system is the role of interlayer interfaces. They are responsible for unification of "language" between layers.

The original features may be measured with use of different techniques and units. Therefore it is necessary to unify them. At the first glance this task may look simple. What can be difficult in changing pixel to centimeters or degrees? But, we have to realise that in most cases subsequent images are affected by various distortions caused by e.g. changing angle of view as UAV hovers over the scene. Compensation and augmentation of various elements in order to pass a unified set of measurements to the next level requires many times the interface to perform task such as pattern recognition, intelligent (adjustable) filtering and so on.

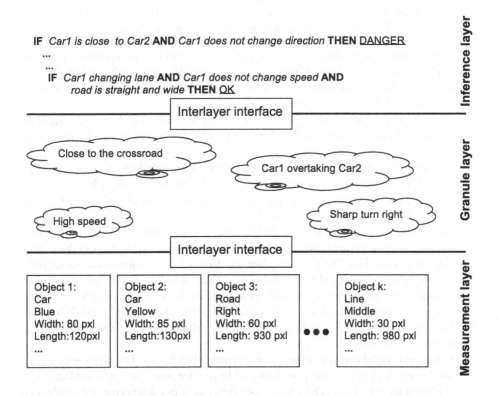

Fig. 2. The layered structure.

The interface between granule and inference layers is responsible for adjusting the output from the process of matching measurements against existing granule concepts. In order to be able to apply the decision rules, we have to express the setting of granules in a particular situation using the terms used by rules. This process includes, among others, the elements of spatio-temporal reasoning, since it is the interface that sends the information about changes in mutual placement of granules in time and about changes in placement of object within granules. This is especially crucial if we intend to make our system adaptive. The ability to avoid another error and to disseminate cases that were improperly treated in the past strongly relies on interface capabilities.

The learning tasks are defined in the same way as in Machine Learning. For example, the identification problem for UAV can be formulated as follows: "*given a collection of situations on roads:* $TRAINING\_SET = \{(S_1, d_1), ..., (S_n, d_n)\}$ *labeled by decision values (verified by experts)*".

The main problem of layered learning is how to design the learning schema consisting of basic learning algorithms that allow to identify properly new situations on the road.

One can see that the natural approach is to decompose the general, complex task into simpler ones. In case of knowledge structure presented in Figure 2 we can propose the following learning schema (Figure 3).

Unfortunately, this simple schema does not work well, because of the occurrence of phenomena characteristic for layered learning, such as:

1. **constrained learning**: It is necessary to determine the training data for particular learning algorithm in the design step. Usually, training data is presented in form of decision table. For example, the decision table for "Granule Learner 1" (see Figure 3) consists of:
   - conditional attributes: *"car speed"*, *"atmosphere condition"*, ... In general, these attributes defined by properties of objects from Measurement Layer.
   - decision attribute: in the simplest case, this attribute has two values: YES – for positive examples and NO – for negative examples.
   - examples (objects, cases) are taken from situations of TRAINING_SET restricted to the conditional attributes.

   the problem comes from the fact that for every situation we have only the global information whether this situation is dangerous or not and which rule can be used to identify this fact, but we do not have information about particular basic concepts thta contribute to that decision. For these situations, we only have some *constrains* between basic concepts. For example, for some situations we can have a constrain for "Granule Learner 1" and "Granule Learner 2" of form: "for this situation, one of those two learners must give a negative answer".

2. **tolerance relation learning**: The learning algorithms (Learners, see Fig. 3) used are not perfect. The possibility of imprecise solution and learning error is their native feature. This raises the question of possible error accumulation in consecutive layers. The problem is to set constrains for error rates of both

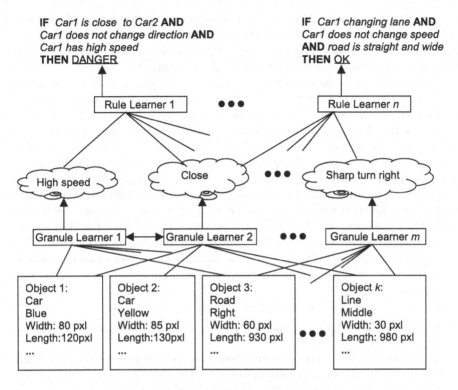

Fig. 3. Learning in layered structure.

individual and ensembles of algorithms in order to achieve acceptable quality of the overall layered learning process. We can apply the general approach for this problem called Rough Mereology (proposed in [4]). This idea is based on determining some standard objects (or *standards* for short) for every concept and corresponding tolerance relation in such a way, that if a new situation is close enough to some basic concepts, this situation will be close enough to higher level concept as well.

4   Construction of Reasoning Engine

Reasoning mechanisms should take into account uncertainties resulting from uncertainties of sensor measurements, missing sensor values, bounds on resources like computation time, robustness with respect to parameter deviation and necessity of adaptation to the changing environment.

In the simplest version the reasoning engine spins around the set of decision rules extracted form background knowledge and observation of learning examples. The output of matching feature measurements against predetermined information granules is fed to that layer thru interlayer interface. The rules that match in a satisfactory degree are contributing to final decision. There are two sub-steps in this process. First is the determination of the degree of matching for a particular rule. This is done with use of techniques known from rough set theory and fuzzy theory. Second is the determination of final decision on the basis of matching and not matching rules. This is almost a classical example of conflict solving. To perform such a task many techniques have been developed within rough set theory itself, as well as in related fields of soft computing.

Since the entire systems operates in very unstable environment it is very likely that the decision is not reached instantly. The inference mechanism requires more information. The request for more information addresses underlying components of the system. To make the decision process faster we should demand the information that not only allows to find matching rules, but also allows elimination of possibly largest set of rules that can be identified as certainly not relevant in a given situation. This is the kind of reasoning scheme resembling medical doctors approach. The physician first tries to eliminate possibilities and then, basing on what remains possible, orders additional examinations that allow to make final diagnosis. The very same mechanism in case of situation identification allows to decrease size of inference system and in consequence, improve effectiveness.

Construction of Reasoning Engine includes:

- Negotiation and dialog methods used for (relevant) concept perception on different layers.
- Spatio-temporal reasoning schemes based on concept perception on different layers with application of soft computing methods (using e.g., rough or/and fuzzy approaches) to reason about the perception results on different layers.

Negotiation and dialog methods are core component of interlayer interfaces, especially while they operate top-down i.e. process the requests send from the inference layer to these below. If in the inference layer no rule match enough in the current situation, the necessity of drawing additional information arises. The catch is to obtain result with minimum possible effort. We can identify within inference layer the set of information granules that, if better described, will be enough to take decision. But, on the other hand, getting this particular, additional information may be complicated from the granule layer point of view. Implicitly, drawing additional information requires also the negotiation between granule and measurement layers.

Within both negotiation processes (inference – granules and granules – measurements) we have to establish the utility function. Such a utility function, if maximised, allow to find the most reasonable set of granules or measurements for further processing. Unfortunately, we may not expect to find an explicit form of this function. Rather we should try to learn its estimation from the set of examples using adaptive approximation methods such as neural networks.

5 Conclusions and Future Researches

We presented a proposition of autonomous system, that can learn how to identify complex objects from examples. Our idea was illustrated by example of dangerous situation identification system which can be integrated for UAV project [?]). We listed some new characteristic aspects for layered learning methods. In next papers weintend to describe a system that learns to make complex decisions. By complex decision we mean e.g. the family of action plans. such a system will not only tell us whats going on below (danger/no danger), but also recommend what urther action should be performed by UAV. This problem is extremely interesting for autonomous systems.

Acknowledgments.

This work has been supported by the Wallenberg Foundation - WITAS Project, by the ESPRIT–CRIT 2 project #20288, and by grant 8T11C 025 19 from the State Committee for Scientific Research (KBN) of the Republic of Poland.

References

1. Fernyhough J., Generation of Qualitative Spatio-Temporal Representations From Visual Input, University of Leeds, 1997.
2. Pal S.K., Ghosh A., Kundu M.K., Soft Computing for Image Processing, Physica–Verlag, Heidelberg, 2000
3. Pawlak Z.: Rough Sets. Theoretical Aspects of Reasoning about Data, Kluwer Academic Publishers, Dordrecht, 1991.
4. Polkowski L., Skowron A.: Towards Adaptive Calculus of Granules, In: [13], vol.1, pp. 201–227.
5. Russ J. C., The Image Processing Handbook, CRC Press LLC & Springer Verlag, Heidelberg, 1999
6. Skowron A., Stepaniuk J.: Tolerance Approximation Spaces, Fundamenta Informaticae, Vol. 27, 1996, 245–253.
7. Skowron A., Stepaniuk J.: Towards Discovery of Information Granules, Lecture Notes in Artificial Intelligence 1704, Springer-Verlag, 1999, pp. 542–547.
8. Skowron A., Stepaniuk J.: Information Granules in Distributed Environment, Lecture Notes in Artificial Intelligence 1711, Springer-Verlag, 1999, pp. 357–365.
9. Stone P. : Layered Learning in Multiagent Systems: A Winning Approach to Robotic Soccer, MIT Press, 2000.
10. WWW SPACENET page: http://agora.scs.leeds.ac.uk/spacenet/.
11. The WITAS project homepage: http://www.ida.liu.se/ext/witas/.
12. Zadeh L.A.: Fuzzy Logic = Computing with Words, IEEE Trans. on Fuzzy Systems Vol. 4, 1996, pp. 103–111.
13. Zadeh L.A., Kacprzyk J. (Eds.): Computing with Words in Information/Intelligent Systems vol.1-2, Physica–Verlag, Heidelberg, 1999.

Rough-Neuro Computing

Lech Polkowski[1], Andrzej Skowron[2]

[1] Polish–Japanese Institute of Information Technology
Koszykowa 86, 02-008 Warsaw, Poland
and
Department of Mathematics and Information Sciences
Warsaw University of Technology
Pl. Politechniki 1,00-650 Warsaw, Poland
e-mail: polkow@pjwstk.waw.pl

[2] Institute of Mathematics
Warsaw University
Banacha 2, 02-097 Warsaw, Poland
e-mail: skowron@mimuw.edu.pl

Abstract. We outline a rough–neuro computing model as a basis for granular computing. Our approach is based on rough sets, rough mereology and information granule calculus.
Keywords: rough sets, neural networks, granular computing, rough mereology

1 Introduction

Rough Mereology [4], [7] is a paradigm allowing for a synthesis of main ideas of two paradigms for reasoning under uncertainty: Fuzzy Set Theory and Rough Set Theory. We present applications of Rough Mereology to the important theoretical idea put forth by Lotfi Zadeh [11], [12], i.e., Granularity of Knowledge by presenting the idea of rough–neuro computing paradigm.

We emphasize an important property of granular computing related to the necessity of lossless compression tuning for complex object constructions. It means that we map a cluster of constructions into one representation. Any construction in the cluster is delivering objects satisfying the specification in satisfactory degree if only objects input to synthesis are sufficiently close to selected standards (prototypes). In rough mereological approach clusters of constructions are represented by the so–called stable schemes (of co–operating agents), i.e., schemes robust to some deviations of parameters of transformed granules. In consequence, the stable schemes are able to return objects satisfying in satisfactory degree the specification not only from standard (prototype) objects but also from objects sufficiently close to them [4], [5]. In this way any stable scheme of complex object construction is a representation of a cluster of similar constructions from clusters of elementary objects.

We extend schemes for synthesis of complex objects (or granules) developed in [7] and [5] by adding one important component. As a result we receive the

W. Ziarko and Y. Yao (Eds.): RSTC 2000, LNAI 2005, pp. 57–64, 2001.

granule construction schemes which can be treated as a generalization of neural network models. The main idea is that granules sent by one agent to another are not, in general, exactly understandable by the receiving agent because these agents are using different languages and usually there is no meaning–preserving translation of formulas of the language of the sending agent to formulas of the receiving agent. Hence, it is necessary to construct some interfaces which will allow to approximately understand received granules. These interfaces can be, in the simplest case, constructed on the basis of exchanged information about agents stored in the form of decision data tables. From these tables the approximations of concepts can be constructed using rough set approach [10]. In our model we assume that for any agent ag and its operation $o(ag)$ of arity n there are approximation spaces $AS_1(o(ag), in), ..., AS_n(o(ag), in)$ which will filter (approximately) the granules received by the agent for performing the operation $o(ag)$. In turn, the granule sent by the agent after performing the operation is filtered (approximated) by the approximation space $AS(o(ag), out)$. These approximation spaces are parameterized with parameters allowing to optimize the size of neighborhoods in these spaces as well as the inclusion relation [8] using as a criterion for optimization the quality of granule approximation. Approximation spaces attached to an operation correspond to neuron weights in neural networks whereas the operation performed by the agent corresponds to the operation realized on the vector of real numbers by the neuron. The generalized scheme of agents is returning a granule in response to input information granules. It can be for example a cluster of elementary granules. Hence our schemes realize much more general computations then neural networks operating on vectors of real numbers. The question, if such schemes can be efficiently simulated by classical neural networks is open.

We would like to call extended schemes for complex object construction *rough-neuro schemes* (for complex object construction). The stability of such schemes corresponds to the resistance to noise of classical neural networks.

In the paper we present in some details the outlined above rough– neuro computing paradigm.

2   Adaptive Calculus of Granules in Distributed Systems

We now present a conceptual scheme for adaptive calculus of granules aimed at synthesizing solutions to problems posed under uncertainty. This exposition is based on our earlier analyzes presented in [4], [7]. We construct a scheme of agents which communicate by relating their respective granules of knowledge by means of transfer functions induced by rough mereological connectives extracted from their respective information systems. We assume the notation of [7] where the reader will find all the necessary information.

We now define formally the ingredients of our scheme of agents.

2.1   Distributed Systems of Agents

We assume that a pair (Inv, Ag) is given where Inv is an *inventory of elementary objects* and Ag is a set of inteligent computing units called shortly *agents*.

We consider an agent $ag \in Ag$. The agent ag is endowed with tools for reasoning about objects in its scope; these tools are defined by components of the agent label. The *label of the agent ag* is the tuple

$$lab(ag) = (\mathcal{A}(ag), M(ag), L(ag), Link(ag), AP\_O(ag), St(ag),$$
$$Unc\_rel(ag), H(ag), Unc\_rule(ag), Dec\_rule(ag))$$

where

1. $\mathcal{A}(ag) = (U(ag), A(ag))$ is an information system of the agent ag; we assume as an example that objects (i.e., elements of $U(ag)$) are granules of the form: $(\alpha, [\alpha])$ where α is a conjunction of descriptors (one may have more complex granules as objects).

2. $M(ag) = (U(ag), [0, 1], \mu_o(ag))$ is a pre - model of L_{rm} with a pre - rough inclusion $\mu_o(ag)$ on the universe $U(ag)$;

3. $L(ag)$ is a set of unary predicates (properties of objects) in a predicate calculus interpreted in the set $U(ag)$; we may assume that formulae of $L(ag)$ are constructed as conditional formulae of logics L_B where $B \subseteq A(ag)$.

4. $St(ag) = \{st(ag)_1, ..., st(ag)_n\} \subset U(ag)$ is the set of *standard objects* at ag;

5. $Link(ag)$ is a collection of strings of the form $t = ag_1 ag_2 ... ag_k ag$; the intended meaning of a string $ag_1 ag_2 ... ag_k ag$ is that $ag_1, ag_2, .., ag_k$ are children of ag in the sense that ag can assemble complex objects (constructs) from simpler objects sent by $ag_1, ag_2, ..., ag_k$. In general, we may assume that for some agents ag we may have more than one element in $Link(ag)$ which represents the possibility of re - negotiating the synthesis scheme.

We denote by the symbol $Link$ the union of the family $\{Link(ag) : ag \in Ag\}$.

6. $AP\_O(ag)$ consists of pairs of the form:

$$(o(ag, t), ((AS_1(o(ag), in), ..., AS_n(o(ag), in)), AS(o(ag), out))$$

where $o(ag, t) \in O(ag)$, n is the arity of $o(ag, t)$, $t = ag_1 ag_2 ... ag_k ag \in Link$, $AS_i(o(ag, t), in)$ is a parameterized approximation space [10] corresponding to the $i - th$ argument of $o(ag, t)$ and $AS(o(ag, t), out)$ is a parameterized approximation space [10] for the output of $o(ag, t)$.

$O(ag)$ is the set of operations at ag; any $o(ag, t) \in O(ag)$ is a mapping of the Cartesian product $U(ag) \times U(ag) \times ... \times U(ag)$ into the universe $U(ag)$; the meaning of $o(ag, t)$ is that of an operation by means of which the agent ag is able to assemble from objects $x_1 \in U(ag_1), x_2 \in U(ag_2), ..., x_k \in U(ag_k)$ the object $z \in U(ag)$ which is an approximation defined by $AS(o(ag, t), out)$ to $o(ag, t)(y_1, y_2, ..., y_k) \in U(ag)$ where y_i is the approximation to x_i defined by $AS_i(o(ag, t), in)$. One may choose here either a lower or an upper approximation.

7. $Unc\_rel(ag)$ is the set of uncertainty relations unc_rel_i of type

$$(o_i(ag, t), \rho_i(ag), ag_1, ..., ag_k, ag, \mu_o(ag_1), ..., \mu_o(ag_k), \mu_o(ag),$$
$$st(ag_1)_i, ..., st(ag_k)_i, st(ag)_i)$$

where $ag_1 ag_2 ... ag_k ag \in Link(ag)$, $o_i(ag, t) \in O(ag)$ and ρ_i is such that

$$\rho_i((x_1, \varepsilon_1), (x_2, \varepsilon_2), .., (x_k, \varepsilon_k), (x, \varepsilon))$$

holds for $x_1 \in U(ag_1), x_2 \in U(ag_2), .., x_k \in U(ag_k)$ and $\varepsilon, \varepsilon_1, \varepsilon_2, .., \varepsilon_k \in [0,1]$ iff $\mu_o(x_j, st(ag_j)_i) = \varepsilon_j$ for $j = 1, 2, .., k$ and $\mu_o(x, st(ag)_i) = \varepsilon$ for the collection of standards $st(ag_1)_i, st(ag_2)_i, .. ., st(ag_k)_i, st(ag)_i$ such that

$$o_i(ag, t)(st(ag_1)_i, st(ag_2)_i, .., st(ag_k)_i) = st(ag)_i.$$

The operation o performed by ag here is more complex then that of [7] as it is composed of three stages: first, approximations to input objects are constructed, next the operation is performed, and finally the approximation to the result is constructed. Relations unc_rel_i provide a global description of this process; in reality, they are composition of analogous relations corresponding to the three stages. As a result, unc_rel_i depend on parameters of approximation spaces. This concerns also other constructs discussed here. It follows that in order to get satisfactory decomposition (similarly, uncertainty and so on) rules one has to search for satisfactory parameters of approximation spaces (this is in analogy to weight tuning in neural computations).

Uncertainty relations express the agents knowledge about relationships a-mong uncertainty coefficients of the agent ag and uncertainty coefficients of its children. The relational character of these dependencies expresses their inten-sionality.

8. $Unc\_rule(ag)$ is the set of uncertainty rules unc_rule_j of type
$$(o_j(ag, t), f_j, ag_1, ag_2, . .., ag_k, ag, st(ag_1), st(ag_2), ..., st(ag_k), st(ag),$$
$$\mu_o(ag_1), ... , \mu_o(ag_k), \mu_o(ag))$$
of the agent ag where $ag_1 ag_2 ... ag_k ag \in Link(ag)$ and $f_j : [0,1]^k \longrightarrow [0,1]$ is a function which has the property that
if $o_j(ag, t)(st(ag_1), st(ag_2), ..., st(ag_k)) = st(ag)$ **and**
$x_1 \in U(ag_1), x_2 \in U(ag_2), ..., x_k \in U(ag_k)$
    satisfy the conditions $\mu_o(x_i, st(ag_i)) \geq \varepsilon(ag_i)$ for $i = 1, 2, .., k$

then $\mu_o(o_j(ag, t)(x_1, x_2, ..., x_k), st(ag)) \geq f_j(\varepsilon(ag_1), \varepsilon(ag_2), .., \varepsilon(ag_k)).$

Uncertainty rules provide functional operators (approximate mereological connectives) for propagating uncertainty measure values from the children of an agent to the agent; their application is in negotiation processes where they inform agents about plausible uncertainty bounds.

9. $H(ag)$ is a strategy which produces uncertainty rules from uncertainty relations; to this end, various rigorous formulas as well as various heuristics can be applied among them the algorithm presented in Section 2.8 of [7].

10. $Dec\_rule(ag)$ is a set of decomposition rules dec_rule_i of type

$$(o_i(ag, t), ag_1, ag_2, ..., ag_k, ag)$$

such that $(\Phi(ag_1), \Phi(ag_2), .., \Phi(ag_k), \Phi(ag)) \in dec\_rule_i$ (where $\Phi(ag_1) \in L(ag_1)$, $\Phi(ag_2) \in L(ag_2)$, ..., $\Phi(ag_k) \in L(ag_k)$, $\Phi(ag) \in L(ag)$ and $ag_1 ag_2 ... ag_k ag \in Link(ag)$) and there exists a collection of standards $st(ag_1), st(ag_2), ..., st(ag_k)$, $st(ag)$ with the properties that

$$o_j(ag, t)(st(ag_1), st(ag_2), .., st(ag_k)) = st(ag),$$

$st(ag_i)$ satisfies $\Phi(ag_i)$ for $i = 1, 2, .., k$ and $st(ag)$ satisfies $\Phi(ag)$.

Decomposition rules are decomposition schemes in the sense that they describe the standard $st(ag)$ and the standards $st(ag_1), ..., st(ag_k)$ from which the standard $st(ag)$ is assembled under o_i in terms of predicates which these standards satisfy.

We may sum up the content of 1 - 10 above by saying that for any agent ag the possible sets of children of this agent are specified and, relative to each team of children, decompositions of standard objects at ag into sets of standard objects at the children, uncertainty relations as well as uncertainty rules, which relate similarity degrees of objects at the children to their respective standards and similarity degree of the object built by ag to the corresponding standard object at ag, are given.

We take rough inclusions of agents as measures of uncertainty in their respective universes. We would like to observe that the mereological relation of being a part is not transitive globally over the whole synthesis scheme because distinct agents use distinct mereological languages.

2.2   Approximate Synthesis of Complex Objects

The process of synthesis of a complex object (signal, action) by the above defined scheme of agents consists in our approach of the two communication stages viz. the top - down communication/negotiation process and the bottom - up communication/assembling process. We outline the two stages here in the language of approximate formulae.

Approximate logic of synthesis For simplicity of exposition and to avoid unnecessarily tedious notation, we assume that the relation $ag' \leq ag$, which holds for agents $ag', ag \in Ag$ iff there exists a string $ag_1 ag_2 ... ag_k ag \in Link(ag)$ with $ag' = ag_i$ for some $i \leq k$, orders the set Ag into a tree. We also assume that $O(ag) = \{o(ag, t)\}$ for $ag \in Ag$ i.e. each agent has a unique assembling operation for a unique t.

The process of synthesis of a complex object (signal, action) by the above defined scheme of agents consists in our approach of the two communication stages viz. the top - down communication/negotiation process and the bottom - up communication process. We outline the two stages here in the language of approximate formulae. To this end we build a logic $L(Ag)$ (cf. [7]) in which we can express global properties of the synthesis process. We recall our assumption that the set Ag is ordered into a tree by the relation $ag' \leq ag$.

Elementary formulae of $L(Ag)$ are of the form $\langle st(ag), \Phi(ag), \varepsilon(ag) \rangle$ where $st(ag) \in St(ag), \Phi(ag) \in L(ag), \varepsilon(ag) \in [0, 1]$ for any $ag \in Ag$. Formulae of $L(ag)$ form the smallest extension of the set of elementary formulae closed under propositional connectives $\vee, \wedge, \neg$ and under the modal operators $[], <> .$

To introduce a semantics for the logic $L(ag)$, we first specify the meaning of satisfaction for elementary formulae. The meaning of a formula $\Phi(ag)$ is defined classically as the set $[\Phi(ag)] = \{u \in U(ag) : u$ has the property $\Phi(ag)\}$; we will denote the fact that $u \in [\Phi(ag)]$ by the symbol $u \models \Phi(ag)$. We extend now the satisfiability predicate $\models$ to approximate formulae: for $x \in U(ag)$, we say that x *satifies* an elementary formula $\langle st(ag), \Phi(ag), \varepsilon(ag) \rangle$, in symbols: $x \models < st(ag), \Phi(ag), \varepsilon(ag) >$, iff (i) $st(ag) \models \Phi(ag)$ and (ii) $\mu_o(ag)(x, st(ag)) \geq \varepsilon(ag)$.

We let

(iii) $x \models \neg \langle st(ag), \Phi(ag), \varepsilon(ag) \rangle$ iff it is not true that $x \models \langle st(ag), \Phi(ag), \varepsilon(ag) \rangle$;

(iv) $x \models \langle st(ag)_1, \Phi(ag)_1, \varepsilon(ag)_1 \rangle \vee \langle st(ag)_2, \Phi(ag)_2, \varepsilon(ag)_2 \rangle$ iff

$$x \models \langle st(ag)_1, \Phi(ag)_1, \varepsilon(ag)_1 \rangle \text{ or } x \models \langle st(ag)_2, \Phi(ag)_2, \varepsilon(ag)_2 \rangle.$$

In order to extend the semantics over modalities, we first introduce the notion of a selection: by a *selection* over Ag we mean a function sel which assigns to each agent ag an object $sel(ag) \in U(ag)$.

For two selections sel, sel' we say that sel induces sel', in symbols $sel \rightarrow_{Ag} sel'$ when $sel(ag) = sel'(ag)$ for any $ag \in Leaf(Ag)$ and

$$sel'(ag) = o(ag, t)(sel'(ag_1), sel'(ag_2), ..., sel'(ag_k))$$

for any $ag_1 ag_2 ... ag_k ag \in Link$.

We extend the satisfiability predicate $\models$ to selections: for an elementary formula $\langle st(ag), \Phi(ag), \varepsilon(ag) \rangle$, we let $sel \models \langle st(ag), \Phi(ag), \varepsilon(ag) \rangle$ iff $sel(ag) \models \langle st(ag), \Phi(ag), \varepsilon(ag) \rangle$.

We now let $sel \models <>< st(ag), \Phi(ag), \varepsilon(ag) >$ when there exists a selection sel' satisfying the conditions:

(i) $sel \rightarrow_{Ag} sel'$; (ii) $sel' \models \langle st(ag), \Phi(ag), \varepsilon(ag) \rangle$.

In terms of logic $L(Ag)$ it is possible to express the problem of synthesis of an approximate solution to the problem posed to the team Ag. We denote by $head(Ag)$ the root of the tree $(Ag, \leq)$.

In the process of top - down communication, a requirement Ψ received by the scheme from an external source (which may be called a *customer*) is decomposed into approximate specifications of the form $\langle st(ag), \Phi(ag), \varepsilon(ag) \rangle$ for any agent ag of the scheme. The decomposition process is initiated at the agent $head(Ag)$ and propagated down the tree.

We are able now to formulate the synthesis problem.

Synthesis problem

Given a formula $\alpha : \langle st(head(Ag)), \Phi(head(Ag)), \varepsilon(head(Ag)) \rangle$ *find a selection sel over the tree* $(Ag, \leq)$ *with the property* $sel \models <> \alpha$.

A solution to the synthesis problem with a given formula

$$\langle st(head(Ag)), \Phi(head(Ag)), \varepsilon(head(Ag)) \rangle$$

is found by negotiations among the agents; negotiations are based on uncertainty rules of agents and their succesful result can be expressed by a top-down recursion in the tree $(Ag, \leq)$ as follows: given a local team $ag_1 ag_2 ... ag_k ag$ with the formula $\langle st(ag), \Phi(ag), \varepsilon(ag) \rangle$ already chosen in negotiations on a higher tree level, it is sufficient that each agent ag_i choose a standard $st(ag_i) \in U(ag_i)$, a formula $\Phi(ag_i) \in L(ag_i)$ and a coefficient $\varepsilon(ag_i) \in [0,1]$ such that

(v) $(\Phi(ag_1), \Phi(ag_2), \ldots, \Phi(ag_k), \Phi(ag)) \in Dec\_rule(ag)$ with standards $st(ag)$, $st(ag_1), \ldots, st(ag_k)$;

(vi) $f(\varepsilon(ag_1), \ldots, \varepsilon(ag_k)) \geq \varepsilon(ag)$ where f satisfies $unc\_rule(ag)$ with $st(ag)$, $st(ag_1), \ldots, st(ag_k)$ and $\varepsilon(ag_1), \ldots, \varepsilon(ag_k), \varepsilon(ag)$.

For a formula α : $\langle st(head(Ag)), \Phi(head(Ag)), \varepsilon(head(Ag)) \rangle$, we call an α - *scheme* an assignment of a formula $\alpha(ag)$: $\langle st(ag), \Phi(ag), \varepsilon(ag) \rangle$ to each $ag \in Ag$ in such manner that (v), (vi) above are satisfied and $\alpha(head(Ag))$ is $\langle st(head(Ag)), \Phi(head(Ag)), \varepsilon(head(Ag)) \rangle$; we denote this scheme with the symbol

$$sch(\langle st(head(Ag)), \Phi(head(Ag)), \varepsilon(head(Ag)) \rangle).$$

We say that a selection *sel* is *compatible* with a scheme

$$sch(\langle st(head(Ag)), \Phi(head(Ag)), \varepsilon(head(Ag)) \rangle)$$

when $\mu_o(ag,t)(sel(ag), st(ag)) \geq \varepsilon(ag)$ for each leaf agent $ag \in Ag$ where $\langle st(ag), \Phi(ag), \varepsilon(ag) \rangle$ is the value of the scheme at ag for any leaf $ag \in Ag$.

Any leaf agent realizes its approximate specification by choosing in the subset $Inv \cap U(ag)$ of the inventory of primitive constructs a construct satisfying the specification.

The goal of negotiations can be summarized now as follows.

Proposition 3.1

For a given a requirement $\langle st(head(Ag)), \Phi(head(Ag)), \varepsilon(head(Ag)) \rangle$ we have:
if a selection *sel* is compatible with a scheme

$$sch(\langle st(head(Ag)), \Phi(head(Ag)), \varepsilon(head(Ag)) \rangle)$$

then $sel \models <> \langle st(head(Ag)), \Phi(head(Ag)), \varepsilon(head(Ag)) \rangle$.

The bottom-up communication consists of agents sending to their parents the chosen constructs. The root agent $root(Ag)$ assembles the final construct.

3   Conclusions

We have outlined a general scheme for rough neuro–computation based on ideas of knowledge granulation by rough mereological tools. An important practical problem is a construction of such schemes (networks) for rough–neuro computing and of algorithms for parameter tuning. We now foresee two possible approaches: the one in which we would rely on new, original decomposition, synthesis and tuning methods in analogy to [7] but in the presence of approximation spaces; the second, in which a rough–neuro computing scheme would be encoded by a neural network in such a way that optimalization of weights in the neural net leads to satisfactory solutions for the rough–neuro computing scheme (cf. [3] for a first attempt in this direction).

Acknowledgments: The research of Lech Polkowski has been supported by the grants: No 8T11C 024 17 from the State Committee for Scientific Research (KBN) of the Republic of Poland and the European ESPRIT Program in CRIT 2 Research Project No 20288. The research of Andrzej Skowron has been supported by the grants of National Committee for Scientific Research, European ESPRIT Program in CRIT 2 Research Project No 20288, and the Wallenberg Foundation.

References

1. Lin, T.Y. Granular Computing on Binary Relations I. Data Mining and Neighborhood Systems (1998), In: [6] **1** pp. 107–121.
2. Z. Pawlak (1991), *Rough Sets – Theoretical Aspects of Reasoning about Data*, Kluwer Academic Publishers, Dordrecht.
3. L. Han, J.F. Peters, S. Ramanna, and R. Zhai (1999), Classifying faults in high voltage power systems: a rough–fuzzy neural computational approach, *Proceedings RSFDGrC'99: 7th International Workshop on Rough Sets, Fuzzy Sets, Data Mining and Granular Soft Computing*, Lecture Notes in Artificial Intelligence **1711**, Springer Verlag, Berlin, pp. 47–54.
4. L. Polkowski, A. Skowron (1996), Rough mereology: A new paradigm for approximate reasoning, *International Journal of Approximate Reasoning* **15/4**, pp. 333–365.
5. L. Polkowski, A. Skowron (1998), Rough mereological foundations for design, analysis, synthesis, and control in distributed systems, *Information Sciences An International Journal* **104/1-2**, Elsevier Science, New York, pp. 129–156.
6. L. Polkowski, A. Skowron (Eds.) (1998), *Rough Sets in Knowledge Discovery* **1-2** Physica-Verlag, Heidelberg 1998.
7. L. Polkowski, A. Skowron (1999), Towards adaptive calculus of granules, In: [13], **1**, pp. 201-227.
8. L. Polkowski, A. Skowron (2000). "Rough Mereology in Information Systems. A Case Study: Qualitative Spatial Reasoning", In: L. Polkowski, T.Y. Lin, S. Tsumoto (Eds.), *Rough sets: New developments*, Studies in Fuzziness and Soft Computing, Physica-Verlag / Springer-Verlag, Heidelberg, 2000 (in print).
9. Skowron, A.; Stepaniuk, J. (1999), Towards Discovery of Information Granules, 3rd European Conference of Principles and Practice of Knowledge Discovery in Databases, September 15–18, 1999, Prague, Czech Republic, Lecture Notes in Artificial Intelligence **1704**, Springer-Verlag, Berlin , pp. 542–547.
10. A. Skowron, J. Stepaniuk, S. Tsumoto (1999), Information Granules for Spatial Reasoning, *Bulletin of the International Rough Set Society* **3/4**, pp. 147-154.
11. L.A. Zadeh (1996), Fuzzy logic = computing with words, *IEEE Trans. on Fuzzy Systems* **4**, pp. 103-111.
12. L.A. Zadeh (1997), Toward a theory of fuzzy information granulation and its certainty in human reasoning and fuzzy logic, *Fuzzy Sets and Systems* **90**, pp. 111-127.
13. Zadeh, L.A., Kacprzyk, J. (eds.): *Computing with Words in Information/Intelligent Systems* vol.1-2, Physica-Verlag, Heidelberg, 1999.

Approximation of Information Granule Sets

Andrzej Skowron[1]   Jaroslaw **Stepaniuk**[2]   James F. **Peters**[3]

[1] Institute of Mathematics, Warsaw University,
Banacha 2, 02-097 Warsaw, Poland,
E-mail: skowron@mimuw.edu.pl
[2] Institute of Computer Science, Bialystok University of Technology,
Wiejska 45A, 15-351 Bialystok, Poland,
E-mail: jstepan@ii.pb.bialystok.pl
[3] Department of Electrical and Computer Engineering, University of Manitoba
15 Gillson Street, ENGR 504,Winnipeg, Manitoba R3T 5V6,Canada
E-mail: jfpeters@ee.umanitoba.ca

Abstract. The aim of the paper is to present basic notions related to granular computing, namely the information granule syntax and semantics as well as the inclusion and closeness (similarity) relations of granules. In particular, we discuss how to define approximation of complex granule sets using the above notions.

1 Introduction

We would like to discuss briefly an example showing a motivation for our work [6]. Let us consider a team of agents recognizing the situation on the road. The aim is to classify a given situation as, e.g., dangerous or *not*. This soft speci-*fication granule* is represented by a family of information granules called case *soft patterns* representing cases, like cars are *too close*. The whole scene (actual situation on the road) is decomposed into regions perceived by local agents. Higher level agents can reason about regions observed by team of their children agents. They can express in their own languages features used by their children. Moreover, they can use new features like attributes describing relations between regions perceived by children agents. The problem is how to organize agents into a team (having, e.g., tree structure) with the property that the information granules synthesized by the team from input granules (being perceptions of local agents from sensor measurements) will identify the situation on the road in the following sense: the granule constructed by the team from input granules representing the situation on the road is sufficiently close to the soft specification granule named dangerous if and only if the situation on the road is really dangerous. We expect that if the team is returning a granule sufficiently close to the soft specification granule dangerous then also a special case of the soft pattern dangerous is identified helping to explain the situation.

The aim of our project is to develop foundations for this kind of reasoning. In particular it is necessary to give precise meaning to the notions like: information granules, soft information granules, closeness of information granules in

W. Ziarko and Y. Yao (Eds.): RSCTC 2000, LNAI 2005, pp. 65–72, 2001.

satisfactory degree, information granules synthesized by team of agents etc. The presented paper realizes the first step toward this goal.

The general scheme is depicted in Figure 1.

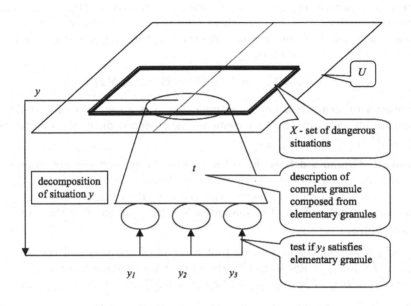

Fig. 1. Illustrative Example

To sum up, we consider a set of agents *Ag*. Each agent is equipped with some approximation spaces (defined using rough set approach [1]). Agents are cooperating to solve a problem specified by a special agent called *customer-agent*. The result of cooperation is a scheme of agents. In the simplest case the scheme can be represented by a tree labeled by agents. In this tree leaves are delivering some information granules (representing of perception in a given situation by leaf agents) and any non-leaf agent $ag \in Ag$ is performing an operation $o(ag)$ on approximations of granules delivered by its children. The root agent returns an information granule being the result of computation by the scheme on granules delivered by leaf agents. It is important to note that different agents use different languages. Thus granules delivered by children agents to their father can be usually perceived by him in an approximate sense before he can perform any operation on delivered granules.

In particular, we point out in the paper a problem of approximation of information granule sets and we show that the first step toward such a notion is similar to the classical rough set approach.

2   Syntax and Semantics of Information Granules

In this section we will consider several examples of information granule constructions. We present now the syntax and semantics of information granules. In the following section we discuss the inclusion and closeness relations for granules.

Elementary granules. In an information system $IS = (U, A)$, elementary granules are defined by $EF_B(x)$, where EF_B is a conjunction of selectors of the form $a = a(x)$, $B \subseteq A$ and $x \in U$. For example, the meaning of an elementary granule $a = 1 \wedge b = 1$ is defined by

$$\|a = 1 \wedge b = 1\|_{IS} = \{x \in U : a(x) = 1 \ \& \ b(x) = 1\}.$$

Sequences of granules. Let us assume that S is a sequence of granules and the semantics $\|\bullet\|_{IS}$ in IS of its elements have been defined. We extend $\|\bullet\|_{IS}$ on S by $\|S\|_{IS} = \{\|g\|_{IS}\}_{g \in S}$.

Example 1. Granules defined by rules in information systems are examples of sequences of granules. Let IS be an information system and let (α, β) be a new information granule received from the rule **if** α **then** β where α, β are elementary granules of IS. The semantics $\|(\alpha, \beta)\|_{IS}$ of (α, β) is the pair of sets $(\|\alpha\|_{IS}, \|\beta\|_{IS})$.

Sets of granules. Let us assume that a set G of granules and the semantics $\|\bullet\|_{IS}$ in IS for granules from G have been defined. We extend $\|\bullet\|_{IS}$ on the family of sets $H \subseteq G$ by $\|H\|_{IS} = \{\|g\|_{IS} : g \in H\}$.

Example 2. One can consider granules defined by sets of rules. Assume that there is a set of rules $Rule\_Set = \{(\alpha_i, \beta_i) : i = 1, \ldots, k\}$. The semantics of $Rule\_Set$ is defined by

$$\|Rule\_Set\|_{IS} = \{\|(\alpha_i, \beta_i)\|_{IS} : i = 1, \ldots, k\}.$$

Example 3. One can also consider as set of granules a family of all granules $(\alpha, Rule\_Set(DT_\alpha))$, where α belongs to a given subset of elementary granules.

Example 4. Granules defined by sets of decision rules corresponding to a given evidence are also examples of sequences of granules. Let $DT = (U, A \cup \{d\})$ be a decision table and let α be an elementary granule of $IS = (U, A)$ such that $\|\alpha\|_{IS} \neq \emptyset$. Let $Rule\_Set(DT_\alpha)$ be the set of decision rules (e.g. in minimal form) of the decision table $DT_\alpha = (\|\alpha\|_{IS}, A \cup \{d\})$ being the restriction of DT to objects satisfying α. We obtain a new granule $(\alpha, Rule\_Set(DT_\alpha))$ with the semantics

$$\|(\alpha, Rule\_Set(DT_\alpha))\|_{DT} = (\|\alpha\|_{IS}, \|Rule\_Set(DT_\alpha)\|_{DT}).$$

This granule describes a decision algorithm applied in the situation characterized by α.

Extension of granules defined by tolerance relation. We present examples of granules obtained by application of a tolerance relation.

Example 5. One can consider extension of elementary granules defined by tolerance relation. Let $IS = (U, A)$ be an information system and let τ be a tolerance relation on elementary granules of IS. Any pair (α, τ) is called a *τ-elementary granule*. The semantics $\|(\alpha, \tau)\|_{IS}$ of (α, τ) is the family $\{\|\beta\|_{IS} : (\beta, \alpha) \in \tau\}$.

Example 6. Let us consider granules defined by rules of tolerance information systems. Let $IS = (U, A)$ be an information system and let τ be a tolerance relation on elementary granules of IS. If **if** α **then** β is a rule in IS then the semantics of a new information granule $(\tau : \alpha, \beta)$ is defined by $\|(\tau : \alpha, \beta)\|_{IS} = \|(\alpha, \tau)\|_{IS} \times \|(\beta, \tau)\|_{IS}$.

Example 7. We consider granules defined by sets of decision rules corresponding to a given evidence in tolerance decision tables. Let $DT = (U, A \cup \{d\})$ be a decision table and let τ be a tolerance on elementary granules of $IS = (U, A)$. Now, any granule $(\alpha, Rule\_Set\,(DT_\alpha))$ can be considered as a representative of information granule cluster $(\tau : (\alpha, Rule\_Set\,(DT_\alpha)))$ with the semantics

$$\|(\tau : (\alpha, Rule\_Set\,(DT_\alpha)))\|_{DT} = \left\{ \|(\beta, Rule\_Set\,(DT_\beta))\|_{DT} : (\beta, \alpha) \in \tau \right\}.$$

Labeled graph granules. We discuss graph granules and labeled graph granules as notions extending previously introduced granules defined by tolerance relation.

Example 8. Let us consider granules defined by pairs (G, E), where G is a finite set of granules and $E \subseteq G \times G$. Let $IS = (U, A)$ be an information system. The semantics of a new information granule (G, E) is defined by $\|(G, E)\|_{IS} = (\|G\|_{IS}, \|E\|_{IS})$, where $\|G\|_{IS} = \{\|g\|_{IS} : g \in G\}$ and $\|E\|_{IS} = \{(\|g\|, \|g'\|) : (g, g') \in E\}$.

Example 9. Let G be a set of granules. Labeled graph granules over G are defined by (X, E, f, h), where $f : X \to G$ and $h : E \to P(G \times G)$. We also assume one additional condition
if $(x, y) \in E$ then $(f(x), f(y)) \in h(x, y)$.
The semantics of labeled graph granule (X, E, f, h) is defined by

$$\{(\|f(x)\|_{IS}, \|h(x, y)\|_{IS}, \|f(y)\|_{IS}) : (x, y) \in E\}.$$

Let us summarize the above presented considerations. One can define the set of granules G as the least set containing a given set of elementary granules G_0 and closed with respect to the defined above operations of new granule construction.

We have the following examples of granule construction rules:

$$\frac{\alpha_1, \dots, \alpha_k\text{- elementary granules}}{\{\alpha_1, \dots, \alpha_k\}\text{- granule}}$$

$$\frac{\alpha_1, \alpha_2\text{- elementary granules}}{(\alpha_1, \alpha_2)\text{- granule}}$$

$$\frac{\alpha\text{- elementary granule },\tau\text{- tolerance relation on elementary granules}}{(\tau:\alpha)\text{- granule}}$$

$$\frac{G\text{- a finite set of granules },E \subseteq G \times G}{(G,E)\text{- granule}}$$

Let us observe that in case of granules constructed with application of tolerance relation we have the rule restricted to elementary granules. To obtain a more general rule like

$$\frac{\alpha\text{- graph granule },\tau\text{- tolerance relation on graph granules}}{(\tau:\alpha)\text{- granule}}$$

it is necessary to extend the tolerance (similarity, closeness) relation on more complex objects. We discuss the problem of closeness extension in the following section.

3   Granule Inclusion and Closeness

In this section we will discuss inclusion and closeness of different information granules introduced in the previous section. Let us mention that the choice of inclusion or closeness definition depends very much on the area of application and data analyzed. This is the reason that we have decided to introduce a separate section with this more subjective part of granule semantics.

The inclusion relation between granules G, G' of degree at least p will be denoted by $\nu_p(G,G')$. Similarly, the closeness relation between granules G, G' of degree at least p will be denoted by $cl_p(G,G')$. By p we denote a vector of parameters (e.g. positive real numbers).

A general scheme for construction of hierarchical granules and their closeness can be described by the following recursive meta-rule: if granules of order $\leq k$ and their closeness have been defined then the closeness $cl_p(G,G')$ (at least in degree p) between granules G, G' of order $k+1$ can be defined by applying an appropriate operator F to closeness values of components of G, G', respectively.

A general scheme of defining more complex granule from simpler ones can be explored using rough mereological approach [2].

Inclusion and closeness of elementary granules. We have introduced the simplest case of granules in information system $IS = (U, A)$. They are defined by $EF_B(x)$, where EF_B is a conjunction of selectors of the form $a = a(x)$, $B \subseteq A$ and $x \in U$. Let $G_{IS} = \{EF_B(x) : B \subseteq A \ \& \ x \in U\}$. In the standard rough set model [1] elementary granules describe indiscernibility classes with respect to some subsets of attributes. In a more general setting see e.g. [3], [5] tolerance (similarity) classes are described.

The crisp inclusion of α in β, where $\alpha, \beta \in \{EF_B(x) : B \subseteq A \ \& \ x \in U\}$ is defined by $\|\alpha\|_{IS} \subseteq \|\beta\|_{IS}$, where $\|\alpha\|_{IS}$ and $\|\beta\|_{IS}$ are sets of objects from IS satisfying α and β, respectively. The non-crisp inclusion, known in KDD, for the case of association rules is defined by means of two thresholds t and t' :

$support_{IS}(\alpha, \beta) = card(\|\alpha \wedge \beta\|_{IS}) \geq t$, and
$accuracy_{IS}(\alpha, \beta) = \frac{support_{IS}(\alpha,\beta)}{card(\|\alpha\|_{IS})} \geq t'$.

Elementary granule inclusion in a given information system IS can be defined using different schemes, e.g., by

$\nu_{t,t'}^{IS}(\alpha, \beta)$ if and only if $support_{IS}(\alpha, \beta) \geq t$ & $accuracy_{IS}(\alpha, \beta) \geq t'$.

The closeness of granules can be defined by

$cl_{t,t'}^{IS}(\alpha, \beta)$ if and only if $\nu_{t,t'}^{IS}(\alpha, \beta)$ and $\nu_{t,t'}^{IS}(\beta, \alpha)$ hold.

Decision rules as granules. One can define inclusion and closeness of granules corresponding to rules of the form **if** α **then** β using accuracy coefficients.

Having such granules $g = (\alpha, \beta)$, $g' = (\alpha', \beta')$ one can define inclusion and closeness of g and g' by $\nu_{t,t'}(g, g')$ if and only if $\nu_{t,t'}(\alpha, \alpha')$ and $\nu_{t,t'}(\beta, \beta')$.

The closeness can be defined by

$cl_{t,t'}(g, g')$ if and only if $\nu_{t,t'}(g, g')$ and $\nu_{t,t'}(g', g)$.

Extensions of elementary granules by tolerance relation. For extensions of elementary granules defined by similarity (tolerance) relation, i.e., granules of the form (α, τ), (β, τ) one can consider the following inclusion measure:

$\nu_{t,t'}^{IS}((\alpha, \tau)(\beta, \tau))$ if and only if

$\qquad \nu_{t,t'}^{IS}(\alpha', \beta')$ for any α', β' such that $(\alpha, \alpha') \in \tau$ and $(\beta, \beta') \in \tau$

and the following closeness measure:

$cl_{t,t'}^{IS}((\alpha, \tau)(\beta, \tau))$ if and only if $\nu_{t,t'}^{IS}((\alpha, \tau)(\beta, \tau))$ and $\nu_{t,t'}^{IS}((\beta, \tau)(\alpha, \tau))$.

Sets of rules. It can be important for some applications to define closeness of an elementary granule α and the granule (α, τ). The definition reflecting an intuition that α should be a representation of (α, τ) sufficiently close to this granule is the following one:

$cl_{t,t'}^{IS}(\alpha, (\alpha, \tau))$ if and only if $cl_{t,t'}(\alpha, \beta)$ for any $(\alpha, \beta) \in \tau$.

An important problem related to association rules is that the number of such rules generated even from simple data table can be large. Hence, one should search for methods of aggregating close association rules. We suggest that this can be defined as searching for some close information granules.

Let us consider two finite sets $Rule\_Set$ and $Rule\_Set'$ of association rules defined by $Rule\_Set = \{(\alpha_i, \beta_i) : i = 1, \ldots, k\}$, and
$Rule\_Set' = \{(\alpha_i', \beta_i') : i = 1, \ldots, k'\}$. One can treat them as higher order information granules. These new granules $Rule\_Set$, $Rule\_Set'$ can be treated as close in a degree at least t (in IS) if and only if there exists a relation rel between sets of rules $Rule\_Set$ and $Rule\_Set'$ such that:

1. For any $Rule$ from the set $Rule\_Set$ there is $Rule'$ from $Rule\_Set'$ such that $(Rule, Rule') \in rel$ and $Rule$ is close to $Rule'$ (in IS) in degree at least t.
2. For any $Rule'$ from the set $Rule\_Set'$ there is $Rule$ from $Rule\_Set$ such that $(Rule, Rule') \in rel$ and $Rule$ is close to $Rule'$ (in IS) in degree at least t.

Another way of defining closeness of two granules G_1, G_2 represented by sets of rules can be described as follows.

Let us consider again two granules $Rule\_Set$ and $Rule\_Set'$ corresponding to two decision algorithms. By $I(\beta_i')$ we denote the set $\{j : cl_p(\beta_j', \beta_i')\}$ for any $i = 1, \ldots, k'$.

Now, we assume $\nu_p\left(Rule\_Set, Rule\_Set'\right)$ if and only if for any $i \in \{1, \ldots, k'\}$ there exists a set $J \subseteq \{1, \ldots, k\}$ such that

$$
cl_p\left(\bigvee_{j\in I(\beta_i')} \beta_j', \bigvee_{j\in J} \beta_j\right) \text{ and } cl_p\left(\bigvee_{j\in I(\beta_i')} \alpha_j', \bigvee_{j\in J} \alpha_j\right)
$$

and for closeness we assume

$cl_p\left(Rule\_Set, Rule\_Set'\right)$ if and only if

$\qquad\nu_p\left(Rule\_Set, Rule\_Set'\right)$ and $\nu_p\left(Rule\_Set', Rule\_Set\right)$.

One can consider a searching problem for a granule $Rule\_Set'$ of minimal size such that $Rule\_Set$ and $Rule\_Set'$ are close.

Granules defined by sets of granules. The previously discussed methods of inclusion and closeness definition can be easily adopted for the case of granules defined by sets of already defined granules. Let G, H be sets of granules.

The inclusion of G in H can be defined by

$\nu_{t,t'}^{IS}(G, H)$ if and only if for any $g \in G$ there is $h \in H$ for which $\nu_{t,t'}^{IS}(g, h)$

and the closeness by $cl_{t,t'}^{IS}(G, H)$ if and only if $\nu_{t,t'}^{IS}(G, H)$ and $\nu_{t,t'}^{IS}(H, G)$.

We have the following examples of inclusion and closeness propagation rules:

$$
\frac{\text{for any } \alpha \in G \text{ there is } \alpha' \in H \text{ such that } \nu_p(\alpha, \alpha')}{\nu_p(G, H)}
$$

$$
\frac{cl_p(\alpha, \alpha'), cl_p(\beta, \beta')}{cl_p((\alpha, \beta), (\alpha', \beta'))}
$$

$$
\frac{\text{for any } \alpha' \in \tau(\alpha) \text{ there is } \beta' \in \tau(\beta) \text{ such that } \nu_p(\alpha', \beta')}{\nu_p((\tau : \alpha), (\tau : \beta))}
$$

$$
\frac{cl_p(G, G') \text{ and } cl_p(E, E')}{cl_p((G, E), cl_p(G', E'))}
$$

where $\alpha, \alpha', \beta, \beta'$ are elementary granules and G, G' are finite sets of elementary granules.

One can also present other discussed cases for measuring the inclusion and closeness of granules in the form of inference rules. The exemplary rules have a general form, i.e., they are true in any information system (under the chosen definition of inclusion and closeness).

4   Approximations of information granule sets

We introduce now the approximation operations for granule sets assuming a given granule system $\mathcal{G}$ specified by syntx, semantics of information granules from the universe U as well as by the relations of inclusion ν_p and closeness cl_q in degrees at least p, q, respectively.

For a given granule g we define its neighborhood $I_p(g)$ to be the set of all information granules from U close to g in degree at least p.

For any subset $X \subseteq U$ we define its lower and upper approximation by

$LOW\,(\mathcal{G}, p, q, X) = \{g \in U : \nu_q\,(I_p\,(g)\,, X)\}$,

$UPP\,(\mathcal{G}, p, t, X) = \{g \in U : \nu_t\,(I_p\,(g)\,, X)\}$, respectively.

where $\nu_q\,(I_p\,(g)\,, X)$ iff for any granule $r \in I_p\,(g)$ the condition $\nu_q\,(r, x)$ holds for some $x \in X$ and $0.5 < t < q$.

Hence it follows that the approximations of sets can be defined analogously to the classical rough set approach. In our next paper we will discuss how to define approximation of complex information granules taking into account their structure (e.g., defined by the relation *to be a part in a degree* [2]).

Conclusions

We have presented the concept of approximation of complex information granule sets. This notion seems to be crucial for further investigations on approximate reasoning based on information granules.

Acknowledgments

This work has been supported by the Wallenberg Foundation, by the ESPRIT–CRIT 2 project #20288, and by grants 8T11C 025 19, 8T11C 009 19 from the State Committee for Scientific Research (KBN) of the Republic of Poland.

References

1. Pawlak Z.: Rough Sets. Theoretical Aspects of Reasoning about Data, Kluwer Academic Publishers, Dordrecht, 1991.
2. Polkowski L., Skowron A.: Towards Adaptive Calculus of Granules, In: [8], vol.1, pp. 201–227.
3. Skowron A., Stepaniuk J.: Tolerance Approximation Spaces, Fundamenta Informaticae, Vol. 27, 1996, pp. 245–253.
4. Skowron A., Stepaniuk J.: Information Granules in Distributed Environment, Lecture Notes in Artificial Intelligence 1711, Springer-Verlag, 1999, pp. 357–365.
5. Stepaniuk J.: Knowledge Discovery by Application of Rough Set Models, ICS PAS Report 887, Institute of Computer Science, Polish Academy of Sciences, Warsaw 1999, and also: L. Polkowski, T.Y. Lin, S. Tsumoto (Eds.), Rough Sets: New Developments, Physica–Verlag, Heidelberg, 2000.
6. WWW SPACENET page: http://agora.scs.leeds.ac.uk/spacenet/.
7. Zadeh L.A.: Fuzzy Logic = Computing with Words, IEEE Trans. on Fuzzy Systems Vol. 4, 1996, pp. 103–111.
8. Zadeh L.A., Kacprzyk J. (Eds.): Computing with Words in Information/Intelligent Systems vol.1–2, Physica–Verlag, Heidelberg, 1999.

Probabilistic Inference and Bayesian Theorem on Rough Sets

Yukari Yamauchi[1] and Masao Mukaidono[1]

Dept. of Computer Science, Meiji University,
Kanagawa, Japan
{yukari, masao}@cs.meiji.ac.jp

Abstract. The concept of (crisp) set is now extended to fuzzy set and rough set. The key notion of rough set is the two boundaries, the lower and upper approximations, and the lower approximation must be inside of the upper approximation. This inclusive condition makes the inference using rough sets complex: each approximation can not be determined independently. In this paper, the probabilistic inferences on rough sets based on two types of interpretation of If-Then rules, conditional probability and logical implication, are discussed. There are some interesting correlation between the lower and upper approximation after probabilistic inference.

1 Introduction

In propositional logic, the truth values of propositions are given either 1(true) or 0(false). Inference based on propositional (binary) logic is done using inference rule : Modus Ponens, shown in Fig.1(a). This rule implies that if an If-Then rule "$A \rightarrow B$" and proposition A are given true(1) as premises, then we come to a conclusion that proposition B is true(1).

The inference rule based on propositional logic is extended to probabilistic inference based on probability theory in order to treat uncertain knowledge. The truth values of propositions are given as the probabilities of events that take any value in the range of $[0, 1]$. Here, U is the sample space (universal set), $A, B \subseteq U$ are events, and the probability of "an event A happens", $\mathbf{P}(A)$ is defined as $\mathbf{P}(A) = |A|/|U|$ ($|U| = 1$, $|A| = a \in [0, 1]$) under the interpretation of randomness. Thus the probabilistic inference rule can be written as Fig.1(b) adapting the style of Modus Ponens.

$$\frac{\begin{array}{c} A \rightarrow B \\ A \end{array}}{B} \qquad \frac{\begin{array}{c} \mathbf{P}(A \rightarrow B) = i \\ \mathbf{P}(A) = a \end{array}}{\mathbf{P}(B) = b \quad {\scriptstyle i,\, a,\, b\, \in\, [0,\, 1]}}$$

Fig. 1. (a)Modus Ponens   (b)Probabilistic Inference

W. Ziarko and Y. Yao (Eds.): RSCTC 2000, LNAI 2005, pp. 73–81, 2001.

If the probability of $A \to B$ and A are given 1 ($i = a = 1$), then b is 1, since the probabilistic inference should be inclusive of modus ponens as a special case. Our focus is to determine the probability of B from the probabilities of $A \to B$ and A that take any value in $[0, 1]$. $A \to B$ is interpreted as "if A is true, then B is true" in meta-language. Traditional Bayes' theorem applied in many probability system adopts conditional probability as the interpretation of If-Then rule. However, the precise interpretation of the symbol "$\to$" is not unique and still under discussion among many researchers.

Nilsson [1] presented a semantical generalization of ordinary first-order logic in which the truth values of sentences can range between 0 and 1. He established the foundation of *probabilistic logic* through a possible-world analysis and probabilistic entailment. However, in most cases, we are not given the probabilities for the different sets of possible worlds, but must induce them from what we are given.

Pawlak [2] discussed the relationship between Bayes' inference rule and decision rules from the rough set perspective, and revealed that two conditional probabilities, called certainty and coverage factors satisfy the Bayes' rule. Related works are done by Yao [6]-[9] on Interval-set and Interval-valued probabilistic reasoning.

Our goal is to deduce a conclusion and its associated probability from given rules and facts and their associated probabilities through simple geometric analysis. The probability of the sentence "if A then B" is interpreted in two ways: conditional probability and the probability of logical implication. We have defined the probabilistic inferences based on the two interpretations of "If-Then" rule, conditional probability and logical implication, and introduce a new variant of Bayes' theorem based on the logical implication [5].

In this paper, analysis on Rough-set based probabilistic inference are done. There are some interesting correlations-relations between the lower and the upper approximation since the lower approximation is inside the upper approximation. This restriction between the lower and upper probabilities is discussed in detail and the traditional Bayes' theorem and a new variant of Bayes' theorem based on the logical implication are applied to the probabilistic inference on rough sets.

2   Probabilistic Inference and Bayes Theorem

2.1   Conditional Probability

Conditional probability, "how often B happens when A is already (or necessary) happens", only deals with the event space that A certainly happens. Thus the sample space changes from U to A.

$$\mathbf{P}(A\overset{c}{\to}B) = \mathbf{P}(B|A) = \mathbf{P}(A \cap B) \div \mathbf{P}(A) \qquad (a \neq 0) \qquad (1)$$

Given $\mathbf{P}(A\overset{c}{\to}B) = i_c$ and $\mathbf{P}(A) = a$, $\mathbf{P}(A \cap B) = i_c \times a$ from Equation(1). Thus the size of the intersection of A and B is fixed. The possible size of B is determined from $A \cap B$ to $(A \cap B) + \sim A$. The probabilistic inference based on the interpretation of if-then rule as the conditional probability determines $\mathbf{P}(B)$ from given $\mathbf{P}(A \overset{c}{\to} B)$ and $\mathbf{P}(A)$ by the following inference style in Fig.2.

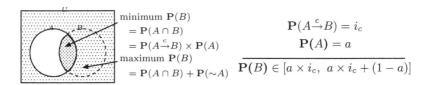

minimum $\mathbf{P}(B)$
$= \mathbf{P}(A \cap B)$
$= \mathbf{P}(A \overset{c}{\to} B) \times \mathbf{P}(A)$
maximum $\mathbf{P}(B)$
$= \mathbf{P}(A \cap B) + \mathbf{P}(\sim A)$

$$\mathbf{P}(A \overset{c}{\to} B) = i_c$$
$$\mathbf{P}(A) = a$$
$$\overline{\mathbf{P}(B) \in [a \times i_c, \ a \times i_c + (1 - a)]}$$

Fig. 2. Conditional Probability

Note, $\mathbf{P}(B)$ can not be determined uniquely from $\mathbf{P}(A \to B)$ and $\mathbf{P}(A)$ thus expressed as the interval probability [4]. When the condition, $a \times i_c = 1 - a(1 - i_c)$ (thus $a = 1$), holds, $\mathbf{P}(B)$ is unique and equal to $\mathbf{P}(A \to B)$.

2.2 Logical Implication

The interpretations of $\to$ (implication) in logics: propositional (binary or Boolean)logic, multi-valued logic, fuzzy logic, etc., are not unique in each logic. However, the most common interpretation of $A \to B$ is $\sim A \vee B$.

$$\mathbf{P}(A \overset{l}{\to} B) = \mathbf{P}(\sim A \cup B)$$
$$= \mathbf{P}(A \cap B) + \mathbf{P}(\sim A). \quad (\mathbf{P}(A) + \mathbf{P}(A \overset{l}{\to} B) \geq 1) \quad (2)$$

Since $\mathbf{P}(A \cap B) = \mathbf{P}(A \overset{l}{\to} B) - \mathbf{P}(\sim A)$ from Equation (2), the possible area of B is determined from $A \cap B$ to $A \cap B + \sim A$ ($A \overset{l}{\to} B$). The probabilistic inference based on the interpretation of if-then rule as the logical implication determines $\mathbf{P}(B)$ as the interval probability from given $\mathbf{P}(A \overset{l}{\to} B)$ and $\mathbf{P}(A)$ as shown in Fig.3.

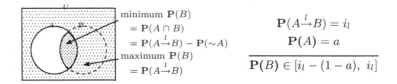

minimum $\mathbf{P}(B)$
$= \mathbf{P}(A \cap B)$
$= \mathbf{P}(A \overset{l}{\to} B) - \mathbf{P}(\sim A)$
maximum $\mathbf{P}(B)$
$= \mathbf{P}(A \overset{l}{\to} B)$

$$\mathbf{P}(A \overset{l}{\to} B) = i_l$$
$$\mathbf{P}(A) = a$$
$$\overline{\mathbf{P}(B) \in [i_l - (1 - a), \ i_l]}$$

Fig. 3. Logical Implication

Similar to the conditional probability, $\mathbf{P}(B)$ is unique and equal to $\mathbf{P}(A \overset{l}{\to} B)$ when the condition $i_l + a - 1 = i_l$ (thus $a = 1$) holds.

2.3   Bayes' Theorem

Bayes' theorem is widespread in application since it is a powerful method to trace a cause from effects. The relationship between a priori probability $\mathbf{P}(A\overset{c}{\to}B)$ and a posteriori probability $\mathbf{P}(B\overset{c}{\to}A)$ is expressed in the following equation by eliminating $\mathbf{P}(A \cap B)$ from the definitions.

$$\mathbf{P}(A\overset{c}{\to}B) = \mathbf{P}(B|A) = \mathbf{P}(A \cap B) \div \mathbf{P}(A),$$
$$\mathbf{P}(B\overset{c}{\to}A) = \mathbf{P}(A|B) = \mathbf{P}(A \cap B) \div \mathbf{P}(B),$$

$$\mathbf{P}(B\overset{c}{\to}A) = \mathbf{P}(A\overset{c}{\to}B) \times \mathbf{P}(A) \div \mathbf{P}(B) \tag{3}$$

Theorem 2.31 *The interpretation of If-Then rule as the logical implication satisfies the following equation.*

$$P(B\overset{l}{\to}A) = P(A\overset{l}{\to}B) + P(A) - P(B) \tag{4}$$

Proof.

$$\begin{aligned}
\mathbf{P}(B\overset{l}{\to}A) &= \mathbf{P}(\sim B \cup A) \\
&= \mathbf{P}(A \cap B) + \mathbf{P}(\sim B) \\
&= \mathbf{P}(A \cap B) + 1 - \mathbf{P}(B) \\
&= \mathbf{P}(A \cap B) + 1 - \mathbf{P}(A) + \mathbf{P}(A) - \mathbf{P}(B) \\
&= \mathbf{P}(A \cap B) + \mathbf{P}(\sim A) + \mathbf{P}(A) - \mathbf{P}(B) \\
&= \mathbf{P}(A\overset{l}{\to}B) + \mathbf{P}(A) - \mathbf{P}(B) \qquad \qquad \square
\end{aligned}$$

Note, the new variant of the Bayes' theorem based on logical implication adopt addition $+$ and subtraction $-$ where the traditional one adopt multiplication $\times$ and division $\div$. This property is quite attractive in operations on multiple-valued domain, and simplicity of calculation.

2.4   Bayes' Inference Based on Conditional Probability

Now, we apply Bayes' theorem as the inference rule, and define $\mathbf{P}(B\overset{c}{\to}A)$ from $\mathbf{P}(A\overset{c}{\to}B)$, $\mathbf{P}(A)$, and $\mathbf{P}(B)$. The inference based on the traditional Bayes' theorem (conditional probability) is shown in Fig.4(a). From $\mathbf{P}(A\overset{c}{\to}B)$ and $\mathbf{P}(A)$, $\mathbf{P}(B)$ is determined as the interval probability by the inference rule, Fig.2 in the previous discussion. Thus $\mathbf{P}(B\overset{c}{\to}A)$ can be determined as follows when $\mathbf{P}(B)$ is unknown.

The condition, $\max((a+b-1),0)/a \leq i_c \leq b/a$, must be satisfied between the probabilities a, b, and i_c. Since $i_c = \mathbf{P}(A \cap B)/a$ thus $\max(a + b - 1, 0) \leq \mathbf{P}(A \cap B) \leq \min(a, b)$.

$$\mathbf{P}(A \overset{c}{\to} B) = i_c$$
$$\mathbf{P}(A) = a$$
$$\mathbf{P}(B) = b$$

$$\overline{\mathbf{P}(B \overset{c}{\to} A) = i_c \times a \div b}$$

$$\mathbf{P}(A \overset{c}{\to} B) = i_c$$
$$\mathbf{P}(A) = a$$
$$\mathbf{P}(B) \in [i_c \times a, \ i_c \times a + (1 - a)]$$

$$\overline{\mathbf{P}(B \overset{c}{\to} A) \in [i_c \times a \div (i_c \times a + (1 - a)), \ 1]}$$

Fig. 4. (a)Bayes' Inference - Conditional   (b)$\mathbf{P}(B)$ unknown

2.5   Bayes' Inference Based on Logical Implication

Similarly, applying the new variant of Bayes' theorem based on logical implication, we get the following inference rule in Fig. 5. $\mathbf{P}(B)$ is determined from $\mathbf{P}(A \overset{l}{\to} B)$ and $\mathbf{P}(A)$ by the inference rule, Fig. 3. Thus $\mathbf{P}(B \overset{l}{\to} A)$ can be determined as follows when $\mathbf{P}(B)$ is unknown.

$$\mathbf{P}(A \overset{l}{\to} B) = i_l$$
$$\mathbf{P}(A) = a$$
$$\mathbf{P}(B) = b$$

$$\overline{\mathbf{P}(B \overset{l}{\to} A) = i_l + a - b}$$

$$\mathbf{P}(A \overset{l}{\to} B) = i_l$$
$$\mathbf{P}(A) = a$$
$$\mathbf{P}(B) \in [i_l + a - 1, i_l]$$

$$\overline{\mathbf{P}(B \overset{l}{\to} A) \in [a, \ 1]}$$

Fig. 5. (a)Bayes' Inference - Logical   (b)$\mathbf{P}(B)$ unknown

The condition, $\max(b, 1 - a) \le i_l \le 1 - a + b$, must be satisfied between the probabilities a, b, and i_l. Since $i_l = \mathbf{P}(A \cap B)/a$ thus $\max(a + b - 1, 0) \le \mathbf{P}(A \cap B) \le \min(a, b)$.

3   Probabilistic inference on Rough Sets

Rough set theory is introduced by Z. Pawlak in 1982 [3], and developed as a mathematical foundation of soft computing. Rough set A is defined by the upper approximation $^*A$ outside the lower approximation $_*A$, as shown in Fig.6.

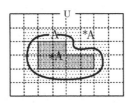

Fig. 6. Rough Set

Thus the probability for a rough set A is given by a paired values, (a_1, a_2). a_1 represents the probability of the lower approximation, and a_2 represents that of the upper approximation.

Given the set of all paired values $\mathcal{P}$,

$$\mathcal{P} = \{(a, b) \mid 0 \le a \le b \le 1\}$$

the paired probability[4] of "A happens" is $\mathbf{P}(A) \in (a_1, a_2)$.

3.1   Inferences on Rough Sets

In the previous section, given $\mathbf{P}(A \to B)$ and $\mathbf{P}(A)$, the size of $\mathbf{P}(A \cap B)$ is determined. Thus $\mathbf{P}(B)$ must be greater than or equal to $\mathbf{P}(A \cap B)$ and less than or equal to $\mathbf{P}(A \cap B) + \mathbf{P}(\sim A)$. This condition holds for both $\mathbf{P}(A \overset{c}{\to} B))$ and $\mathbf{P}(A \overset{l}{\to} B)$. Given $(A \to B)$ and A as rough sets, $\mathbf{P}(A \to B)$ and $\mathbf{P}(A)$ are paired probabilities (i_1, i_2) and (a_1, a_2). The possible probabilities of the lower and upper bound of B, $_*\mathbf{P}(B)$ and $^*\mathbf{P}(B)$, are determined as the same manner in previous discussion.

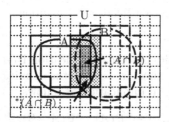

Fig. 7. Probabilistic Inference on Rough Sets

However, since there is an inclusive relation between the lower and upper bounds, they are restricted to each other by certain conditions. For example, if $_*A =^* A$, then $^*\mathbf{P}(B) -_* \mathbf{P}(B)$ $(\mathbf{P}(BndB))$ must be greater than or equal to $^*\mathbf{P}(A \cap B) -_* \mathbf{P}(A \cap B)$ $(\mathbf{P}(BndA \cap B))$.

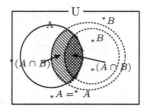

Fig. 8. Probabilistic Inference on Rough Sets - $_*A =^* A$

Conditional probability $\mathbf{P}(A \overset{c}{\to} B)$ is given as $\mathbf{P}(A \cap B) \div \mathbf{P}(A)$. Thus, we assume the definition of the lower and upper conditional probabilities as follows.

$$_*\mathbf{P}(A \overset{c}{\to} B) \equiv \frac{_*\mathbf{P}(A \cap B)}{_*\mathbf{P}(A)} \qquad {}^*\mathbf{P}(A \overset{c}{\to} B) \equiv \frac{^*\mathbf{P}(A \cap B)}{^*\mathbf{P}(A)}$$

Because of the devision in the definition of conditional probability, $_*\mathbf{P}(A \overset{c}{\to} B) \leq {}^*\mathbf{P}(A \overset{c}{\to} B)$ is not always true.

Given $\mathbf{P}(A \overset{c}{\to} B) = (_*i_c, {}^*i_c)$ and $\mathbf{P}(A) = (_*a, {}^*a)$, the minimum ${}^*\mathbf{P}(B)$ is determined as ${}^*\mathbf{P}(A \cap B) = {}^*i_c \times {}^*a$, and the maximum ${}^*\mathbf{P}(B)$ is ${}^*\mathbf{P}(A \cap B) + {}^*\mathbf{P}(\sim A) = {}^*i_c \times {}^*a + (1 - {}^*a)$. Since $_*\mathbf{P}(B) \subseteq {}^*\mathbf{P}(B)$, $_*\mathbf{P}(B)$ is restricted by the sizes of $BndA$ and $Bnd(A \cap B)$.

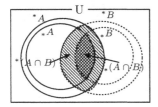

$$\mathbf{P}(A \overset{c}{\to} B) = (_*i_c, {}^*i_c)$$
$$\mathbf{P}(A) = (_*a, {}^*a)$$

$$_*\mathbf{P}(B) \in [_*a \times {}_*i_c, \; _*a \times {}_*i_c + (1 - {}_*a)$$
$$- \max\{0, Bnd(A) - Bnd(A \cap B)\}]$$
$$^*\mathbf{P}(B) \in [{}^*a \times {}^*i_c, \; {}^*a \times {}^*i_c + (1 - {}^*a)]$$

Fig. 9. Rough Set Inference - Conditional Probability

Similarly, the probability of logical implication $\mathbf{P}(A \overset{l}{\to} B)$ is given as $\mathbf{P}(\sim A \cup B) = \mathbf{P}(A \cap B) + \mathbf{P}(\sim A)$. Thus, we assume the definition of the lower and upper probabilities based on logical implication as follows.

$$_*\mathbf{P}(A \overset{l}{\to} B) \equiv_* \mathbf{P}(A \cap B) + 1 - {}^* \mathbf{P}(A) \qquad {}^*\mathbf{P}(A \overset{l}{\to} B) \equiv^* \mathbf{P}(A \cap B) + 1 -_* \mathbf{P}(A)$$

Note, the lower and upper probabilities are not determined independently, since the definition of $\overset{l}{\to}$ includes negation $\sim$.

Given $\mathbf{P}(A \overset{l}{\to} B) = (_*i_l, {}^*i_l)$ and $\mathbf{P}(A) = (_*a, {}^*a)$, the minimum $_*\mathbf{P}(B)$ is determined as ${}^*a - (1 -_* i_l)$, and the maximum $_*\mathbf{P}(B)$ is $_*i_l$. ${}^*\mathbf{P}(B)$ is restricted by the size of $_*\mathbf{P}(B)$ because of the inclusive condition.

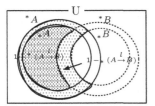

$$\mathbf{P}(A \overset{l}{\to} B) = (_*i_l, {}^*i_l)$$
$$\mathbf{P}(A) = (_*a, {}^*a)$$

$$_*\mathbf{P}(B) \in [{}^*a - (1 -_* i_l), \; _*i_l]$$
$$^*\mathbf{P}(B) \in [\max\{_*a - (1 - {}^*i_l), \; {}^*a - (1 -_* i_l)\}, \; {}^*i_l]$$

Fig. 10. Rough Set Inference - Logical Implication

3.2   Bayes' Theorem on Rough Sets

Now, we apply two types of Bayes' theorem as the rough set inference on paired probabilities.

Conditional probability $\mathbf{P}(A\overset{c}{\to}B)$, $\mathbf{P}(A)$ and $\mathbf{P}(B)$ satisfy Bayes' theorem.

$$_*\mathbf{P}(B\overset{c}{\to}A) = {}_*\mathbf{P}(A\overset{c}{\to}B) \times_* \mathbf{P}(A) \div_* \mathbf{P}(B) \tag{5}$$

$$^*\mathbf{P}(B\overset{c}{\to}A) = {}^*\mathbf{P}(A\overset{c}{\to}B) \times^* \mathbf{P}(A) \div^* \mathbf{P}(B) \tag{6}$$

Given $\mathbf{P}(A\overset{c}{\to}B) = ({}_*i_c, {}^*i_c)$ and $\mathbf{P}(A) = ({}_*a, {}^*a)$ as the paired probabilities, $\mathbf{P}(B)$ is determined in previous discussion. Thus, when $\mathbf{P}(B)$ is unknown, $\mathbf{P}(B\overset{c}{\to}A)$ is determined as follows.

$$_*\mathbf{P}(B\overset{c}{\to}A) \in [\frac{{}_*a \times_* i_c}{{}_*a \times_* i_c + (1 -_* a) - \max\{0, Bnd(A) - Bnd(A \cap B)\}} \,,\, 1]$$

$$^*\mathbf{P}(B\overset{c}{\to}A) \in [\frac{{}^*a \times^* i_c}{{}^*a \times^* i_c + (1 -^* a)} \,,\, 1]$$

Similarly, probability based on logical implication $\mathbf{P}(A\overset{c}{\to}B)$, $\mathbf{P}(A)$ and $\mathbf{P}(B)$ satisfy the new type of Bayes' theorem.

$$_*\mathbf{P}(B\overset{l}{\to}A) = {}^*\mathbf{P}(A\overset{l}{\to}B) +_* \mathbf{P}(A) -^* \mathbf{P}(B) \tag{7}$$

$$^*\mathbf{P}(B\overset{l}{\to}A) = {}_*\mathbf{P}(A\overset{l}{\to}B) +^* \mathbf{P}(A) -_* \mathbf{P}(B) \tag{8}$$

Given $\mathbf{P}(A\overset{l}{\to}B) = ({}_*i_l, {}^*i_l)$ and $\mathbf{P}(A) = ({}_*a, {}^*a)$ as the paired probabilities, $\mathbf{P}(B)$ is determined in previous discussion. Thus, when $\mathbf{P}(B)$ is unknown, $\mathbf{P}(B\overset{l}{\to}A)$ is determined as follows.

$$_*\mathbf{P}(B\overset{l}{\to}A) \in [{}_*a, \, 1 - \max\{0, Bnd(A\overset{l}{\to}B) - Bnd(A)\}]$$

$$^*\mathbf{P}(B\overset{l}{\to}A) \in [{}^*a, \, 1]$$

4   Conclusion

Probabilistic inference on rough sets is discussed. Given the sizes of $\mathbf{P}(A \to B)$ and $\mathbf{P}(A)$, the size of $\mathbf{P}(A \cap B)$ is calculated in both interpretation of "If-then rule: $\to$". Thus $\mathbf{P}(B)$ is determined and applying Bayes' theorem, $\mathbf{P}(B \to A)$ is also determined. However, in rough set inference, the inclusive relation between the lower and upper bound influences each other, thus the lower and upper probabilities are not determined independently. This feature is quite unique and distinguish rough set inference from interval probability and other approximation methods. Farther discussion should be to analyse mathematical aspects of this inference and apply it to knowledge discovery and data mining.

References

1. N.J.Nilsson, "Probabilistic Logic", *Artificial Intelligence,* Vol.28, No.1, 71-78, (1986).
2. Z.Pawlak, "Decision Rules, Bayes' Rule and Rough Sets", N.Zhong, A.Skowron, S.Ohsuga (eds.) *New Directions in Rough Sets, Data Mining, and Granular-Soft Computing* LNAI 1711, 1-9, Springer-Verlag (1999).
3. Z. Pawlak, "Rough Sets", *International Journal of Computer and Information Sciences,* 11, 341-356, (1982).
4. Y.Yamauchi, M.Mukaidono, "Interval and Paired Probabilities for Treating Uncertain Events," IEICE Transactions of Information and Systems, Vol.E82-D, No.5, 955-961, (1999).
5. Y.Yamauchi, M.Mukaidono, "Probabilistic Inference and Bayesian Theorem Based on Logical Implication" N.Zhong, A.Skowron, S.Ohsuga (eds) *New Directions in Rough Sets, Data Mining, and Granular-Soft Computing* LNAI 1711, 334-342, Springer-Verlag (1999).
6. Y.Y.Yao, "A Comparison of Two Interval-valued Probabilistic Reasoning Methods", *Proceedings of the 6th International Conference on Computing and Information,* special issue of *Journal of Computing and Information,* 1, 1090-1105, (1995).
7. Y.Y.Yao, X.Li, "Comparison of Rough-Set and Interval-Set Models for Uncertain Reasoning", *Fundamenta Informaticae,* 27, 289-298, (1996).
8. Y.Y.Yao, Q.Liu, "A Generalized Decision Logic in Interval-Set-Valued Information Tables" N.Zhong, A.Skowron, S.Ohsuga (eds) *New Directions in Rough Sets, Data Mining, and Granular-Soft Computing* LNAI 1711, 285-293, Springer-Verlag (1999).
9. Y.Y.Yao, S.K.M.Wong , "Interval Approaches for Uncertain Reasoning", Z.W.Ras and A.Skowron (eds.) *Foundations of Intelligent Systems,* LNAI 1325, 381-390, (1997).

Interpreting Fuzzy Membership Functions
in the Theory of Rough Sets

Y.Y. Yao and J.P. Zhang

Department of Computer Science, University of Regina
Regina, Saskatchewan, Canada S4S 0A2
E-mail: {yyao,pingz}@cs.uregina.ca

Abstract. A fundamental difficulty with fuzzy set theory is the semantical interpretations of membership functions. We address this issue in the theory of rough sets. Rough membership functions are viewed as a special type of fuzzy membership functions interpretable using conditional probabilities. Rough set approximations are related to the core and support of a fuzzy set. A salient feature of the interpretation is that membership values of equivalent or similar elements are related to each other. Two types of similarities are considered, one is defined by a partition of the universe, and the other is defined by a covering.

1 Introduction

The theory of fuzzy sets is a generalization of the classical sets by allowing partial membership. It provides a more realistic framework for modeling the ill-definition of the boundary of a class [2, 8]. However, a fundamental difficulty with fuzzy set theory is the semantical interpretations of the degrees of membership. The objectives of this paper are to investigate interpretations of fuzzy membership functions in the theory of rough sets, and to establish the connections between core and support of fuzzy sets and rough set approximations.

There are at least two views for interpreting rough set theory [4, 6]. The operator-oriented view treats rough set theory as an *extension* of the classical set theory. Two additional unary set-theoretic operators are introduced, and the meanings of sets and standard set-theoretic operators are unchanged. This view is closely related to modal logics. The set-oriented view treats rough set theory as a *deviation* of the classical set theory. Sets and set-theoretic operators are associated with non-standard interpretations, and no additional set-theoretic operator is introduced. This view is related to many-valued logics and fuzzy sets. A particular set-oriented view is characterized by rough membership functions [3]. The formulation and interpretation of rough membership functions are based on partitions of a universe. By viewing rough membership functions as a special type of fuzzy membership functions, one may be able to provide a sound semantical interpretation of fuzzy membership functions.

The rest of the paper is organized as follows. In Section 2, we examine the interpretation of fuzzy membership functions and show the connections between

W. Ziarko and Y. Yao (Eds.): RSCTC 2000, LNAI 2005, pp. 82–89, 2001.

rough set approximations and the core and support of fuzzy sets, based on rough membership functions defined by a partition of a universe. A unified framework is used for studying both fuzzy sets and rough sets. In Section 3, within the established framework, we apply the arguments to a more general case by extending partitions to coverings of the universe. Three different rough membership functions are introduced. They lead to commonly used rough set approximations.

2   Review and Comparison of Fuzzy Sets and Rough Sets

In this section, some basic issues of fuzzy sets and rough sets are reviewed, examined, and compared by using a unified framework. Fuzzy membership functions are interpreted in terms of rough membership functions. The concepts of core and support of a fuzzy set are related to rough set approximations.

2.1   Fuzzy Sets

The notion of fuzzy sets provides a convenient tool for representing vague concepts by allowing partial memberships. Let U be a finite and non-empty set called universe. A fuzzy subset $\mathcal{A}$ of U is defined by a membership function:

$$\mu_{\mathcal{A}} : U \longrightarrow [0,1]. \tag{1}$$

There are many definitions for fuzzy set complement, intersection, and union. The standard min-max system proposed by Zadeh is defined component-wise by [8]:

$$\mu_{\neg \mathcal{A}}(x) = 1 - \mu_{\mathcal{A}}(x),$$
$$\mu_{\mathcal{A} \cap \mathcal{B}}(x) = \min(\mu_{\mathcal{A}}(x), \mu_{\mathcal{B}}(x)),$$
$$\mu_{\mathcal{A} \cup \mathcal{B}}(x) = \max(\mu_{\mathcal{A}}(x), \mu_{\mathcal{B}}(x)). \tag{2}$$

In general, one may interpret fuzzy set operators using triangular norms (t-norms) and conorms (t-conorms) [2]. Let t and s be a pair of t-norm and t-conorms, we have:

$$\mu_{\mathcal{A} \cap \mathcal{B}}(x) = t(\mu_{\mathcal{A}}(x), \mu_{\mathcal{B}}(x)),$$
$$\mu_{\mathcal{A} \cup \mathcal{B}}(x) = s(\mu_{\mathcal{A}}(x), \mu_{\mathcal{B}}(x)). \tag{3}$$

A crisp set may be viewed as a degenerated fuzzy set. A pair of t-norm and t-conorm reduce to standard set intersection and union when applied to crisp subsets of U. The min is an example of t-norms and the max is an example of t-conorms. An important feature of fuzzy set operators as defined by t-norms and t-conorms is that they are truth-functional operators. In other words, membership functions of complement, intersection, and union of fuzzy sets are defined based solely on membership functions of the fuzzy sets involved [6]. Although they have some desired properties, such as $\mu_{\mathcal{A} \cap \mathcal{B}}(x) \leq \min[\mu_{\mathcal{A}}(x), \mu_{\mathcal{B}}(x)] \leq$

$\max[\mu_\mathcal{A}(x), \mu_\mathcal{B}(x)] \leq \mu_{\mathcal{A}\cup\mathcal{B}}(x)$, there is a lack of a well accepted semantical interpretation of fuzzy set operators.

The concepts of core and support have been introduced and used as approximations of a fuzzy set [2]. The core of a fuzzy set $\mathcal{A}$ is a crisp subset of U consisting of elements with full membership:

$$core(\mathcal{A}) = \{x \in U \mid \mu_\mathcal{A}(x) = 1\}. \tag{4}$$

The support is a crisp subset of U consisting of elements with non-zero membership:

$$support(\mathcal{A}) = \{x \in U \mid \mu_\mathcal{A}(x) > 0\}. \tag{5}$$

With $1 - (\cdot)$ as fuzzy set complement, and t-norms and t-conorms as fuzzy set intersection and union, the following properties hold:

(F1) $core(\mathcal{A}) = \neg(support(\neg\mathcal{A}))$,
         $support(\mathcal{A}) = \neg(core(\neg\mathcal{A}))$,
(F2) $core(\mathcal{A} \cap \mathcal{B}) = core(\mathcal{A}) \cap core(\mathcal{B})$,
         $support(\mathcal{A} \cap \mathcal{B}) \subseteq support(\mathcal{A}) \cap support(\mathcal{B})$,
(F3) $core(\mathcal{A} \cup \mathcal{B}) \supseteq core(\mathcal{A}) \cup core(\mathcal{B})$,
         $support(\mathcal{A} \cup \mathcal{B}) = support(\mathcal{A}) \cup support(\mathcal{B})$,
(F4) $core(\mathcal{A}) \subseteq \mathcal{A} \subseteq support(\mathcal{A})$.

According to (F1), one may interpret *core* and *support* as a pair of dual operators on the set of all fuzzy sets. They map a fuzzy set to a pair of crisp sets. By properties (F2) and (F3), one may say that *core* is distributive over $\cap$ and *support* is distributive over $\cup$. However, *core* is not necessarily distributive over $\cup$ and *support* is not necessarily distributive over $\cap$. These two properties follow from the properties of t-norm and t-conorm. When the min-max system is used, we have equality in (F2) and (F3). Property (F4) suggests that a fuzzy set lies within its core and support.

2.2   Rough Sets

A fundamental concept of rough set theory is indiscernibility. Let $R \subseteq U \times U$ be an equivalence relation on a finite and non-empty universe U. That is, the relation R is reflexive, symmetric, and transitive. The pair $apr = (U, R)$ is called an approximation space. The equivalence relation R partitions the universe U into disjoint subsets called equivalence classes. Elements in the same equivalence class are said to be indistinguishable. The partition of the universe is referred to as the quotient set and is denoted by $U/R = \{E_1, \ldots, E_n\}$.

An element $x \in U$ belongs to one and only one equivalence class. Let

$$[x]_R = \{y \mid xRy\}, \tag{6}$$

denote the equivalence class containing x. For a subset $A \subseteq U$, we have the following well defined rough membership function [3]:

$$\mu_A(x) = \frac{|[x]_R \cap A|}{|[x]_R|},\tag{7}$$

where $|\cdot|$ denotes the cardinality of a set. One can easily see the similarity between rough membership functions and conditional probabilities. As a matter of fact, the rough membership value $\mu_A(x)$ may be interpreted as the conditional probability that an arbitrary element belongs to A given that the element belongs to $[x]_R$.

Rough membership functions may be interpreted as a special type of fuzzy membership functions interpretable in terms of probabilities defined simply by cardinalities of sets. In general, one may use a probability function on U to define rough membership functions [7]. One may view the fuzzy set theory as an uninterpreted mathematical theory of abstract membership functions. The theory of rough set thus provides a more specific and more concrete interpretation of fuzzy membership functions. The source of the fuzziness in describing a concept is the indiscernibility of elements. In the theoretical development of fuzzy set theory, fuzzy membership functions are treated as abstract mathematical functions without any constraint imposed [2]. When we interpret fuzzy membership functions in the theory of rough sets, we have the following constraints:

$$
\begin{array}{ll}
\text{(rm1)} & \mu_U(x) = 1, \\
\text{(rm2)} & \mu_\emptyset(x) = 0, \\
\text{(rm3)} & y \in [x]_R \implies \mu_A(x) = \mu_A(y), \\
\text{(rm4)} & x \in A \implies \mu_A(x) \neq 0 \\
\text{(rm5)} & \mu_A(x) = 1 \implies x \in A, \\
\text{(rm6)} & A \subseteq B \implies \mu_A(x) \leq \mu_B(x).
\end{array}
$$

Property (rm3) is particularly important. It shows that elements in the same equivalence class must have the same degree of membership. That is, indiscernible elements should have the same membership value. Such a constraint, which ties the membership values of individual elements according to their connections, is intuitively appealing. Although this topic has been investigated by some authors, there is still a lack of systematic study [1]. Property (rm4) can be equivalently expressed as $\mu_A(x) = 0 \implies x \notin A$, and property (rm5) expressed as $x \notin A \implies \mu_A(x) \neq 1$.

The constraints on rough membership functions have significant implications on rough set operators. There does not exist a one-to-one relationship between rough membership functions and subsets of U. Two distinct subsets of U may define the same rough membership function. Rough membership functions corresponding to $\neg A$, $A \cap B$, and $A \cup B$ must be defined using set operators and equation (7). By laws of probability, we have:

$$\mu_{\neg A}(x) = 1 - \mu_A(x),$$

$$\mu_{A \cup B}(x) = \mu_A(x) + \mu_B(x) - \mu_{A \cap B}(x),$$
$$\max(0, \mu_A(x) + \mu_B(x) - 1) \le \mu_{A \cap B}(x) \le \min(\mu_A(x), \mu_B(x)),$$
$$\max(\mu_A(x), \mu_B(x)) \le \mu_{A \cup B}(x) \le \min(1, \mu_A(x) + \mu_B(x)). \tag{8}$$

Unlike the commonly used fuzzy set operators, the new intersection and union operators are non-truth-functional. That is, it is impossible to obtain rough membership functions of $A \cap B$ and $A \cup B$ based solely on the rough membership functions of A and B. One must also consider their relationships to the equivalence class $[x]_R$.

In an approximation space, a subset $A \subseteq U$ is approximated by a pair of sets called lower and upper approximations as follows [3]:

$$\underline{apr}(A) = \{x \in U \mid \mu_A(x) = 1\}$$
$$= core(\mu_A),$$
$$\overline{apr}(A) = \{x \in U \mid \mu_A(x) > 0\}$$
$$= support(\mu_A). \tag{9}$$

That is, the lower and upper approximation are indeed the core and support of the fuzzy set μ_A. For any subsets $A, B \subseteq U$, we have:

(R1) $\underline{apr}(A) = \neg(\overline{apr}(\neg A)),$
      $\overline{apr}(A) = \neg(\underline{apr}(\neg A)),$

(R2) $\underline{apr}(A \cap B) = \underline{apr}(A) \cap \underline{apr}(B),$
      $\overline{apr}(A \cap B) \subseteq \overline{apr}(A) \cap \overline{apr}(B),$

(R3) $\underline{apr}(A \cup B) \supseteq \underline{apr}(A) \cup \underline{apr}(B),$
      $\overline{apr}(A \cup B) = \overline{apr}(A) \cup \overline{apr}(B),$

(R4) $\underline{apr}(A) \subseteq A \subseteq \overline{apr}(A),$

By comparing with (F1)-(F4), we can see that rough set approximation operators satisfy the same properties of core and support of fuzzy sets.

Using the equivalence classes $U/R = \{E_1, \ldots, E_n\}$, lower and upper approximations can be equivalently defined as follows:

$$\underline{apr}(A) = \bigcup \{E_i \in U/R \mid E_i \subseteq A\},$$
$$\overline{apr}(A) = \bigcup \{E_i \in U/R \mid E_i \cap A \ne \emptyset\}. \tag{10}$$

The lower approximation $\underline{apr}(A)$ is the union of all the equivalence classes which are subsets of A. The upper approximation $\overline{apr}(A)$ is the union of all the equivalence classes which have a non-empty intersection with A.

3   Generalized Rough Membership Functions based on Coverings of the Universe

In a partition, an element belongs to one equivalence class and two distinct equivalence classes have no overlap. The rough set theory built on a partition, although easy to analyze, may not provide a realistic view of relationships between elements of the universe. One may consider a more realistic model by extending partitions to coverings of the universe [6, 9].

A covering of the universe, $C = \{C_1, \ldots, C_n\}$, is a family of subsets of U such that $U = \bigcup \{C_i \mid i = 1, \ldots, n\}$. Two distinct sets in C may have a non-empty overlap. An arbitrary element x of U may belong to more than one set in C. The family $C(x) = \{C_i \in C \mid x \in C_i\}$ consists of sets in C containing x. The sets in $C(x)$ may describe different types or various degrees of similarity between elements of U. For a set $C_i \in C(x)$, we may compute a value $|C_i \cap A|/|C_i|$ by extending equation (7). It may be interpreted as the membership value of x from the view point of C_i. From the set $C(x)$, we have a family of values $\{|C_i \cap A|/|C_i| \mid x \in C_i\}$. Generalized rough membership functions may be defined by using this family of values. We consider the following three definitions:

$$\text{(minimum)} \quad \mu_A^m(x) = \min \left\{ \frac{|C_i \cap A|}{|C_i|} \mid x \in C_i \right\}, \tag{11}$$

$$\text{(maximum)} \quad \mu_A^M(x) = \max \left\{ \frac{|C_i \cap A|}{|C_i|} \mid x \in C_i \right\}, \tag{12}$$

$$\text{(average)} \quad \mu_A^*(x) = \mathrm{avg} \left\{ \frac{|C_i \cap A|}{|C_i|} \mid x \in C_i \right\}. \tag{13}$$

The minimum, maximum, and average definitions may be regarded as the most permissive, the most optimistic view, and the balanced view in defining rough membership functions. The minimum rough membership function is determined by a set in $C(x)$ which has the smallest overlap with A, and the maximum rough membership function by a set in $C(x)$ which has the largest overlap with A. The average rough membership function depends on every set in $C(x)$. The three rough membership functions are related by:

$$\mu_A^m(x) \leq \mu_A^*(x) \leq \mu_A^M(x). \tag{14}$$

A partition is a special type of coverings. In this case, three rough membership functions reduce to the same rough membership function.

The generalized rough membership functions have the following properties:

(grm1) $\mu_U^m(x) = \mu_U^*(x) = \mu_U^M(x) = 1,$

(grm2) $\mu_\emptyset^m(x) = \mu_\emptyset^*(x) = \mu_\emptyset^M(x) = 0,$

(grm3) $[\forall C_i \in C (x \in C_i \Longleftrightarrow y \in C_i)] \Longrightarrow$
$\qquad\quad [\mu_A^m(x) = \mu_A^m(y), \mu_A^*(x) = \mu_A^*(y), \mu_A^M(x) = \mu_A^M(y)],$

(grm4) $x, y \in C_i \Longrightarrow [\mu_A^m(x) \neq 1 \Longrightarrow \mu_A^m(y) \neq 1, \mu_A^M(x) \neq 0 \Longrightarrow \mu_A^M(y) \neq 0],$

(grm5) $x \in A \Longrightarrow \mu_A^m(x) \neq 0,$

(grm6) $\mu_A^M(x) = 1 \Longrightarrow x \in A,$

(grm7) $A \subseteq B \Longrightarrow [\mu_A^m(x) \leq \mu_B^m(x), \mu_A^*(x) \leq \mu_B^*(x), \mu_A^M(x) \leq \mu_B^M(x)].$

Both (grm3) and (grm4) show the constraints on rough membership functions imposed by the similarity of objects. From the relation $\mu_A^m(x) \leq \mu_A^*(x) \leq \mu_A^M(x)$, we can obtain additional properties. For example, (grm5) implies that $x \in A \Longrightarrow$ $\mu_A^*(x) \neq 0$ and $x \in A \Longrightarrow \mu_A^M(x) \neq 0$. Similarly, (grm6) implies that $\mu_A^m(x) = 1 \Longrightarrow x \in A$ and $\mu_A^*(x) = 1 \Longrightarrow x \in A$.

For set-theoretic operators, one can verify that the following properties:

$$\mu_{\neg A}^m(x) = 1 - \mu_A^M(x),$$
$$\mu_{\neg A}^M(x) = 1 - \mu_A^m(x),$$
$$\mu_{\neg A}^*(x) = 1 - \mu_A^*(x),$$
$$\max(0, \mu_A^m(x) + \mu_B^m(x) - \mu_{A \cup B}^M(x)) \leq \mu_{A \cap B}^m(x) \leq \min(\mu_A^m(x), \mu_B^m(x)),$$
$$\max(\mu_A^M(x), \mu_B^M(x)) \leq \mu_{A \cup B}^M(x) \leq \min(1, \mu_A^M(x) + \mu_B^M(x) - \mu_{A \cap B}^m(x)),$$
$$\mu_{A \cap B}^*(x) = \mu_A^*(x) + \mu_B^*(x) - \mu_{A \cup B}^*(x). \tag{15}$$

We again obtain non-truth-functional rough set operators.

The minimum rough membership function may be viewed as the lower bound on all possible rough membership functions definable using a covering, while the maximum rough membership as the upper bound. The pair $(\mu_A^m(x), \mu_A^M(x))$ may also be used to define an interval-valued fuzzy set [2]. The interval $[\mu_A^m(x), \mu_A^M(x)]$ is the membership value of x with respect to A.

From the three rough membership functions, we define three pairs of lower and upper approximations. For the minimum definition, we have:

$$\begin{aligned}
\underline{apr}^m(A) &= core(\mu_A^m) \\
&= \{x \in U \mid \mu_A^m(x) = 1\} \\
&= \{x \in U \mid \forall C_i \in C(x \in C_i \Longrightarrow C_i \subseteq A)\}, \\
\overline{apr}^m(A) &= spport(\mu_A^m) \\
&= \{x \in U \mid \mu_A^m(x) > 0\} \\
&= \{x \in U \mid \forall C_i \in C(x \in C_i \Longrightarrow C_i \cap A \neq \emptyset)\}.
\end{aligned} \tag{16}$$

For the maximum definition, we have:

$$\begin{aligned}
\underline{apr}^M(A) &= core(\mu_A^M) \\
&= \{x \in U \mid \mu_A^M(x) = 1\} \\
&= \{x \in U \mid \exists C_i \in C(x \in C_i, C_i \subseteq A)\}, \\
&= \bigcup \{C_i \in C \mid C_i \subseteq A\}, \\
\overline{apr}^M(A) &= spport(\mu_A^M) \\
&= \{x \in U \mid \mu_A^M(x) > 0\} \\
&= \{x \in U \mid \exists C_i \in C(x \in C_i, C_i \cap A \neq \emptyset)\} \\
&= \bigcup \{C_i \in C \mid C_i \cap A \neq \emptyset\}.
\end{aligned} \tag{17}$$

The lower and upper approximations in each pair are no longer dual operators. However, $(\underline{apr}^m, \overline{apr}^M)$ and $(\underline{apr}^M, \overline{apr}^m)$ are two pairs of dual operators. The first pair can be derived from the average definition, namely:

$$\underline{apr}^*(A) = \underline{apr}^m(A), \qquad \overline{apr}^*(A) = \overline{apr}^M(A). \tag{18}$$

These approximation operator have been studied extensively in rough set theory. Their connections and properties can be found in a recent paper by Yao [5].

4 Conclusion

Rough membership functions can be viewed as a special type of fuzzy membership functions, and rough set approximations as the core and support of fuzzy sets. This provides a starting point for the interpretation of fuzzy membership functions in the theory of rough sets. We study rough membership functions defined based on partitions and coverings of a universe.

The formulation and interpretation of rough membership functions are inseparable parts of the theory of rough sets. Each rough membership function has a well defined semantical interpretation. The source of uncertainty modeled by rough membership functions is the indiscernibility or similarity of objects. Constraints are imposed on rough membership functions by the relationships between objects. More specifically, equivalent objects must have the same membership value, and similar objects must have similar membership values. These observations may have significant implications for the understanding of fuzzy set theory. The interpretation of fuzzy membership functions in the theory of rough sets provides a more restrictive, but more concrete, view of fuzzy sets. Such semantically sound models may provide possible solutions to the fundamental difficulty with fuzzy set theory regarding semantical interpretations of fuzzy membership functions.

References

1. D. Dubois and H. Prade, Similarity-based approximate reasoning, in: J.M. Zurada, R.J. Marks II, C.J. Robinson (Eds.), *Computational Intelligence: Imitating Life*, IEEE Press, New York, 69-80, 1994.
2. G.J. Klir and B. Yuan, *Fuzzy Sets and Fuzzy Logic: Theory and Applications*, Prentice Hall, New Jersey, 1995.
3. Z. Pawlak and A. Skowron, Rough membership functions, in: R.R. Yager and M. Fedrizzi and J. Kacprzyk (Eds.), *Advances in the Dempster-Shafer Theory of Evidence*, 251-271 1994.
4. Y.Y. Yao, Two views of the theory of rough sets in finite universes, *International Journal of Approximate Reasoning*, **15**, 291-317, 1996.
5. Y.Y. Yao, Relational interpretations of neighborhood operators and rough set approximation operators, *Information Sciences*, **111**, 239-259, 1998.
6. Y.Y. Yao, A comparative study of fuzzy sets and rough sets, *Information Sciences*, **109**, 227-242, 1998.
7. Y.Y. Yao, Generalized rough set models, in: L. Polkowski and A. Skowron (Eds.), *Rough Sets in Knowledge Discovery: Methodology and Applications*, Physica-Verlag, Heidelberg, 286-318, 1998.
8. L.A. Zadeh, Fuzzy sets, *Information and Control*, **8**, 338-353, 1965.
9. W. Zakowski, Approximations in the space (U, Π), *Demonstratio Mathematica*, *XVI*, 761-769, 1983.

Dynamic System Visualization with Rough Performance Maps

James J. Alpigini[1], James F. Peters[2]

[1]Penn State Great Valley School of Graduate Professional Studies
30 East Swedesford Road, Malvern, PA, 19355, USA
jja7@psu.edu
[2]Department of Electrical and Computer Engineering
University of Manitoba, 15 Gillson Street, ENGR 504
Winnipeg, Manitoba R3T 5V6 Canada
jfpeters@ee.umanitoba.ca

Abstract. This paper presents an approach to visualizing a system's dynamic performance with rough performance maps. Derived from the Julia set common in the visualization of iterative chaos, performance maps are constructed using rule evaluation and require a minimum of à priori knowledge of the system under consideration. By the use of carefully selected performance evaluation rules combined with color-coding, they convey a wealth of information to the informed user about dynamic behaviors of a system that may be hidden from all but the expert analyst. A rough set approach is employed to generate an approximation of a performance map. Generation of this new rough performance map allows more intuitive rule derivation and requires fewer system parameters to be observed.

1 Introduction

An approach to visualizing control and other dynamical systems with performance maps based on rough set theory is introduced in this paper. A performance map is derived from the Julia set method common in the study of chaotic equations [1]. Julia sets are fractals with shapes that are generated by iteration of simple complex mappings. The term fractal was introduced by Mandelbrot in 1982 to denote sets with fractured structure [2]. The critical component in a rough performance map is the use of rough set theory [3] in the generation of rules used to evaluate system performance in color-coding system responses. Performance map rules are derived using the method found in [4]. The images resulting from the application of derived rules yield very close approximations of full performance maps. Such maps are called rough performance maps. The contribution of this paper is the application of rough sets in the creation of performance maps in visualizing the performance of dynamical systems.

W. Ziarko and Y. Yao (Eds.): RSCTC 2000, LNAI 2005, pp. 90–97, 2001.

2   Generation of a Performance Map

It has been suggested that the Julia set method can be adapted to visualize control and other dynamic systems [4].   Such a technique addresses the problem of visualizing automatically the state values of some system as its parameters are varied. In effect, the method produces a color-coded performance map (PM), which reflects the dynamic performance of the system across intervals of system parameters.   As with the Julia set, a PM is generated via digital computation, using rules appropriate to the system for which dynamic behavior is being visualized.   The generation of a PM proceeds in a manner similar to a Julia set, namely, with a pixel mapping and color scheme selection.   The pixel mapping for a PM reflects the problem domain, i.e. a pair of system parameters.   Intervals of system parameters, say parameters a and b, are mapped to a computer screen that now represents the parametric plane of the system.   For example, intervals of parameter a and b can be mapped to the x- and y-axis respectively, so each pixel, $p_{x,y}$, represents a unique value of (a,b).   A performance rule is used to evaluate the condition of selected state variables of a system during numerical integration.   Such rules fall into two general types, fire/no fire and qualifying rules.   A fire/no fire rule will fire if a programmed trigger condition is encountered during the simulation whilst a qualifying rule is used to place a qualifying value on an entire system trajectory following a period of integration. Performance rules are typically aggregated to test and detect multiple behaviors, but in any case, their generation is normally intuitive following observations of the dynamic behavior of a system.

3   Simulation for Linear Plant

A performance map can be generated for any system provided that the problem domain can be mapped to the computer screen and that appropriate performance rules can be derived.   In order to demonstrate the utility of this technique, a performance map is now developed for the readily verified second-order, linear, continuous system in (1).

$$Y(s) = \frac{\omega_n^{\,2}}{s \cdot \left(s^2 + 2 \cdot \zeta \cdot \omega_n \cdot s + \omega_n^{\,2} \right)} \tag{1}$$

where ζ =damping ratio and ω_n = natural undamped frequency.

3.1   Generation of Performance Rules

Four qualitative measurements that can be evaluated by fire/no-fire rules, namely, maximum overshoot, O_v, delay time, t_d, rise time, t_r, and settle time, t_s, have been chosen.   While the tuning of parameters ζ and ω_n via classical techniques can set these measurements, it would be most beneficial to be able to assess the parameters to set all measurements at once to meet tuning criteria. To develop an appropriate rule set, it is first necessary to determine the required performance of the system. The rule set will then be used to determine which sets of parameter values $[\omega_n, \zeta]$ affect this performance. The tuning of the system in terms of maximum overshoot, O_v, delay time, t_d, rise time, t_r, and settle time, t_s, are somewhat arbitrary. For the purposes of illustration, the required performance criteria of the system is selected to be: maximum overshoot: $O_v = 0$, maximum delay: $t_d = 0.038$s, maximum rise time: $t_r = 0.068$s and maximum settle time: $t_s = 0.095$s.

The test for maximum overshoot is the most straightforward of the four measurements. This rule will trigger if the magnitude of the system solution $y(t) > 1$ at any time during integration, indicating that an overshoot has occurred, i.e. $O_v > 0$. The test for delay time, t_d, is equally simple, requiring that the magnitude of the system solution be tested at the selected time, $t_d = 0.038$s. This test need only be performed once.

The measurement of rise time, t_r, does not begin until the system solution has reached 10% of its final value. In effect, reaching this value serves to *prime* the rise time rule for trigger detection. Once the rule is primed, then the delay timer is initiated and magnitude of the system is tested at $t_r = 0.068$s. This test is also performed once. Settle time, t_s, is tested in a fashion similar to delay time. Provided the system reaches the programmed settle time, the system solution is tested to determine if it is within 5% of its final value. In order to determine if the system remains within this boundary, this condition is tested following each integration step until the maximum integration time period has elapsed.

3.2   Theoretical Results

Obviously, when the system is stable and damped, the maximum overshoot of the second-order, linear system occurs on the first overshoot, for a unit step input, at a time, t_{max}, which is given by (2).

$$t_{max} = \frac{\pi}{\omega_n \cdot \sqrt{1 - \zeta^2}} \tag{2}$$

The system is critically damped with zero overshoot when $\zeta = 1$, and will exhibit an overshoot whenever $\zeta < 1$. The system is critically damped with zero overshoot when $\zeta = 1$, and will exhibit an overshoot whenever $\zeta < 1$. The rise time t_r and delay time t_d of the system may be approximated (see Fig. 1) as can the settle time t_s (see Fig. 2) [7],[8],[9].

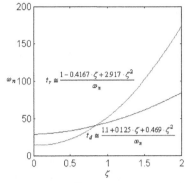 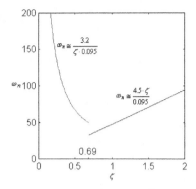

Fig. 1. ω_n vs. ζ for td = 0.038 and tr = 0.068

Fig. 2. Graph of ω_n vs. ζ for $t_s = 0.095$.

3.3   Sample Performance Map

A reasonable expectation of the performance map is that it represents an amalgamation of these graphs, as well as a sharp division at $\zeta = 1.0$ due to overshoot. This is indeed the case as shown by the performance map of Fig. 3.

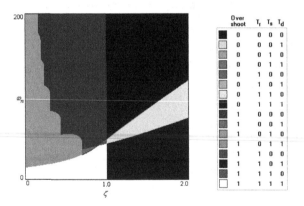

Fig. 3. Performance map of 2nd order linear system

The color key associated with the map shows the pixel colors employed to represent the rules that did and did not fire. In the key, a "1" means that a rule fired, indicating that required performance was not achieved, whilst a "0" indicates that the rule did not fire, and the performance was acceptable for that measurement. The color black, comprising the majority of the upper right quadrant of the map, is used where no rules fired, thus giving an immediate and intuitive visualization of the parameter values, $[\omega_n, \zeta]$, that affect the required system response. Thus, the selection of any pixel within this black region will yield a parameter set that not only provides stable operation, but also operation which is finely tuned for desired performance.

3.4   Validation of Results

The software written to generate the performance map allows the user to select a single pair of parameter values from the map by "clicking" on the image with a computer mouse. This allows the system dynamics to be further explored for these parameters such as by using time graphs, phase portraits or Poincairé maps. A region of the map can also be selected, and a new visualization generated for that set of parameter intervals, effectively rendering a close-up view of a map region. In order to validate the information portrayed by the performance map, time graphs of the system response are considered for parameter sets selected at random from the black and colored regions of the map. Fig. 4 contains a time graph generated for a set of parameters, $[\omega_n, \zeta]$, selected at random from the black region of the map. As stated above, this region reflects the parameters for which the system exhibits a stable, acceptable response. The time graph clearly shows that the plant action is well within the required specifications, i.e. no overshoot, settle time within 0.095s, and so on the required specifications, i.e. no overshoot, settle time within 0.095s, and so on.

Fig. 5 contains a time graph generated for a set of parameters, $[\omega_n, \zeta]$, selected from the white region of the map, where all rules fired, indicating that the maximum overshoot, O_v, delay time, t_d, rise time, t_r, and settle time, t_s, all exceed the limits required. This graph clearly shows that the system response fails to meet these requirements, further validating the performance map technique.

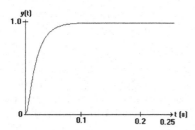

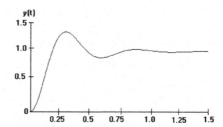

Fig. 4. Time graph of second-order linear system, with $\omega_n = 148.173$ and $\zeta = 1.429$.

Fig. 5. Time graph of second-order linear system, with $\omega_n = 11.296$ and $\zeta = 0.339$.

4   Approximation of Performance Maps

The computational overhead in the generation of a performance map is significant, with a map often requiring as many as 9×10^8 numerical integration steps [4]. In addition, it is possible that some state values required in the formulation of performance rules can be unobservable, negating their value in system evaluation. It is thus desirable develop a means whereby the number of variables required to test a system's performance is reduced. To accomplish this reduction, a rough set to rule derivation is considered [4], [10], [11]. The use of rough rules in the generation of a performance map effectively yields a rough performance map, this being an approximation of the full map (see Table 1).

Table 1. Decision Table and Derived Rules

Ov	Td	Tr	Ts	D	Derived Rules
float(3)	float(3)	float(3)	float(3)	int	
0.000	0.038	0.068	0.095	0	Ov(0.000) AND Ts(0.095) => D(0)
0.050	0.045	0.070	0.090	1	Ov(0.050) AND Ts(0.090) => D(1)
0.050	0.040	0.075	0.090	1	Ov(0.050) AND Ts(0.095) => D(1)
0.050	0.040	0.070	0.095	1	Ov(0.100) AND Ts(0.090) => D(2)
0.100	0.040	0.070	0.090	2	Ov(0.100) AND Ts(0.100) => D(3)
0.100	0.050	0.080	0.100	3	Ov(0.100) AND Ts(0.110) => D(4)
0.100	0.050	0.080	0.110	4	Ov(0.150) AND Ts(0.090) => D(2)
0.100	0.050	0.090	0.100	3	Ov(0.150) AND Ts(0.100) => D(3)
0.150	0.050	0.070	0.090	2	Ov(0.200) AND Ts(0.090) => D(2)
0.150	0.060	0.080	0.100	3	Ov(0.200) AND Ts(0.100) => D(4)
0.200	0.040	0.070	0.090	2	
0.200	0.050	0.080	0.100	4	
0.200	0.040	0.100	0.100	4	

Each tuple in Table 1 represents a set of parameter values and a decision value. This table is filled in an *approximate* manner by assigning a simple scale to reflect the levels of system performance. The scale elements [0, 4] contains decision values where 0 (zero) implies excellent performance and 4 implies very poor system performance. An excellent performance is defined to be the required performance level applied above, namely, maximum overshoot: $O_v = 0$, maximum delay: $t_d = 0.038s$, maximum rise time: $t_r = 0.068s$ and maximum settle time: $t_s = 0.095s$. The remaining performance levels were applied using entirely verbal descriptions. A set of reducts was generated next; the process of which was entirely automated through the use of the Rosetta software package [12]. The result of the reduct generation was that the measurements of delay time t_d and rise time t_r were redundant measurements and were deleted as unnecessary knowledge. The reducts are then maximum overshoot and settle time, $\{O_v, t_s\}$. Finally the set of rough rules shown in Table 1 was generated, which is easily programmed using IF-THEN-ELSE constructs for the generation of a rough performance map.

5   Simulation of Linear Plant

The derived rules in Table 1 are applied to the second-order linear plant described by (1). Derived rules lead to a "rough" performance map, which is an approximation of the full performance map shown in Fig. 3. Being an approximation, there is necessarily some loss of information, however, this loss is acceptable if the ultimate goal of the visualization is to represent the parameter values that yield the required performance. Fig. 6 contains the rough performance map generated for the second-order linear system. The black region of the map, which comprises the majority of the upper right quadrant, reflects the parameter values, $[\omega_n, \zeta]$, that yield the optimum system response.

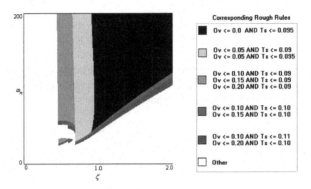

Fig. 6. Rough performance map of the 2nd order linear system

The fractal capacity dimension, also known as the box counting dimension, is based on Shannon's information theory [13] and provides a geometric measurement of an orbit in state space. The capacity dimension is based on a measurement of the number of boxes or cells, of size ε, which are required to fully cover a system trajectory on a phase plane. The capacity dimension model [14] is given in (3).

$$Dc = \lim_{\varepsilon \to 0} \frac{\log N(\varepsilon)}{\log\left(\frac{1}{\varepsilon}\right)} \tag{3}$$

where $N(\varepsilon)$ = number of cells required to fully cover the trajectory on the phase plane ε size of each cell (cell is size $\varepsilon \times \varepsilon$). The capacity dimension is used to quantify the black region of a performance map [4]. Let the map size be $n \times n$ pixels and let the cell size be one pixel. The size of each cell is s = $1/n$. Rather than count the number of cells required to cover a trajectory, count instead the number of colored black pixels. Thus, the capacity dimension of a rough or full performance map Dc_{PM} becomes (4)

$$Dc_{PM} = \lim_{\varepsilon \to 0} \frac{\log N(P_B)}{\log\left(\frac{1}{\varepsilon}\right)} = \frac{\log N(P_B)}{\log n} \tag{4}$$

The fractal or fractional capacity dimension, Dc, is employed to validate the rough where $N(P_B)$ = count of black pixels, and n = width of the performance map in pixels. Using (4), the capacity dimension Dc_{PM} for the full performance map of Fig. 3 is 1.788, while the dimension for the rough performance map of Fig. 6 is 1.794. This provides a validation that the region reflecting rough ruled optimum system performance does indeed provide a good approximation of the parameters $[\omega_n, \zeta]$ that affect the required system response.

6   Concluding Remarks

This paper has presented a method derived from a visualization technique common in the study of chaos, which can be used to visualize the effects of variations in parameters on a system's response. The performance map is generated via digital computation, and with minimal need for rigorous mathematical analysis. Rough sets present a formal and intuitive paradigm whereby a decision table can be reduced to a minimal set of rules using approximate reasoning.

References

[1] G. Julia, Memoire sur l'iteration des fonctions rationnelles.   J. de Math., vol. 8, 1918, 47-245.

[2] B.B. Mandelbrot, Fractal Geometry.   NY: John Wiley & Sons, 1982.

[3] Z. Pawlak, Z, Rough Sets: Theoretical Aspects of Reasoning About Data, Boston, MA: Kluwer Academic Publishers, 1991

[4] J.J. Alpigini, A Paradigm for the Visualization of Dynamic System Performance Using Methodologies Derived from Chaos Theory, Ph.D. Thesis, University of Wales, Swansea, UK, 1999.

[5] R.L. Devaney, Computer Experiments in Mathematics.   MA: Addison-Wesley, 1990, 83-90.

[6] D.W. Russell, J.J. Alpigini, A Visualization Tool for Knowledge Acquisition Using a Julia Set Methodology for Dynamic System Expansion, Proceedings of the 9th International Conference on Software Engineering & Knowledge Engineering (SEKE'97), Madrid, Spain, 18-20 June, 1997, 326-331.

[7] B.C. Kuo, Automatic Control Systems, Seventh Edition, Englewood Cliffs: Prentice Hall International, Inc., 1995, 385-402.

[8] J. DiStefano, A. Stubberud, I Williams, Schaum's Outline Series, Theory and Problems of Feedback and Control Systems, Second Edition, New York: McGraw-Hill, 1995, 114.

[9] J.R. Rowland, Linear Control Systems, Modeling, Analysis, and Design, New York: John Wiley & Sons, 1986, 168-169.

[10] A. Skowron and J. Stepaniuk. Decision rules based on discernability matrices and decision matrices.   In: Proc. Third Int. Workshop on Rough Sets and Soft Computing, San Jose, California, 10-12 November 1994.

[11] J.F. Peters, S. Skowron, Z. Suraj, W. Pedrycz, S. Ramanna, Approximate Real-Time Decision-Making: Concepts And Rough Fuzzy Petri Net Models, Int. Journal of Intelligent Systems, 14 (4), 805-840.

[12] Rosetta software system, http://www.idi.ntnu.no/~aleks/rosetta/

[13] C.E. Shannon, A Mathematical Theory of Communication, The Bell System Technical Journal, V 27, July, October, 1948, 379-423,623-656.

[14] J.J. Alpigini, D.W. Russell, Visualization of Control Regions for Badly Behaved Real-Time Systems, IASTED International Conference on Modeling & Simulation, Pittsburgh, USA, 13-16 May, 1998.

Approximation Spaces of Type-Free Sets

Peter Apostoli and Akira Kanda

Department of Philosophy
University of Toronto
apostoli@cs.toronto.edu

Abstract. We present an approximation space (U, R) which is an inÞnite (hyper-continuum) solution to the domain equation $U \approx C(R)$, the family of elementary subsets of U. Thus U is a universe of type-free sets and R is the relation of indiscernibility with respect to membership in other type-free sets. R thus associates a family $[u]_R$ of elementary subsets with $u \in U$, whence (U, R) induces an *generalized* approximation space $(U, c : U \to U, i : U \to U)$, where $c(u) = \cup[u]_R$ and $i(u) = \cap[u]_R$.

1 Rough Set Theory

The theory of rough sets deals with the approximation of subsets of a universe U of points by two deÞnable or observable subsets called lower and upper approximations ([9]). It extends the theory of point sets by deÞning interior and closure operators over the power set 2^U, typically those of the partition topology associated with an equivalence relation R on U. Approximately given concepts — so-called (*topological*) *rough subsets* — are identiÞed with equivalence classes of subsets under the relation of sharing a common interior and closure ([9], [12]). The study of rough sets thence passes into *abstract set theory*, which studies sets of *sets*. Cattaneo [4] extended the method of point sets to topological rough sets by lowering the type of point sets, representing concepts as points in certain ordered structures, characterizable as modal algebras[1], called abstract approximation spaces. This type lowering process, by which subsets form points, is the foundation of abstract set theory and the cornerstone of FregeÕs attempt to derive mathematics from logic ([2], [3]). We present rough set theory as an extension of the theory of *abstract* sets, and show that (on pain of RussellÕs paradox) the formation of abstract sets in set theoretic comprehension is governed by the method of upper and lower approximations.

Our development marries the fundamental notion of rough set theory — the *approximation space* (U, R) — with that of abstract set theory — a *type-lowering correspondence*[2] from 2^U to U under which subsets form points and vise-versa. The result is a *proximal Frege structure*: an approximation space placed in type-lowering correspondence with its power set. After indicating some of the foundational signiÞcance of such structures, we present a concrete example, called CFG ([1]). The elementary subsets of CFG form a *generalized* approximation space characterizable as a non Kripkean modal

---

[1] As in [13] and [14] where these spaces are called *generalized approximation spaces*.

[2] Terminology established by J. L. Bell in [2].

W. Ziarko and Y. Yao (Eds.): RSCTC 2000, LNAI 2005, pp. 98•105, 2001.
© Springer-Verlag Berlin Heidelberg 2001

algebra. Finally, we propose equivalence classes under R — the basic information granules from which all elementary subsets are composed — as concrete examples of neighborhood systems of deformable subsets as described in advanced computing [7], [8].

1.1 Approximation Spaces (Pawlak, 1982)

Let $U \neq \emptyset$ and $R \subseteq U \times U$ be an equivalence relation. Then (U, R) is said to be an *approximation space*. Pawlak [9] defines the *upper* and *lower approximations* of a set $X \subseteq U$, respectively, as:

$$Cl(X) = \{x \in U \mid (\exists y \in U)(xRy \wedge y \in X)\} = \bigcup \{[u]_R \mid [u]_R \cap X \neq \emptyset\},$$
$$Int(X) = \{x \in U \mid (\forall y \in U)(xRy \rightarrow y \in X)\} = \bigcup \{[u]_R \mid [u]_R \subseteq X\}.$$

R is interpreted as a relation of *indiscernibility* in terms of some prior family of concepts (subsets of U). R-closed subsets of U are called "elementary". Cl is a closure operator over subsets of U, while Int is a projection operator over the subsets of U which is dual to Cl under complementation. Further, $(U, \mathcal{C}(R))$ is a quasi-discrete topological space, where $\mathcal{C}(R)$ is the family of elementary subsets of U. In summary:

$$
\begin{aligned}
&i0 \ Int(X) = U - Cl(U - X) \\
&i1 \ Int(X \cap Y) = Int(X) \cap Int(Y) \\
&i2 \ Int(U) = U \\
&i3 \ Int(A) \subseteq A \\
&i4 \ Int(A) \subseteq Int(Int(A)) \\
&i5 \ Cl(X) = Int(Cl(X)).
\end{aligned}
$$

It may be convenient to break $i1$ into a conjunction of inclusions, e.g.,

$$
\begin{aligned}
&i1(a) \ Int(X) \cap Int(Y) \subseteq Int(X \cap Y) \\
&i1(b) \ Int(X \cap Y) \subseteq Int(X).
\end{aligned}
$$

$i1(b)$ yields the familiar modal Rule of Monotonicity:

$$RM : X \subseteq Y \subseteq U \Rightarrow Int(X) \subseteq Int(Y).$$

Principle $i1(a)$ is called "modal aggregation" by modal logicians and, when expressed in terms of Cl, *additivity* by algebraists.

The quasi-discrete topological spaces are *precisely* the partition topologies, i.e., spaces $(U, \mathcal{C}(R))$ with equivalence relations $R \subseteq U \times U$.

Rough Equality R induces a corresponding *higher-order* indiscernibility relation over subsets of U: X is said to be *roughly equal* to Y ("$X =_R Y$") when $Int(X) = Int(Y)$ and $Cl(X) = Cl(Y)$. Let X, Y be R closed subsets of U. Then $X =_R Y \Rightarrow X = Y$. Thus, the R-closed subsets of U are precisely defined by their respective approximation pairs. Equivalence classes $[X]_{=_R}$ are called *rough* (i.e., vague, tolerant) *subsets* of U. Alternatively, a rough subset of U can be identified with the pair of upper and lower approximations $(In(X), Cl(X))$ that bound $[X]_{=_R}$.

2   Type-Lowering Correspondences and Abstract Rough Sets

2.1   The Set as One and the Set as Many

Frege's doctrine of concepts and their extensions is an attempt to state precisely Cantor's assertion that every "consistent multitude" forms a set. For every concept $X \subseteq U$, Frege postulated an object $\widehat{X} \in U$, "the extension of X," which is an individual of U intended to *represent* X. Frege thus assumed the existence a type-lowering correspondence holding between U and its power set 2^U, i.e., a function $\widehat{\cdot} : 2^U \to U$ mapping[3] higher type entities (subsets) into lower type entities (objects). Frege attempted to govern his introduction of extensions by adopting the principle that the extensions $\widehat{X}$ and $\widehat{Y}$ are strictly identical just in case precisely the same objects (elements of U) fall under the concepts X and Y. This adoption of Basic Law V, as Frege called it, was necessary if extensions where to *represent* concepts in the sense of characterizing precisely which objects fall under them. That this "Basic Law" leads to Russell's paradox regarding the "the class of all classes which do not belong to themselves," and is indeed equivalent in second order logic to the inconsistent principle of naive comprehension, is now well appreciated.

Proximal Frege Structures   Let (U, R) be an approximation space and $\widehat{\cdot} : 2^U \to U$, $\varepsilon : U \to 2^U$ be functions. If in addition, (1) $\widehat{\cdot}$ is a *retraction*, i.e., $\widehat{\varepsilon(u)} = u$ (i.e., $\widehat{\cdot} \circ \varepsilon = 1_U$), and (2) the elementary subsets of U are precisely the $X \subseteq U$ for which $\varepsilon(\widehat{X}) = X$, then $(U, R, \widehat{\cdot}, \varepsilon)$ is called a *proximal Frege structure*. This may be summarized by the equation:

$$\mathcal{C}(R) \approx U \lhd 2^U,$$

where $\mathcal{C}(R)$ is the family of R-closed subsets of U and "$U \lhd 2^U$" indicates that $\widehat{\cdot}$ projects the power set of U onto U.

Let $\mathcal{F} = (U, R, \widehat{\cdot}, \varepsilon)$ be a proximal Frege structure. Writing "$u_1 \in u_2$" for "$u_1 \in \varepsilon(u_2)$", we thus interpret U as a universe of *type-free sets*; ε supports the relation of set membership holding between type-free sets (elements of U); $\widehat{\cdot}$ is Frege's "extension function," mapping concepts to the elements of U that serve as their extensions. $\mathcal{F}$ thus validates the Principle of Naive Comprehension

$$u \in \widehat{X} \Leftrightarrow X(u), \tag{PNC}$$

for *elementary* (R-closed) subsets X of U. When X fails to be elementary, the equivalence also fails and is replaced by a pair of approximating conditionals:

$$(1)\ u \in \widehat{Int(X)} \Rightarrow X(u);$$
$$(2)\ X(u) \Rightarrow u \in \widehat{Cl(X)}.$$

Note we use applicative grammar "$X(u)$" to indicate that u is an element of $X \subseteq U$, reserving "$\in$" for type-free membership. We will write "$\{x : X(x)\}$" to denote $\widehat{X}$ ($X \subseteq U$).

---

[3] See [3], where Bell observes that this function is a *retraction* from 2^U onto U whose section is comprised of precisely Cantor's "consistent multitudes". We characterize the latter as precisely the elementary subsets of U.

Theorem 1. *Let $x, y \in U$. Then, both*
(a) $(\forall u)(u \in x \leftrightarrow u \in y) \leftrightarrow x = y;$
(b) $(\forall u)(x \in u \leftrightarrow y \in u) \leftrightarrow R(x, y).$

Theorem 2. *(Cocchiarella 1972) There are $x, y \in U$ such that $R(x, y)$ but $x \neq y$.*

Assuming the failure of the theorem, Cocchiarella [5] derived Russell's contradiction.

The Boolean Algebra of Type-free Sets Since elements of U represent elementary subsets of U, U forms a complete Boolean algebra under the following definitions of $\cap, \cup$ and $-$: Let $X \subseteq U$, $u \in U$; then

$$\cup X =_{df} \{y : (\exists x)(X(x) \wedge y \in x)\};$$
$$\cap X =_{df} \{y : (\forall x)(X(x) \rightarrow y \in x)\};$$
$$-u =_{df} \{y : y \notin u\}.$$

Then $(\mathcal{C}(R), \cup, \cap, -, \emptyset, U)$ is isomorphic to $(U, \cup, \cap, -, \widehat{\emptyset}, \widehat{U})$ under $\widehat{\cdot} \upharpoonright \mathcal{C}(R)$, the restriction of the type-lowering retraction to elementary subsets of U. We define $u_1 \subseteq u_2$ to be $\varepsilon(u_1) \subseteq \varepsilon(u_2)$, i.e., *inclusion* of type-free sets is the partial ordering naturally associated with the Boolean algebra of type-free sets.

The Discernibility of Disjoint Sets $\mathcal{F}$ is said to have *the Dual Quine property* iff: $x \cap y = \widehat{\emptyset} \Rightarrow \neg R(x, y)$ $(x, y \in U, x \neq \widehat{\emptyset})$.

Theorem 3. *If $\mathcal{F}$ satisfies Dual Quine, then $\mathcal{F}$ provides a model of Peano arithmetic.*

(Translations of) Peano's axioms are derivable in first order logic from the PNC for elementary concepts and Dual Quine. It for this reason that proximal Frege structures provide a generalized model of granular computation.

3 $\mathcal{C}antor \, \mathcal{F}rege \, \mathcal{G}ilmore$ **Set Theory**

3.1   The Algebra of $\mathcal{G}$-sets

Theorem 4. *There is a proximal Frege structure $(M_{max}, \equiv, \widehat{\cdot}, \varepsilon)$, called $\mathcal{CFG}$, satisfying Dual Quine. M_{max} has the cardinality of the hypercontinuum.*

See the Appendix for a sketch of the domain theoretic construction of $\mathcal{CFG}$ given in [1]. Elements of M_{max} are called "$\mathcal{G}$-sets". The complete Boolean algebra

$$(M_{max}, \cup_{max}, \cap_{max}, -_{max}, \widehat{M_{max}}, \widehat{\emptyset}),$$

is called *the algebra of $\mathcal{G}$-sets*, where the subscripted "max" (usually omitted) indicate that these operations are the one naturally associated with type-free sets. We write $\mathbf{U}$ for $\widehat{M_{max}}$ and $\emptyset$ for $\widehat{\emptyset}$. Recall that $\mathcal{C}(\equiv) \approx M_{max} \lhd 2^{M_{max}}$.

Let $a \in M_{max}$; define the *outer penumbra* of a, symbolically, $\Diamond a$, to be the $\mathcal{G}$-set $\cup[a]_{\equiv}$; dually, define the *inner penumbra*, $\Box a$, to be the $\mathcal{G}$-set $\cap[a]_{\equiv}$. These operations, called the *penumbral modalities*, are interpreted using David Lewis' counterpart semantics for modal logic [6]. Your Lewisian counterpart (in a given world) is a person more qualitatively similar to you than any other object (in that world). You are necessarily (possibly) P iff all (some) of your counterparts are P. Similarly, if a and b are indiscernible $\mathcal{G}$-sets, a is more similar to b than any *other* (i.e., discernible!) $\mathcal{G}$-set. Thus we call b a *counterpart* of a whenever $a \equiv b$. Then $\Box a$ ($\Diamond a$) represents the set of $\mathcal{G}$-sets that belong to all (some) of a's counterparts. In this sense we can say that a $\mathcal{G}$-set x *necessarily* (*possibly*) belongs to a just in case x belongs to $\Box a$ ($\Diamond a$).

Plenitude

Theorem 5. *(Plenitude) Let $a, b \in M_{max}$; then $\Box a \equiv a \equiv \Diamond a$. Further, suppose $a \subseteq b$ and $a \equiv b$. Then, for all $c \in M_{max}$, $a \subseteq c \subseteq b \Rightarrow c \equiv b$.*

Corollary 1. *$([a]_\equiv, \cap_{max}, \cup_{max}, \Box a, \Diamond a)$ is a complete lattice with least (greatest) element $\Box a$ ($\Diamond a$).*

3.2   The Penumbral Algebra

We call $(M_{max}, \cup, \cap, -, \mathbf{U}, \emptyset, \Box, \Diamond)$ *the penumbral algebra*. It is a modal algebra satisfying $i0, i2, i3, i4$ and conjectured to be additive, i.e., to satisfy $i1(a)$; in addition,

$$\Diamond m \subseteq \Diamond \Box m,$$

for all $m \in M_{max}$. R. E. Jennings has shown[4] that, on pain of Russell's contradiction, Rule Monotonicity *fails*:

Theorem 6. *There are $a, b \in M_{max}$ such that $a \subseteq b \not\Rightarrow \Box a \subseteq \Box b$,*

whence the converse of [C] does *not* hold in the penumbral algebra.

Closed $\mathcal{G}$-sets, Open $\mathcal{G}$-sets   Let a be a $\mathcal{G}$-set. Then, we say a is *(penumbrally) closed* if $\Diamond a = a$; $\mathbb{C}(M_{max})$ is the set of closed $\mathcal{G}$-sets. Dually, a is *(penumbrally) open* if $\Box a = a$; $\mathbb{O}(M_{max})$ is the set of open $\mathcal{G}$-sets. By Dual Quine and Plenitude, the universe $\mathbf{U}$ and the empty set $\emptyset$ are clopen (penumbrally both open and closed). We conjecture (based upon an analogy with the Rice-Shapiro and Myhill-Shepardson theorems in Recursion Theory) that they are the *only* clopen $\mathcal{G}$-sets.

A Generalized Approximation Space   Since elements of M_{max} represent elementary subsets of M_{max}, $\Box : M_{max} \rightarrow \mathbb{O}(M_{max})$ and $\Diamond : M_{max} \rightarrow \mathbb{C}(M_{max})$ can be interpreted as approximation operations over $\mathcal{C}(\equiv)$. Hence, the penumbral algebra is a generalized approximation space in which elementary subsets are approximated from above by penumbrally closed elementary subsets and from below by penumbrally open elementary subsets: Let $i, c : \mathcal{C}(\equiv) \rightarrow \mathcal{C}(\equiv)$ be maps given by

$$i : X \mapsto \varepsilon(\Box \widehat{X})$$
$$c : X \mapsto \varepsilon(\Diamond \widehat{X}) \qquad (X \in \mathcal{C}(\equiv)).$$

---

[4] In private correspondence.

Then $(\mathcal{C}(\equiv), i, c)$ is an example of a generalized approximation space in *almost* the sense of Y. Y. Yao [14], since it satisfies $i0$, $i2 - i4$, respectively, but fails $i1(b)$. In addition, we conjectured that $(\mathcal{C}(\equiv), i, c)$ is additive, i.e., satisfies $i1(a)$. It is called a *penumbral approximation space*. These approximation maps may be extended to the *full* power set $2^{M_{max}}$, yielding a *non-additive* (see [1]), non-monotonic, generalized approximation space $(2^{M_{max}}, i, c)$ satisfying $i0$ and $i2 - i4$.

Symmetries of the Algebra of Counterparts Let $a \in M_{max}$. The *penumbral border* of a, $\mathbf{B}(a)$, is the $\mathcal{G}$-set $\Diamond a -_{max} \Box a$. Note that $\mathbf{B}(\emptyset) = \mathbf{B}(\mathbf{U}) = \emptyset$. We define the *maximum deformation* of a, a^*, to be the $\mathcal{G}$-set:

$$(a \cup \{x \in \mathbf{B}(a) : x \notin a\}) - \{x \in \mathbf{B}(a) : x \in a\}.$$

By Plenitude, $a^* \equiv a$.

Theorem 7. $([a]_\equiv, \cap_{max}, \cup_{max}, ^*, \Box a, \Diamond a)$ *is a complete Boolean algebra with* * *as complementation. It is called* the algebra of counterparts *of a.*

More generally, let $X \subseteq \mathbf{B}(a)$. We define the X-*transform* of a to be the $\mathcal{G}$-set: $(a \cup \{x \in X : x \notin a\}) - \{x \in X : x \in a\}$. Again, by Plenitude, the X-transform of a belongs to $[a]_\equiv$. Define $f_X : [a]_\equiv \to [a]_\equiv$ by: $f_X(b) =$ the X-transform of b $(b \in [a]_\equiv)$. f_X is called a *deformation* of $[a]_\equiv$. Let $\mathcal{G}_a$ be the family of all f_X for $X \subseteq \mathbf{B}(a)$, i.e., the family of all deformations of a. Note that $f_X(f_X(b)) = b$ $(b \in [a]_\equiv)$.

Lemma 1. $\mathcal{G}_a$ *is a group under the composition of deformations, with $f_\emptyset$ the identity deformation and each deformation its own group inverse.*

Since (trivially) each deformation f_X in $\mathcal{G}_a$ maps congruent (indiscernible) $\mathcal{G}$-sets to congruent $\mathcal{G}$-sets, $\mathcal{G}_a$ may be regarded as a *symmetry group*, every element of which has order 2. Since f_X preserves maximal continuous approximations (see Appendix; in detail, [1]), $\mathcal{G}_a$ is a group of continuous transformations of $\mathcal{G}$-sets.

Neighborhood Systems Neighborhood systems express the semantics of "nearby" or proximity. T. Y Lin [7] has proposed neighborhood systems as a generalization of rough set theory which is useful in characterizing the notion of a "tolerant subset" in soft computing. Let U be a universe of discourse and $u \in U$; a *neighborhood* of u is simply a non-empty subset X of U. A *neighborhood system of u* is family of neighborhoods of u, i.e., simply a family of subsets of U. A *neighborhood system of U* is the assignment of a neighborhood system of u to each $u \in U$.

Admissible Families of Elastically Deformable Subsets The notion of a neighborhood of a *point* can be "lifted" to the notion of the neighborhood of a *subset*. For example, two subsets of the real numbers can be said to be "near" if they differ by a set of measure 0. Lin has proposed interpreting the neighborhood system of a subset as an admissible family of "elastically deformable" characteristic functions:

> *A real world fuzzy set should allow a small amount of perturbation, so it should have an elastic membership function.* [8], p. 1 *An elastic membership function*

can tolerate a small amount of continuous stretching with a limited number of broken points. . . . We need a family of membership functions to express the stretching of a membership function. [7], p. 13. *Mathematically, such an elastic membership function can be expressed by a highly structured subset of the membership function space.* [8], p. 1.

In [7], topological and measure theoretic means, presupposing the system of real numbers, are deployed to make this notion of a "small" amount of perturbation precise. Since indiscernible $\mathcal{G}$-sets transform continuously into one another by swapping boundary elements, they are *almost* co-extensive, disagreeing only upon "exceptional points" or singularities (see [1]). The transformation is *elastic* in that the number of broken points is negligibly small. We therefore offer the algebra of counterparts of a given $\mathcal{G}$-set as a concrete example of an admissible family of "elastically deformable" characteristic functions, i.e., as a tolerant subset in the sense of Lin.

4   Appendix: A Sketch of the Construction of $\mathcal{CFG}$

First, using the theory of SFP objects from Domain Theory [11], we construct a continuum reflexive structure D_∞ satisfying the recursive equation $D_\infty \approx [D_\infty \to T]$, where $\approx$ is order isomorphism, T is the complete partial order (cpo) of truth values

$$
\underset{\nwarrow \qquad \nearrow}{\underline{true} \qquad \underline{false}}
$$
$$
\bot \, ,
$$

(here $\bot$ represents the zero information state, i.e., a classical truth value gap) and $[D_\infty \to T_\infty]$ is the space of all continuous (limit preserving) functions from D_∞ to T, under the information cpo naturally associated with nested partial functions: $f \leq g$ iff $f(d) \leq_T g(d)$ $(d \in D_\infty)$.

The Kleene Strong three valued truth functional connectives are monotone (order preserving) with respect to T, and are trivially continuous. *Unfortunately, $[D_\infty \to T]$ is not closed under the infinitary logical operations* such as arbitrary intersections, whence D_∞ fails to provide a model of partial set theory combining unrestricted abstraction with universal quantification. This is the rock upon which Church's quantified λ-calculus foundered (essentially, Russell's paradox).

However, the characteristic property of SFP objects ensures that each *monotone* function $f : D_\infty \to T$ is maximally approximated by a unique *continuous* function c_f in $[D_\infty \to T]$, whence c_f in D_∞ under representation.

Next, the space M of monotone functions from D_∞ to T is a solution for the reflexive equation $M \approx \langle M \to T \rangle$ where $\langle M \to T \rangle$ is the space of all "hyper continuous" functions from M to T. A monotone function $f : M \to T$ is said to be *hyper-continuous* just in case $c_x = c_y \Rightarrow f(x) = f(y)$ $(x, y \in M)$. The elements of $\langle M \to T \rangle$ correspond to the subsets of M which are closed under the relation of sharing a common maximal continuous approximation. Indeed, $\equiv$ is the restriction of this equivalence relation to maximal elements of M.

We observe that M is closed under arbitrary intersections and unions; more generally it provides a first order model for partial set theory combining unrestricted abstraction with full quantification.

Finally, since $(M, \leq_{info})$ is a complete partial order, M is closed under least upper bounds of $\leq_{info}$-chains. Hence, by Zorn's Lemma, there are $\leq_{info}$-maximal elements of M. Let M_{max} be the set of maximal elements of M. Since

$$(\forall x, y \in M_{max})[x \in y \vee x \notin y],$$

M_{max} is a *classical subuniverse* of M.

Theorem 8. *Let* $a, b \in M_{max}$. *Then* $a \equiv b$ *iff* $c_a = c_b$.

References

1. Apostoli, P., Kanda, A., *Parts of the Continuum: towards a modern ontology of science*, forthcoming in The Poznan Studies in the Philosophy of Science and the Humanities (ed. Leszek Nowak).
2. Bell J. L., "Type Reducing Correspondences and Well-Orderings: Frege's and Zermelo's Constructions Re-examined," J. of Symbolic Logic, **60**, (1995).
3. ———, "The Set as One and the Set as Many," J. of Symbolic Logic, (1999).
4. Cattaneo, G., "Abstract Approximation Spaces for Rough Theories," *Rough Sets in Knowledge Discovery: Methodology and Applications*, eds. L. Polkowski, A. Skowron. Studies in Fuzziness and Soft Computing. Ed.: J. Kacprzyk. Vol. 18, Springer (1998).
5. Cocchiarella, N., "Wither Russell's Paradox of Predication?", Nous, (1972).
6. Lewis, D., "Counterpart Theory and Quantified Modal Logic," J. of Philosophy 65 (1968), pp 113-26. Reprinted in Michael J. Loux [ed.] *The Possible and the Actual* (1979; Cornell U.P.)
7. Lin, T. Y., "Neighborhood Systems: A Qualitative Theory for Fuzzy and Rough Sets," *Advances in Machine Intelligence and Soft Computing*, Volume IV. Ed. Paul Wang, (1997) 132-155.
8. ———, "A Set Theory for Soft Computing. A unified View of Fuzzy Sets via Neighborhoods," *Proceedings of 1996 IEEE International Conference on Fuzzy Systems*, New Orleans, Louisiana, September 8-11, 1996, 1140-1146.
9. Pawlak Z., "Rough Sets," International Journal of Computer and Information Sciences, **11**, 341-350, (1982).
10. ———, "Rough Set Elements," *Rough Sets in Knowledge Discovery: Methodology and Applications*, eds. L. Polkowski, A. Skowron. Studies in Fuzziness and Soft Computing. Ed.: J. Kacprzyk. Vol. 18, Springer (1998).
11. Plotkin G., "A Power Domain Construction," SIAM Journal on Computing, **5**, 452-487, (1976).
12. Wiweger, Antoni, "On Topological Rough Sets," Bulletin of the Polish Academy of Sciences, Mathematics, Vol 37, No. 16, (1989).
13. Yao, Y. Y., "Constructive and Algebraic Methods of the Theory of Rough Sets," Journal of Information Sciences 109, 21 - 47 (1998).
14. ———, "On Generalizing Pawlak Approximation Operators," eds. L. Polkowski and A. Skowron, RSCTC'98, LNAI 1414, pp. 289 - 307, (1998).

RSES and RSESlib - A Collection of Tools for Rough Set Computations

Jan G. Bazan[1] and Marcin Szczuka[2]

[1] Institute of Mathematics, Pedagogical University of Rzeszów
Rejtana 16A, 35-310 Rzeszów, Poland
e-mail: `bazan@univ.rzeszow.pl`
[2] Institute of Mathematics, Warsaw University
Banacha Str. 2, 02-097, Warsaw, Poland
e-mail: `szczuka@mimuw.edu.pl`

Abstract. Rough Set Exploration System - a set of software tools featuring a library of methods and a graphical user interface is presented. Methods, features and abilities of the implemented software are discussed and illustrated with a case study in data analysis.

1 Introduction

Research in decision support systems, classification algorithms in particular those concerned with application of rough sets requires experimental verification. At certain point it is no longer possible to perform every single experiment using software designed for a single purpose. To be able to make thorough, multidirectional practical investigations one have to posess an inventory of software tools that automatise basic operations, so it is possible to focus on the most essential matters. That was the idea behind creation of Rough Set Exploration System, further referred as RSES for short.

First version of RSES and the library RSESlib was released several years ago. The RSESlib is also used in computational kernel of ROSETTA - an advanced system for data analysis (see [16]) constructed at NTNU (Norway) which contributed a lot to RSES development and gained wide recognition. Comparison with other classification systems (see [11]) proves its value.

The RSES software and its computational kernel - the new RSESlib 2.0 library maintains all advantages of previous version. The algorithms from the first incarnation of library are now re-mastered to provide better flexibility, extended functionality and ability to process massive data sets. New algorithms added to the library reflect the current state of our research in classification methods originating in rough sets theory.

The library of functions is not sufficient as an answer to experimenters' demand for helpful tool. Therefore the RSES user interface was constructed. This interface allows to use RSESlib interactively.

W. Ziarko and Y. Yao (Eds.): RSCTC 2000, LNAI 2005, pp. 106–113, 2001.

2   Basic notions

In order to provide clear description further in the paper we bring here some essential definitions

Information system ([12]) is a pair of the form $\mathbf{A} = (U, A)$ where U is a *universe* of *objects* and $A = (a_1, ..., a_m)$ is a set of *attributes* i.e. mappings of the form $a_i : U \rightarrow V_a$, where V_a is called *value set* of the attribute a_i. The decision table is also a pair of the form $\mathbf{A} = (U, A \cup \{d\})$ where the major feature that is different from the information system is the distinguished attribute d. We will further assume that the set of decision values is finite. The $i-$th *decision class* is a set of objects $C_i = \{o \in U : d(o) = d_i\}$, where d_i is the $i-$th decision value taken from decision value set $V_d = \{d_1, ..., d_{rank(d)}\}$

For any subset of attributes $B \subset A$ *indiscernibility relation* $IND(B)$ is defined as follows:

$$xIND(B)y \Leftrightarrow \forall_{a \in B} a(x) = a(y) \tag{1}$$

where $x, y \in U$.

Having indiscernibility relation we may define the notion of reduct. $B \subset A$ is a *reduct* of information system if $IND(B) = IND(A)$ and no proper subset of B has this property. In case of decision tables the *decision reduct* is a set $B \subset A$ of attributes such that it cannot be further reduced and $IND(B) \subset IND(d)$.

Decision rule is a formula of the form

$$(a_{i_1} = v_1) \wedge ... \wedge (a_{i_k} = v_k) \Rightarrow d = v_d \tag{2}$$

where $1 \leq i_1 < ... < i_k \leq m$, $v_i \in V_{a_i}$. Atomic subformulae $(a_{i_1} = v_1)$ are called *conditions.* We say that rule r is *applicable* to object, or alternatively, the object *matches* rule, if its attribute values satisfy the premise of the rule. *Support* denoted as $Supp_{\mathbf{A}}(r)$ is equal to the number of objects from $\mathbf{A}$ for which rule r applies correctly $Match_{\mathbf{A}}(r)$ is the number of objects in $\mathbf{A}$ for which rule r applies in general. Analogously the notion of matching set for a rule or collection of rules may be introduced (see [2], [4]).

By *cut* for an attribute $a_i \in A$, such that V_{a_i} is an ordered set we will denote a value $c \in V_{a_i}$. With the use of cut we may replace original attribute a_i with new, binary attribute which tells as whether actual attribute value for an object is greater or lower than c (more in [8]).

Template of $\mathbf{A}$ is a propositional formula $\bigwedge(a_i = v_i)$ where $a_i \in A$ and $v_i \in V_{a_i}$. A generalised template is the formula of the form $\bigwedge(a_i \in T_i)$ where $T_i \subset V_{a_i}$. An object *satisfies* (matches) a template if for every attribute a_i occurring in the template the value of this attribute on considered object is equal to v_i (belongs to T_i in case of generalised template). The template induces in natural way the split of original information system into two distinct subtables containing objects that do or do not satisfy the template, respectively. Decomposition tree is a binary tree, whose every internal node is labeled by some template and external node (leaf) is associated with a set of objects matching all templates in a path from the root to a given leaf (see [7]).

3   The RSES library v. 2.0

The RSES library (RSESlib) is constructed according to the principles of object oriented programming. The programming language used in implementation is Microsoft Visual C++ compliant with ANSI/ISO C++ (ISO/IEC 14882 standard).

The algorithms that have been implemented in the RSES library fall into two general categories.

First category gathers the algorithms aimed at management and edition of data structures that are present in the library.

The algorithms for performing Rough Set theory based operations on data constitute the second, most essential kind of tools implemented inside RSES library. To give the idea what apparatus is given to the user we describe shortly the most important algorithms.

Reduction algorithms i.e algorithms allowing calculation of the collections of reducts for a given information system (decision table). The exhaustive algorithm for calculation of all reducts is present, however such operation may be time-consuming due to computational complexity of such task (see [13]). Therefore approximate and heuristic solutions such as genetic or Jonhson algorithms were implemented (see [14], [6] for details). The library methods for calculation of reducts allow setting initial conditions for number of reducts to be calculated, required accuracy, coverage and so on. Basing on calculated reduct it is possible to calculate decision rules (see [4]). Procedures for rule calculation allow user to determine some crucial constrains for the set of decision rules. Rules received are accompanied with several coefficients that are further used while the rules are being applied to the set of objects (see [3], [2]). In connection with algorithms for reduct/rule calculation appear the subclass of algorithms allowing shortening of rules and reducts with respect to different requirements (see [3]).

Discretisation algorithms allow to find cuts for attributes. In this way initial decision table is converted to one described with less complex, binary attribute without lose of information about discernibility of objects (see [10], [8], [2]).

Template generation algorithms provide means for calculation of templates and generalised templates. Placed side by side with template generation are the procedures for inducing table decomposition trees (see [7] and [9]).

Classification algorithms used for establishing decision value with use of decision rules and/or templates. Operations for voting among rules with use of different schemes fall into this category (see [3], [9], [4], [2]).

During operation certain functions belonging to RSESlib may read and write information to/from files. Most of the files that can be read or written are regular ASCII text files. They particular sub-types can be distinguished by reviewing the contents or identifying file extensions.

4   The RSES GUI.

To simplify the use of RSES library and make it more intuitive a graphical user interface was constructed. This interface allows interaction with library methods in two modes. First is directed towards ease of use and visual representation of workflow. The other is intended to provide tools for construction of scripts that use RSES functionality

4.1   The project interface.

Project interface window (see Figure 1) consists of two parts. Upper part is the project workspace where icons representing objects occurring during our computation are presented. Lower part is dedicated to messages, status reports, errors and warnings produced during operations. It was designers intention to

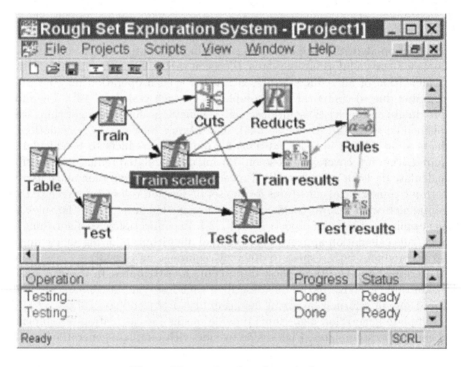

Fig. 1. The project interface window.

simplify the operations on data within project. Therefore, the entities appearing in the process of rough set based computation are represented in the form of icons placed in the upper part of workplace. Such an icon is created every time the data (table, reducts, rules,...) is loaded from the file. User can also place an empty object in the workplace and further fill it with results of operation performed on other objects. The objects that may exist in the workplace are: decision table, collection of reducts, set of rules, decomposition tree, set of cuts and collection of results. Every object (icon) appearing in the project have a set of actions connected with it. By right-clicking on the object the user invokes a context menu for that object. It is also possible to call the necessary action form general pull-down program menu in the main window. Menu choices allow to view and edit objects as well as make them input to some computation. In many cases choice of some option from context menu will cause a new dialog box to open. In this dialog box user can set values of coefficients used in desired calculation, in particular, designate the variable which will store the results of invoked operation. If the operation performed on the object leads to creation of new object or modification of existing one then such a new object is connected with edge originating in object (or objects) which contributed to its current state. Setting of arrows connecting icons in the workspace changes dynamically as new operations are being performed.

The entire project can be saved to file on disk to preserve results and information about current setting of coefficients. That also allows to re-create the entire work on other computer or with other data.

4.2   Scripting interface.

In case we have to perform many experiments with different parameter settings and changing data it is more convenient to plan such a set of operations in advance and then let computer calculate. The idea of simplifying the preparation and execution of compound experiments drove the creation of RSES scripting mechanism.

The mechanism for performing script-based operations with use of RSES library components consists of three major parts. The scripting interface is the part visible to user during script preparation. Other two are behind the scenes and perform simple syntax checking and script execution. We will not describe checking and executing in greater detail.

The user interface for writing RSES based scripts is quite simple. Main window is split into two parts of which upper contains workplace where scripts are being edited and lower contains messages generated during script execution (Figure 2).

The RSES scripting language constructs available to user are:

- Variables. Any variable is inserted to script with predefined type. The type may be either standard (e. g. integer, real) or RSES-specific.
- Functions. User can use all the functions from RSES armory as well as standard arithmetic operations. Function in script always returns a value.
- Procedures. The procedures from RSES library, unlike functions, may not return a value. The procedures correspond to major operations such as: reduct calculation, rule shortening, loading and saving objects.
- Conditional expressions. The expressions of the form **If ... then ... else** may be used. The user defines condition with use of standard operations.
- Loops. Simple loop can be used within RSES script. The user is required to designate loop control variable and set looping parameters.

While preparing a script the user may not freely edit it. He/she can only insert or remove one of the constructs mentioned above using context menu which

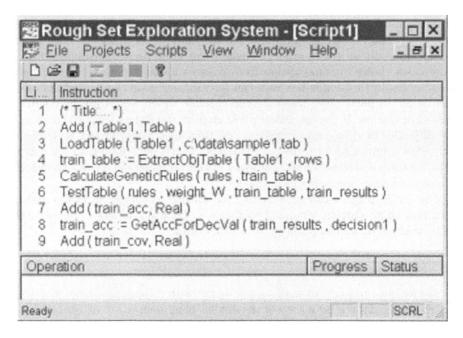

Fig. 2. The script interface window.

appears after right-clicking on the line in the script where insertion/removal is due to occur. The new construct may also be inserted with use of toolbar. The operation of inserting the new construct involves choosing the required operation name and setting all required values. All this is done by easy point-and-click operations within appropriate dialog box supported by pull-down lists of available names. The edition operations are monitored by checker to avoid obvious errors.

Once the script edition is finished it can be executed with menu command **Run**. Before execution the syntax is checked once again. The behaviour of currently running script may be seen in the lower part of interface window.

5   Case study - decomposition by template tree

As already mentioned, the ability of dealing with large data sets is one of key new features of RSESlib 2.0. To deal with such a massive data we use decomposition based on template-induced trees.

Decomposition is a problem of partitioning a large data table into smaller ones. One can use templates extracted from data to partition data table into blocks of objects with common features. We consider here decomposition schemes based on a template tree (see [7]). The main goal of this method is to construct a decomposition tree. Let **A** be a decision table. The algorithm for decomposition tree construction can be presented as follows:

Algorithm 1 *Decomposition by template tree* (see [7])
Step 1 Find the best template T in **A**.
Step 2 Divide **A** onto two subtables: $\mathbf{A}_1$ containing all objects satisfying **T**
and $\mathbf{A}_2 = \mathbf{A} - \mathbf{A}_1$.
Step 3 If obtained subtables are of acceptable size (in the sense of rough set methods)
then stop
else repeat 1-3 for all "too large" subtables.

This algorithm produces a binary tree of subtables with corresponding sets of decision rules for subtables in the leaves of the tree.

The decision tree produced by algorithm presented below can be used to classify a new case to proper decision class. Suppose we have a binary decomposition tree. Let u by a new object and $\mathbf{A}(T)$ be a subtable containing all objects matching template T. We classify object u starting from the root of the tree as follows:

Algorithm 2 *Classification by template tree* (see [7])
Step 1 If u matches template T found for **A**
then: go to subtree related to $\mathbf{A}(T)$
else: go to subtree related to $\mathbf{A}(\neg T)$.
Step 2 If u is at the leaf of the tree then go to 3
else: repeat 1-2 substituting $\mathbf{A}(T)$ (or $\mathbf{A}(\neg T)$) for **A**.
Step 3 Classify u using decision rules for subtable attached to the leaf

This algorithm uses a binary decision tree, however it should not be mistaken for C4.5, ID3. As we told before, in our experiments (see Section 5.1), a rough set methods have been used for classifying algorithm construction in leaves of the decomposition tree (see [2] and [3] for more details).

5.1   Experiments with Forest CoverType data

The Forest CoverType data used in our experiments were obtained from US Forest Service (USFS) Region 2 Resource Information System (RIS) data (see [5] [17]). Data is in rectangle form with 581012 rows and 56 columns (55 + decision). There are 7 decision values. This data has been studied before using Neural Networks and Discriminant Analysis Methods (see [5]).

The original Forest CoverType data was divided into a training set (11340 objects), a validation set (3780 objects) and a testing set (565892 objects) (see [5]). In our experiments we used algorithms presented in Section 5 that are implemented in RSESlib. The decomposition tree for data have been created only by reference to the training set. The validation set was used for adaptation of classifying algorithms which obtained in the leaves of the decomposition tree (ses [2] and [3] for more). The classification algorithm was applied to new cases from the test set. As a measure of classification success we use *accuracy* (see e.g. [7], [5]). The accuracy we define as the ratio of the number of properly classified new cases to the total number of new cases. Three different classification systems applied to the Forest CoverType data have given accuracy of 0.70 (Neural Network), 0.58 (Discriminant Analysis) and 0.73(RSES).

Acknowledgement: In the first place the special tribute should be paid to Professor Andrzej Skowron who, for all these years, is the *spiritus movens* of RSES evolution. We want to stress our gratitude to colleagues who greatly contributed providing their expertise: Adam Cykier, Nguyen Sinh Hoa, Nguyen Hung Son, Jakub Wróblewski, Piotr Synak, Aleksander Ørn.

Development of our software was significantly supported by ESPRIT project 20288, KBN grant 8T11C02511 and Wallenberg Foundation - WITAS project.

References

1. Skowron A., Polkowski L.(ed.), Rough Sets in Knowledge Discovery 1 & 2, Physica Verlag, Heidelberg, 1998
2. Bazan J., A Comparison of Dynamic and non-Dynamic Rough Set Methods for Extracting Laws from Decision Tables, In [1], Vol. 1, pp. 321-365
3. Bazan J., Approximate reasoning methods for synthesis of decision algorithms (in Polish), Ph. D. Thesis, Department of Math., Comp. Sci. and Mechanics, Warsaw University, Warsaw, 1998
4. Bazan J., Son H. Nguyen, Trung T. Nguyen, Skowron A. and J. Stepaniuk, Decision rules synthesis for object classification. In: E. Orłowska (ed.), Incomplete Information: Rough Set Analysis, Physica - Verlag, Heidelberg, 1998, pp. 23-57.
5. Blackard, J.,A., Comparison of Neural Networks and Discriminant Analysis in Predicting Forest Cover Types, Ph. D. Thesis, Department of Forest Sciences, Colorado State University. Ford Collins, Colorado, 1998.
6. Garey M., Johnson D., Computers and Intarctability: A Guide to the Theory of NP-completness, W.H. Freeman&Co., San Francisco, 1998, (twentieth print)
7. Nguyen Sinh Hoa, Data regularity analysis and applications in data mining. Ph. D. Thesis, Department of Math., Comp. Sci. and Mechanics, Warsaw University, Warsaw, 1999
8. Nguyen Sinh Hoa, Nguyen Hung Son, Discretization Methods in Data Mining, In [1], Vol. 1, pp. 451-482
9. Hoa S. Nguyen, A. Skowron and P. Synak, Discovery of data patterns with applications to decomposition and classfification problems. In [1], Vol. 2, pp. 55-97.
10. Nguyen Hung Son, Discretization of real value attributes. Boolean reasoning approach. Ph. D. Thesis, Department of Math., Comp. Sci. and Mechanics, Warsaw University, Warsaw, 1997
11. Michie D., Spiegelhalter D. J., Taylor C. C., Machine Learning, Neural and Statistical Classification, Ellis Horwood, London, 1994
12. Pawlak Z., Rough Sets: Theoretical Aspects of Reasoning about Data, Kluwer, Dordrecht, 1991
13. Rauszer C., Skowron A., The Discernibility Matrices and Functions in Information Systems, In: Słowiński R. (ed.), Intelligent Decision Support, Kluwer, Dordrecht 1992.
14. Wróblewski J., Covering with Reducts - A Fast Algorithm for Rule Generation, Proceeding of RSCTC'98, LNAI 1424, Springer Verlag, Berlin, 1998, pp. 402-407
15. Bazan J., Szczuka M., The RSES Homepage, http://alfa.mimuw.edu.pl/~rses
16. Ørn A., The ROSETTA Homepage, http://www.idi.ntnu.no/~aleks/rosetta
17. Bay, S. D. , The UCI KDD Archive, http://kdd.ics.uci.edu

An Investigation of β-Reduct Selection within the Variable Precision Rough Sets Model

Malcolm Beynon

Cardiff Business School, Cardiff University
Colum Drive, Cardiff, CF10 3EU, Wales, U.K.
BeynonMJ@cardiff.ac.uk

Abstract. The Variable Precision Rough Sets Model (VPRS) is an extension of the original Rough Set Theory. To employ VPRS analysis the decision maker (DM) needs to define satisfactory levels of quality of classification and β (confidence) value. This paper considers VPRS analysis when the DM only defines a satisfactory level of quality of classification. Two criteria for selecting a β-reduct under this condition are discussed. They include the use of permissible β intervals associated with each β-reduct. An example study is given illustrating these criteria. The study is based on US state level data concerning motor vehicle traffic fatalities.

1 Introduction

The Variable Precision Rough Sets Model (VPRS) ([11], [12]), is an extension of the original Rough Set Theory (RST) ([6], [7]). To employ VPRS analysis the decision maker (DM) needs to define satisfactory levels of quality of classification and β (confidence) value. VPRS related research papers (see [1], [5], [11], [12]) do not focus in detail on the choice of β value. There appears to be a presumption this value will be specified by the decision maker (DM). In this paper we consider VPRS analysis when only information on the satisfactory level of quality of classification is known. The lessening of *a priori* assumptions determined by the DM should allow VPRS to be more accessible for analysis of applications.

2 Preliminaries

Central to VPRS analysis is the information system, made up of objects each classified by a set of decision attributes D and characterized by a set of condition attributes C. A value denoting the nature of an attribute to an object is called a descriptor. All descriptor values are required to be in categorical form allowing for certain equivalence classes of objects to be formed. That is, condition and decision classes associated with the C and D sets of attributes respectively. These are used in RST to correctly classify objects.

W. Ziarko and Y. Yao (Eds.): RSCTC 2000, LNAI 2005, pp. 114–122, 2001.

In contrast to RST when an object is classified in VPRS there is a level of confidence (β threshold) in its correct classification. The β value represents a bound on the conditional probability of a proportion of objects in a condition class being classified to the same decision class. Following [1] the value β denotes the proportion of correct classifications, in which case the domain of β is (0.5, 1.0].

For a β value and decision class, those condition classes which have the *largest group proportion* of objects classified to the decision class is at least β, are classified to the decision class. Similarly, there are the condition classes definitely not classified since the proportion of objects in a condition class classified to the decision class does not exceed $1 - \beta$. These sets are known respectively as the β-positive and β-negative regions.[1] The set of condition classes whose proportions of objects classified to the decision class lie between these values $1 - \beta$ and β is referred to as the β-boundary region. Defining the universe U to refer to all the objects in the information system characterised by the set C (with $Z \subseteq U$ and $P \subseteq C$), then:

$$\beta\text{-positive region of the set } Z: \bigcup_{\Pr(Z|X_i) \geq \beta} \{X_i \in EC(P)\} ,$$

$$\beta\text{-negative region of the set } Z: \bigcup_{\Pr(Z|X_i) \leq 1-\beta} \{X_i \in EC(P)\} ,$$

$$\beta\text{-boundary region of the set } Z: \bigcup_{1-\beta < \Pr(Z|X_i) < \beta} \{X_i \in EC(P)\} ,$$

where $EC(\cdot)$ denotes a set of equivalence classes (in this case, condition classes based on a subset of attributes P). Having defined and computed measures relating to the ambiguity of classification, [11] define in VPRS the measure of *quality of classification*, $\gamma^\beta(P, D)$, (with $P \subseteq C$), given by

$$\gamma^\beta(P, D) = \frac{\text{card}\left(\bigcup_{\Pr(Z|X_i) \geq \beta} \{X_i \in EC(P)\} \mid Z \in EC(D)\right)}{\text{card}(U)} ,$$

for a specified value of β. The value $\gamma^\beta(P, D)$, measures the proportion of objects in U for which classification is possible at the specified value of β. This measure is used operationally to define and extract reducts. In RST a reduct is a subset of C offering the same information content as given by C and is used in the rule construction process. Formally in VPRS from [12], an approximate (probabilistic) reduct $RED^\beta(C, D)$, referred to here as a β-reduct, has the twin properties:

1. $\gamma^\beta(C, D) = \gamma^\beta(RED^\beta(C, D), D)$,
2. No proper subset of $RED^\beta(C, D)$, subject to the same β value can also give the same quality of classification.

Since more than one β-reduct may exist, a method of selecting an appropriate (optimum) β-reduct needs to be considered.

---

[1] VPRS model was extended by [3], [4] to incorporate asymmetric bounds on classification probabilities. We restrict our attention here, without loss of generality, to the original VPRS.

3   Explanation of Data

To illustrate the findings (methods of β-reduct selection) in this paper, a real world example is considered. That is, an analysis of the alcohol related traffic fatalities within the 50 US states plus the District of Columbia. More specifically the decision attribute (d_1) is the estimated percent of alcohol related traffic fatalities.[2] Table 1 gives the chosen intervals for d_1 and the number of states in each decision class.

Table 1. Intervals of d_1 defining the decision classes

Interval label	'1'	'2'	'3'
Interval, number of states	[0, 25), 6	[25, 34), 21	[34, 100], 24

To illustrate further the decision classes constructed, Fig. 1 shows geographically the relative positions of the states in each decision class.

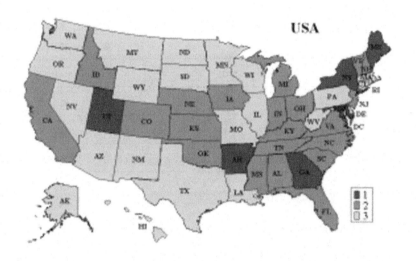

Fig. 1. Geographical illustration of decision classes

Seven condition attributes are used to characterize the motor vehicle accidents (see Table 2) and relate to demographic, behavioural and social-economic factors. They were chosen based on their previous use in the related literature (see [10]).

Each of the condition attributes shown in Table 2 is continuous in nature. Since VPRS requires the data to be in categorical form the Minimum Class Entropy method (see [2]) is used to discretize the data.[3] This is a local supervised discretization method which considers a single condition attribute at a time and utilizes the decision class values. This method requires the DM to choose the number of intervals to

---

[2] That is, motor vehicle accidents which involved a person with a Blood Alcohol Concentration level greater than or equal to 0.1 g/dl.

[3] This discretisation method has been previously used in [9].

categorize a condition attribute. In this study each condition attribute was discretized into two intervals (labeled '0' and '1').

Table 2. Descriptions of condition attributes[4]

Attribute	Description	Function
c_1	Population density	Population (1,000's)/Size of state (sq. miles)
c_2	Youth Cohort	Percentage of state's population aged 15 to 24
c_3	Income	Median household income ($)
c_4	Education	Percentage of population high school graduate or higher
c_5	Speed	Proportion of fatalities speed related
c_6	Seat belt use	Percentage use of seat belts
c_7	Driver's Intensity	Vehicle Miles Travelled (miles)/Licensed Drivers

A visual representation of the boundary values associated with the intervals '0' and '1' for each condition attribute (c_1, .., c_7) are given in Fig. 2.

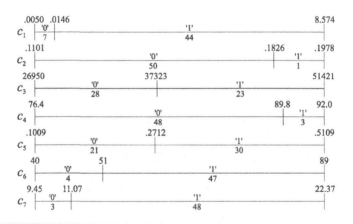

Fig. 2. Intervals (with boundary values) for the condition attributes (c_1, ..., c_7)

In Fig. 2 below each interval label the number of states whose continuous attribute value lies within that interval is given. To illustrate, the c_4 attribute (Education) is described by the intervals [76.4, 89.8) and [89.8, 92.0] which have 48 and 3 states within each interval respectively. The result of the discretization on the condition attributes is an information system denoted here as IS_{US}.

4   VPRS Analysis

The first stage of the VPRS analysis is to calculate the different levels of quality of classification, i.e. $\gamma^\beta(C, D)$ which exist. There are four different levels of $\gamma^\beta(C, D)$

---

[4] Condition attribute data was found from the U.S. Department of Transportation National Highway Safety Administration and the U.S. Census Bureau.

associated with IS_{US}, these are given in Table 3 along with the associated intervals of β values.

Table 3. Quality of classification levels and their associated β intervals

$\gamma^{\beta}(C, D)$	1	37/51	19/51	12/51
β interval	(0.5, 0.643)	[0.643, 0.667)	[0.667, 0.714)	[0.714, 1]

To illustrate the results in Table 3, $\gamma^{\beta}(C, D) = 1$ is only attained for a value of β in the interval (0.5, 0.643). The next stage is to calculate all the β-reducts associated with IS_{US} for the varying levels of $\gamma^{\beta}(C, D)$. Fig. 3 shows all the subsets of attributes with their intervals of permissible β values making them β-reducts, based on the criteria given in section 2 (and [12]).

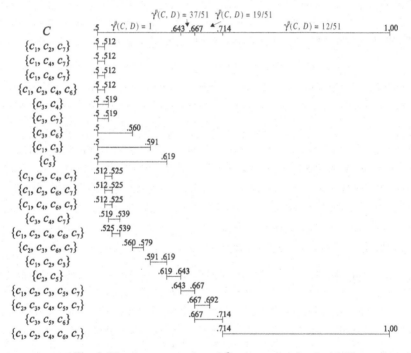

Fig. 3. Visual representation of β-reducts associated with IS_{US}

From Fig. 3, the subset of attributes $\{c_1, c_2, c_7\}$ is a β-reduct for a β value chosen in the interval 0.5 to 0.512. There are 21 β-reducts associated with IS_{US}. Each different level of $\gamma^{\beta}(C, D)$ has a number of associated β-reducts. For $\gamma^{\beta}(C, D) = 1$ there are 17 β-reducts. For the other three levels of $\gamma^{\beta}(C, D)$ there are only one or two associated β-reducts. With all β-reducts identified, the most appropriate (optimum) β-reduct is needed to be selected to continue the analysis. This would require knowledge on what the DM believes are satisfactory levels of $\gamma^{\beta}(C, D)$ and β value. If a β level is known (using Fig. 3) a vertical line can be drawn for that β value and any β-reduct can be chosen for which part (or all) of their permissible β interval lie to the right of the vertical line.[5] This is since a β value is implicitly considered the lower bound on allowed confidence.

In this study no information is given on a satisfactory β value, only a satisfactory level of $\gamma^{\beta}(C, D)$ is known. Hence only those β-reducts are considered which lie in levels of $\gamma^{\beta}(C, D)$ greater than or equal to the satisfactory level given. Under this consideration we offer two methods to the selection of β-reducts.

4.1   Method 1: Largest β Value

The first method chooses a β-reduct which has the highest permissible value of β associated with it from within the intervals of permissible β given by the satisfactory levels of $\gamma^{\beta}(C, D)$. To illustrate, in the case of $\gamma^{\beta}(C, D) = 1$ required, the β interval is (0.5, 0.643). Hence a β-reduct with a permissible β value nearest to 0.643 will be chosen. In this case there is only one β-reduct (see Fig. 3), i.e. $\{c_2, c_5\}$ would be chosen since its upper bound on permissible β is also 0.643.

Using this method there may be more than one β-reduct which have the same largest permissible β value. In this case other selection criteria will need to be used, including least number of attributes etc.. At first sight this would seem an appropriate method to find the optimum β-reduct. However this method is biased towards obtaining a β-reduct associated with the $\gamma^{\beta}(C, D)$ level nearest to the minimum level of satisfactory $\gamma^{\beta}(C, D)$ (a direct consequence of requiring largest β value possible). Where there are a number of different levels of $\gamma^{\beta}(C, D)$ within which to find an optimum β-reduct, this method allows little possibility for a β-reduct to be chosen from the higher $\gamma^{\beta}(C, D)$ levels.

4.2   Method 2: Most Similar β Interval

The second method looks for a β-reduct whose interval of permissible β values is most similar to the β interval for the $\gamma^{\beta}(C, D)$ level it is associated with. To illustrate in the case $\gamma^{\beta}(C, D) = 1$, with $\beta \in (0.5, 0.643)$ the β-reduct $\{c_5\}$ (from Fig. 3) has a permissible interval of β equal to (0.5, 0.619). This interval is most similar to the interval (0.5, 0.643) associated with $\gamma^{\beta}(C, D) = 1$. Using this method, the consideration is at a more general level, i.e. the chosen β-reduct is offering the most similar information content as C. The measure of similarity is not an absolute value but a proportion of the interval of the particular level of $\gamma^{\beta}(C, D)$.

To be more formal defining $\underline{\beta}_{C,\gamma}$ and $\overline{\beta}_{C,\gamma}$ to be the lower and upper bounds on the β interval for a level of quality of classification γ (i.e. $\gamma^{\beta}(C, D)$) based on C. A similar set of boundary points exist for each β-reduct (say R), i.e. $\underline{\beta}_{R,\gamma}$ and $\overline{\beta}_{R,\gamma}$. It

---

[5] β intervals have been considered before in [11] when defining absolute and relatively rough condition classes.

follows the respective ranges of the β intervals can be constructed, for $\gamma^\beta(C, D)$ it is $\left|\beta_{C,\gamma}\right| = \overline{\beta}_{C,\gamma} - \underline{\beta}_{C,\gamma}$. Hence the optimum β-reduct (R) to choose, based on a minimum level of quality of classification γ_m required, is the R, for which $\left|\beta_{R,\gamma}\right| / \left|\beta_{C,\gamma}\right|$ is the largest with $\gamma \geq \gamma_m$. This $\left|\beta_{R,\gamma}\right| / \left|\beta_{C,\gamma}\right|$ value represents the proportion of similarity in β interval sizes between a given β-reduct R and the $\gamma^\beta(C, D)$ level it is associated with. More than one β-reduct may be found using this method, in the same or different $\gamma^\beta(C, D)$ levels

Considering only $\gamma^\beta(C, D) = 1$ demonstrates these two methods can produce different results. Indeed the two measures can be combined to choose a β-reduct.

5   An Example of β-Reduct Selection Method 2

Using the study exposited in section 3, and presuming for example the DMs only requirement is for the VPRS analysis to give classification to at least 66.667% of the states. With at least 34 states needed to be given a classification, the optimum β-reduct can be found from the largest two levels of $\gamma^\beta(C, D)$, i.e. satisfying $\gamma^\beta(C, D) \geq$ 34/51. Inspection of the β-reducts relating to the similarity measure given in method 2 indicates $\{c_1, c_2, c_3, c_5, c_7\}$ (say R) is the optimum β-reduct. This is because it has the same permissible β interval as $\gamma^\beta(C, D) = 37/51$, i.e. $\left|\beta_{R,\gamma}\right| / \left|\beta_{C,\gamma}\right| = 1$. No other β-reduct associated with $\gamma^\beta(C, D) = 1$ or $\gamma^\beta(C, D) = 37/51$ has this level of similarity.

Following the rule construction method given in [1] the minimal set of rules for the β-reduct $\{c_1, c_2, c_3, c_5, c_7\}$ are constructed, see Table 4. It shows there are 6 rules to classify the 37 states.

Table 4. Minimal set of rules associated with β-reduct $\{c_1, c_2, c_3, c_5, c_7\}$

Rule		c_1	c_2	c_3	c_5	c_7		d_1	Strength	Correct	Proportion
1	If		1				then	1	1	1	1
2	If	1		0		1	then	2	7	5	0.714
3	If		0		0		then	2	6	4	0.667
4	If				1	1	then	3	16	12	0.75
5	If	0					then	3	6	6	1
6	If					0	then	3	1	1	1

In Table 4 the strength column relates to how many states the rule classifies (correctly or incorrectly), hence indicates 37 states are given a classification. The correct column indicates how many states are correctly classified by the rule, it totals 29 indicating 78.378% of those states given a classification were correctly classified. The final column gives the proportion of states a rule correctly classified. Fig. 4 gives a geographical illustration of the classification (and non-classification) of all the states by the rules given in Table 4.

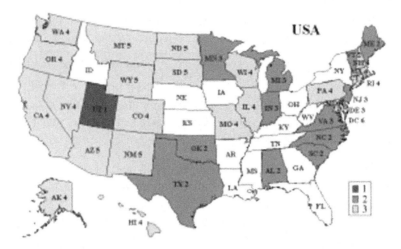

Fig. 4. Geographical representation of classification of states by rules in Table 4

In Fig. 4 each state (which was given a classification) is labelled with the actual rule it is classified by. Further analysis of the rules and their classifications in this study are beyond the focus of this paper. An interesting point to note is that the states Utah (UT) and District of Columbia (DC) are each classified by their own rule (rules 1 and 6 respectively). This follows the reasoning in related literature, which considers these to be outlier states due to their social/behavioral (UT) and demographic (DC) properties.

This study has illustrated the automation of the β-reduct selection process within VPRS analysis. The analysis of permissible intervals of β has enabled a more generalised process of β-reduct selection. However since finding reducts is an NP complete problem in RST [8] it follows the problem is more complicated in VPRS. Further aspects using the principle of permissible β intervals are defined which may help in finding efficiently the optimum β-reduct in VPRS analysis.

- *Spanning set of β-reducts:* A spanning set of β-reducts is the least number of β-reducts whose permissible β intervals cover the full interval of β values (without overlapping) associated with a particular level of $\gamma^\beta(C, D)$. In the case of IS_{US} for when $\gamma^\beta(C, D) = 1$, the β-reducts $\{c_5\}$ and $\{c_2, c_5\}$ form such a spanning set. There may exist more than one spanning set for the same level of quality of classification.
- *Root attributes:* The core of a spanning set of β-reducts forms the root. For IS_{US} when $\gamma^\beta(C, D) = 1$ the set of attributes $\{c_5\}$ is a root.
- *Onto β:* This definition relates to a single quality of classification or information system as a whole. If it is shown a β-reduct exists for any allowable β considered then the system is onto β. This onto principle may effect any discretisation of continuous attributes, since a pre-requisite may be for the discretisation to make sure the information system is onto β. That is the DM is assured they can choose any β value in their analysis and be able to find a β-reduct.

6   Conclusions

An important aspect in this paper is an understanding of the interval of permissible β values which effect the different levels of quality of classification and when a subset of attributes is a β-reduct. This understanding is of particular importance when no knowledge on a specific satisfactory β value is given by the decision maker.

While it may be more favorable for the DM to provide a β value, this paper introduces two methods of selecting a β-reduct without such a known β value. Method 2 in particularly finds the optimum β-reduct which offers an information content most similar to that of the whole set of attributes.

By relaxing some of the *a priori* information a DM needs to consider before VPRS analysis can be undertaken it is hoped the methods discussed in this paper may contribute to VPRS being used more extensively in the future.

References

1. An, A., Shan, N., Chan, C., Cercone, N., Ziarko, W.: Discovering Rules for Water Demand Prediction: An Enhanced Rough-Set Approach. Engng. Applic. Artif. Intell. 9 (1996) 45–653
2. Fayyad, U.M., Irani, K.B.: On the Handling of Continuous-Valued Attributes in Decision Tree Generation. Machine Learning 8 (1992) 87–102
3. Katzberg, J.D., Ziarko, W.: Variable Precision Rough Sets with Asymmetric Bounds. In: Ziarko, W. (ed.): Rough Sets, Fuzzy Sets and Knowledge Discovery (1993) 167–177
4. Katzberg, J.D., Ziarko, W.: Variable precision extension of rough sets. Fundamenta Informaticae 27 (1996) 155–168
5. Kryszkiewicz, M.: Maintenance of Reducts in the Variable Rough Set Model. In: Lin, T.Y. and Cercone, N. (eds.): Rough Sets and Data Mining: Analysis of Imprecise Data, Kluwer Academic Publishers, London, (1997) 355–372
6. Pawlak, Z.: Rough Sets. International Journal of Information and Computer Sciences 11 (1982) 341–356
7. Pawlak, Z.: Rough Sets - Theoretical Aspects of Reasoning About Data. Kluwer Academic Publishers London 1991
8. Skowron, A., Rauszer, C.: The Discernibility Matrices and Functions in Information Systems. In: Slowinski, R. (ed.): Intelligent Decision Support: Handbook of Applications and Advances of Rough Sets Theory, Kluwer Academic Publishers (1992) 331–362
9. Slowinski, R., Zopounidis, C., Dimitras, A.I.: Prediction of company acquisition in Greece by means of the rough set approach. European Journal of Operational Research 100 (1997) 1–15
10. Washington, S., Metarko, J., Fomumung, I., Ross, R., Julian, F., Moran, E.: An interregional comparison: fatal crashes in the Southeastern and non-Southeastern United States: preliminary findings. Accident Analysis and Prevention 31 (1999) 135–146
11. Ziarko, W.: Variable Precision Rough Set Model. Journal of Computer and System Sciences 46 (1993) 39–59
12. Ziarko, W.: Analysis of Uncertain Information in the Framework of Variable Precision Rough Sets. Foundations of Computing and Decision Sciences 18 (1993) 381–396

Application of Discernibility Tables to Calculation of Approximate Frequency Based Reducts

Maciej Borkowski[1] and Dominik Ślęzak[1,2]

[1] Institute of Mathematics, Warsaw University
Banacha 2, 02-097 Warsaw, Poland
[2] Polish-Japanese Institute of Information Technology
Koszykowa 86, 02-008 Warsaw, Poland

Abstract. We provide the unified methodology for searching for approximate decision reducts based on rough membership distributions. Presented study generalizes well known relationships between rough set reducts and boolean prime implicants.

1 Introduction

The notion of a decision reduct was developed within the rough set theory ([2]) to deal with subsets of features being appropriate for description and classification of cases within a given universe. In view of applications, the problem of finding minimal subsets (approximately) determining a specified decision attribute turned out to be crucial. Comparable to the problem of finding minimal (approximate) prime implicants for boolean functions ([1]), it was proved to be NP-hard ([8]). On the other hand, relationship to the boolean calculus provided by the discernibility representation enabled to develop efficient heuristics finding approximately optimal solutions (cf. [6], [7]).

In recent years, various approaches to approximating and generalizing the decision reduct criteria were developed (cf. [6], [9]). An important issue here is to adopt original methodology to be able to deal with indeterminism (inconsistency) in data in a flexible way. Basing on the discernibility characteristics, one can say that a reduct should be meant as an irreducible subset of conditional features, which discerns all pairs of cases behaving too differently with respect to a pre-assumed way of understanding inexact decision information. Another approach is to think about a reduct as a subset, which approximately preserves initial information induced by the whole of attributes.

In this paper we focus on frequency based tools for modeling inconsistencies (cf. [3], [4], [9], [10]). It is worth emphasizing that introduced notion of a (approximate) frequency decision reduct remains in an analogy with the notion of a *Markov boundary*, which is crucial for many applications of statistics and the theory of probability (cf. [5]).

In Section 2 we outline basic notions of rough set based approach to data mining. In Section 3 we recall the relationship between the notion of a rough set decision reduct and a boolean prime implicant. Section 4 contains basic

W. Ziarko and Y. Yao (Eds.): RSCTC 2000, LNAI 2005, pp. 123–130, 2001.

facts concerning the frequency based approach related to the notion of a rough membership function. In Section 5 we generalize the notion of a μ-decision reduct onto the parameterized class of distance based approximations. In Section 6 we illustrate the process of extracting approximate frequency discernibility tables.

2   Decision Tables and Reducts

In the rough set theory ([2]), a sample of data takes the form of an information system $\mathbb{A} = (U, A)$, where each attribute $a \in A$ is identified with function $a : U \to V_a$ from the universe of objects U into the set V_a of all possible values on a. Reasoning about data can be stated as, e.g., a classification problem, where the values of a specified decision attribute are to be predicted under information over conditions. In this case, we consider a triple $\mathbb{A} = (U, A, d)$, called a decision table, where, for the decision attribute $d \notin A$, values $v_d \in V_d$ correspond to mutually disjoint decision classes of objects.

Definition 1. *Let $\mathbb{A} = (U, A, d)$ and ordering $A = \langle a_1, \ldots, a_{|A|} \rangle$ be given. For any $B \subseteq A$, the B-ordered information function over U is defined by*

$$\overrightarrow{Inf}_B(u) = \langle a_{i_1}(u), \ldots, a_{i_{|B|}}(u) \rangle \tag{1}$$

The B-indiscernibility relation is the equivalence relation defined by

$$IND_{\mathbb{A}}(B) = \{(u, u') \in U \times U : \overrightarrow{Inf}_B(u) = \overrightarrow{Inf}_B(u')\} \tag{2}$$

Each $u \in U$ induces a B-indiscernibility class of the form

$$[u]_B = \{u' \in U : (u, u') \in IND_{\mathbb{A}}(B)\} \tag{3}$$

which can be identified with vector $\overrightarrow{Inf}_B(u)$.

Indiscernibility enables us to express global dependencies among attributes:

Definition 2. *Let $\mathbb{A} = (U, A, d)$ be given. We say that $B \subseteq A$ defines d in $\mathbb{A}$ iff*

$$IND_{\mathbb{A}}(B) \subseteq IND_{\mathbb{A}}(\{d\}) \tag{4}$$

or, equivalently, iff for any $u \in U$ $\mathbb{A}$ satisfies the object oriented rule of the form

$$\bigwedge_{a \in B} (a = a(u)) \Rightarrow (d = d(u)) \tag{5}$$

We say that $B \subseteq A$ is a decision reduct iff it defines d and none of its proper subsets does it.

Given $B \subseteq A$ which defines d, we can classify any new case $u_{new} \notin U$ by decision rules of the form (5). The only requirement is that $\mathbb{A}$ must recognize u_{new} with respect to B, i.e., the combination of values observed for u_{new} must fit vector $\overrightarrow{Inf}_B(u)$ for some $u \in U$. Expected degree of the new case recognition is the reason for searching for (approximate) decision reducts, which are of minimal complexity, understood in various ways (cf. [6], [7], [8], [9]).

3   Relationships with Boolean Reasoning

Results provided by the rough set literature state the problems of finding minimal (minimally complex) decision reducts as the NP-hard ones (cf. [6], [8], [9]). It encourages to develop various methods for the effective search for (almost) optimal attribute subsets. These methods are often based on analogies to the optimization problems known from other fields of science. For instance, let us consider the following relationship:

Proposition 1. *([8]) Let $\mathbb{A} = (U, A, d)$ be given. The set of all decision reducts for $\mathbb{A}$ is equivalent to the set of all prime implicants of the boolean discernibility function*

$$f_{\mathbb{A}}(\bar{a}_1, \ldots, \bar{a}_{|A|}) = \bigwedge_{i,j:\, c_{ij} \neq \emptyset} \bigvee_{a \in c_{ij}} \bar{a} \tag{6}$$

where variables $\bar{a}$ correspond to particular attributes $a \in A$, and where for any $i, j = 1, \ldots, |U|$

$$c_{ij} = \begin{cases} \{a \in A : a(u_i) \neq a(u_j)\} \text{ if } d(u_i) \neq d(u_j) \\ \emptyset \quad otherwise \end{cases} \tag{7}$$

The above result enables to adopt heuristics approximating the solutions of the well known problems of boolean calculus (cf. [1]) to the tasks concerning decision reducts (cf. [7], [8]). Obviously, the size of appropriately specified boolean discernibility functions influences crucially the efficiency of adopted algorithms.

Definition 3. *Let $\mathbb{A} = (U, A, d)$ be given. By the discernibility table for $\mathbb{A}$ we mean the collection of attribute subsets defined by*

$$\mathbb{T}_{\mathbb{A}} = \{T \subseteq A : T \neq \emptyset \wedge \exists_{i,j}(T = c_{ij})\} \tag{8}$$

To obtain better compression of the discernibility function, one can apply the absorption law related to the following characteristics.

Definition 4. *Let $\mathbb{A} = (U, A, d)$ be given. By the reduced discernibility table for $\mathbb{A}$ we mean the collection of attribute subsets defined by*

$$\mathbb{T}_{\mathbb{A}}^{\subsetneq} = \{T \in \mathbb{T}_{\mathbb{A}} : \neg \exists_{T' \in \mathbb{T}_{\mathbb{A}}}(T' \subsetneq T)\} \tag{9}$$

Proposition 2. *Let $\mathbb{A} = (U, A, d)$ be given. Subset $B \subseteq A$ defines d in $\mathbb{A}$ iff it intersects with each element of the reduced discernibility table $\mathbb{T}_{\mathbb{A}}^{\subsetneq}$ or, equivalently, iff it corresponds to an implicant of the boolean function*

$$g_{\mathbb{A}}(\bar{a}_1, \ldots, \bar{a}_{|A|}) = \bigwedge_{T \in \mathbb{T}_{\mathbb{A}}^{\subsetneq}} \bigvee_{a \in T} \bar{a} \tag{10}$$

Experiences concerning the performance of boolean-like calculations over discernibility structures (cf. [6], [7], [8]) suggest to pay a special attention to possibility of re-formulation of the above relationship for other types of reducts.

4   Rough Membership Distributions

In applications, we often deal with *inconsistent* decision tables $\mathbb{A} = (U, A, d)$, where there is no possibility of covering the whole of universe by the exact decision rules of the form (5). In case of such a lack of complete conditional specification of decision classes, one has to rely on a kind of representation of initial inconsistency, to be able to measure its dynamics with respect to the feature reduction. We would like to focus on the approach proposed originally in [4], resulting from adopting the frequency based calculus to rough sets.

Definition 5. *Let* $\mathbb{A} = (U, A, d)$, *linear ordering* $V_d = \langle v_1, \ldots, v_r \rangle$, $r = |V_d|$, *and* $B \subseteq A$ *be given. We call a* B-*rough membership distribution the function* $\vec{\mu}_{d/B} : U \rightarrow \triangle_{r-1}$ *defined by*[1]

$$\vec{\mu}_{d/B}(u) = \langle \mu_{d=1/B}(u), \ldots, \mu_{d=r/B}(u) \rangle \tag{11}$$

where, for $k = 1, \ldots, r$, $\mu_{d=k/B}(u) = |\{u' \in [u]_B : d(u') = v_k\}| / |[u]_B|$ *is the rough membership function (cf. [3], [4], [9], [10]) labeling* $u \in U$ *with the degree of hitting the* k-*th decision class with its* B-*indiscernibility class.*

The following is a straightforward generalization of Definition 2:

Definition 6. *Let* $\mathbb{A} = (U, A, d)$ *be given. We say that* $B \subseteq A$ μ-*defines* d *in* $\mathbb{A}$ *iff for each* $u \in U$ *we have*

$$\vec{\mu}_{d/B}(u) = \vec{\mu}_{d/A}(u) \tag{12}$$

We say that $B \subseteq A$ *is a* μ-*decision reduct for* $\mathbb{A}$ *iff it* μ-*defines* d *and none of its proper subsets does it.*

Rough membership distributions can be regarded as a frequency based source of statistical estimation of joint probabilistic distribution over the space of random variables corresponding to $A \cup \{d\}$. From this point of view, the above notion is closely related to the theory of probabilistic conditional independence (cf. [5]): Given $\mathbb{A} = (U, A, d)$, subset $B \subseteq A$ is a μ-decision reduct for $\mathbb{A}$ iff it is a *Markov boundary* of d with respect to A, i.e., iff it is an irreducible subset, which makes d probabilistically independent on the rest of A. This analogy is important for applications of both rough set and statistical techniques of data analysis.

Proposition 3. *(cf. [9]) Let* $\mathbb{A} = (U, A, d)$ *be given. Subset* $B \subseteq A$ μ-*defines* d *iff it intersects with each element of the* μ-*discernibility table defined by*

$$\mathbb{T}_{\mathbb{A}}^{\mu} = \{T \subseteq A : T \neq \emptyset \wedge \exists_{i,j}(T = c_{ij}^{\mu})\} \tag{13}$$

where

$$c_{ij}^{\mu} = \begin{cases} \{a \in A : a(u_i) \neq a(u_j)\} & \text{if } \vec{\mu}_{d/A}(u_i) \neq \vec{\mu}_{d/A}(u_j) \\ \emptyset & \text{otherwise} \end{cases} \tag{14}$$

Proposition 3 relates the task of searching for optimal μ-decision reducts to the procedure of finding minimal prime implicants, just like in the exact case before. Such a relationship enables us to design efficient algorithms approximately solving the NP-hard problem of extracting minimal Markov boundaries.

---

[1] For any $r \in \mathbb{N}$, we denote by $\triangle_{r-1}$ the (r-1)-dimensional simplex of real valued vectors $s = \langle s[1], \ldots, s[r] \rangle$ with non-negative coordinates, such that $\sum_{k=1}^{r} s[k] = 1$.

5 Distance Based Approximations of Rough Memberships

Rough membership information is highly detailed, especially useful if other types of inconsistency representation turn out to be too vague for given data. On the other hand, it is too accurate to handle dynamical changes or noises in data efficiently. To provide a more flexible framework for the attribute reduction, relaxation of criteria for being a μ-decision reduct is needed. The real valued specificity of frequencies enables us to introduce the whole class of intuitive approximations parameterized by the choice of: (1) the way of measuring the distance between distributions, and (2) thresholds up to which we agree to regard close states as practically indistinguishable:

Definition 7. *(cf. [9]) Let $r \in \mathbb{N}$ be given. We say that $\varrho : \triangle_{r-1}^2 \rightarrow [0,1]$ is a normalized distance measure iff for each $s, s', s'' \in \triangle_{r-1}$ we have*

$$\varrho(s, s') = 0 \Leftrightarrow s = s' \qquad \varrho(s, s'') \leq \varrho(s, s') + \varrho(s', s'')$$

$$\varrho(s, s') = \varrho(s', s) \qquad \varrho(s, s') = 1 \Leftrightarrow \exists_{k \neq l}(s[k] = s'[l] = 1) \tag{15}$$

Definition 8. *(cf. [9]) Let $\mathbb{A} = (U, A, d)$, $\varrho : \triangle_{r-1}^2 \rightarrow [0,1]$, $r = |V_d|$, and $\varepsilon \in [0,1)$ be given. We say that $B \subseteq A$ (ϱ, ε)-approximately μ-defines d iff for any $u \in U$*

$$\varrho(\overrightarrow{\mu}_{d/B}(u), \overrightarrow{\mu}_{d/A}(u)) \leq \varepsilon \tag{16}$$

We say that $B \subseteq A$ is a (ϱ, ε)-approximate μ-decision reduct iff it μ-defines d (ϱ, ε)-approximately and none of its proper subsets does it.

Proposition 4. *(cf. [9]) Let $\varrho : \triangle_{r-1}^2 \rightarrow [0,1]$ satisfying (15) and $\varepsilon \in [0,1)$ be given. Then, the problem of finding minimal (ϱ, ε)-approximate μ-decision reduct is NP-hard.*

The above result states that for any reasonable way of approximating conditions of Definition 5 we cannot avoid potentially high computational complexity of the optimal feature reduction process. Still, the variety of possible choices of approximation thresholds and distances enables us to fit data better, by an appropriate tuning. In practice it is more handful to operate with an easily parameterized class of normalized distance measures. Let us consider the following:

Definition 9. *Let $x \in [1, +\infty)$ and $r \in \mathbb{N}$ be given. The normalized x-distance measure is the function $x : \triangle_{r-1}^2 \rightarrow [0,1]$ defined by formula*

$$x(s, s') = \left(\frac{1}{2} \sum_{k=1}^{r} |s[k] - s'[k]|^x \right)^{1/x} \tag{17}$$

One can see that for any $x \in [1, +\infty)$, function (17) satisfies conditions (15). The crucial property of (x, ε)-approximations is the following:

Proposition 5. *Let* $\mathbb{A} = (U, A, d)$, $x \in [1, +\infty)$, $\varepsilon \in [0, 1)$ *and* $B \subseteq A$ *be given. If* B *intersects with each element of the* (x, ε)*-discernibility table*

$$\mathbb{T}_{\mathbb{A}}^{x,\varepsilon} = \{T \subseteq A : T \neq \emptyset \wedge \exists_{i,j}(T = c_{ij}^{x,\varepsilon})\} \tag{18}$$

where

$$c_{ij}^{x,\varepsilon} = \begin{cases} \{a \in A : a(u_i) \neq a(u_j)\} & \text{if } x(\overrightarrow{\mu}_{d/A}(u_i), \overrightarrow{\mu}_{d/A}(u_j)) > \varepsilon \\ \emptyset & \text{otherwise} \end{cases} \tag{19}$$

then it (x, ε)*-approximately* μ*-defines* d *in* $\mathbb{A}$. *If* B *does not intersect with some of elements of the above table, then it cannot* (x, ε')*-approximately* μ*-define* d, *for any* $\varepsilon' \leq \varepsilon/2$.

6   Examples of Discernibility Tables

Proposition 5 provides us with the unified methodology of calculating approximate distance based reducts. We can keep using the procedure introduced for non-approximate decision reducts:

- For $\varepsilon \in [0, 1)$, $x \in [1, +\infty)$, construct the reduced (by absorption) $\mathbb{T}_{\mathbb{A}}^{x,\varepsilon}$;
- Find prime implicants for the corresponding (x, ε)-discernibility function.

It leads to a conclusion that one can apply well known discernibility based algorithms for the decision reduct optimization (cf. [6]) to searching for various types of approximate reducts. Moreover, an appropriate choice of approximation parameters can speed up calculations by reducing the size of a discernibility structure.

For an illustration, let us consider the exemplary decision table in Fig. 1. Since it is enough to focus on (x, ε)-discernibility sets over pairs of objects discernible by A, we present our table in the probabilistic form (cf. [10]), where each record corresponds to an element $u^* \in U/\mathbb{A}$ of the set of $IND_{\mathbb{A}}(A)$-classes.

$U/\mathbb{A}$	$\|[u_i^*]_A\|$	a_1	a_2	a_3	a_4	a_5	$\mu_{d=1/A}(u_i^*)$	$\mu_{d=2/A}(u_i^*)$	$\mu_{d=3/A}(u_i^*)$
u_1^*	10	1	1	0	1	2	0.1	0.5	0.4
u_2^*	10	2	1	1	0	2	1.0	0.0	0.0
u_3^*	10	2	2	2	1	1	0.2	0.2	0.6
u_4^*	10	0	1	2	2	2	0.8	0.1	0.1
u_5^*	10	0	0	0	2	2	0.4	0.2	0.4
u_6^*	10	1	2	0	0	2	0.1	0.2	0.7

Fig. 1. The probabilistic table of A-indiscernibility classes (6 classes) labeled with their: **(1)** object supports (each supported by 10 objects); **(2)** A-ordered information vectors (5 conditional attributes); **(3)** μ-decision distributions (3 decision classes).

In Fig. 2 we present the sizes of reduced (x, ε)-discernibility tables, obtained for constant $x = 2$ under different choices of $\varepsilon \in [0, 1)$. The applied procedure of their extraction looks as follows:

– Within the loop over $1 \leq i < j \leq |U/A|$, find pairs $\langle u_i^*, u_j^* \rangle$ corresponding to distributions remaining too far to each other in terms of inequality

$$x(\overrightarrow{\mu}_{d/A}(u_i^*), \overrightarrow{\mu}_{d/A}(u_j^*)) > \varepsilon \qquad (20)$$

– Simultaneously, store corresponding (x, ε)-discernibility sets $c_{ij}^{x,\varepsilon} \subseteq A$ in a temporary discernibility table, under an online application of absorption.

Fig. 2 contains also basic facts concerning exemplary attribute subsets obtained by an application of simple, exemplary heuristics for searching for prime implicants. One can see that these subsets do satisfy conditions of Definition 8 for particular settings.

x	ε	# pairs	# elts.	# impls.	avg.	found implicants
2	0	15	3	4	2	$\{1,2\},\{2,3\},\{2,4\},\{3,4\}$
2	0.2	12	3	4	2	$\{1,2\},\{2,3\},\{2,4\},\{3,4\}$
2	0.4	7	4	8	2	$\{1,2\},\{1,3\},\{1,4\},\{1,5\}$ $\{2,3\},\{2,4\},\{3,4\},\{3,5\}$
2	0.6	5	3	5	1.8	$\{1,2\},\{1,4\},\{1,5\},\{2,4\}$ $\{3\}$
2	0.8	1	1	3	1	$\{1\},\{2\},\{3\}$

Fig. 2. Absorption-optimized discernibility tables obtained for the above exemplary decision table, under various ε-thresholds and fixed $x = 2$, where: **(1-2)** first two columns refer to (x, ε)-settings; **(3)** The # pairs column presents the number of pairs of A-indiscernibility classes necessary to be discerned; **(4)** The # elts. column presents the number of attribute subsets remaining in a discernibility table after applying the absorption law; **(5-7)** The rest of columns contain the number, average cardinality, and detailed list of attribute subsets found as prime implicants for corresponding boolean discernibility functions.

7   Conclusions

We provide the unified methodology for searching for approximate reducts corresponding to various ways of expressing inexact dependencies in inconsistent decision tables. In particular, we focus on rough membership reducts, which preserve *conditions→decision* frequencies approximately, in terms of the choice of a tolerance threshold and a function measuring distances between frequency distributions.

Presented results generalize well known relationship between rough set reducts and boolean prime implicants onto the whole class of considered approximations. It leads to possibility of using the well known algorithmic framework for searching for minimal decision reducts (cf. [6]) to the approximate μ-decision reduct optimization.

It is also worth emphasizing that introduced tools set up a kind of the rough set bridge between the approximate boolean calculus and the approximate probabilistic independence models. This fact relates our study to a wide range of applications dedicated, in general, to the efficient extraction and representation of data based knowledge.

Finally, described example illustrate how one can influence efficiency of the process of the attribute reduction under inconsistency, by the approximation parameter tuning. Still, further work is needed to gain more experience concerning the choice of these parameters in purpose of obtaining optimal models of the new case classification and data representation.

Acknowledgements

This work was supported by the grants of Polish National Committee for Scientific Research (KBN) No. $8T11C02319$ and $8T11C02519$.

References

1. Brown, E.M.: Boolean Reasoning. Kluwer Academic Publishers, Dordrecht (1990).
2. Pawlak, Z.: Rough sets – Theoretical aspects of reasoning about data. Kluwer Academic Publishers, Dordrecht (1991).
3. Pawlak, Z.: Decision rules, Bayes' rule and rough sets. In: N. Zhong, A. Skowron and S. Ohsuga (eds.), Proc. of the Seventh International Workshop RSFDGrC'99, Yamaguchi, Japan, LNAI **1711** (1999) pp. 1–9.
4. Pawlak, Z., Skowron, A.: Rough membership functions. In: R.R. Yaeger, M. Fedrizzi, and J. Kacprzyk (eds.), Advances in the Dempster Shafer Theory of Evidence, John Wiley & Sons, Inc., New York, Chichester, Brisbane, Toronto, Singapore (1994) pp. 251–271.
5. Pearl, J.: Probabilistic Reasoning in Intelligent Systems: Networks of Plausible Inference. Morgan Kaufmann (1988).
6. Polkowski, L., Skowron, A. (eds.): Rough Sets in Knowledge Discovery, parts **1**, **2**, Heidelberg, Physica-Verlag (1998) pp. 321–365.
7. Skowron, A.: Boolean reasoning for decision rules generation. In: Proc. of the Seventh International Symposium ISMIS'93, Trondheim, Norway, 1993; J. Komorowski, Z. Ras (eds.), LNAI **689**, Springer-Verlag (1993) pp. 295–305.
8. Skowron, A., Rauszer, C.: The discernibility matrices and functions in information systems. In: R. Słowiński (ed.), Intelligent Decision Support. Handbook of Applications and Advances of the Rough Set Theory, Kluwer Academic Publishers, Dordrecht (1992) pp. 311–362.
9. Ślęzak, D.: Various approaches to reasoning with frequency-based decision reducts: a survey. In: L. Polkowski, S. Tsumoto, T.Y. Lin (eds.), Rough Sets in Soft Computing and Knowledge Discovery: New Developments, Physica-Verlag / Springer-Verlag (2000).
10. Ziarko, W.: Decision Making with Probabilistic Decision Tables. In: N. Zhong, A. Skowron and S. Ohsuga (eds.), Proc. of the Seventh International Workshop RSFDGrC'99, Yamaguchi, Japan, LNAI **1711** (1999) pp. 463–471.

Scalable Feature Selection Using Rough Set Theory

Moussa Boussouf, Mohamed Quafafou

IRIN, Université de Nantes, 2 rue de la Houssinière,
BP 92208 - 44322, Nantes Cedex 03, France.
{boussouf,quafafou}@irin.univ-nantes.fr

Abstract. In this paper, we address the problem of feature subset selection using rough set theory. We propose a scalable algorithm to find a set of reducts based on *discernibility function*, which is an alternative solution for the exhaustive approach. Our study shows that our algorithm improves the classical one from three points of view: computation time, reducts size and the accuracy of induced model.

1 Introduction

The irrelevant and redundant features may reduce predictive accuracy, degrade the learner speed (due to the high dimensionality) and reduce the comprehensibility of the induced model. Thus, pruning these features or selecting relevant ones becomes necessary.

In the rough set theory [7][8], the process of feature subset selection is viewed as (relative)reducts computation. In this context, different works have been developed to deal with the problem of feature subset selection. Modrzejewski [5] proposes a heuristic feature selector algorithm, called PRESET. It consists in ordering attributes to obtain an optimal preset decision tree. Kohavi and Frasca in [4] have shown that, in some situations, the *useful* subset does not necessarily contain all the features in the *core* and may be different from a reduct. Using α–RST [9], we have proposed in [10] an algorithm based on wrapper approach to solve this problem. We have shown that we can obtain lower size reducts with higher accuracy than those obtained by classic rough sets concepts. Skowron and Rauszer [13] mentioned that the problem of computing minimal reducts is NP-hard. They proposed an algorithm to find a set of reducts which have the same characteristics as original data. This algorithm needs to compute discernibility between all pairs of objects of training set, i.e., it performs $\frac{N(N-1)}{2}$ comparisons. Consequently, the reducts computation can be time consuming when the decision table has too many objects (or attributes).

In this paper, we study the algorithm of finding reducts based on *discernibility matrix* [13]. We propose a new algorithm based on computing a *minimal discernibility list*, which is an approximative discernibility matrix. We show that our algorithm covers very early the search space with respect to *discernibility function*. It produces lower size reducts, more accurate models and time computation is considerably less consuming comparing with the classical algorithm.

W. Ziarko and Y. Yao (Eds.): RSCTC 2000, LNAI 2005, pp. 131–138, 2001.

2   Finding Reducts in Rough Set Theory

In the rough set theory, an information system has a data table form. Formally, an information system S is a 4-tuple. $S = (U, Q, V, f)$, *where U : is a finite set of objects. Q : is a finite set of attributes. $V = \cup V_q$, where V_q is a domain of attribute q. f is an information function assigning a value for every object and every attribute, i.e., $f : U \times Q \mapsto V$, such that for every $x \in U$ and for every $q \in Q$ $f(x, q) \in V_q$.*

Definition 1. Discernibility Relation: *Let $x, y \in U$ be two distinct objects. The discernibility relation, denoted by DIS, assigns pairs of objects to a subset of Q, i.e., $DIS : U \times U \mapsto \Im(Q)$ where $\Im(Q)$ is all subsets of Q. The discernibility between x and y is defined as follows:*

$$DIS(x, y) = \{q \in Q \mid f(x, q) \neq f(y, q)\}$$

For each pair of objects, the discernibility relation assigns a set of attributes which discern these objects, i.e., it maps a pair of objects with a subset of Q. Consequently, we associate a discernibility matrix, denoted DM, to each information system, where $DIS(i, j)$ is an element of the matrix DM which contains attributes distinguishing an objet i from another object j.

Let two subsets R_1 and R_2 such that $R_1 \subseteq R_2$. We say that R_1 is a reduct of R_2 if, and only if, R_1 has exactly the same characteristics (discrimination power, approximation space, etc.) as R_2 (see [7][8][13] for more details).

Skawron and Rauszer [13] have proposed a finding reducts algorithm, which is based on discernibility matrix. They have proved that the problem of computing reducts in rough set model is transformable to the problem of finding prime implicants of monotonic boolean function called *discernibility function*. Their process of calculating reducts can be summarized in two steps:

- **Step 1:**
  1.1. Computing the discernibility matrix DM: each element of DM contains a set of attributes which discern a pair of objects;
  1.2. Defining the discernibility function: This process leads to a conjunctive form of the discernibility function denoted DF. In fact, each element of DM produces a term represented by a disjunctive form and the conjunction of these terms defines DF;
- **Step 2:** Computing reducts: this process consists of transforming the discernibility function from a conjunctive normal form to a disjunctive one. In fact, this function is reduced performing basically *absorption rules*. This construction produces a reduced disjunctive form. Hence each term of this disjunctive form represents a *reduct*.

Example 1: *Let $Q = \{1, 2, 3, 4, 5\}$ be a set of attributes of an information system with 5 attributes. If $DM = \{\{1\}, \{2, 3\}, \{3, 5\}, \{1, 2\}, \{3, 4, 5\}, \{1, 2\}\}$, then the discernibility function $DF = 1 \wedge (2 \vee 3) \wedge (3 \vee 5)$. The disjunctive form produces two reducts, which are $\{1, 2, 5\}$ and $\{1, 3\}$.*

Extensive researches have been developed to find more refined reducts to deal with real world data and to construct accurate induced models using a machine learning algorithm. Consequently, different extended reduct definitions are introduced: (1) β-reduct defined in the Variable Precision Rough Sets Model [14], (2) α-reduct formalized in the context of a generalized rough set theory, called α-Rough Set Theory [10][11] and (3)*dynamic reducts* [1].

The crucial problem in reducts research is that the mainly used algorithms are based on discernibility matrix and on the work developed by Skowron and Rauszer [13]. However, the calculation of DM is achieved by comparing all pairs of objects, i.e., it needs $\frac{N(N-1)}{2}$ comparisons. The time complexity of the second step is bounded by $p(n)$, where n is the standard code of discernibility function, and p is a polynomial. So, the complexity of finding reducts process is $O(N^2)$. This is prohibitive when we process very large datasets. Using the same algorithm, Bazan et al. [1] search *dynamic reducts*, which are in some sense the most *stable reducts* of a given decision table, i.e., they are the most frequently appearing reducts in subtables created by random samples of a given decision table. To achieve this process, they must perform $\frac{N_i(N_i-1)}{2}$ comparisons for each considered sample. The total number of comparisons is equal to $\sum_{i=1}^{k} \frac{N_i(N_i-1)}{2}$, where N_i is the size of the ith sample. There are many problems faced when applying this approach: first, Kohavi and Frasca in [4] have shown that, in some situations, the *useful* subset does not necessarily contain all the features in the *core* and may be different from a reduct; second, due to the high complexity of the used algorithm, applying this algorithm many times aggravates the problem, especially when the size of each sample is important.

Since the reducts depend on discernibility function of each sample, we say that the stability of reducts depends on stability of discernibility function (step 1.2). Our approach to tackle this problem is different: instead of searching *downstream* the reducts stability, we propose to *stabilize upstream* the discernibility function.

3   Fast Scalable Reducts

We have introduced in the previous section the discernibility relation which assigns pairs of objects to a subset of Q. We denote $\Im(Q)$ the set of all subsets of Q which can be organized as a lattice according to the inclusion operator ($\subset$). Each node of the lattice represents a subset of attributes discerning one pair (or more) of objects. So, each element of DM has a corresponding node in the lattice. But, to define the discernibility function only minimal elements are needed, i.e., big elements are absorbed by the small ones. Consequently, during the step 1 of the algorithm, many comparisons are performed without changing this function. In this context, the problem of finding reducts is viewed as a search for *minimal bound* in the lattice. This bound is represented by Minimal Discernibility List (MDL) and is corresponding to an approximation of the reduced conjunctive form of the discernibility function. This approximation is computed by an incremental and random process following the two steps:

- **Step 1:** Calculate incrementally the minimal discernibility list, MDL. In fact, we select randomly two objects and we determine the subset of attributes P discerning them. The MDL is then updated as follows: if P is minimal, i.e., no element of MDL is included in P, then add P to MDL and remove all $P' \in MDL$ such that $P \subset P'$. This iterative process terminates when the probability to modify the MDL becomes small, (see section 3.1);
- **Step 2:** Calculate reducts from MDL. This step consists in formulating DF from MDL, then transforming the conjunctive form to a reduced disjunctive one. Each final term represents a reduct.

3.1   Probability Evolution

Let Q be the set of M attributes, $\Im(Q)$ the space of all possible subsets that may be represented by a lattice. The cardinality of $\Im(Q)$ is $2^M - 1$ (without considering the empty subset).

Let P be a subset of Q. The number of supersets or coversets (including P) of P, denoted $\mathcal{F}(P)$, is equal to $2^{M-|P|}$, such that $|P|$ denotes the cardinality of P.

Let $MDL = \{P_1, P_2, \dots, P_K\}$ containing K subsets of Q. The total number of supersets, which have no effect on MDL, i.e., which are covered by all elements of MDL, is calculated as follows:

$$\mathcal{F}(MDL) = (-1)^{1-1} \sum_{i_1=1}^{K} \mathcal{F}(P_{i_1})$$

$$+ (-1)^{2-1} \sum_{\substack{i_1=1 \\ i_2 > i_1}}^{K} \mathcal{F}(P_{i_1} \cup P_{i_2})$$

$$+ \dots$$

$$+ (-1)^{j-1} \sum_{\substack{i_1=1 \\ i_j > \dots > i_2 > i_1}}^{K} \mathcal{F}(P_{i_1} \cup P_{i_2} \cup \dots \cup P_{i_j})$$

$$+ \dots$$

$$+ (-1)^{K-1} \sum_{\substack{i_1=1 \\ i_K > \dots > i_j > \dots > i_2 > i_1}}^{K} \mathcal{F}(P_{i_1} \cup P_{i_2} \cup \dots \cup P_{i_j} \cup \dots \cup P_{i_K})$$

such that $\sum_{\substack{i_1=1 \\ i_j > \dots > i_2 > i_1}}^{K} \mathcal{F}(P_{i_1} \cup P_{i_2} \cup \dots \cup P_{i_j})$ represents the number of supersets of all possible unions of j elements of MDL.

For instance, if $MDL = \{P_1, P_2\}$, then $\mathcal{F}(MDL) = \mathcal{F}(P_1) + \mathcal{F}(P_2) - \mathcal{F}(P_1 \cup P_2)$.

The probability for improving the minimal discernibility list, denoted $\mathcal{P}(MDL)$, is calculated as follows:

$$\mathcal{P}(MDL) = 1 - \frac{\mathcal{F}(MDL)}{|\Im(Q)|} = 1 - \frac{\mathcal{F}(MDL)}{2^M - 1}$$

Example 2: *Let $Q = \{1, 2, 3, 4, 5\}$ be a set of attributes of an information system with 5 attributes. Supposing that the current $MDL = \{\{1\}, \{2, 3\}, \{3, 5\}\}$, the reader can check that the probability to improve MDL equals $1 - 22/31 = 0.29$. It means that for the remaining comparisons, we have 71% of chance that two objects are discernible by a subset which is covered by an element of MDL.*

The Figure 1 shows the probability to improve MDL with respect to comparisons number for Pima, Anntyroid and Shuttle datasets (see their characteristics in Table 1). For each dataset and at each iteration, two objects O_1 and O_2 are randomly selected and $DIS(O_1, O_2)$ is calculated. If there is no element in MDL which covers $DIS(O_1, O_2)$ then MDL is improved and $\mathcal{P}(MDL)$ is then calculated. After performing N (which represents the dataset size) comparisons, the mean probability of 10 executions to improve MDL, equals 0.029, 0.038 and 0.032 for Pima, Anntyroid and Shuttle respectively.

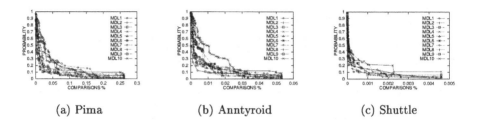

(a) Pima   (b) Anntyroid   (c) Shuttle

Fig. 1. Probability of MDL improvement.

If we consider that this probability is smaller than a given threshold, the MDL represents a good approximation of discernibility function. Thus, only N comparisons are performed instead of $\frac{N(N-1)}{2}$ comparisons of the algorithm based on computing full discernibility matrix. Consequently, only few comparisons are performed to compute MDL. In fact, the percentage of comparisons performed for the three datasets Pima, Anntyroid and Shuttle are respectively 0.261% , 0.053% and 0.0046%. We remark that the more the dataset size is important, the less is the percentage of the performed comparisons.

3.2   FSR Algorithm

Our approach is based on random comparisons between objects of the information system. At each iteration we check the possibility of MDL improvement. The process is stopped when a given number of comparisons is achieved. The reducts are then computed from the resulted MDL. Of course, the best solution is to consider the stopping criteria when the probability to improve MDL exceeds a given threshold. Unfortunately the cost of the probability computation is very high when the MDL size is very large: the complexity of the probability computation is $O(2^{|MDL|})$, where $|MDL|$ represents the MDL cardinality.

The first step of FSR algorithm (Figure 2) consists in computing the minimal discernibility list. After $n = N$, we consider that we have a good approximation of discernibility function. Consequently the complexity of MDL computation is $O(N)$. The second step consists in computing the approximative reducts from MDL. The cost of this step depends on MDL size. The time complexity of

```
Input       IS : Information System of N Examples; n : Comparisons Number;
            D : Condition Attributes; C : Class; /* Q = D ∪ C */
Output      FSR : Fast Scalable Reducts;
STEP 1: for Comparisons:=1 to n
            Random(i,j,N); /* return two objects : 1 ≤ i ≠ j ≤ N */
            R := DIS(IS[i],IS[j]); /* the set of features which discern IS[i] and IS[j] */
            if (C ∈ R) and (∄ E ∈ MDL | E ⊆ R)
                add(R - {C}, MDL); /* Improve MDL */
            endif
        endfor /*end construction of minimal MDL */
STEP 2: FSR=Reducts(MDL); /* Fast Scalable Reducts computation */
```

Fig. 2. FSR: Fast Scalable Reducts algorithm.

reducts construction (by MDL transformation) is bounded by $p(n)$, where n is the standard code of discernibility function, and p is a polynomial.

4   Experimental Results

In order to evaluate candidate scalable feature subsets generated by FSR algorithm, we ran experiments on 18 real-world datasets taken from the UCI Irvine repository, their characteristics are summarized in Table 1. Original datasets are transformed using Fayad and Irani discretization method [3]. Our hybrid approach [2] is used to evaluate and select the best reduct. To achieve this process, we have used C4.5 [12] as an inducer algorithm and Liu&Setiono filter [6] (The allowable inconsistency rate is equal to 3% comparing with the best reduct according to the filter). We stop the MDL computation where the dataset size is exceeded. The seed random function is the same for all datasets.

Our experiments presented in Table 2 show that the classical reducts and those generated with FSR algorithm generally improve the accuracy of C4.5

Table 1. Datasets considered: Size of Train sets, Test protocol, Attributes number, Classes number and the percentage of numeric attributes.

Datasets	Train	Test	Att.	Cla.	Num%	Datasets	Train	test	Att.	Cla.	Num%
Iris	150	5cv	4	3	100	Pima	768	5cv	8	2	100
Wine	178	5cv	13	3	100	Vehicle	846	5cv	18	4	100
Glass	214	5cv	9	6	100	German	1000	5cv	20	2	35
Heart	270	5cv	13	2	46	Segment	2310	5cv	19	7	95
Ecoli	336	5cv	7	8	100	Abalone	3133	1044	8	3	88
Liver	345	5cv	6	2	100	Anntyroid	3772	3428	21	3	29
BreastWD	569	5cv	30	2	100	Pendigits	7494	3498	16	10	100
Australian	690	5cv	14	2	43	Adult	32561	16281	14	2	43
Credit	690	5cv	15	2	210	Shuttle	43500	14500	9	7	100

using original datasets. The FSR algorithm improves the classical reducts algorithm from three points of view:

Table 2. C4.5 accuracy with original datasets, the results of classical and FSR algorithms: C4.5 accuracy, reducts calculation time (in second) and the best reduct size.

Datasets	C4.5 Org.	Classic			FSR		
		C4.5	Time	Size	C4.5	Time	Size
Iris	92.00	92.00	0.03	4	92.00	0.00	2
Wine	80.12	93.28	0.10	7	95.46	0.03	3
Glass	63.84	63.84	0.09	9	65.38	0.00	4
Heart	74.08	77.76	0.18	10	83.72	0.01	6
Ecoli	81.56	81.56	0.19	7	81.56	0.00	5
Liver	69.00	69.00	0.15	6	68.42	0.00	4
BreastWD	94.38	97.00	273.83	11	96.28	2.73	3
Australian	86.68	86.54	1.13	12	86.10	0.01	7
Credit	82.64	82.80	1.05	11	86.32	0.03	5
Pima	78.40	78.40	1.08	8	78.92	0.00	7
Vehicle	71.28	70.10	2.09	17	71.28	0.02	9
German	71.90	74.40	3.27	13	75.00	0.67	7
Segment	93.64	93.66	15.95	13	93.54	0.01	5
Abalone	63.60	63.60	15.38	8	63.50	0.02	7
Anntyroid	94.00	93.80	52.06	20	92.70	0.02	5
Pendigits	91.70	90.30	161.90	16	89.60	0.08	10
Adult	85.40	85.30	1332.13	13	83.50	0.09	9
Shuttle	99.80	99.80	1861.43	9	99.70	0.09	6
MEAN	81.89	82.95	193.58	10.66	83.48	0.21	5.78

1. Reducts size: The size of reducts generated by FSR algorithm is *always* lower than the size of reducts produced by the classical (exhaustive) algorithm. The mean size of reducts is 5.78 for FSR algorithm, whereas it equals 10.66 for the classical one. The most important result is obtained with Breast dataset: among 30 attributes, the FSR algorithm selects only 3 attributes and the accuracy is improved comparing with original data.

2. Accuracy: The accuracy of reducts produced by FSR algorithm generally improves both the accuracy using original datasets and the accuracy of reducts produced by the classical algorithm, differences between average accuracies equal +1.59% and +0.53% respectively. Comparing the accuracy of C4.5 using the best reducts generated by FSR algorithm with the accuracy of the original datasets, the worst accuracy is obtained with Pendigits dataset, it falls by -2.1%, whereas the best one is obtained with Wine dataset, the accuracy improves by +15.34%.

3. Time: The most interesting result is the time of calculating reducts with FSR algorithm. The average time is 0.21s for FSR algorithm and 193.58s for the classical algorithm. So, the algorithm was improved by 921.81 times. The time of FSR algorithm is, in most cases, lower than 1s, except for Breast dataset (which explains that MDL contains many elements, so the second step of FSR algorithm is quietly slow). The best result was obtained with the largest dataset i.e. Shuttle: the FSR was stopped after 0.09s, whereas the classical approach is achieved after 1861.43s, which represents a factor of 20682.

5   Conclusion

In this paper we have studied the problem of feature selection using rough set theory. We have proposed an algorithm to calculate reducts based on finding a minimal discernibility list from an approximative discernibility matrix. In this context, we have shown that the problem of reducts computation using the whole discernibility matrix is reducible to a problem reducts computation based on finding a fast minimal discernibility list, which covers a large space of the discernibility matrix. We claim the use of FSR algorithm instead of the classical one, which is based on computing full comparisons, for three main reasons: (1) the reducts size is lower; (2) the accuracy is generally higher; (3) the FSR time is too lower.

References

1. Bazan G., Skowron A., Synak P.: Dynamic Reducts as a Tool for Extracting Laws from Decisions Tables, ICS Research Report 43/94, Warsaw Univ. of Tech., (1994).
2. Boussouf M.: A Hybrid Approach to Feature Selection. In Proceedings of the 2nd European Symposium, PKDD'98, (1998) 230–238.
3. Fayyad U.M., Irani K.B.: Multi-interval Discretization of Continuous-attributes for classification learning IJCAI'93, (1993) 1022–1027.
4. Kohavi, R., Frasca, B.: Useful feature subset and rough sets reducts. In Proceedings of the 3rd Int. Workshop on Rough Sets and Soft Computing, (1994) 310–317.
5. Modrzejewski, M.: Feature selection using rough sets theory. In Proceedings of the ECML, (1993) 213–226.
6. Liu, H., Setiono, R.: A probabilistic approach for feature selection: A filter solution. In 13th International Conference on Machine Learning (ICML'96), (1996) 319–327.
7. Pawlak, Z.: Rough Sets: Theoretical Aspects of Reasoning About Data. Kluwer Academic Publishers, Dordrecht, The Netherlands (1991).
8. Pawlak, Z.: Rough Sets: present state and the future. Foundations of Computing and Decision Sciences, **18(3-4)**, (1993) 157–166.
9. Quafafou, M.: α-RST: A generalization of rough set theory. In Proceedings in the Fifth International Workshop on Rough Sets and Soft Computing, RSSC'97. (1997).
10. Quafafou, M., Boussouf, M.: Induction of Strong Feature Subsets. In 1st European Symposium, PKDD'97, (1997) 384–392.
11. Quafafou M., Boussouf, M., (1999) : Generalized Rough Sets based Feature Selection. Intelligent Data Analysis Journal, (IDA) 4(1), (1999).
12. Quinlan J. R.: Programs for Machine Learning, Morgan Kaufmann, San Mateo, CA, (1993).
13. Skowron A. and Rauszer C.: The Discernibility Matrice and Functions in Information Systems. Intelligent Decision Support. Handbook of Applications and Advances of Rough Sets Theory. Dordrecht: Kluwer. (1992) 331–362.
14. Ziarko W. : Anaalysis on Uncertain Information in the Framework of Variable Precision Rough Sets. Foundations of Computing and Decision Sciences, **18(3-4)**, (1993) 381–396.

On Algorithm for Constructing of Decision Trees with Minimal Number of Nodes

Igor Chikalov

Faculty of Computing Mathematics and
Cybernetics of Nizhni Novgorod State University
23, Gagarina Av., Nizhni Novgorod, 603600, Russia
Igor.Chikalov@intel.com

Abstract. An algorithm is considered which for a given decision table constructs a decision tree with minimal number of nodes. The class of all information systems (finite and infinite) is described for which this algorithm has polynomial time complexity depending on the number of columns (attributes) in decision tables.

1 Introduction

Decision trees are widely used in different areas of applications. Problem of optimal decision tree constructing is known to be complicated. In the paper an algorithm is considered, which for a given decision table constructs a decision tree with minimal number of nodes. The time of the algorithm work is bounded above by a polynomial on the number of columns and rows in a decision table and on the number of so-called nonterminal separable sub-tables of the table.

Also decision tables over an arbitrary (finite or infinite) information system are considered, and all information systems are described for which the number of rows and the number of nonterminal separable sub-tables are bounded above by a polynomial on the number of columns (attributes) in the table.

The idea of the algorithm is close to so-called descriptive methods of optimization [3, 4, 5, 6]. The obtained results may be useful in test theory the groundwork for which was laid by [1], in rough set theory created in [9, 10] and in their applications.

The algorithm allows generalization on the case of such complexity measures as the depth [8] and the average depth [2] of the decision tree. Similar results were announced in [7]. Also the considered algorithm may be generalized on the case when each column is assigned a weight.

2 Basic Notions

Decision table is a rectangular table filled by numbers from $E_k = \{0, \ldots, k-1\}$, $k \geq 2$, in which rows are pairwise different. Let T be a decision table containing n columns and m rows which are labeled with numbers $1, \ldots, n$ and $1, \ldots, m$ respectively. Denote by $\dim T$ and by $N(T)$ the number of columns and rows

W. Ziarko and Y. Yao (Eds.): RSCTC 2000, LNAI 2005, pp. 139–143, 2001.
© Springer-Verlag Berlin Heidelberg 2001

in T respectively. Let for $i = 1, \ldots, m$, i-th row of the table T is assigned a natural number ν_i, where $0 \leq \nu_i \leq k^n$. The m-tuple $\nu = (\nu_1, \ldots, \nu_m)$ provides the division of the set of the table T rows into classes, in which all rows are labeled with the same number. A two-person game can be associated with the table T. The first player choose a row in the table, and the second one must determine the number the chosen row is labeled. For this purpose he can ask questions to the first player: he can select a column in the table and ask what is the number on the intersection of this column and the chosen row.

Each strategy of the second player can be represented as a *decision tree* that is a marked finite oriented tree with root in which each nonterminal node is assigned a number of column; each edge is assigned a number from E_k; edges starting in a nonterminal node are assigned pairwise different numbers; each terminal node is assigned a number from the set $\{\nu_1, \ldots, \nu_m\}$. A decision tree that represents a strategy of the second player will be called *correct*.

As a complexity measure the number of nodes in decision tree is used. A correct decision tree with minimal number of nodes will be called *optimal*.

Let $i_1, \ldots, i_t \in \{1, \ldots, n\}$ and $\delta_1, \ldots, \delta_t \in E_k$. Denote by $T(i_1, \delta_1) \ldots (i_t, \delta_t)$ the sub-table of the table T containing only such rows, which on intersections with columns labeled by $i_1, \ldots, i_t$ have numbers $\delta_1, \ldots, \delta_t$ respectively. If the considered sub-table differs from T and has at least one row then it will be called *separable sub-table* of the table T. The decision table will be called *terminal* if all rows in the table are labeled with the same number. For a nonterminal table T we denote by $S(T)$ the set of all nonterminal separable sub-tables of the table T including the table T.

3   Algorithm for Constructing of Optimal Decision Trees

In this section the algorithm $\mathcal{A}$ is considered which for a given decision table T constructs an optimal decision tree.

Step 0. If T is a terminal table and all rows of the table T are labeled with the number j, then the result of algorithm's work is the decision tree consisting of one node, which is assigned the number j. Otherwise construct the set $S(T)$ and pass to the first step.

Suppose $t \geq 0$ steps have realized.

Step $(t+1)$. If the table T has been assigned a decision tree then this decision tree is the result of the algorithm $\mathcal{A}$ work. Otherwise choose in the set $S(T)$ a table D satisfying the following conditions:

a) the table D has not been assigned a decision tree yet;

b) for each separable subtable D_1 of the table D either the table D_1 has been assigned a decision tree, or D_1 is a terminal table.

For each terminal separable sub-table D_1 of the table D, assign to the table D_1 the decision tree $\Gamma(D_1)$. Let all rows in the table D_1 be labeled with the number j. Then the decision tree $\Gamma(D_1)$ consists of one node, which is assigned the number j.

For $i \in \{1, \ldots, \dim D\}$ denote by $E(D, i)$ the set of numbers containing in the i-th column of the table D, and denote $I(D) = \{i : i \in \{1, \ldots, \dim D\}, |E(D, i)| \geq 2\}$. For every $i \in I(D)$ and each $\delta \in E(D, i)$ denote by $\Gamma(i, \delta)$ the decision tree assigned to the table $D(i, \delta)$. Let $i \in I(D)$ and $E(D, i) = \{\delta_1, \ldots, \delta_r\}$. Define a decision tree Γ_i. The root of Γ_i is assigned the number i. The root is initial node of exactly r edges $d_1, \ldots, d_r$ which are labeled by the numbers $\delta_1, \ldots, \delta_r$ respectively. The roots of the decision trees $\Gamma(i, \delta_1), \ldots, \Gamma(i, \delta_r)$ are terminal nodes of the edges $d_1, \ldots, d_r$ respectively. Assign to the table D one of the trees Γ_i, $i \in I(D)$, having minimal number of nodes and pass to the next step.

It is not difficult to prove the following statement.

Theorem 1. *For any decision table T filled by numbers from E_k, the algorithm $\mathcal{A}$ constructs an optimal decision tree, and performs exactly $|S(T)| + 1$ steps. The time of the algorithm $\mathcal{A}$ work is bounded below by $c|S(T)|$, where c is a positive constant, and bounded above by a polynomial on $N(T)$, $\dim T$ and $|S(T)|$.*

4 Decision Tables over Information Systems

Let A be a nonempty set, F a nonempty set of functions from A to E_k, and $f \not\equiv \text{const}$ for any $f \in F$. Functions from F will be called *attributes*, and the pair $U = (A, F)$ will be called *k-valued information system*.

For arbitrary attributes $f_1, \ldots, f_n \in F$ and an arbitrary function $\nu : E_k^n \to \{0, 1, \ldots, k^n\}$, we denote by $T(f_1, \ldots, f_n, \nu)$ the decision table with n columns which contains the row $(\delta_1, \ldots, \delta_n) \in E_k^n$ iff the system of equations

$$\{f_1(x) = \delta_1, \ldots, f_n(x) = \delta_n\} \tag{1}$$

is compatible on the set A. This row is assigned the number $\nu(\delta_1, \ldots, \delta_n)$. The table $T(f_1, \ldots, f_n, \nu)$ will be called *a decision table over the information system* U. Denote by $\mathcal{T}(U)$ the set of decision tables over U.

Consider the functions

$$\mathcal{S}_U(n) = \max\{|S(T)| : T \in \mathcal{T}(U), \dim T \leq n\}$$

and

$$\mathcal{N}_U(n) = \max\{N(T) : T \in \mathcal{T}(U), \dim T \leq n\},$$

which characterizes the maximal number of nonterminal separable sub-tables and maximal number of rows respectively depending on the number of columns in decision tables over U.

Using Theorem 1 one can show that for tables over U time complexity of the algorithm $\mathcal{A}$ is bounded above by a polynomial on the number of columns if both the functions $\mathcal{S}_U(n)$ and $\mathcal{N}_U(n)$ are bounded above by a polynomial on n, and time complexity of the algorithm $\mathcal{A}$ has an exponential lower bound if the function $\mathcal{S}_U(n)$ grows expoinentially.

A system of equations of the kind (1) will be called *a system of equations over* U. Two systems of equations are called *equivalent* if they have the same set of solutions. A compatible system of equations will be called *uncancellable* if each its proper subsystem is not equivalent to the system. Let r be a natural number. Information system U will be called r-restricted (restricted) if each uncancellable system of equations over U consists of at most r equations.

Theorem 2. *Let $U = (A, F)$ be k-valued information system. Then the following statements hold:*

a) if U is r-restricted information system then $S_U(n) \leq (nk)^r + 1$ and $N_U(n) \leq (nk)^r + 1$ for any natural n;

b) if U is not restricted information system then $S_U(n) \geq 2^n - 1$ for any natural n.

Proof. a) Let U be r-restricted information system and $T = T(f_1, \ldots, f_n, \nu) \in \mathcal{T}(U)$. One can show that both the values $|S(T)|$ and $N(T)$ do not exceed the number of pairwise nonequivalent compatible subsystems of the system of equations $\{f_1(x) = 0, \ldots, f_n(x) = 0, \ldots, f_1(x) = k - 1, \ldots, f_n(x) = k - 1\}$ including the empty system (the set of solutions of the empty system is equal to A). Each compatible system of equations over U contains an equivalent subsystem with at most r equations. Then $|S(T)| \leq (\dim T)^r k^r + 1$ and $N(T) \leq (\dim T)^r k^r + 1$. Therefore $S_U(n) \leq (nk)^r + 1$ and $N_U(n) \leq (nk)^r + 1$.

b) Let U be not a restricted system and n be a natural number. Then there exists an uncancellable system of equations over U with at least n equations. Evidently, each subsystem of this system is uncancellable. Therefore there exists an uncancellable system over U with n equations. Let it be the system (1), which will be denoted by W. We prove that every two different subsystems W_1 and W_2 of the system W are nonequivalent. Assume the contrary. Then subsystems $W \setminus (W_1 \setminus W_2)$ and $W \setminus (W_2 \setminus W_1)$ are equivalent to W and at least one of them is a proper subsystem of W, which is impossible. Denote $T = T(f_1, \ldots, f_n, \nu)$, where ν is the function, that sets to the the coprrespondence to the n-tuple $\delta \in E_k^n$ the value for which δ is a notation in the system with the radix k. Each proper subsystem $\{f_{i_1}(x) = \delta_1, \ldots, f_{i_t}(x) = \delta_t\}$ of the system W corresponds to the separable subtable $T(f_{i_1}, \delta_1)(f_{i_t}, \delta_t)$ of the table T. Two different subsystems being nonequivalent each other and being nonequivalent to the system W, the subtables corresponding to these subsystems are different and nonterminal. Then $|S(T)| \geq 2^n - 1$. Hence $S_U(n) \geq 2^n - 1$.   □

Example 3. Denote by A the set of all points in the plane. Consider an arbitrary straight line l, which divides the plane into positive and negative open half-planes. Put into correspondence a function $f : A \to \{0, 1\}$ to the straight line l. The function f takes the value 1 if a point is situated in positive half-plane, and f takes the value 0 if a point is situated in negative half-plane or on the line l. Denote by F an infinite set of functions, which correspond to some straight lines in the plane. Consider two cases.

1) Functions from the set F correspond to t infinite classes of parallel straight lines. One can show that the information system U is $2t$-restricted.

2) Functions from the set F correspond to all straight lines in the plane. Then the information system U is not restricted.

Acknowledgments

This work was partially supported by the program "Universities of Russia" (project # 015.04.01.76).

References

1. Chegis, I.A., Yablonskii, S.V.: Logical methods of electric circuit control. Trudy MIAN SSSR **51** (1958) 270–360 (in Russian)
2. Chikalov, I.V.: Algorithm for constructing of decision trees with minimal average depth. Proceedings of The 8th Conference on Information Processing and Management of Uncertainty in Knowledge-Based Systems **1**. July 3-7, Madrid, Spain (2000) 376–379
3. Markov, Al.A.: Circuit complexity of discrete optimization. Diskretnaya Matematika **4**(3) (1992) 29–46 (in Russian)
4. Morzhakov, N.M.: On relationship between complexity of a set description and complexity of problem of linear form minimization on this set. Kombinatorno-algebraicheskiye Metody v Prikladnoi Matematike. Gorky University Publishers, Gorky (1985) 83–98 (in Russian)
5. Morzhakov, N.M.: On complexity of discrete extremal problem solving in the class of circuit algorithms. Matematicheskie Voprosi Kybernetiki **6**. Nauka, Moscow (1996) 215–238 (in Russian)
6. Moshkov, M.Ju.: On problem of linear form minimization on finite set. Kombinatorno-algebraicheskiye Metody v Prikladnoi Matematike. Gorky University Publishers, Gorky (1985) 98–119 (in Russian)
7. Moshkov, M.Ju., Chkalov, I.V.: On efficient algorithms for conditional test constructing. Proceedings of the Twelfth International Conference "Problems of Theoretical Cybernetics" **2**. Nizhni Novgorod, Russia (1999) 165 (in Russian)
8. Moshkov, M.Ju, Chikalov I.V.: On algorithm for constructing of decision trees with minimal depth. Fundamenta Informaticae **41**(3) (2000) 295–299
9. Pawlak, Z.: Rough Sets - Theoretical Aspects of Reasoning about Data. Kluwer Academic Publishers, Dordrecht, Boston, London (1991)
10. Skowron, A., Rauszer, C.: The discernibility matrices and functions in information systems. Intelligent Decision Support. Handbook of Applications and Advances of the Rough Set Theory. Edited by R. Slowinski. Kluwer Academic Publishers, Dordrecht, Boston, London (1992) 331–362

Rough Set-Based Dimensionality Reduction for Multivariate Adaptive Regression Splines

Alexios Chouchoulas and Qiang Shen

Division of Informatics, The University of Edinburgh
{alexios,qiangs}@dai.ed.ac.uk

Abstract. Dimensionality is an obstacle for many potentially powerful machine learning techniques. Widely approved and otherwise elegant methodologies exhibit relatively high complexity. This limits their applicability to real world applications. Friedman's Multivariate Adaptive Regression Splines (MARS) is a function approximator that produces continuous models of multi-dimensional functions using recursive partitioning and multidimensional spline curves that are automatically adapted to the data. Despite this technique's many strengths, it, too, suffers from the dimensionality problem. Each additional dimension of a hyperplane requires the addition of one dimension to the approximation model, and an increase in the time and space required to compute and store the splines. Rough set theory can reduce dataset dimensionality as a preprocessing step to training a learning system. This paper investigates the applicability of the Rough Set Attribute Reduction (RSAR) technique to MARS in an effort to simplify the models produced by the latter and decrease their complexity. The paper describes the techniques in question and discusses how RSAR can be integrated with MARS. The integrated system is tested by modelling the impact of pollution on communities of several species of river algae. These experimental results help draw conclusions on the relative success of the integration effort.

1 Introduction

High dimensionality is an obstacle for many potentially powerful machine learning techniques. Widely approved and otherwise elegant methodologies exhibit relatively high complexity. This places a ceiling on the applicability of such approaches, especially to real world applications, where the exact parameters of a relation are not necessarily known, and many more attributes than necessary are used to ensure all the necessary information is present.

Friedman's Multivariate Adaptive Regression Splines (MARS) [5] is a useful function approximator. MARS employs recursive partitioning and spline Basis functions to closely approximate the application domain [1]. The partitioning and number of Basis functions used are automatically determined by this approach based on the training data. Unlike other function approximators, MARS produces continuous, differentiable approximations of multidimensional functions, due to the use of spline curves. The approach is efficient and adapts itself to

W. Ziarko and Y. Yao (Eds.): RSCTC 2000, LNAI 2005, pp. 144–151, 2001.

the domain and training data. However, it is a relatively complex process, and suffers from the curse of dimensionality [5]. Each dimension of the hyperplane requires one dimension for the approximation model, and an increase in the time and space required to compute and store the splines.

Rough set theory [6] is a methodology that can be employed to reduce the dimensionality of datasets as a preprocessing step to training a learning system on the data. Rough Set Attribute Reduction (RSAR) works by selecting the most information-rich attributes in a dataset, without transforming the data, all the while attempting to lose no information needed for the classification task at hand [2]. The approach is highly efficient, relying on simple set operations, which makes it suitable as a preprocessor for more complex techniques. This paper investigates the application of RSAR to preprocessing datasets for MARS, in an effort to simplify the models produced by the system and decrease their complexity. The integrated system is used to build a model of river algae growth as influenced by changes in the concentration of several chemicals in the water. The success of the application is demonstrated by the reduction in the number of measurements required, in tandem with accuracy that matches very closely that of the original, unreduced dataset.

The paper begins by briefly describing the two techniques in question. Issues pertaining to how RSAR can be used with MARS are also discussed and the integrated system is described. Experiments and their results are then provided and discussed, and conclusions are drawn about the overall success of the integration effort.

2   Background

The fundaments of MARS and RSAR are explained below. The explanations are kept brief, as there already exist detailed descriptions of both techniques in the literature [5, 7].

2.1   Multivariate Adaptive Regression Splines

MARS [5] is a statistical methodology that can be trained to approximate multidimensional functions. MARS uses recursive partitioning and spline curves to closely approximate the underlying problem domain. The partitioning and number of Basis functions used are automatically determined by this approach based on the provided training data.

A spline is a parametric curve defined in terms of control points and a Basis function or matrix, that approximates its control points [4]. Although splines do not generally interpolate their control points, they can approximate them quite closely. The basis function or matrix provides the spline with its continuous characteristics. Two- and 3-dimensional splines are used widely in computer graphics and typography [1].

MARS adapts the general, n-dimensional form of splines for function approximation. It generates a multi-dimensional spline to approximate the shape of the domain's hyper-plane. Each attribute is recursively split into subregions. This partitioning is performed if a spline cannot approximate a region within reasonable bounds. A tree of spline Basis functions is thus built. This allows MARS great flexibility and autonomy in approximating numerous deceptive functions. MARS models may be expressed in the following form, known as ANOVA decomposition [5]:

$$\hat{f}(\mathbf{x}) = a_0 + \sum_{K_m=1} f_i(x_i) + \sum_{K_m=2} f_{ij}(x_i, x_j) + \sum_{K_m=3} f_{ijk}(x_i, x_j, x_k) + \dots,$$

where a_0 is the coefficient of the constant Basis function B_1 [5], f_i is a univariate Basis function of x_i, f_{ij} is a bivariate Basis function of x_i and x_j, and so on. In this context, K_m is the number of variables a Basis function involves. The ANOVA decomposition shows how a MARS model is the sum of Basis functions, each of which expresses a relation between a subset of the variables of the entire model. As an example, a univariate Basis function f_i is defined as:

$$f_i(x_i) = \sum_{K_m=1} a_m B_m(x_i), \quad \text{where} \quad B_m^q(\mathbf{x}) = \prod_{k=1}^{K_m} \left[s_{km} \cdot \left(x_{km} - t_{km} \right) \right]_+^q.$$

Here, a_m is the coefficient of Basis function B_m which only involves variable x_i. B_m is the Basis function in question, involving K_m ordinates x_{km} of point $\mathbf{x}$ ($1 \leq k \leq K_m$); q is the order of the multivariate spline, with $q \geq 1$; $s_{km} = \pm 1$; and t_{km} is ordinate m of the control point $\mathbf{t}_m$.

MARS uses recursive partitioning to adjust the Basis functions' coefficients ($\{a_m\}_1^M$, for each Basis function B_m, $1 \leq m \leq M$), and to partition the universe of discourse into a set of these disjoint regions $\{R_m\}_1^M$. A region R is split into two subregions if and only if a Basis function cannot be adjusted to fit the data in R within a predefined margin [5].

Unlike many other function approximators, MARS produces continuous, differentiable approximations of multidimensional functions, thanks to the use of splines. MARS is particularly efficient and produces good results. The continuity of the resultant approximative models is one of the most desirable results if statistical analysis is to be performed. However, MARS is relatively complex, and suffers from the curse-of-dimensionality problem. Each dimension of the hyperplane requires one dimension for the approximation model, and an increase in the time and space required to compute and store the splines. The time required to perform predictions increases exponentially with the number of dimensions. Further, MARS is very sensitive to outliers. Noise may mar the model by causing MARS to generate a much more complex model as it tries to incorporate the noisy data into its approximation. A technique that simplified the produced models and did away with some of the noise would thus be very desirable. This forms the very reason that the Rough Set-based Attribute Reduction technique is adopted herein to build an integrated approach to multivariate regression with reduced dimensionality.

2.2   The Rough Set Attribute Reduction Method (RSAR)

Rough set theory [6] is a flexible and generic methodology. Among its many uses is dimensionality reduction of datasets. RSAR, a technique utilising Rough set theory to this end, removes redundant input attributes from datasets of nominal values, making sure that very little or no information essential for the task at hand is lost. In fact, the technique can *improve* the information content of data: by removing redundancies, learning systems can focus on the useful information, perhaps even producing better results than when run on unreduced data.

RSAR works by maximising a quantity known as *degree of dependency*. The degree of dependency $\gamma_P(X)$ of a set Y of decision attributes on a set of conditional attributes X provides a measure of how important that set of conditional attributes is in classifying the dataset examples into the classes in Y. If $\gamma_X(Y) = 0$, then classification Y is independent of the attributes in X, hence the conditional attributes are of no use to this classification. If $\gamma = 1$, then Y is completely dependent on X, hence the attributes are indispensable. Values $0 < \gamma_X(Y) < 1$ denote partial dependency. To calculate $\gamma_X(Y)$, it is necessary to define the *indiscernibility* relation. Given a subset of the set of attributes, $P \subseteq A$, two objects x and y in a dataset $\mathbb{U}$ are *indiscernible* with respect to P if and only if $f(x, Q) = f(y, Q) \ \forall \ Q \subseteq P$ (where $f(\alpha, B)$ is the classification function represented in the dataset, returning the classification of object α using the conditional attributes contained in the set B). The indiscernibility relation for all $P \in A$ is written as $\text{IND}(P)$ (derived fully elsewhere). The upper approximation of a set $P \subseteq \mathbb{U}$, given an equivalence relation $\text{IND}(P)$, is defined as $\overline{P}Y = \bigcup\{X : X \in \mathbb{U}/\text{IND}(P), X \cap Y \neq \emptyset\}$. Assuming equivalence relations P, Q in $\mathbb{U}$, it is possible to define the *positive region* $\text{POS}_P(Q)$ as $\text{POS}_P(Q) = \bigcup_{X \in Q} \underline{P}X$. Based on this,

$$\gamma_P(Q) = \frac{\| \text{POS}_P(Q) \|}{\| \mathbb{U} \|},$$

where $\| Set \|$ is the cardinality of *Set*. The naïve version of the RSAR algorithm evaluates $\gamma_P(Q)$ for all possible subsets of the dataset's conditional attributes, stopping when it either reaches 1, or there are no more combinations to investigate. This is clearly not guaranteed to produce the minimal reduct set of attributes. Indeed, given the complexity of this operation, it becomes clear that naïve RSAR is intractable for large dimensionalities.

The QUICKREDUCT Algorithm [2] escapes the NP-hard nature of the naïve version by searching the tree of attribute combinations in a best-first manner. It starts off with an empty subset and adds attributes one by one, each time selecting the attribute whose addition to the current subset will offer the highest increase of $\gamma_P(Q)$. The algorithm stops on satisfaction of one of three conditions: a $\gamma_P(Q)$ of 1 is reached; adding another attribute does not increase γ; or all attributes have been added. As a result, the use of QUICKREDUCT makes RSAR very efficient. It can be implemented in a relatively simple manner, and a number of software optimisations further reduce its complexity in terms of both space and time. Additionally, it is evident that the RSAR will not compromise with a set of conditional attributes that contains large part of the information of the initial set — it will *always* attempt to reduce the attribute set without losing *any* information significant to the classification at hand.

3   An Application Domain

Concern for environmental issues has increased greatly in the last decade [3]. The production of waste, toxic and otherwise, from a vast number of different manufacturing processes and plants is one of the most important issues. It influences directly the future of humanity's food and water supply. It has become clear that even changes in farming and sewage water treatment can affect the river, lake and sea ecologies.

The alga, an ubiquitous single-celled plant, is the most successful coloniser of any ecology on the planet. There are numerous different species of algae, and most of them respond rapidly to environmental changes. Wild increases in summer algae population in recent years are an indirect result of nearby human activities. Booming algae communities are detrimental to water clarity, river life and human activities in such areas, since algae growth is associated with toxic effects. Biologists are attempting to isolate the chemical parameters that control such phenomena.

The aim of this application of the MARS and RSAR techniques is to predict the concentration of seven species of river alga, based on a set of parameters. Samples were taken from European rivers over the period of one year, and analysed to measure the concentrations of eight chemicals. The pH of the water was also measured, as well as the season, river size and flow rate. Population distributions for each of the species were determined in the samples. It is relatively easy to locate relations between one or two of these quantities and a species of algae, but the process of identifying such relations requires well-trained personnel with expertise in Chemistry and Biology and involves microscopic examination that is difficult to Thus, such a process becomes expensive and slow, given the number of quantities involved here.

The dataset includes 200 instances [3]. The first three attributes of each instance (season, river size and flow rate) are represented as linguistic variables. Chemical concentrations and algae population estimates are represented as continuous quantities. The dataset includes a few missing values. To prepare the dataset for use by this technique, the linguistic values were mapped to integers.

RSAR is relatively easy to interface to MARS. The only obstacle is the fact that RSAR works better with discrete values. Massaging the data into a suitable representation is therefore necessary. The dataset's first three conditional attributes are already discrete values. The chemical concentrations exhibit an exponential distribution (as shown in figure 1). These were transformed by $\lfloor \log(x+1) \rfloor$, where x is the attribute value. This was only necessary for RSAR to make its selection of attributes. The transformed dataset was *not* fed to MARS.

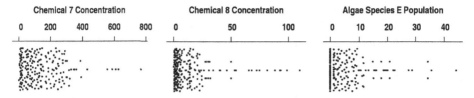

Fig. 1. Density plots for three of the algae dataset attributes.

The seven decision attributes were converted to a logarithmic scale, then quantised into four regions each to signify four exponentially increasing levels of abundance. This processing was performed for the sake of both RSAR and MARS. Although other preprocessing techniques [7] may also be employed to implement this kind of conversion, the use of this particular quantisation is reasonable in a real-world context because of the way the algae population 'counts' are obtained. It is assumed that the river's water is perfectly homogeneous and that any sample of the water, no matter how small, is statistically representative. A few drops of each sample are examined visually via microscope and the number of algae are counted. This allows for human errors in determining the population, as well as statistical inaccuracies. Quantisation alleviates this problem. In addition, if the aim is to predict the behaviour of algae communities, it is far more intuitive to provide linguistic estimates of the population like 'normal', 'lower' and 'higher'.

4   Experimental Results

This paper claims that the application of RSAR to MARS can reduce the time and space required to store MARS models, and that the accuracy of the models does not drop significantly. To test these claims, two series of experiments were performed: one produced MARS models based on the original, unreduced data; the other employed RSAR to reduce the dimensionality of the data and invoked MARS to produce models. The Algae dataset was split randomly (using a 50% split ratio) into training and test datasets, both transformed as described earlier. 100 runs were performed for each experiment series.

For the second experiment, RSAR was run on the suitably preprocessed algae dataset. The reduction algorithm selected seven of the eleven conditional attributes. This implies that the dataset was reasonably information-rich before reduction, but not without redundancies.

For convenience, each of the seven algae species (one for each decision attribute) were processed separately in order to provide seven different MARS models for each experiment run. This simplified assessing the results. For each species, the overall root-mean-square of the difference between the known labelling of a datum and the corresponding MARS prediction was obtained. Although the input conditional attributes are integer due to their quantisation, MARS predicted values are, of course, continuous values. It was preferred not to quantise these in determining the error, so as to perceive the situation more accurately. To provide

Table 1. Experimental results, showing RMS errors.

Alga	Before		After	
	Min	Max	Min	Max
Species A	0.923	1.639	0.924	1.642
Species B	0.893	1.362	0.932	1.389
Species C	0.822	1.202	0.856	1.206
Species D	0.497	0.748	0.595	0.768
Species E	0.723	1.210	0.768	1.219
Species F	0.762	1.158	0.892	1.259
Species G	0.669	0.869	0.689	0.872

a clearer picture, errors for both quantised-quantised and continuous-quantised results are shown. They help provide a minimum and maximum for the expected errors of the experiments.

The results are shown on table 1. Minimum and maximum RMS errors are shown separately for each alga species. It is clear from these results that the implications of employing RSAR as a preprocessor for MARS are minimal. The slight drops in accuracy exhibited after the dimensionality reduction indicates that the process has removed some of the necessary information. However, a fuller investigation of the domain reveals that this is due to the quantisation process employed for this domain, rather than the RSAR methodology itself. Despite this accuracy reduction, however, the MARS models obtained from the low-dimensionality data are simpler by at least a factor of 2^4. This is based on a conservative assumption that each of the four removed attributes is split into only two subregions by MARS. Given the relative complexity of even small MARS models, this reduction in model size is particularly welcome. Processing time required by MARS decreases similarly, although the algorithm's efficiency is such that time requirements are not as important as space requirements.

The advantages of reducing the dimensionality extend to the runtime of the system: performing predictions with the simplified MARS models is much faster and easier. The drop in dataset dimensionality allows for fewer measured variables, which is important for dynamic systems where observables are often restricted, or where the cost of obtaining more measurements is high. Reducing the number of measurements to be made significantly enhances the resultant system, with minimal impact on accuracy.

5   Conclusion

Most attempts to build function approximators and learning systems of all types stumble on the curse of dimensionality. This enforces a ceiling on the applicability of many otherwise elegant methodologies, especially when applied to real world applications, where the exact parameters of a relation are not necessarily known, and many more attributes than necessary are sometimes used to ensure that all the information is present.

MARS is a function approximator that generates continuous models based on spline curves and divide-and-conquer recursive partitioning techniques. Rough set theory can be employed to reduce the dimensionality of datasets. RSAR selects the most information rich attributes in a dataset, while attempting to lose no information needed for the classification task. This paper has presented an approach that integrates RSAR and MARS. RSAR helps reduce the dimensionality of the domain with which MARS has to cope. The RSAR algorithm has proved to be both generalisable and useful in stripping datasets of insignificant information, while retaining more important conditional attributes.

MARS is sensitive to outliers and noise and is particularly prone to high-dimensionality problems. Employing RSAR as a preprocessor to MARS provides accurate results, emphasising the strong points of MARS, and allowing it to be applied to datasets consisting of a moderate to high number of conditional attributes. The resultant MARS model becomes smaller and is processed faster by real-world application systems.

Neither of the two techniques are, of course, perfect. Improvements to the integration framework and further research into the topic remain to be done. For instance, investigating a means for RSAR to properly deal with unknown values would be helpful in a smoother co-operation between RSAR and MARS. Also, it would be interesting to examine the effects of the use of the integrated system in different application domains. Ongoing work along this direction is being carried out at Edinburgh.

References

1. Bartels, R., Beatty, J. and Barsky, B. *Splines for Use in Computer Graphics and Geometric Modelling.* Morgan Kaufmann (1987).
2. Chouchoulas, A. and Shen, Q. Rough Set-Aided Rule Induction for Plant Monitoring. *Proceedings of the 1998 International Joint Conference on Information Science (JCIS'98)*, **2** (1998) 316–319.
3. ERUDIT, European Network for Fuzzy Logic and Uncertainty Modelling in Information Technology. *Protecting Rivers and Streams by Monitoring Chemical Concentrations and Algae Communities (Third International Competition)* http://www.erudit.de/erudit/activities/ic-99/problem.htm (1999).
4. Foley, J. D., van Dam, A., Feiner, S. K., Hughes, J. F. and Philips, R. L. *Introduction to Computer Graphics.* Addison-Wesley (1990).
5. Friedman, J. H. Multivariate Adaptive Regression Splines. *Annals of Statistics*, **19** (1991) 1–67.
6. Pawlak, Z. *Rough Sets: Theoretical Aspects of Reasoning About Data.* Kluwer Academic Publishers, Dordrecht (1991).
7. Shen, Q. and Chouchoulas, A. A modular approach to generating fuzzy rules with reduced attributes for the monitoring of complex systems. *Engineering Applications of Artificial Intelligence* **13** (2000) 263–278.
8. Shen, Q. and Chouchoulas, A. Combining Rough Sets and Data-Driven Fuzzy Learning. *Pattern Recognition*, **32** (1999) 2073–2076.

Rough, and Fuzzy-Rough Classification Methods Implemented in RClass System

Grzegorz Drwal

Institute of Mathematics
Silesian Technical University
Kaszubska 23, 44-100 Gliwice
gdrwal@euler.matfiz.polsl.gliwice.pl

Abstract. This paper presents the RClass system, which was designed as a tool for data validation and the classification of uncertain information. This system uses rough set theory based methods to allow handling uncertain information. Some of proposed classification algorithms also employ fuzzy set theory in order to increase a classification quality. The knowledge base of the RClass system is expressed as a deterministic or non-deterministic decision table with quantitative or qualitative values of attributes, and can be imported from standard databases or text files.

Keywords: Rough sets, fuzzy sets, classifier

1. Introduction

In many real situations, man can choose the proper decision on the basis of uncertain (imprecise, incomplete, inconsistent) information. The imitation of this ability with automatic methods requires the understanding of how humans collect, represents, process and utilize information. Although this problem is far from being solved, several promising formalisms have appeared, like fuzzy set theory proposed by L.A. Zadeh [11] and rough set theory proposed by Z. Pawlak [7]. Fuzzy set theory allows the utilization of uncertain knowledge by means of fuzzy linguistic terms and their membership functions, which reflects human's understanding of the problem. Rough set theory enables to find relationships between data without any additional information (like prior probability, degree of membership), only requiring knowledge representation as a set of if-then rules.

Rough set theory based classification methods have been implemented in the RClass system [3], which will be described in this paper. This system also uses hybrid fuzzy-rough methods in order to improve the classification of quantitative data. Both these approaches - rough and fuzzy-rough are discussed and compered below.

The paper is divided into six sections. Section 1 contains introductory remarks. Sections 2 and 3 introduce the theoretical basis of the RClass system: - the rough classification method is presented in section 2, the so-called fuzzy-rough method is proposed in section 3. The architecture of the system is briefly described in section 4, section 5 presents simple numerical example and, finally, the concluding remarks are detailed in section 6.

W. Ziarko and Y. Yao (Eds.): RSCTC 2000, LNAI 2005, pp. 152–159, 2001.

2. The Rough Classification Method

The knowledge base of the RClass system is represented by a decision table [5], denoted as DT=<U, A, B, V, f>, where U is a finite set of objects, A - a finite set of condition attributes, B - a finite set of decision attributes, $V = \bigcup_{q \in Q = A \cup B} V_q$, where V_q is a domain of the attribute q, and f: U×Q→V is called the information function. For MISO (Multiple Input Single Output) type systems, like RClass, we can assume that B={d} and A={$a_1, a_2, ... a_N$}. The decision table is deterministic (consistent) if A→B. In other cases it is non-deterministic (inconsistent).

System RClass is an example of a maximum classifier, which decides:

$$\text{observation } x' \in \text{class } y_k \Leftrightarrow B'(x', y_k) = \max_{y_i \in V_d} \{B'(x', y_i)\}. \tag{1}$$

Function $B'(x', y_i)$, called the decision function, gives the certainty of classification observation x' to class y_k. We propose to define this function, by analogy to the fuzzy compositional rule of inference [12], as:

$$B'(x', y) = *_{\substack{S \\ r \in U}} \left(\mu^P(T_A(x'), I_A(r)) *_T \mu^P(I_A(r), T_B(y))\right), \tag{2}$$

where, respectively, T_A, T_B, I_A are tolerance sets containing observation x', decision y_i and rule r. Operators $*_T, *_S$ denote t-norms (e.g. min) or t-conorms (e.g. max) and symbol μ^P stands for rough inclusion [8].

Tolerance sets T_A, T_B, I_A can be regarded as a set of all objects, with similar values of respective attributes:

$$T_P(x') = \{r_i \in U: \text{Des}_P(r) \, \tilde{P} \, x'\}; \tag{3}$$

$$I_P(r) = \{r_i \in U : \text{Des}_P(r) \, \tilde{P} \, \text{Des}_P(r_i)\}, \tag{4}$$

in which symbol $\tilde{P}$ denotes a tolerance relation defined for a set of attributes P. The description $\text{Des}_P(r)$ of object r∈U in terms of values of attributes from P is known as:

$$\text{Des}_P(r) = \{(q, v): f(r, q) = v, \forall q \in P\}. \tag{5}$$

Employing tolerance approximation spaces [9,10] allows classification of unseen objects and objects with quantitative values of attributes without discretization. The definition of tolerance relation in RClass system is based on concept of similarity function:

$$\forall r, r' \in U \qquad \text{Des}_P(r) \, \tilde{P} \, \text{Des}_P(r') \Leftrightarrow p_P(r, r') \geq \Gamma, \tag{6}$$

where p_P stands for similarity function, $\Gamma \in [0,1]$ denotes the arbitrary chosen threshold value of this similarity function.

The similarity function measures the degree of a resemblance between two objects r, r'∈ U and can be defined, on the basis of the set of attributes P, as follows:

$$p_P(r,r') = \underset{q_i \in P}{X} \left(w_{q_i} * s_{q_i} \left(f(r,q_i), f(r',q_i) \right) \right),$$

(7)

where: symbol $\underset{q_i \in P}{X}$ stands for any t-norm (e.g. min, prod) or average operator,

w_{q_i} represents the weight of attribute q_i,

s_{q_i} denotes the measure of similarity between values of attribute q_i. As this measure, according to the types of values of the attribute, can be used one of functions presented in the table 1.

Table 1. Similarity measures for different types of attributes

Type of attribute q_i	Similarity measure
Quantitative (e.g. age=18,23, ...)	$s_{q_i}(a_s,a_t) = 1 - \dfrac{|a_s - a_t|}{max_{q_i} - min_{q_i}}$
Qualitative (e.g. color = red, green, ...)	$s_{q_i}(a_s,a_t) = \begin{cases} 0, \text{if } a_s = a_t \\ 1, \text{if } a_s \neq a_t \end{cases}$
Ordered qualitative (e.g. pain = weak, medium, ...)	$s_{q_i}(a_s,a_t) = 1 - \dfrac{|s-t|}{card(V_{q_i}) - 1}$

Symbols max_{q_i} and min_{q_i}, used in the table 1, denotes respectively maximal or minimal value of attribute q_i.

In systems containing only qualitative attributes, the tolerance relation defined above becomes an equivalence relation and therefore the equation (2) can be simplified to following formula:

$$B'(x',y_i) = \mu^P(T_A(x'), T_B(y_i))$$

(8)

This formula stands for the problem of choosing the best decision rule:

$$R_i: T_A(x') \Rightarrow T_B(y_i)$$

(9)

Such decision rule is deterministic (consistent or certain) in DT if $T_A(x') \subseteq T_B(y_i)$, and R_i is non-deterministic (inconsistent or possible) in DT, for other cases. In RClass system, decision rules are evaluated using the concept of rough inclusion. Here are examples of rough inclusions, which are implemented in this system:

- $\mu_1^P(X, Y) = \begin{cases} 1, & \text{if } X \subseteq Y \\ 0, & \text{otherwise} \end{cases};$ (10)

- $\mu_2^P(X, Y) = \begin{cases} \dfrac{\text{card}(X \cap Y)}{\text{card}(X)}, & \text{jeúli } X \neq \varnothing \\ 1, & \text{jeúli } X = \varnothing \end{cases};$ (11)

- $\mu_3^P(X, Y) = \dfrac{\text{card}\big((U \setminus X) \cup Y\big)}{\text{card}(U)};$ (12)

- $\mu_4^P(X, Y) = \begin{cases} 1, & X \subseteq Y \\ \dfrac{\text{card}(Y)}{\text{card}(U)}, & X \not\subset Y. \end{cases}$ (13)

Inclusion 1 is a typical inclusion from set theory and allows only refining deterministic rules from non-deterministic. Inclusion 2 is well known [8] as a standard rough inclusion. For systems with specified strength factors and specificity factors [2] the definition of cardinality operator „card" can be extended as follows:

$$\text{card}^P(X) = \sum_{r_i \in U} \text{Strength\_factor}(r_i) * \text{Specificity\_factor}(r_i) * \chi_x(r_i),\quad (14)$$

where χ_x denotes the characteristic function of set X.

3. The Fuzzy-Rough Classification Method

Applying rough set theory based analysis to the knowledge, which contains measurable, quantitative data, requires using specialized discretization methods or employing a tolerance relation instead of an equivalence relation (as it was proposed above). The disadvantage of both these solutions is that they group similar objects to classes without remembering differences between objects of the same class. We think that this lost, additional information would improve the classification process and therefore for such measurable data we suggest using the fuzzy-rough classification method that was implemented in RClass system. This method, which is a modification of presented above rough method, allow taking into account not only the fact of similarity between two objects, but also the degree of this similarity.

In the fuzzy-rough classification method, we defined classes of tolerance relation as fuzzy sets - the membership functions of these sets determine the degree of similarity of objects, which belong to the same class. This means that tolerance sets $T_P(x)$, $I_P(r)$ from the formula (2) were replaced by fuzzy sets $RT_P(x)$, $RI_P(r)$ with the following membership functions:

$$\mu_{RT_P(x)}(r) = \mathop{\mathrm{X}}_{q_i \in P} \big(w_{q_i} * s_{q_i} \big(f(r, q_i), x_i \big) \big);\quad (15)$$

$$\mu_{RI_P(r')}(r) = \underset{q_i \in P}{\mathbf{X}} \left(w_{q_i} * s_{q_i} \left(f(r,q_i), f(r',q_i) \right) \right). \tag{16}$$

Due to the use of fuzzy sets, we had to apply fuzzy set theory's operators instead of classical set theory's operators - e.g. the following cardinality operator:

$$\forall X \subseteq U \quad card^{RP}(X) = \sum_{r_i \in U} \mu_X(r_i), \tag{17}$$

Employing fuzzy sets operators to definition of a rough inclusion, we introduced the concept of a fuzzy-rough inclusion as an extension of a rough inclusion. After all these changes the final formula for decision function of the fuzzy-rough classifier can be written as:

$$B'(y) = \underset{r \in U}{*_S} \left(\mu^{RP} \left(RT_A(x'), RI_A(r) \right) *_T \mu^{RP} \left(RI_A(r), RT_B(y) \right) \right), \tag{18}$$

where μ^{RP} denotes the any fuzzy-rough inclusion.

It can be easily noticed that, when we have knowledge, which contains only qualitative information, the presented above fuzzy-rough method is equal to the rough method (because there are no differences between object of the same class and therefore all fuzzy sets are crisp sets).

4. Architecture of the RClass System

The most important parts of the RClass system are the classification mechanism, knowledge base and user interface. The classification mechanism, based on rough and fuzzy-rough methods, is the kernel of the RClass system. The knowledge base contains a decision table represented by a set of MISO rules. The user interface is the standard interface for Windows applications (see figure 1). These modules, which enable basic functions of the RClass system, are completed by a knowledge base manager, knowledge analysing tools and an internal editor. The knowledge base manager is responsible for reading and writing knowledge bases from/to external files. It also allows importation of files from many well-known applications. Knowledge analysing tools consist of several basic functions of data analysis based on rough set theory, such as weights calculation, a reduct finding, and the quality of classification calculation. The internal editor enables simple knowledge base modification (for creating knowledge bases and deep modification, an external editor is suggested). The architecture of the RClass system is described in Fig.2.

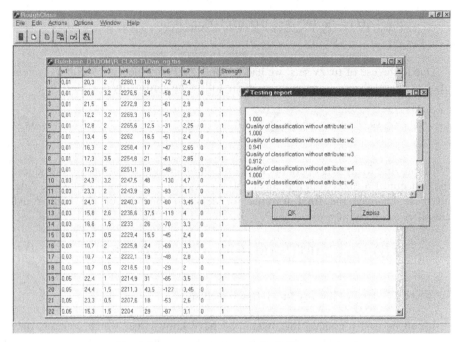

Fig. 1. The user interface of the RClass system

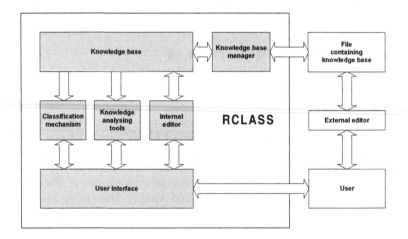

Fig. 2. The architecture of the RClass system

System RClass is a 32-bits object-oriented application working in a Windows95/NT environment. The object-oriented structure [4] makes later development straightforward and allows use of a variety of object-oriented software libraries. 32-bits architecture seems to be more stable and faster than the 16-bits one. The Delphi 3.0 32-bits object-oriented software environment with its Visual Component Library (VCL), was chosen to implement the RClass system, mainly because it can offer a diversity of data base management functions.

5. Numerical Example

This example presents results obtained in RClass system on well-known Iris data set. This data set made by E. Anderson [1] contains 150 samples belonging to one of three classes (Iris Setosa, Iris Versicolor, and Iris Virginica) with 50 samples each. The instances describe iris plant using four input features: sepal length (x_1), sepal width (x_2), petal length (x_3), and petal width (x_4). This data set is very simple – the first class is linearly separable from other two classes, our rough set analysis proved the strong relationships between decision and attributes x_3, x_4 and a big excess of used information. We successfully applied our system to solve much more complicated classification tasks, but we decided to present the Iris problem, because it is often used as a benchmark. The Iris data set is also easily available on Internet network and can be achieved e.g. from address http://www.ics.uci.edu/~mlearn/MLRepository.html. In order that classification task not to be so trivial we used original data set reduced to two most significant input variables x_3, x_4. For our experiments, we prepared training set employing first 25 instances of each class, the remaining 75 samples of original data set were used as a testing set.

Table 2. The knowledge base of fuzzy-rough classifier

x_3	x_4	dec	Strength of rule
1.4	0.2	1	25
4.3	1.3	2	25
5.2	1.7	3	12
6	2.2	3	13

Table 3. The knowledge base of rough classifier

x_3	x_4	dec	Strength of rule
1.5	0.2	1	25
4.3	1.3	2	25
5.7	2.1	3	10
4.9	1.8	3	5
4.9	1.7	3	5
5	1.5	3	5

From the training data set we extracted 4 rules (see table 2) as a database of fuzzy-rough classifier and 6 rules (see table 3) as a database of rough classifier. In spite of such significant knowledge reduction the results achieved for both presented solutions were satisfactory - 2.7% errors for fuzzy-rough classification method and 5.3% errors for rough classification method. The obtained results are similar to results achieved in other systems on the same reduced data sets - e.g. neuro-fuzzy system NEFCLASS, presented in paper [6], came out with 4%

incorrectly classified objects. The better result of fuzzy–rough method shows that even if we employ tolerance relation instead of equivalence relation it can cause the loss of some necessary information and therefore the lower classification quality. It proves that although rough set theory's based analysis can be used for both qualitative and quantitative information, it is more specialized for processing qualitative data.

6. Conclusions

The rough classification system RClass presented above seems to be a universal tool for handling imprecise knowledge. The system is realised as a shell system, which potentially can be applied to various areas of human activity. There are many knowledge validation methods (weights calculation, data reduction and rule extraction) and two classification methods (rough and fuzzy-rough) implemented in the RClass system, therefore it can be used both as a classifier and as a pre-processor of knowledge.

References

[1] Anderson E. „*The IRISes of the Gaspepenisula*" Bull. Amer. IRIS Soc., 59, 2-5, 1935;

[2] Grzymala-Busse J.W. „*Classification of Unseen Examples under Uncertainty*", Fundamenta Informaticae 30, 1997;

[3] Drwal G., Mrózek A. „*System RClass - Software Implementation of the Rough Classifier*" 7th International Symposium on Intelligent Information Systems, pp. 392-395, Malbork 1998;

[4] Mayer B. „*Object-Oriented Software Construction*", Engewood Clif, New York, Prentice Hall, 1988;

[5] Mrózek A. „*Use of Rough Sets and Decision Tables for Implementing Rule-based Control of Industrial Process*", Bulletin of the Polish Academy of Sciences T.S., 34, 1986;

[6] Nauck D., Kruse R. „*A Neuro-Fuzzy Method to Learn Fuzzy Classification Rules from Data*" Fuzzy Sets and Systems 89, pp. 277-288, Elsevier Science, 1997;

[7] Pawlak Z. „*Rough Sets*" International Journal of Computer and Information Sciences, 1982;

[8] Polkowski L., Skowron A. „*Rough Mereological Foundations for Design, Analysis, Synthesis and Control in Distributed Systems*", Information Sciences , vol. 104, Numbers 1-2, 1998;

[9] Polkowski L., Skowron A., Zytkow J. „*Tolerance Based Rough Sets*", Proc. of Third International Workshop on Rough Sets and Soft Computing, San Jose State University, CA, 1994;

[10] Stepaniuk J. „*Rough Sets Similarity Based Learning*", Proc. of 5th European Congress on Intelligent Techniques & Soft Computing, 1634-1638, Aachen, 1997.

[11] Zadeh L.A. „*Fuzzy Sets*", Information and Control, Vol. 8, 1965;

[12] Zadeh L.A. „Outline of a New Approach to the Analysis of Complex Systems and Decision Processes" IEEE Transaction on Systems, Man, and Cybernetics SMC-3, 1973.

Rough Set Approach to Decisions under Risk

S. Greco[1], B. Matarazzo[1], and R. Slowinski[2]

[1]Faculty of Economics, University of Catania, 95129 Catania, Italy

[2]Institute of Computing Science, Poznan University of Technology, 60965 Poznan, Poland

Abstract: In this paper we open a new avenue for applications of the rough set concept to decision support. We consider the classical problem of decision under risk proposing a rough set model based on stochastic dominance. We start with the case of traditional additive probability distribution over the set of states of the world, however, the model is rich enough to handle non-additive probability distributions and even qualitative ordinal distributions. The rough set approach gives a representation of decision maker's preferences in terms of *"if..., then..."* decision rules induced from rough approximations of sets of exemplary decisions.

1.   Introduction

Decisions under risk have been intensively investigated by many researchers (for a comprehensive review see [2]). In this paper, we present an approach to this problem based on the rough sets theory   ([4]). Since decisions under risk involve data expressed on preference-ordered domains (larger outcomes are preferable to smaller outcomes), we consider the **D**ominance-based **R**ough **S**et **A**pproach (DRSA) [3]. The paper has the following plan. Section 2 recalls basic principles of DRSA. Section 3 introduces the rough sets approach to decision under risk. Section 4 presents a didactic example and section 5 contains conclusions.

2.   Dominance-Based Rough Set Approach (DRSA)

For algorithmic reasons, knowledge about objects is represented in the form of an information table. The rows of the table are labelled by *objects*, whereas columns are labelled by *attributes* and entries of the table are *attribute-values*, called *descriptors*.

Formally, by an *information table* we understand the 4-tuple $S=<U,Q,V,f>$, where U is a finite set of objects, Q is a finite set of *attributes*, $V = \bigcup_{q \in Q} V_q$ and V_q is a domain of the attribute q, and $f:U \times Q \to V$ is a total function such that $f(x,q) \in V_q$ for

W. Ziarko and Y. Yao (Eds.): RSCTC 2000, LNAI 2005, pp. 160–169, 2001.

every $q \in Q$, $x \in U$, called an *information function* [4]. The set Q is, in general, divided into set C of *condition attributes* and set D of *decision attributes*. In general, the notion of condition attribute differs from that of *criterion* because the scale (domain) of a criterion has to be ordered according to a decreasing or increasing preference, while the domain of the condition attribute does not have to be ordered.

Assuming that all condition attributes $q \in C$ are criteria, let S_q be an *outranking relation* [5] on U with respect to criterion q such that $x S_q y$ means "x is at least as good as y with respect to criterion q". We suppose that S_q is a total preorder, i.e. a strongly complete and transitive binary relation, defined on U on the basis of evaluations $f(\cdot, q)$.

Furthermore, assuming that the set of decision attributes D (possibly a singleton $\{d\}$) makes a partition of U into a finite number of classes, let $\mathbf{Cl} = \{Cl_t, \ t \in T\}$, $T = \{1, ..., n\}$, be a set of these classes such that each $x \in U$ belongs to one and only one $Cl_t \in \mathbf{Cl}$. We suppose that the classes are ordered, i.e. for all $r, s \in T$, such that $r > s$, the objects from Cl_r are preferred (strictly or weakly [5]) to the objects from Cl_s. More formally, if S is a *comprehensive outranking relation* on U, i.e. if for all $x, y \in U$, $x S y$ means "x is at least as good as y", we suppose: $[x \in Cl_r, \ y \in Cl_s, \ r > s] \Rightarrow [x S y$ and *not* $y S x]$. The above assumptions are typical for consideration of a *multiple-criteria sorting problem*.

The sets to be approximated are called *upward union* and *downward union* of classes, respectively:

$$Cl_t^{\geq} = \bigcup_{s \geq t} Cl_s , \quad Cl_t^{\leq} = \bigcup_{s \leq t} Cl_s , \quad t = 1, ..., n.$$

The statement $x \in Cl_t^{\geq}$ means "x belongs at least to class Cl_t", while $x \in Cl_t^{\leq}$ means "x belongs at most to class Cl_t".

Let us remark that $Cl_1^{\geq} = Cl_n^{\leq} = U$, $Cl_n^{\geq} = Cl_n$ and $Cl_1^{\leq} = Cl_1$. Furthermore, for $t = 2, ..., n$, we have:

$$Cl_{t-1}^{\leq} = U - Cl_t^{\geq} \quad \text{and} \quad Cl_t^{\geq} = U - Cl_{t-1}^{\leq} .$$

The key idea of rough sets is approximation of one knowledge by another knowledge. In **C**lassical **R**ough **S**et **A**pproach (CRSA) [4] the knowledge approximated is a partition of U into classes generated by a set of decision attributes; the knowledge used for approximation is a partition of U into elementary sets of objects that are indiscernible by a set of condition attributes. The elementary sets are seen as "granules of knowledge" used for approximation.

In DRSA, where condition attributes are criteria and classes are preference-ordered, the knowledge approximated is a collection of upward and downward unions of classes and the "granules of knowledge" are sets of objects defined using dominance relation instead of indiscernibility relation. This is the main difference between CRSA and DRSA.

We say that x *dominates* y with respect to $P \subseteq C$, denoted by xD_Py, if xS_qy for all $q \in P$. Given $P \subseteq C$ and $x \in U$, the "granules of knowledge" used for approximation in DRSA are:

- a set of objects dominating x, called *P-dominating set*, $D_P^+(x)=\{y \in U: yD_Px\}$,
- a set of objects dominated by x, called *P-dominated set*, $D_P^-(x)=\{y \in U: xD_Py\}$.

For any $P \subseteq C$ we say that $x \in U$ belongs to Cl *without any ambiguity* if $x \in Cl$ and for all the objects $y \in U$ dominating x with respect to P, we have $y \in Cl$, i.e. $D_P^+(x) \subseteq Cl$. Furthermore, we say that $y \in U$ *could belong* to Cl if there would exist at least one object $x \in Cl_t^{\geq}$ such that y dominates x with respect to P, i.e. $y \in D_P^+(x)$.

Thus, with respect to $P \subseteq C$, the set of all objects belonging to $Cl_t^{\geq}$ without any ambiguity constitutes the *P-lower approximation* of Cl, denoted by $\underline{P}(Cl_t^{\geq})$, and the set of all objects that could belong to $Cl_t^{\geq}$ constitutes the *P-upper approximation* of $Cl_t^{\geq}$, denoted by $\overline{P}(Cl_t^{\geq})$:

$$\underline{P}(Cl_t^{\geq})=\{x \in U: D^+(x) \subseteq Cl_t^{\geq}\}, \quad \overline{P}(Cl_t^{\geq})=\bigcup_{x \in Cl_t^{\geq}} D_P^+(x), \quad \text{for } t=1,\ldots,n.$$

Analogously, using $D^-(x)$ one can define *P-lower approximation* and *P-upper approximation* of Cl_t:

$$\underline{P}(Cl_t^{\leq})=\{x \in U: D^-(x) \subseteq Cl_t^{\leq}\}, \quad \overline{P}(Cl_t^{\leq})=\bigcup_{x \in Cl_t^{\leq}} D_P^-(x), \quad \text{for } t=1,\ldots,n.$$

The *P-boundaries* (*P-doubtful regions*) of $Cl_t^{\geq}$ and $Cl_t^{\leq}$ are defined as:

$$Bn_P(Cl_t^{\geq})=\overline{P}(Cl_t^{\geq}) - \underline{P}(Cl_t^{\geq}), \quad Bn_P(Cl_t^{\leq})=\overline{P}(Cl_t^{\leq}) - \underline{P}(Cl_t^{\leq}), \quad \text{for } t=1,\ldots,n.$$

Due to complementarity of the rough approximations [3], the following property holds:

$$Bn_P(Cl_t^{\geq})=Bn_P(Cl_{t-1}^{\leq}), \quad \text{for } t=2,\ldots,n, \quad \text{and} \quad Bn_P(Cl_t^{\leq})=Bn_P(Cl_{t+1}^{\geq}), \quad \text{for } t=1,\ldots,n-1.$$

For every $t \in T$ and for every $P \subseteq C$ we define the *quality of approximation of partition* **Cl** by set of attributes P, or in short, *quality of sorting*:

$$\gamma_P(Cl)=\frac{\text{card}\left(U - \left(\bigcup_{t \in T} Bn_P\left(Cl_t^{\leq}\right)\right)\right)}{\text{card}(U)}=\frac{\text{card}\left(U - \left(\bigcup_{t \in T} Bn_P\left(Cl_t^{\geq}\right)\right)\right)}{\text{card}(U)}.$$

The quality expresses the ratio of all P-correctly sorted objects to all objects in the table.

Each minimal subset $P \subseteq C$ such that $\gamma_P(Cl) = \gamma_C(Cl)$ is called a *reduct* of Cl and denoted by RED_{Cl}. Let us remark that an information table can have more than one reduct. The intersection of all reducts is called the *core* and denoted by $CORE_{Cl}$.

The dominance-based rough approximations of upward and downward unions of classes can serve to induce a generalized description of objects contained in the information table in terms of "*if...*, *then...*" decision rules. For a given upward or downward union of classes, $Cl_t^{\geq}$ or $Cl_s^{\leq}$, the decision rules induced under a hypothesis that objects belonging to $\underline{P}(Cl_t^{\geq})$ or $\underline{P}(Cl_s^{\leq})$ are *positive* and all the others *negative*, suggest an assignment to "at least class Cl_t" or to "at most class Cl_s", respectively; on the other hand, the decision rules induced under a hypothesis that objects belonging to the intersection $\overline{P}(Cl_s^{\leq}) \cap \overline{P}(Cl_t^{\geq})$ are *positive* and all the others *negative*, are suggesting an assignment to some classes between Cl_s and Cl_t ($s<t$).

Assuming that for each $q \in C$, $V_q \subseteq \mathbf{R}$ (i.e. V_q is quantitative) and that for each $x, y \in U$, $f(x,q) \geq f(y,q)$ implies xS_qy (i.e. V_q is preference-ordered), the following three types of decision rules can be considered:

1) $D_{\geq}$-*decision rules* with the following syntax:

   if $f(x,q_1) \geq r_{q1}$ and $f(x,q_2) \geq r_{q2}$ and $...f(x,q_p) \geq r_{qp}$, then $x \in Cl_t^{\geq}$,

   where $P = \{q_1,...,q_p\} \subseteq C$, $(r_{q1},...,r_{qp}) \in V_{q1} \times V_{q2} \times ... \times V_{qp}$ and $t \in T$;

2) $D_{\leq}$-*decision rules* with the following syntax:

   if $f(x,q_1) \leq r_{q1}$ and $f(x,q_2) \leq r_{q2}$ and $... f(x,q_p) \leq r_{qp}$, then $x \in Cl_t^{\leq}$,

   where $P = \{q_1,...,q_p\} \subseteq C$, $(r_{q1},...,r_{qp}) \in V_{q1} \times V_{q2} \times ... \times V_{qp}$ and $t \in T$;

3) $D_{\geq \leq}$-*decision rules* with the following syntax:

   if $f(x,q_1) \geq r_{q1}$ and $f(x,q_2) \geq r_{q2}$ and $... f(x,q_k) \geq r_{qk}$ and $f(x,q_{k+1}) \leq r_{qk+1}$ and $...$ $f(x,q_p) \leq r_{qp}$, then $x \in Cl_s \cup Cl_{s+1} \cup ... \cup Cl_t$,

   where $O' = \{q_1,...,q_k\} \subseteq C$, $O'' = \{q_{k+1},...,q_p\} \subseteq C$, $P = O' \cup O''$, O' and O'' not necessarily disjoint, $(r_{q1},...,r_{qp}) \in V_{q1} \times V_{q2} \times ... \times V_{qp}$, $s, t \in T$ such that $s<t$;

As it is possible that $\{q_1,...,q_k\} \cap \{q_{k+1},...,q_p\} \neq \emptyset$, in the condition part of a $D_{\geq \leq}$-decision rule we can have "$f(x,q) \geq r_q$" and "$f(x,q) \leq r'_q$", where $r_q \leq r'_q$, for some $q \in C$. Moreover, if $r_q = r'_q$, the two conditions boil down to "$f(x,q) = r_q$".

Since each decision rule is an implication, by a *minimal* decision rule we understand such an implication that there is no other implication with an antecedent of at least the same weakness and a consequent of at least the same strength.

3.   DRSA for Decision under Risk

To apply rough sets theory to decision under risk, we consider the following basic elements:

- a set $S=\{s_1, s_2, ..., s_n\}$ of states of the world, or simply *states*, which are supposed to be mutually exclusive and collectively exhaustive,

- an a priori probability distribution P over the states of the world: more precisely, the probabilities of states s_1, s_2, ..., s_n are p_1, p_2, ..., p_n, respectively ($p_1+ p_2+ ...+ p_n=1$, $p_i\geq 0$, i=1,...n),

- a set $A=\{A_1, A_2, ..., A_m\}$ of *acts*,

- a set $X=\{x_1, x_2, ..., x_r\}$ of consequences or outcomes that for the sake of simplicity we suppose to be expressed in monetary terms and therefore $X\subseteq R$,

- a function g: $A\times S\rightarrow X$ assigning to each act-state pair $(A_i, s_j)\in A\times S$ a consequence $x_h\in X$,

- a set of classes $Cl=\{Cl_1, Cl_2, ..., Cl_t\}$, such that $Cl_1\cup Cl_2\cup ...\cup Cl_t=A$, $Cl_p\cap Cl_q=\varnothing$ for each $p,q\in\{1,2...,t\}$ with $p\neq q$; the classes of Cl are preference-ordered according to an increasing order of their indices, in the sense that for each $A_i,A_j\in A$, if $A_i\in Cl_p$ and $A_j\in Cl_q$ with $p>q$, then A_i is preferred to A_j,

- a function e: $A\rightarrow Cl$ assigning each act $A_i\in A$ to a class $Cl_j\in Cl$.

In this context, two different types of dominance can be considered:

1) (classical) *dominance*: given $A_i,A_j\in A$, A_i dominates A_j iff for each possible state of nature act A_i gives an outcome at least as good as act A_j. More formally, $g(A_i, s_k)\geq g(A_j, s_k)$, for each $s_k\in S$,

2) *stochastic dominance*: given $A_i,A_j\in A$, for each outcome $x\in X$, act A_i gives an outcome at least as good as x with a probability at least as large as act A_j.

Case 1) corresponds to the case in which the utility is state dependent (see e.g. [6]) while case 2) corresponds to a model of decision under risk proposed by Allais [1]. In this paper we consider this second case.

On the basis of an a priori probability distribution P, we can assign to each subset of states of world $W\subseteq S$ ($W\neq\varnothing$) the probability P(W) that one of the states in W is verified, i.e. $P(W) = \sum_{i:s_i\in W} p_i$, and then to build up the set Π of all the possible values P(W), i.e. $\Pi = \{\pi\in [0,1]: \pi=P(W), W\subseteq S\}$.

We define the following function z: $A\times S\rightarrow\Pi$ assigning to each act-state pair $(A_i, s_j)\in A\times S$ a probability $\pi\in\Pi$ as follows:

$$z(A_i, s_j) = \sum_{r:g(A_i,s_r)\geq g(A_i,s_j)} p_r .$$

Therefore, $z(A_i, s_j)$ represents the probability of obtaining an outcome whose value is at least $g(A_i, s_j)$ by act A_i.

On the basis of function $z(A_i, s_j)$ we can define the function $\rho\colon A \times \Pi \to X$ as follows: $\rho(A_i, \pi) = \min\limits_{j:z(A_i,s_j)\geq\pi} g(A_i, s_j)$.

Thus $\rho(A_i, \pi)=x$ means that by act A_i we can gain *at least* x with a given probability π.

Using the function $z(A_i, s_j)$, we can also define the function $\rho'\colon A \times \Pi \to X$ as $\rho'(A_i, \pi) = \max\limits_{j:z(A_i,s_j)\leq\pi} g(A_i, s_j)$.

$\rho'(A_i, \pi)=x$ means that by act A_i we can gain *at most* x with a given probability π.

If the elements Π, $0=\pi_{(1)}, \pi_{(2)} , ..., \pi_{(d)}=1$ ($d=card(\Pi)$), are reordered in such a way that $\pi_{(1)}\leq\pi_{(2)}\leq ... \leq\pi_{(d)}$, then we have $\rho(A_i, \pi_{(j)})= \rho'(A_i, 1-\pi_{(j-1)})$.

Therefore, $\rho(A_i, \pi_{(j)})\leq x$ is equivalent to $\rho'(A_i, 1-\pi_{(j-1)})\geq x$, $A_i\in A$, $\pi_{(j)}\in \Pi$, $x\in X$.

Given $A_i, A_j\in A$, A_i stochastically dominates A_j if and only if $\rho(A_i, \pi)\geq\rho(A_j, \pi)$ for each $\pi\in \Pi$. This is equivalent to say: given $A_i, A_j\in A$, A_i stochastically dominates A_j if and only if $\rho'(A_i, \pi)\leq\rho'(A_j, \pi)$ for each $\pi\in \Pi$.

We can apply DRSA in this context considering as set of objects U the set of acts A, as set of attributes (criteria) Q the set $\Pi\cup\{\mathbf{cl}\}$, where $\mathbf{cl}$ is an attribute representing the classification of acts from A into classes from $\mathbf{Cl}$, as set V the set $X\cup\mathbf{Cl}$, as information function f a function $\mathbf{f}$ such that $\mathbf{f}(A_i, \pi)=\rho(A_i, \pi)$ and $\mathbf{f}(A_i,\mathbf{cl})=e(A_i)$. With respect to the set of attributes Q, the set C of condition attributes corresponds to the set Π and the set of decision attributes D corresponds to $\{\mathbf{cl}\}$.

The aim of this rough set approach to decision under risk is to explain the preferences of the decision maker represented by the assignments of the acts from A to the classes of $\mathbf{Cl}$ in terms of stochastic dominance expressed by means of function ρ.

4.   A Didactic Example

The following example illustrates the approach. Let us consider

- a set $S=\{s_1, s_2, s_3\}$ of states of the world,

- an a priori probability distribution P over the sates of the world defined as follows: $p_1=.25, p_2=.35, p_3=.40$,

- a set $A=\{A_1, A_2, A_3, A_4, A_5, A_6\}$ of acts

- a set $X=\{0, 10, 15, 20, 30\}$ of consequences

- a set of classes $Cl=\{Cl_1, Cl_2, Cl_3\}$, where Cl_1 is the set of bad acts, Cl_2 is the set of medium acts, Cl_3 is the set of good acts,

- a function $g:A\times S\to X$ assigning to each act-state pair $(A_i, s_j)\in A\times S$ a consequence $x_h\in X$ and a function e: $A\to Cl$ assigning each act $A_i\in A$ to a class $Cl_j\in Cl$ presented in the following Table 1.

Table 1 Acts, consequences and assignment to classes from **Cl**

	p_i	A_1	A_2	A_3	A_4	A_5	A_6
s_1	.25	30	0	15	0	20	10
s_2	.35	10	20	0	15	10	20
s_3	.40	10	20	20	20	20	20
cl		good	medium	medium	bad	medium	good

Table 2 shows the values of function $\rho(A_i, \pi)$.

Table 2 Acts, values of function $\rho(A_i, \pi)$ and assignment to classes from **Cl**

	A_1	A_2	A_3	A_4	A_5	A_6
.25	30	20	20	20	20	20
.35	10	20	20	20	20	20
.40	10	20	20	20	20	20
.60	10	20	15	15	20	20
.65	10	20	15	15	20	20
.75	10	20	0	15	10	20
1	10	0	0	0	10	10
Cl	good	medium	medium	bad	medium	good

Table 2 is the data table on which the DRSA is applied. Let us give some examples of the interpretation of the values in Table 2. If we consider the column of act A_3 we have that by act A_3

• the value 20 in the row corresponding to .25 means that the outcome is at least 20 with a probability of at least .25.

• the value 15 in the row corresponding to .65 means that the outcome is at least 15 with a probability of at least .65.

• the value 0 in the row corresponding to .75 means that the outcome is at least 0 with a probability of at least .75.

If we consider the row corresponding to .65, then

• the value 10 relative to A_1, means that by act A_1 the outcome is at least 10 with a probability of at least .65,

• the value 20 relative to A_2, means that by act A_2 the outcome is at least 20 with a probability of at least .65,

and so on.

Applying rough set approach we approximate the following *upward union* and *downward union* of classes:

$Cl_2^\geq = Cl_2 \cup Cl_3$, i.e. the set of the acts at least medium,

$Cl_3^\geq = Cl_3$, i.e. the set of the acts (at least) good,

$Cl_1^\leq = Cl_1$, i.e. the set of the acts (at most) bad,

$Cl_2^\leq = Cl_1 \cup Cl_2$, i.e. the set of the acts at most medium.

The **first result** of the DRSA approach was a discovery that the data table (Table 2) is *not consistent*. Indeed, Table 2 shows that act A_4 stochastically dominates act A_3, however act A_3 is assigned to a better class (medium) than act A_4 (bad). Therefore, act A_3 cannot be assigned without doubts to the set of the class of the at least medium acts as well as act A_4 cannot be assigned without doubts to the set of the class of the (at most) bad acts. In consequence, lower approximation and upper approximation of $Cl_2^\geq$, $Cl_3^\geq$ and $Cl_1^\leq$, $Cl_2^\leq$ are equal, respectively, to

$$\underline{C}(Cl_2^\geq) = \{A_1,A_2,A_5,A_6\} = Cl_2^\geq - \{A_3\}, \quad \overline{C}(Cl_2^\geq) = \{A_1,A_2,A_3,A_4,A_5,A_6\} = Cl_2^\geq \cup \{A_4\},$$

$$\underline{C}(Cl_3^\geq) = \{A_1,A_6\} = Cl_3^\geq, \quad \overline{C}(Cl_3^\geq) = \{A_1,A_6\} = Cl_3^\geq,$$

$$\underline{C}(Cl_1^\leq) = \varnothing = Cl_1^\leq - \{A_4\}, \quad \overline{C}(Cl_1^\leq) = \{A_3,A_4\} = Cl_1^\leq \cup \{A_3\},$$

$$\underline{C}(Cl_2^\leq) = \{A_2,A_3,A_4,A_5\} = Cl_2^\leq, \quad \overline{C}(Cl_2^\leq) = \{A_2,A_3,A_4,A_5\} = Cl_2^\leq.$$

Since there are two inconsistent acts on a total of six acts (A_3,A_4), then the quality of approximation (quality of sorting) is equal to 4/6.

The **second discovery** was one **reduct** of condition attributes (criteria) ensuring the same quality of sorting as the whole set Π of probabilities: $RED_{Cl}^1 = \{.25, .75, 1\}$. This means that we can explain the preferences of the decision maker using only the probabilities in RED_{Cl}^1. RED_{Cl}^1 is also the core because no probability value in RED_{Cl}^1 can be removed without deteriorating the quality of sorting.

The **third discovery** was a set of minimal decision rules describing the decision maker's preferences *[within parentheses there is a verbal interpretation of corresponding decision rule]* (within parentheses there are acts supporting the corresponding rule):

1) if $\rho(A_i, .25) \geq 30$, *then* $A_i \in Cl_3^\geq$,

[if the probability of gaining at least 30 is at least .25, then act A_i is (at least) good] (A_1),

2) *if* $\rho(A_i, .75) \geq 20$ *and* $\rho(A_i, 1) \geq 10$, *then* $A_i \in Cl_3^\geq$,

[if the probability of gaining at least 20 is at least .75 and the probability of gaining at least 10 is (at least) 1 (i.e. for sure the gaining is at least 10) , then act A_i is (at least) good] (A_6),

3) *if* $\rho(A_i, 1) \geq 10$, *then* $A_i \in Cl_2^\geq$,

[if the probability of gaining at least 10 is (at least) 1 (i.e. for sure the gaining is at least 10) , then act A_i is at least medium] (A_1, A_5, A_6),

4) *if* $\rho(A_i, .75) \geq 20$, *then* $A_i \in Cl_2^\geq$,

[if the probability of gaining at least 20 is at least .75 , then act A_i is at least medium] (A_2, A_6),

5) *if* $\rho(A_i, .25) \leq 20$ *(i.e.* $\rho'(A_i, 1) \geq 20)$ *and* $\rho(A_i, .75) \leq 15$ *(i.e.* $\rho'(A_i, .35) \geq 15)$, *then* $A_i \in Cl_2^\leq$,

[if the probability of gaining at most 20 is (at least) 1 (i.e. for sure you gain at most 20) and the probability to gain at most 15 is at least .35, then act A_i is at most medium] (A_3, A_4, A_5),

6) *if* $\rho(A_i, 1) \leq 0$ *(i.e.* $\rho'(A_i, .25) \geq 0)$, *then* $A_i \in Cl_2^\leq$,

[if the probability of gaining at most 0 is at least .25, then act A_i is at most medium] (A_2, A_3, A_4),

7) *if* $\rho(A_i, 1) \geq 0$ *and* $\rho(A_i, 1) \leq 0$ (i.e. $\rho(A_i, 1) = 0$) *and* $\rho(A_i, .75) \leq 15$ *(i.e.* $\rho'(A_i, .35) \geq 10)$, *then* $A_i \in Cl_1 \cup Cl_2$,

[if the probability of gaining at least 0 is 1 (i.e. for sure the outcome is at least 0) and the probability of gaining at most 15 is at least .35, then act A_i is bad or medium, without enough information to assign A_i to only one of the two classes] (A_3, A_4).

Minimal sets of minimal decision rules represent the most concise and non-redundant knowledge contained in Table 1 (and, consequently, in Table 2). The above minimal set of 7 decision rules uses 3 attributes (probability .25, .75 and 1) and 11 elementary conditions, i.e. 26% of descriptors from the original data table (Table 2). Of course this is only a didactic example: representation in terms of decision rules of larger sets of exemplary acts from real applications are more synthetic in the sense of the percentage of used descriptors from the original data table.

5.   Conclusions

We introduced the rough sets theory of decisions under risk using the idea of stochastic dominance. The results are quite encouraging. Let us observe that we considered an additive probability distribution, but an extension to non-additive probability, and even to a qualitative ordinal probability, is straightforward. Furthermore, in case of the elements of set Π are numerous (like in real case application), a subset $\Pi' \subset \Pi$ or a set of the most significant probability values (e.g. 0, .1, .2, ..., .9, 1) can be considered.

Acknowledgement The research of the first two authors has been supported by the Italian Ministry of University and Scientific Research (MURST). The third author wishes to acknowledge financial support of the KBN research grant from State Committee for Scientific Research.

References

1. Allais, M., "The so-called Allais paradox and rational decision under uncertainty", in *Expected Utility Hypotheses and the Allias Paradox*, edited by M. Allais and O. Hagen, pp. 437-681, Reidel, 1979.

2. Fishburn, P. C., *Nonlinear Preferences and Utility Theory*, The John Hopkins University Press, 1988.

3. Greco, S., Matarazzo, B., Slowinski, R., "The use of rough sets and fuzzy sets in MCDM", in *Advances in Multiple Criteria Decision Making*, edited by T. Gal, T. Hanne and T. Stewart, chapter 14, pp. 14.1-14.59,Kluwer Academic Publishers, 1999.

4. Pawlak, Z., "*Rough Sets. Theoretical Aspects of Reasoning about Data*", Kluwer Academic Publishers, 1991.

5. Roy, B., "*Méthodologie Multicritère d'Aide à la Décision*", Paris, Economica, 1985.

6. Wakker, P.P., Zank, H., State Dependent Expected Utility for Savage's State Space, *Mathematics of Operations Research*, 24-1, 1999.

Variable Consistency Model of Dominance-Based Rough Sets Approach

S. Greco[1], B. Matarazzo[1], R. Slowinski[2], and J. Stefanowski[2]

[1]Faculty of Economics, University of Catania, 95129 Catania, Italy
[2]Institute of Computing Science, Poznan University of Technology, 60965 Poznan, Poland

Abstract. Consideration of preference-orders requires the use of an extended rough set model called Dominance-based Rough Set Approach (DRSA). The rough approximations defined within DRSA are based on consistency in the sense of dominance principle. It requires that objects having not-worse evaluation with respect to a set of considered criteria than a referent object cannot be assigned to a worse class than the referent object. However, some inconsistencies may decrease the cardinality of lower approximations to such an extent that it is impossible to discover strong patterns in the data, particularly when data sets are large. Thus, a relaxation of the strict dominance principle is worthwhile. The relaxation introduced in this paper to the DRSA model admits some inconsistent objects to the lower approximations; the range of this relaxation is controlled by an index called consistency level. The resulting model is called variable-consistency model (VC-DRSA). We concentrate on the new definitions of rough approximations and their properties, and we propose a new syntax of decision rules characterized by a confidence degree not less than the consistency level. The use of VC-DRSA is illustrated by an example of customer satisfaction analysis referring to an airline company.

1. Introduction

Rough sets theory introduced by Pawlak [6] is an approach for analysing information about objects described by attributes. It is particularly useful to deal with inconsistencies of input information caused by its granularity. The original rough set approach does not consider, however, the attributes with preference-ordered domains, i.e. *criteria*. Nevertheless, in many real-life problems the *ordering properties* of the considered attributes play an important role. For instance, such features of objects as product quality, market share, debt ratio are typically treated as criteria in economical problems. Motivated by this observation, Greco, Matarazzo and Slowinski [1,3] proposed a generalisation of the rough set approach to problems where ordering properties should be taken into account. Similarly to the original rough sets, this approach is based on approximations of partitions of the objects into pre-defined categories, however, differently to the original model, the categories are ordered from the best to the worst and the approximations are constructed using a *dominance relation* instead of an indiscernibility relation. The considered dominance relation is

W. Ziarko and Y. Yao (Eds.): RSCTC 2000, LNAI 2005, pp. 170–181, 2001.

built on the basis of the information supplied by criteria. The new *Dominance-based Rough Set Approach* (DRSA) was applied to solve typical problems of *Multiple-Criteria Decision Aiding* (MCDA), i.e. choice, ranking and sorting (see e.g. [1,3]).

In this paper, we consider a variant of DRSA used to *multiple-criteria sorting problems*, which concerns an assignment of objects evaluated by a set of criteria to some pre-defined and preference-ordered decision classes. In this variant of DRSA, the sets to be approximated with the dominance relation are, so-called, *upward* and *downward unions of decision classes*. There are known encouraging results of its applications, e.g. to evaluation of bankruptcy risk [2].

The analysis of large real-life data tables shows, however, that for some multiple-criteria sorting problems the application of DRSA identifies large differences between lower and upper approximations of the unions of decision classes and, moreover, rather weak decision rules, i.e. supported by few objects from lower approximations. The reason is that inconsistency, in the sense of dominance principle, between objects x and y assigned to very distant classes, h and t, respectively, (x dominates y, while class h is worse than t) causes inconsistency (ambiguity) also with all objects belonging to intermediate classes (from h to t) and dominated by x. In such cases it seems reasonable to relax the conditions for assignment of objects to lower approximations of the unions of decision classes. Classically, only non-ambiguous objects can be included in lower approximations. The relaxation will admit some ambiguous objects as well; the range of this ambiguity will be controlled by an index called *consistency level*. The aim of this article is to present a generalization of DRSA to variable consistency model (VC-DRSA).

This kind of relaxation has been already considered within the classical indiscernibility-based rough set approach, by means of so-called *variable precision rough set model* (VPRS) [11]. VPRS allows defining lower approximations accepting a limited number of counterexamples controlled by pre-defined level of certainty.

The paper is organized as follows. In section 2, main concepts of VC-DRSA are introduced, including rough approximations, approximation measures and decision rules. An illustrative example presented in section 3 refers to a real problem of customer satisfaction analysis in an airline company. The final section groups conclusions.

2. Variable Consistency Dominance-Based Rough Set Approach (VC-DRSA)

For algorithmic reasons, information about objects is represented in the form of an information table. The rows of the table are labelled by *objects*, whereas columns are labelled by *attributes* and entries of the table are *attribute-values*. Formally, by an *information table* we understand the 4-tuple $S=<U,Q,V,f>$, where U is a finite set of objects, Q is a finite set of *attributes*, $V = \bigcup_{q \in Q} V_q$ and V_q is a domain of the attribute q, and $f:U \times Q \rightarrow V$ is a total function such that $f(x,q) \in V_q$ for every $q \in Q$, $x \in U$, called an *information function* [6]. The set Q is, in general, divided into set C of *condition attributes* and set D of *decision attributes*.

Assuming that all condition attributes $q \in C$ are *criteria*, let $\succeq_q$ be an *weak preference relation* on U with respect to criterion q such that $x \succeq_q y$ means "x is at least as good as y with respect to criterion q". We suppose that $\succeq_q$ is a total preorder, i.e. a strongly complete and transitive binary relation, defined on U on the basis of evaluations $f(\cdot, q)$.

Furthermore, assuming that the set of decision attributes D (possibly a singleton $\{d\}$) makes a partition of U into a finite number of decision classes, let $Cl = \{Cl_t, t \in T\}$, $T = \{1, \dots, n\}$, be a set of these classes such that each $x \in U$ belongs to one and only one class $Cl_t \in Cl$. We suppose that the classes are preference-ordered, i.e. for all $r, s \in T$, such that $r > s$, the objects from Cl_r are preferred to the objects from Cl_s. The above assumptions are typical for consideration of a *multiple-criteria sorting problem*.

The sets to be approximated are called *upward union* and *downward union* of classes, respectively:

$$Cl_t^{\geq} = \bigcup_{s \geq t} Cl_s, \quad Cl_t^{\leq} = \bigcup_{s \leq t} Cl_s, \quad t = 1, \dots, n.$$

The statement $x \in Cl_t^{\geq}$ means "x belongs at least to class Cl_t", while $x \in Cl_t^{\leq}$ means "x belongs at most to class Cl_t".

Let us remark that $Cl_1^{\geq} = Cl_n^{\leq} = U$, $Cl_n^{\geq} = Cl_n$ and $Cl_1^{\leq} = Cl_1$. Furthermore, for $t = 2, \dots, n$, we have:

$$Cl_{t-1}^{\leq} = U - Cl_t^{\geq} \quad \text{and} \quad Cl_t^{\geq} = U - Cl_{t-1}^{\leq}.$$

The key idea of rough sets is approximation of one knowledge by another knowledge. In classical rough set approach (CRSA), the knowledge approximated is a partition of U into classes generated by a set of decision attributes; the knowledge used for approximation is a partition of U into elementary sets of objects that are indiscernible by a set of condition attributes. The elementary sets are seen as "*granules of knowledge*" used for approximation.

In DRSA approach, where condition attributes are criteria and classes are preference-ordered, the knowledge approximated is a collection of *upward* and *downward unions of classes* and the "granules of knowledge" are sets of objects defined using a dominance relation instead of an indiscernibility relation. This is the main difference between CRSA and DRSA. Let us define now the dominance relation.

We say that x *dominates* y with respect to $P \subseteq C$, denoted by $x D_P y$, if $x \succeq_q y$ for all $q \in P$. Given $P \subseteq C$ and $x \in U$, the "granules of knowledge" used for approximation in DRSA are:

- a set of objects dominating x, called *P-dominating set*, $D_P^+(x) = \{y \in U: y D_P x\}$,

- a set of objects dominated by x, called *P-dominated set*, $D_P^-(x) = \{y \in U: x D_P y\}$.

For any $P \subseteq C$ we say that $x \in U$ belongs to $Cl_t^{\geq}$ with *no ambiguity at consistency level* $l \in (0, 1]$, if $x \in Cl_t^{\geq}$ and at least $l*100\%$ of all objects $y \in U$ dominating x with respect to P also belong to $Cl_t^{\geq}$, i.e.

$$\frac{card\left(D_P^+(x)\cap Cl_t^{\geq}\right)}{card\left(D_P^+(x)\right)}\geq l.$$

The level l is called *consistency level* because it controls the degree of consistency between objects qualified as belonging to $Cl_t^{\geq}$ without any ambiguity. In other words, if $l<1$, then $(1-l)*100\%$ of all objects $y\in U$ dominating x with respect to P do not belong to $Cl_t^{\geq}$ and thus contradict the inclusion of x in $Cl_t^{\geq}$.

Analogously, for any $P\subseteq C$ we say that $x\in U$ belongs to $Cl_t^{\leq}$ *with no ambiguity at consistency level* $l\in(0, 1]$, if $x\in Cl_t^{\leq}$ and at least $l*100\%$ of all the objects $y\in U$ dominated by x with respect to P also belong to $Cl_t^{\leq}$, i.e.

$$\frac{card\left(D_P^-(x)\cap Cl_t^{\leq}\right)}{card\left(D_P^-(x)\right)}\geq l.$$

Thus, for any $P\subseteq C$, each object $x\in U$ is either ambiguous or non-ambiguous at consistency level l with respect to the upward union $Cl_t^{\geq}$ ($t=2,...,n$) or with respect to the downward union $Cl_t^{\leq}$ ($t=1,...,n-1$).

The concept of non-ambiguous objects at some consistency level l leads naturally to the definition of P-lower approximations of the unions of classes $Cl_t^{\geq}$ and $Cl_t^{\leq}$.

$$\underline{P}^l\left(Cl_t^{\geq}\right)=\{x\in Cl_t^{\geq} : \frac{card\left(D_P^+(x)\cap Cl_t^{\geq}\right)}{card\left(D_P^+(x)\right)}\geq l\}, \quad \underline{P}^l\left(Cl_t^{\leq}\right)=\{x\in Cl_t^{\leq} : \frac{card\left(D_P^-(x)\cap Cl_t^{\leq}\right)}{card\left(D_P^-(x)\right)}\geq l\}.$$

Given $P\subseteq C$ and consistency level l, we can define the P-*upper approximations* of $Cl_t^{\geq}$ and $Cl_t^{\leq}$, denoted by $\overline{P}^l\left(Cl_t^{\geq}\right)$ and $\overline{P}^l\left(Cl_t^{\leq}\right)$, by complementation of $\underline{P}^l\left(Cl_{t-1}^{\leq}\right)$ and $\underline{P}^l\left(Cl_{t+1}^{\geq}\right)$ with respect to U:

$$\overline{P}^l\left(Cl_t^{\geq}\right)=U-\underline{P}^l\left(Cl_{t-1}^{\leq}\right), \quad \overline{P}^l\left(Cl_t^{\leq}\right)=U-\underline{P}^l\left(Cl_{t+1}^{\geq}\right).$$

$\overline{P}^l\left(Cl_t^{\geq}\right)$ can be interpreted as the set of all the objects belonging to $Cl_t^{\geq}$, *possibly ambiguous* at consistency level l. Analogously, $\overline{P}^l\left(Cl_t^{\leq}\right)$ can be interpreted as the set of all the objects belonging to $Cl_t^{\leq}$, *possibly ambiguous* at consistency level l. The P-*boundaries* (P-doubtful regions) of $Cl_t^{\geq}$ and $Cl_t^{\leq}$ are defined as:

$$Bn_P(Cl_t^{\geq})=\overline{P}^l\left(Cl_t^{\geq}\right)-\underline{P}^l\left(Cl_t^{\geq}\right), \quad Bn_P(Cl_t^{\leq})=\overline{P}^l\left(Cl_t^{\leq}\right)-\underline{P}^l\left(Cl_t^{\leq}\right), \quad \text{for } t=1,...,n.$$

The *variable consistency* model of the dominance-based rough set approach provides some degree of flexibility in assigning objects to lower and upper approximations of the unions of decision classes. It can easily be demonstrated that for $0<l'<l\leq 1$ and $t=2,...,n$,

$$\underline{P}^l\left(Cl_t^{\geq}\right)\subseteq\underline{P}^{l'}\left(Cl_t^{\geq}\right) \quad \text{and} \quad \overline{P}^{l'}\left(Cl_t^{\geq}\right)\subseteq\overline{P}^l\left(Cl_t^{\geq}\right).$$

The *variable consistency* model is inspired by Ziarko's model of the *variable precision* rough set approach [11,12], however, there is a significant difference in the definition of rough approximations because $\underline{P}^l(Cl_t^{\geq})$ and $\overline{P}^l(Cl_t^{\geq})$ are composed of non-ambiguous and ambiguous <u>objects</u> at consistency level l, respectively, while Ziarko's $\underline{P}^l(Cl_t)$ and $\overline{P}^l(Cl_t)$ are composed of P-indiscernibility <u>sets</u> such that at least $l*100\%$ of these sets are included in Cl_t or have an non-empty intersection with Cl_t, respectively. If one would like to use Ziarko's definition of variable precision rough approximations in the context of multiple-criteria sorting, then the P-indiscernibility sets should be substituted by P-dominating sets $D_P^+(x)$, however, then the notion of ambiguity that naturally leads to the general definition of rough approximations (see [9]) looses its meaning. Moreover, bad side effect of a direct use of Ziarko's definition is that a lower approximation $\underline{P}^l(Cl_t^{\geq})$ may include objects y assigned to Cl_h, where h is much less than t, if y belongs to $D_P^+(x)$ that was included in $\underline{P}(Cl^{\geq})$. When the decision classes are preference ordered, it is reasonable to expect that objects assigned to far worse classes than the considered union are not counted to the lower approximation of this union.

Furthermore, the following properties can be proved:

1) $$\overline{P}^l(Cl_t^{\geq})=Cl_t^{\geq}\cup\{x\in Cl_{t-1}^{\leq}:\frac{card\left(D_P^-(x)\cap Cl_{t-1}^{\leq}\right)}{card\left(D_P^-(x)\right)}<l\}=$$

$$=Cl_t^{\geq}\cup\{x\in Cl_t^{\leq}:\frac{card\left(D_P^-(x)\cap Cl_t^{\geq}\right)}{card\left(D_P^-(x)\right)}>1-l\},$$

$$\overline{P}^l(Cl_t^{\leq})=Cl_t^{\leq}\cup\{x\in Cl_{t+1}^{\geq}:\frac{card\left(D_P^-(x)\cap Cl_{t+1}^{\geq}\right)}{card\left(D_P^-(x)\right)}<l\}=$$

$$=Cl_t^{\leq}\cup\{x\in Cl_t^{\geq}:\frac{card\left(D_P^+(x)\cap Cl_t^{\leq}\right)}{card\left(D_P^+(x)\right)}>1-l\},$$

2) $\underline{P}(Cl_t)\subseteq Cl_t^{\geq}\subseteq\overline{P}^l(Cl_t^{\geq})$, $\underline{P}(Cl_t^{\leq})\subseteq Cl_t^{\leq}\subseteq\overline{P}^l(Cl_t^{\leq})$.

Due to complementarity of the rough approximations [3], also the following property holds:

$Bn_P(Cl_t^{\geq})=Bn_P(Cl_{t-1}^{\leq})$, for $t=2,...,n$, and $Bn_P(Cl_t^{\leq})=Bn_P(Cl_{t+1}^{\geq})$, for $t=1,...,n-1$.

For every $t\in T$ and for every $P\subseteq C$ we define the *quality of approximation of partition* Cl by set of criteria P, or in short, *quality of sorting*:

$$\gamma_P(Cl) = \frac{card\left(U - \left(\bigcup_{t\in T} Bn_P\left(Cl_t^{\leq}\right)\right)\right)}{card\,(U)} = \frac{card\left(U - \left(\bigcup_{t\in T} Bn_P\left(Cl_t^{\geq}\right)\right)\right)}{card\,(U)}.$$

The quality expresses the ratio of all P-correctly sorted objects to all objects in the table.

Each minimal subset $P \subseteq C$ such that $\gamma_P(Cl) = \gamma_C(Cl)$ is called a *reduct* of Cl and denoted by RED_{Cl}. Let us remark that an information table can have more than one reduct. The intersection of all reducts is called the *core* and denoted by $CORE_{Cl}$.

Let us remind that the dominance-based rough approximations of upward and downward unions of classes can serve to induce a generalized description of objects contained in the information table in terms of "*if..., then...*" decision rules. For a given upward or downward union of classes, $Cl_t^{\geq}$ or $Cl_s^{\leq}$, the decision rules induced under a hypothesis that objects belonging to $\underline{P}^l\left(Cl_t^{\geq}\right)$ (or $\underline{P}^l\left(Cl_s^{\leq}\right)$) are *positive* and all the others *negative*, suggest an assignment to "at least class Cl_t" (or to "at most class Cl_s"). They are called D$_{\geq}$- (or D$_{\leq}$) *certain decision rules* because they assign objects to classes without any ambiguity. Next, if upper approximations differ from lower approximations, another kind of decision rules can be induced under the hypothesis that objects belonging to $\overline{P}\left(Cl_t^{\geq}\right)$ (or to $\overline{P}\left(Cl_s^{\leq}\right)$) are *positive* and all the others *negative*. These rules are called D$_{\geq}$- (or D$_{\leq}$) *possible decision* rules suggesting that an object *could belong* to "at least class $Cl_t^{\geq}$" (or "at most class $Cl_s^{\leq}$"). Yet another option is to induce D$_{\geq\leq}$-approximate decision rules from the intersection $\overline{P}^l(Cl_s^{\leq}) \cap \overline{P}^l(Cl_t^{\geq})$ instead of possible rules. For more discussion see [8].

Within VC-DRSA, decision rules are induced from examples belonging to extended approximations. So, it is necessary to assign to each decision rules an additional parameter α, called *confidence of the rule*. It controls the discrimination ability of the rule.

Assuming that for each $q \in C$, $V_q \subseteq \mathbf{R}$ (i.e. V_q is quantitative) and that for each $x, y \in U$, $f(x,q) \geq f(y,q)$ implies $x \succeq_q y$ (i.e. V_q is preference-ordered), the following two basic types of variable-consistency decision rules can be considered:

1)   D$_{\geq}$-*decision rules* with the following syntax:

if $f(x,q_1) \geq r_{q1}$ and $f(x,q_2) \geq r_{q2}$ and $...f(x,q_p) \geq r_{qp}$, then $x \in Cl_t^{\geq}$ with confidence α,

where $P = \{q_1,...,q_p\} \subseteq C$, $(r_{q1},...,r_{qp}) \in V_{q1} \times V_{q2} \times ... \times V_{qp}$ and $t \in T$;

2)   D$_{\leq}$-*decision rules* with the following syntax:

if $f(x,q_1) \leq r_{q1}$ and $f(x,q_2) \leq r_{q2}$ and $... f(x,q_p) \leq r_{qp}$, then $x \in Cl_t^{\leq}$ with confidence α,

where $P = \{q_1,...,q_p\} \subseteq C$, $(r_{q1},...,r_{qp}) \in V_{q1} \times V_{q2} \times ... \times V_{qp}$ and $t \in T$;

We say that an object *supports* a decision rule if it matches both condition and decision parts of the rule. On the other hand, an object is *covered* by a decision rule if it matches the condition part of the rule. More formally, given a $D_\geq$-rule ρ : *if* $f(x,q_1) \geq r_{q1}$ *and* $f(x,q_2) \geq r_{q2}$ *and* $...f(x,q_p) \geq r_{qp}$, *then* $x \in Cl_t^\geq$, an object $y \in U$ supports decision rule ρ iff $f(y,q_1) \geq r_{q1}$ and $f(y,q_2) \geq r_{q2}$ and $... f(y,q_p) \geq r_{qp}$ and $y \in Cl_t^\geq$, while y is covered by ρ iff $f(y,q_1) \geq r_{q1}$ and $f(y,q_2) \geq r_{q2}$ and $... f(y,q_p) \geq r_{qp}$. Similar definitions hold for $D_\leq$-decision rules.

Let *Cover(ρ)* denote the set of all objects covered by the rule ρ. Thus, the confidence α of $D_\geq$-decision rule ρ is defined as: $\dfrac{card\left(Cover(\rho) \cap Cl_t^\geq\right)}{card\left(Cover(\rho)\right)}$. For $D_\leq$-decision rule the confidence is defined in a similar way.

Let us remark that the decision rules are induced from *P*-lower approximations whose composition is controlled by user-specified consistency level *l*. In consequence, the value of confidence α for the rule should be constrained from the bottom. It seems reasonable to require that the smallest accepted confidence of the rule should not be lower than the currently used consistency level *l*. Indeed, in the worst case, some objects from the *P*-lower approximation may create a rule using all criteria from *P* thus giving a confidence $\alpha \geq l$. The user may have a possibility of increasing this lower bound for confidence of the rule but then decision rules may not cover all objects from the approximations.

Moreover, we require that each decision rule is minimal. Since a decision rule is an implication, by a *minimal* decision rule we understand such an implication that there is no other implication with an antecedent of at least the same weakness (in other words, rule using a subset of elementary conditions or/and weaker elementary conditions) and a consequent of at least the same strength (in other words, rule assigning objects to the same union or sub-union of classes) with a not worse confidence $\alpha \geq l$.

Consider a $D_\geq$-decision rule "*if* $f(x,q_1) \geq r_{q1}$ *and* $f(x,q_2) \geq r_{q2}$ *and* $...f(x,q_p) \geq r_{qp}$, *then* $x \in Cl_t^\geq$ " with confidence α. If there exists an object $y \in \underline{P}^l\left(Cl_t^\geq\right)$, $P = \{q_1, q_2, ..., q_p\}$ and $l \leq \alpha$, such that $f(y,q_1) = r_{q1}$ and $f(y,q_2) = r_{q2}$ and $...f(y,q_p) = r_{qp}$, then y is called *basis* of the rule. Each $D_\geq$-decision rule having a basis is called *robust* because it is "founded" on an object existing in the data table. Analogous definition of robust decision rules holds for $D_\leq$-decision rules.

The induction of variable-consistency decision rules can be done using properly modified algorithms proposed for DRSA. Let us remind that in DRSA, decision rules should have confidence equal to 1. The key modification of rule induction algorithms for VC-DRSA consists in accepting as rules such conjunctions of elementary conditions that yield confidence $\alpha \geq l$. Let us also notice that different strategies of rule induction could be used [10]. For instance, one can wish to induce a minimal and complete set of rules covering all input examples, or all minimal rules, or a subset of rules satisfying some user's pre-defined requirements, e.g. generality or support. The details of one of the rule induction algorithms for VC-DRSA can be found in [4].

3. Illustrative Example

Let us illustrate the above concepts on a didactic example. The example refers to a real problem of customer satisfaction analysis [7] in an airline company. The company has diffused a questionnaire to its customers in order to get opinion about the quality of its services. Among the questions of the questionnaire there are three items concerning specific aspects of the aircraft comfort: space for hand luggage (q_1), seat width (q_2) and leg room (q_3). Moreover, there is also a question about an overall evaluation of the aircraft comfort (d). A customer's answer on each of these questions gives an evaluation on a three grade ordinal scale: poor, average, good.

The data table contains 50 objects (questionnaires) described by the set $C=\{q_1, q_2, q_3\}$ of 3 criteria corresponding to the considered aspects of the aircraft comfort and the overall evaluation $D=\{d\}$. All criteria are to be maximized. The scale of criteria is number-coded: 1=poor, 2=average, 3=good. The overall evaluation d creates three decision classes, which are preference-ordered according to increasing class number, i.e. Cl_1=poor, Cl_2=average, Cl_3=good. The analysed data are presented in Table 1.

Table 1. Customer satisfaction data table

Cust.	q_1	q_2	q_3	d	Cust.	q_1	q_2	q_3	d
1	1	3	2	2	26	2	2	1	2
2	1	3	1	1	27	1	2	2	1
3	3	3	1	1	28	3	2	2	2
4	3	3	2	3	29	1	3	2	2
5	3	1	3	1	30	2	3	1	2
6	2	3	1	2	31	1	1	1	1
7	2	1	2	2	32	1	2	2	2
8	1	1	3	2	33	3	1	1	1
9	2	3	3	3	34	2	2	1	2
10	3	3	1	1	35	3	2	1	2
11	1	3	3	2	36	2	2	2	2
12	2	1	1	2	37	1	1	2	1
13	1	1	1	1	38	3	1	2	2
14	1	3	3	3	39	3	3	1	1
15	1	1	1	1	40	1	1	1	1
16	2	2	3	2	41	3	1	1	1
17	2	1	1	2	42	1	2	1	1
18	1	2	3	2	43	1	3	2	2
19	3	2	1	3	44	3	1	2	1
20	2	2	2	1	45	2	2	3	1
21	2	3	3	3	46	3	2	2	1
22	1	2	3	2	47	2	1	3	1
23	3	2	2	2	48	1	2	2	3
24	2	2	1	2	49	2	2	3	3
25	1	3	1	1	50	3	3	3	3

The marketing department of the airline company wants to analyse the influence of the three specific aspects on the overall evaluation of the aircraft comfort. Thus, a sample of questionnaires was analysed using VC-DRSA. As the decision classes are ordered, the following downward and upward unions of classes are to be considered:

at most poor: $Cl_1^\leq$ = {2,3,5,10,13,15,20,24,25,27,31,33,37,39,40,41,42,44,45,46,47},

at most average: $Cl_2^\leq$ ={1,2,3,5,6,7,8,10,11,12,13,15,16,17,18,20,22,23,24,25,26,27, 28,29,30,31, 32,33,34,35,36,37,38,39,40,41,42,43,44,45,46,47};

at least average: $Cl_2^\geq$ ={1,4,6,7,8,9,11,12,14,16,17,18,19,21,22,23,26,28,29,30,32, 34,35,36,38,43,48,49,50},

at least good - $Cl_3^\geq$ ={4,9,14,19,21,48,49,50}.

Let us observe that in the data table there are several inconsistencies. For instance, object #3 dominates object #6, because its evaluations on all criteria q_1, q_2, q_3 are not worse, however, it is assigned to the decision class Cl_1 worse than Cl_2 to which belongs object #6. This means that the customer #3 gave an evaluation for all the considered aspects not worse than the evaluation given by customer # 6 and, on another hand, customer #3 gave an overall evaluation of the aircraft comfort worse than the overall evaluation of customer #6. There are 99 inconsistent pairs in the data table violating the dominance principle in this way.

The data table has been analysed by VC-DRSA assuming the confidence level l=0.8. In this case, the approximations of upward and downward unions of decision classes are the following (the objects present in the lower approximations obtained for confidence level l=1 are in bold):

$\underline{C}^{0.8}(Cl_1^\leq)$={**2,13,15,25,31,37,40,42**},

$\overline{C}^{0.8}(Cl_1^\leq)$ ={2,3,5,6,7,8,10,12,13,15,17,19,20,24,25,26,27,30,31,33,34,35,36,37,38, 39,40,41,42,43,44,45,46,47},

$Bn_C^{0.8}(Cl_1^\leq)$={3,5,6,7,8,10,12,17,19,20,24,26,27,30,33,34,35,36,38,39,40,41,43,44, 45,46,47};

$\underline{C}^{0.8}(Cl_2^\leq)$={1,2,3,**5,6,7,8**,10,11,**12,13,15**,16,**17**,18,20,22,23,**24,25,26**,27,28,29,**30,31**, 32,**33,34**,35,36,**37,38**,39,**40,41**, 42,43,**44**,45,46,**47**},

$\overline{C}^{0.8}(Cl_2^\leq)$={1,2,3,5,6,7,8,10,11,12,13,15,16,17,18,19,20,22,23,24,25,26,27,28,29,30, 31,32,33,34,35,36,37,38,39,40,41,42,43,44,45,46,47,48,49},

$Bn_C^{0.8}(Cl^\leq)$={19,48,49};

$\underline{C}^{0.8}(Cl_2^\geq)$={1,**4,9,11,14**,16,18,**21**,22,23,28,**29**,32,48,49,**50**},

$\overline{C}^{0.8}(Cl_2^\geq)$={1,3,4,5,6,7,8,9,10,11,12,14,16,17,18,19,20,21,22,23,24,26,272,28,29,30, 32,33,34,35,36,38,39,40,41,43, 44,45,46,47, 48,49,50},

$Bn_C^{0.8}(Cl_2^\geq)$={3,5,6,7,8,12,17,19,20,24,26,27,30,33,34,35,36,38,39,40,41,43, 44,45,46,47};

$\underline{C}^{0.8}(Cl_3^\geq)$={**4,9**,14,**21,50**},

$\overline{C}^{0.8}(Cl_3^\geq)$={4,9,14,19,21,48,49,50},

$Bn_C^{0.8}(Cl_3^\geq)$={19,48,49}.

The set of all robust decision rules having a confidence level $\alpha \geq 0.8$ was induced from the above approximations. Let us remark that rules having confidence $\alpha = 1$ are the same as obtained with the DRSA rule induction algorithm. The induced rules are listed below:

Rule 1. *if* $(f(x,q_1) \leq 1)$ *and* $(f(x,q_2) \leq 1)$, *then* $x \in Cl_1^{\leq}$ $[\alpha = .83]$

Rule 2. *if* $(f(x,q_1) \leq 1)$ *and* $(f(x,q_3) \leq 1)$, *then* $x \in Cl_1^{\leq}$ $[\alpha = 1]$

Rule 3. *if* $(f(x,q_1) \leq 1)$ *and* $(f(x,q_2) \leq 1)$ *and* $(f(x,q_3) \leq 2)$, *then* $x \in Cl_1^{\leq}$ $[\alpha = 1]$

Rule 4. *if* $((f(x,q_1) \leq 1)$, *then* $x \in Cl^{\leq}$ $[\alpha = 0.89]$

Rule 5. *if* $((f(x,q_1) \leq 2)$, *then* $x \in Cl^{\leq}$ $[\alpha = 0.85]$

Rule 6. *if* $((f(x,q_2) \leq 2)$, *then* $x \in Cl^{\leq}$ $[\alpha = 0.91]$

Rule 7. *if* $((f(x,q_2) \leq 1)$, *then* $x \in Cl^{\leq}$ $[\alpha = 1]$

Rule 8. *if* $(f(x,q_1) \leq 2)$ *and* $(f(x,q_2) \leq 2)$, *then* $x \in Cl^{\leq}$ $[\alpha = 0.92]$

Rule 9. *if* $((f(x,q_3) \leq 2)$, *then* $x \in Cl^{\leq}$ $[\alpha = 0.92]$

Rule 10. *if* $((f(x,q_3) \leq 1)$, *then* $x \in Cl^{\leq}$ $[\alpha = 0.95]$

Rule 11. *if* $(f(x,q_1) \leq 2)$ *and* $(f(x,q_3) \leq 1)$, *then* $x \in Cl^{\leq}$ $[\alpha = 1]$

Rule 12. *if* $(f(x,q_1) \leq 2)$ *and* $(f(x,q_3) \leq 2)$, *then* $x \in Cl^{\leq}$ $[\alpha = 0.96]$

Rule 13. *if* $(f(x,q_2) \leq 2)$ *and* $(f(x,q_3) \leq 2)$, *then* $x \in Cl^{\leq}$ $[\alpha = 0.93]$

Rule 14. *if* $(f(x,q_2) \geq 2)$ *and* $(f(x,q_3) \geq 2)$, *then* $x \in Cl$ $[\alpha = 0.82]$

Rule 15. *if* $(f(x,q_2) \geq 3)$ *and* $(f(x,q_3) \geq 2)$, *then* $x \in Cl$ $[\alpha = 1]$

Rule 16. *if* $(f(x,q_2) \geq 2)$ *and* $(f(x,q_3) \geq 3)$, *then* $x \in Cl$ $[\alpha = 0.9]$

Rule 17. *if* $(f(x,q_1) \geq 3)$ *and* $(f(x,q_2) \geq 2)$ *and* $(f(x,q_3) \geq 2)$, *then* $x \in Cl$ $[\alpha = 0.83]$

Rule 18. *if* $(f(x,q_2) \geq 3)$ *and* $(f(x,q_3) \geq 3)$, *then* $x \in Cl$ $[\alpha = 0.8]$

Rule 19. *if* $(f(x,q_1) \geq 3)$ *and* $(f(x,q_2) \geq 3)$ *and* $(f(x,q_3) \geq 2)$, *then* $x \in Cl$ $[\alpha = 1]$

Rule 20. *if* $(f(x,q_1) \geq 2)$ *and* $(f(x,q_2) \geq 3)$ *and* $(f(x,q_3) \geq 3)$, *then* $x \in Cl$ $[\alpha = 1]$

Managers of the airline company appreciated the easy verbal interpretation of the above rules. For example rule 16 says that if seat width is at least average and leg room is (at least) good, then the overall evaluation of the comfort is at least average with a confidence of 90%, independently of the space for hand luggage. This rule covers 10 examples, i.e. 20% of all questionnaires. Let us remark that such a strong pattern would not be discovered by DRSA because of one negative example (#45) being inconsistent with four other positive examples (#16,18,22,49).

Let us remark that relaxation of the confidence level has at least the following two positive consequences:
1) it enlarges lower approximations, permitting to regain many objects that were inconsistent with some marginal objects from outside of the considered union of

classes: for instance six objects, #16,18,22,32,48,49, inconsistent with object #45 entered the lower approximation $\underline{C}^{0.8}(Cl_2^{\geq})$;

2) it discovers strong rule patterns that did not appear when the dominance was strictly observed in rule induction: for instance rule 16 above.

The above aspects are very useful when dealing with large real data sets, which is usually the case of customer satisfaction analysis.

4. Conclusions

The relaxation of the dominance principle introduced in the dominance-based rough set approach results in a more flexible approach insensitive to marginal inconsistencies encountered in data sets. The variable-consistency model thus obtained maintains all basic properties of the rough sets theory, like inclusion, monotonicity with respect to supersets of criteria and with respect to the consistency level. The rough approximations resulting from this model are the basis for construction of decision rules with a required confidence. The variable-consistency model is particularly useful for analysis of large data sets where marginal inconsistencies may considerably reduce the lower approximations and prevent discovery of strong rule patterns.

Acknowledgement
The research of S. Greco and B. Matarazzo has been supported by the Italian Ministry of University and Scientific Research (MURST). R. Slowinski and J. Stefanowski wish to acknowledge financial support of the KBN research grant no. 8T11F 006 19 from the State Committee for Scientific Research.

References

1. Greco S., Matarazzo B., Slowinski R., *Rough Approximation of Preference Relation by Dominance Relations*, ICS Research Report 16/96, Warsaw University of   Technology, Warsaw, 1996. Also in *European Journal of Operational Research* 117 (1999) 63-83.

2. Greco S., Matarazzo B., Slowinski R., A new rough set approach to evaluation of bankruptcy risk. In C. Zopounidis (eds.), *Operational Tools in the Management of Financial Risk*, Kluwer Academic Publishers, Dordrecht, Boston, 1998, 121-136.

3. Greco S., Matarazzo B., Slowinski R., The use of rough sets and fuzzy sets in MCDM. In T. Gal, T. Stewart and T. Hanne (eds.) *Advances in Multiple Criteria Decision Making*, chapter 14, Kluwer Academic Publishers, Boston, 1999, 14.1-14.59.

4. Greco S., Matarazzo B., Slowinski R., Stefanowski J., An algorithm for induction of decision rules consistent with dominance principle. In: *Proc. 2nd Int. Conference on Rough Sets and Current Trends in Computing*, Banff, October 16-19, 2000 (to appear).

5. Pawlak, Z., Rough sets, *International Journal of Information & Computer Sciences* 11 (1982) 341-356.

6. Pawlak, Z., *Rough Sets. Theoretical Aspects of Reasoning about Data*, Kluwer Academic Publishers, Dordrecht, 1991.

7. Siskos Y., Grigoroudis E., Zopounidis C., Sauris O., Measuring customer satisfaction using a collective preference disaggregation model, *Journal of Global Optimization*, 12 (1998) 175-195.

8. Slowinski R., Stefanowski J, Greco, S., Matarazzo, B., Rough sets processing of inconsistent information. *Control and Cybernetics* 29 (2000) no.1, 379-404.

9. Slowinski R., Vanderpooten D., A generalized definition of rough approximations based on similarity. *IEEE Transactions on Data and Knowledge Engineering*, 12 (2000) no. 2, 331-336.

10. Stefanowski J., On rough set based approaches to induction of decision rules. In Polkowski L., Skowron A. (eds.) *Rough Sets in Data Mining and Knowledge Discovery*, vol. 1, Physica-Verlag, Heidelberg, 1998, 500-529.

11. Ziarko W. Variable precision rough sets model. *Journal of Computer and Systems Sciences* 46 (1993) no. 1, 39-59.

12. Ziarko W. Rough sets as a methodology for data mining. In Polkowski L., Skowron A. (eds.) *Rough Sets in Data Mining and Knowledge Discovery*, vol. 1, Physica-Verlag, Heidelberg, 1998, 554-576.

Approximations and Rough Sets Based on Tolerances

Jouni Järinen

Turku Centre for Computer Science (TUCS)
Lemminkäisenkatu 14 A, FIN-20520 Turku, Finland
jjarvine@cs.utu.fi

Abstract In rough set theory it is supposed that the knowledge about objects is limited by an indiscernibility relation. Commonly indiscernibility relations are assumed to be equivalences interpreted so that two objects are equivalent if we cannot distinguish them by their properties. However, there are natural indiscernibility relations which are not transitive, and here we assume that the knowledge about objects is restricted by a tolerance relation R. We study R-approximations, R-definable sets, R-equalities, and investigate briefly the structure of R-rough sets.

1 Indiscernibility

The rough set theory introduced by Z. Pawlak in the early eighties [15] deals with incomplete information about objects. More precisely, it considers situations in which objects may be indiscernible by their properties. A primary application area has been data mining.

In rough set theory it is assumed that the knowledge about objects is restricted by an indiscernibility relation. Usually indiscernibility relations are supposed to be equivalences such that two objects are equivalent if we cannot distinguish them by our information. Since such an equivalence E induces a partition whose blocks are the equivalence classes, the objects of the given universe U can be divided into three classes with respect to any subset $X \subseteq U$:

1. the objects, which surely are in X;
2. the objects, which are surely not in X;
3. the objects, which possibly are in X.

The objects in class 1 form the lower E-approximation of X, and the objects of type 1 and 3 form together its upper E-approximation. The E-boundary of X consists of objects in class 3. Subsets of U which are identical to both of their approximations are called E-definable.

In this work we assume that the indiscernibility relations are tolerances. Namely, transitivity is not an obvious property of indiscernibility relations. For example, transitive relations fail to capture the indiscernibility involved in the concept of heap as it appears in the *Eubulide's paradox*: "What is the smallest number of grains making a heap of grains?" (see [8], for example).

In Pawlak's information systems [14] indiscernibility relations which are equivalences arise naturally when one considers a given set of attributes: two objects are

W. Ziarko and Y. Yao (Eds.): RSCTC 2000, LNAI 2005, pp. 182–189, 2001.
© Springer-Verlag Berlin Heidelberg 2001

equivalent when their values of all attributes in the set are the same. However, some of the natural indiscernibility relations encountered in nondeterministic information systems are not necessarily transitive (see [13], for example).

Example 1. Suppose that the set $U = \{1 \ldots 7\}$ consists of seven patients called 1 2 ... 7, respectively. Their body temperatures (temp), blood pressures (BP), and hemoglobin values (Hb) are given in Table 1.

	temp	BP	Hb
1	39.3	103/65	125
2	39.1	97/60	116
3	39.2	109/71	132
4	37.1	150/96	139
5	37.3	145/93	130
6	37.8	143/95	121
7	36.7	138/83	130

Table 1.

Let us define an indiscernibility relation R so that two patients are R-related if their values for body temperature, blood pressure and hemoglobin are so close to each other that the differences are insignificant. The tolerance R is represented graphically by the following graph.

For example, patients 1 and 2 are R-related since their values for body temperature, blood pressure, and hemoglobin are essentially the same.

We end this section by noting that most of the results issued in this work are presented in the author's doctoral dissertation [6] where all proofs and many further facts can be found. We also remark that all our lattice theoretical notions and results appear in [1], for example.

2   Approximations

First we study approximations determined by tolerance relations. B. Konikowska [7] and J. A. Pomykała [18] considered approximation operations defined by strong similarity relations of nondeterministic information systems. Also J. Nieminen [9] has studied approximations induced by tolerances but his definition is not the same as ours. Furthermore, J. A. Pomykała [17] and W. Żakowski [21] have investigated related notions of approximations defined by covers. In [19] A. Skowron and J. Stepaniuk introduced tolerance approximation spaces and they presented in this context several results concerning particularly attribute reduction. J. Järvinen considered in [5] dependence spaces induced by preimage relations. In this setting attribute dependencies induced by reflexive and symmetric indiscernibility relations can be studied.

If a binary relation R on a set U is reflexive and symmetric, it is called a *tolerance relation* on U. The set of all tolerance relations on U is denoted by $\mathrm{Tol}(U)$. For all $R \in \mathrm{Tol(U)}$ and $a \in U$, the set $a/R = \{b \in U \mid aRb\}$ is the *R-neighborhood* of a.

Definition 2. Let U be a set of *objects* and let R be a tolerance on U. The *lower R-approximation* and the *upper R-approximation* of a set $X \subseteq U$ are defined by

$$X_R = \{x \in U \mid x/R \subseteq X\};$$
$$X^R = \{x \in U \mid x/R \cap X \neq \emptyset\},$$

respectively. The set $B_R(X) = X^R - X_R$ is called the *R-boundary* of X.

The set X_R (resp. X^R) consists of elements which surely (resp. possibly) belong to X in view of the knowledge provided by R. The R-boundary is the actual area of uncertainty. It consists of elements whose membership in X cannot be decided when R-related objects cannot be distinguished from each other.

Let us consider the tolerance R defined in Example 1. The upper R-approximation of $X = \{4, 5, 6\}$ is $X^R = \{4, 5, 6, 7\}$ and its lower R-approximation is $X_R = \{4, 6\}$. Hence, the R-boundary of X is $B_R(X) = \{5, 7\}$. For example, 5 is in $B_R(X)$ since both in X and in $X^\complement$ there is an object which is indiscernible from 5.

Our first proposition gives some basic properties of approximations. Note that the set of all subsets of a set U is denoted by $\wp(U)$, and for any $X \subseteq U$, $X^\complement = U - X$ is the *complement* of X.

Proposition 3. *If $R \in \mathrm{Tol}(U)$ and $X, Y \subseteq U$, then*
  (a) $\emptyset_R = \emptyset^R = \emptyset$ *and* $U_R = U^R = U$;
  (b) $X_R \subseteq X \subseteq X^R$;
  (c) $(X_R)^\complement = (X^\complement)^R$ *and* $(X^R)^\complement = (X^\complement)_R$;
  (d) $B_R(X) = B_R(X^\complement)$;
  (e) $X \subseteq Y$ *implies* $X^R \subseteq Y^R$ *and* $X_R \subseteq Y_R$.

Note that Proposition 3(d) means simply that if we cannot decide when an object is in X, we obviously cannot decide whether it is in $X^\complement$ either.

Definition 4. Let $\mathcal{P} = (P, \leq)$ be an ordered set. A pair $(\blacktriangleright, \blacktriangleleft)$ of maps $\blacktriangleright : P \to P$ and $\blacktriangleleft : P \to P$ (which we refer to as the *right map* and the *left map*, respectively) is called a *dual Galois connection* on $\mathcal{P}$ if $\blacktriangleright$ and $\blacktriangleleft$ are order-preserving and $p^{\blacktriangleleft\blacktriangleright} \leq p \leq p^{\blacktriangleright\blacktriangleleft}$ for all $p \in P$.

It is now obvious that a dual Galois connection on $\mathcal{P} = (P, \leq)$ is a Galois connection between $\mathcal{P}$ and $\mathcal{P}^\partial = (P, \geq)$. The following proposition presents some basic properties of dual Galois connections, which follow easily from the properties of Galois connections (see [1, 12], for example). But before that we define closure and interior operators.

Let $\mathcal{P} = (P, \leq)$ be an ordered set. Then a map $c : P \to P$ is a *closure operator* if for all $x, y \in P$, $x \leq c(x)$, $x \leq y$ implies $c(x) \leq c(y)$, and $c(c(x)) = c(x)$. An element $x \in P$ is called *closed* if $c(x) = x$. The set of all c-closed elements is denoted by P_c. It is well-known that $P_c = \{c(x) \mid x \in P\}$. If $i : P \to P$ is a closure operator on $\mathcal{P}^\partial = (P, \geq)$, it is an *interior operator* on $\mathcal{P}$.

Proposition 5. *Let* $(\blacktriangleright, \blacktriangleleft)$ *be a dual Galois connection on a complete lattice* $\mathcal{P}$.

(a) *For all* $p \in P$, $p^{\blacktriangleright \blacktriangleleft \blacktriangleright} = p^{\blacktriangleright}$ *and* $p^{\blacktriangleleft \blacktriangleright \blacktriangleleft} = p^{\blacktriangleleft}$.

(b) *The map* $c\colon P \to P$, $p \mapsto p^{\blacktriangleright \blacktriangleleft}$ *is a closure operator on* $\mathcal{P}$ *and the map* $k\colon P \to P$, $p \mapsto p^{\blacktriangleleft \blacktriangleright}$ *is an interior operator on* $\mathcal{P}$.

(c) *If* c *and* k *are the mappings defined in* (b), *then restricted to the sets of* c-*closed elements* P_c *and* k-*closed elements* P_k, *respectively,* $\blacktriangleright$ *and* $\blacktriangleleft$ *yield a pair* $\blacktriangleright\colon P_c \to P_k$, $\blacktriangleleft\colon P_k \to P_c$ *of mutually inverse order-isomorphisms between the complete lattices* $(P_c, \leq)$ *and* $(P_k, \leq)$.

(d) *If* $A \subseteq P$, *then* $(\bigvee A)^{\blacktriangleright} = \bigvee \{p^{\blacktriangleright} \mid p \in A\}$ *and* $(\bigwedge A)^{\blacktriangleleft} = \bigwedge \{p^{\blacktriangleleft} \mid p \in A\}$.

(e) *The kernel of the map* $\blacktriangleright\colon P \to P$ *is a congruence on* $(P, \vee)$ *such that the greatest element in the congruence class of any* $p \in P$ *is* $p^{\blacktriangleright \blacktriangleleft}$.

(f) *The kernel of the map* $\blacktriangleleft\colon P \to P$ *is a congruence on* $(P, \wedge)$ *such that the least element in the congruence class of any* $p \in P$ *is* $p^{\blacktriangleleft \blacktriangleright}$.

We can now write the following proposition.

Proposition 6. *If* $R \in \text{Tol}(U)$, *then* $(^{R}, _{R})$ *is a dual Galois connection on* $(\wp(U), \subseteq)$.

Let $\blacktriangleright$ and $\blacktriangleleft$ be maps on $\wp(U)$. We say that $(\blacktriangleright, \blacktriangleleft)$ is a *an approximation pair*, if there exists an $R \in \text{Tol}(U)$ such that $X^{\blacktriangleright} = X^{R}$ and $X^{\blacktriangleleft} = X_{R}$ for all $X \subseteq U$. In [6] J. Järvinen characterized the dual Galois connections on $(\wp(U), \subseteq)$ which are approximation pairs.

We end this section by presenting a corollary of Propositions 5 and 6.

Corollary 7. *Let* $R \in \text{Tol}(U)$, $X \subseteq U$, *and* $\mathcal{H} \subseteq \wp(U)$.

(a) $((X^{R})_{R})^{R} = X^{R}$ *and* $((X_{R})^{R})_{R} = X_{R}$;

(b) *the map* $X \mapsto (X^{R})_{R}$ *is a closure operator and the map* $X \mapsto (X_{R})^{R}$ *is an interior operator;*

(c) $(\bigcup \mathcal{H})^{R} = \bigcup \{X^{R} \mid X \in \mathcal{H}\}$ *and* $(\bigcap \mathcal{H})_{R} = \bigcap \{X_{R} \mid X \in \mathcal{H}\}$;

(d) $(\bigcap \mathcal{H})^{R} \subseteq \bigcap \{X^{R} \mid X \in \mathcal{H}\}$ *and* $(\bigcup \mathcal{H})_{R} \supseteq \bigcup \{X_{R} \mid X \in \mathcal{H}\}$.

We note that if E is an equivalence relation, then $(X_E)_E = (X_E)^E = X_E$ and $(X^E)^E = (X^E)_E = X^E$ (cf. Corollary 7(a)). Note also that the inclusions in Corollary 7(d) can be proper, and this holds for approximations defined by equivalences as well.

3   Definable Sets

In this section we consider R-definable sets, where R is a tolerance.

Definition 8. *Let* $R \in \text{Tol}(U)$. *A set* $X \subseteq U$ *is* R-*definable if* $X_R = X^R$.

We denote by $\text{Def}(R)$ the set of all R-definable sets. It is obvious that a set X is R-definable if and only if its R-boundary $B_R(X)$ is empty.

In Example 1, $\text{Def}(R) = \{\emptyset, \{1, 2, 3\}, \{4, 5, 6, 7\}, U\}$. For instance, the R-definable set $\{1, 2, 3\}$ consists of the patients who probably have certain influenza.

By Lemma 9(a), to show that a set is definable requires only half as much work as the definition suggests. Lemma 9(b) presents another interesting property of definable sets (cf. Corollary 7(d)).

Lemma 9. *Let $R \in \mathrm{Tol}(U)$ and $X, Y \subseteq U$.*
  (a) *$X \in \mathrm{Def}(R)$ iff $X^R = X$ iff $X_R = X$;*
  (b) *if $X \in \mathrm{Def}(R)$, then $(X \cup Y)_R = X_R \cup Y_R$ and $(X \cap Y)^R = X^R \cap Y^R$.*

Next we present some notions which we shall need. The set of all equivalences on U is denoted by $\mathrm{Eq}(U)$. We say that $X \subseteq U$ is *saturated* by an equivalence $E \in \mathrm{Eq}(U)$, if X is the union of some equivalence classes of E or $X = \emptyset$. The set of all sets saturated by E is denoted by $\mathrm{Sat}(E)$. A family $\mathcal{F} \subseteq \wp(U)$ is called a *complete field of sets* if $\emptyset, U \in \mathcal{F}$, $X^{\complement} \in \mathcal{F}$ for all $X \in \mathcal{F}$ and $\bigcup \mathcal{H}, \bigcap \mathcal{H} \in \mathcal{F}$ for all $\mathcal{H} \subseteq \mathcal{F}$. For $R \in \mathrm{Tol}(U)$, we denote by R^e the smallest equivalence on U, which includes R. It is well-known and obvious that $R^e = \bigcap \{E \in \mathrm{Eq}(U) \mid R \subseteq E\}$.

Proposition 10. *If $R \in \mathrm{Tol}(U)$, then*
  (a) *$\mathrm{Def}(R) = \mathrm{Sat}(R^e)$;*
  (b) *$\mathrm{Def}(R)$ is a complete field of sets.*

Note also that X^{R^e} is the least R-definable set including X, and that X_{R^e} is the greatest R-definable set included in X.

Let $E \in \mathrm{Eq}(U)$. By Proposition 10(a), the E-definable sets are the unions of some (or none) E-classes, and this actually is Pawlak's original definition of E-definable sets [15]. We also mention that the sets X^E and X_E are E-definable for all $X \subseteq U$, but the approximations induced by tolerances are not necessarily definable. For instance in Example 1, $\{2\}^R = \{1, 2\}$ and $\{1, 2\}_R = \{2\}$.

4   Rough Equalities

First we define different types of equalities based on approximations. For equivalence relations the corresponding notions were defined in [10, 11], where M. Novotný and Z. Pawlak also characterized all three types of rough equalities on finite sets. M. Steinby [20] has generalized these characterizations by omitting the assumption of finiteness, and in [6] J. Järvinen characterized rough equalities defined by tolerances.

Definition 11. Let $R \in \mathrm{Tol}(U)$. We define in $\wp(U)$ the *lower R-equality* $\approx_R$, the *upper R-equality* $\approx^R$, and the R-equality $\equiv_R$ by the following conditions:

$$X \approx_R Y \iff X_R = Y_R;$$
$$X \approx^R Y \iff X^R = Y^R;$$
$$X \equiv_R Y \iff X_R = Y_R \text{ and } X^R = Y^R.$$

Because the pair $(^R, _R)$ is a dual Galois connection on $(\wp(U), \subseteq)$, the relation $\approx^R$ is a congruence on $(\wp(U), \cup)$ such that the greatest element in the $\approx^R$-class of any $X \subseteq U$ is $(X^R)_R$ by Proposition 5. Analogously, the relation $\approx_R$ is a congruence on $(\wp(U), \cap)$ such that the least element in the $\approx_R$-class of any $X \subseteq U$ is $(X_R)^R$. The relation $\equiv_R$ is an equivalence on $\wp(U)$, but it is not usually a congruence on $(\wp(U), \cup)$ or on $(\wp(U), \cap)$, and this holds even if R is an equivalence.

Let us denote by $\mathcal{G}(R)$ the set of the greatest elements of the $\approx^R$-classes and $\mathcal{L}(R)$ denotes the set of the least elements of the $\approx_R$-classes. Now the next proposition holds.

Proposition 12. *For any $R \in \mathrm{Tol}(U)$,*

(a) $\mathcal{G}(R) = \{X_R \mid X \subseteq U\}$ *and* $\mathcal{L}(R) = \{X^R \mid X \subseteq U\};$

(b) *the complete lattices $(\mathcal{G}(R), \subseteq)$ and $(\mathcal{L}(R), \subseteq)$ are order-isomorphic.*

For an equivalence E (see e.g. [20]), $\mathcal{G}(E) = \mathcal{L}(E) = \mathrm{Sat}(E) = \mathrm{Def}(E)$. Furthermore, if E_1 and E_2 are different equivalences, then obviously $\approx_{E_1}$ differs from $\approx_{E_2}$, $\approx^{E_1}$ differs from $\approx^{E_2}$, and $\equiv_{E_1}$ differs from $\equiv_{E_2}$. We end this section by showing that different tolerances may define the same lower and upper equality.

Example 13. Let $U = \{a, b, c, d\}$ and let R, S, and T be tolerances on U such that

$$a/R = c/S = b/T = \{a, b, c\}; \qquad b/R = d/S = a/T = \{a, b, d\};$$
$$c/R = a/S = d/T = \{a, c, d\}; \qquad d/R = b/S = c/T = \{b, c, d\}.$$

Now the relations $\approx_R$, $\approx_S$, and $\approx_T$ are equal, and they have the six congruence classes $\{\emptyset, \{a\}, \ldots, \{c, d\}\}, \{\{a, b, c\}\}, \{\{a, b, d\}\}, \{\{a, c, d\}\}, \{\{b, c, d\}\}$, and $\{U\}$. Similarly, the relations $\approx^R$, $\approx^S$ and $\approx^T$ are identical and they have six congruence classes $\{\emptyset\}, \{\{a\}\}, \{\{b\}\}, \{\{c\}\}, \{\{d\}\}$, and $\{\{a, b\}, \ldots, U\}$. It is obvious that also $\equiv_R, \equiv_S$, and $\equiv_T$ are the same. They have 11 equivalence classes.

5   Rough Sets

In the classical setting a rough set may be defined as an equivalence class of sets which look the same in view of the of the knowledge restricted by the given indiscernibility relation, i.e., as a class of sets having the same lower approximation and the same upper approximation. Rough sets defined by a tolerance can be defined the same way.

Definition 14. Let $R \in \mathrm{Tol}(U)$. The equivalence classes of the R-equality $\equiv_R$ are called *R-rough sets.*

For studying the structure of the system of R-rough sets it is convenient to adopt a formulation used (for equivalence relations) by T. B. Iwiński [4]. It is based on a fact that R-rough sets can be equivalently viewed as pairs (X_R, X^R), where $X \subseteq U$, since each $\equiv_R$-class $\mathcal{C}$ is uniquely determined by the pair (X_R, X^R), where X is any member of $\mathcal{C}$.

Let $R \in \mathrm{Tol}(U)$. For any $X \subseteq U$, the pair $R\langle X \rangle = (X_R, X^R)$ is called the *R-approximation* of X. The set of all R-approximations of the subsets of U is $\mathcal{A}(R) = \{R\langle X \rangle \mid X \subseteq U\}$.

Let R be the tolerance on $U = \{1, \ldots, 7\}$ considered in Example 1. For example, the sets $\{1, 5\}$ and $\{2, 3, 4, 6, 7\}$ belong to the same R-rough set. Note that these sets are the complements of each other! This rough set can be represented by the pair $(\emptyset, U)$. The other members of it are $\{2, 3, 5\}$ and $\{1, 4, 6, 7\}$.

There is a canonical order-relation $\leq$ on $\wp(U) \times \wp(U)$ defined by

$$(X_1, X_2) \leq (Y_1, Y_2) \iff X_1 \subseteq Y_1 \text{ and } X_2 \subseteq Y_2.$$

Because $\mathcal{A}(R) \subseteq \wp(U) \times \wp(U)$ for all $R \in \mathrm{Tol}(U)$, the set $\mathcal{A}(R)$ may be ordered by $\leq$. M. Gehrke and E. Walker have shown in [3] that if E is an equivalence, then $(\mathcal{A}(E), \leq)$

is a complete Stone lattice isomorphic to $(\mathbf{2}^I \times \mathbf{3}^J, \leq)$, where $I = \{a/E : |a/E| = 1\}$ and $J = \{a/E : |a/E| > 1\}$. Here $\mathbf{n}$ denotes the set $\{0, 1, \ldots, n-1\}$ ordered by $0 < 1 \cdots < n - 1$.

The rough set systems defined by tolerances differ essentially from the ones defined by equivalences. Our final example shows that $(\mathcal{A}(R), \leq)$, where $R \in \mathrm{Tol}(U)$, is not necessarily even a semilattice when $U \geq 5$.

Example 15. Let $U = \{1, 2, 3, 4, 5\}$ and let R be a tolerance on U such that

$$1/R = \{1, 2\}, 2/R = \{1, 2, 3\}, 3/R = \{2, 3, 4\}, 4/R = \{3, 4, 5\}, 5/R = \{4, 5\}.$$

The Hasse diagram of $(\mathcal{A}(R), \leq)$ is given in Figure 1. Note that $(\mathcal{A}(R), \leq)$ is not a join-semilattice because, for instance, the elements $(1, 123)$ and $(\emptyset, 1234)$ do not have a supremum. Similarly, $(\mathcal{A}(R), \leq)$ is not a meet-semilattice since the elements $(12, 1234)$ and $(1, U)$ do not have an infimum.

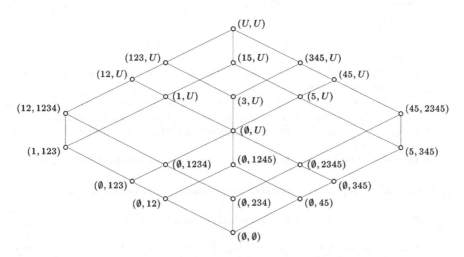

Figure 1.

On the other hand, $(\mathcal{A}(R), \leq)$ is always a lattice, if $R \in \mathrm{Tol}(U)$ and $|U| \leq 4$.

Conclusions. We have studied approximations and rough sets based on tolerances. The motivation of this work was that there are natural indiscernibility relations which are not transitive. The main contribution of this paper is the observation that the properties of approximations and rough sets defined by tolerances differ from the classical ones defined by equivalences. The biggest differences are that the tolerance approximations are not necessarily definable, that different tolerances may induce the same rough equalities, and that the ordered sets of tolerance rough sets are not necessarily even semilattices. In the end of each section these differences were briefly discussed. The future works are to study how our work can be applied for example to Pawlak's information systems and to contexts in the sense of Wille [2].

Acknowledgement. The author thanks Professor Magnus Steinby for the careful reading of the manuscript and for his many valuable comments and suggestions.

References

1. B. A. DAVEY, H. A. PRIESTLEY, *Introduction to Lattices and Order*, Cambridge University Press, Cambridge, 1990.
2. B. GANTER, R. WILLE, *Formal Concept Analysis. Mathematical Foundations* (translated from the 1996 German original by CORNELIA FRANZKE), Springer-Verlag, Berlin, 1999.
3. M. GEHRKE, E. WALKER, *On the Structure of Rough Sets*, Bulletin of the Polish Academy of Sciences, Mathematics 40 (1992), pp 235–245.
4. T. B. IWIŃSKI, *Algebraic Approach to Rough Sets*, Bulletin of the Polish Academy of Sciences, Mathematics 35 (1987), pp 673–683.
5. J. JÄRVINEN, *Preimage Relations and Their Matrices*, in [16], pp 139–146.
6. J. JÄRVINEN, *Knowledge Representation and Rough Sets*, Doctoral Dissertation, University of Turku, March 1999 (available at http://www.cs.utu.fi/jjarvine/thesis).
7. B. KONIKOWSKA, *A Logic for Reasoning about Relative Similarity*, Studia Logica 58 (1997), pp 185–226.
8. S. MARCUS, *The Paradox of the Heap of Grains in Respect to Roughness, Fuzziness and Negligibility*, in [16], pp 19–23.
9. J. NIEMINEN, *Rough Tolerance Equality and Tolerance Black Boxes*, Fundamenta Informaticae 11 (1988), pp 289–296.
10. M. NOVOTNÝ, Z. PAWLAK, *Characterization of Rough Top Equalities and Rough Bottom Equalities*, Bulletin of the Polish Academy of Sciences, Mathematics 33 (1985), pp 92–97.
11. M. NOVOTNÝ, Z. PAWLAK, *On Rough Equalities*, Bulletin of the Polish Academy of Sciences, Mathematics 33 (1985), pp 98–113.
12. O. ORE, *Galois Connexions*, Transactions of American Mathematical Society 55 (1944), pp 493–513.
13. E. ORŁOWSKA, *Logic of Nondeterministic Information*, Studia Logica XLIV (1985), pp 93–102.
14. Z. PAWLAK, *Information Systems. Theoretical Foundations*, Informations Systems 6 (1981), pp 205–218.
15. Z. PAWLAK, *Rough Sets*, International Journal of Computer and Information Sciences 5 (1982), pp 341–356.
16. L. POLKOWSKI, A. SKOWRON (eds.), *Rough Sets and Current Trends in Computing*, Lecture Notes in Artificial Intelligence 1424, Springer–Verlag, Berlin/Heidelberg, 1998.
17. J. A. POMYKAŁA, *On Definability in the Nondeterministic Information System*, Bulletin of the Polish Academy of Sciences, Mathematics 36 (1988), pp 193–210.
18. J. A. POMYKAŁA, *On Similarity Based Approximation of Information*, Demonstratio Mathematica XXVII (1994), pp 663–671.
19. A. SKOWRON, J. STEPANIUK, *Tolerance Approximation Spaces*, Fundamenta Informaticae 27 (1996), pp 245–253.
20. M. STEINBY, *Karkeat joukot ja epätäydellinen tieto* (*Rough Sets and Incomplete Knowledge*, in Finnish). In J. PALOMÄKI, I. KOSKINEN (eds.), *Tiedon esittämisestä*, Filosofisia tutkimuksia Tampereen yliopistosta 62, Tampere, pp 1–21, 1997.
21. W. ŻAKOWSKI, *Approximations in the Space* (U, Π), Demonstratio Mathematica 16 (1983), pp 761–769.

A Rough Set Approach to Inductive Logic Programming

Herman Midelfart and Jan Komorowski

Department of Computer and Information Science,
Norwegian University of Science and Technology,
N-7491 TRONDHEIM,
Norway,
{herman, janko}@idi.ntnu.no

Abstract. We investigate a Rough Set approach to treating imperfect data in Inductive Logic Programming. Due to the generality of the language, we base our approach on neighborhood systems. A first-order decision system is introduced and a greedy algorithm for finding a set of rules (or clauses) is given. Furthermore, we describe two problems for which it can be used.

1 Introduction

Inductive Logic Programming (ILP) has been defined as the intersection between (inductive) Machine Learning and Logic Programming [9]. The field is concerned with finding hypotheses or sets of rules from examples, and first-order logic is used as representation language. Compared to traditional Machine Learning (ML) and Rough Sets (RS), ILP has some advantages. Firstly, a more expressive representation is used. This means that more complex concepts (e.g., that two attributes are equal) may be found. Secondly, ILP allows the use of background knowledge in a natural and compact way since such knowledge can be expressed as predicates. Thirdly, problems that are not handled well by traditional ML such as the multiple-instance problem [1] may be represented and solved in this framework.

There have been some attempts to combine ILP and RS already. Siromoney and Inoue [12] introduce and formalize a method where two sets of clauses are found – one for the positive examples (or the positive region), and one for the negative examples (or the negative region). The sets are found by an ordinary ILP-system such as Progol [8]. Their approach is based on ILP, but not confined to ILP. It could easily be adopted to propositional problems as well, provided that there are only two decision classes. Stepaniuk [14] uses the same transformational scheme as in LINUS [3] and introduces a method for reducing the background knowledge by removing irrelevant clauses. The method assumes that the background knowledge is extensional (i.e., it consists of only ground atoms). Liu and Zhong [6] notice that imperfect data (such as uncertain, inconsistent, missing and noisy data) is not handled well in the ILP. They suggest several problem settings for RS in ILP, however, without offering any solutions.

W. Ziarko and Y. Yao (Eds.): RSCTC 2000, LNAI 2005, pp. 190–198, 2001.

In this paper, we describe a framework based on RS for handling imperfect data in ILP, and give a greedy algorithm for finding a set of clauses. We assume that the reader is familiar with RS and Logic Programming. RS notions and notation follow Komorowski et al. [2] which is also a good introduction to RS.

2   Inductive Logic Programming

2.1   The Normal Setting

In ILP, the goal is to find a hypothesis H from background knowledge $\mathcal{B}$ and examples E where $\mathcal{B}$, E, and H usually are sets of definite clauses. The set E can be divided further into two sets, the positive examples E^+, and the negative examples, E^-. In the normal setting[1] the following conditions must hold: $\forall e \in E^- : \mathcal{B} \not\models e$, $\forall e \in E^- : \mathcal{B} \cup H \not\models e$, $\exists e \in E^+ : \mathcal{B} \not\models e$, and $\forall e \in E^+ : \mathcal{B} \cup H \models e$.

The normal setting is the main setting of ILP, but the definition given here is rather general. Additional restrictions are usually set. For example, only the definition of a *single* predicate is found, and this predicate is called the *target* predicate. Furthermore, the clauses are often assumed *function-free* since clauses with functions can be turned into function-free ones by means of flattening [11].

In this paper, we will use these restrictions. In addition, we will assume that H is a set of *normal non-recursive* clauses. The clauses in E will be represented as tuples (a, b) where a and b are ground. b corresponds to the body and a to the head. An example clause can be put in this form by first skolemizing it and then breaking the head and the body apart.

2.2   Declarative Bias

Since the language of clauses is infinite, all ILP systems restrict the language in some way. Most ILP systems employ a declarative bias so the restrictions can be set by the user. Mode declarations originate from Prolog, and are declarations of the kind of the terms that are allowed in a predicate. They have been used in several ILP systems including Progol [8]. We adopt a very similar scheme here. By position, we mean the place where a term occurs in an atom. So, term t_i has position i $(1 \leq i \leq n)$ in atom $p(t_1, \ldots, t_n)$.

Definition 1. *A mode declaration is a declaration of a predicate in either the head* modeh $(pred(t_1, \ldots, t_n))$ *or the body* modeb $(recall, pred(t_1, \ldots, t_n))$. *Each t_i defines the type and the kind of term allowed at the corresponding position in the predicate. A t_i can have one of the following forms:*

> $+t$ - *the term must be an input variable of type t.*
> $-t$ - *the term must be an output variable of type t.*
> $\#t$ - *the term must be a ground constant of type t.*

The recall (modeb only) restricts the number of times that a predicate with the same combination of input variables may appear in a clause.

For example, $\text{modeb}(2, owns(+person, -stuff))$, declares the first term of $owns$ to be an input variable of type $person$ and the second term to be an output variable of type $stuff$. For each combination of input variables, the predicate can be repeated twice. So $owns(X, B), owns(X, C)$ is permitted, but $owns(X, B)$, $owns(X, C), owns(X, D)$ is not.

Let $m(l)$ be a function associating a literal l with the mode declaration defining it. Given atom $a = p(s_1, \ldots, s_n)$, $In(a)$ denotes the set of the input variables in a and their types, i.e., $In(a) = \{\langle x, t\rangle \mid m(a) = \text{mode}(\text{b}|\text{h})(R, p(t_1, \ldots, t_n))$ and $s_i = x$ and $t_i = +\text{t}, 1 \le i \le n\}$. $Out(a)$, and $Cons(a)$ are defined similarly for output variables and constants. For a set of atoms B, the sets of input variables, output variables, and constants are the unions of the corresponding sets for each atom in B, e.g., $In(B) = \bigcup_{a \in B} In(a)$.

3   Towards Rough Inductive Logic Programming

Rough Sets Theory is based on propositional logic, and makes assumptions which are not true in first-order logic. When going from propositional logic to first-order these assumptions have be dealt with.

Firstly, the indiscernibility relation assumes that all attributes can be considered separately. This is not true in ILP. Attributes correspond to atoms, and atoms cannot generally be considered separately. For example, if P and Q are true, then so is $P \wedge Q$, but the truth of $\exists (P(x, y))$ and $\exists (Q(y, x))$ does not imply $\exists (P(x, y) \wedge Q(y, x))$. Thus we have to consider the whole body of a clause. This corresponds to comparing information vectors in RS.

Secondly, each object in the universe U (i.e., each example) satisfies one information vector or rule condition. This is not generally true in ILP. For each atom, we can create a positive and a negative literal, but they are not always complementary. So, an example may satisfy several bodies. For example, if $B = \{Q(a, b), Q(a, c), R(b)\}$ then both $c_1 = P(X) \leftarrow Q(X, Y) \wedge R(Y)$ and $c_2 = P(X) \leftarrow Q(X, Y) \wedge \neg R(Y)$ imply example $P(a)$. However, the set $\{c_1, c_2\}$ is complete in the sense that any example which follows from $P(X) \leftarrow Q(X, Y)$ also follows from at least one of them. Thus we need a more general relation than indiscernibility relation since the resulting clauses are not always complementary.

4   A First-Order Decision System

We are now ready to define a first-order equivalent of a decision system.

Definition 2. *A first-order decision system I is a tuple $\langle B, E, M, M_h\rangle$. B is a set of definite clauses defining the background knowledge. E is a set of ground examples, $e = (a, b)$, where b corresponds to the body and a to the head. M is a set of modeb declarations. M_h is a modeh declaration for the target predicate which occurs as the head in all of the examples in E. The target predicate is allowed in neither the bodies of the examples nor the background knowledge.*

We will assume that C is a set containing the constants in $\mathcal{B}$ and E, and V is a countable set of variables. h denotes an atom of the target predicate defined by M_h and is instantiated from the variables in V.

From M, C and V, we can define an atom set A, but not every possible atom should be allowed in A. For example, if $mode(p(+t))$ is a mode declaration, but $p(X) \in A$ is not connected to the head (i.e., X does not occur as an input variable in h or an output variable in A), $p(X)$ will violate the mode declaration. The reason is that an input variable has to be instantiated. We define an atom set inductively.

Definition 3. *(Atom set) A is an atom set if $A = \emptyset$ or $A = C \cup B$, B is an atom set, and $C \subseteq \rho(B)$ where $\rho(B)$ is defined as:*

1. *$p(s_1, \dots, s_n) \in \rho(B)$ if $modeb(r, p(t_1, \dots, t_n))$ is in M and for $1 \le i \le n$:*
   (a) $s_i = x$ if $t_i = +t$, $\langle x, t \rangle \in Out(B) \cup In(h)$,
   (b) $s_i = x$ if $t_i = -t$ and x is a new variable which does not occur in B.
   (c) $s_i = c$ if $t_i = \#t$ and $c \in C$ is of type t
   and p occurs with the same combination of input variables at most $r - 1$ times in B
2. *$(s_1 = s_2) \in \rho(B)$ if $m = modeb(r, t_1 = t_2)$ and for $i = 1, 2$:*
   (a) $s_i = x$ if $t_i = +t$, $\langle x, t \rangle \in Out_M(B) \cup In_M(h)$,
   (b) $s_i = x$ if $t_i = -t$, $\langle x, t \rangle \in Out(h)$
   (c) $s_i = x$ if $t_i = \#t$ and x is a new variable which does not occur in B.

$\rho(B)$ defines a mapping from atom sets to atoms and is similar to a refinement operator in ILP. The predicate $= /2$ is treated specially for several reasons. Firstly, we want the language of attribute-value pairs (used in ML an RS) to be a sub-language of ours. Such pairs may be declared by mode declarations like $modeb(1, +\text{person} = \#\text{person})$, but the constants can only be assigned when bodies are created from the atom set. The reason is that we need to distinguish between different constant values rather than the truth values of the predicate. So, a position corresponding to a $\#t$ definition contains a variable in the atom set. Secondly, the output variables in an atom are new variables. It must be possible to unify or test the equality of such variables. In particular, it must be possible to unify them with the output variables in the head. This is done by the $= /2$ predicate which is the only predicate that is allowed to contain output variables from the head.

The atoms in the atom set cannot be considered separately, but must be considered together with the rest of the atoms in the set. So, to evaluate how well each atom distinguishes between the examples, we need to create bodies from the atom set.

Definition 4. *(Body) Let A be an atom set. c is a body created from A if it is a conjunction of literals such that for each a in A, l_a is in c where $l_a = a\sigma$ if $m(a) = \text{modeb}(r, t_1 = t_2)$ and either t_1 or t_2 are equal to $\#t$ and σ is a substitution assigning x a constant of type t, where $\langle x, t \rangle \in Cons(a)$. Otherwise l_a is either a or $\neg a$.*

When we use negation-as-failure, the new variables introduced by a negative literal are universally quantified. A clause such as $p(X) \leftarrow q(X, Y) \wedge \neg r(Y, Z)$ is an abbreviation of $\forall X (p(X) \leftarrow \exists Y \forall Z (q(X, Y) \wedge \neg r(Y, Z)))$. However, variable Z will not be instantiated by SLDNF-resolution if $\neg r(Y, Z)$ succeeds. This means that the output variables of a negative literal cannot be used as inputs variables in other literals. Thus each literal in the body must be connected to the head through a chain of positive literals. Given a body, we call the maximal subset satisfying this property for an admissible body.

Definition 5. *(Admissible body) Let c be a body created from an atom set. Then c' is an admissible body if c' is the maximal subset of c such that for each literal l in c', $In(l) \subseteq (In(h) \cup Out(pos(c')))$ where $pos(c')$ contains the positive literals in c'.*

Definition 6. *(Body set) Let A be an atom set. A body set $BS(A)$ is the set of all admissible bodies created from A.*

5   A Neighborhood System for ILP

The coverage $Cov(r)$ of a rule r is the set of objects which satisfies the condition of the rule. In a propositional decision system, each object satisfies only the condition of one rule and two objects are indiscernible if they satisfy the same condition. Thus the coverage of a rule is equal to the elementary set of objects satisfying its condition. As mention in Section 3, an object may satisfy several clause bodies when only a single atom is negated. Therefore we have no such relationship between elementary sets and coverages in our situation. Moreover, it is not possible to define an equivalence relation since the coverages may overlap. However, we may define a neighborhood system [4, 5] which is a generalization of RS.

A neighborhood system (NS) is a mapping from a set V to collections of subsets of $U : p \in V \rightarrow \{N_p\} \subseteq 2^{2^U}$ where the N_p's are subsets of U which are associated with p. In our problem both U and V correspond to the set of examples E, N_p corresponds to the coverage of some body which is satisfied by example p. We can define a neighborhood system as follows.

Definition 7. *(Coverage) Let c be a body, and h the target atom. The coverage of c is $Cov(c) = \{(a, b) \in E \mid M(\mathcal{B} \cup b) \models \exists(c\theta) \text{ and } a = h\theta\}$.*

Definition 8. *A neighborhood system $NS_{BS(A)}(e)$ is a mapping from examples to coverages, $NS_{BS(A)}(e) = \{Cov(c) \mid e \in Cov(c) \text{ and } c \in BS(A)\}$.*

The upper and lower approximations of a neighborhood system are:

$$\underline{A}X = \{e \in E \mid \exists Cov(c) \in NS(e) : Cov(c) \subseteq X\}$$
$$\overline{A}X = \{e \in E \mid \forall Cov(c) \in NS(e) : Cov(c) \cap X \neq \emptyset\}$$

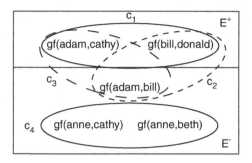

Fig. 1. Coverages of the body set in Example 1.

Example 1. We would like to find a definition of grandfather (denote by $gf/2$) from the predicates $father/2$ and $parent/2$ given the following background knowledge and examples:

$$\mathcal{B} = \begin{cases} parent(X,Y) \leftarrow father(X,Y), \\ parent(X,Y) \leftarrow mother(X,Y), \\ mother(anne,bill), & father(adam,bill), \\ mother(anne,beth), & father(adam,beth), \\ mother(cathy,donald), & father(bill,cathy), \\ mother(cathy,sue) \end{cases}$$

$$E^+ = \begin{cases} gf(adam,cathy) & (e_1) \\ gf(bill,donald) & (e_2) \end{cases} \qquad E^- = \begin{cases} gf(anne,cathy) & (e_3) \\ gf(anne,beth) & (e_4) \\ gf(adam,bill) & (e_5) \end{cases}$$

Assume that the target atom is $h = gf(X,Y)$, and that we have found the following atom set $A = \{father(X,V), parent(V,W), W = Y\}$. Then the body set is:

$$BS(A) = \begin{cases} father(X,Z) \wedge parent(Z,W) \wedge W = Y & (c_1) \\ father(X,Z) \wedge parent(Z,W) \wedge W \neq Y & (c_2) \\ father(X,Z) \wedge \neg parent(Z,W) & (c_3) \\ \neg father(X,Z) & (c_4) \end{cases}$$

This body set has the coverages depicted in Fig. 1. The neighborhood system constructed from them is: $NS(e_1) = \{\{e_1,e_2\},\{e_1,e_3\}\}$, $NS(e_2) = \{\{e_1,e_2\}, \{e_2,e_3\}\}$, $NS(e_3) = \{\{e_1,e_3\},\{e_2,e_3\}\}$, $NS(e_4) = \{\{e_4,e_5\}\}$, and $NS(e_5) = \{\{e_4,e_5\}\}$. The lower and upper approximation for E^+ and E^- are: $\underline{A}E^+ = \{e_1,e_2\}$, $\overline{A}E^+ = \{e_1,e_2,e_3\}$, $\underline{A}E^- = \{e_4,e_5\}$, and $\overline{A}E^- = \{e_3,e_4,e_5\}$.

6   Two Problems of Finding Clauses

Given a first-order decision system there are basically two problems that can be solved depending whether we want find a definition of a predicate or a function.

6.1   The Predicate Problem

Usually, the ILP problem has two decision classes – the positive and the negative examples. Hence, we may define a function $d(e)$ which denotes the decision class as $d(e) = +$, if $e \in E^+$, and $d(e) = -$, if $e \in E^-$. Since we have only two classes, it is not necessary to find clauses for both of them. We can create clauses that predict one of the classes and assume that an instance is in the other class if it is not predicted by the clauses. So, there are two ways to generate clauses:

1. Find one set of clauses which explicitly predicts both positive and negative instances of the predicate.
2. Find two sets of clauses which correspond to the upper and lower approximation of the positive examples. The negative instances can be predicted as not following from these sets, i.e., by means of the closed world assumption (CWA).

6.2   The Function Problem

In the problem above, there are just two decision classes, but a decision system $\mathbf{A} = (U, A \cup \{d\})$ has usually more than two classes. Moreover, each information vector has only one associated decision class, if the system is consistent. This corresponds to learning a functional predicate where the set A is a set of input variables and d is an output variable. A predicate is functional if each binding of the input variables has at most one binding of the output variables. The problem of finding such a functional predicate can be formulated as follows: Find a hypothesis H such that $\forall (a, b) \in E : \mathcal{B} \cup b \cup H \models a$ and $\forall (a, b) \in E \ \forall \sigma :$ $\mathcal{B} \cup b \cup H \models h\theta_{in}\sigma$ implies $h\theta_{in}\sigma = a$. Here, h is an instance of the target predicate with new variables at each position such that, $h\theta = a$, and θ_{in} is a subset of θ containing the substitution for the input variables. σ denotes some substitution binding all variables in h that correspond to constants and output variables.

This problem can be solved in our framework. The decision class of an example $e = (a, b)$ is just the substitution corresponding to the output variables and the constants in a. Hence, the decision class of e is $d(e) = \theta \backslash \theta_{in}$, and $E^i = \{e \in E \mid d(e) = i\}$ contains the examples with class i.

7   A Greedy Algorithm for Finding a Set of Clauses

Finding a reduct is NP-hard [13], and this is a subproblem of finding a set of rules. Since a (propositional) decision system can be mapped into our first-order decision system, finding a set of clauses must be NP-hard, as well. Thus we have to use an approximation algorithm.

A greedy algorithm for finding a set of clauses is given in Algorithm 1. It maintains an atom set B and a body set BS and iteratively adds the atom that discerns the most examples, to the atom set. The body set is extended correspondingly. Since an atom set can be infinite, a maximum depth bound k is set on the atoms.

Algorithm 1 A greedy algorithm for finding a set of clauses

Input: A first-order decision system $\langle \mathcal{B}, E, M, M_h \rangle$, and a max. depth k.
Output: A set of clauses H.

1: $h = p(X_1, \ldots, X_n)$ is defined by M_h and the X_i are distinct variables
2: $R = \{\langle e_i, e_j \rangle \in E^2 \mid d(e_i) \neq d(e_j)\}$), $B = \emptyset$, $BS = \emptyset$
3: **while** $R \neq \emptyset$ **do**
4:    $\forall q \in \rho(B)$ if $Depth(B \cup \{q\}, h) \leq k$ then compute $R(q) = R \cap Dis(BS_q))$ where
        $BS_q = \{q \wedge b, \neg q \wedge b \mid b \in BS$ and $(\neg)q \wedge b$ is admissible$\}$
5:    pick the p with the highest $|R(p)|$
6:    **if** $R(p) = 0$ **then** break the while loop
7:    $BS = \{b \in BS_p \mid b$ has the highest accuracy among the bodies covering $e \in E\}$
8:    $B = B \cup \{p\}$, $R = R \backslash Dis(BS)$
9: $H = \{head(b) \leftarrow b \mid b \in BS\}$ where $head(b) = \bigvee\{h(d(e)) \mid e \in Cov(b)\}$ and
    $h(x) = h$ if $x = +$, $h(x) = \neg h$ if $x = -$, and $h(x) = h\theta$ if x is a substitution θ

Definition 9. *(Depth) Let h be a target atom. The depth of variable v in atom set A, $Depth(v, A, h)$ is*

1. *0, if v occurs in h.*
2. *$\max_{\langle u, t \rangle \in In(a)} Depth(u, C, h) + 1$ if $\langle v, t' \rangle \in Out(a)$ and $In(a) \subseteq (Out(C) \cup In(h))$ for some subset C of A.*
3. *∞, otherwise.*

The depth of atom set A is the maximum depth of its variables, $Depth(A, h) = \max_{\langle v, t \rangle \in Var(C)} Depth(v, A, h)$ where $Var(C) = In(C) \cup Out(C)$.

For example, $Depth(\{q(X, Y), q(Y, Z)\}, p(X)) = 2$ and variable Y has depth 1, given the following mode declarations: $\mathtt{modeh}(q(+h))$ and $\mathtt{modeb}(1, q(+h, -h))$.

An example e_i is discerned from example e_j if there is no body which is satisfied by both or there is some body that is satisfied only by e_i and examples of the same decision class as e_i, but not e_j. More formally, $\langle e_i, e_j \rangle \in Dis(BS)$ iff $d(e_i) \neq d(e_j)$ and

1. $NS_{BS}(e_i) \cap NS_{BS}(e_j) = \emptyset$, or
2. there is an $Cov(c) \in NS_{BS}(e_i) \backslash NS_{BS}(e_j)$ such that $Cov(c) \subseteq E^{d(e_i)}$

To compute the discernibility $Dis(BS(B))$ of an atom set B, the coverage of each body in the body set $BS(B)$ has to be found. However, the body set contains at worst $2^{|B|}$ bodies, and is not bounded by the examples since the same example may satisfy several bodies. Thus the body set (BS) is reduced in line 7 such that only the best body for each example is kept.

In RS, we have an initial set of condition attributes A and the goal is to find a subset B such that they have the same positive region. However, in a first-order decision system there are no initial atom sets like A. So, instead all examples are initially assumed indiscernible and the algorithm terminates when all examples have been discerned ($R = \emptyset$), or R cannot be reduced any further ($R(p) = 0$ for all $p \in \rho(B)$). In this latter case the examples that cannot be discerned by the corresponding body set are assumed indiscernible. Further details can be found in [7].

8   Conclusion and Future Work

We have given a framework for ILP based on Rough Set Theory, and a greedy algorithm for finding a set of clauses. Our approach should be useful when the examples are indistinguishable due to sparse or missing background knowledge. In the future, we will investigate applications of this framework. We will also pursue other approaches to rough inductive logic programming. Especially, finding a reduct seems like an interesting problem. This is not possible with the algorithm presented here since the body set is reduced greedily. This means that there may be other bodies that discern better between the examples than those in the body set found by the algorithm. So even if we create an initial atom set (e.g. the set of all atoms with depth $\leq k$), we cannot guarantee that the positive regions of this set and the set found by the algorithm, will be exactly identical (though in practice they will be close). However, we believe that finding a reduct is achievable if another representation is chosen for the body set and are currently investigating this.

References

1. T.G. Dietterich, R.H. Lathrop, and T. Lozano-Pérez. Solving the multiple-instance problem with axis-parallel rectangles. *Artificial Intelligence*, 89(1-2):31–71, 1997.
2. J. Komorowski, Z. Pawlak, L. Polkowski, and A. Skowron. A rough set perspective on data and knowledge. In Pal and Skowron (Eds), *Rough Fuzzy Hybridization*, pp. 107–121. Springer-Verlag, 1999.
3. N. Lavrač and S. Džeroski. *INDUCTIVE LOGIC PROGRAMMING: Techniques and Applications*. Ellis Horwood, 1994.
4. T.Y. Lin. Granular computing on binary relations I: Data mining and neighborhood systems. In Polkowski and Skowron [10], pp. 107–121.
5. T.Y. Lin. Granular computing on binary relations II: Rough set representations and belief functions. In Polkowski and Skowron [10], pp. 122–140.
6. C. Liu and N. Zhong. Rough problems settings for inductive logic programming. In Zhong et al. [15], pp. 168–177.
7. H. Midelfart and J. Komorowski. Rough inductive logic programming. Technical report, Dept. of Comp. and Info. Science, Norwegian Univ. of Sci. and Tech., 2000.
8. S. Muggleton. Inverse entailment and Progol. *New Generation Computing*, 13(3/4): 245–286, 1995.
9. S. Muggleton and L. De Raedt. Inductive logic programming: Theory and methods. *Journal of Logic Programming*, 19, 20:629–679, 1994.
10. L. Polkowski and A. Skowron (Eds). *Rough Sets in Knowledge Discovery 1: Methodology and Applications*, Physica-Verlag, 1998.
11. C. Rouveirol. Flattening and saturation: Two representation changes for generalization. *Machine Learning*, 14:219–232, 1994.
12. A. Siromoney and K. Inoue. The generic rough set inductive logic programming model and motifs in strings. In Zhong et al. [15], pp. 158–167.
13. A. Skowron and C. Rauszer. The discernibility matrices and functions in information systems. In *Intelligent decision support: Handbook of Applications and Advances of Rough Sets Theory*, pp. 331–362, 1992.
14. J. Stepaniuk. Rough sets and relational learning. In H. J. Zimmermann, editor, *Proceedings of EUFIT'99*, 1999.
15. N. Zhong, A. Skowron, and S. Ohsuga, editors. *Proceedings of RSFDGrC'99*, LNAI 1711. Springer-Verlag, 1999.

Classification of Infinite Information Systems

Mikhail Moshkov

Research Institute for Applied Mathematics and
Cybernetics of Nizhny Novgorod State University
10, Uljanova St., Nizhny Novgorod, 603005, Russia
moshkov@nnucnit.unn.ac.ru

Abstract. In the paper infinite information systems are investigated
which are used in pattern recognition, discrete optimization, computational geometry. The depth and the size of deterministic and nondeterministic decision trees over such information systems are studied. A
partition of the set of all infinite information systems on two classes is
considered. Systems from the first class are near to the best from the
point of view of deterministic and nondeterministic decision tree time
and space complexity. Decision trees for systems from the second class
have in the worst case large time or space complexity. In proofs (which
are too long and not included to the paper) methods of test theory [1,
2] and rough set theory [8, 9] are used.

1 Introduction

Decision rules and deterministic decision trees are widely used in different applications for problem solving and for knowledge representation. One can interpret
a complete decision rule system (a system which is applicable for any input) as
a nondeterministic decision tree. In this paper both deterministic and nondeterministic decision trees are studied.

In rough set theory finite information systems are considered usually. However the notion of infinite information system is helpful in pattern recognition,
discrete optimization, computational geometry [3, 6]. In the paper arbitrary infinite information systems are considered.

Rough set theory allows to describe problems of different nature such that
the solution of the problem may be obtained exactly or approximately if we know
values of attributes from a finite set which forms the description of a problem.
The efficiency of decision trees for such problem solving depends on their time
and space complexity. From [4, 5, 7] and from results of this paper follows that
for an arbitrary infinite information system in the worst case

- the minimal depth of deterministic decision tree (as a function on the number of attributes in a problem description) either is bounded from below by
logarithm and from above by logarithm to the power $1 + \epsilon$, where ϵ is an arbitrary positive constant, or grows linearly;

- the minimal depth of nondeterministic decision tree (as a function on the
number of attributes in a problem description) either is bounded from above by
a constant or grows linearly;

W. Ziarko and Y. Yao (Eds.): RSCTC 2000, LNAI 2005, pp. 199–203, 2001.
© Springer-Verlag Berlin Heidelberg 2001

- the minimal number of nodes in deterministic and nondeterministic decision trees (as a function on the number of attributes in a problem description) has either polynomial or exponential growth.

In the paper a classification of infinite information systems is considered. All such systems are divided on two classes.

For information systems from the first class

- there exist deterministic decision trees whose depth grows almost as logarithm, and the number of nodes grows almost as a polynomial on the number of attributes in a problem description;

- there exist nondeterministic decision trees whose depth is bounded from above by a constant, and the number of nodes grows almost as a polynomial on the number of attributes in a problem description.

The information systems from the first class are near to the best from the point of view of decision tree use. From stated above follows that for these systems the growth of the depth of decision trees is almost minimal, and the growth of the number of nodes in decision trees is almost minimal too. In the case when the number of attributes in a problem description is relatively small the considered decision trees may be used in practice. Note also that for information systems from the first class there exist complete decision rule systems for which the length of rules is bounded from above by a constant and the number of rules grows almost as a polynomial on the number of attributes in a problem description.

For an arbitrary information system from the second class in the worst case

- the minimal depth of deterministic decision tree (as a function on the number of attributes in a problem description) grows linearly;

- nondeterministic decision trees have at least linear growth of the depth or have at least exponential growth of the number of nodes (depending on the number of attributes in a problem description).

2   Basic Notions

Let A be a nonempty set, B be a finite nonempty set with at least two elements, and F be a nonempty set of functions from A to B. Functions from F will be called *attributes* and the triple $U = (A, B, F)$ will be called *an information system*.

The set A may be interpreted as the set of inputs for problems over the information system U. A *problem over* U is an arbitrary $(n + 1)$-tuple $z = (\nu, f_1, \ldots, f_n)$ where $\nu : B^n \to \mathbf{N}$, $\mathbf{N}$ is the set of natural numbers, and $f_1, \ldots, f_n \in F$. The number $\dim z = n$ will be called *the dimension* of the problem z. The problem z may be interpreted as a problem of searching for the value $z(a) = \nu(f_1(a), \ldots, f_n(a))$ for an arbitrary $a \in A$. Different problems of pattern recognition, discrete optimization, fault diagnosis and computational geometry can be represented in such form.

As algorithms for problem solving we will consider decision trees. A *decision tree over* U is a marked finite tree with the root in which the root and the edges starting in the root are assigned nothing; each terminal node is assigned a number from $\mathbf{N}$; each node which is neither the root nor terminal (such nodes are called *working*) is assigned an attribute from F; each edge is assigned an element from B. A decision tree is called *deterministic* if the root is initial node of exactly one edge, and edges starting in a working node are assigned pairwise different elements.

Let Γ be a decision tree over U. A *complete path* in Γ is an arbitrary sequence $\xi = v_0, d_0, v_1, d_1, \ldots, v_m, d_m, v_{m+1}$ of nodes and edges of Γ such that v_0 is the root, v_{m+1} is a terminal node, and v_i is the initial and v_{i+1} is the terminal node of the edge d_i for $i = 0, \ldots, m$. Now we define a subset $\mathcal{A}(\xi)$ of the set A associated with ξ. If $m = 0$ then $\mathcal{A}(\xi) = A$. Let $m > 0$, the attribute f_i be assigned to the node v_i and δ_i be the element from B assigned to the edge d_i, $i = 1, \ldots, m$. Then $\mathcal{A}(\xi) = \{a : a \in A, f_1(a) = \delta_1, \ldots, f_m(a) = \delta_m\}$.

We will say that a decision tree Γ over U solves a problem z over U *nondeterministically* if for each $a \in A$ there exists a complete path ξ in Γ such that $a \in \mathcal{A}(\xi)$, and for each $a \in A$ and each complete path ξ in Γ such that $a \in \mathcal{A}(\xi)$ the terminal node of the path ξ is assigned the number $z(a)$.

We will say that a decision tree Γ over U solves a problem z over U *deterministically* if Γ is a deterministic decision tree which solves the problem z nondeterministically.

As time complexity measure we will consider *the depth* of a decision tree which is the maximal number of working nodes in a complete path in the tree. As space complexity measure we will consider the number of nodes in a decision tree. We denote by $h(\Gamma)$ the depth of a decision tree Γ, and by $L(\Gamma)$ we denote the number of nodes in Γ.

3   Classification

Consider an information system $U = (A, B, F)$. If F is an infinite set then U is called *an infinite* information system.

We will say that the information system U has *infinite independence dimension* (or, in short, *infinite I-dimension*) if the following condition holds: for each $t \in \mathbf{N}$ there exist attributes $f_1, \ldots, f_t \in F$ and two-element subsets $B_1, \ldots, B_t$ of the set B such that for arbitrary $\delta_1 \in B_1, \ldots, \delta_t \in B_t$ the system of equations

$$\{f_1(x) = \delta_1, \ldots, f_t(x) = \delta_t\} \tag{1}$$

is compatible (has solution) on the set A. If the considered condition does not hold then we will say that the information system U has *finite I-dimension*.

Now we consider the condition of decomposition for the information system U. Let $t \in \mathbf{N}$. A nonempty subset D of the set A will be called (t, U)-set if D coincides with the set of solutions on A of a system of the kind (1), where $f_1, \ldots, f_t \in F$ and $\delta_1, \ldots, \delta_t \in B$.

We will say that the information system U satisfies *the condition of decomposition* if there exist numbers $m, p \in \mathbf{N}$ such that every $(m + 1, U)$-set is a union of p sets each of which is (m, U)-set.

The considered classification divides the set of infinite information systems on two classes: C_1 and C_2. The class C_1 consists of all infinite information systems each of which has finite I-dimension and satisfies the condition of decomposition. The class C_2 consists of all infinite information systems each of which has infinite I-dimension or does not satisfy the condition of decomposition.

4   Bounds on Time and Space Complexity

In the following theorem time and space complexity of deterministic decision trees are considered.

Theorem 1. *Let $U = (A, B, F)$ be an infinite information system. Then the following statements hold:*

a) if $U \in C_1$ then for any ϵ, $0 < \epsilon < 1$, there exists a constant $c \in \mathbf{N}$ such that for each problem z over U there exists a decision tree Γ over U which solves the problem z deterministically and for which $h(\Gamma) \leq c(\log_2 n)^{1+\epsilon} + 1$ and $L(\Gamma) \leq |B|^{c(\log_2 n)^{1+\epsilon}+2}$ where $n = \dim z$;

b) if $U \in C_1$ then there is no $n \in \mathbf{N}$ such that for each problem z over U with $\dim z = n$ there exists a decision tree Γ over U which solves the problem z deterministically and for which $h(\Gamma) < \log_{|B|}(n+1)$;

c) if $U \in C_2$ then there is no $n \in \mathbf{N}$ such that for each problem z over U with $\dim z = n$ there exists a decision tree Γ over U which solves the problem z deterministically and for which $h(\Gamma) < n$.

In the following theorem time and space complexity of nondeterministic decision trees are considered.

Theorem 2. *Let $U = (A, B, F)$ be an infinite information system. Then the following statements hold:*

a) if $U \in C_1$ then for any ϵ, $0 < \epsilon < 1$, there exist constants $c_1, c_2 \in \mathbf{N}$ such that for each problem z over U there exists a decision tree Γ over U which solves the problem z nondeterministically and for which $h(\Gamma) \leq c_1$ and $L(\Gamma) \leq |B|^{c_2(\log_2 n)^{1+\epsilon}+2}$ where $n = \dim z$;

b) if $U \in C_2$ then there is no $n \in \mathbf{N}$ such that for each problem z over U with $\dim z = n$ there exists a decision tree Γ over U which solves the problem z nondeterministically and for which $h(\Gamma) < n$ and $L(\Gamma) \leq 2^n$.

So the class C_1 is very interesting from the point of view of different applications. The following example characterizes both the wealth and the boundedness of this class.

Example 3. Let $m, t \in \mathbf{N}$. We denote by $Pol(m)$ the set of all polynomials which have integer coefficients and depend on variables $x_1, \ldots, x_m$. We denote by $Pol(m, t)$ the set of all polynomials from $Pol(m)$ such that the degree of each polynomial is at most t. We define information systems $U(m)$ and $U(m, t)$ as follows: $U(m) = (\mathbf{R}^m, E, F(m))$ and $U(m, t) = (\mathbf{R}^m, E, F(m, t))$ where $\mathbf{R}$

is the set of real numbers, $E = \{-1, 0, +1\}$, $F(m) = \{\text{sign}(p) : p \in Pol(m)\}$ and $F(m, t) = \{\text{sign}(p) : p \in Pol(m, t)\}$. One can prove that $U(m) \in \mathcal{C}_2$ and $U(m, t) \in \mathcal{C}_1$. Note that the system $U(m)$ has infinite I-dimension.

5 Conclusion

In the paper a classification of infinite information systems is considered. The class of information systems is described which are near to the best from the point of view of deterministic and nondeterministic decision tree time and space complexity. This classification may be useful for the choice of information systems for investigation of problems of pattern recognition, discrete optimization, computational geometry.

Acknowledgments

This work was partially supported by Russian Foundation of Fundamental Research (project 99-01-00820), by Federal Program "Integration" (project A0110) and by Program "Universities of Russia" (project 015.04.01.76).

References

1. Chegis, I.A., Yablonskii, S.V.: Logical methods of electric circuit control. Trudy MIAN SSSR **51** (1958) 270–360 (in Russian)
2. Moshkov, M.Ju.: Conditional tests. Problemy Kybernetiki **40**. Edited by S.V. Yablonskii. Nauka Publishers, Moscow (1983) 131–170 (in Russian)
3. Moshkov, M.Ju.: Decision Trees. Theory and Applications. Nizhni Novgorod University Publishers, Nizhni Novgorod (1994) (in Russian)
4. Moshkov, M.Ju.: Comparative analysis of deterministic and nondeterministic decision tree complexity. Global approach. Fundamenta Informaticae **25**(2) (1996) 201–214
5. Moshkov, M.Ju.: Unimprovable upper bounds on time complexity of decision trees. Fundamenta Informaticae **31**(2) (1997) 157–184
6. Moshkov, M.Ju.: On time complexity of decision trees. Rough Sets in Knowledge Discovery 1. Methodology and Applications (Studies in Fuzziness and Soft Computing **18**). Edited by L. Polkowski and A. Skowron. Phisica-Verlag. A Springer-Verlag Company (1998) 160–191
7. Moshkov, M.Ju.: Deterministic and nondeterministic decision trees for rough computing. Fundamenta Informaticae **41**(3) (2000) 301–311
8. Pawlak, Z.: Rough Sets - Theoretical Aspects of Reasoning about Data. Kluwer Academic Publishers, Dordrecht, Boston, London (1991)
9. Skowron, A., Rauszer, C.: The discernibility matrices and functions in information systems. Intelligent Decision Support. Handbook of Applications and Advances of the Rough Set Theory. Edited by R. Slowinski. Kluwer Academic Publishers, Dordrecht, Boston, London (1992) 331–362

Some Remarks on Extensions and Restrictions of Information Systems

Zbigniew Suraj[*]

Institute of Mathematics
Pedagogical University
Rejtana 16A, 35-310 Rzeszów, Poland
e-mail: zsuraj@univ.rzeszow.pl

Abstract. Information systems (data tables) are often used to represent experimental data [4],[5],[8]. In [16] it has been pointed out that notions of extension and restriction of information system are crucial for solving different class of problems [6],[7],[10]-[11],[12]-[15]. The intent of this paper is to present some properties of an information system extension (restriction) and methods of their verification.

Keywords: information systems, indiscernibility relation, discernibility function, minimal rules, information system extension.

1 Introduction

In the rough set theory, the information systems [5] and the rules extracted from information systems are the most common form of representing knowledge. In [16] it has been pointed out that notions of extension and restriction of information system are crucial for solving different class of problems, among others: (i) the synthesis problem of concurrent systems specified by information systems [10],[15], (ii) the problem of discovering concurrent data models from experimental tables [13], (iii) the re-engineering problem for cooperative information systems [12],[14], (iv) the real-time decision making problem [6],[11], (v) the control design problem for discrete event systems [7].

The main idea of an information system extension can be explained as follows: A given information system S defines an extension S' of S created by adding to S all new objects corresponding to known attribute values. If an extension S' of S is consistent with all rules true in S (i.e., any object u from a set of objects U matching the left hand side of the rule also matches its right hand side and there is an object u matching the left hand side of the rule) and S' is the largest (w.r.t. the number of objects in S) extension of S with that property then the system S' is called a maximal consistent extension of S.

Maximal consistent extensions are used in the design of concurrent system specified by information systems [10],[15]. If an information system S specifies a concurrent system then the maximal consistent extension of S represents the

---

[*] This work has been partially supported by the grant No. 8T11C 025 19 from the State Committee for Scientific Research in Poland.

W. Ziarko and Y. Yao (Eds.): RSCTC 2000, LNAI 2005, pp. 204–211, 2001.

largest set of global states of the concurrent system consistent with all rules true in S. The set of global states can include new states of the given concurrent system consistent with all rules true in S. Moreover, if we are interested in determining a minimal (w.r.t. the number of objects) description of a given concurrent system then we can compute a minimal consistent restriction of an information system which specifies that concurrent system.

The rest of the paper is structured as follows. A brief presentation of the basic concepts underlying the rough set theory is given in Section 2. The basic definitions and procedures for computing extensions are presented in Section 3.

2   Preliminaries of Rough Set Theory

2.1 Information Systems

An *information system* is a pair $S = (U, A)$, where U - is a non-empty, finite set called the *universe*, A - is a non-empty, finite set of *attributes*, i.e., $a : U \to V_a$ for $a \in A$, where V_a is called the *value set* of a. Elements of U are called *objects*.

The set $V = \bigcup_{a \in A} V_a$ is said to be the *domain* of A.

Example 1. Consider an information system $S = (U, A)$ with $U = \{u_1, u_2, u_3\}$, $A = \{a, b\}$ and the values of the attributes are defined as in Table 1.

U/A	a	b
u_1	0	1
u_2	1	0
u_3	0	0

Table 1. An example of an information system

Let $S = (U, A)$ be an information system and let $B \subseteq A$. A binary relation $ind(B)$, called an *indiscernibility relation*, is defined by $ind(B) = \{(u, u') \in U \times U$ for every $a \in B, a(u) = a(u')\}$. Any information system $S = (U, A)$ determines an *information function* $Inf_A : U \to P(A \times V)$ defined by $Inf_A(u) = \{(a, a(u)) : a \in A\}$ where $V = \bigcup_{a \in A} V_a$ and $P(X)$ denotes the powerset of X. The set $\{Inf_A(u) : u \in U\}$ is denoted by $\text{INF}(S)$. The values of an information function will be sometimes represented by vectors of the form $(v_1, ..., v_m), v_i \in V_a$ for $i = 1, ..., m$ where $m = \text{card}(A)$. Such vectors are called *information vectors* (over V and A).

If $S = (U, A)$ is an information system then the *descriptors* of S are expressions of the form (a, v) where $a \in A$ and $v \in V_a$. Instead of (a, v) we also write $a = v$ or a_v. If τ is a Boolean combination of descriptors then by $\| \tau \|$ we denote the meaning of τ in the information system S.

2.2 Rules in Information Systems

Rules express some of the relationships between values of the attributes described in the information systems.

Let $S = (U, A)$ be an information system and let V be the domain of A.

A *rule* over A and V is any expression of the following form: (1) $a_{i_1} = v_{i_1} \wedge ... \wedge a_{i_r} = v_{i_r} \Rightarrow a_p = v_p$ where $a_p, a_{i_j} \in A, v_p, v_{i_j} \in V_{a_{i_j}}$ for $j = 1, ..., r$.

A rule of the form (1) is called *trivial* if $a_p = v_p$ appears also on the left hand side of the rule. The rule (1) is *true in* S if $\emptyset \neq \| a_{i_1} = v_{i_1} \wedge ... \wedge a_{i_r} = v_{i_r} \| \subseteq \| a_p = v_p \|$.

The fact that the rule (1) is true in S is denoted in the following way: $a_{i_1} = v_{i_1} \wedge ... \wedge a_{i_r} = v_{i_r} \underset{S}{\Longrightarrow} a_p = v_p$. By $D(S)$ we denote the set of all rules true in S.

Let $R \subseteq D(S)$. An information vector $\mathbf{v} = (\mathbf{v_1}, ..., \mathbf{v_m})$ is *consistent* with R iff for any rule $a_{i_1} = v_{i_1} \wedge ... \wedge a_{i_r} = v_{i_r} \underset{S}{\Longrightarrow} a_p = v_p$ in R if $\mathbf{v}_{i_j} = v_{i_j}$ for $j = 1, ..., r$ then $v_p = \mathbf{v_p}$.

2.3 Discernibility Matrix

The *discernibility matrix* and the *discernibility function* [9] help to compute minimal forms of rules w.r.t. the number of attributes on the left hand side of the rules.

Let $S = (U, A)$ be an information system and let $U = \{u_1, ..., u_n\}$, $A = \{a_1, ..., a_m\}$. By $M(S)$ we denote an $n \times n$ matrix (c_{ij}), called the *discernibility matrix* of S, such that $c_{ij} = \{a \in A : a(u_i) \neq a(u_j)\}$ for $i, j = 1, ..., n$.

A *discernibility function* $f_{M(S)}$ for an information system S is a Boolean function of m propositional variables $a_1^*, ..., a_m^*$ (where $a_i \in A$ for $i = 1, ..., m$) defined as the conjunction of all expressions $\bigvee c_{ij}^*$ where $\bigvee c_{ij}^*$ is the disjunction of all elements of $c_{ij}^* = \{a^* : a \in c_{ij}\}$ for $1 \leq j < i \leq n$ and $c_{ij} \neq \emptyset$. In the sequel we write a instead of a^*.

2.4 Minimal Rules in Information Systems

Now we recall a method for generating the minimal (i.e., with minimal left hand sides) form of rules in information systems [10],[15]. The method is based on the idea of Boolean reasoning [1] applied to discernibility matrices defined in [9].

Let $S = (U, A)$ be an information system and $B \subset A$. For every $a \notin B$ we define a function $d_a^B : U \to P(V_a)$ such that $d_a^B(u) = \{v \in V_a : \text{there exists } u' \in U \ u' \ ind(B) \ u \text{ and } a(u') = v\}$ where $P(V_a)$ denotes the powerset of V_a.

Let $S = (U, A)$ be an information system. We are looking for all minimal rules in S of the form: $a_{i_1} = v_{i_1} \wedge ... \wedge a_{i_r} = v_{i_r} \underset{S}{\Longrightarrow} a = v$ where $a \in A, v \in V_a, a_{i_j} \in A$ and $v_{i_j} \in V_{a_{i_j}}$ for $j = 1, ..., r$.

The above rules express functional dependencies between the values of the attributes of S. These rules are computed from systems of the form $S' = (U, B \cup \{a\})$ where $B \subset A$ and $a \in A - B$.

First, for every $v \in V_a, u_l \in U$ such that $d_a^B(u_l) = \{v\}$ a modification $M(S'; a, v, u_l)$ of the discernibility matrix is computed from $M(S')$. By $M(S'; a, v, u_l) = (c_{ij}^*)$ (or M, in short) we denote the matrix obtained from $M(S')$ in the following way:

> $\quad$ if $i = l$ then $c_{ij}^* = \emptyset$;
> $\quad$ if $c_{lj} \neq \emptyset$ and $d_a^B(u_j) \neq \{v\}$ then $c_{lj}^* = c_{lj} \cap B$
> $\quad$ else $c_{lj}^* = \emptyset$.

Next, we compute the discernibility function f_M and the prime implicants [**]
[17] of f_M taking into account the non-empty entries of the matrix M (when all
entries c_{ij}^* are empty we assume f_M to be always true).

Finally, every prime implicant $a_{i_1} \wedge \ldots \wedge a_{i_r}$ of f_M determines a rule $a_{i_1} = v_{i_1} \wedge \ldots \wedge a_{i_r} = v_{i_r} \overset{\Longrightarrow}{S} a = v$ where $a_{i_j}(u_l) = v_{i_j}$ for $j = 1, \ldots, r$, $a(u_l) = v$.

The set of all rules constructed in the above way for any $a \in A$ is denoted
by $\mathrm{OPT}(S, a)$. We put $\mathrm{OPT}(S) = \bigcup \{ \mathrm{OPT}(S, a) : a \in A \}$.

We compute all minimal rules true in $S' = (U, B \cup \{a\})$ of the form $\tau \Rightarrow a = v$, where τ is a term in disjunctive form over B and $V_B = \bigcup_{a \in B} V_a$, with a
minimal number of descriptors in any disjunct. To obtain all possible functional
dependencies between the attribute values it is necessary to repeat this process
for all possible values of a and for all remaining attributes from A.

3   Extensions and Restrictions of Information Systems

Let $S = (U, A)$ be an information system. For $S = (U, A)$, a system $S' = (U', A')$
such that $U \subseteq U'$, $A' = \{a' : a \in A\}$, $a'(u) = a(u)$ for $u \in U$ and $V_a = V_{a'}$ for
$a \in A$ will be called an *extension* of S. S is then called a *restriction* of S'.

The number of the extensions of a given information system determines

Proposition 1. *Let $S = (U, A)$ be an information system, $k = card(U)$, $n = card(V_{a_{i_1}} \times \ldots \times V_{a_{i_l}})$ where $a_{i_j} \in A$ for $j = 1, \ldots, l$ and $l = card(A)$. Then the
number of extensions of S is equal to $2^{n-k} - 1$.*

It follows from the following formula: $C_{n-k}^1 + C_{n-k}^2 + \ldots + C_{n-k}^{n-k}$ where C_j^i denotes
the number of i-element combinations of a set with j-elements.

Let $S = (U, A)$ be an information system and let U'' denotes the set of objects
corresponding to all admissible global states of S which do not appear into S,
i.e., U'' equals the difference between the cartesian product of the value sets for
all attributes $a \in A$ and those from $\mathrm{INF}(S)$. We say that an information system
$S' = (U', A)$ is a *maximal extension* of S iff $U' = U \cup U''$.

Proposition 2. *Let $S = (U, A)$ be an information system. There exists only one
maximal extension S' of S.*

Let $S = (U, A)$ be an information system. An information system $S' = (U', A)$
is called a *minimal restriction* of S iff S' is a restriction of S and any restriction
S'' of S is an extension of S'.

Proposition 3. *A given information system S has at least one a minimal restriction S' of S (different from S).*

[**] An implicant of a Boolean function f is any conjunction of literals (variables or
their negations) such that if the values of these literals are true under an arbitrary
valuation v of variables then the value of the function f under v is also true. A prime
implicant is a minimal implicant. Here we are interested in implicants of monotone
Boolean functions only, i.e. functions constructed without negation.

Example 2. Consider the information system S from Example 1. The system S_1 represented by Table 2 is its extension, but the system S_2 obtained from S_1 by deleting the object u_3 is not. S_1 is the maximal extension of S. However, the system S_3 obtained from S_1 by deleting objects u_3 and u_4 is the minimal restriction of S.

U_1/A_1	a	b
u_1	0	1
u_2	1	0
u_3	0	0
u_4	1	1

Table 2. The information system S_1

3.1 Maximal Consistent Extensions of Information Systems

The notion of a maximal consistent extension of a given information system has been introduced in [10]. In this subsection, we present some properties of this notion and procedures for computing the maximal consistent extension.

Let $S' = (U', A)$ be an extension of $S = (U, A)$. We say that S' is a *consistent extension* of S iff $D(S) \subseteq D(S')$. S' is called a *maximal* consistent extension of S iff S' is a consistent extension of S and any consistent extension S'' of S is a restriction of S'.

From the above definition follows

Proposition 4. *Let $S = (U, A)$ be an information system. There exists only one maximal consistent extension S' of S.*

PROCEDURE for computing maximal consistent extension S' of S:

Input: An information system $S = (U, A)$ and the set $\mathrm{OPT}(S)$ of all rules constructed as in subsection 2.4 for S.

Output: The maximal consistent extension S' of S.

Step 1. Compute all admissible global states of S which do not appear in S.

Step 2. Verify (using the set $\mathrm{OPT}(S)$ of rules) which global states of S obtained in *Step 1* are consistent with rules true in S.

It is known that, in general, the set $\mathrm{OPT}(S)$ of all rules constructed as described in subsection 2.4 can be exponential complexity (w.r.t. the number of attributes). Nevertheless, there are several methodologies allowing to deal with this problem in practical applications (see e.g. [2],[3],[4] pages 3-97).

Proposition 5. *Let $S = (U, A)$ be an information system and S' its maximal consistent extension. The set $\mathrm{OPT}(S)$ of all minimal rules of S defined in subsection 2.4 is empty iff S' is equal to the space of all possible values of attributes from A.*

In order to decide if a given information system has the maximal consistent extension the following proposition can be useful.

Proposition 6. *Let $S =(U, A)$ be an information system. If S' is the maximal consistent extension of S different from S then for at least two attributes $a, b \in A$ $card(V_a) > 2$ and $card(V_b) > 2$.*

Proposition 6 constitutes a necessary condition for the existence of the maximal consistent extension of an information system different from that system.

Proposition 7. *Let $S =(U, A)$ be an information system and S' its maximal consistent extension. If there exists such combination of attribute values $(v_{i_1}, v_{i_2}, ..., v_{i_n})$ where $v_{i_j} \in V_{a_j}$ for $j = 1, 2, ..., n$ and $n = card(A)$ that there no exist a functional dependency between any two attribute values from this combination in S, then S' is different from S.*

Proposition 8. *Let $S =(U, A)$ be an information system and S' its maximal consistent extension. S' is the information system in which all new added objects to S have the property mentioned in Proposition 7.*

Example 3. Consider an information system S represented by Table 3. Applying to S the method for generating the minimal form of rules described in subsection 2.4 we obtain the following set $\mathrm{OPT}(S)$ of rules: $a_1 \vee a_2 \underset{S}{\Longrightarrow} b_0$, $b_1 \vee b_2 \underset{S}{\Longrightarrow} a_0$.

U/A	a	b
u_1	0	1
u_2	1	0
u_3	0	2
u_4	2	0

Table 3. An example of an information system S

After running the procedure for computing maximal consistent extension of S we obtain the system S' including all objects of the system S and new object u_5 such that $a(u_5) = b(u_5) = 0$.

PROCEDURE for finding a description of maximal consistent extension S' of S:

Input: An information system $S =(U, A)$, the set $\mathrm{OPT}(S)$ of all minimal rules of S defined in subsection 2.4, and V - the domain of A.

Output: A description of maximal consistent extension S' of S in the form of Boolean formula constructed from descriptors over A and V.

Step 1. Rewrite each rule from $\mathrm{OPT}(S)$ to the form of Boolean formula.

Step 2. Construct the conjunction of formulas obtained in *Step* 1.

Step 3. Compute prime implicants of the formula obtained in *Step* 2.

In order to find the description of all new elements of maximal consistent extension S' of S (i.e., those from outside of the set of objects U) it is sufficient to execute a procedure presented below.

PROCEDURE for finding a description of all new elements of maximal consistent extension S' of S:

Input: As for the procedure presented above.

Output: A description of new elements of maximal consistent extension S' of S in the form of Boolean formula constructed from descriptors over A and V.

Step 1. Execute the procedure for finding a description of maximal consistent extension of S.

Step 2. Rewrite each row from the system S to the form of Boolean formula.

Step 3. Construct the negation of the formula obtained in *Step* 2.

Step 4. Construct the conjunction of formulas obtained in *Steps* 2 and 3.

Step 5. Compute prime implicants of the formula obtained in *Step* 4.

Example 4. Consider again the information system S and OPT(S) described in Example 3. Now by applying to S the procedure for finding a description of its maximal consistent extension we obtain the following Boolean formula: $(\neg((a = 1) \vee (a = 2)) \vee (b = 0)) \wedge (\neg((b = 1) \vee (b = 2)) \vee (a = 0))$. After simplifications (using Boolean theory laws) we get the formula of the form: $(a = 0) \vee (b = 0)$. This formula is matching to all objects of S' from Example 3.

In order to find a description of new elements of maximal consistent extension of S represented by Table 3 we can perform the above procedure.

As result we obtain the following Boolean formula: $\neg(((a = 0) \wedge (b = 1)) \vee ((a = 1) \wedge (b = 0)) \vee ((a = 0) \wedge (b = 2)) \vee ((a = 2) \wedge (b = 0))) \wedge ((a = 0) \vee (b = 0))$. After simplifications we get the result formula of the form: $(a = 0) \wedge (b = 0)$. It is matching to the object u_5 of the system S' from Example 3.

Let $S = (U, A)$ be a restriction of $S' = (U', A)$. We say that S is a *consistent restriction* of S' iff $D(S) \subseteq D(S')$. S is a *minimal consistent restriction* of S' iff S is a consistent restriction of S' and any consistent restriction S'' of S' is an extension of S.

Remark 1. The information system S from Example 3 is the minimal consistent restriction of the system S' considered in Example 3.

4   Concluding Remarks

The extensions (restrictions) of information systems appear in many investigations related to the rough set methods for solving different class of problems [16]. The presented approach can be treated as a constructive method of the information system extension to the largest data table including the same knowledge as the original information system. Maximal (minimal) consistent extensions (restrictions) provide a basis for modeling concurrent systems using rough set methods.

Acknowledgment

I am grateful to Professor A. Skowron for stimulating discussions and interesting suggestions about this work.

References

1. Brown, E.M.: Boolean Reasoning. Kluwer, Dordrecht (1990)
2. Michalski, R., Bratko, I., Kubat, M. (eds.): Machine Learning and Data Mining. Methods and Applications. Wiley, New York (1998)
3. Nadler, M., Smith, E.P.: Pattern Recognition Engineering. Wiley, New York (1993)
4. Pal, S.K., and Skowron, A. (eds.): Rough-Fuzzy Hybridization. A New Trend in Decision Making, Springer-Verlag, Singapore (1999)
5. Pawlak, Z.: Rough sets - Theoretical Aspects of Reasoning about Data. Kluwer, Dordrecht (1991)
6. Peters, J.F., Skowron, A., Suraj, Z., Pedrycz, W., Ramanna, S.: Approximate Real-Time Decision Making: Concepts and Rough Fuzzy Petri Net Models. International Journal of Intelligent Systems 14-4 (1998) 4-37
7. Peters, J.F., Skowron, A., Suraj, Z.: An Application of Rough Set Methods in Control Design. Fundamenta Informaticae (to appear)
8. Polkowski, L., Skowron, A. (eds.): Rough Sets in Knowledge Discovery 2. Applications, Case Studies and Software Systems. Physica-Verlag, Heidelberg (1998)
9. Skowron, A., Rauszer, C.: The discernibility matrices and functions in information systems. In: Słowiński, R. (ed.): Intelligent Decision Support: Handbook of Applications and Advances of Rough Sets Theory, Kluwer, Dordrecht (1992) 331-362
10. Skowron, A., Suraj, Z.: Rough Sets and Concurrency. Bulletin of the Polish Academy of Sciences 41-3 (1993) 237-254
11. Skowron, A., Suraj, Z.: A Parallel Algorithm for Real-Time Decision Making: A Rough Set Approach. Journal of Intelligent Information Systems 7, Kluwer, Dordrecht (1996) 5-28
12. Suraj, Z.: An Application of Rough Set Methods to Cooperative Information Systems Re-engineering. In: Tsumoto, S., Kobayashi, S., Yokomori, T., Tanaka, H., Nakamura, A. (eds.): Proceedings of the Fourth International Workshop on Rough Sets, Fuzzy Sets and Machine Discovery (RSFD-96), Tokyo (1996) 364-371
13. Suraj, Z.: Discovery of Concurrent Data Models from Experimental Tables: A Rough Set Approach. Fundamenta Informaticae 28-3,4 (1996) 353-376
14. Suraj, Z.: Reconstruction of Cooperative Information Systems under Cost Constraints: A Rough Set Approach. Journal of Information Sciences 111 (1998) 273-291
15. Suraj, Z.: The Synthesis Problem of Concurrent Systems Specified by Dynamic Information Systems. In: [8] 418-448
16. Suraj, Z.: Rough Set Methods for the Synthesis and Analysis of Concurrent Processes, ICS PAS Reports 893 (1999). Also in: Polkowski, L. (eds.): Studies in Fuzziness and Soft Computing, Physica-Verlag, Heidelberg (to appear)
17. Wegener, I.: The complexity of Boolean functions. Wiley and B.G. Teubner, Stuttgart (1987)

Valued Tolerance and Decision Rules

Jerzy Stefanowski[1] and Alexis Tsoukiàs[2]

[1] Institute of Computing Science
Poznan University of Technology, 60-965, Poznan, Poland
`Jerzy.Stefanowski@cs.put.poznan.pl`
[2] LAMSADE - CNRS, Université Paris Dauphine
75775 Paris Cedex 16, France
`tsoukias@lamsade.dauphine.fr`

Abstract. The concept of valued tolerance is introduced as an extension of the usual concept of indiscernibility (which is a crisp equivalence relation) in rough sets theory. Some specific properties of the approach are discussed. Further on the problem of inducing rules is addressed. Properties of a "credibility degree" associated to each rule are analysed and its use in classification problems is discussed.

1 Introduction

Rough sets theory hes been introduced by Pawlak to deal with a vague description of objects. The starting point of this theory is an observation that objects having the same description are *indiscernible* (similar) with respect to the available information. Although original rough sets theory has been used to face several problems, the use of the indiscernibility relation may be too rigid in some real situations. Therefore several generalisations of this theory have been proposed. Some of them ([1, 12]) extend the basic idea to a fuzzy context while others use more general similarity or tolerance relations instead of classical indiscernibility relation (see e.g. [8, 9]). There are also combinations of both extensions by [2, 3] where lower and upper approximations are fuzzy sets based on a fuzzy similarity relation. Properties of extended binary relations were studied in [13].

In this paper we introduce the concept of valued tolerance relation as a new extension of rough sets theory. A functional extension of the concepts of upper and lower approximation is introduced so that to any subset of the universe a degree of lower (upper) approximability can be associated. In other terms, any subset of the universe can be lower (upper) approximation of a given set, but to a different degree. Further on, such a functional extension enables to compute a credibility degree for any rule induced from the input information table. Such an idea first appeared in our previous work on incomplete information tables [10].

The paper is organised as follows. In section 2, we discuss motivations for using valued tolerance relation. Then, in section 3 we introduce formally the concept of valued tolerance. Some specific properties of this approach are also examined. In section 4, problems of inducing decision rules and computing credibility of rule are discussed. Results are summarised in section 5.

W. Ziarko and Y. Yao (Eds.): RSCTC 2000, LNAI 2005, pp. 212–219, 2001.

2   Why Valued Tolerance?

Original rough sets theory is based on the concept of indiscernibility relation which is a crisp equivalence relation (complete, reflexive, symmetric and transitive relation valued in $\{0, 1\}$). Practically speaking, two objects, described by a set of attributes, are indiscernible iff they have identical values. However, real life suggests that this is a very strong assumption. Objects may be practically indiscernible without having identical values. The idea of substituting indiscernibility with the concept of similarity has already been studied in e.g. [9, 8]. Moreover, it could be the case that objects can be "more or less similar" depending on the particular information available. Consider the following two examples.

	c_1 c_2 c_3 c_4
x_1	A B B C
x_2	A B * C
x_3	A * * *
Example 1	

	c_1 c_2 c_3 c_4
x_1	90 20 50 80
x_2	91 21 51 81
x_3	85 15 45 75
Example 2	

Example 1. Three objects x_1, x_2, x_3 and four attributes c_1, c_2, c_3, c_4 are given, each attribute equipped with a discrete nominal scale A,B,C,D. Besides the above information table is provided, where * is representing the "unknown" value of attribute:

If any similarity is to be considered among the three objects it is easy to suggest that "is more possible that x_2 is similar to x_1 than x_3 to x_1". Conventional rough sets theory simply does not apply in this case and its usual extensions handling unknown values will consider the three objects as completely different or totally identical ([4, 6]). However, being able to give a value to the possibility that objects are similar could open interesting operational directions (see [10]).

Example 2. Three objects x_1, x_2, x_3 and four attributes c_1, c_2, c_3, c_4 are given, each attribute equipped with an interval scale in the interval [0,100]. Besides, the above information table is provided:

If any similarity is to be considered the reader might agree that is reasonable to consider that "is more possible that x_2 is similar to x_1 than x_3 to x_1", while x_2 and x_3 are not similar at all. This is an effect of the existence of a discrimination threshold. In such a model (see [7]) objects are different only if they have a difference of more than the established threshold (in the example such a threshold is 5). However, the threshold by itself does not solve the problem, for the same discrimination problem could be considered near the threshold: i.e. why a difference of 4 is not significant and a difference of 5 it is? It is more natural to consider that the possibility that two objects are similar decreases as the difference of value of the two objects increases. The use of a valued tolerance appears to be more appropriate in this case also.

From the above considerations it is clear (for us) the opportunity of introducing a valued tolerance when comparing objects, whatever the purpose of the comparison is. Our aim is to introduce such a concept. In the following we will restrict ourselves to symmetric (possibly valued) similarity relations which we denote as (possibly valued) tolerance relations.

3   Rough Sets and Valued Tolerance

In the following we introduce an approach already discussed for handling incomplete information tables in [10]. Given a valued tolerance relation on the set A we can define a "tolerance class", that is a fuzzy set with membership function the "possibility of tolerance" to a reference object $x \in A$. The problem is to define the concepts of upper and lower approximation of a set Φ. The approach we will adopt in this paper considers, coherently with the rest, approximation as a *continuous valuation*. Given a set Φ to describe and a set $Z \subseteq A$ we will try to define the degree by which Z approximates from the top or from the bottom the set Φ. Technically we will try to give the functional correspondent of the concepts of lower and upper approximation. Some researchers [1–3] have similar concerns and explored the idea of combining fuzzy and rough sets, but under our perspective lower and upper approximations are not fuzzy sets to which elements from the universe of discourse may more or less belong. Each subset of A may be a lower or upper approximation of Φ, but to different degrees (such an approach has been inspired by the work of Kitainik [5]).

For this purpose we need to translate in a functional representation the usual logical connectives of negation, conjunction etc. ($x, y, z \cdots$ represent in the following membership degrees). For this purpose we consider negation functions $N(x) : N(x) = 1 - x$, T-norms $T(x, y) = \min(x, y)$ or xy or $\max(x + y - 1, 0)$ and T-conorms $S(x, y) = \max(x, y)$ or $x + y - xy$ or $\min(x + y, 1)$. If $S(x, y) = N(T(N(x), N(y)))$ we call the triplet $\langle N, T, S \rangle$ a De Morgan triplet. $I(x, y)$, the degree by which x may imply y is again a function that could satisfy the following (almost) incompatible properties $I(x, y) = S(N(x), y)$ or $x \leq y \Leftrightarrow I(x, y) = 1$.

Coming back to our lower and upper approximations we know that, given a set $Z \subseteq A$, a subset of attributes $B \subseteq C$ and a set Φ, the usual definitions are: $Z = \Phi_B \Leftrightarrow \forall z \in Z$, $\Theta_B(z) \subseteq \Phi$ and $Z = \Phi^B \Leftrightarrow \forall z \in Z$, $\Theta_B(z) \cap \Phi \neq \emptyset$ where $\Theta_B(z)$ is the tolerance class of element z created on the basis of the subset of attributes B. The functional translation of such definitions is: $\forall x \ \phi(x) =_{def} T_x \phi(x)$; $\exists x \ \phi(x) =_{def} S_x \phi(x)$; $\Phi \subseteq \Psi =_{def} T_x(I(\mu_\Phi(x), \mu_\Psi(x)))$; $\Phi \cap \Psi \neq \emptyset =_{def} \exists x \ \phi(x) \wedge \psi(x) =_{def} S_x(T(\mu_\Phi(x), \mu_\Psi(x)))$ we get: $\mu_{\Phi_B}(Z) = T_{z \in Z}(T_{x \in \Theta_B(z)}(I(R_B(z, x), \hat{x})))$, $\mu_{\Phi^B}(Z) = T_{z \in Z}(S_{x \in \Theta_B(z)}(T(R_B(z, x), \hat{x})))$ where: $\mu_{\Phi_B}(Z)$ is the degree for set Z to be a B-lower approximation of Φ; $\mu_{\Phi^B}(Z)$ is the degree for set Z to be a B-upper approximation of Φ; $\Theta_B(z)$ is the tolerance class of element z; T, S, I are the functions previously defined; as far as $I(x, y)$ is concerned we will always choose to satisfy De Morgan law ($I(x, y) = S(N(x), y)$). This is due to the particular case of $\mu_{\Phi_B}(Z)$ where $\hat{x} \in \{0, 1\}$. If we choose any other representation then lower approximability collapse to $\{0, 1\}$. $R_B(z, x)$ is the membership degree of element x in the tolerance class of z (at the same time is the valued tolerance relation between elements x and z for attribute set B; in our case $R_B(z, x) = T_{j \in B} R_j(z, x)$); $\hat{x}$ is the membership degree of element x in the set Φ ($\hat{x} \in \{0, 1\}$).

In the following we provide some formal properties that such an approach fulfill. The reader should remark that in the following Φ^c denotes the complement of set Φ with respect to the universe.

Proposition 1. *If T, S, I fulfill the De Morgan law and R_B is a valued tolerance relation then $\forall Z \in A \;\; \mu_{\Phi_B}(Z) \leq \mu_{\Phi^B}(Z)$.*

Proof: In order to demonstrate the proposition we observe that both the lower and the upper approximability are T-norms on the same set Z. It is sufficient now to demonstrate that $\forall z \in Z$ the argument of the T-norm defining the lower approximability is less or equal to the argument of the T-norm defining the upper approximability. Thus we have to demonstrate that: $T_{x \in \Theta_B(z)}(S(1 - R_B(z, x), \hat{x})) \leq S_{x \in \Theta_B(z)}(T(R_B(z, x), \hat{x}))$. Since min is the largest T-norm and max is the smallest T-conorm is sufficient to demonstrate that: $\min_{x \in \Theta_B(z)}(\max(1 - R_B(z, x), \hat{x})) \leq \max_{x \in \Theta_B(z)}(\min(R_B(z, x), \hat{x}))$. We distinguish two cases:
1. Consider $z = x_k \in \Phi^c$. Then $\hat{x_k} = 0$. Therefore when $x = x_k$ we have $\max(1 - R_B(x_k, x_k), \hat{x_k}) = 0$ so that $\min_{x \in \Theta_B(z)}(\max(1 - R_B(z, x), \hat{x})) = 0$. At the same time: $\forall x \in \Phi^c \; \hat{x} = 0$ and $\min(R_B(z, x), \hat{x}) = 0$ and $\forall x \in \Phi \; \hat{x} = 1$ and $\min(R_B(z, x), \hat{x}) = R_B(z, x)$ so that $\max_{x \in \Theta_B(z)}(\min(R_B(z, x), \hat{x})) = \max_{x \in \Phi}(R_B(z, x)) \geq 0$. Therefore if $z \in \Phi^c$ we get $\mu_{\Phi_B}(z) = 0 \leq \mu_{\Phi^B}(z)$.
2. Consider $z = x_k \in \Phi$. Then $\hat{x_k} = 1$. Therefore when $x = x_k$ we have $\min(R_B(x_k, x_k), \hat{x_k}) = 1$ so that $\max_{x \in \Theta_B(z)}(\min(R_B(z, x), \hat{x})) = 1$. At the same time: $\forall x \in \Phi \; \hat{x} = 1$ and $\max(1 - R_B(z, x), \hat{x}) = 1$ and $\forall x \in \Phi^c \; \hat{x} = 0$ and $\max(1 - R_B(z, x), \hat{x}) = 1 - R_B(z, x) \leq 1$ so that $\min_{x \in \Theta_B(z)}(\max(1 - R_B(z, x), \hat{x})) = \min_{x \in \Phi^c}(1 - R_B(z, x)) \leq 1$. Therefore if $z \in \Phi$ we get $\mu_{\Phi_B}(z) \leq 1 = \mu_{\Phi^B}(z)$. We immediately obtain the following corollary.

Corollary 1.
If $z \in \Phi$ then $\mu_{\Phi_B}(z) = \min_{x \in \Phi^c}(1 - R_B(z, x)) \leq \mu_{\Phi^B(z)} = 1$
If $z \in \Phi^c$ then $\mu_{\Phi_B}(z) = 0 \leq \mu_{\Phi^B(z)} = \max_{x \in \Phi}(R_B(z, x))$

Proposition 2. *If T, S, I respect the De Morgan law and R_B is a valued tolerance relation then $\forall z \; \mu_{\Phi_B}(z) = 1 - \mu_{(\Phi^c)^B}(z)$.*

Proof: Denote by $\hat{x^c}$ the membership of x to Φ^c. Clearly $\hat{x^c} = 1 - \hat{x}$. We then have: $\mu_{\Phi_B}(z) = T_{x \in \Theta_B(z)}(S(1 - R_B(z, x), \hat{x})) = T_{x \in \Theta_B(z)}(S(1 - R_B(z, x), 1 - \hat{x^c})) = T_{x \in \Theta_B(z)}(1 - T(R_B(z, x), \hat{x^c})) = 1 - S_{x \in \Theta_B(z)}(T(R_B(z, x), \hat{x^c})) = 1 - \mu_{(\Phi^c)^B}(z)$. We immediately obtain the following corollary.

Corollary 2.
$\mu_{\Phi_B}(Z) = 1 - S_{z \in Z}(\mu_{(\Phi^c)^B}(z)), \; \mu_{\Phi^B}(Z) = 1 - S_{z \in Z}(\mu_{(\Phi^c)_B}(z))$

Finally we can show the following result.

Proposition 3. $\forall Z \subset A \;\; \hat{B} \subset B \Rightarrow \mu_{\Phi_{\hat{B}}}(Z) \leq \mu_{\Phi_B}(Z)$

Proof: Since $R_B(z, x) = T_{j \in B} R_j(z, x)$, if $\hat{B} \subseteq B$ then $\forall x \; R_{\hat{B}}(z, x) \geq R_B(z, x)$ and therefore $\forall x \; 1 - R_{\hat{B}}(z, x) \leq 1 - R_B(z, x)$. Then by definition of lower approximability the proposition holds.

The consequence of the above results is that in order to compute the lower (upper) approximability of any subset $\Phi \subseteq A$ is sufficient to compute the upper (lower) approximability of each single element of A of the sets Φ and Φ^c. Operationally we can fix a threshold k (l) for the lower (upper) approximability and then add elements to the empty set by decreasing order of their lower (upper) approximability. Consider the following example.

Example 3. A set of 12 objects $x_1, \cdots x_{12}$, four attributes c_1, c_2, c_3, c_4 and a decision attribute d are given, each attribute equipped with an interval scale in the interval $[0,100]$. Besides, the following decision table is provided:

	x_1	x_2	x_3	x_4	x_5	x_6	x_7	x_8	x_9	x_{10}	x_{11}	x_{12}
c_1	89	85	80	79	74	70	68	63	59	57	55	25
c_2	91	87	83	80	76	71	70	64	64	59	56	32
c_3	95	90	92	85	83	79	74	70	69	57	56	15
c_4	87	86	80	77	71	70	66	62	60	56	54	48
d	Φ	Φ	Ψ	Φ	Ψ	Ψ	Φ	Ψ	Ψ	Φ	Ψ	Φ

A constant threshold of $k = 10$ applies in order to consider two objects as surely not different (similar) to any of the four attributes. Following an approach recently introduced [11] we consider for each attribute c_j a valued tolerance relation as follows:

$$R_j(x,y) = \frac{\max(0, \min(c_j(x), c_j(y)) + k - \max(c_i(x), c_j(y)))}{k}$$

where: k is the discrimination threshold. It is easy to observe that: $\forall x, y \in A$ $R_j(x,y) = 1$ iff $c_j(x) = c_j(y)$, $\forall x, y \in A$ $R_j(x,y) \in]0,1[$ iff $| c_j(x) - c_j(y) | < k$, $\forall x, y \in A$ $R_j(x,y) = 0$ iff $| c_j(x) - c_j(y) | \geq k$. If $R_j(x,y)$ represents the necessity that x is similar to y then, in presence of several attributes, a way to evaluate the necessity that x is comprehensively similar to y is to take the T-norm of the different similarities. We get: $R(x,y) = \min_j(R_j(x,y))$. Applying this formula to Example 3 we get the comprehensive valued relation on the set A. For instance, for object x_1 $R(x_1, x_1) = 1$, $R(x_1, x_2) = 0.5$, $R(x_1, x_3) = 0.1$ and for the rest we have $R(x-1, y) = 0$. Using the above information we can compute the lower and upper approximability for each element of set A. For instance, $\mu_{\Phi_B}(x_1) = 0.9$ $\mu_{\Phi^B}(x_1) = 1$ $\mu_{\Psi_B}(x_1) = 0$ and $\mu_{\Psi^B}(x_1) = 0.1$.

4   Decision Rules

In order to induce classification rules from the decision table on hand we may accept now rules with a "credibility degree" derived from the fact that objects may be similar to the conditional part of the rule only to a certain degree, besides the fact the implication in the decision part is also uncertain. More formally we give the following representation for a rule: $\rho_i =_{def} \bigwedge_{c_j \in B} (c_j(x) = v) \rightarrow (d = \phi)$ where: $B \subseteq C$, v is the value of conditional attribute c_j, ϕ is the value of decision attribute d. We may use the valued relation $s_B(x, \rho_i)$ in order to indicate that element x "supports" rule ρ_i or that, x is similar to some extend to the conditional part of rule ρ_i on attributes B. The relation s is a valued tolerance relation defined exactly as relation R. We denote as $S(\rho_i) = \{x : s_B(x, \rho_i) > 0\}$ and as $\Phi = \{x : d(x) = \phi\}$. In a case of crisp relation ρ_i is a classification rule iff: $\forall x \in S(\rho_i) : \Theta_B(x) \subseteq \Phi$. Shifting in the valued case we can compute a credibility degree for any rule ρ_i calculating the credibility for the previous formula which can be rewritten as: $\forall x, y \; s_B(x, \rho_i) \rightarrow (R_B(x, y) \rightarrow \Phi(y))$. We get: $\mu(\rho_i) = T_{x \in S(\rho_i)}(I(s_B(x, \rho_i), T_{y \in \Theta_B(x)}(I(\mu_{\Theta_B(x)}(y), \mu_\Phi(y)))))$; where: $\mu_{\Theta_B(x)}(y) = R_B(x, y)$ and $\mu_\Phi(y) \in \{0, 1\}$.

Finally it is necessary to check whether B is a non-redundant set of conditions for rule ρ_i, i.e. to look if it is possible to satisfy the condition: $\exists \hat{B} \subset B : \mu(\rho_i^{\hat{B}}) \geq \mu(\rho_i^B)$. We can equivalently state that if there is no $\hat{B}$ satisfying the condition then B is a "non redundant" set of attributes for rule ρ_i. Before we continue the presentation of our approach in rule induction, is important to state the following result.

Proposition 4. *Consider a rule ρ_i classifying objects to a set $\Phi \subseteq A$ under a set of attributes B. If T, S, I satisfy the De Morgan law and R_B is a valued tolerance, the credibility $\mu(\rho_i)$ of the rule is upper bounded by the lower approximability of set Φ by the element x_k whose description (under attributes B) coincides with the conditional part of the rule.*

Proof: Consider the definition of rule credibility.
$T_{y \in \Theta_B(x)}(I(\mu_{\Theta_B(x)}(y), \mu_\Phi(y))) = \mu_{\Phi_B}(x)$. Considering that $I(x, y) = S(N(x), y)$ and that $s_B = R_B$ we can rewrite: $\mu(\rho_i) = T_{x \in S(\rho_i)}(S(1 - R_B(x, \rho_i), \mu_{\Phi_B}(x)))$
We distinguish four cases.
1) It exists an element $x_k \in A$ whose description (under attributes B) coincides with the conditional part of rule ρ_i. We have $R_B(x, \rho_i) = 1$. Therefore $S(1 - R_B(x, \rho_i), \mu_{\Phi_B}(x)) = \mu_{\Phi_B}(x)$ in this case.
2) For all x for which $R_B(x, \rho_i) = 0$ we get $S(1 - R_B(x, \rho_i), \mu_{\Phi_B}(x)) = 1$.
3) For all x for which $1 - R_B(x, \rho_i) > \mu_{\Phi_B}(x)$ we get
$S(1 - R_B(x, \rho_i), \mu_{\Phi_B}(x)) \geq 1 - R_B(x, \rho_i)$ since max is the smallest T-conorm.
4) For all x for which $1 - R_B(x, \rho_i) < \mu_{\Phi_B}(x)$ we get
$S(1 - R_B(x, \rho_i), \mu_{\Phi_B}(x)) \geq \mu_{\Phi_B}(x)$ since max is the smallest T-conorm.
Denoting x_k, x_l, x_i, x_j the x for the four cases respectively we obtain: $\mu(\rho_i) = T_{x \in S(\rho_i)}(\mu_{\Phi_B}(x_k), \{\forall x_l : 1\}, \{\forall x_i : 1 - R_B(x_i, \rho_i)\}, \{\forall x_j : \mu_{\Phi_B}(x_j)\})$ Since by definition $T(x, y) \leq \min(x, y) \leq x$, $\mu_{\Phi_B}(x_k)$ is an upper bound for $\mu(\rho_i)$.

Operationally, the user should fix a credibility threshold for the induced rules in order to prevent proliferation of rules considered as "unsafe" for the classification purposes. A sensitivity analysis could be performed around such a threshold to find accepted rules.

In general, elementary conditions of the induced rules are created using the description of objects in the decision table. Assuming that the user has defined a credibility threshold at level λ, it is possible to use the result of Proposition 4.1. to induce decision rules, i.e. when choosing objects as candidates for inducing a classification rules for class Φ, it is sufficient to choose only objects with lower approximability of a set Φ not worse than λ. Other objects could be skipped; Further on it is necessary to search for the non-reduced sets of conditions (in general it corresponds to the problem of looking for local reducts); Finally, given credibility threshold λ one can relax the previous requirements to non-redundant rule, i.e. it may be accepted that from rule ρ_i with credibility $\mu(\rho_i)$ new rules could be generated with shortest condition part but with lower credibility however still over the allowed threshold. Let us notice that the problem of inducing all rules with accepted credibility from examples in the information table has exponential complexity in the worst case. However, fixing sufficient high value of credibility threshold may reduce the search space.

Continuation of Example 3: Consider the information table of Example 3. We choose min as the T-norm and max as T-conorm. Let us consider creating a decision rule basing on description of the object x_1, i.e. $\rho_1 : (c_1 = 89) \wedge (c_2 = 91) \wedge (c_3 = 95) \wedge (c_4 = 87) \rightarrow (d = \Phi)$. Three objects (x_1, x_2, x_3) are similar to its condition part with $R_C(x_1, \rho_i) = 1$, $R_C(x_2, \rho_i) = 0.5$ and $R_C(x_3, \rho_i) = 0.1$. So $S(\rho_i) = \{x_1, x_2, x_3\}$. We can compute credibility of the rule according to formula: $\mu(\rho_i) = \min_{x \in S(\rho_i)}(\max(1 - R_B(x, \rho_i), \mu_{\Phi_B}(x)))$. So, taking values of lower approximability we have $\mu(\rho_i) = \min(\max(1 - 1, 0.9), \max(1 - 0.5, 0.6), \max(1 - 0.1, 0)) = 0.6$. However, this is a rule with a redundant set of conditions. As one can check, it can be reduced to the much simpler form $\rho_1 : (c_4 = 87) \rightarrow (d = \Phi)$ which is still supported by objects $S(\rho_i) = \{x_1, x_2, x_3\}$ with credibility degree $= 0.6$. Proceeding in a similar way we can induce other decision rules.

Let us now consider the use of induced decision rules to classify new unclassified objects. The problem is to assign such objects to a-priori known sets (decision classes) on the basis of their tolerance to the conditional part of the already induced rules. We have a double source of uncertainty. First, the new object will be similar to a certain extend to the conditional part of a given rule. Second the rule itself has a credibility (classification is not completely sure any more). In general a new unclassified object will be more or less similar to more than one decision rule and such rules may indicate different decision classes. Therefore an unclassified object can be assigned to several different classes. In order to choose one class the following procedure is proposed:

1. For each decision rules ρ_i in the set of induced rules we calculate the tolerance of the new object z to its condition part, $R_B(z, \rho_i)$.

2. Then we compute the membership degree of object z to each decision class Φ_i as $\mu_{\Phi_i}(z) = T(R_B(z, \rho_i), \mu(\rho_i))$. Then we choose the class with the maximum membership degree.

3. If a tie occurs (the same membership for different classes) choose the rule with the highest number of supporting objects $S(\rho_i)$.

5   Conclusion

In the paper we develop the idea that valued tolerance relations (symmetric valued similarity relations) can be more suitable when objects are compared for classification purposes. Particularly when rough sets are used, the classic indiscernibility relation (which is a crisp equivalence relation) can be a too strong assumption with respect to the available information.

The main contribution of the paper consists in considering that any subset of the universe of discourse can be considered as a lower (upper) approximation of set Φ, but to a different degree, due to the existence of a valued tolerance relation among the elements of the universe. A number of formal properties of this approach are demonstrated and discussed in the paper. Further on, the availability of a lower (upper) approximability degree for each set with respect to a decision class Φ enables to compute classification rules equipped with a credibility degree. A significant result obtained in the paper consists in demonstrating that a rule credibility is upper bounded by the lower approximability of the set whose elements description coincides with the conditional part of the rule. Finally, two sources of uncertainty during classification of new objects were discussed.

References

1. Dubois D., Prade H.: Rough Fuzzy Sets and Fuzzy Rough Sets. International Journal of General Systems, **17**, (1990), 191–209.
2. Greco S., Matarazzo B. Slowinski R.: Fuzzy similarity relation as a basis for rough approximations. In Polkowski L., Skowron A. (eds.), Proc. of the RSCTC-98, Springer Verlag, Berlin, LNAI 1424, (1998), 283–289.
3. Greco S., Matarazzo B. Slowinski R.: Rough set processing of vague information using fuzzy similarity relations. In Calude C.S., Paun G. (eds), Finite vs infinite: contributions to an eternal dilemma, Springer Verlag, Berlin, (2000), 149–173.
4. Grzymala-Busse J.W.: On the unknown attribute values in learning from examples. Proc. of Int. Symp. on Methodologies for Intelligent Systems, (1991), 368–377.
5. Kitainik L.: Fuzzy Decision Procedures with Binary Relations, Kluwer Academic, Dordrecht, (1993).
6. Kryszkiewicz M.: Properties of incomplete information systems in the framework of rough sets. In Polkowski L., Skowron A. (eds.), Rough Sets in Data Mining and Knowledge Discovery, Physica-Verlag, Heidelberg, (1998), 422–450.
7. Luce R.D.: Semiorders and a theory of utility discrimination, Econometrica, **24**, (1956), 178–191.
8. Skowron A., Stepaniuk J.: Tolerance approximation spaces, Fundamenta Informaticae, **27**, (1996), 245–253.
9. Słowiński R., Vanderpooten D.: Similarity relation as a basis for rough approximations, In Wang P. (ed.), Advances in Machine Intelligence and Soft Computing, vol. IV., Duke University Press, (1997), 17–33.
10. Stefanowski J., Tsoukiàs A.: On the extension of rough sets under incomplete information, in N. Zhong, A. Skowron, S. Ohsuga, (eds.), New Directions in Rough Sets, Data Mining and Granular-Soft Computing, Springer Verlag, LNAI 1711, Berlin, (1999), 73–81.
11. Tsoukiàs A., Vincke Ph.: A characterization of PQI interval orders, to appear in *Discrete Applied Mathematics*, (2000).
12. Yao Y.: Combination of rough sets and fuzzy sets based on α-level sets. In Lin T.Y., Cercone N. (eds.), Rough sets and data mining, Kluwer Academic, Dordrecht, (1996), 301–321.
13. Yao Y., Wang T.: On rough relations: an alternative fromulation. In N. Zhong, A. Skowron, S. Ohsuga, (eds.), New Directions in Rough Sets, Data Mining and Granular-Soft Computing, Springer Verlag, LNAI 1711, Berlin, (1999), 82–90.

A Conceptual View of Knowledge Bases
in Rough Set Theory

Karl Erich Wolff

University of Applied Sciences Darmstadt, Department of Mathematics
Schöfferstr. 3, D-64295 Darmstadt, Germany
ERNSTSCHRÖDERCENTER FOR CONCEPTUAL KNOWLEDGE PROCESSING
Research Group Concept Analysis at Darmstadt University of Technology
E-mail: wolff@mathematik.tu-darmstadt.de

Abstract. Basic relationships between Rough Set Theory (RST) and Formal Concept Analysis (FCA) are discussed. Differences between the "partition oriented" RST and the "order oriented" FCA concerning the possibility of knowledge representation are investigated. The fundamental connection between RST and FCA is that the knowledge bases of RST and the scaled many-valued contexts of FCA are shown to be nearly equivalent.

1 Introduction: Rough Sets and Formal Concepts

The purpose of this paper is to discuss some basic relationships between Rough Set Theory (RST) and Formal Concept Analysis (FCA). This discussion is a part of the more general investigation of the purposes, the underlying philosophical ideas, the formalizations, and the technical tools of knowledge theories as, for example, data analysis, data base theory, evidence theory, formal languages, automata theory, system theory, and Fuzzy Theory. Their relations to classical knowledge theories like geometry, algebra, logic, statistics, probability theory and physics also should be studied.

Both RST and FCA were introduced, independently, in 1982, by Z. Pawlak [7]and R. Wille [10]respectively. Roughly speaking, both theories formalize in some meaningful way the concept of "concept"; more careful explanation is given in section 2. Both theories are used to investigate parts of "reality" described by some "measurement protocol" from which "data" are obtained, technically represented mainly in data tables. Both theories use data tables as the central tool for the development of decision aids. Both theories have been widely applied in science and also in industry. But until now only a few personal contacts between the RST and FCA communities have occured. This paper should open up detailed discussion.

Despite their many similarities there are several differences between the two communities. I believe that the underlying philosophical ideas are quite different. That seems to influence strongly the research aims and the strategies in research development. That should be discussed through personal interaction where the differences between

W. Ziarko and Y. Yao (Eds.): RSCTC 2000, LNAI 2005, pp. 220–228, 2001.

the formalizations of basic concepts in the theories can be discussed more clearly and easily.

2 Formal Concept Analysis

Formal Concept Analysis (FCA) was introduced by Wille [10] and developed in the research group Concept Analysis of the mathematical department at Darmstadt University of Technology (Germany). It is now the mathematical basis for research in Conceptual Knowledge Processing. For the mathematical foundations the interested reader is referred to Ganter, Wille [4,5]. Two elementary introductions were written by Wolff [13] and Wille [12]. From the latter we quote:

Formal Concept Analysis is based on the philosophical understanding that a concept is constituted by two parts: its extension which consists of all objects belonging to the concept, and its intension which comprises all attributes shared by those objects. For formalizing this understanding it is necessary to specify the objects and attributes which shall be considered in fulfilling a certain task. Therefore, Formal Concept Analysis starts with the definition of a formal context ... .

2.1 Formal Contexts and Concept Lattices

The following definition of a *formal context* was motivated by the observation that the specific meaning of concepts in human thinking and communication is always determined by contexts. For the description of contexts we use the most simple verbal utterance which states that an object has an attribute. Therefore a *formal context* is defined as a triple (G,M,I) of sets where I is a binary relation between G and M, i.e., $I \subseteq G \times M$. The set G is called the set of *objects* (Gegenstände), the set M is called the set of *attributes* (Merkmale) and the statement that the pair $(g,m) \in I$ is read "object g has attribute m".

Why is this simple definition important for a formal theory on concepts? It is important since it is possible to define for each formal context $\mathbf{K}$ a very meaningful conceptual hierarchy $(\mathbf{B}(\mathbf{K}),\leq)$, whose elements, the *formal concepts of* K, represent units of thought consisting of two parts, the extension and the intension, just as it is understood in philosophical investigations dating back to Arnauld, Nicole ([1],1685).

A *formal concept* of $\mathbf{K} = (G,M,I)$ is defined as a pair (A,B) where $A \subseteq G$, $B \subseteq M$ and $A^{\uparrow} = B$ and $B^{\downarrow} = A$ where $A^{\uparrow}$ is the set of common attributes of A, formally described as $A^{\uparrow} := \{m \in M \mid \forall g \in A \ g \ I \ m \}$ and $B^{\downarrow}$ is the set of common objects of B, $B^{\downarrow} := \{g \in G \mid \forall m \in B \ g \ I \ m \}$. A is called the *extent* and B the *intent* of (A,B).

The set of all formal concepts of $\mathbf{K}$ is denoted by $\mathbf{B}(\mathbf{K})$. The conceptual hierarchy among concepts is defined by set inclusion: For $(A_1, B_1), (A_2, B_2) \in \mathbf{B}(\mathbf{K})$ let $(A_1, B_1) \leq (A_2, B_2) :\Leftrightarrow A_1 \subseteq A_2$ (which is equivalent to $B_2 \subseteq B_1$).

An important role is played by the *object concepts* $\gamma(g) := (\{g\}^{\uparrow\downarrow}, \{g\}^{\uparrow})$ for $g \in G$ and dually the *attribute concepts* $\mu(m) := (\{m\}^{\downarrow}, \{m\}^{\downarrow\uparrow})$ for $m \in M$.

The pair $(\mathbf{B(K)},\leq)$ is an ordered set, i.e., $\leq$ is reflexive, antisymmetric, and transitive on $\mathbf{B(K)}$. It has some important properties:

$(\mathbf{B(K)},\leq)$ is a complete lattice, called the *concept lattice of K*, and any complete lattice is isomorphic to a concept lattice; $(\mathbf{B(K)},\leq)$ contains the entire information in $\mathbf{K}$, i.e., $\mathbf{K}$ can be reconstructed from $\mathbf{B(K)}$. If $\mathbf{B(K)}$ is finite it can be drawn as a line diagram in the plane, such that $\mathbf{K}$ can be reconstructed.

Line diagrams of concept lattices can be drawn automatically by computer programs (Wille [11]) and serve as an important communication tool for the representation of multidimensional data (Wolff [14]).

It is clear that binary relations and therefore formal contexts are used in nearly all branches of mathematics and in many applications; therefore Formal Concept Analysis is very useful in many situations, even if the formal contexts are not finite. One of the most famous infinite examples is the context $(\mathbf{Q}, \mathbf{Q}, \leq_Q)$ of the rational numbers $\mathbf{Q}$ with the usual rational ordering $\leq_Q$. The concept lattice $\mathbf{B}(\mathbf{Q}, \mathbf{Q}, \leq_Q)$ is isomorphic to the complete lattice of all real numbers including ∞ and $-\infty$ with the usual ordering on this set. This conceptual construction of the real numbers shows that Formal Concept Analysis covers not only finite structures.

Since each complete lattice is isomorphic to a concept lattice, and complete lattices, closure systems, and closure operators are mathematically equivalent, Formal Concept Analysis enriches the application of these theories by a strong communicational component, which stems from the contextual meaning of the objects and attributes and the rich possibilities for visualizing multidimensional data by line diagrams of concept lattices.

2.2 Conceptual Scaling

The word 'scaling' is understood here in the sense of 'embedding something in a certain (usually well-known) structure', called a scale: for example, embedding some objects according to the values of measurements of their temperature into a temperature scale. Another example is the embedding of conference talks in the time schedule, which is usually a direct product of two time chains, one for hours and one for days. More generally, in conceptual scaling objects or values are embedded in the concept lattice of some formal context, called a conceptual scale.

Conceptual Scaling Theory was developed by Ganter and Wille [3]. The general process in conceptual scaling starts with the representation of knowledge in a data table with arbitrary values and possibly missing values. These data tables are formally described by *many-valued contexts (G,M,W,I)*, where G is a set of 'objects', M is a set of 'many-valued attributes', W is a set of 'values' and I is a ternary relation, $I \subseteq G{\times}M{\times}W$, such that for any $g \in G$, $m{\in} M$ there is at most one value w satisfying $(g,m,w) \in I$. Therefore, a many-valued attribute m can be understood as a (partial) function and we write $m(g) = w$ iff $(g,m,w) \in I$. A many-valued attribute m is called *complete* if it is a function. (G,M,W,I) is called *complete* if each $m \in M$ is complete.

The central granularity-choosing process in conceptual scaling theory is the construction of a formal context $\mathbf{S}_m = (W_m, M_m, I_m)$ for each $m{\in} M$ such that W_m

$\supseteq mG := \{m(g) \mid g \in G\}$. Such formal contexts, called *conceptual scales*, represent a contextual language about the set of values of m. Usually one chooses W_m as the set of all 'possible' values of m with respect to some purpose. Each attribute $n \in M_m$ is called a *scale attribute*. The set $n^{\downarrow} = \{w \mid w\ I_m\ n\}$ is the extent of the attribute concept of n in the scale S_m. Hence, the choice of a scale induces a selection of subsets of W_m - describing the granularity of the contextual language about the possible values. The set of all intersections of these subsets constitutes just the closure system of all extents of the concept lattice of S_m.

The granularity of the language about the possible values of m induces in a natural way a granularity on the set G of objects of the given many-valued context, since each object g is mapped via m onto its value m(g) and m(g) is mapped via the object concept mapping γ_m of S_m onto $\gamma_m(m(g))$: $g \to m(g) \to \gamma_m(m(g))$.

Hence the set of all object concepts of S_m plays the role of a frame within which each object of G can be embedded.

For two attributes m, m´ $\in$ M each object g is mapped onto the corresponding pair:
$g \to (m(g), m´(g)) \to (\gamma_m(m(g)), \gamma_m(m´(g))) \in B(S_m) \times B(S_{m´})$.

The standard scaling procedure, called *plain scaling,* constructs from a *scaled many-valued context* $((G,M,W,I), (S_m \mid m \in M))$, consisting of a many-valued context (G,M,W,I) and a scale family $(S_m \mid m \in M)$, the *derived context*, denoted by
$K := (G, \{(m,n) \mid m \in M, n \in M_m\}, J)$, where
$g\ J\ (m,n)$ iff $m(g)\ I_m\ n$ ($g \in G$, $m \in M$, $n \in M_m$).

The concept lattice $B(K)$ can be (supremum-)embedded into the direct product of the concept lattices of the scales (Wille [10], Ganter, Wille [4,5]). That leads to a very useful visualization of multidimensional data in so-called *nested line diagrams,* which is implemented in the program TOSCANA (Vogt, Wille [9]).

Scaled many-valued contexts are essentially the same as information channels in the sense of (Barwise, Seligman [2]), which was shown by the author (Wolff [18]).

Finally we mention that Fuzzy Theory, introduced by Zadeh [20], also developed some notion of a scale, namely the linguistic variables (Zadeh [21]). It was shown by Wolff ([15,16]) that Fuzzy Theory can be extended (by replacing the unit interval in the definition of the membership function by an arbitrary ordered set (L,≤)) to so-called L-Fuzzy Theory, which allows for developing, analogously as with Formal Concept Analysis, a Fuzzy Scaling Theory which is equivalent to Conceptual Scaling Theory.

3 Looking at Rough Sets Conceptually

In this paper I do not repeat the basic notions in Rough Set Theory systematically. I just describe some important differences between RST and FCA.

3.1 Partitioning or Ordering the Universe ?

Pawlak ([8], p.2) starts his definition of a *knowledge base* with "a finite set $U \neq \varnothing$ (the universe) of objects we are interested in." The set U plays the same role in RST as the set G of objects (which may be infinite or empty) in FCA. Furthermore, Pawlak writes "Any subset $X \subseteq U$ of the universe will be called a *concept* or a *category* in U...". This notion of "concept" corresponds extensionally to the notion of "extent of an attribute concept" and intensionally to "attribute" in FCA, which is also indicated by the name "category". Therefore the classical distinction between extents and intents, which goes back to the "Port Royal Logic" of Arnauld and Nicole ([1], 1685) is not represented in the "extensionally oriented" descriptions in RST.

It is remarkable that Pawlak does not introduce the notion of an arbitrary system of "overlapping" subsets of U, defined as a pair (U, **S**), where **S** is a subset of the power set **P**(U). The reason is that Pawlak ([8], p.3) "is mainly interested in this book with concepts which form a partition (classification) of a certain universe U...".

This decision, to use partitions of the universe, is an expression of a certain (often successful) mode of thinking in nominal structures. Therefore the set inclusion does not play the same prominent role in RST as in FCA, where the conceptual hierarchy is defined by the inclusion of the extents (or, equivalently, the inverse inclusion of the intents) of formal concepts. The background for this "ordinal thinking" in FCA is the successful application of ordinal structures, used for example in the conceptual thinking of Aristotle, in the subspace structures in spatial geometry and in lattices in logics. At first glance, the nominal approach in RST and the ordinal approach in FCA seem to be very different and incomparable. But each partition, for example the partition {M,F} of the universe of all people with the class M of male and the class F of female people, can be described by a formal context, for example $(M \cup F, \{M,F\}, \in)$. The formal context of a partition is a *nominal scale*, defined as a formal context (G,M,I) where the relation I is a function from G to M. The concept lattice of a nominal scale is isomorphic to an antichain together with a top and a bottom element.

On the other side, from any formal context $\mathbf{K} = (G,M,I)$ one can obtain a partition of the object set G by taking the inverse images of the object-concept mapping $\gamma : G \to \mathbf{B}(\mathbf{K})$. This partition $\mathbf{p}(\gamma) := \{\gamma^{-1}\gamma(g) \mid g \in G \}$ is called the *partition of γ*. The classes of $\mathbf{p}(\gamma)$ are just the equivalence classes of the relation $\sim$, where $g \sim h$ is defined by $g^{\uparrow} = h^{\uparrow}$ (for $g, h \in G$). That means, that g and h are indiscernible in the sense, that g and h have exactly the same attributes in **K**. The classes of $\mathbf{p}(\gamma)$ are called the *contingents* of γ (or the *object contingents of K*).

3.2 Knowledge Bases and Scaled Many-Valued Contexts

Pawlak ([8], p.3) defines a *knowledge base* as a pair (U,**R**) where **R** is a family of equivalence relations on the finite, non-empty set U, the universe of the knowledge base. The FCA conceptual counterpart of a knowledge base (U,**R**) is a scaled many-valued context $((G,M,W,I), (\mathbf{S}_m \mid m \in M))$. In the following we show how to construct a suitable scaled many-valued context from a knowledge base. The main idea is to take the equivalence relations $R \in \mathbf{R}$ as many-valued attributes and the equivalence class $[x]_R$ as the value of R at the object $x \in U$. Then nominal scaling of the equivalence classes yields a derived context which has as contingents just the indiscernibility classes of $(U,\mathbf{R})$, where the indiscernibility relation $IND(\mathbf{R})$ is the intersection of all equivalence relations in $\mathbf{R}$. The formal description is given in Theorem 1.

Theorem 1:
Let $(U,\mathbf{R})$ be a knowledge base. Then the scaled many-valued context $\mathbf{sc}(U,\mathbf{R}) := ((U,\mathbf{R},W,I), (S_R \mid R \in \mathbf{R}))$ is defined by: $W := \{ [x]_R \mid x \in U, R \in \mathbf{R}\}$ and $(x,R,w) \in I :\Leftrightarrow w = [x]_R$ and the nominal scale $S_R := (U/R, U/R, =)$ for each many-valued attribute $R \in \mathbf{R}$. Then the indiscernibility classes of $(U,\mathbf{R})$ are exactly the contingents of the derived context $\mathbf{K}$ of $\mathbf{sc}(U,\mathbf{R})$.

Proof: The derived context of $\mathbf{sc}(U,\mathbf{R})$ is by definition $\mathbf{K} = (U, \{(R, [x]_R) \mid R \in \mathbf{R}, [x]_R \in U/R \}, J)$, where the relation J is defined by $x J (R, [y]_R) :\Leftrightarrow [x]_R = [y]_R$ (for $x, y \in U$ and $R \in \mathbf{R}$). Let γ denote the object-concept mapping of $\mathbf{K}$. Then we have to prove that for all $x, y \in U : \gamma(x) = \gamma(y) \Leftrightarrow (\forall R \in \mathbf{R}\ (x,y) \in R)$. Let $x, y \in U$, then $(\forall R \in \mathbf{R}\ (x,y) \in R) \Leftrightarrow (\forall R \in \mathbf{R}\ [x]_R = [y]_R) \Leftrightarrow (\forall R \in \mathbf{R}\ x J (R, [y]_R)) \Leftrightarrow \gamma(x) = \gamma(y)$.

The construction of a knowledge base from a scaled many-valued context is described in Theorem 2.

Theorem 2:
Let $\mathbf{SC} := ((G,M,W,I), (S_m \mid m \in M))$ be a scaled many-valued context, and $\mathbf{K} := (G, \{(m,n) \mid m \in M, n \in M_m \}, J)$ its derived context. Then the knowledge base $\mathbf{kb}(\mathbf{SC})$ is defined by $\mathbf{kb}(\mathbf{SC}) := (G, \mathbf{R})$, where $\mathbf{R} := \{R_m \mid m \in M\}$ and for $m \in M$ $R_m := \{(g,h) \in G \times G \mid \gamma_m(g) = \gamma_m(h) \}$ and γ_m is the object-concept mapping of the m-part of $\mathbf{K}$; clearly, the m-part of K is the formal context $(G, \{(m,n) \mid n \in M_m \}, J_m)$ where $J_m := \{(g, (m,n)) \in J \mid n \in M_m \}$. Then the indiscernibility classes of $\mathbf{kb}(\mathbf{SC})$ are exactly the contingents of the derived context $\mathbf{K}$ of $\mathbf{SC}$.

Proof: Let $g, h \in G$, then the object-concept mapping γ of $\mathbf{K}$ satisfies: $\gamma(g) = \gamma(h) \Leftrightarrow \forall m \in M\ \gamma_m(g) = \gamma_m(h) \Leftrightarrow \forall m \in M\ (g,h) \in R_m \Leftrightarrow (g,h) \in IND(\mathbf{R})$.

If we combine the two operations sc and kb we conclude that the nominal scaling in the definition of $\mathbf{sc}$ is "compatible" with the operation $\mathbf{kb}$ in the sense of the following theorem:

Theorem 3:
For any knowledge base $(U, \mathbf{R})$: $\mathbf{kb}(\mathbf{sc}(U, \mathbf{R})) = (U, \mathbf{R})$.

Proof: Theorem 3 follows from Theorem 1 and Theorem 2.

Clearly, for a scaled many-valued context $SC := ((G,M,W,I), (S_m \mid m \in M))$ we can not expect that $sc(kb(SC))$ equals SC. The reason is that the ordinal structures of the concept lattices of the m-parts of the derived context K of SC are not represented in the knowledge base $kb(SC)$.

This investigation uncovers another aspect of RST, which is in a way not evident from the viewpoint of theory construction: Pawlak ([8], p.5) introduces *basic categories* as "set theoretical intersections of elementary categories", where elementary categories are the classes of the equivalence relations in (U, R). The set of all basic categories of (U, R) is exactly the closure system of the extents of the derived context K of $sc(U, R)$. That the basic category "red and triangular" describes a subconcept (in RST and in FCA) of the basic category "red" shows that the ordinal structure of the conceptual hierarchy is used, expressed with the lattice operation "infimum", which corresponds to the intersection of extents according to the Basic Theorem on Concept Lattices (Ganter, Wille [5], p.20). The same theorem states that the supremum in the concept lattice corresponds not necessarily to the union of the extents, but to the closure of the union of the extents. Therefore, the construction of all unions of basic categories leads to a usually much larger lattice than the corresponding concept lattice.

The above-studied relationship between scaled many-valued contexts and knowledge bases should be seen also in connection with information channels, which are essentially the same as scaled many-valued contexts (Wolff [18]).

4 Conclusion and Future Perspectives

This short comparison between some fundamental notions in RST and FCA can be extended further to discuss, for example, the role of granularity in RST, Fuzzy Theory, and FCA on the basis of "L-Fuzzy Scaling Theory" being equivalent to Conceptual Scaling Theory (Wolff [15,16]. Another interesting field is the reduction of knowledge and the conceptual role of reducts. For practical purposes several kinds of dependencies should be carefully compared. Numerous applications where the set of variables of a data table is divided into two (or more) parts of independent and dependent variables, of input and output variables, of time and space variables can be described using the decomposition and embedding techniques in FCA. Examples are decision tables, switching circuits, automata and the representation of systems in Mathematical System Theory (Mesarovic, Takahara [6]). The role of "time" in the formal description of systems was conceptually investigated by the author (Wolff [17]); conceptual time systems, consisting of a many-valued context with an object set of "time granules", a "time part" and a "space part" lead to a conceptual system theory in which state and phase spaces are introduced as concept lattices. This approach can be used for a conceptual description of arbitrary automata, including an interpretation of the edges of the directed graph of the automaton as implications of a suitable formal context.

References

1. Arnauld, A., Nicole, P.: La Logique ou L'Art de penser, outre les Règles communes, plusieurs observations nouvelles, propres à former le jugement. Sixième Édition revue et de nouveau augmentée. A Amsterdam, Chez Abraham Wolfgang 1685.German translation by Christos Axelos: Die Logik oder die Kunst des Denkens. Wissenschaftliche Buchgesellschaft, Darmstadt, 1994.

2. Barwise, J., Seligman, J.: Information Flow – The Logic of Distributed Systems. Cambridge Tracts in Theoretical Computer Science 44, 1997.

3. Ganter, B., R. Wille: *Conceptual Scaling*. In: F. Roberts (ed.) Applications of combinatorics and graph theory to the biological and social sciences,139-167. Springer Verlag, New York, 1989.

4. Ganter, B., R. Wille: Formale Begriffsanalyse: Mathematische Grundlagen. Springer-Verlag, Berlin-Heidelberg 1996.

5. Ganter, B., R. Wille: Formal Concept Analysis: mathematical foundations. (translated from the German by Cornelia Franzke) Springer-Verlag, Berlin-Heidelberg 1999.

6. Mesarovic, M.D., Takahara, Y.: General Systems Theory: Mathematical Foundations.Academic Press, London, 1975.

7. Pawlak, Z.; Rough Sets.International Journal of Computer and Information Sciences, 11, 1982, 341-356.

8. Pawlak, Z.; Rough Sets: Theoretical Aspects of Reasoning About Data. Kluwer Academic Publishers, 1991.

9. Vogt, F., R.Wille; TOSCANA - a graphical tool for analyzing and exploring data. In: R.Tamassia, I.G.Tollis (eds.); Graph Drawing. Springer-Verlag, Heidelberg, 1994, 193-205.

10. Wille, R.; Restructuring lattice theory: an approach based on hierarchies of concepts. In: Ordered Sets (ed. I.Rival). Reidel, Dordrecht-Boston, 1982, 445-470.

11. Wille, R.: Lattices in data analysis: how to draw them with a computer. In: I.Rival (ed.): Algorithms and order. Kluwer, Dordrecht-Boston 1989, 33-58.

12. Wille, R.; Introduction to Formal Concept Analysis. In: G. Negrini (ed.): Modelli e modellizzazione. Models and modelling. Consiglio Nazionale delle Ricerche, Instituto di Studi sulli Ricerca e Documentatione Scientifica, Roma 1997, 39-51.

13. Wolff, K.E.; A first course in Formal Concept Analysis - How to understand line diagrams. In: Faulbaum, F. (ed.): SoftStat '93, Advances in Statistical Software 4, Gustav Fischer Verlag, Stuttgart 1994, 429-438.

14. Wolff, K.E.: Comparison of graphical data analysis methods. In: Faulbaum, F. & Bandilla, W. SoftStat '95 Advances in Statistical Software 5, Lucius&Lucius, Stuttgart 1996, 139-151.

15. Wolff, K.E.; Conceptual Interpretation of Fuzzy Theory. In: Zimmermann, H.J.: EUFIT'98, 6th European Congress on Intelligent Techniques and Soft Computing, Aachen 1998, Vol. I, 555-562.

16. Wolff, K.E.; Concepts in Fuzzy Scaling Theory: Order and Granularity. Proceedings of the 7th European Conference on Intelligent Techniques and Soft Computing, EUFIT'99. Preprint Fachhochschule Darmstadt, 1999a. To appear in Fuzzy Sets and Systems.

17. Wolff, K.E.; Concepts, States, and Systems. In: Dubois, D.M. (ed.): Computing Anticipatory Systems. CASYS'99 - Third International Conference, Liège, Belgium, August 9-14, 1999 American Institute of Physics Conference Proceedings 517 (selected papers), 2000, p. 83-97.

18. Wolff, K.E.; Information Channels and Conceptual Scaling. In: Stumme G. (ed.): Working with Conceptual Structures – Contributions to ICCS 2000, (8th International Conference on

Conceptual Structures: Logical, Linguistic, and Computational Issues). Shaker Verlag, Aachen, 277-283.

19. Zadeh, L.A.; The Concept of State in System Theory. In: M.D. Mesarovic: Views on General Systems Theory. John Wiley & Sons, New York, 1964, 39-50.

20. Zadeh, L.A.; Fuzzy sets. Information and Control **8**, 1965, 338 - 353.

21. Zadeh, L.A.; The concept of a linguistic variable and its application to approximate reasoning. Part I: Inf. Science **8**,199-249; Part II: Inf. Science **8**, 301-357; Part III: Inf. Science **9**, 43-80, 1975.

Computer Vision Using Fuzzy Logic
for Robot Manipulator
Lab. Implementation

Adriano Breunig[1], Haroldo R. de Azevedo[2], Edilberto P. Teixeira[2], and
Elmo B. de Faria[3]

[1] Escola Técnica Federal de Mato Grosso
Depto. de Informática, Cuiabá, 78.005-390/MT, Brasil
`abreunig@uol.com.br`
[2] Universidade Federal de Uberlândia, Depto. de Engenharia Elétrica
Uberlândia, 38.400-902/MG, Brasil
`{haroldo,edilbert}@ufu.br`
[3] Instituto Luterano de Ensino Superior de Itumbiara, Depto. de Informática
Itumbiara, 75.523-200/GO, Brasil
`ebfaria@ufu.br`

Abstract. This paper presents a Laboratory implementation of a computer vision using fuzzy logic techniques for the detection of boundaries in images obtained through a camera installed in a robotic arm. Here, these images were captured, digitized and analysed using fuzzy algorithms, in order to used for the control of the robotic arm carrying the camera. The image treatment process uses the method proposed by Pal and Majumder (1987), which employs image enhancement techniques through the use of fuzzy set concepts. The image recognition process generates the control signal necessary to move the robotic arm in a given specific pathway, as well as to determine the next action to be taken at the end of the task.

1 Introduction

The artificial Vision System tecnology is increasingly becoming more important. Its aplications may be found in several areas, such as in industries, in medicine, in robots, etc. In the case of robots the vision is indispensable. The vision systems for use with robots should be designed following to basic criterion (Groover, Weiss, Nagel and Odrey, 1989):

1st - Relatively Low Cost;
2nd - Relatively Fast Response Time.

Usually there are two methods for the implementation of vision systems. One takes the image of the object and tries to reduce it to contour lines that form the profile of that object. This method utilises filters to obtain the image information and uses contrast enhancement to turn all parts of the image into either black or white (thresholding).The formed image is usually called a **binary image**.

W. Ziarko and Y. Yao (Eds.): RSCTC 2000, LNAI 2005, pp. 229–237, 2001.

The advantage of binary images is that they supply well defined limits, easily recognized by simple algorithms. This type of implementation is used by systems for the processing of two dimensional images.

The other method for the implementation of vision systems tries to give to the computer an image closer to what humans perceive. This method gives information about the brightness of the image to the computer. This allows the computer to obtain two important image caracterics, which are not possible to obtain with the technique of high image contrast: surfaces and shadows. This can be used to obtain three dimensional image information and to solve conflicts when one object partially blocks an other. This method is usually used for three dimensional vision systems.

In this work we shall use the first method, as we will have a controlled working environment, utilizing images with suficient clear shapes.

2   Vision System

The images is obtained with a camera connected straight to the parallel port of a computer. The image is digitized with a resolution of 64×64 pixels and 256 levels of grey. The image will be analyzed through the techniques of Image Enhancement and Histogram, using for this purpose, Fuzzy Algorithms. The image analysis will generate a control signal for the movement of the six joints robotic arm Fig. 1.

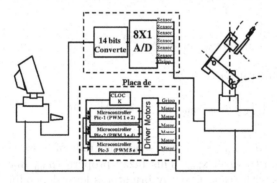

Fig. 1. General scheme of the robot system

3   Pre-processing

The analogical video signal obtained with the camera is sampled and quantized in order to be suitable for computer processing.

3.1   Image Representation - Matrix Form

A monocromatic image may be described as a mathematical function $f(x,y)$ of the light intensity, where its value at a point (x, y) is proportional to the level of grey at that point.

This function $f(x,y)$, Eq. 1, is represented by a $M \times N$ matrix with "l" levels of grey, each one representing a pixel.

3.2   Image Representation - Fuzzy Set Theory

A image F, where $F = f(x, y)$ (Eq. 2), with dimensions $M \times N$ and "l" levels of grey may be considered as a matrix of singletons fuzzy, each one with an $f(x,y)$ value that indicates the relative brightness value of grey level "l", where $1 = 0, 1, 2, 3...L - 1$. Applying the fuzzy set theory we may write (Pal and Majumder, 1987):

$$f(x,y) = \begin{bmatrix} f(0,0) & f(0,1) & \cdots & f(0, N-1) \\ f(1,0) & f(1,1) & \cdots & f(1, N-1) \\ \vdots & \vdots & \vdots & \vdots \\ f(M-1,0) & f(M-1,1) & \cdots & f(M-1, N-1) \end{bmatrix} \tag{1}$$

$$F = \begin{bmatrix} m_{11}/x_{11} & m_{12}/x_{12} & \cdots & m_{1n}/x_{1n} & \cdots & m_{1N}/x_{1N} \\ m_{21}/x_{21} & m_{22}/x_{22} & \cdots & m_{2n}/x_{2n} & \cdots & m_{2N}/x_{1N} \\ \vdots & \vdots & \vdots & \vdots & \vdots & \vdots \\ m_{m1}/x_{m1} & m_{m2}/x_{m2} & \cdots & m_{mn}/x_{mn} & \cdots & m_{mM}/x_{mN} \\ \vdots & \vdots & \vdots & \vdots & \vdots & \vdots \\ m_{M1}/x_{M1} & m_{M2}/x_{M2} & \vdots & m_{Mn}/x_{Mn} & \vdots & m_{MN}/x_{MN} \end{bmatrix} \tag{2}$$

or, in union form (Eq. 2):

$$x = \bigcup_m \bigcup_n \mu_{mn}/x_{mn}, m = 1, 2, ...M \quad \text{and} \quad n = 1, 2, ...N$$

where μ_{mn}/x_{mn}, $(0 \le \mu_{mn} \le 1)$ represents the grade of possessing some property μ_{mn} by the (m,n)th pixel x_{mn}. This fuzzy property μ_{mn} may be defined in a number of ways with respect to any brightness level depending on the problem at hand.

3.3   Digitizing

The signal acquired through the camera must be sampled and quantized in order to be suitable for computer processing. An increase in the valuess of M and N in (2) means a better image resolution. Quantization rounds up the values of each pixel and places, them in the range of 0 a $2^n - 1$, where the larger the value of n the greater the number of grey levels present in the digitized image.

The digitized signal is a converted analogic to digital signal, in which the number of samples per unit of time gives the sampling rate, and the number of bits of the analogical/digital converter determines the number of levels of grey.

It should be decided which values of M, N and l are suitable, talking into account image quality and the number of bits necessary for storage. We may calcule the quantity of bytes utilized by means of the expression: $N \times M \times l/8$ (Marques and Vieira, 1999). It should be remember that image quality is a subjective concept and it is strougally dependent on the aplication itself.

4   Image Treatment

The captured image is made up of two parts: the object to be identified and the background. In order to separate these two parts it is necessary to have a good contrast, that is, a significant difference in light intensity between points of the object and point of the background must exist.

Image treatment is based on the principal of thresholding (Gonzalez. e Woods, 1992), that is, to assign the value 0 for pixels above a given value X (determined by the Histogram analises), and a value 1 for pixels below or at most equal to X. The image enhancement algorithm utilized in this work was suggested by Pal e Majumder (1987), and show in Fig. 2. This technique modifics the pixels using the properties of the fuzzy set. This procedure involves a preliminary image enhancement in block "E", followed by a smoothing in block "S" and then a further enhancement. The fuzzy operator INT (contrast intensification) is used in both enhancements. The pupose of the image smoothing is to blur the image before the second enhancement.

Fig. 2. Block diagram of the enhancement model

The algorithm used in the image enhancement (Fig. 2) is explained in the pages that follow.

4.1   Histogram

The histogram gives na idea of the image quality in terms of contrast and relative proportions of white and black in the image.

The histogram of an image is usually a grafic representation of bars, which indicates the l levels of greys and the quantity of pixels in the image. The vertical axis represents the amountly of pixels for a given l levels of grey. The horizontal axis represents the l levels of grey.

Here, the histogram is used to give na inicial idea of the image quality for future comparison with the image obtained after blocks "E" and "S". These will be explained in sub-section 4.2 and 4.3.

4.2 Image Enhancement through the Concepts of Fuzzy Sets

Image enhancement is acomplished through the INT operator. The operation of contrast intensification (INT) upon a fuzzy set $\mathbf{A}$ criates another fuzzy set $\mathbf{A'} = \mathbf{INT}(\mathbf{A})$, where the membership function, is given by:

$$\mu_{A'}(x) = \mu_{INT(A)}(x) = \begin{cases} 2[\mu_A(x)]^2, 0 \leq \mu_A(x) \leq 0.5 & (3a) \\ [1 - 2(1 - \mu_A(x))]^2, 0.5 \leq \mu_A(x) \leq 1 & (3b) \end{cases}$$

This operation reduces the fuzziness of a set $\mathbf{A}$ by increasing the values of $\mu_{\mathbf{A}}(x)$ which are above 0.5 and decreasing those which are below it. Let us now define operation (3) by transformation T_1 of the membership function $\mu(x)$ (Pal e King, 1981). In general, each μ_{mn} in F (Eq. 2) may be modified to μ'_{mn} to enhance the image F in the property domain by a transformation function T_r, where:

$$\mu'_{mn} = T_r(\mu_{mn}) = \begin{cases} T'_r(\mu_{mn}), 0 \leq \mu_{mn} \leq 0.5 & (4a) \\ T'_r(\mu_{mn}), 0.5 \leq \mu_{mn} \leq 1 & (4b) \end{cases}$$

and, $r = 1, 2, ...$

The transformation function T_r is defined as successive applications of T_1 by the recursive relationship:

$$T_s(\mu_{mn}) = T_1\{T_{s-1}(\mu_{mn})\}, s = 1, 2, \cdots \qquad (5)$$

and $T_1(\mu_{mn})$ represents the operator INT defined in Eq.3. This is shown that, as r increases, the curve tends to be steeper because of the successive application of INT. In the limiting case, as $r \to \infty$, T_r produces a two-level (binary) image.

4.3 Smoothing Algorithm - Averaging

This method is based on averaging the intensities within neighbors and is usually used to remove "pepper and salt" noise. The smoothed (Block "S" - Fig. 2) (m,n)th pixel intensity is:

$$x'_{mn} = 1/N_1 \sum_{Q_1} x_{ij}, (i, j) \neq (m, n), (i, j) \in Q_1 \qquad (6)$$

The higher the values of Q_1 the greater is the degree of blurring.

The smoothing algorithm described above blur the image by attenuating the high spatial frequency components associated with edges and other abrupt changes in grey levels.

5   Hardware System

In this section, the hardware system, developed to implement the control strategy presented in this paper, is described. There are six DC motors to drive the joints and six potentiometers to measure the joint angles.

The driving circuit includes high frequency transistors, switched by pulse width (PWM). The end effector is actuated using a pneumatic transmitter. The three base joints are free to turn up to 270°, and the upper three joints can only rotate 180°.

To test the strategies presented in this paper, a specific circuit had to be developed, since the manufacturer's driving circuit does not allow for most of the requirements of the approach developed in this research project. Therefore, only the original mechanism, the motors, and the potentiometers are used in this work. All the electronic circuits and software programs have been developed and implemented. Three Microchip^{TM} PIC micro controllers are used (Microchip, 1997).

They have been chosen because they are cheap and relatively ease to implement. They use RISC technology of 14 bits. Its possible to deal with a large number of interruption schemes and to generate PWM signals independently of the CPU synchronization command.

Figure 1 presents a general scheme of the robot system used in this work. Basically, the system consists of two hierarchical levels. In the highest level there is a Pentium computer working as a host. In the second level, there are three micro controllers of the type PIC16C73A.

6   The Fuzzy Controller

The use of a fuzzy controller for trajectory tracking of such a robot arm is very appropriate, considering the nonlinear behavior of the system. There are some propositions for the application of fuzzy logic to robot control (Luh, 1983). The main reason for the application of fuzzy control for this kind of application is, basically, the nonlinear characteristic of robots. There are many other approaches for controlling robot arms (Bonitz, 1996, Moudgal, Passino, & Yurkovich, 1995, Rocco, 1996). Some of them demand a lot of computer effort. In some cases, where very fast response is required, the computational burden could be prohibiting. The use of fuzzy logic in such a case is very attractive since it requires a minimum of computer time. On the other hand, the powerful of fuzzy logic is such that, the coupling effects of the joints can be completely compensated using the appropriate fuzzy rules. The approach presented in this paper differs in some extend to the methods found in the literature (Young, & Shiah, 1997, Kawasaki, Bito, & Kanzaki, 1996, Cox, E., 1994) since there is a combination of individual fuzzy controllers for each joint and a master fuzzy controller used to generate the set points. Therefore, the couplings found in the general robot dynamic equation are completely compensated. There are also two fuzzy controllers for each joint: one for speed control and another for position control.

They work in such a way that, speed controller actuates in the first part of the trajectory. At the end of it, the position controller takes place leading to a very smooth positioning. The decision for the switching of all those controllers is also accomplished by the master fuzzy controller. A simplified scheme is presented in Fig.3. In Fig. 3, **Q** and •**Q** are the desired position and speed of the joint, respectively. The position controller uses the error and error variation as input. On the other hand, the inputs for the speed controller are the speed error and error variation. Each fuzzy controller has the following characteristics:

- Gaussian membership functions
- Defuzzification using the center of area
- The use of simple product as the minimum operation
- Concentration operator for the output function.

The universe of discourse of position for each one of the base joints is $[-270°$ to $270°]$ and for the upper joints is $[-180°$ to $180°]$. The universe of discourse of the output, for each joint is $[-192°$ to $192°]$, corresponding to the PWM input of the motor drives. The rule base was defined using cardinality equal to 7, for every controller, although smaller rule basis could also be tested with excellent results. The rules were implemented based on the experience acquired from the observation of the robot behavior.

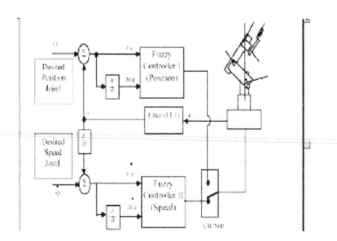

Fig. 3. Control system for join 1

7   Practical Results

The result of the enhancement method (Fig. 2) is shown in Figures 5 a 10.

The results shown in figures 5 to 10, were obtained directly from the camera acquisition system. The Fig. 5 and 8 are the original image and histogram of

Fig. 5. Original Image

Fig. 6. Image after first enhance

Fig. 7. Image after smoothing

Fig. 8. Image after second enhance

the original image. In the next step the original image is Enhance, smoothed Fig. 5(b) and them a second enhancement Fig. 5(c). Finally the binary image Fig. 5(c) and 5(e) is obtained and applied in the control system of the robot manipulator.

The results shown in figures 5(a) to 5(f), were obtained directly from the camera acquisition system. The figures 6 and 9 are the original image and histogram of the original image. In the next step the original image is Enhance figure 6, smoothed figure 7 and them a second enhancement figure 8. Finally the binary image figures 8 and 10 is obtained and applied in the control system of the robot manipulator.

The results shown in figures 11 to 16, were obtained directly from the robot acquisition system. In the test, each joint started from the 50°. The desired final angle, for each joint is showed as a continuous line, in figure. Some very small overshoots could be seen.

But they could be very well compensated with a more careful adjustment of the rule base. The switching from the speed to the position controller is accomplished by the master fuzzy controller. Figure 17 explains how the switching idone.

8   Conclusions

The results obtained shown that the method applied is effective for controlled environment. The model Fig.3, shown good possibilities for improvement since the method used in block "S", may be altered, by means of, for instance a maxmim filter, and in this way improve the results sent to the second block "E".

The study also shows that although the averaging filter had been used, with only two interactions in each block "E", we have obtained a binary image.

The results obtained for image enhancement using fuzzy logic, show us effective, very speed and of the simple implementation.

References

1. Bonitz, R. G., and Hsia, T. C., February (1996). Internal Force-Based Impedance Control for Cooperating Manipulators, IEEE Transactions on Robotics and Automation, vol. 12, no. 1.
2. Cox, E., (1994).The Fuzzy Systems Handbook: A practitioner's Guide to Building, Using, and Maintaining Fuzzy Systems, Academic Press.
3. Gonzalez, R. C. and Woods, R. E. (1992). Digital Image Processing, Addison-Wesley Publishing Company.
4. Kawasaki, H., Bito, T. and Kanzaki, K., June (1996). An Efficient Algorithm for the Model-Based Adaptive Control of Robotic Manipulators, IEEE Transactions on Robotics and Automation, vol. 12, no. 3.
5. Luh, J. Y. S. (1983). Conventional controller design for industrial robots - A tutorial. IEEE Trans. Syst. Man. Cyber.,vol. SMC-13, n° 3,Mai.
6. Microchip Technology Incorporated (1997). PIC16C7X Data Sheet, DS30390E, USA.
7. Moudgal, V. G., Kwong, W. A., Passino, K. M. and Yurkovich, S., May (1995). Fuzzy Learning Control for a Flexible-Link Robot, IEEE Transactions Fuzzy Systems, vol. 3, no. 2.
8. Pal, Sankar K. and Majumder, Dwijesh K. Dutta. (1987). Fuzzy Mathematical Approach to Pattern Recognition, John Wiley & Sons.
9. Rocco, P., August (1996). Stability of PID Control for Industrial Robot Arms, IEEE Transactions on Robotics and Automation, vol. 12, no. 4.
10. Young, K. Y., & Shiah, S. J., November (1997). An Approach to Enlarge Learning Space Coverage for Robot Learning Control, IEEE Transactions on Fuzzy Systems, vol. 5, no. 4.

On Fuzzy Lattice

Kankana Chakrabarty

School of Mathematical and Computer Sciences
The University of New England
Armidale 2351, New South Wales, Australia
kankanac@turing.une.edu.au

Abstract. The notion of lattice has a wide variety of applications in the areas of physical sciences, communication systems, and information analysis. Fuzzy set theory, with its proximity to a a multitude of areas closely related to AI, provides an alternative to the traditional concept of set membership. Nanda[4] proposed the concept of fuzzy lattice, using the notion of fuzzy partial ordering. But after a critical study, it has been observed that his definition contains some redundancy. As a consequence of this observation, we present a modified definition of fuzzy lattice in this paper. ...

1 Introduction

Lattice structure has been found to be extremely important in the areas of quantum logic, ergodic theory, Reynold's operators, communication systems and information analysis systems. Some system models often include excessive complexity of the situation which in turn may lead to consequences where it is difficult to formulate the model or the model is too complicated to be used in practice. This practical inconvenience is caused by our inability to differentiate events in real situations exactly and thus to define instrumental notions in precise form.

The concept of human cognition and interaction with the outer world involves objects that are not sets in the classical sense, but fuzzy sets, which imply classes with unsharp boundaries in which the transition from membership to non-membership is gradual rather than abrupt.

Nanda[4] proposed the notion of fuzzy lattice using the concept of fuzzy partial ordering. After performing a critical study, we observe some amount of redundency present in his definition and his definition is found to be incomplete. As a consequence of this redundancy, we modify the definition of fuzzy lattice in the present paper.

W. Ziarko and Y. Yao (Eds.): RSCTC 2000, LNAI 2005, pp. 238–242, 2001.
© Springer-Verlag Berlin Heidelberg 2001

2   Preliminaries

Let Ω be any set and let $\tilde{R}$ be a fuzzy relation defined over Ω. Then, $\tilde{R}$ is said to be

- **Max-min transitive** if $\tilde{R} \circ \tilde{R} \subseteq \tilde{R}$ or more explicitly, if $\forall (x, y, z) \in \Omega^3$,

$$\mu_{\tilde{R}(x,z)} \geq \min\{\mu_{\tilde{R}(x,y)}, \mu_{\tilde{R}(y,z)}\}$$

- **Reflexive** if $\forall x \in \Omega$,

$$\mu_{\tilde{R}(x,x)} = 1$$

- **Perfect antisymmetry** if

$$\forall (x, y) \in \Omega^2, x \neq y, \mu_{\tilde{R}(x,y))} > 0 \Rightarrow \mu_{\tilde{R}(y,x)} = 0$$

where $\mu_{\tilde{R}(x,y)}$ represents the membership value of the pair (x, y) in $\tilde{R}$.

A fuzzy relation $\tilde{P}$ defined over a set Ω is said to be **fuzzy partial ordering** if and only if it is reflexive, max-min transitive and perfectly antisymmetric. A set Ω along with a fuzzy partial ordering $\tilde{P}$ defined on it, is called a **fuzzy partially ordered set**.

Let Ω be a fuzzy partially ordered set with a fuzzy partial order $\tilde{P}$ defined over it. With each $x \in \Omega$, we associate two fuzzy sets

- The **dominating class**, $\tilde{P} \geq (x)$, defined by

$$\tilde{P} \geq (x)(y) = \tilde{P}(y, x).$$

- The **dominated class**, $\tilde{P} \leq (x)$, defined by

$$\tilde{P} \leq (x)(y) = \tilde{P}(x, y).$$

Let $\mathcal{M}$ be a non-fuzzy subset of Ω. Then the **fuzzy upper bound** of $\mathcal{M}$, denoted by $U_{\phi(\mathcal{M})}$ is a fuzzy set defined by

$$U_{\phi(\mathcal{M})} = \bigcap_{x \in \mathcal{M}} \tilde{P} \geq (x).$$

The **fuzzy lower bound** of $\mathcal{M}$, denoted by $L_{\phi(\mathcal{M})}$ is a fuzzy set defined by

$$L_{\phi(\mathcal{M})} = \bigcup_{x \in \mathcal{M}} \tilde{P} \leq (x).$$

3   Comment

In his paper [4], Nanda defined the notion of fuzzy lattice in the following way.

3.1   Definition

A fuzzy partially ordered set X (a non-fuzzy set X with a fuzzy partial order defined on it) is called a fuzzy lattice if every two-element non-fuzzy set (i.e. every pair of elements) in X has fuzzy lower bound and fuzzy upper bound.

3.2   Observation

After a critical study, we analysed the above definition as proposed by Nanda[4], and observed that in case of any fuzzy partially ordered set X, every two element non-fuzzy set in it has fuzzy lower bound and fuzzy upper bound which are nothing but two fuzzy subsets of X. Hence we conlude that this definition is incomplete in the sense that according to this definition, every fuzzy partially ordered set becomes a fuzzy lattice. Hence we modify the concept and redefine the notion of fuzzy lattice.

4   Modified Definition

The modified definition of fuzzy lattice is as follows:

4.1   Definition

Let X be a fuzzy partially ordered set and let $\tilde{A}$ be a fuzzy subset of X. Then $\tilde{A}$ is said to be a fuzzy lattice in X if every pair of elements in X has a fuzzy lower bound L_ϕ and a fuzzy upper bound U_ϕ (where both L_ϕ and U_ϕ are fuzzy subsets of X) satisfying the following two conditions:

- $\mu_{\max\{U_\phi\}}(x) \geq \mu_{\tilde{A}}(x) \quad \forall x \in X.$
- $\mu_{\min\{L_\phi\}}(x) \geq \mu_{\tilde{A}}(x) \quad \forall x \in X.$

4.2   An Example

Let $X = \{\xi_1, \xi_2, \xi_3\}$ be a set and $\tilde{P}$ be a fuzzy partial order defined over X as below:

$\tilde{P}$	ξ_1	ξ_2	ξ_3
ξ_1	1	1	.8
ξ_2	0	1	0
ξ_3	0	.7	1

Also let $\tilde{A} = \{\xi_1/0, \xi_2/0, \xi_3/.6\}$ be any fuzzy subset of X.

In this case, we have

$$U_{\phi(\xi_1,\xi_2)} = \{\xi_1/1, \xi_2/0, \xi_3/0\}$$
$$L_{\phi(\xi_1,\xi_2)} = \{\xi_1/1, \xi_2/1, \xi_3/.8\}$$
$$U_{\phi(\xi_1,\xi_3)} = \{\xi_1/.8, \xi_2/0, \xi_3/0\}$$
$$L_{\phi(\xi_1,\xi_3)} = \{\xi_1/1, \xi_2/1, \xi_3/1\}$$
$$U_{\phi(\xi_2,\xi_3)} = \{\xi_1/0, \xi_2/0, \xi_3/.7\}$$
$$L_{\phi(\xi_2,\xi_3)} = \{\xi_1/0, \xi_2/1, \xi_3/1\}$$

Therefore,

$$\max\{U_\phi\} = \{\xi_1/1, \xi_2/0, \xi_3/.7\}$$

and

$$\min\{L_\phi\} = \{\xi_1/0, \xi_2/1, \xi_3/.8\}$$

Hence for all $\xi \in X$, we have

$$\mu_{\max\{U_\phi\}}(\xi) \geq \mu_{\tilde{A}}(\xi),$$

$$\mu_{\min\{L_\phi\}}(\xi) \geq \mu_{\tilde{A}}(\xi).$$

Thus $\tilde{A}$ is a fuzzy lattice in X.
But in this case, if we consider $\tilde{B} = \{\xi_1/.5, \xi_2/0, \xi_3/.9\}$, then $\tilde{B}$ is not a fuzzy lattice in X.

4.3   A Trivial Property

Let $\tilde{A}$ be a fuzzy lattice in the fuzzy partially ordered set X and let $\tilde{B}$ be any fuzzy subset of X such that $\mu_{\tilde{B}}(x) \leq \mu_{\tilde{A}}(x) \ \ \forall x \in X$. Then $\tilde{B}$ is also a fuzzy lattice in X.

Proof: Let any two element non-fuzzy set in X have the fuzzy lower bound L_ϕ and the fuzzy upper bound U_ϕ. Since $\tilde{A}$ is a fuzzy lattice in X, hence for all $x \in X$, we have

$$\mu_{\max\{U_\phi\}}(x) \geq \mu_{\tilde{A}}(x),$$

$$\mu_{\min\{L_\phi\}}(x) \geq \mu_{\tilde{A}}(x).$$

Also we have here,

$$\mu_{\tilde{B}}(x) \leq \mu_{\tilde{A}}(x) \ \ \forall x \in X.$$

Hence, for all $x \in X$,

$$\mu_{\max\{U_\phi\}}(x) \geq \mu_{\tilde{B}}(x),$$

$$\mu_{\min\{L_\phi\}}(x) \geq \mu_{\tilde{B}}(x).$$

This implies $\tilde{B}$ is also a fuzzy lattice and hence the property.

5 Conclusion

The most important semantics associated with the use of fuzzy sets is the expression of proximity and representation of incomplete states of information. In view of the possible areas of application of lattice structure in case of soft computing, we studied the notion of fuzzy lattice as proposed by Nanda. Some redundancies in this definion has been observed and consequently a modified definition of fuzzy lattice has been proposed.

6 Acknowledgement

The author is extremely grateful to the referees for their valuable comments which helped in the modification of this paper.

References

1. Birkoff, G.: Lattice theory. 3rd edn. Amer. Math. Soc. Colloquium Pub.**25** (1984)
2. Chakrabarty, K.: On Bags and Fuzzy Bags. Advances in Soft Computing. Soft Computing Techniques and Applications. Physica-Verlag, Heidelberg New York (2000) 201-212
3. Dubois, D., Prade, H.: Fuzzy Sets and Systems - Theory and Applications. Academic Press. (1980)
4. Nanda, S.: Fuzzy Lattices, Bulletin Calcutta Math.Soc. **81** (1989) 1-2
5. Zadeh, L.A.: Fuzzy Sets, Inform.Con. **8** (1965) 338-353

Fuzzy-Logic-Based Adaptive Control for a Class of Nonlinear Discrete-Time System

Hugang Han[1], Chun-Yi Su[2], and Shuta Murakami[3]

[1] Hiroshima Prefectural University
562 Nanatsuka-cho, Shobara-shi, Hiroshima 727-0023, Japan
kan@bus.hiroshima-pu.ac.jp
[2] Concordia University
1455 de Maisonneuve Blvd. W., Montreal, Quebec, Canada H3G 1M8
cysu@me.concordia.ca
[3] Kyushu Institute of Technology
1-1 Sensui-cho, Tobata-ku, Kitakyushu, Fukuoka 804-8550, Japan
murakami@comp.kyutech.ac.jp

Abstract. Recently various continuous adaptive fuzzy control schemes have been proposed to deal with nonlinear systems with poorly understood dynamics by using parameterized fuzzy approximators. However, practical applications call for discrete-time adaptive fuzzy controller design because almost all these controllers are implemented on digital computers. To meet such a demand, in this paper a discrete-time adaptive fuzzy control scheme is developed. The strategy ensures the global stability of the resulting closed-loop system in the sense that all signals involved are uniformly bounded and tracking error will be asymptotically in decay.

1 Introduction

The application of fuzzy set theory to control problems has been the focus of numerous studies [1]. The motivation is often that the fuzzy set theory provides an alternative to the traditional modeling and design of control systems when system knowledge and dynamic models in the traditional sense are uncertain and time varying. Without any doubt, the stability is the most important requirement for any control system. It is highly desirable to design a fuzzy controller which guarantees the stability. Recently, some researches have been focused on use of the Lyapunov synthesis approach to construct stable adaptive fuzzy control system[2–6]. The most fundamental idea may refer to the creative works in [2, 3]. Namely, to deal with unknown control systems, the fuzzy model would be considered as an approximation model to approximate unknown linear or nonlinear functions in the plant, where the fuzzy model is expressed as a serious of fuzzy basis function (FBF) expansion. Ultimately, instead of using the unknown functions, the fuzzy model is utilized directly to construct the control inputs based on the Lyapunov synthesis approach.

However, thus designed fuzzy control schemes are usually implemented on digital

W. Ziarko and Y. Yao (Eds.): RSCTC 2000, LNAI 2005, pp. 243–250, 2001.
© Springer-Verlag Berlin Heidelberg 2001

computers because where computers are the manipulators in practical applications, so there has existed a gap between the designed and realizable control algorithms. With the increasing applications of advanced computer technologies, it is much more meaningful to design control schemes in discrete-time. So far, however, all of the adaptive fuzzy control systems have been designed only for a continue-time system except using the so-called T-S type fuzzy model[7]. In this paper, we will introduce a stable discrete-time adaptive fuzzy control algorithm for a class of unknown sampled-data nonlinear systems. The control scheme considered here is an integration of fuzzy control component, in which the FBF expansion can be considered as an universal approximator, with sliding control component of the variable structure control with a sector[8, 9]. The developed controller guarantees the global stability of the resulting closed-loop system in the sense that all signals involved are uniformly bounded and tracking error will be asymptotically in decay.

2   System Description

We are interested in the single-input/single-output nonlinear discrete-time system

$$x(k+1) + f\left(x(k-n+1), x(k-n+2), \ldots, x(k)\right) = bu(k) \qquad (1)$$

where k is current time, x is the output, and u is the input in the system. $f(\cdot)$ is nonlinear function with n being system order. It is assumed that the order n is known but the nonlinear function $f(\cdot)$ is unknown. It should be noted that more general classes of nonlinear systems could be transformed into this structure[10]. The control objective is to force the state vector $X(k) = [x_1(k), x_2(k), \ldots, x_n(k)]^T$ $= [x(k-n+1), x(k-n+2), \ldots, x(k)]^T$ to follow a specified desired trajectory, $X_d(k) = [x_d(k-n+1), x_d\ (k-n+2), \ldots, x_d(k)]^T$. Defining the traking error vector, $\tilde{X}(k) = X(k) - X_d(k)$, the problem is thus to design a control $u(k)$ which ensures that $\tilde{X}(k) \to 0$, as $k \to \infty$. For simplicity in this initial discussion, we take $b = 1$ in the subsequent development.

3   Fuzzy Approximator

We consider a fuzzy system for which there are four principal elements in such a fuzzy system: fuzzifier, fuzzy rule base, fuzzy inference engine, and defuzzifier. Let input space $X \in R^n$ be a compact product space. Assume that there are N rules in the rule base and each of which has the following form:

$$R_j : IF\ x_1(k)\ is\ A_j^1\ and\ x_2(k)\ is\ A_j^2\ and$$

$$\ldots\ and\ x_n(k)\ is\ A_j^n,\ THEN\ z(k)\ is\ B_j$$

where $j = 1, 2, \ldots, N, x_i(k)(i = 1, 2, \ldots, n)$ are the input variables to the fuzzy system at current time k, $z(k)$ is the output variable of the fuzzy system at current time k, and A_j^i and B_j are linguistic terms characterized by fuzzy membership functions $\mu_{A_j^i}(x)$ and $\mu_{B_j}(z)$, respectively.

As in [11], we consider a subset of the fuzzy systems with *singleton fuzzifier*, *product inference*, and *Gaussian membership function*. Hence, such a fuzzy system can be written as

$$h(X) = \frac{\sum_{j=1}^{N} \omega_j(k) \left(\prod_{i=1}^{n} \mu_{A_j^i}(x_i(k)) \right)}{\sum_{j=1}^{N} \prod_{i=1}^{n} \mu_{A_j^i}(x_i(k))} \tag{2}$$

where $h: U \subset R^n \to R$, $\omega_j(k)$ is the point in R at which $\mu_{B_j}(\omega_j(k)) = 1$, named as the connection weight; $\mu_{A_j^i}(x_i(k))$ is the *Gaussian membership function*, defined by

$$\mu_{A_j^i}(x_i(k)) = \exp\left[-\left(\frac{x(k) - \xi_j^i(k)}{\sigma_j^i(k)} \right)^2 \right] \tag{3}$$

where $\sigma_j^i(k)$ and $\xi_j^i(k)$ are real-valued parameters, in which $\xi_j^i(k)$ indicates the position and $\sigma_j^i(k)$ indicates the variance of the membership function. Here, if we take $\dfrac{\prod_{i=1}^{n} \mu_{A_j^i}(x_i(k))}{\sum_{j=1}^{N} \prod_{i=1}^{n} \mu_{A_j^i}(x_i(k))}$ in (2) as basis functions and $\omega_j(k)$ as coefficients, $h(X)$, then, can be viewed as a linear combination of the basis functions. Therefore, we give a definition regarding the basis function as follows.

Definition 1 *Define fuzzy basis functions (FBF) as*

$$g_j(X) = \prod_{i=1}^{n} \mu_{A_j^i}(x_i(k)), \quad j = 1, 2, \ldots, N \tag{4}$$

where $\mu_{A_j^i}(x_i(k))$ is the Gaussian membership functions defined in (3).

Then, the fuzzy system (2) is equivalent to a FBF expansion

$$h(X) = \sum_{j=1}^{N} \omega_j(k) \cdot g_j(X) \triangleq W^T(k) \cdot G(X(k)) \tag{5}$$

where, $W(k) = [\omega_1(k), \omega_2(k), \cdots, \omega_N(k)]^T$, $G(X(k)) = [g_1(X(k)), g_2(X(k)), \cdots, g_N(X(k))]^T$, and notation $\triangleq$ means a definition. For convenience, throughout this paper, the notation $(\cdot)$ is sometimes omitted when no confusion is likely to arise. For example we will express $G(X(k))$ as $G(k)$ with k being current time. We now show an important property of FBF expansion[3].

Theorem 1 *For any given real continuous function f on the compact set $U \in R^n$ and arbitrary ε, there exists optimal FBF expansion $h^*(X(k)) = W^*(k) \cdot G(X(k))$ such that*

$$\sup_{X \in U} |f(X(k)) - h^*(X(k))| < \varepsilon \tag{6}$$

This theorem states that the FBF expansion (5) is universal approximator on a compact set. Herein, we use terms *fuzzy universal approximator* or *fuzzy approximator* to refer to the FBF expansion. Since the fuzzy universal approximator is characterized by parameter vectors W, the optimal h^* does contains an optimal vector W^*.

4   Adaptive Fuzzy Control System

In this paper, we adopt the variable structure theory to construct our adaptive fuzzy control system. An error metric is firstly defined as

$$s(k) = (q^{-1} + \lambda)^{n-1}\tilde{x}(k) \quad \text{with } \lambda > 0 \tag{7}$$

where q^{-1} is the delay operator, and λ defines the bandwidth of the error dynamics of the system. The error metric above can be rewritten as $s(k) = \Lambda^T \tilde{X}(k)$ with $\Lambda^T = [\lambda^{n-1}, (n-1)\lambda^{n-2}, \cdots, 1]$. The equation $s(k) = 0$ defines a time-varying hyperplane in R^n on which the tracking error vector $\tilde{X}(k)$ decays exponentially to zero, so that perfect tracking can be asymptotically obtained by maintaining this condition [13]. In this case the control objective becomes the design of controller to force $s(k) = 0$.

An increment $\Delta s(k + 1)$ can be written as

$$\begin{aligned}\Delta s(k+1) &= s(k+1) - s(k) \\ &= \Lambda_1 \tilde{X}(k) + \lambda^{n-1}\tilde{x}(k+1)\end{aligned} \tag{8}$$

where $\Lambda_1^T = [(n-1)\lambda^{n-2} - \lambda^{n-1}, \cdots, 1 - (n-1)\lambda, -1]$. Substituting $\tilde{x}(k+1) = x(k+1) - x_d(x+1)$ into (8) and combining (1) yield,

$$\frac{1}{\lambda^{n-1}}\Delta s(k+1) = u(k) - h(k) \tag{9}$$

where $h(k)$ denotes

$$h(k) = f(X) + x_d(k+1) - \frac{1}{\lambda^{n-1}}\Lambda_1^T \tilde{X}(k) \tag{10}$$

It naturally suggests that when $h(X)$ is known, a control input of form

$$u(k) = -k_d s(k) + h(k), \qquad k_d > 0 \tag{11}$$

leads to a closed-loop system $\Delta s(k) = -k_d s(k)$, and hence, $\tilde{X}(k) \to 0$ as $k \to \infty$. The problem is how $u(k)$ can be determined when $h(X)$, which concludes an unknown functions $f(X)$ and b, are unknown. Therefore we have to approximate them to achieve control objective. Here the fuzzy approximator described in the previous section is used. Let us denote $h^*(X(k)) = W_h^{*T}G_h(k)$ to be the optimal fuzzy approximator of the unknown function $h(X(k))$. However, we have no idea to know the optimal parameter vector W_h^* in the optimal fuzzy approximator. Generally, the estimate, denoted $\hat{h}(X(k)) = \hat{W}_h^T G_h(k)$, is adopted instead of the optimal fuzzy approximator $h^*(X(k))$. Regarding the optimal fuzzy approximator, we make following assumptions.

Assumption 1 *The optimal fuzzy approximator error, $d_h(k) = h^*(X(k)) - h(X(k))$, satisfy the following inequality, where ε_h^* is known small positive value.*

$$\sup_{X \in U} |d_h(k)| < \varepsilon_h^* \tag{12}$$

Assumption 2 *In defined subspace $C_{w_h} \ni W_h^*, C_{\varepsilon_h} \ni \varepsilon_h^*$, there are some positive constants $c_{w_h}, c_{\varepsilon_h}$, that satisfies the following inequalities.*

$$\|\hat{W}_h(k) - W_h^*\|_2 \le c_{w_h}, \qquad \forall \hat{W}_h(k) \in C_{w_h} \tag{13}$$

$$|\hat{\varepsilon}_h(k)| + |\varepsilon_h^*| \le c_{\varepsilon_h}, \qquad \forall \hat{\varepsilon}_h(k) \in C_{\varepsilon_h} \tag{14}$$

where $\hat{\varepsilon}_h(k)$ is the estimate of ε_f^*, as well as $\hat{W}_h(k)$.

Remarks:

1. Base the Theorem 1, the assumption 1 is reasonable to provide.
2. It seems the assumption 2 is difficult to satisfy. However, in the each defined subspace where the optimal parameter (or vector) is contained, such an inequality is easy to check. Therefore, the problem is how to force the each estimate to enter the each defined subspace. At the upcoming algorithm, the projection algorithm is adopted to ensure each estimate is within the each defined subspace.

Inspired by the above control structure in (11), using the fuzzy approximator $\hat{h}(X)$, our adaptive control law is now described below:

$$u(k) = u_{fu}(k) + u_{vu}(k) \tag{15}$$

where $u_{fu}(k)$, expressed by,

$$\begin{aligned} u_{fu}(k) &= \hat{h}(k) - \hat{\varepsilon}_h(k)sgn(s(k)) \\ &= \hat{W}_h^T(k)G_h(k) - \hat{\varepsilon}_h(k)sgn(s(k)) \end{aligned} \tag{16}$$

is fuzzy component of control law which will attempt to recover or cancel the unknown function h. And the adaptive component is synthesized by

$$\hat{W}_h(k) = \wp\left\{\hat{W}_h(k-1) - \Gamma_h G_h(k)s(k)\right\} \tag{17}$$

$$\hat{\varepsilon}_h(k) = \wp\left\{\hat{\varepsilon}_h(k-1) + \gamma_h|s(k)|\right\} \tag{18}$$

where $\hat{W}_h$, and $\hat{\varepsilon}_h$ are the estimates of W_h^*, and ε_h^*, respectively; $\wp$ represents the projection operator necessary to ensure that $\hat{W}_h(k) \in C_{w_h}$, and $\hat{\varepsilon}_h(k) \in C_{\varepsilon_h}$ for $\forall k$ [14]; $\Gamma_h \in R^{N \times N}$, and $\gamma_h > 0$ determine the rates of adaptation in which Γ_h is a symmetric positive define matrix and selected to be satisfy the following inequality:

$$G_h^T(k)\Gamma_h G_h(k) \le \frac{2}{\lambda^{n-1}} \tag{19}$$

It should note that the term like $\hat{\varepsilon}_h(k)sgn(s(k))$ in (16) actually reflects the component for compensation of the approximating error $(\hat{h}(k) - h(k))$. If the fuzzy approximator provide good description of the unknown function $h(k)$, then the term $\hat{\varepsilon}_h(k)$ should be small as well as the optimal fuzzy approximator error $d_h(k)$. Conversely, if the fuzzy approximator is poor, $\hat{\varepsilon}_h(k)$ will rise automatically to the necessary level, ensuring the stability of the overall system.

$u_{vu}(k)$ is the variable structure control component that is expressed by

$$u_{vu}(k) = \beta_{h1} c_{w_h} \|G_h(k)\| + \beta_{h2} c_{\varepsilon_h} \tag{20}$$

The variable structure control component only works outside of the sector $\Omega(k)$[8] defined by

$$\Omega(k) = \Omega_{h1}(k) \cup \Omega_{h2}(k) \tag{21}$$

with

$$\Omega_{h1}(k) = \{s(k)| \ \|\|G_h(k)\|s(k)\| c_{w_h} \leq \rho_{h1}\} \tag{22}$$

$$\Omega_{h2}(k) = \{s(k)| \ |s(k)|c_{\varepsilon_h} \leq \rho_{h2}\} \tag{23}$$

$$\rho_{h1} = \frac{\lambda^{n-1}(\ell+1)^2 c_{w_h}}{2\ell} \|G_h(k)\| A(k) \tag{24}$$

$$\rho_{h2} = \frac{\lambda^{n-1}(\ell+1)^2 c_{\varepsilon_h}}{2\ell} A(k) \tag{25}$$

$$A(k) = c_{w_h}\|G_h(k)\| + c_{\varepsilon_h} \tag{26}$$

where ℓ is a positive constant chosen to guarantee the system robustness. And the switching type coefficients in (20) are determined as follows.

$$\beta_{h1} = \begin{cases} \ell, & \text{for } c_{w_h}\|G_h(k)\|s(k) < -\rho_{h1} \\ 0, & \text{for } c_{w_h} \|\|G_h(k)\|s(k)\| \leq \rho_{h1} \\ -\ell, & \text{for } c_{w_h}\|G_h(k)\|s(k) > \rho_{h1} \end{cases} \tag{27}$$

$$\beta_{h2} = \begin{cases} \ell, & \text{for } c_{\varepsilon_h}s(k) < -\rho_{h2} \\ 0, & \text{for } c_{\varepsilon_h}|s(k)| \leq \rho_{h2} \\ -\ell, & \text{for } c_{\varepsilon_h}s(k) > \rho_{h2} \end{cases} \tag{28}$$

Remarks:

1. In contrast with the continuous-time adaptive fuzzy control system[6], the term of variable structure control component $u_{vu}(k)$ corresponds to the sliding component $u_{su}(t)$. In [6], via a modulate $m(t)$ the $u_{su}(t)$ only works on the specified region A_d^C where the fuzzy approximator could not effectively approximate the unknown function in a sense. Similarly, here $u_{vu}(k)$ only works outside of the sector $\Omega(k)$ which is defined on the tracking error metric $s(k)$ via the switching type coefficients (27~ 28).

2. Compared with the other adaptive schemes given in [8] [9], there is an important difference. In [8] [9], the convergence of tracking error metric depends on a assumption like $\sup_j \tilde{\omega}_j(k) \leq d$, where d is a known constant, which may not be easy to check, because it could not ensure that $\hat{\omega}_j$ is bounded. On the other hand, in our scheme, by using a projection algorithm we can easily realize $\hat{W}(k) \in \mathcal{C}_w$ so that $\|\tilde{W}(k)\| < c_\omega$. This will enhance the flexibility of the system.

The stability of the closed-loop system described by (1) and (15-28) is established in the following theorem.

Theorem 2 *Under the assumptions 1 and 2, if the plant (1) is controlled by (15), (16), (20), (27-28), and the adaptive component is synthesized by (17-18), then the tracking error metric of the system will stably enter the sector defined by (21-26). When the system tracking error metric is driven inside the sector, $|s(k+1)|$ is usually in a small magnitude such that it can be assumed that $|\Delta s(k+1)| \leq ((\gamma_h - k_d)\lambda^{n-1})^{1/2}|s(k)|$ with $0 < k_d < \gamma_h$. Then, $\tilde{X}(k) \to 0$ as $k \to \infty$ and the all signals are bounded.*

We have no enough room to give the stability proof and present the simulation results which confirm the correctness of the proposed control algorithm.

5   Conclusion

In this paper, we proposed a discrete-time adaptive fuzzy control scheme for a class of unknown sampled-data nonlinear systems. The developed controller guarantees the global stability of the resulting closed-loop system in the sense that all signals involved are uniformly bounded and tracking error would be asymptotically in decay.

References

1. I. J. Leontaritis and S. A. Billings, "Input-output parametric models for nonlinear systems. Part I and II," *International Journal of Control*, vol. 41, pp. 303-344, 1985.
2. L.-X. Wang, "Stable adaptive fuzzy control of nonlinear system," *IEEE Trans. on Fuzzy Systems*, vol. 1, pp. 146-155, 1993.
3. C.-Y. Su and Y. Stepanenko, "Adaptive control of a class of nonlinear systems with fuzzy logic," *IEEE Trans. on Fuzzy Systems*, vol. 2, pp. 258-294, 1994.
4. Y.-G. Piao, H.-G. Zhang and Z. Bien, "Design of fuzzy direct controller and stability analysis for a class of nonlinear system," *Proceedings of the ACC*, pp. 2274-2275, USA, 1998.
5. T. Chai and S. Tong, "Fuzzy direct adaptive control for a class of nonlinear systems," *Fuzzy Sets and Systems*, vol. 103, no. 3, pp. 379-384, 1999.
6. H. Han, C.-Y. Su and S. Murakami, "Further results on adaptive control of a class of nonlinear systems with fuzzy logic," *Journal of Japan Society for Fuzzy Theory and System*, vol. 11, no. 5, pp. 501-509, 1999.

7. S.-J. Wu and C.-T. Lin, "Global stability analysis of discrete time-invariant optimal fuzzy control systems," *International Journal of Fuzzy Systems*, vol. 1, no. 1, pp. 60-68, 1999.

8. Katsuhisa Furuta, "VSS type self-tuning control," *IEEE Trans. on Industrial Electronics*, vol. 40, no. 1, pp. 37-44, 1993.

9. F. Sun, Z. Sun, and P.-Y. Woo, "Stable neural-network-based adaptive control for sampled-data nonlinear systems," *IEEE Trans. on Neural Networks*, vol. 9, no. 5, pp. 956-968, 1998.

10. S. Sastry and M. Bodson, *Adaptive Control*, Prentice Hall, 1989.

11. L.-X. Wang and J. M. Mendel, "Back-propagation fuzzy system as nonlinear dynamic system identifiers," *Proc. of IEEE Conf. on Fuzzy Systems*, pp. 1409-1418, San Diego, 1992.

12. C.-J. Chien and L.-C. Fu, "Adaptive control of interconnected systems using neural network design," *Proc. of 37th IEEE Conf. on Decision and Control*, Tampa, TL. 1998.

13. J.-J. Slotine and J. A. Coetsee, "Adaptive sliding controller synthesis for nonlinear systems," *Int. J Control*, vol. 43, pp. 1631-1651, 1986.

14. Changyun Wen, "A robust adaptive controller with minimal modification for discrete time-varying systems," *IEEE Trans. on Auto. Control*, vol. 39, no.5, pp. 987-991, 1994.

15. J.-J.E.Slotine and W. Li, *Applied nonlinear control*, Prentice Hall, 1991.

Conditional Probability Relations in Fuzzy Relational Database

Rolly Intan and Masao Mukaidono

Meiji University, Kanagawa-ken, Japan

Abstract. In 1982, Buckles and Petry [1] proposed fuzzy relational database for incorporating non-ideal or fuzzy information in a relational database. The fuzzy relational database relies on the spesification of *similarity relation* [8] in order to distinguish each scalar domain in the fuzzy database. These relations are reflexive, symmetric, and max-min transitive. In 1989, Shenoi and Melton extended the fuzzy relational database model of Buckles and Petry to deal with *proximity relation* [2] for scalar domain. Since reflexivity and symmetry are the only constraints placed on proximity relations, proximity relation is considered as generalization of similarity relation.

In this paper, we propose design of fuzzy relational database to deal with *conditional probability relation* for scalar domain. These relations are reflexive and not symmetric. We show that naturally relation between fuzzy information is not symmetric. In addition, we define a notion of redundancy which generalizes redundancy in classical relational database. We also discuss partitioning of domains with the objective of developing equivalence class.

1 Introduction

Fuzzy relational database which is proposed by Buckles and Petry(1982) [1], as in classical relational database theory, consists of a set of tuples, where t_i represents the i-th tuple and if there are m domains D, then $t_i = (d_{i1}, d_{i2}, ..., d_{im})$. In the classical relational database, each component of tuples d_{ij} is an atomic crisp value of tuple t_i with the restriction to the domain D_j, where $d_{ij} \in D_j$. However, in fuzzy relational database, each component of tuples d_{ij} is not limited to atomic crisp value; instead, $d_{ij} \subseteq D_j$ $(d_{ij} \neq \emptyset)$ as defined in the following definition.

Definition 1. *[1]. A **fuzzy database relation**, R, is a subset of the set of cross product*

$$2^{D_1} \times 2^{D_2} \times \cdots \times 2^{D_m}, \quad \text{where } 2^{D_j} = 2^{D_j} - \emptyset.$$

A fuzzy tuple is a member of a fuzzy database relation as follows.

Definition 2. *[1]. Let $R \subseteq 2^{D_1} \times 2^{D_2} \times \cdots \times 2^{D_m}$ be a fuzzy database relation. A **fuzzy tuple** t (with respect to R) is an element of R.*

W. Ziarko and Y. Yao (Eds.): RSCTC 2000, LNAI 2005, pp. 251–260, 2001.

Even though the fuzzy relational database considers components of tuples as set of data values from the corresponding domains, by applying the concept of equivalence classes, it is possible to define a notion of redundancy which is similar to classical relational database theory. The starting point is the definition of an interpretation of a fuzzy tuple as shown in the following definition.

Definition 3. *[1]. Let* $t_i = (d_{i1}, d_{i2}, ..., d_{im})$ *be a fuzzy tuple. An* **interpreta-tion** *of* t_i *is a tuple* $\theta = (a_1, a_2, ..., a_m)$ *where* $a_j \in d_{ij}$ *for each domain* D_j.

It is important to note that each component of an interpretation is an atomic value. In classical relational database, each tuple is the same as its interpretation because every component of tuples is an atomic crisp value. The process of determining redundancy in the term of interpretations of fuzzy tuples is defined as follows.

Definition 4. *[1]. Two tuples* t_i *and* t_j *are redundant if and only if they possess an identical interpretation.*

The definition of redundancy above is a generalization of the concept of redundancy in classical database theory. Specifically, the absence of redundant tuples in classical relational database means there are no multiple occurrences of the same interpretations. Similarly, in fuzzy relational database, there should be no more than one tuple with a given interpretation.

The fuzzy relational database of Buckles and Petry relies on the specification of *similarity relation* for each distinct scalar domain in the fuzzy database. A similarity relation, s_j, for a given domain, D_j, maps each pair of elements in the domain to an element in the closed interval $[0,1]$.

Definition 5. *[8] A* **similarity relation** *is a mapping,* $s_j : D_j \times D_j \rightarrow [0,1]$, *such that for* $x, y, z \in D_j$,

$$s_j(x,x) = 1 \quad \text{(reflexivity)},$$
$$s_j(x,y) = s_j(y,x) \quad \text{(symmetry)},$$
$$s_j(x,z) \geq \max\{\min[s_j(x,y), s_j(y,z)]\} (\max - \min \text{ transitivity})$$

The same with similarity relation in fuzzy relational database of Buckles and Petry, a specific type of similarity relation known as the *identity relation* is used in classical relational database as defined in the following definition.

Definition 6. *A* **identity relation** *is a mapping,* $s_j : D_j \times D_j \rightarrow \{0,1\}$, *such that for* $x, y \in D_j$,

$$s_j(x,y) = \begin{cases} 1 \ if \ x = y, \\ 0 \ otherwise. \end{cases}$$

There is considerable criticism about the use of similarity relation, especially for the point of max-min transitivity in fuzzy relational database (see e.g., [7]). To easily understand this criticism, a simple illustration is given by an example. Let us suppose that *one* is similar to *two* to a level of 0.8, and *two* is similar to *three* to a level of 0.8. According to max-min transitivity, the similarity between *one* and *three* must be no less than 0.8. Therefore max-min transitivity is considered as a very restrictive constraint. Considering this reason, in 1989, Shenoi and Melton extended fuzzy relational database model of Buckles and Petry to deal with *proximity relation* for scalar domain as defined in the following definition.

Definition 7. *[2] A **proximity relation** is a mapping,* $s_j : D_j \times D_j \to [0,1]$, *such that for* $x, y \in D_j$,

$$s_j(x, x) = 1 \quad \text{(reflexivity)},$$
$$s_j(x, y) = s_j(y, x) \quad \text{(symmetry)},$$

In their other paper [6], Shenoi and Melton introduced a notion of α-redundant tuples as defined in the following definition.

Definition 8. *Two tuples,* t_i *and* t_j, *are* α-redundant, *denote by* $t_i \sim_\alpha t_j$, *where* $\alpha = (\alpha_1, ..., \alpha_m)$, *whenever*

$$d_{i1} \sim_{\alpha_1} d_{j1}, ..., d_{im} \sim_{\alpha_m} d_{jm}.$$

In general, α *is a subset of levels associated with* t_i *and* t_j.

However, every word naturally has different range of meaning where there are some words have more general meaning than the others. For simple example, talking about hair color, the word 'Brown' is more general and broader, while 'Light Brown' is narrower and more specific. The word 'Brown' can cover a wider range of meaning in color than the word 'Light Brown'. So the range of meaning in color is different in these two words. In our sentences, it is correct to say that 'Light Brown is Brown', but not 'Brown is Light Brown'. Moreover, we can say that similarity level of 'Brown' given 'Light Brown' and similarity level of 'Light Brown' given 'Brown' are different. Therefore, we consider the relation of similarity between two scalar domain should be not symmetric rather than symmetric. Considering this reason, in this paper, we propose design of fuzzy relational database to deal with *conditional probability relation* for scalar domain. These relations are reflexive and not symmetric. In addition, related to this method, we define a notion of redundancy which generalizes redundancy in classical relational database and even the concept of redundancy which was proposed by Shenoi and Melton [6]. We also discuss partitioning of domains with the objective of developing equivalence class by using not only α-level to determine the degree of non-ideality, but also r as reference data for clustering.

2   Fuzzy Relational Database with Conditional Probability Relation

In this section, we propose conditional probability relation as a basis of dealing with scalar domains in fuzzy relational database. A conditional probability relation is defined in the following definition.

Definition 9. *A **conditional probability relation** is a mapping, $s_j : D_j \times D_j \to [0,1]$, such that for $x, y \in D_j$,*

$$s_j(x|y) = \frac{|x \cap y|}{|y|} \tag{1}$$

where $s_j(x|y)$ means similarity level of x given y.

Considering our example in Section 1, it is possible to set

$$s_{hair}(Brown|LightBrown) = 0.9,$$

$$s_{hair}(LightBrown|Brown) = 0.4.$$

The expression means that similarity level of 'Brown' given 'Light Brown' is 0.9 and similarity level of 'Light Brown' given 'Brown' is 0.4, respectively. These conditions can be easily understood by imaging every scalar domain represented in set. In our example, size of set 'Brown' is bigger than size of set 'Light Brown', as shown in the following figure.

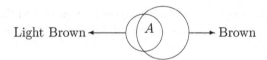

Fig. 1. Intersection: 'Brown' and 'Light Brown'

Area A which is the intersection's area between 'Brown' and 'Light Brown', represents similarity area of them. We can calculate the similarity level of 'Brown' given 'Light Brown' and the similarity level of 'Light Brown' given 'Brown' denoted by $s_{hair}(Brown|LightBrown)$ and $s_{hair}(LightBrown|Brown)$, respectively using (1) as follows.

$$s_{hair}(Brown|LightBrown) = \frac{|A|}{|LightBrown|},$$

$$s_{hair}(LightBrown|Brown) = \frac{|A|}{|Brown|},$$

where $|Brown|$ and $|LightBrown|$ represent size of set 'Brown' and 'Light Brown', respectively. Corresponding to Figure 1, $|Brown| > |LightBrown|$ implies

$$s_{hair}(Brown|LightBrown) > s_{hair}(LightBrown|Brown).$$

Furthermore, ideal information or crisp data and non-ideal information or fuzzy information can be represented by using fuzzy sets. In that case, fuzzy set can be used as a connector to represent imprecise data from *total ignorance* (the most imprecise data) to *crisp* (the most precise data) as follows.

Definition 10. *[3] Let U be universal set, where $U = \{u_1, u_2, ..., u_n\}$.* **Total ignorance(TI)** *over U and* **crisp** *of $u_i \in U$ are defined as*

$$\text{TI over } U = \{1/u_1, ..., 1/u_n\},$$
$$\text{Crisp}(u_i) = \{0/u_1, ..., 0/u_{i-1}, 1/u_i, 0/u_{i+1}, ..., 0/u_n\},$$

respectively.

Similarity level of two fuzzy informations which are represented in two fuzzy sets can be approximately calculated by using conditional probaaility of two fuzzy sets [3]. In that case, $|y| = \sum_u \chi_u^y$, where χ_u^y is membership function of y over u, and intersection is defined as minimum.

Definition 11. *Let $x = \{\chi_1^x/u_1, ..., \chi_n^x/u_n\}$ and $y = \{\chi_1^y/u_1, ..., \chi_n^y/u_n\}$ are two fuzzy sets over $U = \{u_1, u_2, ..., u_n\}$. $s_j : D_j \times D_j \to [0, 1]$, such that for $x, y \in D_j$,*

$$s_j(x|y) = \frac{\sum_{i=1}^n \min\{\chi_i^x, \chi_i^y\}}{\sum_{i=1}^n \chi_i^y},$$

where $s_j(x|y)$ means level similarity of x given y.

Example 1. Let us suppose that two fuzzy sets, 'Warm' and 'Rather Hot' which represent condition of temperature, are arbitrarily given in the following membership functions.

$$Warm = \{0.2/22, 0.5/24, 1/26, 1/28, 0.5/30, 0.2/32\}$$

$$Rather Hot = \{0.5/30, 1/32, 1/34, 0.5/36\}$$

By Definition 11, we calculate similarity level of 'Warm' given 'Rather Hot' and similarity level of 'Rather Hot' given 'Warm', respectively as follows.

$$s_{temperature}(Warm|RatherHot) = \frac{\min(0.5, 0.5) + \min(0.2, 1)}{0.5 + 1 + 1 + 0.5} = \frac{0.7}{3},$$

$$s_{temperature}(RatherHot|Warm) = \frac{\min(0.5, 0.5) + \min(0.2, 1)}{0.2 + 0.5 + 1 + 1 + 0.5 + 0.2} = \frac{0.7}{3.4}.$$

Calculation of similarity level based on conditional probability relation implies some conditions in the following theorem.

Theorem 1. *Let $s_j(x|y)$ be similarity level of x given y and $s_j(y|x)$ be similarity level of y given x, such that for $x, y, z \in D_j$,*

1. *if $s_j(x|y) = s_j(y|x) = 1$ then $x = y$,*
2. *if $s_j(y|x) = 1$ and $s_j(x|y) < 1$ then $x \subset y$,*
3. *if $s_j(x|y) = s_j(y|x) > 0$ then $|x| = |y|$,*
4. *if $s_j(x|y) < s_j(y|x)$ then $|x| < |y|$,*
5. *if $s_j(x|y) > 0$ then $s_j(y|x) > 0$,*
6. *if $s_j(x|y) \geq s_j(y|x) > 0$ and $s_j(y|z) \geq s_j(z|y) > 0$ then $s_j(x|z) \geq s_j(z|x)$.*

The conditional probability relation in Definition 9 is defined through a model or an interpretation based on conditional probability. From Theorem 1, we can define two other interesting mathematical relations, *resemblance relation* and *conditional transitivity relation*, based on their constraints represented by axioms in representing similarity level of a relation as follows, although we do not clarify their properties in this paper.

Definition 12. *A **resemblance relation** is a mapping,* $s_j : D_j \times D_j \to [0,1]$, *such that for* $x, y \in D_j$,

$$s_j(x, x) = 1 \quad \text{(reflexivity)},$$

$$\text{if } s_j(x, y) > 0 \text{ then } s_j(y, x) > 0.$$

Definition 13. *A **conditional transitivity relation** is a mapping,* $s_j : D_j \times D_j \to [0,1]$, *such that for* $x, y \in D_j$,

$$s_j(x, x) = 1 \quad \text{(reflexivity)},$$

$$\text{if } s_j(x, y) > 0 \text{ then } s_j(y, x) > 0,$$

$$\text{if } s_j(x, y) \ge s_j(y, x) > 0 \text{ and } s_j(y, z) \ge s_j(z, y) > 0 \text{ then } s_j(x, z) \ge s_j(z, x).$$

Since a relation which satisfies conditional transitivity relation must satisfy resemblance relation, it is clearly seen that resemblance relation is more general than conditional transitivity relation. On the other hand, conditional transitivity relation generalizes proximity relation, because proximity relation is just a spesific case of conditional transitivity relation. Related to Theorem 1, we also conclude that conditional probability relation which is defined in Definition 9 is also a specific example of conditional transitivity relation. Generalization level of identity relation, similarity relation, proximity relation, conditional probability, conditional transitivity relation and resemblance relation is clearly shown in Figure 2.

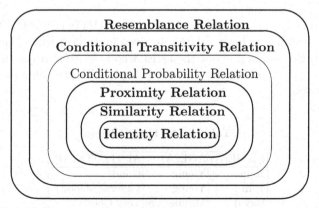

Fig. 2. Generalization level

3   Redundancy in Tuples

In this section, we discuss and define a notion of redundancy which generalizes redundancy in classical relational database and even the concept of redundancy in fuzzy relational database proposed by Shenoi and Melton [6].

Based on the Theorem 1, we define a notion of redundancy tuples as follows.

Definition 14. *Tuple* $t_i = (d_{i1}, d_{i2}, ..., d_{im})$ *is α-redundant in relation R if there is a tuple* $t_j = (d_{j1}, d_{j2}, ..., d_{jm})$ *which covers all information of* t_i *with the degree of non-ideality* α*, where* $α = (α_1, ..., α_m)$*, whenever*

$$\forall x \in d_{ik}, \exists y \in d_{jk}, \ s_k(y|x) \geq \alpha_k$$

for $k = 1, 2, ..., m$

In classical relational database, all of its scalar domain are atomic values and each distinct scalar domain is disjoint. It is clear that the identity relation is used for the treatment of ideal information where a domain element may have no similarity to any other elements of the domain; each element may be similar only unto itself. Consequently, a tuple is redundant if it is exactly the same as another tuple. However, in fuzzy relational database, a domain element may have similarity level to any other elements of the domain. Moreover, considering the range of meaning, a fuzzy information may cover any other fuzzy information (i.e., 'Brown' covers 'Light Brown') with the certain degree of non-ideality. Therefore, we define the concept of redundant tuple in fuzzy relational database as defined in Definition 14 where components of tuples may not be single value as proposed in the fuzzy relational database model of Buckles and Petry. Compared to Definition 4 and 8 which also define redundant tuples, Definition 14 appeals to be more general as a consequence that symmetry is just a special case in conditional probability relation.

Example 2. Let us consider the relation scheme ARTIST(NAME, AGE, APTITUDE). An instance of fuzzy relation is given in Table 1 where each tuple represents someone's opinion about the artist who is written in the tuple. Domain NAME is unique which means that every tuple with the same name indicates the same person.

Table 1. ARTIST Relation

NAME(N)	AGE(A)	APTITUTE(AP)
John	Young	Good
Tom	Young	{Average,Good}
David	Middle Age	Very Good
Tom	[20,25]	Average
David	About-50	Outstanding

Table 2. Similarity Level of AGE

	Young	[20,25]	Midle Age	About-50
Young	1.0	0.8	0.3	0.0
[20, 25]	0.4	1.0	0.2	0.0
Middle Age	0.3	0.4	1.0	0.9
About-50	0.0	0.0	0.5	1.0

Table 3. Similarity Level of APTITUDE

	Average	Good	Very Good	Outstanding
Average	1.0	0.6	0.3	0.0
Good	0.6	1.0	0.8	0.6
Very Good	0.15	0.4	1.0	0.9
Outstanding	0.0	0.3	0.9	1.0

Let us suppose that similarity level of scalar domain AGE and APTITUDE are given in Table 2 and 3.

From Table 2, similarity level of $Young$ given $[20, 25]$, $s_A(Young|[20, 25])$, is equal to 0.8, on the other hand, similarity level of $[20, 25]$ given $Young$, $s_A([20, 25]|Young)$ is equal to 0.4. In that case, $Young$ covers a wider range of meaning in AGE than $[20, 25]$. Now, we want to remove redundant tuples in Table 1 with arbitrary $\alpha = (1.0, 0.0.7, 0.8)$ which coresponds to N, A, and AP, where $\alpha_N = 1.0, \alpha_A = 0.7, \alpha_{AP} = 0.8$. We must set α to 1.0 especially for domain NAME, because domain NAME is crisp domain and each distinct scalar domain indicates different person.

By applying the formula in Definition 14, there are two redundant tuples, $\langle Tom, [20, 25], Average \rangle$ and $\langle David, About - 50, Outstanding \rangle$, which are covered by $\langle Tom, Young, \{Average, Good\} \rangle$ and $\langle David, Middle\ Age, Very\ Good \rangle$, respectively. Table 4 shows the final result after removing the two redundant tuples.

Table 4. ARTIST Relation (free redundancy)

NAME(N)	AGE(A)	APTITUTE(AP)
John	Young	Good
Tom	Young	{Average,Good}
David	Middle Age	Very Good

4   Clustering in Scalar Domains

In fuzzy relational database model of conditional probability relation, there are two parameters, α-cut and r which are required to produce partitioning scalar domains into equivalence classes or disjoint clusters. In that case, α corresponds to the degree of non-ideality and r means *reference* which is used as reference data to be compared to in the process of partitioning scalar domain.

Definition 15. *If* $s : D \times D \to [0, 1]$ *is a conditional probability relation, then the equivalence class (disjoint partition) on D with α-cut and reference r is denoted by E_α^r and is given by*

$$E_\alpha^r = \{x \in D | s(x|r) \geq \alpha\}.$$

Example 3. Given conditional probability relation for domain *Hair Color*, is shown in Table 5.

Table 5. Similarity Level of Hair Color

	Bk	DB	A	R	LB	Bd	Bc
Black(Bk)	1.0	0.8	0.7	0.5	0.2	0.0	0.0
Dark Brown(DB)	0.8	1.0	0.8	0.6	0.4	0.2	0.0
Auburn(A)	0.7	0.8	1.0	0.9	0.6	0.4	0.0
Red(R)	0.5	0.7	0.9	1.0	0.8	0.6	0.3
Light Brown(LB)	0.1	0.2	0.3	0.4	1.0	0.8	0.6
Blond(Bd)	0.0	0.1	0.2	0.3	0.8	1.0	0.9
Bleached(Bc)	0.0	0.0	0.0	0.1	0.4	0.6	1.0

Let us suppose that we want to cluster the data with $r = Red$; this gives rise to the following clusters for various values of α.

$$\alpha \in (0.7, 1.0] : \{Auburn, Red\}$$
$$\alpha \in (0.4, 0.7] : \{Dark\ Brown, Black\}$$
$$\alpha \in [0.0, 0.4] : \{Blond, Bleached, Light\ Brown\}$$

5   Conclusions

In this paper, we extended fuzzy relational database to deal with conditional probability relations for scalar domain. These relations are reflexive and not symmetric. We showed that naturally similarity level of two fuzzy informations or non-ideal informations is not symmetric. Similarity level of two fuzzy informations may be symmetric if and only they have the same range of meaning. Therefore, conditional probability relation generalizes proximity relation which is proposed by Shenoi and Melton in 1989. Moreover, we proposed two other relations related to the conditional probability relation, resemblance relation and conditional transitivity relation which are useful for the treatment of imprecise informations. We also defined a notion of redundancy which also generalizes the concept of redundancy in classical relational database. Finally, process of clustering scalar domain was given by using not only α-level to determine the degree of non-ideality, but also r as reference data for clustering. In our next paper, we will show that this process of clustering scalar domain is very applicable in the application of approximate data querying.

References

1. Buckles, B.P., Petry, F.E., 'A Fuzzy Representation of Data for Relational Database', *Fuzzy Sets and Systems, 5*, (1982), pp. 213-226.
2. Dubois, D., Prade, H., *Fuzzy Sets and Systems: Theory and Applications*, (Academic Press, New York, 1980).
3. Intan, R., Mukaidono, M., 'Application of Conditional Probability in Constructing Fuzzy Functional Dependency (FFD)', *Proceedings of AFSS'00*, (2000) to be appeared.
4. Intan, R., Mukaidono, M., 'Fuzzy Functional Dependency and Its Application to Approximate Querying', *Proceedings of IDEAS'00*, (2000) to be appeared.
5. Shenoi, S., Melton, A., 'Proximity Relations in The Fuzzy Relational Database Model', *Fuzzy Sets and Systems, 31*, (1989), pp. 285-296
6. Shenoi, S., Melton, A., Fan, L.T., 'Functional Dependencies and Normal Forms in The Fuzzy Relational Database Model', *Information Science, 60*, (1992), pp.1-28
7. Tversky, A., 'Features of Similarity', *Psychological Rev. 84(4)*, (1977), pp. 327-353.
8. Zadeh, L.A., 'Similarity Relations and Fuzzy Orderings ', *Inform. Sci.3(2)*, (1970), pp. 177-200.

A New Class of Necessity Measures and Fuzzy Rough Sets Based on Certainty Qualifications

Masahiro Inuiguchi and Tetsuzo Tanino

Department of Electronics and Information Systems
Graduate School of Engineering, Osaka University
2-1, Yamada-Oka, Suita, Osaka 565-0871, Japan
{inuiguti, tanino}@eie.eng.osaka-u.ac.jp
http://vanilla.eie.eng.osaka-u.ac.jp

Abstract. In this paper, we propose a new class of necessity measures which satisfy (R1) $N_A(B) > 0 \Leftrightarrow \exists \varepsilon > 0$; $[A]_{1-\varepsilon} \subseteq (B)_\varepsilon$, (R2) $\exists h^* \in (0,1)$; $N_A(B) \geq h^* \Leftrightarrow A \subseteq B$ and (R3) $N_A(B) = 1 \Leftrightarrow (A)_0 \subseteq [B]_1$. It is shown that such a necessity measure is designed easily by level cut conditioning approach. A simple example of such a necessity measure is given. The proposed necessity measure is applied to fuzzy rough set based on certainty qualifications. It is demonstrated that the proposed necessity measure gives better upper and lower approximations of a fuzzy set than necessity measures defined by S-, R- and reciprocal R-implications.

1 Introduction

In [3], we showed that a necessity measure can be obtained by specifying a level cut condition. However, the usefulness of this result was not clearly demonstrated. In [4], we showed that fuzzy rough sets based on certainty qualifications give better approximations than previous fuzzy rough sets [1]. We have not yet discuss about the selection of the necessity measure to define fuzzy rough sets.

In this paper, we demonstrate that, by the level cut conditioning approach, we can obtain a new and interesting class of necessity measures which satisfy

(R1) $N_A(B) > 0$ if and only if there exists $\varepsilon > 0$ such that $[A]_{1-\varepsilon} \subseteq (B)_\varepsilon$,
(R2) there exists $h^* \in (0,1)$ such that $N_A(B) \geq h^*$ if and only if $A \subseteq B$,
(R3) $N_A(B) = 1$ if and only if $(A)_0 \subseteq [B]_1$,

where $[A]_h = \{x \in X \mid \mu_A(x) \geq h\}$, $(A)_h = \{x \in X \mid \mu_A(x) > h\}$, $\mu_A : X \to [0,1]$ is a membership function of a fuzzy set A and X is the universal set. Moreover, we demonstrate that the proposed necessity measures provide better lower and upper approximations in fuzzy rough sets based certainty qualifications than often used necessity measures.

2 Necessity Measures and Level Cut Conditions

Given a piece of information about unknown variable x that x is in a fuzzy set A, the certainty degree of the event that x is in a fuzzy set B is evaluated by

$$N_A(B) = \inf_x I(\mu_A(x), \mu_B(x)), \tag{1}$$

W. Ziarko and Y. Yao (Eds.): RSCTC 2000, LNAI 2005, pp. 261–268, 2001.

where $I : [0,1]^2 \to [0,1]$ is an implication function, i.e., a function such that

(I1) $I(0,0) = I(0,1) = I(1,1) = 1$ and $I(1,0) = 0$.

$N_A(B)$ of (1) is called a necessity measure. Such necessity measures are used in approximate reasoning, information retrieval, fuzzy mathematical programming, fuzzy data analysis and so on.

The quality of reasoning, decision and analysis in methodologies based on necessity measures depends on the adopted implication function. There are a lot of implication functions. The implication which defines the necessity measure has been selected by proverbiality and by tractability. The selected implication function should be fit for the problem setting. The authors [3] proposed the implication function selection by specifying two modifier functions with single parameters, g^m and $g^M : [0,1]^2 \to [0,1]$ which are roughly corresponding to concentration and dilation modifiers such as linguistic expressions 'very' and 'roughly', respectively. In other words, from the viewpoint that a necessity measure relates to an inclusion relation, we suppose that $N_A(B) \geq h$ is equivalent to an inclusion relation between fuzzy sets A and B with a parameter h, i.e.,

$$N_A(B) \geq h \Leftrightarrow m_h(A) \subseteq M_h(B), \tag{2}$$

where the inclusion relation between fuzzy sets is defined normally, i.e., $A \subseteq B$ if and only if $\mu_A(x) \leq \mu_B(x)$, for all $x \in X$. $m_h(A)$ and $M_h(B)$ are fuzzy sets defined by $\mu_{m_h(A)}(x) = g^m(\mu_A(x), h)$ and $\mu_{M_h(B)}(x) = g^M(\mu_B(x), h)$. From its meaning and technical reason, g^m and g^M are imposed to satisfy

(g1) $g^m(a, \cdot)$ is lower semi-continuous and $g^M(a, \cdot)$ upper semi-continuous for all $a \in [0,1]$,

(g2) $g^m(1,h) = g^M(1,h) = 1$ and $g^m(0,h) = g^M(0,h) = 0$ for all $h > 0$,

(g3) $g^m(a,0) = 0$ and $g^M(a,0) = 1$ for all $a \in [0,1]$,

(g4) $h_1 \geq h_2$ implies $g^m(a,h_1) \geq g^m(a,h_2)$ and $g^M(a,h_1) \leq g^M(a,h_2)$ for all $a \in [0,1]$,

(g5) $a \geq b$ implies $g^m(a,h) \geq g^m(b,h)$ and $g^M(a,h) \geq g^M(b,h)$ for all $h \leq 1$,

(g6) $g^m(a,1) > 0$ and $g^M(a,1) < 1$ for all $a \in (0,1)$.

Under the assumption that g^m and g^M satisfy (g1)–(g6), we proved that there exists a necessity measure which satisfies (2) and defined by the following implication function (see [3]):

$$I^L(a,b) = \sup_{0 \leq h \leq 1} \{h \mid g^m(a,h) \leq g^M(b,h)\}. \tag{3}$$

There are a lot of implication functions which can be represented by (3).

Table 1 shows pairs (g^m, g^M) with respect to S-implications, R-implications and reciprocal R-implications (see [2]) defined by continuous Archimedean t-norms and strong negations. A continuous Archimedean t-norm is a conjunction function which is represented by $t(a,b) = f^*(f(a) + f(b))$ with a continuous and strictly decreasing function $f : [0,1] \to [0,+\infty)$ such that $f(1) = 0$ (see [2]), where $f^* : [0,+\infty) \to [0,1]$ is a pseudo-inverse defined by $f^*(r) = \sup\{h \mid$

Table 1. g^m and g^M for I^S, I^R and I^{r-R} with continuous Archimedean t-norms

I	$f(0)$	$g^m(a,h)$, when $h>0$	$g^M(a,h)$, when $h>0$
I^S	—	$\max(0, 1 - f(a)/f(n(h)))$	$\min(1, f(n(a))/f(n(h)))$
I^R	$<+\infty$	$\max(0, 1 - f(a)/(f(0)-f(h)))$	$\min(1, (f(0)-f(a))/(f(0)-f(h)))$
	$=+\infty$	a	$f^{-1}(\max(0, f(a)-f(h)))$
I^{r-R}	$<+\infty$	$(f(n(a))-f(h))/(f(0)-f(h))$	$\min(1, f(n(a))/(f(0)-f(h)))$
	$=+\infty$	$n(f^{-1}(\max(0, f(n(a))-f(h))))$	a
all	—	$g^m(a,h)=0$, when $h=0$	$g^M(a,h)=1$, when $h=0$

$f(h) \geq r\}$. A strong negation is a bijective strictly decreasing function $n : [0,1] \to [0,1]$ such that $n(n(a)) = a$. Given a t-norm t and a strong negation n, the associated S-implication function I^S, R-implication function I^R and reciprocal R-implication I^{r-R} are defined by $I^S(a,b) = n(t(a,n(b)))$, $I^R(a,b) = \sup_{0 \leq h \leq 1}\{h \mid t(a,h) \leq b\}$ and $I^{r-R}(a,b) = I^R(n(b),n(a))$.

3   A New Class of Necessity Measures

Dienes, Gödel, reciprocal Gödel and Łukasiewitz implications are often used to define necessity measures. S-implications, R-implications and reciprocal R-implications are also considered since they are more or less generalized implication functions defined by t-norms and strong negations. The t-norms often used are a minimum operation and continuous Archimedean t-norms.

None of those often used implication functions satisfies three conditions (R1)–(R3). Indeed, checking a sufficient condition to (R2), i.e., there exists $h \in (0,1)$ such that $g^m(\cdot, h) = g^M(\cdot, h)$ for R- and reciprocal R-implications, the equality holds when $h = 1$ but never holds when $h \in (0,1)$. For S-implications, the equality never holds for $h \in [0,1]$. Those facts can be confirmed from Table 1 in case of continuous Archimedean t-norms.

We propose a new class of necessity measures which satisfy (R1)–(R3). (R1)–(R3) can be rewritten by using g^m and g^M as shown in the following theorem.

Theorem 1. *(R1)–(R3) can be rewritten as follows by using g^m and g^M:*

(R1') $\lim_{h \to +0} g^m(a,h) = 0$ and $\lim_{h \to +0} g^M(a,h) = 1$ for all $a \in (0,1)$,
(R2') there exists $h^* \in (0,1)$ such that $g^m(\cdot, h) = g^M(\cdot, h)$ and $g^m(\cdot, h)$ is bijective if and only if $h = h^*$,
(R3') one of the following two assertions holds:
   (i) $g^m(a,1) = 1$ for all $a \in (0,1)$ and $g^M(b,1) < 1$ for all $b \in (0,1)$,
   (ii) $g^m(a,1) > 0$ for all $a \in (0,1)$ and $g^M(b,1) = 0$ for all $b \in (0,1)$.

Theorem 1 shows that there are a lot of necessity measures which satisfy (R1)–(R3). An example is given in the following example.

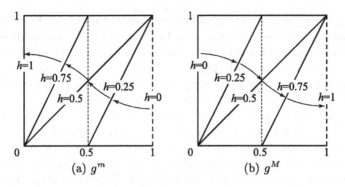

Fig. 1. Transition of g^m and g^M form $h = 0$ to $h = 1$

Example 1. The following (g^m, g^M) satisfies (g1)–(g6) as well as (R1')–(R3') with $h^* = 0.5$:

$$g^m(x, h) = \begin{cases} \max\left(0, \dfrac{x + 2h - 1}{2h}\right), & \text{if } h \in [0, 0.5), \\[2mm] \min\left(1, \dfrac{x}{2 - 2h}\right), & \text{if } h \in [0.5, 1]. \end{cases} \tag{4}$$

$$g^M(x, h) = \begin{cases} \min\left(1, \dfrac{x}{2h}\right), & \text{if } h \in [0, 0.5), \\[2mm] \max\left(0, \dfrac{x + 1 - 2h}{2 - 2h}\right), & \text{if } h \in [0.5, 1], \end{cases} \tag{5}$$

where all denominators are treated as $+0$ when they become 0. Those g^m and g^M are illustrated in Figure 1. From (3), the implication function associated with this pair (g^m, g^M) is obtained as

$$I^L(a, b) = \begin{cases} 1, & \text{if } a = 0 \text{ or } b = 1, \\[2mm] \dfrac{1 - a + b}{2}, & \text{otherwise}. \end{cases} \tag{6}$$

The necessity measure is defined by using this implication function, which satisfies (2) with g^m and g^M defined by (4) and (5).

4   Fuzzy Rough Sets Based on Certainty Qualifications

Rough sets have been known as theory to deal with uncertainty mainly caused by indiscernibility between objects and have been applied to reduction of information tables, data mining and expert systems. As a generalization of rough sets, fuzzy rough sets have been proposed in [1]. In fuzzy rough sets, the equivalence relation is extended to a similarity relation and lower and upper approximations of fuzzy sets are defined by necessity and possibility measures, respectively. The authors [4] has proposed fuzzy rough sets based on certainty qualifications. It is shown that, by using certainty qualifications, we obtain better lower and upper approximations than the previous ones. In the previous fuzzy rough sets, lower and upper approximations cannot be obtained when a fuzzy partition is given instead of a similarity relation. On the other hand, fuzzy rough sets based on certainty qualifications give lower and upper approximations even in this case.

In this section, we introduce fuzzy rough sets based on certainty qualifications proposed in [4]. A certainty qualification is a restriction of possible candidates for a fuzzy set A by $N_A(B) \geq q$, where q and a fuzzy set B is given. In many cases, the family of fuzzy sets A's which satisfy $N_A(B) \geq q$ is characterized by the greatest element in the sense of set-inclusion, i.e., $A_1 \subseteq A_2$ if and only if $\mu_{A_1}(x) \leq \mu_{A_2}(x)$, for all $x \in X$. To ensure this, we assume

(I2) $I(c,b) \leq I(a,d)$, $0 \leq a \leq c \leq 1$, $0 \leq b \leq d \leq 1$,
(I3) I is upper semi-continuous.

Under the assumptions (I2) and (I3), we obtain the greatest element $\hat{A}$ as $\mu_{\hat{A}}(x) = \sigma[I](h, \mu_B(x))$, where the implication function I defines the necessity measure and a functional σ is defined by $\sigma[I](a,b) = \sup_{0 \leq h \leq 1}\{h \mid I(h,b) \geq a\}$. All fuzzy sets A's such that $A \subseteq \hat{A}$ satisfy $N_A(B) \geq q$.

A converse certainty qualification is also conceivable. A converse certainty qualification is a restriction of possible candidates for a fuzzy set B by $N_A(B) \geq q$, where q and A are given. The family of fuzzy sets B's which satisfy $N_A(B) \geq q$ is characterized by the smallest element in the sense of set-inclusion. Under the assumptions (I2) and (I3), the smallest element $\check{B}$ is obtained as $\mu_{\check{B}}(x) = \xi[I](\mu_A(x), q)$, where the implication function I defines the necessity measure and a functional ξ is defined by $\xi[I](a,b) = \inf_{0 \leq h \leq 1}\{h \mid I(a,h) \geq b\}$. All fuzzy sets B's such that $B \supseteq \check{B}$ satisfy $N_A(B) \geq q$.

Based on certainty and converse certainty qualifications, two kinds of lower and upper approximations can be defined. One is based on certainty qualifications and the other based on converse certainty qualifications. As selected in [4], lower and upper approximations based on converse certainty qualification are adopted also in this paper since they are simpler than the others.

Given a fuzzy binary relation R which is reflexive, i.e., $\mu_R(x,x) = 1$, for all $x \in X$, a fuzzy rough set of a fuzzy set A is a pair $(R_\square(A), R^\diamond(A))$ defined by

$$\mu_{R_\square(A)}(x) = \sup_y \xi[I](\mu_{[y]_R}(x), N_{[y]_R}(A)), \quad \mu_{R^\diamond(A)}(x) = n\left(\mu_{R_\square(A^c)}(x)\right) \quad (7)$$

where $n : [0,1] \to [0,1]$ is a strong negation. The complement of A, A^c is defined by $\mu_{A^c}(x) = n(\mu_A(x))$. $[y]_R$ is a fuzzy equivalent class defined by $\mu_{[y]_R}(x) = \mu_R(x,y)$.

It has been shown that we always have $R_\square(A) \subseteq A \subseteq R^\diamond(A)$ which ensures that $R_\square(A)$ and $R^\diamond(A)$ are lower and upper approximations of A (see [4]). Moreover, $R_\square(A)$ is a better lower approximation than the one previously defined by a necessity measure and $R^\diamond(A)$ is a better upper approximation than the one previously defined by a possibility measure. Basic properties of rough sets are preserved in fuzzy rough sets based on certainty qualifications under additional assumptions on the implication function I (see [4]).

An interesting result of fuzzy rough sets based on certainty qualification is the fact that we can define lower and upper approximations even when a fuzzy partition $\Phi = \{F_1, F_2, \ldots, F_n\}$ is given instead of a reflexive fuzzy relation R. A fuzzy partition $\Phi = \{F_1, F_2, \ldots, F_n\}$ is a family of fuzzy sets F_i's satisfy

(P1) $\inf_x \max_{i=1,2,\ldots,n} \mu_{F_i}(x) > 0$,

(P2) $\sup_x \min(\mu_{F_i}(x), \mu_{F_j}(x)) < 1$, for all $i, j \in \{i = 1, 2, \ldots, n\}$, $i \neq j$.

When such a fuzzy partition is given, lower and upper approximations, $\Phi_\square(A)$ and $\Phi^\diamond(A)$ are defined as follows:

$$\mu_{\Phi_\square(A)}(x) = \max_{i=1,2,\ldots,n} \xi[I](\mu_{F_i}(x), N_{F_i}(A)), \quad \mu_{\Phi^\diamond(A)}(x) = n\left(\mu_{\Phi_\square(A^c)}(x)\right) \quad (8)$$

5   Necessity Measures and Fuzzy Rough Sets

Let us discuss about necessity measures for fuzzy rough sets defined by certainty qualifications. In order to obtain $R_\square(A) \neq \emptyset$ or $\Phi_\square(A) \neq \emptyset$ (resp. $R^\diamond(A) \neq X$ or $\Phi^\diamond(A) \neq X$), $N_{[x]_R}(A)$ or $N_{F_i}(A)$ (resp. $N_{[x]_R}(A^c)$ or $N_{F_i}(A^c)$) should be positive (resp. less than one). From this point of view, we discuss the conditions for $N_A(B) > 0$ and $N_A(B) = 1$ so that we know the variety of pairs of fuzzy sets (A, B) such that $N_A(B) \in (0, 1)$.

First we have the following theorem.

Theorem 2. *Let a t-norm t satisfies*

$$t(\varepsilon_1, \varepsilon_2) > 0, \text{ for all } \varepsilon_1, \varepsilon_2 > 0. \quad (9)$$

Then we have the following assertions:

(i) For necessity measures $N_A(B)$ defined by S-implications, we have

$$N_A(B) > 0 \Leftrightarrow \text{ there exists } \varepsilon > 0 \text{ such that } [A]_{1-\varepsilon} \subseteq (B)_\varepsilon. \quad (10)$$

(ii) For necessity measures $N_A(B)$ defined by R-implications, we have

$$N_A(B) > 0 \Leftrightarrow \text{ there exists } \varepsilon > 0 \text{ such that } \atop (A)_{\varepsilon'} \subseteq (B)_{\varepsilon'} \text{ for all } \varepsilon' \in (0, \varepsilon]. \quad (11)$$

(iii) For necessity measures $N_A(B)$ defined by reciprocal R-implications, we have

$$N_A(B) > 0 \Leftrightarrow \text{ there exists } \varepsilon > 0 \text{ such that } \atop [A]_{1-\varepsilon'} \subseteq [B]_{1-\varepsilon'} \text{ for all } \varepsilon' \in (0, \varepsilon]. \quad (12)$$

Equation (9) holds when t is a minimum operation, an arithmetic product or a continuous Archimedean t-norm with $f(0) = +\infty$. Theorem 2 shows that the condition for $N_A(B) > 0$ is quite restrictive when $N_A(B)$ is defined by R- and reciprocal R-implications with t-norms satisfy (9). Thus, t-norms which violate (9), such as a bounded product, should be adopted for necessity measures defined by R- and reciprocal R-implications. For $N_A(B) = 1$, we have the following theorem.

Theorem 3. *The following assertions are valid:*

(i) When (9) holds, for necessity measures $N_A(B)$ defined by S-implications, we have $N_A(B) = 1 \Leftrightarrow (A)_0 \subseteq [B]_1$.

(ii) For necessity measures $N_A(B)$ defined by R- and reciprocal R-implications with continuous t-norms, we have $N_A(B) = 1 \Leftrightarrow A \subseteq B$.

From Theorem 3, we know that the domain of A that $N_A(B) \in (0,1)$ is large when $N_A(B)$ is defined by S-implications with t-norm satisfies (9). On the other hand, the necessity measures defined by R- and reciprocal R-implications with continuous t-norms cannot evaluate the difference between two different inclusion relations, e.g., $A \subseteq B$ and $(A)_0 \subseteq [B]_1$ (in both cases, we have $N_A(B) = 1$). Thus, in those necessity measures, the information $N_A(B) = 1$ is less effective to estimate B under known A.

From the discussions above, among the often used necessity measures, S-implications with t-norms satisfying (9) seem to be good for applications to fuzzy rough sets based on certainty qualifications. However, those implications have the following disadvantage.

Theorem 4. *Consider necessity measures defined by S-implications with t-norms satisfying (9). For fuzzy rough sets based on certainty qualification under a fuzzy partition $\Phi = \{F_1, F_2, \ldots, F_n\}$, we have*

(i) $\Phi_\square(A)$ is normal only when there exists F_i such that $N_{F_i}(A) = 1$, where a fuzzy set A is said to be normal if there exists $x \in X$ such that $\mu_A(x) = 1$.

(ii) There exists $x \in X$ such that $\mu_{\Phi^\circ(A)}(x) = 0$ only when there exists F_i such that $N_{F_i}(A^c) = 1$.

From Theorem 3(i), the necessary and sufficient condition of $N_{F_i}(A) = 1$ is $(F_i)_0 \subseteq [A]_1$ and this condition is quite restrictive so that, in many cases, $(F_i)_0 \subseteq [A]_1$ does not hold. Thus, Theorem 4 implies that, in many cases, $\Phi_\square(A)$ is not normal and that $\mu_{\Phi^\circ(A)}(x) > 0$ for all $x \in X$ holds when an S-implication with t-norm satisfying (9) is adopted to define the necessity measure.

Now let us consider necessity measures satisfy (R1)–(R3). From (R1) and (R3), we know that the domain of A that $N_A(B) \in (0,1)$ is large in the proposed necessity measures. Moreover, we have the following theorem.

Theorem 5. *Consider necessity measures satisfy (R1)–(R3). For fuzzy rough sets based on certainty qualification under a fuzzy partition $\Phi = \{F_1, F_2, \ldots, F_n\}$, we have*

(i) $\Phi_\square(A)$ is normal if there exists F_i such that $N_{F_i}(A) \geq h^*$.

(ii) There exists $x \in X$ such that $\mu_{\Phi^\circ(A)}(x) = 0$ if there exists F_i such that $N_{F_i}(A^c) \geq h^*$.

From Theorem 5, the proposed necessity measures seem to be better than the often used necessity measures in application to fuzzy rough sets based on certainty qualifications. By the proposed measure, we may obtain better lower and upper approximations of a fuzzy set. Let us examine this conjecture in the following example.

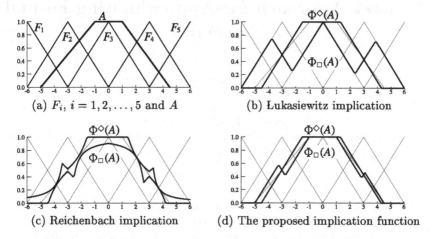

(a) F_i, $i = 1, 2, \ldots, 5$ and A (b) Łukasiewitz implication

(c) Reichenbach implication   (d) The proposed implication function

Fig. 2. Lower and upper approximations of A by three implication functions

Example 2. Let $\Phi = \{F_1, F_2, \ldots, F_5\}$ and A be a fuzzy partition and a fuzzy set given in Figure 2(a). We compare lower and upper approximations $\Phi_\Box(A)$ and $\Phi^\Diamond(A)$ with necessity measures defined by three different implication functions. We consider Łukasiewitz implication I^L ($I^L(a, b) = \min(1 - a + b, 1)$), Reichenbach implication I^{Rei} ($I^{\mathrm{Rei}}(a, b) = 1 - a + ab$) and a proposed implication function defined by (6). Łukasiewitz implication is an R-implication, a reciprocal R-implication and, at the same time, an S-implication defined by a strong negation $n(a) = 1 - a$ and a t-norm, more concretely, a bounded product $t(a, b) = \max(0, a+b-1)$ which does not satisfy (9). On the other hand, Reichenbach implication is an S-implication defined by a strong negation $n(a) = 1 - a$ and a t-norm, more concretely, an arithmetic product $t(a, b) = ab$ which satisfies (9). The lower and upper approximations $\Phi_\Box(A)$ and $\Phi^\Diamond(A)$ with respect to the three implication functions are shown in Figure 2(b)–(d). From these figures, the proposed implication function gives the best approximations among three implication functions.

References

1. Dubois, D., Prade, H.: Putting Rough Sets and Fuzzy Sets Together. In: Słowinski, R. (ed.), Intelligent Decision Support: Handbook of Applications and Advances of the Rough Sets Theory, Kluwer Academic Publishers, Dordrecht (1992) 203–232
2. Fodor, J., Roubens, M.: Fuzzy Preference Modeling and Multicriteria Decision Making. Kluwer Academic Publishers, Dordrecht (1994)
3. Inuiguchi, M., Tanino, T.: Level Cut Conditioning Approach to the Necessity Measure Specification. In: Zhong, N., Skowron, A., Ohsuga, S. (eds.): New Directions in Rough Sets, Data Mining, and Granular-Soft Computing: Proceedings of 7th International Workshop, RSFDGrC'99, Springer, Berlin (1999) 193–202
4. Inuiguchi, M., Tanino, T.: Fuzzy Rough Sets Based on Certainty Qualifications. Proceedings of AFSS 2000 (2000) 433–438

A Fuzzy Approach for Approximating Formal Concepts

Jamil Saquer[1] and Jitender S. Deogun[2] [*]

[1] jms481f@mail.smsu.edu, Computer Science Department, Southwest Missouri State
University, Springfield, MO 65804, USA
[2] deogun@cse.unl.edu, Computer Science & Engineering Department, University of
Nebraska, Lincoln, NE 68588, USA

Abstract. In this paper we present a new approach for approximating
concepts in the framework of formal concept analysis. We investigate two
different problems. The first, given a set of features B (or a set of objects
A), we are interested in finding a formal concept that approximates B
(or A). The second, given a pair (A, B), where A is a set of objects
and B is a set of features, we are interested in finding a formal concept
that approximates (A, B). We develop algorithms for implementing the
approximation techniques presented. The techniques developed in this
paper use ideas from fuzzy sets. The approach we present is different
and simpler than existing approaches which use rough sets.

1 Introduction

Formal concept analysis (FCA) is a mathematical framework developed by Wille
and his colleagues at Darmstadt/Germany that is useful for representation and
analysis of data [11]. A pair consisting of a set of objects and a set of features common to these objects is called a concept. Using the framework of FCA, concepts
are structured in the form of a lattice called the concept lattice. The concept
lattice is a useful tool for knowledge representation and knowledge discovery [4].
Formal concept analysis has also been applied in the area of conceptual modeling
that deals with the acquisition, representation and organization of knowledge [6].
Several concept learning methods have been implemented in [1, 4, 5] using ideas
from formal concept analysis.

Not every pair of a set of objects and a set of features defines a concept
[11]. Furthermore, we might be faced with a situation where we have a set of
features (or a set of objects) and need to find the best concept that approximates
these features (or objects). For example, when a physician diagnosis a patient, he
finds a disease whose symptoms are the closest to the symptoms that the patient
has. In this case we can think of the symptoms as features and the diseases as
objects. Another example is in the area of information retrieval where user's

---

[*] This research was supported in part by the Army Research Office, Grant No.
DAAH04-96-1-0325, under DEPSCoR program of Advanced Research Projects
Agency, Department of Defense.

W. Ziarko and Y. Yao (Eds.): RSCTC 2000, LNAI 2005, pp. 269–276, 2001.

query can be understood as a set of features and the answer to the query can be understood as the set of objects that possess these features. It is therefore of fundamental importance to be able to find concept approximations regardless how little information is available.

In this paper we present a general approach for approximating concepts that uses ideas from fuzzy set theory. We first show how a set of features (or objects) can be approximated by a concept. We then extend our approach for approximating a pair of a set of objects and a set of features. Based on our approach, we present efficient algorithms for concept approximation.

The notion of concept approximation was first introduced in [7, 8] and further investigated in [9, 10]. All these approaches use rough sets as the underlying approximation model. In this paper, we use fuzzy sets as the approximation model. This approach is simpler and the approximation is presented in terms of a single formal concept as compared to two in terms of lower and upper approximations [7–10]. Moreover, the concept approximation algorithms that result from using a fuzzy set approach are simpler.

The organization of this paper is as follows. In Section 2 we give an overview of FCA results that we need for this paper. In Section 3, we show how to approximate a set of features or a set of objects. In Section 4, we show how to approximate a pair of a set of objects and a set of features. A numerical example explaining the approximation ideas is given in Section 5. Finally, a conclusion is drawn in Section 6.

2   Background

Relationships between objects and features in FCA is given in a *context* which is defined as a triple (G, M, I), where G and M are sets of objects and features (also called attributes), respectively, and $I \subseteq G \times M$. An example of a context is given in Table 1 where an "X" is placed in the ith row and jth column to indicate that the object at row i possesses the feature at column j. If object g possesses feature m, then $(g, m) \in I$ which is also written as gIm. The set of features common to a set of objects A is denoted by $\beta(A)$ and defined as $\{m \in M \mid gIm \ \forall g \in A\}$. Similarly, the set of objects possessing all the features in a set $B \subseteq M$ is denoted by $\alpha(B)$ and given by $\{g \in G \mid gIm \ \forall m \in B\}$. A *formal concept* (or simply a *concept*) in the context (G, M, I) is defined as a pair (A, B) where $A \subseteq G$, $B \subseteq M$, $\beta(A) = B$ and $\alpha(B) = A$. A is called the *extent* of the concept and B is called its *intent*. For example, the pair (A, B) where $A = \{4, 5, 8, 9, 10\}$ and $B = \{c, d, f\}$ is a formal concept. On the other hand, the pair (A, B) where $A = \{2, 3, 4\}$ and $B = \{f, h\}$ is not formal concept because $\alpha(B) \neq A$. A pair (A, B) where $A \subseteq G$ and $B \subseteq M$ which is not a formal concept is called a *non-definable concept* [10]. The Fundamental Theorem of FCA states that the set of all formal concepts on a given context with the ordering $(A_1, B_1) \leq (A_2, B_2)$ iff $A_1 \subseteq A_2$ is a complete lattice called the *concept lattice* of the context [11]. The concept lattice of the context given in Table 1 is shown in Figure 1 where concepts are labeled using *reduced labeling* [2]. The extent of a concept C in Figure 1 consists of the objects at C and the objects at the concepts that can be reached from C going downward following descending paths towards the bottom concept C_1. Similarly, the intent of C consists of the features at C and the features at the concepts that can be reached from C going upwards following ascending paths to the top concept C_{23}. The extent and intent of each concept in Figure 1 are also given in Table 2.

Table 1. Example of a context

	a	b	c	d	e	f	h	i	j	k	l	x
1				X								
2				X		X	X					
3						X	X					
4	X	X	X	X	X	X	X			X		
5		X	X	X	X	X	X					X
6						X	X	X	X	X	X	X
7						X	X	X	X			X
8		X	X	X	X	X						
9		X	X		X					X		
10		X	X	X	X					X		

3 Approximating a Set of Features or a Set of Objects

Since approximating a set of objects works analogous to approximating a set of features, we only show how to approximate a set of features. Let $B \subseteq M$ be a set of features. Our goal is to find a formal concept C intent of which is as similar to B as possible. The concept C is then said to approximate B. Define a membership function $f_C(B)$ that gives a measure of how well C approximates B as follows [1]

$$f_C(B) = \frac{\frac{|B \cap \text{Intent}(C)|}{|B \cup \text{Intent}(C)|} + \frac{|\alpha(B) \cap \text{Extent}(C)|}{|\alpha(B) \cup \text{Extent}(C)|}}{2}.$$

The range of $f_C(B)$ is the interval $[0,1]$. $f_C(X) = 0$ when B and $\alpha(B)$ are disjoint from the intent and extent of C, respectively. $f_C(B) = 1$ when $B = \text{Intent}(C)$ and therefore, $\alpha(B) = \text{Extent}(C)$. In general, the closer the value of $f_C(B)$ to 1, the greater the similarity between B and the intent of C. Conversely, the closer the value of $f_C(B)$ to 0, the less the similarity between B and the intent of C.

To approximate a set of features B, we find a formal concept C such that the value of $f_C(B)$ is the closest to 1. If we find more than one concept satisfying the approximation criteria, we choose only one such concept. In this case, we say

---

[1] One can, as well, think of $f_C(B)$ as a similarity measure between B and C.

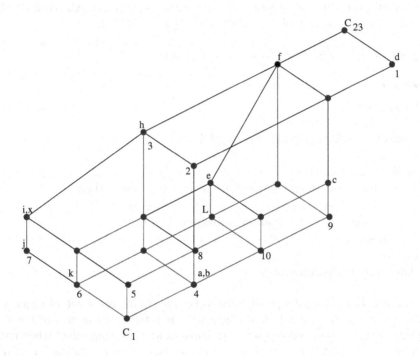

Fig. 1. Concept lattice for the context given in Table 1 with reduced labeling

that these concepts *equally approximate* B. In some applications, for example medical diagnosis, we may need to present the user with all the concepts that equally approximate B.

The pseudo-code for the algorithm for approximating a set of features is given in Algorithm 1. The input to this algorithm is the set of all formal concepts on a given context (G, M, I), which we denote by L, and a set of features $B \subseteq M$. [2] L_i denotes the ith concept in L. The output is a formal concept C that approximates B and the value of $f_C(B)$. The idea of the algorithm is similar to that of finding a maximal element in a set. Finding the value of $f_C(B)$ requires evaluating $\alpha(B)$ which requires time equals to $|B||G|$ [2]. [3] The value of $\alpha(B)$ is assigned to the variable Obj outside the do loop to improve the efficiency of Algorithm 1. The running time complexity of Algorithm 1 is $O(|L| + |B||G|)$.

Algorithm 1. *Approximate a set B of features*

$$C \leftarrow L_1$$
$$\text{Obj} \leftarrow \alpha(B)$$
// Assign $f_C(B)$ to maxvalue
maxvalue $\leftarrow$ Evaluate-Membership(Obj,B,C)
$$n \leftarrow |L|$$
for $(i \leftarrow 2; i \leq n; i++)$
    if (Evaluate-Membership(Obj,B,L_i) > maxvalue) then
        $C \leftarrow L_i$
        maxvalue $\leftarrow$ Evaluate-Membership(Obj,B,C)
    end if
end for
Answer $\leftarrow$ C and maxvalue

The function *Evaluate-Membership* takes as arguments a set of objects A, a set of features B, and a formal concept C. It returns the degree of membership or similarity between the set of features B and the concept C when called with arguments $\alpha(B)$, B and C as is done in Algorithm 1. Likewise, *Evaluate-Membership* returns the degree of similarity between the set of objects A and the concept C when called with arguments A, $\beta(A)$ and C. It also returns the degree of similarity between the pair (A, B) and the concept C when called with arguments A, B and C.

Algorithm 2. *Evaluate-Membership(A,B,C)*

$$Return \quad \frac{\dfrac{|A \cap Extent_{(C)}|}{|A \cup Extent_{(C)}|} + \dfrac{|B \cap Intent_{(C)}|}{|B \cup Intent_{(C)}|}}{2}$$

---

[2] The most efficient algorithm for finding all formal concepts of a context is called the Next algorithm [3]. This algorithm is also described in [2].
[3] $|X|$ is the cardinality of the set X which is the number of elements of X.

The code for the function *Evaluate-Membership* requires the evaluation of set intersections and set unions which can be implemented very efficiently, in constant time, using bit vectors and bit operations.

4   Approximating a Pair of a Set of Objects and a Set of Features

Suppose that we are given a set of objects A and a set of features B. We call the process of finding a concepts C such that the extent of C is as similar to A as possible and intent of C is as similar to B as possible *concept approximation*. We also say that the concept C approximates the pair (A, B). Define a membership function $f_C(A, B)$ that indicates how well the formal concept C approximates the pair (A, B) as follows

$$f_C(A, B) = \frac{\frac{|A \cap \text{Extent}(C)|}{|A \cup \text{Extent}(C)|} + \frac{|B \cap \text{Intent}(C)|}{|B \cup \text{Intent}(C)|}}{2}.$$

The expression $|A \cap \text{Extent}(C)|/|A \cup \text{Extent}(C)|$ indicates how similar A is to Extent(C) and the expression $|B \cap \text{Intent}(C)|/|B \cup \text{Intent}(C)|$ indicates how similar B is to Intent(C). It is also easy to see that the range of $f_C(A, B)$ is the interval $[0, 1]$. $f_C(A, B) = 0$ when C and (A, B) do not have any element in common and $f_C(A, B) = 1$ when (A, B) is equal to the formal concept C. The closer the value of $f_C(A, B)$ to 1, the greater the similarity between the pair (A, B) and the formal concept C. Conversely, the closer the value of $f_C(A, B)$ to 0, the less the similarity between (A, B) and C.

Algorithm 3 gives the pseudo-code for approximating a pair (A, B). The input to Algorithm 3 is the set L of all formal concepts on a given context (G, M, I), a set of objects A and a set of features B. The output is a concept C that approximates (A, B) and the value of $f_C(A, B)$ which is used as an indication of how well C approximates (A, B). The idea of Algorithm 3 is similar to that of Algorithm 1. The running time complexity of Algorithm 3 is $O(|L|)$.

Algorithm 3. *Approximate a pair (A, B) of a set of objects and a set of features*

```
C ← L₁
// Assign f_C(A, B) to maxvalue
maxvalue ← Evaluate − Membership(A, B, C)
n ← |L|
for (i ← 2; i ≤ n; i++)
    if (Evaluate − Membership(A, B, Lᵢ) > maxvalue ) then
        C ← Lᵢ
        maxvalue ← Evaluate − Membership(A, B, C)
    end if
end for
Answer ← C and maxvalue
```

5   Numerical Example

In this section we give a numerical example of the approximation ideas discussed in Sections 3 and 4.

Consider the context (G, M, I) given in Table 1 which gives information about 10 objects and 12 features that the objects can have. This context has 23 formal concepts which were generated using the algorithm Next described in [2]. Table 2 gives details about executing Algorithm 1 on the set of features $B = \{f,h,i\}$ and about executing Algorithm 3 on the pair $(A_1, B_1) = (\{4, 6, 9\}, \{e,f,h\})$ which is a non-definable concept because $\alpha(B_1) \neq A_1$. The first column in each row is a label of the concept under consideration. The second and third columns give the extent and intent of the concept. The fourth column contains the value of $f_C(\{f,h,i\})$ which is the degree of similarity between $\{f,h,i\}$ and C. Finally, the fifth column gives the value of $f_C(A_1, B_1)$ the degree of similarity between the pair (A, B) and C.

Considering the values of $f_C(\{f,h,i\})$ in the fourth column, Algorithm 1 returns $(\{5, 6, 7\}, \{f,h,i,x\})$ as the formal concept approximating the set of features $\{f,h,i\}$ with similarity value of 0.8750. Similary, considering the values in the fifth column, Algorithm 3 returns the formal concept $(\{4, 6\}, \{e, f, h, l\})$ as a result of approximating the non-definable concept $(\{4, 6, 9\}, \{e,f,h\})$ with similarity value of 0.7083.

6   Conclusion

This paper presents a new approach for approximating concepts in the framework of formal concept analysis. This approach is based on a fuzzy set theoretic background and is different from the previous approaches which use ideas from rough set theory. We gave algorithms for approximating a set B of features and for approximating a pair (A, B) of a set of objects and a set of features. The time complexity for the algorithm for approximating a set of features is $O(|L| + |B||G|)$ where L is the set of all formal concepts and G is the set of all objects. The time complexity for our algorithm for approximating a pair (A, B) is $O(|L|)$.

References

1. C. Carpineto, G. and Romano, A lattice Conceptual Clustering System and its Application to Browsing Retrieval, Machine Learning, 10, 95-122, 1996.
2. B. Ganter and R. Wille, Formal Concept Analysis: Mathematical Foundations, (Springer, Berlin, 1999).
3. B. Ganter, Two Basic Algorithms in concept analysis. FB4-Preprint No. 831, TH Darmstadt, 1984.
4. R. Godin, and R. Missaoui, An Incremental Concept Formation for Learning from Databases, Theoretical Computer Science, 133, 387-419, 1994.
5. T. B. Ho, An Approach to Concept Formation Based on Formal Concept Analysis, IEICE Trans. Information and Systems, E78-D, 553-559, 1995.

Table 2. Results of Running Algorithm 1 on {f,h,i} and Algorithm 3 on ({4,6,9},{e,f,h}).

Concept	Extent	Intent	$f_C(\{f,h,i\})$	$f_C(A_1, B_1)$
C_1	{}	{a,b,c,d,e,f,h,i,j,k,l,x}	0.1250	0.1250
C_2	{6}	{e,f,h,i,j,k,l,x}	0.3542	0.3542
C_3	{6,7}	{f,h,i,j,x}	0.6333	0.2917
C_4	{5}	{c,d,e,f,h,i,x}	0.3810	0.2143
C_5	{5,6}	{e,f,h,i,x}	0.6333	0.4250
C_6	{5,6,7}	{f,h,i,x}	0.8750	0.3000
C_7	{4}	{a,b,c,d,e,f,h,l}	0.1111	0.3542
C_8	{4,10}	{c,d,e,f,l}	0.0714	0.2917
C_9	{4,9,10}	{c,d,f,l}	0.0833	0.3333
C_{10}	{4,6}	{e,f,h,l}	0.3250	0.7083
C_{11}	{4,6,10}	{e,f,l}	0.2000	0.5000
C_{12}	{4,6,9,10}	{f,l}	0.2083	0.5000
C_{13}	{4,5,8}	{c,d,e,f,h}	0.2667	0.4000
C_{14}	{4,5,8,10}	{c,d,e,f}	0.1667	0.2833
C_{15}	{4,5,8,9,10}	{c,d,f}	0.1714	0.2667
C_{16}	{4,5,6,8}	{e,f,h}	0.4500	0.7000
C_{17}	{4,5,6,8,10}	{e,f}	0.2917	0.5000
C_{18}	{2,4,5,8}	{d,f,h}	0.3333	0.3333
C_{19}	{2,4,5,8,9,10}	{d,f}	0.1875	0.2679
C_{20}	{2,3,4,5,6,7,8}	{f,h}	0.5476	0.4583
C_{21}	{2,3,4,5,6,7,8,9,10}	{f}	0.3333	0.3333
C_{22}	{1,2,4,5,8,9,10}	{d}	0.0556	0.1250
C_{23}	{1,2,3,4,5,6,7,8,9,10}	{}	0.1500	0.1500

6. H. Kangassalo, On the concept of concept for conceptual modeling and concept deduction, in Information Modeling and Knowledge Bases III, Ohsuga et al. (eds.), IOS Press, 17-58, 1992.

7. R. Kent, Rough Concept Analysis: A Synthesis of Rough Set and Formal Concept Analysis, Fundamenta Informaticae 27 (1996) 169-181.

8. R. Kent, Rough Concept Analysis, In Rough Sets, Fuzzy Sets and Knowledge Discovery, (Springer-Verlag, 1994) 245-253.

9. J. Saquer and J. Deogun, Formal Rough Concept Analysis, in: Proc. RSFDGrC'99, Lecture Notes in Computer Science, vol. 1711, (Springer-Verlag, 1999) 91-99.

10. J. Saquer and J. Deogun, Concept Approximations for Formal Concept Analysis, to appear in Proc. ICCS'00, Darmstadt, Germany, August 14-18, 2000.

11. R. Wille 1982, Restructuring Lattice Theory: an Approach Based on Hierarchies of Concepts, in: Ivan Rivali, ed., Ordered sets, (Reidel, Dordecht-Boston, 1982) 445-470.

12. L. A. Zadeh, Fuzzy Sets, Information and Control 8: pp. 338-353, 1965.

On Axiomatic Characterizations of Fuzzy Approximation Operators
I. The Fuzzy Rough Set Based Case⋆

Helmut Thiele

Department of Computer Science I,
University of Dortmund,
D–44221 Dortmund, Germany
Phone: +49 231 755 6152
Fax: +49 231 755 6555
E-mail: `thiele@ls1.cs.uni-dortmund.de`
WWW: `http://ls1-www.informatik.uni-dortmund.de`

Abstract. In a previous paper we have developed an axiomatic characterization of approximation operators defined by the classical diamond and box operator of modal logic. The paper presented contains the analogous results of approximation operators which are defined by using the concepts of fuzzy rough sets.

Keywords. Rough sets, fuzzy rough sets, approximation operators.

1 Introduction and Fundamental Definitions

In the paper presented we shall use the well-known concepts of the "classical" crisp and fuzzy set theory, respectively.

For definiteness we recall the following definitions.

Let U be an arbitrary non-empty crisp set. The power set of U and the twofold Cartesian product of U by itself are denoted by $\mathbb{P}U$ and $U \times U$, respectively. Binary relations on U are sets of the form $R \subseteq U \times U$. We denote $[x, y] \in R$ also by xRy.

For an arbitrary binary relation R on U we define

Definition 1. *1. R is said to be reflexive on U $=_{\text{def}} \forall x(x \in U \to xRx)$.*

2. R is said to be transitive on U $=_{\text{def}} \forall x \forall y \forall z(xRy \wedge yRz \to xRz)$.

3. R is said to be symmetric on U $=_{\text{def}} \forall x \forall y(xRy \to yRx)$.

4. R is said to be an equivalence relation on U $=_{\text{def}} R$ is reflexive, transitive, and symmetric on U.

For mappings $\Phi : \mathbb{P}U \to \mathbb{P}U$ we define

Definition 2. *1. Φ is said to be embedding on $\mathbb{P}U$*
$=_{\text{def}} \forall X(X \subseteq U \to X \subseteq \Phi(X))$.

⋆ This research was supported by the Deutsche Forschungsgemeinschaft as part of the Collaborative Research Center "Computational Intelligence(531)"

W. Ziarko and Y. Yao (Eds.): RSCTC 2000, LNAI 2005, pp. 277–285, 2001.

2. Φ *is said to be closed on* $\mathbb{P}U$
   $=_{\text{def}} \forall X(X \subseteq U \to \Phi(\Phi(X)) \subseteq \Phi(X))$.
3. Φ *is said to be monotone on* $\mathbb{P}U$
   $=_{\text{def}} \forall X \forall Y(X \subseteq Y \subseteq U \to \Phi(X) \subseteq \Phi(Y))$.
4. Φ *is said to be symmetric on* U
   $=_{\text{def}} \forall x \forall y(x, y \in U \land y \in \Phi(\{x\}) \to x \in \Phi(\{y\}))$.
5. Φ *is said to be a closure and a symmetric closure operator on* $\mathbb{P}U$
   $=_{\text{def}} \Phi$ fulfils the items 1, 2, 3 and 1, 2, 3, 4, respectively.
6. Φ *is said to be strongly compact on* $\mathbb{P}U$
   $=_{\text{def}} \forall X \forall y(X \subseteq U \land y \in \Phi(X) \to \exists x_0(x_0 \in X \land y \in \Phi(\{x_0\})))$.

Fuzzy sets F on U are mappings of the form $F : U \to \langle 0, 1 \rangle$ where $\langle 0, 1 \rangle$ denotes the set of all real numbers r with $0 \leq r \leq 1$. We define $\mathbb{F}\mathbb{P}U =_{\text{def}} \{F | F : U \to \langle 0, 1 \rangle\}$ and call $\mathbb{F}\mathbb{P}U$ the fuzzy power set of U. For arbitrary fuzzy sets F and G on U we put $F \sqsubseteq G =_{\text{def}} \forall x(x \in U \to F(x) \leq G(x))$.

In section 3 of this paper we shall consider binary fuzzy relations $S : U \times U \to \langle 0, 1 \rangle$ and (crisp-fuzzy) mappings $\Psi : \mathbb{P}U \to \mathbb{F}\mathbb{P}U$. To formulate the results of section 3 we have to modify definitions 1 and 2 as follows. For defining some kinds of fuzzy transitivity we fix two arbitrary real functions $\tau, \pi : \langle 0, 1 \rangle \times \langle 0, 1 \rangle \to \langle 0, 1 \rangle$.

Definition 3. *1. S is said to be fuzzy reflexive on* U
   $=_{\text{def}} \forall x(x \in U \to S(x, x) = 1)$.
2. *S is said to be fuzzy standard transitive on* U
   $=_{\text{def}} \forall x \forall y \forall z(x, y, z \in U \to \min(S(x, y), S(y, z)) \leq S(x, z))$.
3. *S is said to be fuzzy conjunction-like τ-transitive on* U
   $=_{\text{def}} \forall x \forall y \forall z(x, y, z \in U \to \tau(S(x, y), S(y, z)) \leq S(x, z))$.
4. *S is said to be fuzzy implication-like π-transitive on* U
   $=_{\text{def}} \forall x \forall y \forall z(x, y, z \in U \to S(x, y) \leq \pi(S(y, z), S(x, z)))$.
5. *S is said to be fuzzy symmetric on* U
   $=_{\text{def}} \forall x \forall y(x, y \in U \to S(x, y) = S(y, x))$.
6. *S is said to be a fuzzy standard equivalence relation on* U
   $=_{\text{def}} S$ fulfils the items 1, 2, and 5.

Remark 1. The items 3 and 4 lead to further kinds of fuzzy equivalence relations. These cases will be considered in a following paper, in particular, with respect to the investigations in section 3.

Now, we are going to modify definition 2. Assume $\Psi : \mathbb{P}U \to \mathbb{F}\mathbb{P}U$. For expressing the closedness of Ψ we need the iteration $\Psi(\Psi(X))$ where $X \subseteq U$ and $z \in U$.

Definition 4. $\Psi(\Psi(X))(z) =_{\text{def}} \sup\{\min(\Psi(X)(y), \Psi(\{y\})(z)) | y \in U\}$.

Definition 5. *1. Ψ is said to be fuzzy embedding on* $\mathbb{P}U$
   $=_{\text{def}} \forall X \forall y(X \subseteq U \land y \in X \to \Psi(X)(y) = 1)$.
2. *Ψ is said to be fuzzy closed on* $\mathbb{P}U$
   $=_{\text{def}} \forall X(X \subseteq U \to \Psi(\Psi(X)) \sqsubseteq \Psi(X))$.

3. Ψ is said to be *fuzzy monotone on* $[\mathbb{P}U, \mathbb{F}\mathbb{P}U]$
   $=_{\text{def}} \forall X \forall Y (X \subseteq Y \subseteq U \to \Psi(X) \sqsubseteq \Psi(Y))$.
4. Ψ is said to be *fuzzy symmetric on* U
   $=_{\text{def}} \forall x \forall y (x, y \in U \to \Psi(\{x\})(y) = \Psi(\{y\})(x))$.
5. Ψ is said to be a *fuzzy closure* and a *fuzzy symmetric closure operator
   on* $[\mathbb{P}U, \mathbb{F}\mathbb{P}U] =_{\text{def}} \Psi$ fulfils the items 1, 2, 3 and 1, 2, 3, 4, respectively.
6. Ψ is said to be *fuzzy strongly compact on* $[\mathbb{P}U, \mathbb{F}\mathbb{P}U]$
   $=_{\text{def}} \forall X \forall y (X \subseteq U \wedge y \in U \to \exists x_0 (x_0 \in X \wedge \Psi(X)(y) \leq \Psi(\{x_0\})(y)))$.

2 An Axiomatic Characterization of Rough Set Based Approximation Operators

For a better understanding the definitions and results of section 3 we recall the fundamental definitions and main results of [13] in the present section where we also correct some slight mistakes.

In the paper [13] we started our investigations by recalling the definitions of the upper and the lower rough approximation $\langle R \rangle X$ and $[R]X$, respectively, where R is an equivalence relation on U and $X \subseteq U$.

Because of lacking space we considered only the construct $\langle R \rangle X$. By using concepts of modal logic the definition of $\langle R \rangle X$ inspires to introduce the following approximation operator $\text{OPER}(R)$ where R is an arbitrary binary relation on U and $X \subseteq U$.

Definition 6. $\text{OPER}(R)(X) =_{\text{def}} \{y | y \in U \wedge \exists x (x \in X \wedge [x, y] \in R)\}$.

Obviously, we have $\text{OPER}(R) : \mathbb{P}U \to \mathbb{P}U$. For the following investigations it will be very important to describe the images $\text{OPER}(R)$ without using the mapping OPER. To this end we generate a binary relation $\text{REL}(\Phi)$ where $\Phi : \mathbb{P}U \to \mathbb{P}U$.

Definition 7. $\text{REL}(\Phi) =_{\text{def}} \{[x, y] | x, y \in U \wedge y \in \Phi(\{x\})\}$.

Then without any assumption to $R \subseteq U \times U$ we obtain the following

Theorem 1. *For every* $R \subseteq U \times U$, $\text{REL}(\text{OPER}(R)) = R$.

This theorem means that R can be uniquely reconstructed from $\text{OPER}(R)$. In other words, OPER is an injection from $\{R | R \subseteq U \times U\}$ into $\{\Phi | \Phi : \mathbb{P}U \to \mathbb{P}U\}$.

Now, we are going to describe the set $\{\text{OPER}(R) | R \subseteq U \times U\}$.

Lemma 1. *For every* $R \subseteq U \times U$, $\text{OPER}(R)$ *is monotone and strongly compact on* $\mathbb{P}U$.

For formulating the following two lemmas and the following theorem we start our constructions with a mapping $\Phi : \mathbb{P}U \to \mathbb{P}U$, in contrast to theorem 1.

Lemma 2. *If* Φ *is monotone on* $\mathbb{P}U$ *then*
      for every $X \subseteq U$, $\text{OPER}(\text{REL}(\Phi))(X) \subseteq \Phi(X)$.

Lemma 3. *If Φ is strongly compact on $\mathbb{P}U$ then*
$$for\ every\ X \subseteq U,\ \Phi(X) \subseteq \mathrm{OPER}(\mathrm{REL}(\Phi))(X).$$

Theorem 1, lemma 2, and lemma 3 imply the following final theorem where $\Phi : \mathbb{P}U \to \mathbb{P}U$.

Theorem 2.

1. *If Φ is monotone and strongly compact on $\mathbb{P}U$ then for every $X \subseteq U$, $\mathrm{OPER}(\mathrm{REL}(\Phi))(X) = \Phi(X)$.*
2. *The mapping OPER is a bijection from the set of all binary relations on U onto the set of all monotone and strongly compact operators on $\mathbb{P}U$.*
3. *The mapping REL is the inversion of the mapping OPER and vice versa.*

Furthermore, in [13] we investigated how properties of binary relations on U are translated into properties of operators on $\mathbb{P}U$ by the mapping OPER and vice versa by the mapping REL. Here we present the following slightly corrected and complemented results, respectively.

Theorem 3. *1. R is reflexive on U iff $\mathrm{OPER}(R)$ is embedding on $\mathbb{P}U$.*
2. If Φ is embedding on $\mathbb{P}U$ then $\mathrm{REL}(\Phi)$ is reflexive on U.
3. If $\mathrm{REL}(\Phi)$ is reflexive on U and Φ is monotone and strongly compact on $\mathbb{P}U$ then Φ is embedding on $\mathbb{P}U$.

Theorem 4. *1. R is transitive on U iff $\mathrm{OPER}(R)$ is closed on $\mathbb{P}U$.*
2. If Φ is closed and monotone on $\mathbb{P}U$ then $\mathrm{REL}(\Phi)$ is transitive on U.
3. If $\mathrm{REL}(\Phi)$ is transitive on U and Φ is monotone and strongly compact on $\mathbb{P}U$ then Φ is closed on $\mathbb{P}U$.

Theorem 5. *1. R is reflexive and transitive on U iff $\mathrm{OPER}(R)$ is a closure operator on $\mathbb{P}U$.*
2. If Φ is a closure operator on $\mathbb{P}U$ then $\mathrm{REL}(\Phi)$ is reflexive and transitive on U.
3. If $\mathrm{REL}(\Phi)$ is reflexive and transitive on U and Φ is monotone and strongly compact on $\mathbb{P}U$ then Φ is a closure operator on $\mathbb{P}U$.

Theorem 6. *1. R is an equivalence relation on U iff $\mathrm{OPER}(R)$ is a symmetric closure operator on $\mathbb{P}U$.*
2. If Φ is a symmetric closure operator on $\mathbb{P}U$ then $\mathrm{REL}(\Phi)$ is an equivalence relation on U.
3. If $\mathrm{REL}(\Phi)$ is an equivalence relation on U and Φ is monotone and strongly compact on $\mathbb{P}U$ then Φ is a symmetric closure operator on $\mathbb{P}U$.

3   An Axiomatic Characterization of Fuzzy Rough Set Based Approximation Operators

Assume
$$S : U \times U \to \langle 0, 1 \rangle \text{ and } X \subseteq U.$$

The upper and lower S-fuzzy approximation $\langle S \rangle X$ and $[S]X$, respectively, of the crisp set X are defined as follows where $y \in U$ (see also [1–9]).

Definition 8.   *1.* $((\langle S \rangle X)(y) =_{\text{def}} \sup\{\min(S(x, y), \gamma_X(x))| x \in U\}$,
2. $([S]X)(y) =_{\text{def}} \inf\{\max(1 - S(x, y), \gamma_X(x))| x \in U\}$.

In literature sometimes the constructs $\langle S \rangle X$ and $[S]X$ are called Fuzzy Rough Sets, in particular, if S is a fuzzy standard equivalence relation on U.

Because the functions min and max are mutually dual we obtain for every $S : U \times U \to \langle 0, 1 \rangle$, $X \subseteq U$, and $y \in U$,

$$(\langle S \rangle X)(y) = 1 - ([S]\overline{X})(y)$$

and

$$([S]X)(y) = 1 - (\langle S \rangle \overline{X})(y)$$

i.e. $\langle S \rangle X$ and $[S]X$ are also mutually dual.

Assume

$$\tau, \pi : \langle 0, 1 \rangle \times \langle 0, 1 \rangle \to \langle 0, 1 \rangle .$$

Generalizing ideas presented in [10] we define

Definition 9.   *1.* $((\langle S, \tau \rangle X)(y) =_{\text{def}} \sup\{\tau(S(x, y), \gamma_X(x))| x \in U\}$,
2. $([S, \pi]X)(y) =_{\text{def}} \inf\{\pi(S(x, y), \gamma_X(x))| x \in U\}$.

In the papers [10, 11] we have stated some properties of $\langle S, \tau \rangle X$ and $[S, \pi]X)$, in particular, if τ is a t-norm and π is a certain kind of implication. Furthermore, we underlined that $\langle S, \tau \rangle X$ and $[S, \pi]X)$ are mutually dual if τ and π fulfil the equation

$$\forall r \forall s (r, s \in \langle 0, 1 \rangle \to \pi(r, s) = 1 - \tau(r, 1 - s)).$$

Because of lacking space, in the following we shall consider only the construct $\langle S \rangle X$ and the fuzzy approximation operator which can be generated by $\langle S \rangle X$. The remaining cases described above will be investigated in a following paper.

Definition 8 inspires to introduce the following "Fuzzy-Rough-Like" approximation operator FROPER as a mapping from $\mathbb{P}U$ into $\mathbb{F}\mathbb{P}U$.

Assume $S : U \times U \to \langle 0, 1 \rangle$, $X \subseteq U$, and $y \in U$.

Definition 10. $\text{FROPER}(S)(X)(y) =_{\text{def}} (\langle S \rangle X)(y)$.

The following lemma simplifies the definition of FROPER.

Lemma 4. $\text{FROPER}(S)(X)(y) = \sup\{S(x, y)| x \in X\}$.

Proof. Applying the definition of the characteristic function γ_X of the crisp set $X \subseteq U$ and the equation $\min(r, 0) = 0$ for every $r \in \langle 0, 1 \rangle$.   □

Obviously, we have $\text{FROPER}(S) : \mathbb{P}U \to \mathbb{F}\mathbb{P}U$. As in section 2 we ask the question how to describe the images $\text{FROPER}(S)$ without using the mapping OPER. To this end by an arbitrary mapping $\Psi : \mathbb{P}U \to \mathbb{F}\mathbb{P}U$ we generate the binary fuzzy relation $\text{FREL}(\Psi)$ on U as follows where $x, y \in U$.

Definition 11. $\text{FREL}(\Psi)(x, y) =_{\text{def}} \Psi(\{x\})(y)$.

Then without any assumption we get the following theorem.

Theorem 7. *1. For every $S : U \times U \to \langle 0, 1 \rangle$, FREL(FROPER($S$)) = S.*

2. FROPER is an injection from the set of all binary fuzzy relations on U into the set $\{\Psi | \Psi : \mathbb{P}U \to \mathbb{F}\mathbb{P}U\}$.

3. FREL is the inversion of FROPER with respect to $\{\text{FROPER}(S) | S : U \times U \to \langle 0, 1 \rangle\}$ and vice versa.

This theorem means that S can be uniquely reconstructed from FROPER(S). Now, we are going to describe the set $\{\text{FROPER}(S) | S : U \times U \to \langle 0, 1 \rangle\}$.

Lemma 5. *For every $S : U \times U \to \langle 0, 1 \rangle$, the mapping FROPER($S$) is fuzzy monotone on $[\mathbb{P}U, \mathbb{F}\mathbb{P}U]$.*

Proof. We have to show

(1) $\forall X \forall Y (X \subseteq Y \subseteq U \to \text{FROPER}(S)(X) \sqsubseteq \text{FROPER}(S)(Y))$.

From $X \subseteq Y \subseteq U$ we get

$$\{S(x,y) | x \in X\} \subseteq \{S(x,y) | x \in Y\},$$

hence (1) holds.   □

Now, we are going to prove that FROPER(S) is fuzzy strongly compact on $[\mathbb{P}U, \mathbb{F}\mathbb{P}U]$. To show this we additionally need the concept of *submodality* which we have already introduced in [12], but here in the following slightly modified form. Assume $S : U \times U \to \langle 0, 1 \rangle$.

Definition 12. *S is said to be submodal with respect to its first argument* $=_{\text{def}} \forall X \forall y (X \subseteq U \wedge y \in U \to \exists x_0 (x_0 \in X \wedge \sup\{S(x,y) | x \in X\} = S(x_0, y)))$.

Lemma 6. *If S is submodal with respect to its first argument then FROPER(S) is fuzzy strongly compact on $[\mathbb{P}U, \mathbb{F}\mathbb{P}U]$.*

Proof. Assume

(2) $X \subseteq U$ and $y \in U$.

We have to prove

(3) $\exists x_0 (x_0 \in X \wedge \text{FROPER}(S)(X)(y) \leq \text{FROPER}(S)(\{x_0\})(y))$.

Because S is submodal with respect to its first argument, we have

(4) $\exists x_0 (x_0 \in X \wedge \sup\{S(x,y) | x \in X\} = S(x_0, y))$,

hence by definition of FROPER

(5) $\text{FROPER}(X)(y) = S(x_0, y)$.

Furthermore, by definition of FROPER we get

(6) $\text{FROPER}(\{x_0\})(y) = \sup\{S(x,y) | x \in \{x_0\}\} = S(x_0, y)$,

hence (5) and (6) imply (3).   □

By the following lemmas 7 and 8 we solve the problem how the set $\{\text{FROPER}(S)|S : U \times U \rightarrow \langle 0, 1 \rangle\}$ can be characterized without using the mapping FROPER.

Lemma 7. *If Ψ is a fuzzy monotone mapping on $[\mathbb{P}U, \mathbb{F}\mathbb{P}U]$ then for every $X \subseteq U$, $\text{FROPER}(\text{FREL}(\Psi))(X) \sqsubseteq \Psi(X)$.*

Proof. We have to prove that for every $y \in U$,

$$(7) \qquad \text{FROPER}(\text{FREL}(\Psi))(X)(y) \leq \Psi(X)(y).$$

By lemma 4 we have

$$(8) \qquad \text{FROPER}(\text{FREL}(\Psi))(X)(y) = \sup\{\text{FREL}(\Psi)(x, y)|x \in U\},$$

hence by definition of $\text{FREL}(\Psi)$

$$(9) \qquad \text{FROPER}(\text{FREL}(\Psi))(X)(y) = \sup\{\Psi(\{x\})(y)|x \in X\}.$$

Obviously, it is sufficient to show

$$(10) \qquad \Psi(\{x\})(y) \leq \Psi(X)(y) \text{ if } x \in X.$$

But (10) holds because $\{x\} \subseteq X$ and Ψ is a fuzzy monotone mapping on $[\mathbb{P}U, \mathbb{F}\mathbb{P}U]$, hence lemma 7 holds. □

Lemma 8. *If Ψ is fuzzy strongly compact on $[\mathbb{P}U, \mathbb{F}\mathbb{P}U]$ then for every $X \subseteq U$, $\Psi(X) \sqsubseteq \text{FROPER}(\text{FREL}(\Psi)(X)$.*

Proof. We have to prove that for every $y \in U$,

$$(11) \qquad \Psi(X)(y) \leq \text{FROPER}(\text{FREL}(\Psi))(X)(y).$$

Because Ψ is fuzzy strongly compact on $[\mathbb{P}U, \mathbb{F}\mathbb{P}U]$ we get

$$(12) \qquad \exists x_0(x_0 \in X \wedge \Psi(X)(y) \leq \Psi(\{x_0\})(y)).$$

To prove (11) it is sufficient to show

$$(13) \qquad \Psi(\{x_0\})(y) \leq \text{FROPER}(\text{FREL}(\Psi))(\{x_0\})(y)$$

because the mapping $\text{FROPER}(\text{FREL}(\Psi))$ is fuzzy monotone on $[\mathbb{P}U, \mathbb{F}\mathbb{P}U]$ (see lemma 5).

Furthermore, by definition of FROPER we have

$$(14) \qquad \begin{aligned} \text{FROPER}(\text{FREL}(\Psi))(\{x_0\})(y) \\ = \sup\{\text{FREL}(\Psi)(x, y)|x \in \{x_0\}\} \\ = \text{FREL}(\Psi)(x_0, y), \end{aligned}$$

hence by definition of FREL,

$$(15) \qquad \text{FREL}(\Psi)(x_0, y) = \Psi(\{x_0\})(y),$$

hence (13) holds. □

Theorem 8. *1. If Ψ is fuzzy monotone and fuzzy strongly compact on $[\mathbb{P}U, \mathbb{F}\mathbb{P}U]$ then for every $X \subseteq U$, $\mathrm{FROPER}(\mathrm{FREL}(\Psi))(X) = \Psi(X)$.*

2. FROPER is a bijection from the set of all binary fuzzy relations on U onto the set of all mappings Ψ which are fuzzy monotone and fuzzy strongly compact on $[\mathbb{P}U, \mathbb{F}\mathbb{P}U]$.

3. FREL is the inversion of FROPER and vice versa.

Now, analogously to section 2 we investigate how special properties of binary fuzzy relations S on U will be translated into properties of operators $\Psi : \mathbb{P}U \to \mathbb{F}\mathbb{P}U$ by the mapping FROPER and vice versa by the mapping FREL.

Remark 2. Because of lacking space we omit the proofs of all the following theorems.

Theorem 9. *1. S is fuzzy reflexive on U iff $\mathrm{FROPER}(S)$ is fuzzy embedding on $\mathbb{P}U$.*

2. If Ψ is fuzzy embedding on $\mathbb{P}U$ then $\mathrm{FREL}(\Psi)$ is fuzzy reflexive on U.

3. If $\mathrm{FREL}(\Psi)$ is fuzzy reflexive on U and Ψ is fuzzy strongly compact on $[\mathbb{P}U, \mathbb{F}\mathbb{P}U]$ then Ψ is fuzzy embedding on $\mathbb{P}U$.

Theorem 10. *1. S is fuzzy standard transitive on U iff $\mathrm{FROPER}(S)$ is fuzzy closed on $[\mathbb{P}U, \mathbb{F}\mathbb{P}U]$.*

2. If Ψ is fuzzy closed and fuzzy monotone on $[\mathbb{P}U, \mathbb{F}\mathbb{P}U]$ then $\mathrm{FREL}(\Psi)$ is fuzzy standard transitive on U.

3. If $\mathrm{FREL}(\Psi)$ is fuzzy standard transitive on U and Ψ is fuzzy monotone and fuzzy strongly compact on $[\mathbb{P}U, \mathbb{F}\mathbb{P}U]$ then Ψ is fuzzy closed on $[\mathbb{P}U, \mathbb{F}\mathbb{P}U]$.

Theorem 11. *1. S is fuzzy reflexive and fuzzy standard transitive on U iff $\mathrm{FROPER}(S)$ is a fuzzy closure operator on $[\mathbb{P}U, \mathbb{F}\mathbb{P}U]$.*

2. If Ψ is a fuzzy closure operator on $[\mathbb{P}U, \mathbb{F}\mathbb{P}U]$ then $\mathrm{FREL}(\Psi)$ is fuzzy reflexive and fuzzy standard transitive on $[\mathbb{P}U, \mathbb{F}\mathbb{P}U]$.

3. If $\mathrm{FREL}(\Psi)$ is fuzzy reflexive and fuzzy standard transitive on $[\mathbb{P}U, \mathbb{F}\mathbb{P}U]$ then Ψ is a fuzzy closure operator on $[\mathbb{P}U, \mathbb{F}\mathbb{P}U]$.

Theorem 12. *1. S is a fuzzy standard equivalence relation on U iff $\mathrm{FROPER}(S)$ is a fuzzy symmetric closure operator on $[\mathbb{P}U, \mathbb{F}\mathbb{P}U]$.*

2. If Ψ is a fuzzy symmetric closure operator on $[\mathbb{P}U, \mathbb{F}\mathbb{P}U]$ then $\mathrm{FREL}(\Psi)$ is a fuzzy standard equivalence relation on U.

3. If $\mathrm{FREL}(\Psi)$ is a fuzzy standard equivalence relation on U and Ψ is fuzzy monotone and fuzzy strongly compact on $[\mathbb{P}U, \mathbb{F}\mathbb{P}U]$ then Ψ is a fuzzy symmetric fuzzy closure operator on $[\mathbb{P}U, \mathbb{F}\mathbb{P}U]$.

Acknowledgements. The author wishes to thank Claus-Peter Alberts for his help in preparing the paper.

References

1. Biswas, R.: On rough fuzzy sets. Bull. Polish Acad. Sci., Math, 42:351–355, 1994.
2. Biswas, R.: On rough sets and fuzzy rough sets. Bull. Polish Acad. Sci., Math, 42:345–349, 1994.
3. Dubois, D., Prade, H.: Rough fuzzy sets and fuzzy rough sets. International Journal of General Systems, 17:191–209, 1990.
4. Nakumura, A.: Fuzzy rough sets. In: Note on Multiple Valued Logic in Japan, Volume 9, 1–8, 1988.
5. Nanda, S., Majumdar, S.: Fuzzy rough sets. Fuzzy Sets and Systems, 45:157–160, 1992.
6. Novotný, M., Pawlak, Z.: Characterisation of rough top equalities and rough bottom equalities. Bull. Polish Acad. Sci., Math, 33:91–97, 1985
7. Novotný, M., Pawlak, Z.: On rough equalities. Bull. Polish Acad. Sci., Math, 33:99–104, 1985
8. Pawlak, Z.: Rough sets, basic notations. Technical report, ICS PAS, Report 486, 1981.
9. Pawlak, Z.: Rough sets. International Journal of Information and Computer Science, 11:341–356, 1982.
10. Thiele, H.: On the definitions of modal operators in fuzzy-logic. In: 23rd International Symposium on Multiple-Valued Logic, 62–67, Sacramento, California, May 24-27, 1993.
11. Thiele, H.: Fuzzy rough sets versus rough fuzzy sets – an interpretation and a comparative study using concepts of modal logics. In: 5th European Congress on Intelligent Techniques and Soft Computing (EUFIT'97), Volume1, 159–167, Aachen, Germany, September 8-11, 1997.
12. Thiele, H.: On Closure Operators in Fuzzy Deductive Systems and Fuzzy Algebras. Paper, 28th International Symposium on Multiple-Valued Logic, Proceedings, 304–309, Fukuoka, Japan, May 27-29, 1998. Extended Version: On Closure Operators in Fuzzy Algebras and Fuzzy Deductive Systems. Technical Report CI-34/98, 23 pages, University of Dortmund, Collaborative Research Center 531 (Computational Intelligence), April 1998.
13. Thiele, H.: On Axiomatic Characterisations of Crisp Approximation Operators. Paper, 7th International Conference on Fuzzy Theory and Technology, Proceedings 56–59, Atlantic City, New Jersey, USA, February 27 - March 3, 2000. To be published in "Information Sciences".

LEM3 Algorithm Generalization Based on Stochastic Approximation Space

M.C. Fernández-Baizán[1], C. Pérez-Llera[2], J. Feito-García[3], and A. Almeida[2]

[1]Univ. Politécnica Madrid, Dpt. Lenguajes, E-28660 Boadilla del Monte, Spain
cfbaizan@fi.upm.es
[2]Univ. Oviedo, Dpt. Informática, E-33271 Gijón, Spain
cpllera@etsiig.uniovi.es
[3]Univ. León, Dpt. Dirección y Economía de la Empresa, E- León, Spain
ddejfg@unileon.es

Abstract. This work introduces a generalization of the algorithm LEM3, an incremental learning system of production rules from examples, based on the Boolean Approximation Space introduced by Pawlak. The generalization is supported in the Stochastic Approximation Space introduced by Wong and Ziarko. In this paper, *stochastic limits* in the precision of the upper and lower approximations of a class are addressed. These allow the generation of *certain rules* with a certainty level β ($0.5 \leq \beta \leq 1$). Also the modifications in LEM3 necessary in order to handle examples with missing attribute values are introduced.

1 Introduction

The main characteristics of the LEM3 system [4] are:
- It is inductive, supervised, incremental and learns with full memory (using all previous examples for later learning).
- Learned knowledge is expressed in classification rules formed by conjunctions of attribute-value (*a-v*) pairs.
- The rule learning strategy follows a nonincremental learning program, LEM2, in generating minimal rules. LEM2 solves the inconsistency using the Rough Sets theory introduced by Pawlak [1,3]. Two **examples** are **inconsistent** if they have the same values in all the condition attributes, but have different classes. When a class, X, contains inconsistent examples, it means that X is not definable by the set of all condition attributes. The basic idea is to replace X by its upper and lower approximations generated by the set of all condition attributes. For sets of inconsistent examples, two types of rules can be generated: **possible rules** learned from the upper approximation of each class, and **certain rules** learned from the lower approximation of each class. One feature that differentiates LEM3 from LEM2, is the use of the *Global Data Structure* to capture knowledge learned from previous examples. In LEM3, this structure is proposed to support the incremental updating of upper and lower approximations, based on Boolean Approximation Space (**BoolAS**).

W. Ziarko and Y. Yao (Eds.): RSCTC 2000, LNAI 2005, pp. 286–290, 2001.
© Springer-Verlag Berlin Heidelberg 2001

The main contribution of this paper is the generalization, based on the Stochastic Approximation Space (**StocAS**), of the algorithm for incremental updating of upper and lower approximations proposed in LEM3. The **StocAS** [2] allows the use of stochastic limits in determining the upper and lower approximations. Thus classification rules with possible or certain ownership to a class, supported by a certainty level β can be generated ($0.5 <= \beta <= 1$). The use of the **StocAS** is proposed in [4] as future work.

2 Global Data Structure (GDS) in Boolean Approximation Space (BoolAS)

A **BoolAS**, A, is an ordered pair (U,R), where U is a non-empty set called **universe** and R is an equivalence relation on U called **indiscernibility relation**. For each subset X in U, X is characterized by a pair of sets, the **upper approximation** and **lower approximation** of X in A, which are defined as: $\overline{R}X = \{x \in U | [x]_R \cap X \neq \varnothing\}$ and $\underline{R}X = \{x \in U | [x]_R \subseteq X\}$ where $[x]_R$ denotes the equivalence class of R containing x.

The **GDS** stores information learned from previous training examples such as: consistency of a-v pairs, instances denoted by a-v pairs, a-v pairs relative to each class, and an instance-count for each class (which can be used to calculate conditional probabilities that determine the measurement of "goodness" of a-v pairs). The **GDS** consists of three tables: the Block Table (**BT**), the Relevant a-v Pair Table (**Ra-vT**) and the Lower and Upper Approximation Tables (**L&UAT**).

In **BT** examples represented by integers are stored. The table indices are a-v pairs and classes. A **block** is a set of examples indexed by an a-v pair. The **BT** is used to store consistencies of a-v pairs and examples presented to the learning algorithm. An **a-v pair** is said to be **consistent** if it is associated with only one class, otherwise, it is **inconsistent**. A block is consistent if all the examples in the block are of the same class.

The **Ra-vT** contains a-v pairs that are relevant to each class. An **a-v pair** is **relevant** to a class if it can describe at least one instance of the class. For each class, a **low-list** and an **up-list** of relevant a-v pairs are stored in this table. An a-v pair that is relevant and consistent with a class is stored in the low-list of the class. An a-v pair that is relevant and inconsistent with a class is stored in up-list of the class. The lists are ordered by the "goodness" of relevant a-v pairs. The **goodness** of an a-v pair, p, with respect to a class, X, is defined as the conditional probability of an instance, e, being in class X, given that e is in the block denoted by p. The lists are used to minimise the search space of rule-generating procedure. The **BT** and **Ra-vT** tables are updated with each example. For each a-v pair in a new example, there are three steps in updating the tables: 1) Checking the consistency of an a-v pair; 2) Inserting the example into the **BT** with the index (attribute,value,class) and marking the block as consistent or inconsistent, based on the result of the first step; 3) Inserting the a-v pair into the proper a-v pair list of the **Ra-vT**.

The **L&UAT** contain three sets of examples for each class: the set of examples that belong to the class, and two sets corresponding to the $\overline{R}X$ and $\underline{R}X$ of the class. Every time a new example is presented to the learning system, is added to the set corresponding to its class. Also the $\overline{RX}$ and $\underline{RX}$ of all the classes are updated with the information provided by the new example. If the new example is consistent, it is only necessary to update the $\overline{RX}$ and $\underline{RX}$ of the class associated with the new example. If however, the new example is inconsistent, it is necessary to update all the $\overline{RX}$ and/or $\underline{RX}$ of the classes which this inconsistency influences.

The following problems found in [4] are addressed in this study: updating the goodness of an a-v pair, and managing unknown values of attributes.

3 Generalization of LEM3 to the Stochastic Approximation Space (StocAS)

A **StocAS** is a triplet $A=(U,R,P)$, where U is a universe, R is an equivalence relation on U, and P is a probability measurement of subsets of U. The lower and upper approximation of a subset X in U are defined by using the concept of stochastic approximation with a certainty level β $(0.5 \leq \beta \leq 1)$. The **R-upper** and **R-lower** β-**approximations** of X in A are defined in [4] as: $\overline{RX}=\{x \in U | \beta > P(X|[x]) >= 0.5\}$ and $\underline{RX}=\{x \in U | P(X|[x]) >= \beta\}$ where $P(X|[x])$ is the conditional probability defined as $P(X \cap [x])/P([x])$.

In the $\overline{RX}$ definition we use the parameter α instead of the constant 0.5: $\overline{RX}=\{x \in U | P(X|[x]) >= \alpha\}$, and we establish $\alpha=(1-\beta)$ to maintain coherence with the meaning of the certainty level β. In [4], Chan also considers a superior limit β in the precision of the $\overline{RX}$: $\overline{RX}=\{x \in U | \beta > P(X | [x]) >= 0.5\}$. In this work, this superior limit is not used, so as to conserve coherence with the definition of $\overline{RX}$ according to the Rough Sets theory. The region that defines the previous expression is in fact, the *boundary region* [3]. The generation of possible rules from the boundary region examples, instead of from the $\overline{RX}$ examples, has the disadvantage that the rules obtained only classify the boundary region examples, so that generalization is lost. On the other hand it has the advantage that the rules generated do not contain rules that have already been generated from the $\underline{RX}$, thereby eliminating the later process of rules simplification.

In **StocAS**, an **example**, e_i, is **consistent** with a class, X, if the conditional probability $P(X|[e_i]) >= \beta$, where $[e_i]$ is the intersection of the blocks denoted by the a-v pairs of the example. The case of $\beta=1$, is the consistency condition in the **BoolAS**. In **BoolAS**, an example can be either consistent or inconsistent. In **StocAS** an example may be consistent with one class, yet inconsistent with another; or it could be inconsistent with several classes.

In **StocAS**, an **a-v pair** is **consistent** with a class, X, if the conditional probability $P(X|[(a,v)]) >= \beta$, where $[(a,v)]$ is the block denoted by the a-v pair. The **Ra-vT** updating procedure, in LEM3, was modified to include this new concept of consistency of an a-v pair with a class, and to insert the relevant a-v pairs of a class, X, in the low-list of that class. In **BoolAS**, an a-v pair is consistent if it is associated with only one class. In **StocAS**, an a-v pair can begin by being consistent with a class, X, stop being consistent with this class, become consistent with a different class. It is necessary to consider that in **BoolAS** the consistency concept of an a-v pair is used to simplify the updating process of the $\overline{R}X$ and $\underline{R}X$ of a class. In **BoolAS**, the fact that an example contains a consistent pair with its class is reason enough to affirm that the example is also consistent. In **StocAS** this is not so easy; it is necessary to verify that all the pairs of the example are consistent with the class. Finally, this modification of the algorithm was discarded as it did not contribute any major simplification to the approximations updating process, and added great complexity to the **Ra-vT** updating process. The new updating procedure of $\overline{R}X$ and $\underline{R}X$ is the following.

```
Procedure UPDATE_STOCHASTIC_APPROXIMATIONS
/* Input: an example e_i of class X */
/* Output: updated lower and upper β-approximation of class X
and of the rest of classes. */
begin
   Add e_i to examples set of class X;
   Y=Intersection of all blocks denoted by a-v pairs in e_i,
   including examples with an unknown pair;
   If P(X|Y) >= β
   then
      Add Y to lower β-approximation of X;
      Add Y to upper β-approximation of X;
      for each class X' ≠ X do
        If P(X'|Y) < β y P(X'|Y) >= α
        then
           Delete Y from lower β-approximation of X';
           Add Y to upper β-approximation of X';
        else if P(X'|Y) < α
              then
                 Delete Y from lower β-approximation of X';
                 Delete Y from upper β-approximation of X';
              endif
        endif
      endfor
   else     /* P(X|Y) < β */
      Add Y to upper β-approximation of X;
      For each class X' ≠ X do
        If P(X'|Y) < β and P(X'|Y) >= α
        then
           Delete Y from lower β-approximation of X';
           Add Y to upper β-approximation of X';
        else if P(X'|Y) < α
              then
                 Delete Y from lower β-approximation of X';
                 Delete Y from upper β-approximation of X';
              endif
        endif
      endfor
   endif
End;  /* Procedure UPDATE_STOCHASTIC_APPROXIMATIONS */
```

4 Conclusions

A generalization of algorithm LEM3 for learning production rules from consistent and inconsistent examples, including the treatment of attributes with unknown values has been introduced. This extension, based in the Stochastic Approximation Space introduced by Wong and Ziarko, is suggested in [4] as future work. The R-upper and R-lower β-approximations of X are defined in [4] as: $\overline{R}X=\{x\in U| \ \beta>P(X|[x])>=0.5\}$ and $\underline{R}X=\{x\in U|P(X|[x])>=\beta\}$. In the $\overline{R}X$ definition we use the parameter α instead of the constant 0.5, and $\alpha=(1-\beta)$ in order to maintain coherence with the certainty level β. In [4] is also considered a superior limit β in the precision of $\overline{R}X$. This work has been done without this superior limit β in order to conserve the coherence with the definition of $\overline{R}X$, according to the Rough Sets theory.

A value $\alpha>0$ means the elimination of the inconsistent examples of some upper approximations. In other words, the elimination of rules that have the least conditional probability. This can be considered as a vertical purification of the rules set. If α increases, the total number of rules diminishes. A value of $\alpha=0.5$ means the inclusion of the inconsistent examples only in the $\overline{R}X$ of the class to which at least half of the examples belong. If there is no class that fulfills this condition, the examples are not included in any $\overline{R}X$. When parameter β diminishes, the inconsistent examples whose consistency with a class is greater than β are considered in the $\underline{R}X$ of that class. In other words, **certain examples** with a **certainty level** β are considered. This makes the rules generated from $\underline{R}X$ more general because they include more examples. Thus, the rules are compacted, and generally the number of rules diminishes.

References

1. Pawlak, Z.: Information Systems, Theoretical Foundations. Inf. Syst. **6** 3 (1981) 205–218
2. Wong, S.K.M., Ziarko, W.: INFER-An Adaptative Decision Support System Based on the Probabilistic Approximate Classsification. In Proc. 6th Int. Workshop on Expert Systems and their Applications (1986) 713-7262
3. Pawlak, Z.: Theoretical Aspects of Reasoning about Data. Kluwer Academic Publishers. Dordrecht, Boston, London (1991)
4. Chan, C.C.: Incremental Learning of Production Rules from Examples under Uncertainty: A Rough Set Approach. Int. J. of Software Engineering and Knowledge Engineering **1** 4 (1991) 439-461
5. Grzymala-Busse, J.W.: A New Version of the Rule Induction System LERS. Fundamenta Informaticae 31 (1997) 27-39
6. Grzymala-Busse, J.W.: Knowledge acquisition under Uncertainty: A Rough Sets Approach. J. Intell. Robotic Syst. 1 (1998) 3-16

Using the Apriori Algorithm to Improve Rough Sets Results

María C. Fernández-Baizán[1], Ernestina Menasalvas Ruiz[1]
José M. Peña Sánchez[1], Juan Francisco Martínez Sarrías[1], Socorro Millán[2]
{cfbaizan, emenasalvas, jpena}@fi.upm.es, juanfran@pegaso.ls.fi.upm.es,
millan@borabora.edu.co, *

[1] Departamento de Lenguajes y Sistemas Informáticos e Ingeniería del Software,
Facultad de Informática, U.P.M., Campus de Montegancedo, Madrid
[2] Universidad del Valle, Cali. Colombia

Abstract. Ever since Data Mining first appeared, a considerable amount of algorithms, methods and techniques have been developed. As a result of research, most of these algorithms have proved to be more effective and efficient. For solving problems different algorithms are often compared. However, algorithms that use different approaches are not very often applied jointly to obtain better results. An approach based on the joining of a predictive model (rough sets) together with a link analysis model (the Apriori algorithm) is presented in this paper.

Keywords: Data Mining models joining, Rough Sets, Association rules.

1 Introduction

The Rough Set methodology provides a way to generate decision rules. Some condition values may be unnecesary in a decision rule. Thus it is always desirable to reduce the amount of information required to describe a concept. A reduced number of condition attributes results in a set of rules with higher support. On the other hand, this kind of rules are easier to understand. The concept of *reduct* is used when there is a need for reducing the number of attributes, but this is a computational expensive process.

One way to construct a simpler model computed from data, easier to understand and with more predictive power, is to create a set of simplified rules [11]. A simplified rule (also refered to as minimal rule or kernel rule) is one in which the number of conditions in its antecedent is minimal. Thus, when dealing with decision rules, some condition values can be unnecessary and can be dropped to generate a simplified rule preserving essential information. In [7] an approach to simplify decision tables is presented. Such an approach consists of three steps: 1) Computation of reducts of condition attributes; 2) Elimination of duplicate

---

* This work has been partially supported by UPM under project "Design of a Data Warehouse to be integrated with a Data Mining system"

W. Ziarko and Y. Yao (Eds.): RSCTC 2000, LNAI 2005, pp. 291–295, 2001.

rows; 3) Elimination of superfluous values of attributes. This approach to the problem is not very useful because both the computation of reducts and the superfluous equivalence classes are NP-Hard.

Many algorithms and methods have been proposed and developed to generate minimal decision rules, some based on inductive learning [6], [8], [3] and some other based on Rough Sets theory [11], [10], [12], [2], [9]. Rough sets theory provides a sound basis for the extraction of qualitative knowledge (dependecies) from very large relational databases.

Shan [11] proposes and develops a systematic method for computing all minimal rules, called maximally general rules, based on decision matrices.

Based on rough sets and boolean reasoning, Bazan [2] proposes a method to generate decision rules using dynamic reducts, stable reducts of a given decision table that appear frequently in random samples of a decision table.

Skowron [12] proposes a method that when applied over consistent decisions tables make it possible to obtain minimal decision rules. Based on the relative discernibility matrix notion.

An incremental learning algorithm for computing a set of all minimal decision rules based on the decision matrix method is proposed in [10].

On the other hand, different algorithms have been proposed to calculate reducts based on Rough Sets Theory. However, finding the minimal reduct is a NP-hard problem [13], so its computational complexity makes application in large databases imposible. In [4] a heuristic algorithm to calculate a reduct of the decision table is proposed. The algorithm is based on two matrices that are calculated using information from the Positive Region. In Chen and Lin [5] a modified notion of reducts is introduced.

In this paper we propose to execute prior to Rough Set methodology the Apriori algorithm in order to discover strong dependencies that can, in general, be useful to reduce the original set of attributes.

Observe that this approach wil not generate a minimal reduct. Nevertheless it is important to note that as a side effect it is possible to obtain strong rules to classify the concept that will be refined using the rough set methodology.

The rest of the paper is organized as follows.

Section 2: Rough Sets and association rules introduction.

Section 3: describes the new approach.

Section 4: results discussion and future work.

2   Preliminaries

2.1   Rough Sets Theory

The original Rough Set model was proposed by Pawlak [7]. This model is concerned with the analysis of deterministic data dependencies. According to Ziarko [14] Rough Set Theory is the discovery representation and analysis of data regularities. In this model, the objects are classified into indiscernibility classes based on pairs (attribute, values).

The following are the basic concepts of the rough set model.

Let OB be a non-empty set called the universe, and let IND be an equivalence relation over the universe OB, called an indiscernibility relation which represents a classification of the universe into classes of objects which are indiscernible or identical in terms of the knowledge provided by the given attributes. The main notion in Rough Sets Theory is that of the approximation space which is formally defined as $A = (OB, IND)$.

Equivalence classes of the relation are also called elementary sets. Any finite union of elementary sets is refered to as a definable set. Let's take $X \subseteq OB$ which represents a concept. It is not always the case that X can be defined exactly as the union of some elementary sets. That is why two new sets are defined: $\underline{Apr}(X) = \{o \in OB/[o] \subseteq X\}$ will be called the **lower approximation** $\overline{Apr}(X) = \{o \in OB/[o] \cap X \neq \emptyset\}$ will be called the **upper approximation**. Any set defined in terms of its lower and upper approximations is called a **rough set**.

2.2   Information Systems

The main computational effort in the process of data analysis in rough set theory is associated with the determination of attribute relationships in information systems. An Information System is a quadruple: $S = (OB, AT, V, f)$ where:

- OB is a set of objects
- AT is a set of attributes
- $V = \bigcup V_a$ being V_a the values of attribute a
- $f : OB \times AT \to V$

2.3   Decision Tables

Formally, a decision table S is a quadruple $S = (OB, C, D, V, f)$. All the concepts are defined similarly to those of information systems; the only difference is that the set of attributes has been divided into two sets, C and D, which are conditions and decision respectively.

Let P be a non empty subset of $C \cup D$, and let x, y be members of OB. x, y are indiscernible by P in S if and only if $f(x, p) = f(y, p)$ for all $p \in P$. Thus P defines a partition on OB. This partition is called a classification of OB generated by P. Then for any subset P of $C \cup D$, we can define an approximation space, and for any $X \subset OB$ the lower approximation of X in S and the upper approximation of X in S will be denoted as $\underline{P}(X)$ and $\overline{P}(X)$, respectively.

2.4   Association Rules

The purpose of association discovery is to find items that imply the presence of other items. An association rule is formally described as follows:

Let $I = \{i_1, i_2 \dots, i_n\}$ be a set of literals called items.

Let D be a set of transactions, each transaction $T \subset I$.

An association rule is an implication of the form $X \to Y$ where $X \subset I$ and

$Y \subset I$ and $X \cap Y = \emptyset$. The rule $X \to Y$ holds in the transaction set D with confidence c if $c\%$ of transactions in D that contain X also contain Y. The rule $X \to Y$ holds in the transaction set D with support s if $s\%$ of transactions in D contain $X \cup Y$.

Given a set of transaction D the problem of mining association rules is to generate all association rules that have support and confidence greater that the user-specified minimum support ($minsup$) and minimum confidence ($minconf$). In order to derive the association rules two steps are required: 1) Find the large itemsets for a given $minsup$; 2) Compute rules for a given $minconf$ based on the itemsets obtained before.

3   The Cooperative Algorithm

Algorithm input:

- T the input decision data table. Note that the set of attributes AT will be divided in condition (eventually antecedent) and decision (described attribute)
- $minconf$ the minimun confidence for rules
- n maximun number of condition allowed in the antecendent of the rules

We will assume that input table (T) is one that is either discrete or has been discretized in a pre-processing stage. We will also assume that the input table has been binarized. Formally expressed:

Let AT be a set of attributes, let $A \in AT$ be an attribute. Let $V = \{a_1, a_2, \dots, a_n\}$ be the set of values of A.

The binarization of A will yield n attributes $A_1, A_2, \dots, A_n$ such that for any $o \in OB$ if $f(o, A) = a_i$ then $f(o, A_j) = 1$ if $i = j$ $f(o, A_j) = 0$ if $i \neq j$

The process that will be performed is as follows:

1. Execute the Apriori algorithm being T the input data table. Internally the algorithm will calculate the best $minsupport$ and k (size of large itemsets) depending on the nature of T. We will call this new set of rules T_{Assoc}.
2. Delete all those rules in T_{Assoc} in which the decision attribute occurs as part of their antecedent.
3. If $\forall i \exists k$ such that $A_k \to D_i$ then A_k is a superfluos attribute so that it can be removed from AT. Remove also such rules from T_{Assoc}.
4. Analysis of the rules in the new T_{Assoc}. This set of rules contains two kind of rules that we have called *strong classification rules* and *meta-rules*. The former is composed by all those rules in T_{Assoc} in which the consequent is some D_i with confidence $\geq minconf$. The latter is a set containing rules that will allow us to reduce the number of condition attributes.
   - Include *strong classification rules* in the *output data mining rule set*
   - Those associations rules in *meta-rules* that only contain condition attributes have to be taken into account as they highlight dependencies among condition attributes. We will call this set of rules T_{red}
5. Reduce AT taking into account the rules in T_{red}
6. Execute the Positive Region algorithm to obtain a set of rules that will be included in *output data mining rule set*.

4 Conclusions

This approach provides a sound basis for the definition of a new cooperative algorithm to obtain comprehensible rules, while avoiding the computational complexity of *classical* predictive methods.

5 Acknowledgments

This work has been partially supported by the Universidad Politécnica de Madrid under grant A9913 for the project entitled "Design of a Data Warehouse to be integrated with a Data Mining system" and under grant AL200-1005-2.38 for project of collaboration with the University of Valle (Colombia).

References

1. Agrawal R., Imielinski T., Swami A. Mining association rules between sets of Item in large Databases, Proceedings of ACM SIGMOD, pages 207-216, May 1993.
2. J. Bazan Dynamic Reducts and statistical inference IPMU'96 1147-1151
3. Grzymala-Busse J. W. LERS: A system for learning from examples based on Rough Sets In Slowinski, R. (ed.) Intelligent Decision Support: Handbook of applicactions and advances of Rough Set theory. Kluwer Netherlands pp. 3-18
4. Fernandez C.,Menasalvas E., J.Pena J., Castano M. A new approach for efficient calculus of reducts in large databases Joint Conference of Information Sciences, 1997, pp.340-343
5. Chen R., Lin T. Y. Finding Reducts in Very Large DataBases Joint Conference of Information Sciences, 1997, pp.350-352
6. Michalski R., Carbonell J. Mitchell T. M. Machine Learning: An Artificial Intelligence Approach, vol. 1, Tioga Publishing, Palo Alto CA. 1986
7. Z. Pawlak, Rough Sets - Theoretical Aspects of Reasoning about Data, Kluwer, 1991.
8. J.R. Quinlan *Induction of decision Trees* Machine Learning 1: 81-106, 1986, Kluwer Academic Publisher
9. Shan N. W. Ziarko, An incremental learning algorithm for constructing decision rules Rough Sets, Fuzzy Sets and Knowledge Discovery. W. Ziarko ed. Springer Verlag 1994 pp. 326-334
10. Shan N. W. Ziarko, Data-Base acquisition an incremental modification of classification rules pp. 357-368 Computational Intelligence, vol. 11 n. 2, 1995.pp. 357-369.
11. Shan N., Rule Discovery from Data using decision matrices Master degree Thesis. University of Regina. 1995
12. Skowron A., Extracting laws from decision tables: A Rough Set approach Computational Intelligence, vol. 11 n. 2, 1995.pp. 371-387
13. Skowron A., The discernibility matrices and functions in Information Systems Decision Support by Experience, R. Slowinski (Ed.) Kluwer Academic Publishers, 1992
14. W. Ziarko The discovery, analysis, and representation of data dependencies in databases KDD-93 pp. 195-209
15. W. Ziarko Rough Sets and Knowledge Discovery: An Overview In Rough Sets, Fuzzy Sets and knowledge Discovery. W. Ziarko (ed.). Springer-Verlag 1994 pp.11-15

An Application for Knowledge Discovery Based on a Revision of VPRS Model

Juan F. Gálvez[1], Fernando Diaz[1], Pilar Carrión[1] and Angel Garcia[2]

[1] Dpto. de Lenguajes y Sistemas Informáticos
Universidad de Vigo, Campus Las Lagunas, 32004 Ourense, Spain
{galvez,fdiaz,pcarrion}@uvigo.es
[2] Dpto. de Organización y Estructura de la Información
Universidad Politécnica de Madrid, Spain
agarcia@ei.upm.es

Abstract. In this paper, we present a particular study of the negative factors that affect the performance of university students. The analysis is carried out using the CAI (*Conjuntos Aproximados con Incertidumbre*) model that is a new revision of the VPRS (*Variable Precision Rough Set*) model. The major contribution of the CAI model is the approximate equality among knowledge bases. This concept joined with the revision of the process of knowledge reduction (concerning both attributes and categories), allow a significant reduction in the number of generated rules and the number or attributes per rule as it is showed in the case of study.

1 Introduction

One of the first approaches for extracting knowledge from data, without statistical background, was the Rough Sets Theory (RS), introduced by Z. Pawlak [1, 2]. It arose from the necessity of having a formal framework to manage imprecise knowledge originated from empirical data. One of advantages of this model is the lack of necessity about preliminary or additional knowledge about the data analysed.

Tools development based on this approach and its further application to real problems has shown some limitations of the model, being the most remarkable the incapability of managing uncertain information. An extension to this model was proposed by W. Ziarko [3, 4]. Derived from it, without any additional hypotheses, it incorporates uncertainties management and is called *Variable Precision Rough Sets* (VPRS).

The CAI (*Conjuntos Aproximados con Incertidumbre*) model [5, 6, 7] is derived from the VPRS model. As the VPRS model, the CAI model is also suited to deal with uncertain information but with the aim of improve the classification power in order to induce stronger rules. Approximate equality of different knowledge bases is defined to reach this goal and is an attempt for introducing uncertainty at two different levels: the constituting blocks of knowledge (elementary categories) and the overall knowledge.

The paper is organised as follows. Sections 2 and 3 briefly introduce basic concepts on RS and VPRS, respectively. Section 4 introduces the approximate

W. Ziarko and Y. Yao (Eds.): RSCTC 2000, LNAI 2005, pp. 296–303, 2001.

equality concept and the knowledge reduction process under the CAI model. In Section 5, a case of study is presented to show the improvement of the CAI model, in terms of a significant reduction in the number of generated rules and the number of attributes per rule. Concretely, the application is about the analysis of negative factors that affect the performance of students in a Databases subject at University of Vigo. Finally, the conclusions of this work are presented.

2   Rough Sets Fundamentals

The Rough Set framework allows the treatment of imprecise knowledge. Imprecision is defined from the fact that knowledge granularity influencing the reference universe definition originates indiscernibility [1, 2, 8]. Basics notions are the knowledge base and the lower and upper-approximation.

Let U be, a finite and not empty set representing a reference universe and let be R an equivalence relation where $R \subseteq U \times U$, then we call knowledge base K to the pair $K = (U, R)$.

The knowledge base K establishes a partition of the universe into disjoint categories (i.e. a classification), denoted by U/R and where their elements, $[x]_R$, are equivalence classes of R. Objects that belonging to the same class are indistinguishable under relation R, called indiscernibility relation. Since these classes represent elementary properties of the universe expressed by means of knowledge K, they are also called elementary categories.

Objects belonging to the same category are not distinguishable, which means that their membership status with respect to an arbitrary subset of the domain may not always be clearly definable. This fact leads to the definition of a set in terms of lower and upper approximations. The lower approximation is a description of the domain objects which are known with certainty to belong to the subset of interest, whereas the upper approximation is a description of the objects which possibly belong to the subset. Any subset defined through its lower and upper approximations is called a rough set.

Formally, given a knowledge base $K = (U, R)$, and being $X \subseteq U$, the R-lower approximation and R-upper approximation (denoted by RX and $\overline{R}X$, respectively), are defined as follows:

$$RX = \bigcup\{[x]_R \in U/R : [x]_R \subseteq X\}$$
$$\overline{R}X = \bigcup\{[x]_R \in U/R : [x]_R \cap X \neq \emptyset\}$$

Once defined these approximations of X, the reference universe U is divided in three different regions: the positive region $POS_R(X)$, the negative region $NEG_R(X)$ and the boundary region $BND_R(X)$, defined as follows:

$$POS_R(X) = RX$$
$$NEG_R(X) = U - \overline{R}X$$
$$BND_R(X) = \overline{R}X - RX$$

The positive region or lower approximation of X includes those objects that could be unmistakably classified as X members using knowledge R. Similarly, the negative region include those objects that belong to $-X$ (the complementary set of X). Finally, the boundary region is the indiscernible area of the universe U.

3   Variable Precision Rough Sets

The basic concept introduced in the VPRS model is the relationship of *majority inclusion*. Its definition lies on a criterion $c(X, Y)$ called relative classification error, defined as follows:

$$c(X, Y) = 1 - \frac{card(X \cap Y)}{card(X)} \tag{1}$$

being X and Y two subsets of the universe U.

According to this, the traditional inclusion relationship between X and Y is defined in the following way:

$$X \subseteq Y \quad \text{iff} \quad c(X, Y) = 0 \tag{2}$$

Rough inclusion arises by the relaxation of the traditional inclusion when an admissible error level is permitted in classification. This error is explicitly expressed as β. Then, the rough inclusion relationship between X and Y, is defined as:

$$X \subseteq_\beta \quad \text{iff} \quad c(X, Y) \le \beta \tag{3}$$

Ziarko established as a requisite that at least the 50% of the elements have to be common elements, then being $0 \le \beta < 0.5$.

Under this assumption, Ziarko redefines the concepts of lower and upper approximation. Let be a knowledge base $K = (U, R)$ and a subset $X \subseteq U$, the β-lower approximation of X, (denoted by $R_\beta X$) and the β-upper approximation (denoted by $R_\beta X$) are defined as follows:

$$R_\beta X = \bigcup \{[x]_R \in U/R : [x]_R \subseteq_\beta X\}$$

$$R_\beta X = \bigcup \{[x]_R \in U/R : c([x]_R, X) < 1 - \beta\}$$

Alike in the Pawlaks model, the reference universe U could be divided in three different regions. These are the positive region $POS_{R,\beta}(X)$, the negative region $NEG_{R,\beta}(X)$ and the boundary region $BND_{R,\beta}(X)$, defined as follows:

$$POS_{R,\beta}(X) = R_\beta X$$

$$NEG_{R,\beta}(X) = U - R_\beta X$$

$$BND_{R,\beta}(X) = R_\beta X - R_\beta X$$

These new definitions originate a reduction of the boundary region allowing more objects to be classified as X members. The relationship among the VPRS model and the RS model is established considering that the RS model is a particular case within the VPRS when $\beta = 0$.

4   The CAI Model

As mentioned above, the major contribution of the CAI model is the approximate equality and its use in the definition of knowledge equivalence [5, 6, 7].

In the RS theory two different knowledge bases $K = (U, P)$ and $K' = (U, Q)$ are equivalent, denoted by $K \approx K'$, if $U/P = U/Q$, where U/P and U/Q are the classifications induced by knowledge P and Q, respectively. So, P and Q are equivalent if their constituting blocks are identical.

In the CAI model, uncertainty is introduced at two different levels: the constituting blocks of knowledge (elementary categories) and the overall knowledge, through the relationship of majority inclusion. So that, two different knowledge bases P and Q are equivalent or approximately equal, and denoted by $P \approx_\beta Q$, if the majority of their constituting blocks are similar.

Approximate equality forces to redefine the concepts implied in the knowledge reduction. As the Rough Set theory, the reduct and core are the basics concepts of the process of knowledge reduction.

In order to introduce these concepts it is necessary to define the indiscernibility relation $IND(\mathbf{R})$. If $\mathbf{R}$ is a family of equivalence relationships, then $\cap \mathbf{R}$ is also an equivalence relationship, denoted by $IND(\mathbf{R})$.

Formally, the concept of approximate equality is defined as follows. Given two different knowledge bases $K = (U, \mathbf{P})$ and $K' = (U, \mathbf{Q})$, we say that knowledge $\mathbf{P}$ is a β-specialization of knowledge $\mathbf{Q}$, denoted by $\mathbf{P} \subset_\beta \mathbf{Q}$, iff

$$1 - \frac{card(\{[x]_\mathbf{P} \in IND(\mathbf{P}) \, / \, \exists [x]_\mathbf{Q} \in IND(\mathbf{Q}) : [x]_\mathbf{P} \subseteq_\beta [x]_\mathbf{Q}\})}{card(IND(\mathbf{P}))} \leq \beta \quad (4)$$

Finally, we say that the two knowledge bases are approximately equal if the following property is held,

$$\mathbf{P} \approx_\beta \mathbf{Q} \Leftrightarrow \mathbf{P} \subset_\beta \mathbf{Q} \wedge \mathbf{Q} \subset_\beta \mathbf{P} \quad (5)$$

Let $\mathbf{R}$ be a family of equivalence relations and let $R \in \mathbf{R}$. We will say that R is β-dispensable in $\mathbf{R}$ if $IND(\mathbf{R}) \approx_\beta IND(\mathbf{R} - \{R\})$, otherwise R is β-indispensable in $\mathbf{R}$. The family $\mathbf{R}$ is β-independent if each $R \in \mathbf{R}$ is β-indispensable in $\mathbf{R}$; otherwise $\mathbf{R}$ is β-dependent.

Proposition 1. *If* $\mathbf{R}$ *is* β-*independent and* $\mathbf{P} \subseteq \mathbf{R}$, *then* $\mathbf{P}$ *is also* β-*independent.*

$\mathbf{Q} \subseteq \mathbf{P}$ is a β-reduct of $\mathbf{P}$ if $\mathbf{Q}$ is β-independent and $IND(\mathbf{Q}) \approx_\beta IND(\mathbf{P})$. The set of all β-indispensable relations in $\mathbf{P}$ will be called the β-core of $\mathbf{P}$, and will be denoted by $CORE_\beta(\mathbf{P})$.

Proposition 2. $CORE_\beta(\mathbf{P}) = \cap RED_\beta(\mathbf{P})$, *where* $RED_\beta(\mathbf{P})$ *is the family of all* β-*reducts of* $\mathbf{P}$.

The β-indispensable relations, β-reducts, and β-core can be similarly defined relative to an specific indiscernibility relation $\mathbf{Q}$, or elementary category. For example, the relative β-core is the union of all β-indispensable relations respect to the relation $\mathbf{Q}$.

An important kind of knowledge representation system is the decision table formalism, which is broadly used in many applications. A decision table specifies what decisions (actions) should be undertaken when some conditions are satisfied. Most decision problems can be formulated employing decision tables and therefore, this tool is particularly useful in decision making applications.

The concepts defined in CAI model allow us to confront the reduction of an information system (decision table) of the similar way at Pawlak's methodology [2] with the exception in the treatment of the inconsistencies. In the CAI model the inconstancies are necessaries because they provide useful information in the generation of rules. Therefore, we follow two steps [6]:

- Computation of β-reducts of condition attributes, which is equivalent to eliminate some columns from the decision table.
- Computation of β-reducts of categories, which is equivalent to eliminate superfluous values of attributes.

In this methodology there are not elimination of duplicate rows neither elimination of inconsistencies, because this information is useful for the rule. In consequence, and due to the inconsistency treatment, the induced decision rules may have an associate error in any case no greater than the admissible error β.

5   Case of Study

This study deals with the analysis of negative factors that affect the students performance in a Databases subject. The data were obtained from Computer Science students of the University of Vigo [6, 9]. The purpose of this paper is to show the classification power of CAI model.

We analyse 16 condition attributes (statistic and personal data) and 1 decision attribute (student's mark) relating to 118 students. If the decision attribute value is 0, the student fails.

Firstly, we apply the method of Sect. 4, searching the reducts for the condition attributes and reduction of superfluous values of attributes for each reduct. The result of this process is the generation of decision rules.

We need to select the β value for generating decision rules. This is an empirical process. In order to select the rules we use the following criteria:

- Rules that cover more objects.
- Rules with less number of condition attributes.

Under these criteria the obtained results, considering $\beta = 0.25$, are:

```
IF (Family Environment(Rural) AND (Access Mode (FPII))
   THEN Failed [The rule covers 15 objects with an error of 0.20].
IF (Access Mode (FPII) AND Father Studies (Primary))
   THEN Failed [The rule covers 11 objects with an error of 0.18].
IF (Father Studies (University))
   THEN Failed [The rule covers 8 objects with an error of 0.25].
IF (Family Environment(Rural) AND Father Studies (Bachelor's degree))
   THEN Failed [The rule covers 8 objects with an error of 0.12].
```

We would like to emphasise that only three of the analysed attributes take part in the strongest rules. The following conclusions can be derived from these rules:

- The negative factors that affect the students performance in a Databases subject are:
  - The family environment of the student if it is rural.
  - The access mode of the student if it is the FP-II access. This access mode is reserved in Spain for students which do not study bachelor's degree.
- Father Studies influence in the result too, however it is not conclusive, probably due to the lack of analysed objects and the high number of attributes.

The Table 1 shows the results that have been obtained by the CAI model with three different values of β. Analysing the results, we can notice the following

Table 1. Results obtained by the CAI Model

β values	Rules	Number of attrs.	Number of objects	Error
$\beta = 0.0$	1476	1 attr.: 1.08% 2 attr.: 27.23% $\geq$3 attr.: 71.69%	1 obj.: 75.0% 2 obj.: 16.0% 3 obj.: 4.8% $\geq$4 obj.: 4.2%	Error=0.0: 100%
$\beta = 0.25$	174	1 attr.: 10.5% 2 attr.: 77.5% $\geq$3 attr.: 12.0%	1 obj.: 74.15% 2 obj.: 11.5% 3 obj.: 2.3% $\geq$4 obj.: 12.05%	Error=0.0: 88.5% 0<Error< β: 11.5%
$\beta = 0.35$	102	1 attr.: 25.5% 2 attr.: 68.6% 3 attr.: 5.9%	1 obj.: 62.6% 2 obj.: 9.9% 3 obj.: 13.8% $\geq$4 obj.: 13.7%	Error=0.0: 75.5% 0<Error< β: 24.5%

advantages of the CAI model:

- Reduction of the number of the condition attributes in the rules. With $\beta > 0$ the majority of rules have 2 attributes.
- The number of generated rules decreases when β parameter grows. With $\beta = 0$, 1476 rules are generated, with $\beta > 0$ the number of rules is reduced over a 90%.
- Generation of rules with a higher classification power. The strongest rules were obtained with $\beta > 0$.
- We can control the uncertainty degree that can be introduced in our classification.

These advantages are agree with the perspectives of the Rough Set model presented by Pawlak in [10].

6   Conclusions

In this paper we present a new model based on VPRS model, referred to as the CAI model. As starting point, a revision of knowledge equivalence, and the definition of the concept of approximate equality are introduced. These concepts have led to a new redefinition of the knowledge reduction process. The consequences of this redefinition are the generation of stronger rules, in the sense, that the induced rules are more significant and with a less number of attributes in theirs antecedents.

Moreover, we have done a first approach to the study of negative factors that affect the performance of university students. We want to emphasise the few studies about this problem which have been carried out in Spain. We believe that CAI model can be a good starting point for further and deeper studies.

Finally, our research group have developed a software library that allows us the use of CAI model in practical cases. Now, we are interested in mechanisms for an automatic generation of Bayesian networks from data sets based on CAI model, using detection of dependence-independence attributes to build the network [11].

References

1. Pawlak, Z.: Rough Sets. *International Journal of Computer and Information Sciences*, **11**, pp. 341–346, 1982.
2. Pawlak, Z.: *Rough Sets: Theoretical Aspects of Reasoning about Data*. Kluwer Academic Publishers, Dordrecht, 1991.
3. Ziarko, W.: Variable Precision Rough Set Model. *Journal of Computer and System Sciences*, **46**, pp. 39–59, 1993.
4. Ziarko, W.: Analysis of Uncertain Information in the Framework of Variable Precision Rough. *Foundations of Computing and Decision Sciences*, **18**, no. 3–4, pp. 381–396, 1993.
5. Gálvez, J.F., Garcia, A.: A Revision of Knowledge Equivalence Characterisation under the Variable Precision Rough Set Model. In *Proceedings of the World Multiconference on Systemics, Cybernetics and Informatics. ISAS'98*, vol. 2, pp. 270–273, Orlando, FL (USA), 1998.
6. Gálvez, J.F.: *Definición de un Modelo de Adquisición del conocimiento en Sistemas Causales basado en Conjuntos Aproximados de Precisión Variable*. Doctoral Thesis, University of Vigo, 1999.
7. Gálvez, J.F., Diaz, F., Carrión, Garcia, A.: A new model based on VPRS Theory for improving classification. In *Proceedings of the eighth International Conference on Information Processing and Management of Uncertainty in Knowledge-based systems. IPMU'2000*, vol. 3, pp. 1953–1956, Madrid (Spain), 2000.
8. Pawlak, Z.: Granularity of Knowledge, Indiscernibility and Rough Sets. *IEEE*, pp. 106–110, 1998.
9. Gálvez, J.F., Garcia, A.: Using Rough Sets to Analyse Lack of Success in Academic Achievement. In *Proceedeings of the World Multiconference on Systemics, Cybernetics and Informatics. ISAS'97*, 1997.

10. Pawlak, Z.: Rough Sets. Present State and Perspectives. In *Proceedings of the sixth International Conference on Information Processing and Management of Uncertainty in Knowledge-based systems. IPMU'1996*, pp. 1137–1145, Granada (Spain), 1996.
11. Diaz, F., Corchado, J.M.: Rough Set based Learning for Bayesian networks. *International Workshop on Objective Bayesian Methodology*, Valencia (Spain), 1999.

An Algorithm for Induction of Decision Rules Consistent with the Dominance Principle

Salvatore **Greco**[1], Benedetto **Matarazzo**[1], Roman **Slowinski**[2], Jerzy **Stefanowski**[2]

[1]Faculty of Economics, University of Catania, Corso Italia 55, 95 129 Catania, Italy
[2]Institute of Computing Science, Poznan University of Technology,
3A Piotrowo Street, 60-965 Poznan, Poland,

Abstract. Induction of decision rules within the dominance-based rough set approach to the multiple-criteria sorting decision problem is discussed in this paper. We introduce an algorithm called DOMLEM that induces a minimal set of generalized decision rules consistent with the dominance principle. An extension of this algorithm for a variable consistency model of dominance based rough set approach is also presented.

1. Introduction

The key aspect of Multiple-Criteria Decision Analysis (MCDA) is consideration of objects described by multiple criteria representing conflicting points of view. Criteria are attributes with preference-ordered domains. For example, if decisions about cars are based on such characteristics as price and fuel consumption, these characteristics should be treated as criteria because a decision maker usually considers lower price as better than higher price and moderate fuel consumption more desirable than higher consumption. Regular attributes, such as e.g. colour and country of production are different from criteria because their domains are not preference-ordered.

As pointed out in [1,6] the Classical Rough Set Approach (CRSA) cannot be applied to *multipl-criteria decision problems*, as it does not consider criteria but only regular attributes. Therefore, it cannot discover another kind of *inconsistency* concerning violation of the *dominance principle,* which requires that objects having better evaluations (or at least the same evaluations) cannot be assigned to a worse class. For this reason, Greco, Matarazzo and Slowinski [1] have proposed an extension of the rough sets theory, called Dominance-based Rough Set Approach (DRSA), that is able to deal with this inconsistency typical to exemplary decisions in MCDA problems. This innovation is mainly based on substitution of the *indiscernibility relation* by a *dominance relation.* In this paper we focus our attention on one of the major classes of MCDA problems which is a counterpart of multiple-attribute classification problem within MCDA: it is called *multiple-criteria sorting problem.* It concerns an assignment of some objects evaluated by a set of criteria into some pre-defined and preference-ordered decision classes (categories).

Within DRSA, due to preference-order among decision classes, the sets to be approximated are, so-called, *upward* and *downward unions* of decision classes. For

W. Ziarko and Y. Yao (Eds.): RSCTC 2000, LNAI 2005, pp. 304–313, 2001.

each decision class, the corresponding upward union is composed of this class and all better classes. Analogously, the downward union corresponding to a decision class is composed of this class and all worse classes. The consequence of considering criteria instead of regular attributes is the necessity of satisfying the dominance principle, which requires a change of the approximating items from indiscernibility sets to dominating and dominated sets. Given object x, dominating set is composed of all objects evaluated not worse than x on all considered criteria, while dominated set is composed of all objects evaluated not better than x on all considered criteria. Moreover, the syntax of DRSA decision rules is different from CRSA decision rules. In the condition part of these rules, the elementary conditions have the form: "evaluation of object x on criterion q is at least as good as a given level" or "evaluation of object x on criterion q is at most as good as a given level". In the decision part of these rules, the conclusion has the form: "object x belongs (or possibly belongs) to at least a given class" or " object x belongs (or possibly belongs) to at most a given class".

The aim of this paper is to present an algorithm for inducing DRSA decision rules. This algorithm, called DOMLEM, is focused on inducing a minimal set of rules that cover all examples in the input data. Moreover, we will show how this algorithm can be extended to induce decision rules in a generalization of DRSA, called Variable Consistency DRSA model (VC-DRSA). This generalization accepts a limited number of counterexamples in rough approximations and in decision rules [2].

The paper is organized as follows. In the next sections, the main concepts of DRSA are briefly presented. In section 3, the DOMLEM algorithm is introduced and illustrated by a didactic example. Extensions of the DOMLEM algorithm for VC-DRSA model are discussed in section 4. Conclusions are grouped in final section.

2. Dominance-based Rough Set Approach

Basic concepts of DRSA are briefly presented (for more details see e.g. [1]). It is assumed that examplary decisions are stored in a *data table*. By this table we understand the 4-tuple $S=<U,Q,V,f>$, where U is a finite set of objects, Q is a finite set of *attributes*, $V = \bigcup_{q \in Q} V_q$ and V_q is a domain of the attribute q, and $f\colon U \times Q \to V$ is a total function such that $f(x,q) \in V_q$ for every $q \in Q$, $x \in U$. The set Q is, in general, divided into set C of *condition attributes* and set D of *decision attributes*.

Assuming that all condition attributes $q \in C$ are criteria, let S_q be an *outranking relation* on U with respect to criterion q such that $xS_q y$ means "x is at least as good as y with respect to criterion q". Furthermore, assuming that the set of decision attributes D (possibly a singleton $\{d\}$) makes a partition of U into a finite number of classes, let $Cl=\{Cl_t,\ t \in T\}$, $T=\{1,..., n\}$, be a set of these classes such that each $x \in U$ belongs to one and only one $Cl_t \in Cl$. We suppose that the classes are ordered, i.e. for all $r,s \in T$, such that $r>s$, the objects from Cl_r are preferred to the objects from Cl_s. The above assumptions are typical for consideration of a *multiple-criteria sorting problem*.

The sets to be approximated are *upward union* and *downward union* of classes, respectively: $Cl_t^{\geq} = \bigcup_{s \geq t} Cl_s$, $Cl_t^{\leq} = \bigcup_{s \leq t} Cl_s$, $t=1,...,n$.

Then, the indiscernibility relation is substituted by a *dominance relation*. We say that x *dominates* y with respect to $P \subseteq C$, denoted by $xD_P y$, if $xS_q y$ for all $q \in P$. The dominance relation is reflexive and transitive. Given $P \subseteq C$ and $x \in U$, the "granules of knowledge" used for approximation in DRSA are:

- a set of objects dominating x, called *P-dominating set*, $D_P^+(x) = \{y \in U: yD_P x\}$,

- a set of objects dominated by x, called *P-dominated set*, $D_P^-(x) = \{y \in U: xD_P y\}$.

Using $D_P^+(x)$ sets, *P-lower* and *P-upper approximation* of $Cl_t^\geq$ are defined as:

$$\underline{P}(Cl_t^\geq) = \{x \in U: D_P^+(x) \subseteq Cl \}, \quad \overline{P}(Cl_t^\geq) = \bigcup_{x \in Cl_t^\geq} D_P^+(x), \quad \text{for } t=1,\ldots,n.$$

Analogously, *P-lower* and *P-upper approximation* of $Cl_t^\leq$ are defined as:

$$\underline{P}(Cl_t^\leq) = \{x \in U: D_P^-(x) \subseteq Cl_t^\leq \}, \quad \overline{P}(Cl_t^\leq) = \bigcup_{x \in Cl_t^\leq} D_P^-(x), \quad \text{for } t=1,\ldots,n.$$

The *P-boundaries* (*P*-doubtful regions) of $Cl_t^\geq$ and $Cl_t^\leq$ are defined as:

$$Bn_P(Cl_t^\geq) = \overline{P}(Cl_t^\geq) - \underline{P}(Cl), \quad Bn_P(Cl_t^\leq) = \overline{P}(Cl_t^\leq) - \underline{P}(Cl^\leq), \quad \text{for } t=1,\ldots,n.$$

These approximations of upward and downward unions of classes can serve to induce generalized "*if... then...*" decision rules. For a given upward or downward union $Cl_t^\geq$ or $Cl_s^\leq$, $s,t \in T$, the rules induced under a hypothesis that objects belonging to $\underline{P}(Cl)$ or to $\underline{P}(Cl_s^\leq)$ are *positive* and all the others *negative*, suggest an assignment of an object to "at least class Cl_t" or to "at most class Cl_s", respectively. They are called *certain* $D_\geq$- (or $D_\leq$)-*decision rules* because they assign objects to unions of decision classes without any ambiguity. Next, if upper approximations differ from lower ones, *approximate* $D_{\geq \leq}$- *decision rules* can be induced under a hypothesis that objects belonging to the intersection $\overline{P}(Cl_s^\leq) \cap \overline{P}(Cl_t^\geq)$ ($s<t$) are *positive* and all the others *negative*. They suggest an assignment of objects to some classes between Cl_s and Cl_t. Yet another option is to induce $D_\geq$- (or $D_\leq$)-*possible decision rules* instead of approximate ones under the hypothesis that objects belonging to $\overline{P}(Cl_t^\geq)$ or to $\overline{P}(Cl_t^\leq)$ are *positive* and all the others *negative*. These rules suggest that an object *could belong* to "at least class Cl_t" or "at most class Cl_s", respectively.

Assuming that for each criterion $q \in C$, $V_q \subseteq \mathbf{R}$ (i.e. V_q is quantitative) and that for each $x,y \in U$, $f(x,q) \geq f(y,q)$ implies $xS_q y$ (i.e. V_q is preference-ordered), the following five types of decision rules can be considered:

1) *certain* $D_\geq$-*decision rules* with the following syntax:

  if $f(x,q_1) \geq r_{q1}$ and $f(x,q_2) \geq r_{q2}$ and $\ldots f(x,q_p) \geq r_{qp}$, then $x \in Cl$,

2) *possible* $D_\geq$-*decision rules* with the following syntax:

  if $f(x,q_1) \geq r_{q1}$ and $f(x,q_2) \geq r_{q2}$ and $\ldots f(x,q_p) \geq r_{qp}$, then x could belong to Cl ,

3) *certain* $D_\leq$-*decision rules* with the following syntax:

  if $f(x,q_1) \leq r_{q1}$ and $f(x,q_2) \leq r_{q2}$ and $\ldots f(x,q_p) \leq r_{qp}$, then $x \in Cl^\leq$,

4) *possible* $D_\leq$-*decision rules* with the following syntax:

if $f(x,q_1){\leq}r_{q1}$ and $f(x,q_2){\leq}r_{q2}$ and $...$ $f(x,q_p){\leq}r_{qp}$, then x could belong to $Cl^{\leq}$,
where $P{=}\{q_1,...,q_p\}{\subseteq}C$, $(r_{q1},...,r_{qp}){\in}V_{q1}{\times}V_{q2}{\times}...{\times}V_{qp}$ and $t{\in}T$;

5) *approximate* $D_{\geq\leq}$-*decision rules* with the following syntax:

if $f(x,q_1){\geq}r_{q1}$ and $f(x,q_2){\geq}r_{q2}$ and $...$ $f(x,q_k){\geq}r_{qk}$ and $f(x,q_{k+1}){\leq}r_{qk+1}$ and $...$ $f(x,q_p){\leq}r_{qp}$,
then $x{\in}Cl_s{\cup}Cl_{s+1}...{\cup}Cl_t$,

where $O'{=}\{q_1,...,q_k\}{\subseteq}C$, $O''{=}\{q_{k+1},...,q_p\}{\subseteq}C$, $P{=}O'{\cup}O''$, O' and O'' not necessarily disjoint, $(r_{q1},...,r_{qp}){\in}V_{q1}{\times}V_{q2}{\times}...{\times}V_{qp}$, $s,t{\in}T$ such that $s{<}t$. As it is possible that $\{q_1,...,q_k\}{\cap}\{q_{k+1},...,q_p\}{\neq}\varnothing$, in the condition part of a $D_{\geq\leq}$-decision rule we can have "$f(x,q){\geq}r_q$" and "$f(x,q){\leq}r'_q$", where $r_q{\leq}r'_q$, for some $q{\in}C$. Moreover, if $r_q{=}r'_q$, the two conditions boil down to "$f(x,q){=}r_q$".

The rules of type 1) and 3) represent certain knowledge extracted from the data table, while the rules of type 2), 4) represent possible knowledge, and rules of type 5) represent ambiguous knowledge.

Moreover, each decision rule should be minimal. Since a decision rule is an implication, by a *minimal* decision rule we understand such an implication that there is no other implication with an antecedent of at least the same weakness (in other words, rule using a subset of elementary conditions or/and weaker elementary conditions) and a consequent of at least the same strength (in other words, rule assigning objects to the same union or sub-union of classes).

Consider a $D_{\geq}$-decision rule "if $f(x,q_1){\geq}r_{q1}$ and $f(x,q_2){\geq}r_{q2}$ and $...$ $f(x,q_p){\geq}r_{qp}$, then $x{\in}Cl$ ". If there exists an object $y{\in}\underline{P}(Cl$) such that $f(y,q_1){=}r_{q1}$ and $f(y,q_2){=}r_{q2}$ and $...$ $f(y,q_p){=}r_{qp}$, then y is called *basis* of the rule. Each $D_{\geq}$-decision rule having a basis is called *robust* because it is "founded" on an object existing in the data table. Analogous definition of robust decision rules holds for the other types of rules.

We say that an object *supports* a decision rule if it matches both condition and decision parts of the rule. On the other hand, an object is *covered* by a decision rule if it matches the condition part of the rule.

A set of certain and approximate decision rules is *complete* if three following conditions are fulfilled: each $y{\in}\underline{C}(Cl_t^{\geq})$ supports at least one certain $D_{\geq}$-decision rule whose consequent is $x{\in}Cl_r$ ", with $r,t{\in}\{2,...,n\}$ and $r{\geq}t$; each $y{\in}\underline{C}(Cl_t^{\leq})$ supports at least one certain $D_{\leq}$-decision rule whose consequent is $x{\in}Cl_u^{\leq}$ ", with $u,t{\in}\{1,...,n{-}1\}$ and $u{\leq}t$; and each $y{\in}\overline{C}(Cl_s^{\leq}){\cap}\overline{C}(Cl$) supports at least one approximate $D_{\geq\leq}$-decision rule whose consequent is $x{\in}Cl_v{\cup}Cl_{v+1}{\cup}...{\cup}Cl_z$", with $s,t,v,z{\in}T$ and $s{\leq}v{<}z{\leq}t$.

In simple words, complete means that the set of rules is able to cover all objects from the data table in such a way that consistent objects are re-assigned to their original classes and inconsistent objects are assigned to clusters of classes referring to this inconsistency. An analogous definition of completeness can be formulated for a set of possible decision rules.

We call *minimal* each set of minimal decision rules that is complete and non-redundant, i.e. exclusion of any rule from this set makes it non-complete.

3. DOMLEM algorithm

Various algorithms have been proposed for induction of decision rules within CRSA (see e.g. [4,7,3] for review). Many of these algorithms tend to generate a minimal set of rules with the smallest number of rules. It is an NP-hard problem, so it is natural to use heuristic algorithms for rule induction, like LEM2 algorithm proposed by Grzymala [3]. In this paper, we approach the same problem with respect to DRSA.

The proposed rule induction algorithm, called DOMLEM, is built on the idea of MODLEM algorithm [8]. The latter, inspired by LEM2 [3], was designed to handle directly numerical attributes during rule induction.

The main procedure of DOMLEM is iteratively repeated for all lower or upper approximations of the upward (downward) unions of decision classes. Depending on the type of the approximation we are getting the corresponding type of decision rules,: e.g. of type1) from lower approximation of upward unions of classes, and of type 2) from upper approximation of upward unions of classes.

Moreover, taking into account the preference-order of decision classes and the requirement of minimality of decision rules, the procedure is repeated starting from the strongest union of classes, e.g. for type 1) decision rules the lower approximations of upward unions of classes should be considered in the decreasing order of the classes.

In the algorithm, $P \subseteq C$ and E denotes a complex (conjunction of elementary conditions e) being a candidate for a condition part of the rule. Moreover, $[E]$ denotes a set of objects matching the complex E. Complex E is accepted as a condition part of the rule iff $\varnothing \neq [E] = \bigcap_{e \in E} [e] \subseteq B$, where B is the considered approximation. For the sake of simplicity, in the following we present the general scheme of the DOMLEM algorithm only for a case of type 1) decision rules.

Procedure DOMLEM
(**input:** L_{upp} – a family of lower approximations of upward unions of decision classes:

$\{ \underline{P}(Cl), \underline{P}(Cl_{t-1}^{\geq}), \ldots \underline{P}(Cl_2^{\geq}) \}$; **output:** $R_{\geq}$ set of $D_{\geq}$-decision rules);

begin
   $R_{\geq} := \varnothing$;
   for each $B \in L_{upp}$ **do**
   begin
      E:=**find_rules**(B);
      for each rule $E \in \mathbf{E}$ **do**
      if E is a minimal rule **then** $R_{\geq} := R_{\geq} \cup E$;
   end
end.

Function find_rules
(**input:** a set B; **output:** a set of rules $\mathbf{E}$ covering set B);
begin
   $G := B$; {a set of objects from the given approximation}
   $\mathbf{E} := \varnothing$;

while $G \neq \varnothing$ **do**
begin
   $E := \varnothing$; {starting complex}
   $S := G$; {set of objects currently covered by E}
   while $(E = \varnothing)$ **or not** $([E] \subseteq B)$ **do**
   begin
     $best := \varnothing$; {best candidate for elementary condition}
     for each criterion $q_i \in P$ **do begin**
      $Cond := \{(f(x,q_i) \geq r_{qi}) : \exists x \in S \ (f(x,q_i) = r_{qi})\}$;
      {for each positive object from S create an elementary condition}
      for each $elem \in Cond$ **do**
     if $evaluate(\{elem\} \cup E)$ is_better_than $evaluate(\{best\} \cup E)$ **then** $best := elem$;
     end; {for}
     $E := E \cup \{best\}$; {add the best condition to the complex}
     $S := S \cap [best]$;
   end; {while not $([E] \subseteq B)$}
   for each elementary condition $e \in E$ **do**
     if $[E - \{e\}] \subseteq B$ **then** $E := E - \{e\}$;
   create a rule on the basis of $\mathbf{E}$;
   $\mathbf{E} := \mathbf{E} \cup \{E\}$; {add the induced rule}
   $G := B - \cup_{E \in \mathbf{E}}[E]$; {remove examples covered by the rule}
end; {while $G \neq \varnothing$}
end {function}

Let us comment the choice of a best condition using function *evaluate(E)*. A candidate E for a condition part of a rule could be evaluated by various measures. In the current version of DOMLEM the complex E with the highest ratio $\big|[E] \cap G\big| / \big|[E]\big|$ is chosen. In case of a tie, the complex E with the highest value of $\big|[E] \cap G\big|$ is chosen.

In the case of other types of decision rules, the above scheme works with corresponding approximations and elementary conditions. For example, in the case of type 3) rules, the corresponding approximations are the lower approximations of the downward unions of classes, considered in the increasing order of preference, and the elementary conditions are of the form $f(x,q_i) \leq r_{qi}$. In the case of type 5) rules, there are considered intersections of upper approximations of upward and downward unions of classes $\overline{P}(Cl_s^{\geq}) \cap \overline{P}(Cl_t^{\leq})$, $s < t$, and the elementary conditions have the form $f(x,q) \geq r_q$ and $f(x,q') \leq r'_{q'}$ for $q, q' \in C$; if $q = q'$, then $r_q \leq r'_q$. Furthermore, because of testing minimality of rules, in the case of type 5) rules, it is useful to discover in a given intersection $K = \overline{P}(Cl_s^{\geq}) \cap \overline{P}(Cl_t^{\leq})$, $s < t$, two subsets of objects, called "lower edge" and "upper edge" defined respectively as: the set of objects from K that do not dominate any other object from K having different evaluation on considered criteria, and the set of objects from K that are not dominated by any other object from K having different evaluation on considered criteria. Then combinations of conditions based on object from the "lower edge" with conditions based on objects from the "upper edge" are the only candidates for entering a complex.

Notice that requirement for inducing robust decision rules restricts the search space as only conjunctions of elementary conditions with thresholds referring to the same basis objects are allowed. Let us shortly discuss the computation complexity of the DOMLEM algorithm. We assume that the basic operation is checking which examples are covered by a complex (condition). Let m denotes a number of attributes, n is a number of objects. In the worst case each rule covers a single object using all criteria. In this case inducing robust rules requires at most $n(nm+3m-2)/2$ operations. On the other hand, while looking for non-robust rules one cannot restrict the search to conditions based on basic object only. Thus (assuming that each criterion is on average chosen once) we need at most $nm(n+1)(m+1)/4$ operations. So, the complexity of the algorithm is polynomial.

Illustrative example:

Consider the following example (see Table 1.). A set of 17 objects is described by the set of 3 criteria $C=\{q_1, q_2, q_3\}$ – all are to be maximized according to preference. The decision attribute d classifies objects into three decision classes Cl_1, Cl_2, Cl_3 which are preference-ordered according to increasing class number.

Table 1. Illustrative data table

Object	q_1	q_2	q_3	d
1	1.5	3	12	Cl_2
2	1.7	5	9.5	Cl_2
3	0.5	2	2.5	Cl_1
4	0.7	0.5	1.5	Cl_1
5	3	4.3	9	Cl_3
6	1	2	4.5	Cl_2
7	1	1.2	8	Cl_1
8	2.3	3.3	9	Cl_3
9	1	3	5	Cl_1
10	1.7	2.8	3.5	Cl_2
11	2.5	4	11	Cl_2
12	0.5	3	6	Cl_2
13	1.2	1	7	Cl_2
14	2	2.4	6	Cl_1
15	1.9	4.3	14	Cl_2
16	2.3	4	13	Cl_3
17	2.7	5.5	15	Cl_3

The downward and upward unions of classes are the following $Cl_1^{\leq}=\{3,4,7,9,14\}$, $Cl_2^{\leq}=\{1,2,3,4,6,7,9,10,11,12,13,14,15\}$, $Cl_2^{\geq}=\{1,2,5,6,8,10,11,12,13,15,16,17\}$, $Cl_3^{\geq}=\{5,8,16,17\}$. There are 5 inconsistent objects violating the dominance principle, i.e. 6,8,9,11,14. For instance, object # 9 dominates object # 6, because it is better on all criteria q_1, q_2, q_3, however, it is assigned to the decision class Cl_1 worse than Cl_2 to which belongs object # 6. So, the C approximations of upward and downward unions of decision classes are: $\underline{C}(Cl_1^{\leq})=\{3,4,7\}$, $\overline{C}(Cl_1^{\leq})=\{3,4,6,7,9,14\}$, $Bn_C(Cl_1^{\leq})=\{6,$

9,14}, $\underline{C}(Cl_2^\leq)=\{1,2,3,4,6,7,9,10,12,13,14,15\}$, $\overline{C}(Cl_2^\leq)=\{1,2,3,4,6,7,\ 8,9,10,11,12,$
13,14,15}, $Bn_C(Cl_2^\leq)=\{8,11\}$, $\underline{C}(Cl_2^\geq)=\{1,2,5,8,10,11,12,13,15,16,17\}$, $\underline{C}(Cl_2^\geq)=$
{1,2,5,6,8,9,10,11,12,13,14,15,16,17}, $Bn_C(Cl_2^\geq)=\{6,9,14\}$, $\underline{C}(Cl_3^\geq)=\{5,16,17\}$,
$\overline{C}(Cl_3^\geq)=\{5,8,11,\ 16,17\}$, $Bn_C(Cl_3^\geq)=\{8,11\}$.

Let us illustrate in detail the induction of certain D_2-decision rules for the upward union $Cl_3^\geq$. The lower approximation $\underline{C}(Cl_3^\geq)$ is an input set B to the DOMLEM function *find_rules*. The elementary conditions for objects {5,16,17} are as follows (reported elements mean: the condition e_i, the set of objects satisfying the condition e_i, the first evaluation measure $|[e_i]\cap G|/|[e_i]|$, the second evaluation measure $|[e_i]\cap G|$):

$e_1=(f(x,q_1)\geq 2.3)$, {5,8,11,16,17}, 0.6, 3; $e_5=(f(x,q_2)\geq 5.5)$, {17}, 1.0, 1;
$e_2=(f(x,q_1)\geq 2.7)$, {5,17}, 1.0, 2; $e_6=(f(x,q_3)\geq 9)$, {1,2,5,8,11,15,16,17}, 0.38, 3;
$e_3=(f(x,q_2)\geq 4)$, {2,5,11,15,16,17}, 0.5, 3; $e_7=(f(x,q_3)\geq 13)$, {15,16,17}, 0.67, 2;
$e_4=(f(x,q_2)\geq 4.3)$, {2,5,15,17}, 0.5, 2; $e_8=(f(x,q_3)\geq 15)$, {17}, 1.0, 1;

The condition e_2 is found the best because its first measure is the highest and it covers more positive examples than e_5 and e_8. Moreover, as e_2 satisfies the inclusion $[e_2]\subseteq B$, it can be used to create a rule covering two objects # 5 and 17. They are removed from G and the last remaining positive example to be covered is 16. Now, there are available three elementary conditions: $e_9=(f(x,q_1)\geq 2.3)$, {8,11,16}, 0.33, 1; $e_{10}=(f(x,q_2)\geq 4)$ {2,11,15,16}, 0.25, 1; $e_{11}=(f(x,q_3)\geq 13)$, {15,16}, 0.5, 1.

The condition $e_{11}=(f(x,q_3)\geq 13)$ is chosen due the highest first evaluation measure. On the other hand, it is not sufficient to create a rule using only this condition because it covers object # 15 which is a negative example. So, in the next iteration one has to consider complexes $E=e_9\wedge e_{11}$ and $E=e_{10}\wedge e_{11}$. As the complex $E=e_9\wedge e_{11}$ has a higher first evaluation measure e_9 is chosen. Notice that $(f(x,q_3)\geq 13)$ and $(f(x,q_1)\geq 2.3)$ can be now accepted for the condition part of a rule as it covers objects # 16 and 17. Proceeding in this way one obtains finally the minimal set of decision rules:

if $(f(x,q_3)\leq 2.5)$, then $x\in Cl_1^\leq$ {3, 4}

if $(f(x,q_2)\leq 1.2)$, and $(f(x,q_1)\leq 1.0)$ then $x\in Cl_1^\leq$ {4, 7}

if $(f(x,q_1)\leq 2.0)$, then $x\in Cl_2^\leq$ {1, 2, 3, 4, 6, 7, 9, 10, 12, 13, 14, 15}

if $(f(x,q_1)\geq 2.7)$, then $x\in Cl_3^\geq$ {5, 17}

if $(f(x,q_3)\geq 13.0)$ and $(f(x,q_1)\geq 2.3)$, then $x\in Cl_3^\geq$ {16, 17}

if $(f(x,q_1)\geq 1.2)$ and $(f(x,q_3)\geq 7.0)$, then $x\in Cl_2^\geq$ {1, 2, 5, 8, 11, 13, 15, 16, 17}

if $(f(x,q_2)\geq 2.8)$ and $(f(x,q_3)\geq 6.0)$, then $x\in Cl_2^\geq$ {1, 2, 5, 8, 11, 12, 15, 16, 17}

if $(f(x,q_2)\geq 2.8)$ and $(f(x,q_1)\geq 1.7)$, then $x\in Cl_2^\geq$ {2, 5, 8, 10, 11, 15, 16, 17}

if $(f(x,q_1)\geq 2.3)$ and $(f(x,q_2)\leq 3.3)$, then $x\in Cl_2\cup Cl_3$ {8}
if $(f(x,q_1)\geq 2.5)$ and $(f(x,q_2)\leq 4.0)$, then $x\in Cl_2\cup Cl_3$ {11}
if $(f(x,q_3)\leq 6.0)$ and $(f(x,q_1)\geq 2.0)$, then $x\in Cl_1\cup Cl_2$ {14}
if $(f(x,q_3)\geq 4.5)$ and $(f(x,q_3)\leq 5.0)$, then $x\in Cl_1\cup Cl_2$ {6, 9}

4. Decision rules in Variable Consistency model of Dominance-based Rough Set Approach

In [2] we proposed a generalization of DRSA to variable consistency model (VC-DRSA). It allows to define lower approximations of the unions of decision classes accepting limited number of negative examples controlled by pre-defined level of consistency $l \in (0, 1]$. Within VC-DRSA, given $P \subseteq C$ and consistency level l, the P-lower and P-upper approximations of the upward unions of classes are the following:

$$\underline{P}^l(Cl_t^\geq) = \{x \in Cl_t^\geq : \frac{card(D_P^+(x) \cap Cl_t^\geq)}{card(D_P^+(x))} \geq l\},$$

$$\overline{P}^l(Cl_t^\geq) = Cl_t^\geq \cup \{x \in Cl_{t-1}^\leq : \frac{card(D_P^-(x) \cap Cl_{t-1}^\leq)}{card(D_P^-(x))} < l\}.$$

The definitions of approximations for downward unions are analogic – see [2]. These approximations are used for induction of decision rules having the same syntax as in DRSA. In the VC-DRSA context each decision rule is characterized by an additional parameter α called *confidence of the rule*. It is the ratio of the number of objects supporting the rule and the number of objects covered by the rules.

The induction of such rules can be done after simple modifications of the DOMLEM algorithm. First, the inputs B of the algorithm are the new P^l-approximations of upward or downward unions of decision classes. Notice that in DRSA the complex E was accepted as a condition part of a rule iff $[E] \subseteq B$. This corresponds to the requirement that $|[E] \cap B|/|[E]|$, should be equal to 1. The keypoint of VC-DRSA is a relaxation of this requirement permitting to build a rule based on a complex E having a confidence α not worse than the consistency level l. The rest of the algorithm remains unchanged.

Continuation of the example. Let us assume that the user considers only criteria $P = \{q_1, q_2\}$ and is interested in analysing upward union $Cl_2^\geq$. The DRSA leads to $\underline{P}(Cl_2^\geq) = \{1,2,5,8,10,11,15,16,17\}$ and boundary $Bn_P(Cl_2^\geq) = \{6,9,12,13,14\}$. The two following decision rules are induced to describe objects from $\underline{P}(Cl_2^\geq)$:

if $(f(x,q_1) \geq 1.7)$ and $(f(x,q_2) \geq 2.8)$, then $x \in Cl_2^\geq$, $\{2,5,8,10,11,15,16,17\}$

if $(f(x,q_1) \geq 1.5)$ and $(f(x,q_2) \geq 3)$, then $x \in Cl_2^\geq$, $\{1,2,5,8,11,15,16,17\}$.

Let us assume now that the user works with VC-DRSA accepting consistency level l equal to 0.75. As P-dominating sets of objects # 6, 12 and 13 are contained in $Cl_2^\geq$ with a degree greater than the consistency level (0.83, 0,9 and 0.91, respectively) they can be added to the lower approximation $\underline{P}^{0.75}(Cl_2^\geq)$. The boundary region is now composed of only two objects # 9 and 14. Further on, the following rules are induced from the lower approximation $\underline{P}^{0.75}(Cl_2^\geq)$ (within parentheses there are objectss supporting the corresponding rules and objects only covered by the corresponding rules but not satisfying their decision parts - the latter are marked by "*"):

if $(f(x,q_1) \geq 1.2)$, *then* $x \in Cl_2^{\geq}$ with confidence 0.91, $\{1,2,5,8,10,11,13,14^*,15,16,17\}$

if $(f(x,q_2) \geq 2)$, *then* $x \in Cl_2^{\geq}$ with confid. 0.79, $\{1,2,3^*,5,6,8,9^*,10,11,12,14^*,15,16,17\}$

One can also notice that these rules are supported by more examples (i.e. 10 and 11, respectively) than the previous ones (8 in both).

6. Conclusions

The paper addressed the important issue of inducing decision rules for multicriteria sorting problems. As none of already known rule induction algorithms can be directly applied to multicriteria sorting problems, we introduced a specific algorithm called DOMLEM. It produces a complete and non-redundant, i.e. minimal, set of decision rules. It heuristically tends to minimize the number of generated rules. It was also extended to produce decision rules accepting a limited number of negative examples within the variable consistency model of the dominance rough sets approach.

Acknowledgement The research of S.Greco and B.Matarazzo has been supported by the Italian Ministry of University and Scientific Research (MURST). R. Slowinski and J. Stefanowski wish to acknowledge support of the KBN research grant 8T11F 006 19 from the State Committee for Scientific Research.

References

1. Greco, S., Matarazzo B., Slowinski R., The use of rough sets and fuzzy sets in MCDM. In T. Gal, T. Stewart and T. Hanne (eds.) *Advances in Multiple Criteria Decision Making*, chapter 14, pp. 14.1-14.59. Kluwer Academic Publishers 1999.
2. Greco, S., Matarazzo B., Slowinski R., Stefanowski J., Variable consistency model of dominance-based rough set approach; submitted to *RSCTC'2000 conference*, Banff 2000.
3. Grzymala-Busse, J.W., LERS - a system for learning from examples based on rough sets, In Slowinski, R., (ed.) *Intelligent Decision Support. Handbook of Applications and Advances of the Rough Sets Theory*, Kluwer Academic Publishers, Dordrecht, 1992, 3-18.
4. Komorowski J., Pawlak Z., Polkowski L. Skowron A., Rough Sets: tutorial. In Pal S.K., Skowron A. (eds.) *Rough Fuzzy Hybridization. A new trend in decision making*, Springer Verlag, Singapore, 1999, 3-98.
5. Pawlak, Z., *Rough Sets. Theoretical Aspects of Reasoning about Data*, Kluwer Academic Publishers, Dordrecht, 1991.
6. Slowinski R., Stefanowski J, Greco, S., Matarazzo B., Rough sets processing of inconsistent information. *Control and Cybernetics* 29 (2000) no. 1, 379-404.
7. Stefanowski J., On rough set based approaches to induction of decision rules. In Polkowski L., Skowron A. (eds.) *Rough Sets in Data Mining and Knowledge Discovery*, Physica-Verlag, vol. 1, 1998, 500-529.
8. Stefanowski J., Rough set based rule induction techniques for classification problems, In *Proc. 6th European Congress on Intelligent Techniques and Soft Computing* vol. 1, Aachen Sept. 7-10, 1998, 109-113.

Data Mining for Monitoring Loose Parts in Nuclear Power Plants

J. W. Guan, D. A. Bell

School of Information and Software Engineering
University of Ulster at Jordanstown
BT37 0QB, Northern Ireland, U.K.
{J.Guan,DA.Bell}@ulst.ac.uk

Abstract. Monitoring the mechanical impact of a loose (detached or drifting) part in the reactor coolant system of a nuclear power plant is one of the essential functions for operation and maintenance of the plant. Large data tables are generated during this monitoring process. This data can be "mined " to reveal latent patterns of interest to operation and maintenance. Rough set theory has been applied successfully to data mining. It can be used in the nuclear power industry and elsewhere to identify classes in datasets, finding dependencies in relations and discovering rules which are hidden in databases. This paper can be considered as one of a series, the earlier ones being summarized in Guan & Bell (2000a). These methods can be used to understand and control aspects of the causes and effects of loose parts in nuclear power plants. So in this paper we illustrate the use of our data mining methods by means of a running example using Envelope Rising Time data ERT on monitoring loose parts in nuclear power plants.

Introduction

A significant percentage of the world's electricity is now produced by nuclear power plants. Monitoring the mechanical impact of a loose (detached or drifting) part in the reactor coolant system of a nuclear power plant is one of the essential functions for operation and maintenance of the plant. One way of contributing to the solutions of problems in this area is to gain clear insights into causes and effects of loose parts using *data mining*. This is the computer-based technique of discovering interesting, useful, and previously unknown patterns from massive databases — such as those generated in nuclear energy operation. Our approach to data mining is to use rough set theory (Pawlak, 1991; Pawlak, Grzymala-Busse, Slowinski, & Ziarko 1995).

In many applications rough set theory has been applied successfully to rough classification and knowledge discovery. We have previously presented results of such application to nuclear power generation operation and control (Guan & Bell 1998, 2000b). Methods for using rough sets to identify classes in datasets, finding dependencies in relations and discovering rules which are hidden in databases have been developed. We use these methods again but here we apply them to the

W. Ziarko and Y. Yao (Eds.): RSCTC 2000, LNAI 2005, pp. 314–321, 2001.

loose part monitoring problem. The data to be mined here is the Envelope Rising Time data ERT, expressing the features of loose parts. This data is analyzed as a running example throughout this paper, and is used to illustrate our algorithms as they are presented.

The shape of the paper follows the usual pattern for this series of papers on the nuclear power generation, and the algorithms are those used for other safety applications in previous papers. In section 1, we discuss decision tables. A decision table consists of two finite sets: a universe U and an attribute set $A = C \cup D$, where $C \cap D = \emptyset$, and attributes in C are called *condition attributes*, and attributes in D are called *decision attributes*. We have previously proposed algorithms for discovery of knowledge based on rough analysis (Bell Guan 1998). Section 2 illustrates a simple classification for an attribute, and we introduce rough subsets and support subsets. In section 3, we discuss the support degree to a decision attribute from many condition attributes. Section 4 deals with the significance of a condition attribute a in C. We deal with *attribute significancy* on one subset of condition attributes. These methods are also applied to the ERT data. Finally, in section 5, an algorithm to discover knowledge is used for the ERT data.

1   Information Systems and Decision Tables

An *information system* $\mathcal{I}$ is a system $< U, A >$, where

1. $U = \{u_1, u_2, ..., u_i, ..., u_{|U|}\}$ is a finite non-empty set, called the *universe* or *object space*; elements of U are called *objects*;

2. $A = \{a_1, a_2, ..., a_j, ..., a_{|A|}\}$ is also a finite non-empty set; elements of A are called *attributes*;

3. for every $a \in A$ there is a mapping a from U into some space $a : U \rightarrow a(U)$, and $a(U) = \{a(u) \mid u \in U\}$ is called the *domain* of attribute a.

We want to find dependencies in relations and to discover rules which are hidden in databases. We can consider some attributes are condition attributes and some others are decision ones. Then we can discover the relation between condition and decision, predict decision from condition. Thus, an information system $< U, A >$ is called a *decision table*, if we have $A = C \cup D$ and $C \cap D = \emptyset$, where attributes in C are called *condition attributes* and *attributes* in D are called *decision attributes*.

Example 1.1. The Oh ERT data — Envelope Rising Time on monitoring loose parts

rule	TR1	TR2	TR3	Confid.	rule	TR1	TR2	TR3	Confid.
u_1	Short	Short	Middle	High	u_7	Middle	Short	Middle	Middle
u_2	Short	Middle	Middle	High	u_8	Middle	Middle	Middle	Middle
u_3	Short	Middle	Long	High	u_9	Middle	Middle	Short	Low
u_4	Short	Short	Short	Middle	u_{10}	Long	Any	Any	Low
u_5	Short	Short	Long	Middle	u_{11}	Any	Long	Any	Low
u_6	Short	Middle	Short	Middle					

In the table above $TR1, TR2, TR3$ are the Envelope Rising Time of the first, second, third arrived signal, respectively.

Oh (Oh 1996) made the fuzzy rule base for the evaluation of confidence level of input signals based on the relationship between the wave arrival sequences and envelope pattern changes.

Note that this table is an information system, where $A = \{TR1, TR2, TR3, \text{Confidence}\}$. In fact, it is also a database relation.

To consider it as a decision table, let $C = \{TR1, TR2, TR3\}$ and $D = \{y\}$, $y = \text{Confidence}$. Then the information system becomes a decision table.

2   Rough Subsets and Support Subsets

Let $< U, A >$ be an information system. For every attribute $a \in A$ we can introduce a classification U/a in universe U as follows: two objects $u, v \in U$ are in the same class if and only if $a(u) = a(v)$.

Let W be a subset of U. For a classification U/a, the *rough subset* of subset W from U/a is a pair of subsets $(W^{(U/a)^-}, W^{(U/a)^+})$, where

1. $W^{(U/a)^+} = \cup_{V \in U/a, V \cap W \neq \emptyset} V$ is called the *upper approximation* to W from U/a. It is not used in this paper;

2. $W^{(U/a)^-} = \cup_{V \in U/a, V \subseteq W} V$, also denoted by $S_a(W)$, is called the *lower approximation* to W from U/a. Subset $S_a(W)$ is also said to be the *support subset* to W from attribute a, and $spt_a(W) = |S_a(W)|/|U|$ is said to be the *support degree* to W from attribute a.

Briefly, $S_x(W)$ means that rule "$x = x(S_x(W))$ implies $y = y(S_x(W))$" has *strength* $spt_x(W) = |S_x(W)|/|U|$ (Grzymala-Busse 1991, Guan Bell 1991).

When $spt_a(W) = 0$ there is an *inconsistence*.

Let $y \in D$ be a decision attribute in a decision table $< U, A >$, where $A = C \cup D, C \cap D = \emptyset$. We now consider an overall decision for a decision attribute $y \in D$ rather than a local decision for a "decision subset" $W \in U/y$.

The *support subset to decision attribute* $y \in D$ from condition attribute $a \in C$ is subset $S_a(y) = \cup_{W \in U/y} W^{(U/a)^-} = \cup_{W \in U/y}(\cup_{V \in U/a, V \subseteq W} V)$, and $spt_a(y) = |\cup_{W \in U/y} W^{(U/a)^-}|/|U|$ is called the *support degree* to y from a.

If $U/y = U/\delta$, where δ is the "universal" partition $U/\delta = \{U\}$, then we have $S_a(y) = U, spt_a(y) = 1$ for all $a \in C$.

Example 2.1. For the ERT data, we have classsifications as follows.

$U/y = \{W_1, W_2, W_3\}$, where $W_1 = \{u_1, u_2, u_3\}$,

$W_2 = \{u_4, u_5, u_6, u_7, u_8\}$, $W_3 = \{u_9, u_{10}, u_{11}\}$;

$U/TR1 = \{\{u_1, u_2, u_3, u_4, u_5, u_6\}, \{u_7, u_8, u_9\}, \{u_{10}\}, \{u_{11}\}\}$,

$U/TR2 = \{\{u_1, u_4, u_5, u_7\}, \{u_2, u_3, u_6, u_8, u_9\}, \{u_{10}\}, \{u_{11}\}\}$, and

$U/TR3 = \{\{u_1, u_2, u_7, u_8\}, \{u_3, u_5\}, \{u_4, u_6, u_9\}, \{u_{10}, u_{11}\}\}$.

Also, we have

$S_{TR1}(W_1) = S_{TR1}(W_2) = \{\}, S_{TR1}(W_3) = \{u_{10}, u_{11}\}$;

$S_{TR2}(W_1) = S_{TR2}(W_2) = \{\}, S_{TR2}(W_3) = \{u_{10}, u_{11}\}$;

$S_{TR3}(W_1) = S_{TR3}(W_2) = \{\}, S_{TR3}(W_3) = \{u_{10}, u_{11}\}$.

For example, $S_{TR2}(W_3) = \{u_{10}, u_{11}\}$ means the tuples $u = u_{10}, u_{11}$ in $W_3 = \{u_9, u_{10}\}$ support a rule which states that condition $TR2(u) = Any, Long$ implies decision $y(u) = Low$ with strength $spt_{TR2}(W_3) = 2/11$.

Also, $S_{TR2}(W_1) = \{\}$ means there is an inconsistence.

For example, let us consider a class $\{u_1, u_4, u_5, u_7\}$ in $U/TR2$.

On one hand, a tuple $u = u_1$ in the class and in $W_1 = \{u_1, u_2, u_3\}$ supports a rule which states that condition $TR2(u) = Short$ implies decision $y(u) = High$.

On the other hand, another tuple $u = u_4$ in the class but not in $W_1 = \{u_1, u_2, u_3\}$ supports a rule which states that condition $TR2(u) = Short$ implies decision $y(u) = Middle$.

These two rules are *inconsistent*.

Finally, we have $S_{TR1}(y) = S_{TR2}(y) = S_{TR3}(y) = \{u_{10}, u_{11}\}$.

3   The Support Degree from Many Conditions

Let $< U, A >$ be an information system. For two attributes $a, b \in A$ we need to compute the following classification U/ab in universe U: two objects $u, v \in U$ are in the same class if and only if $a(u) = a(v)$ and $b(u) = b(v)$.

For a set $X \subseteq A$ of attributes we define classification U/X in universe U as follows: two objects $u, v \in U$ are in the same class if and only if $a(u) = a(v)$ for every $a \in X$.

We also define $U/\emptyset = U/\delta$, where δ is the "universal classification" $U/\delta = \{U\}, |U/\delta| = 1$.

Let $W \subseteq U$ be a subset of universe U. We now consider support to the subset W from a set of condition attributes.

For a condition attribute set $X \subseteq C$ in the decision table $< U, C \cup D >$, the *support subset* to W from condition attributes X is subset $S_X(W) = W^{(U/X)^-} = \cup_{V \in U/X, V \subseteq W} V$, and $spt_X(W) = |W^{(U/X)^-}|/|U|$ is called the support degree to W from X.

Briefly, $S_X(W)$ means that rule "$X = X(S_X(W))$ implies $Y = Y(S_X(W))$" has *strength* $spt_X(W) = |S_X(W)|/|U|$.

When $spt_a(W) = 0$ there is an *inconsistence*.

Let $Y \subseteq D$ be a subset of some decision attributes in $< U, C \cup D >$. We now consider support to the subset Y from a condition attribute set $X \subseteq C$, the *support subset* to attributes Y from condition attributes X is subset

$$S_X(Y) = \cup_{W \in U/Y} W^{(U/X)^-} = \cup_{W \in U/Y}(\cup_{V \in U/X, V \subseteq W} V),$$

and $spt_X(Y) = | \cup_{W \in U/Y} W^{(U/X)^-} |/|U|$ is called the support degree to Y from X.

Example 3.1. For the ERT data, we find that $U/TR1 \wedge TR2$ is
$\{\{u_1, u_4, u_5\}, \{u_2, u_3, u_6\}, \{u_7\}, \{u_8, u_9\}, \{u_{10}\}, \{u_{11}\}\}$; $U/TR1 \wedge TR3$ is
$\{\{u_1, u_2\}, \{u_3, u_5\}, \{u_4, u_6\}, \{u_7, u_8\}, \{u_9\}, \{u_{10}\}, \{u_{11}\}\}$; $U/TR2 \wedge TR3$ is
$\{\{u_1, u_7\}, \{u_2, u_8\}, \{u_3\}, \{u_4\}, \{u_5\}, \{u_6, u_9\}, \{u_{10}\}, \{u_{11}\}\}$;
$U/TR1 \wedge TR2 \wedge TR3$ is

$\{\{u_1\}, \{u_2\}, \{u_3\}, \{u_4\}, \{u_5\}, \{u_6\}, \{u_7\}, \{u_8\}, \{u_9\}, \{u_{10}\}, \{u_{11}\}\}.$
Also, we have
$S_{TR1 \wedge TR2}(W_1) = \{\}, S_{TR1 \wedge TR2}(W_2) = \{u_7\}, S_{TR1 \wedge TR2}(W_3) = \{u_{10}, u_{11}\}.$

For example, $S_{TR1 \wedge TR2}(W_2) = \{u_7\}$ means the tuple $u = u_7$ in $W_2 = \{u_4, u_5, u_6, u_7, u_8\}$ supports a rule which states that condition $TR1(u) = Middle$ and $TR2(u) = Short$ implies decision $y(u) = Middle$.

Also, $S_{TR1 \wedge TR2}(W_1) = \{\}$ means there is an inconsistence.

For example, let us consider a class $\{u_1, u_4, u_5\}$ in $U/TR1 \wedge TR2$.

On one hand, a tuple $u = u_1$ in the class and in $W_1 = \{u_1, u_2, u_3\}$ supports a rule which states that conditions $TR1(u) = Short$ and $TR2(u) = Short$ implies decision $y(u) = High$.

On the other hand, another tuple $u = u_4$ in the class but not in $W_1 = \{u_1, u_2, u_3\}$ supports a rule which states that condition $TR1(u) = Short$ and $TR2(u) = Short$ implies decision $y(u) = Middle$.

These two rules are *inconsistent*.

Finally, we have $S_{TR1 \wedge TR3}(W_1) = \{u_1, u_2\}, S_{TR1 \wedge TR3}(W_2) = \{u_4, u_6, u_7, u_8\},$
$S_{TR1 \wedge TR3}(W_3) = \{u_9, u_{10}, u_{11}\}; S_{TR2 \wedge TR3}(W_1) = \{u_3\},$
$S_{TR2 \wedge TR3}(W_2) = \{u_4, u_5\}, S_{TR2 \wedge TR3}(W_3) = \{u_{10}, u_{11}\};$
$S_{TR1 \wedge TR2 \wedge TR3}(W_1) = W_1 = \{u_1, u_2, u_3\},$
$S_{TR1 \wedge TR2 \wedge TR3}(W_2) = W_2 = \{u_4, u_5, u_6, u_7, u_8\},$
$S_{TR1 \wedge TR2 \wedge TR3}(W_3) = W_3 = \{u_9, u_{10}, u_{11}\}.$

4   Significance, Significant Subsets of Attributes

Definition 4.1. Let X be a non-empty subset of C: $\emptyset \subset X \subseteq C$. Let Y be a subset of D: $Y \subseteq D$ such that $Y \neq \emptyset, U/Y \neq U/\delta = \{U\}$. Given an attribute $x \in X$, we say that x is *significant* (for Y) in X if $S_X(Y) \supset S_{X-\{x\}}(Y)$; and that x is *not significant* or *nonsignificant* (for Y) in X if $S_X(Y) = S_{X-\{x\}}(Y)$.

Definition 4.2. Let X be a non-empty subset of C: $\emptyset \subset X \subseteq C$. Let Y be a subset of D: $Y \subseteq D$ such that $Y \neq \emptyset, U/Y \neq U/\delta = \{U\}$. Given an attribute $x \in X$, we define the *significance* of x (for Y) in X as $sig^Y_{X-\{x\}}(x) = \frac{|S_X(Y)| - |S_{X-\{x\}}(Y)|}{|U|}$.

In the special case where Y is a singleton, $Y = \{y\}$, we also denote $sig^Y_{X-\{x\}}(x)$ by $sig^y_{X-\{x\}}(x)$: $sig^Y_{X-\{x\}}(x) = \frac{|S_X(y)| - |S_{X-\{x\}}(y)|}{|U|}$.

In the special case where X is a singleton, $X = \{x\}$, we also denote $sig^Y_\emptyset(x)$ by $sig^Y(x)$: $sig^Y(x) = sig^Y_\emptyset(x) = \frac{|S_x(Y)| - |S_\delta(Y)|}{|U|} = \frac{|S_x(Y)|}{|U|}$.

So we always have $sig^Y(x) > 0$ unless $S_x(Y) = \emptyset$.

Also, in the special case where X contains two attributes, $X = \{x_1, x_2\}$, we denote $sig^Y_{\{x_2\}}(x_1)$ by $sig^Y_{x_2}(x_1)$: $sig^Y(x) = sig^Y_{x_2}(x_1) = \frac{|S_X(Y)| - |S_{x_2}(Y)|}{|U|}$.

Definition 4.3. Let $\mathcal{D} = < U, C \cup D >$ be a decision table, where C is the condition attribute set and D is the decision attribute set. Let Y be a subset of D: $Y \subseteq D$ such that $Y \neq \emptyset, U/Y \neq U/\delta = \{U\}$. Let X be a non-empty subset of

C: $\emptyset \subset X \subseteq C$. The non-empty subset X is said to be *significant* or independent (for Y) if each $x \in X$ is significant (for Y) in X; otherwise X is *nonsignificant* (for Y). An empty set $\emptyset$ is said to be *significant* (for Y).

To check whether or not X is significant, we can computes $|X|$ significances $sig^Y_{X-\{x_i\}}(x_i)$ for $i = 1, 2, ..., |X|$ to check whether or not they are greater than 0.

Example 4.1. For the ERT data, we have the following.

(1) From example 2.1 we find that

$sig^y(TR1) = spt_{TR1}(y) = 2/11$; $sig^y(TR2) = spt_{TR2}(y) = 2/11$;
$sig^y(TR3) = spt_{TR3}(y) = 2/11$.

(2) From example 3.1, we find that

$sig^y_{TR1}(TR2) = 1/11$; $sig^y_{TR2}(TR1) = 1/11$; $sig^y_{TR1}(TR3) = 7/11$;
$sig^y_{TR3}(TR1) = 7/11$; $sig^y_{TR3}(TR2) = 3/11$; $sig^y_{TR2}(TR3) = 3/11$;
$sig^y_{TR1 \wedge TR3}(TR2) = 2/11$; $sig^y_{TR1 \wedge TR2}(TR3) = 8/11$; $sig^y_{TR2 \wedge TR3}(TR1) = 6/11$.

Also, we know the significancy (for y) of the following subsets of $X = \{TR1, TR2, TR3\}$.

(1) $\emptyset$ is significant.

(2) Singletons $\{TR1\}, \{TR2\}, \{TR3\}$ are significant since
$sig^y(TR1), sig^y(TR2), sig^y(TR3) > 0$.

(3) $\{TR1, TR2\}$ is significant since $sig^y_{TR2}(TR1) > 0$ and so $TR1$ is significant in $\{TR1, TR2\}$.

$\{TR1, TR3\}$ is significant since $sig^y_{TR1}(TR3) = 7/11 > 0$ and $sig^y_{TR3}(TR1) = 7/11 > 0$.

$\{TR3, TR2\}$ is significant since $sig^y_{TR3}(TR2) = 3/11 > 0$ and $sig^y_{TR2}(TR3) = 3/11 > 0$.

$\{TR1, TR2, TR3\}$ is significant since $sig^y_{TR2 \wedge TR3}(TR1) = 6/11 > 0$ and so $TR1$ is significant in $\{TR1, TR2, TR3\}$, $sig^y_{TR1 \wedge TR3}(TR2) = 2/11 > 0$ and so $TR2$ is significant, $sig^y_{TR1 \wedge TR2}(TR3) = 8/11 > 0$ and so $TR3$ is significant.

5   An Algorithm to Discover Knowledge

Definition 5.1. In $\mathcal{D} =< U, C \cup D >$, let Y be a subset of D: $Y \subseteq D$ such that $Y \neq \emptyset, U/Y \neq U/\delta = \{U\}$. Let X be a non-empty or empty subset $X \subseteq C$ of C. A subset X_0 of X is said to be a *key* of X (for Y) if X_0 satisfies

(1) $S_{X_0}(Y) = S_X(Y)$;
(2) if $X' \subset X_0$ then $S_X(Y) \supset S_{X'}(Y)$.

The empty subset $\emptyset$ has key $\emptyset$ (for Y).

Applying the significance measure, we can design an algorithm to discover knowledge as follows (Grzymala-Busse, Slowinski et al 1992).

Let $\mathcal{D} =< U, C \cup D >$ be a decision table, where C is the condition attribute set and D is the decision attribute set.

Let X be a non-empty subset of C: $\emptyset \subset X \subseteq C$.

Let Y be a subset of D: $Y \subseteq D$ such that $Y \neq \emptyset, U/Y \neq U/\delta = \{U\}$.

Algorithm D. *This algorithm finds one key of X (for Y) and discovers knowledge.*

Step 1. Let $X = \{x_1, x_2, ..., x_j, ..., x_J\}$. Compute $U/x_1, U/x_2, ..., U/x_J, U/y$ for $y \in Y$ and

$U/Y = \{W_1, W_2, ..., W_i, ..., W_I\}$.

Compute U/X and $S_X(W_I), S_X(W_I), S_X(W_I), S_X(W_I)$.

Compute $sig^Y(x_j)$ for $j = 1, 2, ..., J$.

Choose x_{j_1} such that $|S_{x_{j_1}}(W_i)|$ for some i is a maximum (when there are more than one possibilities it can be run in parallel mode concurrently for all possibilities).

For every i, j such that $S_{x_j}(W_i) \neq \emptyset$ discover the following rules:
$(u \in S_{x_j}(W_i))$ **If** (x_j) x_j is $x_j(u)$ **then** (y) y is $y(u)$ **with strength** $spt_{x_j}(W_i)$.

If $\mathcal{W}_1 = \{W_i \in U/Y | S_{x_j}(W_i) \subset S_X(W_i)$ for all $j\}$ is empty or $J = 1$ then the algorithm is completed. Now $(\neg \exists_i \forall_j (S_{x_j}(W_i) \subset S_X(W_i)) = \forall_i \exists_j (S_{x_j}(W_i) = S_X(W_i)))$ for each $W_i \in U/Y$ there is an x_j such that $S_{x_j}(W_i) = S_X(W_i)$. The collection of these x_j's may be a key of X (for Y).

Otherwise, go to step 2.

Step 2. Compute $U/x_{j_1}x_j$ and $sig^Y_{x_{j_1}}(x_j)$ for $j \neq j_1$.

Choose an x_{j_2} such that $sig^Y_{x_{j_1}}(x_{j_2})$ is a maximum.

For $S_{x_{j_1}x_{j_2}}(W_i) - S_{x_{j_1}}(W_i) - S_{x_{j_2}}(W_i) \neq \emptyset$, where $W_i \in \mathcal{W}_1$, we can discover the following rules: $(u \in S_{x_{j_1}x_{j_2}}(W_i) - S_{x_{j_1}}(W_i) - S_{x_{j_2}}(W_i))$

If (x_{j_1}) x_{j_1} is $x_{j_1}(u)$, (x_{j_2}) x_{j_2} is $x_{j_2}(u)$, **then** (y) y is $y(u)$ **with strength** $spt_{x_{j_1}x_{j_2}}(W_i)$.

If $\mathcal{W}_2 = \{W_i \in \mathcal{W}_1 | S_{x_{j_1}x_{j_2}}(W_i) \subset S_X(W_i)\}$ is empty or $J = 2$ then the algorithm is completed and $\{x_{j_1}, x_{j_2}\}$ may be a key of X (for Y).

Otherwise, go to step 3.

...

Step $|X|$. Compute $U/x_{j_1}x_{j_2}...x_{j_{|X|-1}}x_j$ and

$sig^Y_{x_{j_1}x_{j_2}...x_{j_{|X|-1}}}(x_j)$ for $j \neq j_1, j_2, ..., j_{|X|-1}$.

Let $x_{j_{|X|}} = x_j$.

For $S_X(W_i) - S_{X-\{x_{j_{|X|}}\}}(W_i) - S_{x_{j_{|X|}}}(W_i) \neq \emptyset$ where $W_i \in \mathcal{W}_{|X|-1}$, we can discover the following rules: $(u \in S_X(W_i) - S_{X-\{x_{j_{|X|}}\}}(W_i) - S_{x_{j_{|X|}}}(W_i))$

If (x_{j_1}) x_{j_1} is $x_{j_1}(u)$, (x_{j_2}) x_{j_2} is $x_{j_2}(u)$,..., $(x_{j_{|X|-1}})$ $x_{j_{|X|-1}}$ is $x_{j_{|X|-1}}(u)$, $(x_{j_{|X|}})$ $x_{j_{|X|}}$ is $x_{j_{|X|}}(u)$ **then** (y) y is $y(u)$ **with strength** $spt_{x_{j_1}x_{j_2}...x_{j_{|X|}}}(W_i)$.

Then the algorithm is completed and $\{x_{j_1}, x_{j_2}, ..., x_{j_{|X|-1}}, x_{j_{|X|}}\}$ may be a key of X (for Y).

Example 5.1. By using this algorithm, the decision table of example 1.1 can be reduced to the following:

RULES	TR1	TR2	TR3	Confidence	Strength
u_1, u_2	Short		Middle	High	2/11
u_3	Short	Middle	Long	High	1/11
u_4, u_6	Short		Short	Middle	2/11
u_5	Short	Short	Long	Middle	1/11
u_7, u_8	Middle		Middle	Middle	2/11
u_9	Middle		Short	Low	1/11
u_{10}, u_{11}			Any	Low	2/11

Now this is an illustrative example. However, if it is scaled up, it can be appreciated that condensed rules of thumb can be distilled from large tables of monitoring data. The final rule on the table is interesting. If we take 'any' to be the same as blank ("doesn't matter") it says "confidence is low".

6   Summary and Future Work

In this paper we apply some algorithms with relatively low computation times which are helpful in the distillation of rules from data on loose parts based on rough sets. The results show how condensed rules which may be useful for safety and general control can be derived from large collections of data.

References

1. Agrawal, R., Imielinski, T., & Swami, A. (1993). Database mining: a performance perspective, *IEEE Transactions on Knowledge and Data Engineering (Special issue on Learning and Discovery in Knowledge-Based Databases)*, 5(6), 914-925.
2. Grzymala-Busse, J. (1991). Managing uncertainty in expert systems. Kluwer Academic Publishers.
3. Guan, J.W., & Bell, D.A. (1991). *Evidence theory and its applications*, Vol.1 (Studies in Computer Science and Artificial Intelligence 7). Elsevier.
4. Guan, J.W., & Bell, D.A. (2000a). Rough knowledge discovery for nuclear safety, *International Journal of General Systems*, Vol.29(2000), No.2, 231-249.
5. Guan, J.W., & Bell, D.A. (2000b). Knowledge discovery for controlling nuclear power plants, *Proc. of the 4th International FLINS Conference on Intelligent Techniques and Soft Computing in Nuclear Science and Engineering* (World Scientific, Singapore), August 28-30, 2000, Bruges, Belgium. (http://www.sckcen.be/flins2000)
6. Oh, Y. G. et al (1996). Fuzzy logic utilization for the diagnosis of metallic loose part impact in nulear power plant, *Intelligent System and Soft Computing for Nuclear Science and Industry*, Proceedings of the 2nd International FLINS Workshop, Mol, Belgium, September 25-27, 1996, 372-378.
7. Pawlak, Z. (1991). *Rough sets: theoretical aspects of reasoning about data*. Kluwer.
8. Pawlak, Z., Grzymala-Busse, J., Slowinski, R., & Ziarko, W. (1995). Rough sets. CACM, 38(11), 89-95.
9. Polkowski, L.; Skowron, A. 1998a, *Rough sets in knowledge discovery, Vol.1, Methodology and Applications*, Physica-Verlag, A Springer-Verlag Company, 1998.
10. Polkowski, L.; Skowron, A. 1998b, *Rough sets in knowledge discovery, Vol.2, Applications, Case Studies, and Software Systems*, Physica-Verlag, A Springer-Verlag Company, 1998.
11. Yao, Y. Y., & Wong S. K. M. (1992). A decision theoretic framework for approximating concepts. International Journal of Man-machine Studies, 37, 793-809.
12. Yao, Y. Y., & Lin T. Y. (1996). Generalization of rough sets using modal logics, Intelligent automation and soft computing, Vol.2(1996), No.2, 103-120.
13. Ziarko, W. (1991). The discovery, analysis, and representation of data dependencies in databases. In G. Piatetsky-Shapiro, & W. J. Frawley (Eds.), Knowledge Discovery in Databases (pp. 177-195). AAAI Press/MIT Press.

Supervised Learning: A Generalized Rough Set Approach

Jianchao Han[1], Xiaohua Hu[2], and Nick Cercone[1]

[1] Department of Computer Science, University of Waterloo,
Waterloo, Ontario, N2L 3G1 Canada
[2] Knowledge Stream Partner, 148 State St., Boston, MA 02109

Abstract. Classification rules induction is a central problem addressed by machine learning and data mining. Rough sets theory is an important tool for data classification. Traditional rough sets approach, however, pursuits the fully correct or certain classification rules without considering other factors such as uncertain class labeling, importance of examples, as well as the uncertainty of the final rules. A generalized rough sets model, GRS, is proposed and a classification rules induction approach based on GRS is suggested. Our approach extends the variable precision rough sets model and attempts to reduce the influence of noise by considering the importance of each training example and handling the uncertain class labels. The final classification rules are also measured with the uncertainty factor.

Keywords: Rough set theory, supervised learning, classification, rule induction.

1 Introduction

Supervised learning or classification is an important research topic in machine learning and data mining [1]. Rough sets theory can be used to induce rules from large data sets [5]. It is complementary to statistical methods and provides the necessary framework to conduct data analysis and knowledge discovery from imprecise and ambiguous data. A number of algorithms and systems for learning classifiers have been developed based on this theory [2, 4, 8].

The original rough sets approach pursuits the fully correct and certain classifications within the available information. Unfortunately, the available information usually allows only for partial classification. As a result, classification with a controlled degree of uncertainty, or a classification error rate, is outside the realm of rough set theory. The variable precision rough set model (VP-model) [7] presents the concept of the majority inclusion relation. Rules which are almost always correct, called *strong* rules, can be extracted with the VP-model. Such *strong* rules are useful for decision support in a rule-based expert system. However, these approaches have limitations. The following are some of them.

(1) All tuples are treated with equal importance [2, 3]. Usually, the original data are generalized into concise form by finding attribute reduct. The tuples with the same values of attributes in the reduct are combined together and

W. Ziarko and Y. Yao (Eds.): RSCTC 2000, LNAI 2005, pp. 322–329, 2001.

a "vote" field is attached for the count of the combined tuples. In real-world applications, different tuples may have different degrees of importance to the decision attributes, thus should have different contributions to the "vote".

(2) All training examples m ust be crisply labeled, otherwise the final result will be incorrect and lead to big classification error. In some actual applications, however, it is very expensive and/or risky to make a yes-no decision. In such cases, it could be helpful to associate some uncertainty factor with the class label.

(3) The lower and upper approximations of a concept are defined based on the strict set inclusion operation, which has no tolerance to the noise data in the classification [6]. For example, assume $X = \{x_1, x_2, \ldots, x_{99}, \ldots, x_{500}\}$, and $E_1 = \{x_1, x_2, \ldots, x_{99}, x_{501}\}$, and $E_2 = \{x_{500}, x_{501}, \ldots, x_{599}\}$ are two equivalent classes. All examples in E_1 are in X except x_{501}, and all examples in E_2 are not in X except x_{500}. In the traditional model, both E_1 and E_2 are treated equally and thus put in the boundary region. However, in practice, x_{501} may be noise in E_1 and x_{500} may be noise in E_2. It seems reasonable to put E_1 in the positive region and E_2 in the negative region.

We propose an approac h for learning classification rules based on a new generalized rough set model, GRS, which extends the concept of the variable precision rough set model. Our new approach will deal with the situations where uncertain objects may exist, different objects may have different degrees of importance attached, and different classes may have different noise ratios. The original rough sets model and the VP-model of rough sets [7] become a special case of this model. The primary advantage of the GRS model is that it extends the traditional rough sets model to work well in noisy environments.

This paper is organized as follows. A generalized rough set model, GRS, is suggested in Section 2 to overcome the above limitations. A supervised learning approach based on GRS is developed in Section 3. In Section 4, an illustration is investigated using the generalized model to learn classification rules from noise data. Finally, Section 5 is concluding remarks.

2   Generalized Rough Sets Model

In order to overcome the limitations discussed in the previous section, we propose a generalized rough set model which is developed from the traditional model and the VP-model to deal with the importance of tuples and uncertain ty class labels, respectively. For simplicity, we only consider binary classifications, but it can be easily extended to multiple classifications.

For our purpose, the information system IS is extended to the uncertain information system UIS as follows:

$$UIS = <U, C, D, \{V_a\}_{a \in C}, f, g, d>,$$

where $U = \{u_1, u_2, \ldots, u_n\}$ is a non-empty set of tuples, C is a non-empty set of condition attributes, D is a binary decision attribute with possible values 1 and 0, where 1 represents the positive class, while 0 the negative class. V_a is the domain of attribute "a" with at least two elements. f is a function: $U \times C \to V = \bigcup_{a \in C} V_a$, which maps each pair of tuple and attribute to an attribute value. g is a function: $U \to [0, 1]$, which maps each tuple to a value between 0 and 1, indicating the certainty of being positive example. d is a function: $U \to [0, 1]$, which assigns each tuple an importance factor to represents how important the tuple is for the classification task. The importance factor corresponds to the condition attribute set C, while the class label certainty corresponds to the decision attribute D. $d \times u$ contributes to the positive class, while $d \times (1 - u)$ contributes to the negative class.

We adapt the concept of relative classification error introduced in [8] to deal with the noise data. The main idea is to put some boundary examples into the positive region or negative region. Therefore, the *strong* rules which are almost always correct are obtained. In actual applications, positive class and negative class may contain different kinds of noise and have different noise tolerance degrees. Two classification factors P_β and N_β $(0.0 \leq P_\beta, N_\beta \leq 1.0)$ are introduced to solve this problem, which can be determined by estimating noise degree in the positive region and the negative region, respectively.

Let E be a non-empty equivalent class in the approximation space. The classification ratios of E with respect to the positive class P_{class} and negative class N_{class} are defined as

$$C_P(E) = \sum_{x \in E}(d(x) \times g(x))/\sum_{x \in E} d(x),$$
$$C_N(E) = \sum_{x \in E}(d(x) \times (1 - g(x)))/\sum_{x \in E} d(x)$$
$$= 1 - C_P(E),$$

respectively. $C_P(E)$ is the certainty to classify E to the positive region, while $C_N(E)$ is the certainty to classify E to the negative region. If tuples in E are classified to positive class, the classification error rate is $1 - C_P(E)$. If tuples in E are classified to negative class, the classification error rate is $1 - C_N(E)$.

For the pre-specified precision threshold P_β and N_β, E is classified to the positive class if $C_P(E) \geq P_\beta$, or to the negative class if $C_N(E) \geq N_\beta$. Otherwise, E is put to the boundary region.

The concepts of set approximation in IS can be extended for UIS according to the classification factors P_β and N_β. Let $R_{P,N}$ be the indiscernibility relation based on a set of condition attributes B and $R^*_{P,N} = \{E_1, E_2, \ldots, E_n\}$ be the collection of equivalent classes of $R_{P,N}$. Assume $X \subseteq U$, then the positive lower approximation and upper approximation of X with respect to precision P_β and N_β, denoted $POS_{P_\beta}(X|B)$ and $NEG_{N_\beta}(X|B)$ respectively, are defined as

$$POS_{P_\beta}(X|B) = \bigcup \{E \in R^*_{P,N} : C_P(E) \geq P_\beta\},$$

$$NEG_{N_\beta}(X|B) = \bigcup \{E \in R^*_{P,N} : C_N(E) \geq N_\beta\}.$$

Similarly, the boundary region of X, $BND_{P,N}(X)$, is composed of those elementary sets which are neither in the positive region nor in the negative region of X,

$$BND_{P_\beta,N_\beta}(X|B) = \bigcup \{E \in R^*_{P,N} : C_P(E) < P_\beta, C_N(E) < N_\beta\}.$$

Clearly, the difference of the extended concepts from the original ones is that the elementary sets are divided into positive or negative region in terms of their classification ratio instead of their inclusion in the target concept. P_β and N_β can be adjusted for different data sets with different noise levels. The positive region and negative region shrink while the boundary region expends as P_β and N_β increase. On the other hand, as P_β and N_β decrease, the boundary area shrinks and the positive region and negative region will expand.

Assume B is a subset of condition attribute set, $B \subseteq C$, and D is the decision attribute. Let B^* be a family of equivalent class of the indiscernibility relation $IND_{P,N}(D) = \{P_{class}, N_{class}\}$. For the given classification factors P_β and N_β, the dependency degree of the decision attribute D on the condition attribute set B is defined as

$$\gamma(B, D, P_\beta, N_\beta) = \frac{\sum_{x \in POS_{P_\beta}(P_{class}|B)} d(x) + \sum_{x \in NEG_{N_\beta}(P_{class}|B)} d(x)}{\sum_{x \in U} d(x)}.$$

Simply, the dependency degree $\gamma(B, D, P_\beta, N_\beta)$ of the decision attribute D on the condition attribute set B at precision level P_β and N_β is the proportion of the tuples in U that can be classified into positive or negative class with an error rate less than the pre-specified threshold $(1 - P_\beta)$ and $(1 - N_\beta)$.

Finally, the concept of attribute reduct in UIS is proposed. By substituting the functional dependency degree in the traditional reduct definition with the dependency degree $\gamma(B, D, P_\beta, N_\beta)$ defined as above, the attribute reduct can be generalized to allow for a further reduction of attributes.

For the given classification factors P_β, N_β:

1. an attribute $a \in B$ is redundant in B if

$$\gamma(B - \{a\}, D, P_\beta, N_\beta) = \gamma(B, D, P_\beta, N_\beta);$$

   otherwise a is indispensable;
2. If all attributes in B are indispensable, then B is called *orthogonal*;
3. B is called a *reduct* of the condition attribute set C in UIS if and only if B is *orthogonal* and $\gamma(B, D, P_\beta, N_\beta) = \gamma(C, D, P_\beta, N_\beta)$.

Thus, an attribute reduct is such a subset of condition attributes that the decision attribute has the same dependency degree on it as that on the entire set of condition attributes, and no attribute can be eliminated from it without affecting the dependency degree.

The concept of reduct is very useful in those applications where it is necessary to find the most important collection of condition attributes responsible for a cause-and-effect relationship and also useful for eliminating noise attributes from the information system. Given an information system, there may exist more than one reduct. Each reduct can be used as an alternative group of attributes which could represent the original information system with the classification factors P_β and N_β. An open question is how to select an optimal reduct. It certainly depends on the optimality criteria. The computational procedure for verifying a single reduct is very straightforward, but finding all reducts is hard.

3   Learning Classification Rules

Based on the generalized rough set model, we can design a procedure for learning classification rules as follows.

Assume an uncertain information system $UIS =< U, C, D, \{V_a\}_{a \in C}, f, g, d >$ and classification precision factors P_β and N_β. The rule induction procedure consists of following three steps.

Step 1: Compute the dependency degree of the decision attribute D on the condition attribute set C:

- Calculate the discernibility equivalent classes of the training set U with respect to the condition attribute set C. Generally, each distinct tuple forms an equivalent class. Let X_C be the set of all such equivalent classes;
- For each equivalent class $X \in X_C$, calculate its classification ratios $C_P(X)$ and $C_N(X)$;
- Calculate the positive and negative regions of U with respect to C, $POS_{P_\beta}(U|C)$ and $NEG_{N_\beta}(U|C)$, in terms of the classification ratios above;
- Calculate the dependency degree of D on C according to the positive and negative regions, that is, $\gamma(C, D, P_\beta, N_\beta)$.

Step 2: Find the generalized attribute reducts of the condition attribute set C according to the attribute dependency:

For each non-empty subset B of C, $B \subseteq C$,
- Calculate the discernibility equivalent classes of U with respect to B. Let X_B be the set of all such equivalent classes;
- For each equivalent class $X \in X_B$, calculate its classification ratios $C_P(X)$ and $C_N(X)$;
- Calculate the positive and negative regions of U with respect to B, $POS_{P_\beta}(U|B)$ and $NEG_{N_\beta}(U|B)$;
- Calculate the dependency degree of D on B, that is, $\gamma(B, D, P_\beta, N_\beta)$.
- Compare the dependency degree of D on C with that of D on B. If $\gamma(C, D, P_\beta, N_\beta) = \gamma(B, D, P_\beta, N_\beta)$, then B is an attribute reduct of C; otherwise it is not.

Let $RED(C, D, P_\beta, N_\beta)$ be the set of all generalized attribute reducts of C.

Step 3: Construct classification rules with certainty factors. For each attribute reduct $B \in RED(C, D, P_\beta, N_\beta)$ of C, a set of rules can be achieved as follows with each rule corresponding to an equivalent class with respect to B.

Assume B consists of m attributes, $B = \{B_1, B_2, \ldots, B_m\}$, and X is an equivalent class with respect to B, $X \in X_B$. According to the definition of discernibility relation, all tuples in X have the same attribute values for all attributes in B, assuming $B_1 = b_1, B_2 = b_2, \ldots, B_m = b_m$. A classification rule for the positive class can be constructed according to X.

$Rule_P(X)$:   **If** $B_1 = b_1, B_2 = b_2, \ldots, B_m = b_m$
        then D=*positive class* $(CF = CF_P(X))$,

where the certainty factor $CF_P(X)$ is defined as $CF_P(X) = \min_{x \in X} g(x)$.

Similarly, a classification rule for the negative class can be obtained as

$Rule_N(X)$: **If** $B_1 = b_1, B_2 = b_2, \ldots, B_m = b_m$
    then $D=negative\ class$ $(CF = CF_N(X))$,

where the certainty factor $CF_N(X)$ is defined as $CF_N(X) = \min_{x \in X}(1 - g(x))$.

For the dual rules $Rule_P(X)$ and $Rule_N(X)$, they have the same condition part but opposite decision labels with different certainty factors, and the sum of their certainty factors is not 1. This shows that believing that an example belongs to the positive class with certainty of α does not mean believing it belonging to the negative class with certainty of $1 - \alpha$. It can be proved that $CF(Rule_P(X)) + CF(Rule_N(X)) \leq 1$, which means that there may exist a certainty boundary for which we know nothing.

Thus, for reduct B, we achieve two sets of classification rules, one for the positive class and the other for the negative class, and each set consists of $card(X_B)$ rules with different certainty factors.

4   An Illustration

In this section, we consider an example using the GRS model to learn classification rules. Table 1 illustrates a set of training examples, $U = \{e_i\}$, ($i = 1, 2, \ldots, 6$). The set of condition attributes is $C = \{C_1, C_2\}$ and their domains are $V_{C_1} = \{0, 1\}$ and $V_{C_2} = \{0, 1, 2\}$, respectively. The binary decision attribute is $D = \{0, 1\}$. Each tuple in the table is labeled as the positive class 1 with a certainty (column g), and assigned a importance degree (column d).

Table 1. An example of an uncertain information system

U	C_1	C_2	D	g	d
e_1	0	0	1	0.95	1
e_2	0	1	1	0.67	0.75
e_3	0	2	1	0.15	1
e_4	1	0	1	0.85	1
e_5	1	1	1	0.47	0.75
e_6	1	2	1	0.10	1

Assume $P_\beta = 0.85$, $N_\beta = 0.80$. Initially, because all tuples are distinct, each of them forms an equivalent class with respect to the discernibility relation. Thus, we have six elementary sets, which are $X_1 = \{e_1\}, X_2 = \{e_2\}, \ldots$, and $X_6 = \{e_6\}$. Compute their classification ratio as follows:

	X_1	X_2	X_3	X_4	X_5	X_6
C_P	0.95	0.67	0.15	0.85	0.47	0.10
C_N	0.05	0.33	0.85	0.15	0.53	0.90

Since $C_P(X_1) \geq P_\beta$, $C_P(X_4) \geq P_\beta$ and $C_N(X_3) \geq N_\beta$, $C_N(X_6) \geq N_\beta$, taking the concept as the entire table (because all tuples in U are labeled as the positive class, though, with uncertainty factor), we have $POS_{0.85}(U|C) = \{X_1, X_4\}$, and $NEG_{0.80}(U|C) = \{X_3, X_6\}$. The boundary region is composed of other examples, $BND_{0.85, 0.80}(U|C) = \{X_2, X_5\}$.

If we want the positive and negative regions to be more "pure", we can increase P_β and N_β. Suppose $P_\beta = 0.9$, $N_\beta = 0.9$, then we have $C_P(X_1) \geq P_\beta$, and $C_N(X_6) \geq N_\beta$. Hence, $POS_{0.90}(U|C) = \{X_1\}$, $NEG_{0.90}(U|C) = \{X_6\}$, and $BND_{0.90, 0.90}(U|C) = \{X_2, X_3, X_4, X_5\}$. The equivalent classes X_3 and X_4 are no longer good enough to be in the positive region, so they are put in the boundary and the positive region and negative region shrink.

We can calculate the degree of dependency between the condition attribute set C and the decision attribute D with classification factors $P_\beta = 0.85$ and $N_\beta = 0.80$. According to above results, the degree of dependency between C and D is, $\gamma(C, D, 0.80, 0.85) = 0.73$.

Dropping the condition variable C_1 from C, we get a subset $C' = \{C_2\}$. Assume $P_\beta = 0.85$ and $N_\beta = 0.80$ again. The discernibility equivalent classes are $X_1 = \{e_1, e_4\}$, $X_2 = \{e_2, e_5\}$ and $X_3 = \{e_3, e_6\}$ according to the condition attribute set C'. Compute C_P and C_N for each equivalent class as follows: $C_P(X_1) = 0.90$, $C_N(X_1) = 0.10$, $C_P(X_2) = 0.57$, $C_N(X_2) = 0.43$, $C_P(X_3) = 0.125$, and $C_N(X_3) = 0.875$.

It is easy to see that $POS_{0.85}(U|C') = \{X_1\}$ and $NEG_{0.80}(U|C') = \{X_3\}$. Thus, we have $\gamma(C', D, 0.80, 0.85) = 0.73$.

From above, we know that $\gamma(C', D, 0.80, 0.85) = \gamma(C, D, 0.80, 0.85)$, so $C' = \{C_2\}$ is a reduct of C on D.

Therefore, three classification rules can be achieved from the three equivalent classes X_1, X_2, and X_3, respectively.

Rule 1: If $C_2 = 0$ then $D = 1$ *(positive class)* with $CF = 0.85$.
Rule 2: If $C_2 = 1$ then $D = 1$ *(positive class)* with $CF = 0.47$.
Rule 3: If $C_2 = 2$ then $D = 1$ *(positive class)* with $CF = 0.10$.

Similarly, the three converse classification rules for the negative class can be obtained.

5   Concluding Remarks

In this paper we analyzed the limitations of the traditional rough set theory and extended it to a generalized rough sets model for modeling the classification process in the noise environment. We developed an approach for classification rules induction based on this generalized model. The illustration shows that this approach works well for dealing with unprecise training examples, especially the uncertain class labeling. This is crucial for the cases in which it is expensive or risk to correctly and precisely label the training examples. We also considered the importance degrees of training examples. Different training examples may have different contributions to the classification. In real world applications, this is helpful for users to specify which are important examples and which are not. Generally, the typical examples recognized by domain experts are always important, while the examples automatically collected ma y not be so important.

There are two questions which need to be addressed in our future research. W e only discussed the simple binary classification in which the decision attribute has only two values of *positive* or *negative* class. For the multiple classifications we must solve such problems that how the uncertainty factor is attached to each label and how the final rules are built. In addition, we assumed that condition attributes are all discrete valued. For the numerical attributes the discretization must be performed before the induction procedure starts [3].

Another question is about the calculation of attribute reducts. As mentioned in the paper, verifying a reduct is straightforward, but finding all reducts is hard. There have been some approaches for attacking this problem in literature [2, 4, 5]. The results obtained are significant and encouraging. This will be one of our next research tasks.

Ac kno wledgmen: t The authors are mem bers of the Institute for Robotics and Intelligent Systems (IRIS) and wish to acknowledge the support of the Networks of Centres of Excellence of the Government of Canada, the Natural Sciences and Engineering Research Council, and the participation of PRECARN Associates Inc.

References

1. Fayyad, U., Piatetsky-Shapiro, G., Smyth, P. and Uthurusamy, R. (1996) Advances in Knowledge Discovery and Data Mining, AAAI Press/MIT Press.
2. Hu, X. and Cercone, N. (1996) Mining Knowledge Rules from Databases: An Rough Set Approach, in Proc. of the 12th International Conf. on Data Engineering, 96-105.
3. Hu, X. and Cercone, N. (1999) Data Mining via Generalization, Discretization and Rough Set Feature Selection, Knowledge and Information System: An International Journal 1(1), 33-66.
4. Lin, T. Y. and Cercone, N. (1997) Applications of Rough Sets Theory and Data Mining, Kluwer Academic Publishers.
5. Pawlak, Z. (1991) Rough Sets: Theoretical Aspects of Reasoning About Data, Kluwer Academic Publishers.
6. Wong, S. K. M., Ziarko, W. and Le, R. L. (1986). Comparison of Rough-set and Statistical Methods in Inductive Learning. International Journal of Man-Machine Studies 24, 52-72, 1986.
7. Ziarko, W. (1993) V ariable Precision Rough Set Model, Journal of Computer and System Sciences 46(1), 39-59.
8. Ziarko, W. (1994) Rough Sets, F uzzy Sets and Kno wledge Discovery, Springer-Verlag.

Unification of Knowledge Discovery and Data Mining using Rough Sets Approach in A Real-World Application

Julia Johnson, Mengchi Liu, Hong Chen

Department of Computer Science
University of Regina
Regina, SK S4S 0A2 Canada
e-mail: {johnson, mliu, hongc}@cs.uregina.ca

Abstract. In today's fast paced computerized world, many business organizations are overwhelmed with the huge amount of fast growing information. It is becoming difficult for traditional database systems to manage the data effectively. Knowledge Discovery in Databases (KDD) and Data Mining became popular in the 1980s as solutions for this kind of data overload problem. In the past ten years, Rough Sets theory has been found to be a good mathematical approach for simplifying both the KDD and Data Mining processes. In this paper, KDD and Data Mining will be examined from a Rough Sets perspective. Based on the Rough Sets research on KDD that has been done at the University of Regina, we will describe the attribute-oriented approach to KDD. We will then describe the linkage between KDD and Rough Sets techniques and propose to unify KDD and Data Mining within a Rough Sets framework for better overall research achievement. In the real world, the dirty data problem is a critical issue exists on many organizations. In this paper, we will describe in detail how this KDD with Rough Sets approach framework will be applied to solve a real world dirty data problem.

1. Introduction

Many businesses in today's world are overwhelmed with the huge amount of fast growing information. It is becoming more difficult for traditional systems to manage the data effectively. KDD and Data Mining became very popular in the 1980s in discovering useful information from data. The Rough Sets Theory is a mathematical approach for simplifying KDD. KDD and Data Mining have been adopted as solutions for better data management. In the real world, dirty data is a common problem existing in many organizations. Data cleaning is an important application in the KDD application areas.

The rest of the paper is organized as follows. In Section 2, we overview Knowledge Discovery and Data Mining based on paper [1]. In Section 3, we discuss the research on KDD within a Rough Sets approach that has been done at the University of Regina as presented in papers [8] and [9]. In Section 4, we propose the idea of unification of knowledge discovery and data mining using the Rough Sets approach. In Section 5, we describe a real-world KDD application on data cleaning. Finally, in Section 6, we summarize the main ideas from the observations of KDD, Data Mining, and Rough Sets concepts, as well as present the conclusions.

2. Definition of Knowledge Discovery and Data Mining

The definitions for KDD, Data Mining, and the KDD Process ([1], p. 83) are given below:

W. Ziarko and Y. Yao (Eds.): RSCTC 2000, LNAI 2005, pp. 330–337, 2001.

"**Knowledge Discovery in Databases** is the non-trivial process of identifying valid, novel, potentially useful, and ultimately understandable patterns in data."

"**Data Mining** is a step in the KDD process consisting of applying data analysis and discovery algorithms that, under acceptable computational efficiency limitations, produce a particular enumeration of patterns over the data."

> **KDD Process** is the *process* of using the database along with any required selection, preprocessing, subsampling, and transformations of it; to apply data mining methods (algorithms) to enumerate patterns from it; and to evaluate the products of data mining to identify the subset of the enumerated patterns deemed "knowledge".

3. Overview of Rough Sets and Knowledge Discovery

The Rough Sets Theory was first introduced by Zdzislaw Pawlak in the 1980s and this theory provides a tool to deal with vagueness or uncertainty. A Rough Set is defined as a pair of sets, the lower approximation and upper approximation that approximate an arbitrary subset of the domain. The lower approximation consists of objects that are sure to belong to the subset of interest, where the upper approximation consists of objects possibly belonging to the subset [11].

In the past decade Rough Sets theory has become very popular in the research field of Knowledge Discovery and Data Mining. Rough Sets theory has been considered as a good mathematical approach for simplifying the KDD process. The Rough Sets model has been used in various KDD research areas such as marketing research [9], industrial control [9], medicine, drug research [11], stock market analysis and others.

In paper [10], Dr. Ziarko points out that the key idea in the Rough Sets approach is that the imprecise representation of data helps uncover data regularities. Knowledge discovered from data may be often incomplete or imprecise [10]. By using the Rough Sets model, data can be classified more easily using the lower approximation and upper approximation of a set concept [7]. Thus, data can be induced with the best description of the subset of interest, representing uncertain knowledge, identifying data dependencies, and using the discovered patterns to make inferences [7].

Various research on KDD within a Rough Sets approach has been done at the University of Regina. Paper [8] describes the use of Rough Sets as a tool for Knowledge Discovery. The key approach to KDD in this paper is the use of the attribute-oriented Rough Sets method. This method is based on the generalization of the information, which includes the examination of the data at different abstraction levels, followed by discovering, analyzing, and simplifying the relationships for the data. Rough Sets method of the reduction of knowledge is applied to eliminate the irrelevant attributes in the database. During the data analysis and reduction stages, Rough Sets techniques help to analyze the dependencies of the attributes and help to identify the irrelevant attributes during the information reduction process. Thus, the Rough Sets approach helps to generate a minimum subset of the generalized attributes as well as helping the generalized information system to derive the rules about the data dependencies within the database.

Rough Sets theory has been applied in various application areas in the Knowledge Discovery area. Some Rough Sets based software packages such as Datalogic have also appeared on the market [9]. In the real world, a number of systems have been developed and implemented for Data Mining purposes. DBMiner [3] is a real-world data mining and data warehouse system. A model of Rough Sets Data Miner (RSDM) implementation has also been proposed in paper [2] to unify the framework between the Data Mining and Rough Sets techniques.

4. Unification of Knowledge Discovery and Data Mining using Rough Sets Approach

Data Mining is an important step in the KDD process. In paper [1], the authors discuss the challenges of the KDD research field and conclude that there would be an advantage to unify the KDD process and Data Mining. The authors take a first step towards unification by providing a clear overview of KDD and Data Mining. Data Mining methods and the components of the Data Mining algorithms are also described. Following their concept, we would like to take the next step towards the unification of KDD and Data Mining. The next step that we are going to do is applying the KDD application using a Rough Sets approach. The attribute-oriented Rough Sets approach helps to simplify the data reduction process as well as to derive rules about data dependencies in the database. Applications described in paper [9] have also proved that Rough Sets theory can be used for identifying data regularities in the decision tables. RSDM and Datalogic software tool are two examples of this unification of KDD and Data Mining within a Rough Sets framework. Unification of these approaches has a great potential for the overall research achievement in KDD, Data Mining, and Rough Sets research areas. In the next section, we use a real-world KDD application in the area of data cleaning to discuss in more detail a specific KDD application using a Rough Set approach.

5. KDD, Data Mining with Rough Sets Approach in the Real-World Application

5.1 Background of XYZ and ABC & Overview of the ABC Computer System

XYZ is a global consulting firm with its headquarters in Honolulu, Hawaii. XYZ provides the service in developing the pension software systems for its clients. ABC is XYZ's largest client. ABC is a shoes company with its headquarters in Wailuku, Maui. ABC consists of two main divisions: Administration and Benefits.

Prior to the year 1996, ABC Administration and Benefits division used two IBM mainframe systems running independently for its daily operations. At that time, the

Benefits system only supported the Health and Welfare administration and a payment system for the Health and Welfare payment. The pension administration and short-term disability administration used manual processing. In late 1996, ABC requested XYZ to move their existing mainframe based systems to a new consolidated system under the PC Windows95 environment using Visual Basic as the development tool and the Microsoft SQL Server as the database system. Steps for restructuring the ABC systems are below:

(1). Convert the Administration system and the Benefits Health and Welfare systems from IBM mainframe to Visual Basic.

(2). Enhance the payment system, develop the Pension subsystem, Payee subsystem, and Weekly Indemnity (WI) subsystem, and consolidate them with the Health and Welfare subsystem as the new ABC Benefits system.

(3). Merge the Administration and Benefits systems as one consolidated system.

5.2 Description of the ABC Data problems

During the development and implementation for the ABC systems, various data problems arose. The first problem came from the data conversion from IBM mainframe system to Microsoft SQL Server database systems due to the inconsistent data from old systems. The second ABC data problem is due to lack of accurate information. For example, a temporary social security number (i.e. a member's last name) will often be used for the new member when real SSN is not available. Other ABC data problems come from users' typo errors, as well as data errors generated from the new systems due to the new systems software bugs.

The Personal table is the most important table in the ABC Administration and Benefits database. The Personal table is the parent table of many child tables such as Hw_Membership (Health and Welfare Membership) table. When there are duplicated personal records in the Personal table, this might cause two duplicated Hw_Membership records for this member, as well as more data duplication errors for this member's dental and medical claim payments. Thus, fixing the data in the Personal table is the most important starting point for cleaning the ABC data.

5.3 Rough Sets Approach for Solving the ABC data problems

As mentioned above, data cleaning for the Personal table is the key data cleaning issue for the ABC data problem. The Personal table has over 60 attributes and contains over 20,000 records. It is a time-consuming task to check through every single record. The patterns for duplicated personal records are also uncertain. Rough Sets theory will be a good approach for dealing with this Personal data duplication uncertainty. By adopting the attribute-oriented Rough Sets approach [4, 8], the most relevant attributes will be key in deciding the features of possible duplicated personal records. For example, the SSN, last_name, and birth_date are important attributes in determining the uniqueness of a person, however, the cell_phone_number attribute is considered irrelevant attributes in determining the uniqueness of a person. Thus, we eliminate the irrelevant Personal attributes and only keep the relevant Personal attributes for further examination.

Using the Rough Sets theory to analyze the ABC data problem, an information table is constructed as below. The SSN, last_name, first_name, birth_date, gender, address are treated as the condition attributes. An abbreviation of "s" indicates two or more records contain the same values (i.e. same last name). The abbreviation of "d" indicates that two or more records contain different values (i.e. different birth date). The decision attribute "duplicated?" with y (yes) indicates that the examined personal records are referring to the same person and n (no) indicates different persons.

Table 1: Information Table

I.e.	Condition						Decision
	SSN	LName	FName	Birth_Date	Gender	Address	Duplicated?
e1	d	s	s	s	s	d	y
e2	d	d	s	d	d	d	n
e3	s	s	d	s	d	s	n
e4	s	s	d	s	d	s	y
e5	s	s	d	s	d	s	n
e6	s	s	d	s	s	d	y
e7	s	s	d	d	s	d	y
e8	s	s	d	d	s	d	n

To summarize, the indiscernability classes for the information with respect to condition attributes SSN, last_name, first_name, birth_date, gender, address are {e1}, {e2}, {e3, e4, e5}, {e6}, {e7, e8}. The indiscernability classes for the information table with respect to the attribute address deleted are also {e1}, {e2}, {e3, e4, e5}, {e6}, {e7, e8}. When the indiscernability classes are the same for a set of condition attributes and a superset of those condition attributes, then any attribute which is a member of the superset, but not the subset, is redundant.

Inconsistent information occurs in the above information table. For example, e3, e4, e5 have the same values for condition attribute address, but with different outcomes -- e3 and e5 have a No value for the decision and e4 has a Yes value. Also, we see from the table that e7 and e8 belong to the same indiscernability class, but differ in the value of their decision attribute. All the examples in a given indiscernability class have the same values for their condition attributes. The reasons for the indiscernability classes {e3, e4, e5} and {e7, e8}, that is the reasons why the individuals within each of these classes are indiscernible, are explained below:

(1). The reason for e3 is that the data from the old system were stored incorrectly for a deceased member and his widow. Due to lack of personal information about the deceased member, the widow's SSN and birth date were mistakenly stored for the deceased member. The reason for e4 is that two personal records for the same person came from two different subsystems. One subsystem contains the wrong first name and gender information for the person. The reason for e5 is that the data from the old system mixed up both the father's and daughter's personal information when both father and daughter are ABC members.

(2). The reason for e7 is also that two personal records for the same person came from two different subsystems. One subsystem has the wrong first name and birth date entered for the person. However, the reason for e8 is that two persons happen to have the same last name and with the same temporary SSN stored as last name in the system.

The following rules can be deduced based on the information table.
Certain rules: (generated from example e1, e2, and e6)

(SSN, d) ∧ (last_name, s) ∧ (first_name, s) ∧ (birth_date, s) ∧ (gender, s) ⇒ (Duplicated, y)

(SSN, d) ∧ (last_name, d) ∧ (first_name, s) ∧ (birth_date, d) ∧ (gender, d) ⇒ (Duplicated, n)

(SSN, s) ∧ (last_name, s) ∧ (first_name, d) ∧ (birth_date, s) ∧ (gender, s) ⇒ (Duplicated, y)

Possible rules:

(1). (SSN, s) ∧ (last_name, s) ∧ (first_name, d) ∧ (birth_date, s) ∧ (gender, d) ⇒ (Duplicated, y)

(2). (SSN, s) ∧ (last_name, s) ∧ (first_name, d) ∧ (birth_date, s) ∧ (gender, d) ⇒ (Duplicated, n)

(3). (SSN, s) ∧ (last_name, s) ∧ (first_name, d) ∧ (birth_date, d) ∧ (gender, s) ⇒ (Duplicated, y)

(4). (SSN, s) ∧ (last_name, s) ∧ (first_name, d) ∧ (birth_date, d) ∧ (gender, s) ⇒ (Duplicated, n)

Rules (1) and (2) are deduced from example e3, e4, and e5. Rules (3) and (4) are deduced from example e7 and e8. Following the method suggested in [6], the strength of a rule for the uncertain rules is:

$$\frac{\text{\# of positive examples covered by the rule}}{\text{\# of examples covered by the rule (including both positive and negative)}}$$

Based on this definition, rule 1 has a probability of 1/3. Example e4 is a positive example covered by this rule. Example e3 and e5 are negative examples covered by this rule. Similarly, the probability for rule 2 is 2/3, and for both rule 3 and 4 is ½.

5.4 Algorithm

Our algorithm was inspired by the arguments in favor of an equational theory [5] to allow equivalence of individuals in the domain to be determined rather than simply value or string equivalence (i.e., We want to compare actual persons rather than comparing attributes of persons). A KDD approach to data cleaning is taken in work [5] where the data is processed multiple times with different keys on each pass to assist in identifying individuals independent of their attribute values. Separate rules are used in algorithm [5] for name comparison, SSN comparison and address comparison. The results of each comparison are accumulated with the results of the next rule to be applied but all of the rules are precise rules.

In contrast, we have combined uncertain rules within the general framework of algorithm [5] to solve the duplicated ABC Personal records problem. Algorithm [5] refers to an experimentally generated Personnel and Claims multi-database the fields of which bear much similarity to those of our ABC Personal records database. The rules that we have provided in Section 6.4.1 facilitate decision making when cleaning the duplicated personal records. Our algorithm is provided below:

While scan through the whole table, for all possible duplicated personal records,
    /* Part 1: For certain rules. */

Case a: (SSN, d) ^ (last_name, s) ^ (first_name, s) ^ (birth_date, s) ^ (gender, s)
  Merge the Personal records
  Purge the duplicated Personal records
Case b: (SSN, s) ^ (last_name, s) ^ (first_name, d) ^ (birth_date, s) ^ (gender, s)
  Merge the Personal records
  Purge the duplicated Personal records
/* Part II: For possible rules */
Case c: (SSN, s) ^ (last_name, s) ^ (first_name, d) ^ (birth_date, s) ^ (gender, d)
  Insert the record into the exception report
Case d: (SSN, s) ^ (last_name, s) ^ (first_name, d) ^ (birth_date, d) ^ (gender, s)
  Insert the record into the exception report
EndWhile

Those personal records that satisfy certain rules are duplicated records, and thus will be merged. Those personal records that satisfy possible rules will be printed into an exception report. The exception report will be sent to the users to get further detail checking. For those records in the exception report, there is a 33% chance of duplicated records for Case c situation, and there is a 50% chance of duplicated records for Case d situation.

User intervention is required for the data to which the uncertain rules apply. Users usually have to review the exception report and make a decision about data that may or may not be duplicated. However, in many situations, even users find it hard to distinguish whether or not the personal records are duplicated. In addition, the software often encounters bugs that prevent further normal operation of the system because of the dirty data. Data cleaning is a necessary task in order for the system to continue running smoothly. On the other hand, data recovery is a much more time-consuming and costly task. If the data are deleted by mistake, it will be much more difficult to recover the data than to clean the data. The strength of the rule will be useful in helping both the users and the consultants to make a decision about the pros and cons for cleaning the data. The strength of the rule for the rules 1, 2, 3, and 4 is 33%, 66%, 50%, 50% respectively. The precedence of the strength of rule is Rule 2 > Rule 3 = Rule 4 > Rule 1. Thus, when proceeding with data cleaning for fixing the software bugs, we can make a decision to apply rule 2 first. If this fixes the software bug, we do not need to apply rules 1, 3, and 4. If the problem still cannot be solved, we will then apply rule 3 or 4. In the worst case, if the problem still exists after applying rule 3 or 4, then rule 1 will be applied. In conclusion, data cleaning using Rough Sets approach will help clean up the data more accurately, as well as help users make better decisions. The overall data quality will be better improved with less cost.

6. Summary and Conclusions

Researchers often do not distinguish between KDD and Data Mining. At the University of Regina, as we have seen in Section 3, research is being conducted that applies Rough Sets to KDD. Data Mining is a step within the KDD process for finding patterns in a reduced information table. It has been observed in paper [1] that it would be beneficial to clarify the relationship between Data Mining and KDD.

Although the Data Mining step of searching for patterns of interest in the data has received the most attention in the literature, in this paper we addressed additional steps in the KDD process. We have provided a common framework based on a Rough Sets approach for understanding the relationship between KDD and Data Mining. A real-world KDD application of the data cleaning was described in detail in a Rough Sets approach. Our conclusions are as follows:

- From the literature [1], we conclude that Knowledge Discovery in Database and Data Mining are not the same thing
- From the research done at the University of Regina, we have observed that KDD within a Rough Sets approach has great advantages for simplifying the KDD process
- Unifying KDD and Data Mining within a Rough Set approach will benefit the knowledge discovery research overall achievement
- KDD within a Rough Set approach has advantages for a real-world organization data cleaning problem

Besides these advantages of the Rough Sets approach in the areas of KDD and Data Mining, there is still a lot of potential usage to be discovered in KDD and Data Mining areas from the Rough Sets approach. We are sure that more and more researchers have realized the importance of unifying KDD, Data Mining, and Rough Sets.

References

[1] Fayyad, U., Piatetsky-Shapiro, and G., Smyth, P. 1996. Knowledge Discovery and Data Mining: Towards a Unifying Framework. *Proceedings KDD-96*, pp. 82-88.

[2] Fernandez-Baizan, M., Ruiz, E., and Wasilewska, A., 1998. A Model of RSDM Implementation. *Rough Sets and Current Trends in Computing*, Springer, pp. 186-193.

[3] Han, J., Fu, Y., Wang, W., and etc. 1996. DBMiner: A System for Mining Knowledge in Large Relational Databases. *Proceedings KDD-96*, pp. 250-255.

[4] Hu, XiaoHua, 1995. *Knowledge Discovery in Database: An Attribute-Oriented Rough Set Approach*. University of Regina.

[5] Hernandez, M., and Stolfo, S., Real-world Data is Dirty: Data Cleansing and The Merge/Purge Problem. URL:
http://www.cs.columbia.edu/~sal/merge-purge.html

[6] Pawlak, Z., Grzymala-Busse J., Slowinski, R., Ziarko, W., Rough Sets. *Communications of the ACM*. vol38, no11, November 1995.

[7] Piatetsky-Shapiro, G., *An Overview of Knowledge Discovery in Database: Recent Progress and Challenges*, in Ziarko, W.(ed.), Rough Sets, Fuzzy Sets and Knowledge Discovery, Springer-Verlag, pp. 1-10.

[8] Shan, N., Ziarko, W., Hamilton, H, and Cercone, N. 1995. Using Rough Sets as Tools for Knowledge Discovery. *Proceedings KDD-95*, pp.263-268.

[9] Ziarko, W., Knowledge Discovery by Rough Sets Theory. *Communications of the ACM*. vol. 42, no. 11, 1999.

[10] Ziarko, W., *Rough Sets and Knowledge Discovery: An Overview*, in Ziarko, W. (ed.), Rough Sets, Fuzzy Sets and Knowledge Discovery, Springer-Verlag, pp. 11-14.

[11] A Brief Introduction to Rough Sets, Electronic Bulletin of the Rough Set Community, URL: http://cs.uregina.ca/~roughset/rs.intro.txt

The Application of a Distance Learning Algorithm in Web-Based Course Delivery

Aileen H. Liang, Wojciech Ziarko, and Brien Maguire

Department of Computer Science, University of Regina, Canada S4S 0A2
{hliang,ziarko,rbm}@cs.uregina.ca

Abstract. This paper discusses the implementation of the Distance Learning Algorithm (DLA), which is derived from the Rough Set Based Inductive Learning Algorithm proposed by Wong and Ziarko in 1986. Rough Set Based Inductive Learning uses Rough Set theory to find general decision rules. Because this algorithm was not designed for distance learning, it was modified into the DLA to suit the distance learning requirements. In this paper, we discuss implementation issues.

1 Introduction

As distance education over the World Wide Web (WWW) becomes more and more popular, the lack of contact and feedback between online students and the instructor, inherent to the course delivery mode, becomes a growing concern. For this reason, the Distance Learning Algorithm (DLA) has been proposed to overcome some of the problems involved [2]. DLA is derived from the Rough Set Based Inductive Learning Algorithm [4]. Inductive Learning is a research area in Artificial Intelligence. It has been used to model the knowledge of human experts by using a carefully chosen sample of expert decisions to infer decision rules. Rough Set Based Inductive Learning uses Rough Set theory to compute general decision rules. Because the Rough Set-Based Inductive Learning algorithm was not designed for distance learning, we have modified it into DLA to fit distance learning situations. Furthermore, we have implemented it using Java to make it more portable in a distance delivery environment.

In this paper, we discuss the implementation of the Distance Learning Algorithm. We illustrate how a Decision Tree can be used to find the reduced information table. This paper is organized as follows. Section 2 gives an overview of distance education. Section 3 introduces Rough Sets and Inductive Learning. Section 4 describes the Distance Learning Algorithm. Section 5 discusses the implementation of DLA. Section 6 concludes the paper and points out future work.

2 Overview of Distance Education

Distance Education is a form of education that is perhaps easiest to classify by describing it as the sum of delivery methodologies that have moved away from

W. Ziarko and Y. Yao (Eds.): RSCTC 2000, LNAI 2005, pp. 338–345, 2001.

the traditional classroom environment. From the simple correspondence course, broadcast TV with reverse audio, to specialized video-conferencing tools, such as Proshare or Flashback, and web-based course [7], distance-based education has helped many people obtain college credits, complete training, or update knowledge to adapt to the new information society. Distance Education differs from traditional education in many ways. It offers educational programs to facilitate a learning strategy without day-to-day contact teaching. It provides interactive study materials and decentralized learning facilities where students seek academic and other forms of educational assistance when they need it. With the recent popularity of the WWW, web-based distance education has become more and more popular due to its convenience, low cost, suitability to delivery at a distance, and flexible time scheduling features.

3   Rough Sets and Inductive Learning

Rough Set theory was introduced by Zdzislaw Pawlak in the early 1980s [?] as a mathematical tool for approximate modeling of classification problems. It is based on an information table that is used to represent those parts of reality that constitute our domain of interest. Given an information table of examples U with distinct attributes, Rough Set theory allows us to classify the information table in two different ways: by the condition attributes and by a decision attribute in the information table to find equivalence classes called indiscernability classes $U = \{X_1, ..., X_n\}$. Objects within a given indiscernability class are indistinguishable from each other on the basis of those attribute values. Each equivalence class based on the decision attribute defines a concept. We use $Des(X_i)$ to denote the description, i.e., the set of attribute values, of the equivalence class X_i.

Rough Set theory allows a concept to be described in terms of a pair of sets, lower approximation and upper approximation of the class.

Let Y be a concept. The lower approximation $\underline{Y}$ and the upper approximation $\overline{Y}$ of Y are defined as

$$\underline{Y} = \{e \in U \mid e \in X_i \text{ and } X_i \subseteq Y\}$$
$$\overline{Y} = \{e \in U \mid e \in X_i \text{ and } X_i \cap Y \neq \emptyset\}$$

In other words, the lower approximation is the intersection of all those equivalence classes that are contained by Y and the upper approximation is the union of all those equivalence classes that have a non-empty intersection with Y.

If an element is in $\overline{Y} - \underline{Y}$, we cannot be certain if it is in Y. Therefore, the notion of a discriminant index of Y has been introduced to measure the degree of certainty in determining whether or not elements in U are members of Y [5]. The discriminant index of Y is defined as follows:

$$\alpha_{Q_1}(Y) = 1 - \frac{|\overline{Y} - \underline{Y}|}{|U|}$$

For the information table, not all condition attributes are necessary to classify the information table. There exists a minimal subset of condition attributes that suffices for the classification. The process used to obtain such a subset is called knowledge reduction [3], which is the essential part of Rough Set theory. The minimal set of such essential condition attributes is called the reduct of the information table.

To describe this more clearly, let us first define a concept of a positive region of a classification with respect to another classification. For mathematical reasons, we use equivalence relations here instead of classifications.

Let R be an equivalence relation over U. We use U/R to denote the family of all equivalence classes of R. Let $\mathbf{R}$ be a family of equivalence relations over U. Then $\cap\mathbf{R}$, the intersection of all equivalence relations belonging to $\mathbf{R}$, is also an equivalence relation and is denoted by $IND(\mathbf{R})$. Then $U/IND(\mathbf{R})$ is the family of all equivalence classes of the equivalence relation $IND(\mathbf{R})$. Let R be an equivalence relation in $\mathbf{R}$. Then R is indispensable if $U/IND(\mathbf{R} - R) \neq U/IND(\mathbf{R})$. If every $R \in \mathbf{R}$ is indispensable, then $U/IND(\mathbf{R})$ is a reduct of U.

Let Q be the equivalence relation over U and P a concept, which is an equivalence class based on the decision attribute. The *P-positive* region of Q, denoted $POS_P(Q)$, is defined as follows:

$$POS_P(Q) = \bigcup_{X \in U/Q} X$$

In other words, the P-positive region of Q is the set of all objects of the universe U which can be classified to classes of U/Q employing knowledge expressed by the concept P.

Let $\mathbf{P}$ be a family of equivalence relations over U and Q an equivalence relation. Then an equivalence relation $R \in \mathbf{P}$ is *Q-dispensable* in $\mathbf{P}$ if

$$POS_{IND(\mathbf{P})}(Q) = POS_{IND(\mathbf{P}-\{R\})}(Q)$$

Otherwise, R is Q-indispensable in $\mathbf{P}$.

If every R in $\mathbf{P}$ is Q-dispensable in P, we say that $\mathbf{P}$ is Q-independent (or $\mathbf{P}$ is independent with respect to Q).

The family $S \subseteq P$ is called a Q-reduct of P, if and only if S is the Q-independent subfamily of P, and $POS_S(Q) = POS_P(Q)$.

Inductive Learning, a research area in Artificial Intelligence, is used to model the knowledge of human experts by using a carefully chosen sample of expert decisions and by inferring decision rules automatically, independent of the subject of interest. Rough set based Inductive Learning uses Rough Set theory to compute general decision rules [4, 6]. Their closeness determines the relationship between the set of attributes and the concept.

4   Distance Learning Algorithm

The Inductive Learning Algorithm proposed by Wong and Ziarko [4] was not designed for a distance learning situation for the following reasons. First, it only outputs deterministic rules at the intermediate level. For distance education, nondeterministic rules at the intermediate step can inform online students about important information useful in guiding their distance learning. Second, for distance education, we are primarily interested in one concept Y, such that $Des(Y) = \{Fail\}$, i.e., the failure concept. We want to find out what causes online students to fail. In contrast, the Inductive Learning Algorithm covers multiple concepts. Thus, we have adapted the Inductive Learning Algorithm to a distance education environment and the result is the DLA [2]. Unlike the Inductive Learning Algorithm, DLA calculates the reduct as well.

The DLA algorithm has four steps. Step 1 computes the reduct. Step 2 initializes variables. As long as the number of condition attributes is not zero, Step 3, the while loop, is executed. Inside the while loop, the discriminant indices of condition attributes are computed first, then the highest index value is stored into result and this condition attribute is removed from the condition attributes. Next, the algorithm outputs the deterministic or non-deterministic decision rules and determines a new domain. If the new domain is empty, the algorithm executes Step 4 that outputs all the deterministic or non-deterministic decision rules [2].

5   Implementation of the Distance Learning Algorithm

Before discussing the implementation, we first introduce the problem. There is a text file that contains students' records including assignments, quizzes, and one final examination. Failure in the final examination means failure in the course. Therefore, the main issue here is that we want to find out what material students failed to understand as shown by their results in assignments and quizzes. The assumption is that the lack of understanding of key areas of the course material results in failure on the final exam. Thus, we use a table of results from course work to determine the rules associated with failure on the final exam. Then we can inform students in subsequent courses of the core sections of the course and provide guidance for online students. The following table is from a Java class at the University of Regina in which students use the Web for their course material. There are 115 students, 6 quizzes, and one final examination:

	Quiz 1	Quiz 2	Quiz 3	Quiz 4	Quiz 5	Quiz 6	FINAL
s_1	98	100	90	89	91	85	90
s_2	100	90	30	45	55	32	40
s_3	68	70	80	89	91	85	85
s_4	88	100	80	69	75	85	81
s_5	76	65	50	70	46	49	46
..	..	..	..	..	..	..	..
s_{115}	58	70	90	60	78	97	87

We chose the Java programming language to do the implementation. The reasons for choosing this programming language included the following factors. First, Java is becoming more and more popular. Second, it has various utilities that can simplify the programming task. Third, Java aims to become a universal programming language running on a wide variety of machines. Finally, it can easily be used on the web applications. To implement this work, we divide the task into five classes. The five classes are: RoughSet class, ReadData class, Reduct class, Inductive class, and Strength class.

RoughSet Class This class imports the ReadData class, Reduct class, Inductive class, and Strength class. It contains the main method that makes this class the executable class. It defines several two dimensional arrays to hold the converted information table (students' records) and calls the methods that are defined in the imported classes.

ReadData Class This class reads the student information needed from any text file, converts percentage marks into pass or fail marks, combines entries with the same values, and generates a small table. The StringTokenizer class predefined in the Java language is used to process the text file. Using Table 1 as our example, the ReadData class reads Table 1, combines the students, and derives the collapsed student information table that contains the eight sample classes show in the table below. The total number of students in Table 2 is 104. The other 11 students received failure scores in all areas and do not need to be considered here.

	Quiz 1	Quiz 2	Quiz 3	Quiz 4	Quiz 5	Quiz 6	FINAL	Total
e_1	p	p	p	p	p	p	p	76
e_2	p	p	f	f	p	f	f	4
e_3	p	f	f	p	p	f	f	3
e_4	p	p	f	p	p	p	p	6
e_5	p	p	f	p	p	f	f	1
e_6	p	p	p	f	f	f	f	5
e_7	f	p	p	p	p	p	p	8
e_8	p	p	p	p	p	f	p	10

Reduct Class The ReadData class has successfully obtained the collapsed table, but it may contain redundant condition attributes, which means that some condition attributes may not determine the decision at all. Therefore, we need to find the reduct and remove the redundant attributes.

To implement this step, linear searching and sorting algorithms are used to accomplish this task and a three dimensional array is used to hold the equivalence relations; that is, the family $\mathbf{R} = \{Q1, Q2, Q3, Q4, Q5, Q6, Final\}$. The **positive region** of R is held in an one dimensional array. $POS_{\mathbf{R}}(F) = \{e_1, e_2, e_3, e_4, e_5, e_6, e_7, e_8\}$

In order to compute the reduct of R with respect to Final, we then have to find out whether the family R is F-dependent or not. We use a while loop to remove Q1, Q2, Q3, Q4, Q5, and Q6 respectively and check the positive region in the mean time. If the positive region of each quiz is different from the positive region of Final, we keep this condition attribute in order to find the reduct. The following is the partial result:

$$U/IND(\mathbf{R} - \{Q1\}) = \{\{e_1, e_7\}, \{e_2\}, \{e_3\}, \{e_4\}, \{e_5\}, \{e_6\}, \{e_8\}\}$$
$$POS_{\mathbf{R}-\{Q1\}}(F) = \{e_1, e_2, e_3, e_4, e_5, e_6, e_7, e_8\} = POS_{\mathbf{R}}(F)$$
$$U/IND(\mathbf{R} - \{Q2\}) = \{\{e_1\}, \{e_2\}, \{e_3, e_5\}, \{e_4\}, \{e_6\}, \{e_7\}, \{e_8\}\}$$
$$POS_{\mathbf{R}-\{Q2\}}(F) = \{e1, e_2, e_3, e_4, e_5, e_6, e_7, e_8\} = POS_{\mathbf{R}}(F) \ldots$$
$$U/IND(\mathbf{R} - \{Q6\}) = \{\{e_1, e_7, e_8\}, \{e_2, e_3, e_4, e_5\}, \{e_6\}\}$$
$$POS_{\mathbf{R}-\{Q6\}}(F) = \{e_1, e_6, e_7, e_8\} \neq POS_{\mathbf{R}}(F)$$

Because each positive region of Quiz 3, Quiz 5, and Quiz 6 does not equal to the positive region of Final, $Quiz3, Quiz5$ and $Quiz6$ are indispensable. Therefore, we obtain the reduct in a two-dimensional array as follow:

	Quiz 3	Quiz 5	Quiz 6	FINAL	Total
e_1	p	p	p	p	76
e_2	f	p	f	f	4
e_3	f	p	f	f	3
e_4	f	p	p	p	6
e_5	f	p	f	f	1
e_6	p	f	f	f	5
e_7	p	p	p	p	8
e_8	p	p	f	p	10

Inductive Class The remaining work required is to find the decision rules. To find the decision rules, Inductive Class uses several arrays repeatedly to hold the domain, concept, indiscernibility classes, lower and upper approximations, and discriminant indices in a while loop. During the computation inside the loop, linear searching and sorting algorithms are used many times in order to obtain the correct result. The routine then shows the alpha values, i.e, the discriminant indices. The following is the outcome:

Domain: $U = \{e_1, e_2, e_3, e_4, e_5, e_6, e_7, e_8\}$
Failure Concept: $Y = \{e_2, e_3, e_5, e_6\}$

The indiscernibility classes based on Quiz 3 are:

$Y_1 = \{e_1, e_6, e_7, e_8\}$ $Des(Y_1) = \{Quiz3 = P\}$
$Y_2 = \{e_2, e_3, e_4, e_5\}$ $Des(Y_2) = \{Quiz3 = F\}$
upper approximation: $\overline{Y} = \{e_1, e_6, e_7, e_8, e_2, e_3, e_4, e_5\}$
lower approximation: $\underline{Y} = \{\}$
discriminant index: $\alpha_{Q_3} = 1 - |\overline{Y} - \underline{Y}|/|U| = 0.00$

The rest are computed similarly and we obtain the results shown in the following table:

Attribute	α value
Quiz 3	0.00
Quiz 5	0.125
Quiz 6	0.375

Because Quiz 6 has the highest discriminant index that best determines its membership in the concept Y, we obtain the first decision rule.

$$Rule1 : \{Quiz6 = f\} \Rightarrow \{Final = f\}$$

The decision tree below is used here to find the new domain which is the same as the formula mentioned in the Distance Learning Algorithm:

$$E' = E' - [(E' - \overline{Y}) \cup \underline{Y}]$$

{e1:p,e4:p,e7:p}   {e2:f,e3:f,e5:f,e6:f,e8:p}

Based on the decision tree, we obtain the horizontally reduced table as follows:

	Quiz 3	Quiz 5	Quiz 6	FINAL	Total
e_2	f	p	f	f	4
e_3	f	p	f	f	3
e_5	f	p	f	f	1
e_6	p	f	f	f	5
e_8	p	p	f	p	10

We then merge Quiz 6 with the rest of the condition attributes to find the highest discriminant indices. That produces two combinations: Quiz 3 and Quiz 6, Quiz 5 and Quiz 6. The following table shows the result of the second round of the program:

Attribute	α value
Quizzes 3 and 6	0.60
Quizzes 5 and 6	0.20

Because Quizzes 3 and 6 have the highest discriminant index, we thus obtain the second decision rule:

$$Rule2 : \{Quiz3 = f, Quiz6 = f\} \Rightarrow \{Final = f\}$$

The resultant decision tree is shown in the following figure:

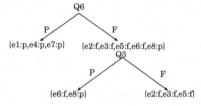

By applying the same method, we finally obtain the last rule:

$$Rule3 : \{Quiz5 = f, Quiz6 = f\} \Rightarrow \{Final = f\}$$

Strength Class We have the rules available now, but how strongly should online students believe in these rules? The Strength class answers that question. It finds the strength of each rule by using the following formula:

$$\frac{\#\ of\ positive\ cases\ covered\ by\ the\ rule}{\#\ of\ cases\ covered\ by\ the\ rule\ (including\ both\ positive\ and\ negative)}$$

To implement the last step, we use the linear search algorithm again. The strength of the rules is finally held in an one-dimensional array and at the end of the computation, the program posts this information about the rules, as seen in the following table:

Rules	Strength
R1: {Quiz 6 = f} $\Rightarrow$ {Final = f}	56.52%
R2: {Quiz 3 = f, Quiz 6 = f} $\Rightarrow$ {Final = f}	100.0%
R2: {Quiz 5 = f, Quiz 6 = f} $\Rightarrow$ {Final = f}	100.0%

The information shown in the above table tells readers that the rules measure previous online students' performance. It guides repeating and new online students in focusing their studies and provides information to the course instructor about whether the online materials need to be modified or reordered. The first rule has a probability of 56.52%.h That means if a student fails Quiz 6, he or she has a 56.52% possibility of failing the final. The Second and third rules say that if a student fails to understand the material related to Quizzes 3 and 6, or Quizzes 5 and 6, then the student has a 100% possibility of failing the course. Therefore, the materials related to Quiz 3, 5, and 6 are the core materials for this online class. As the quiz 6 related materials are so important, the instructor, therefore, might wish to provide additional examples to reinforce the understandability of the material.

6   Conclusion

This paper demonstrates that by applying Rough Sets to distance education over the World Wide Web, the problem of limited feedback can be improved. This makes distance education more useful. Rough Sets base distance learning allows decision rules to be induced that are important to both students and instructors. It thus guides students in their learning. For repeating students, it specifies the areas they should focus on according to the rules applied to them. For new students, it tells them which sections need extra effort in order to pass the course. Rough Sets based distance education can also guide the instructor about the best order in which to present the material. Based on the DLA results, the instructor may reorganize or rewrite the course notes by providing more examples and explaining more concepts. Therefore, Rough Sets based distance education improves the state-of-the-art of Web learning by providing virtual student/teacher feedback and making distance education more effective.

References

1. P. L. Brusilovsky. A Framework for Intelligent Knowledge Sequencing and Task Sequencing. In *Proceedings of the 2nd International Conference on Intelligent Tutoring Systems*, pages 499-506, Montreal, Canada, June 10-12, 1992. Springer-Verlag LNCS 608.
2. A. H. Liang, B. Maguire, and J. Johnson. Rough Set Based WebCT Learning. In *Proceedings of the 1st International conference on Web-Age Information Management*, Shanghai, P.R. China, June 21-23, 2000. Springer-Verlag LNCS 1846.
3. Z. Pawlak, Rough Sets. *International Journal of Computation Information Science*, Vol. 11, pages 341-356, 1982.
4. S. K. M. Wong and W. Ziarko, Algorithm for Inductive Learning. *Bulletiun of the Polish Academy of Sciences*, Vol. 34, No. 5-6, pages 241-372, 1986.
5. S.K.M. Wong, W. Ziarko, and R.L. Ye. Comparison of rough-set and statistical methods in inductive learning. *International Journal of Man-Machine Studies*, 24:52–72, 1986.
6. N. Shan, W. Ziarko, H.J. Hamilton, and N. Cercone. Using rough sets as tools for knowledge discovery. In *Proceedings of the International Conference on Knowledge Discovery and Data Mining (KDD'95)*, pages 263–268. AAAI Press, 1995.
7. Kurt Maly, Hussein Abedl-Wahab, C. Michael Overstreet, J. Christian Wild, Ajay K. Gupta, Alaa Youssef, Emilia Stoica, and Ehab S. Al-Shaer. INTERACTIVE DISTANCE LEARNING OVER INTRANETS, In *IEEE Internet Computing*, pages 60-71. v.1, no.1, Febryary 1997.

Searching Decision Rules in Very Large Databases Using Rough Set Theory

Tsau Young ("T.Y.") Lin[1,2] and Hui Cao[1]

[1] Department of Mathematics and Computer Science
San Jose State University, San Jose, California 95192
[2] Berkeley Initiative in Soft Computing
University of California,Berkeley, California 94720
tylin@cs.sjsu.edu; tylin@cs.berkely.edu

1 Introduction

This paper applies rough set theory to find decision rules using ORACLE RDBMS. The major steps include elimination of redundant attributes and of redundant attribute values for eac h tuple. In this paper three algorithms to extract high frequency decision rules from v ery large decision tables are presented. One algorithm uses pure SQL syntax. The other tw o use sorting algorithm and binary tree data structure respectively. The performances among these methods are ev aluated and compared. The use of binary tree structure impro es the computational time tremendously. So pure SQL results indicate some major change to query optimizer may be desirable if it will be used for data mnining.

2 Rough Sets Methodology

A decision table is a variation of a relation [3], in which we distinguish two kinds of v ariables: condition attributes (denoted as $COND$) and decision attributes (denoted as $DECS$); each row represents a decision rule; see Table 1.

	Condition attributes					Decision attributes
Rule #	TEST	LOW	HIGH	NEW	CASE	RESULT
R1	1	0	0	11	2	1
R2	4	0	1	11	6	3
R3	0	1	1	10	2	1
R4	0	1	1	12	20	10
R5	1	1	0	12	20	10
R6	1	1	0	23	60	30
R7	1	0	0	23	60	88

T able 1. Decision table

The most important concept in rough set theory is reduct. A relativ e reduct is a minimal subset of condition attributes that can at least classify the table

W. Ziarko and Y. Yao (Eds.): RSCTC 2000, LNAI 2005, pp. 346–353, 2001.

into the same equivalence classes as decision attributes[1]. For example, the set {LOW, CASE} is one of relative reducts of Table 1.

Applying rough set methodology for data mining is about the reduction or simplification of a decision table:

1. Step 1: elimination of redundant columns. This is to search for relative reducts
2. Step 2: elimination of redundant values for each rule. This is to search for value reducts.

Step 1: search relative reducts

Finding relative reducts is to find the minimal subsets S of $COND$ which decision attributes are (full) functional dependent on, i.e. $S ==> DECS$[3]. However, most databases contain noise, so partial dependency with degree k, is more practical approach. Therefore, the basic algorithm to search for relative reducts is to compute the number of consistent rules with respect to S, cs, then compute the degree of dependency, $k' = cs/$(total rows of database). If $k' >= k$, S is a relative reduct. Otherwise it is not.

Since for each entity, we have to scan the whole database once to decide whether it is a consistent rule or not, the running time for the above algorithm is always $O(n^2)$, where n is the number of entities of the database. Here, we present a faster method. For a partial dependency with degree k, the maximal inconsistent rules allowed, called E, is $E = n*(1-k)$. The number of inconsistent rules actually appearing inside the database is denoted as $exCount$. Obviously when $exCount > E$ the subset S to be checked is not a relative reduct; we can stop checking the rest entities.

Step 2: search value reducts.

To illustrate the idea, we will use Table 1 and one of its relative reduct {TEST, LOW, NEW}. For rule R1, the minimal condition is {TEST, NEW}. That is its value reduct. We present all value reducts of relative reduct {TEST, LOW, NEW} in the Table 2. In other words, Table 2 represents the minimal conditions for each rule.

Rule #	Condition attributes					Decision attributes
	TEST	LOW	HIGH	NEW	CASE	RESULT
R1	1			11		1
R2	4					3
R3				10		1
R4				12		10
R5				12		10
R6		1		23		30
R7		0		23		88

Table 2. Value reducts of Table 1

We summarize the features of value reducts of a rule as follows.

1. For a value subset, if it is unique in this rule, it is a value reduct of this rule. For example, {TEST=1, NEW=11} is unique for rule R1. Using this only can determine rule R1; no other attributes are needed.
2. For a value subset, if it is not unique in this rule, but it always implies the same decision attributes values, it is also a value reduct. For example, in both rule R4 and R5, we have {NEW=12} ==> {RESULT=10}. Therefore, it is the value reduct for both R4 and R5.

The value reducts (or decision rules) we discuss so far have features:

1. all rules are included even though they only appear once in the database;
2. all rules are strictly consistent. That is, if {NEW=23} ==> {RESULT=30} it can not imply other values of RESULT.

This kind of decision rules only work for very clean databases. Actual databases usually contain noises or errors. In the subsequent parts of this paper, we will consider a more useful definition of decision rules, which has the following features.

1. Those rules that appear just a few times are not considered. Only high frequency rules are considered, that is, the number of times it appears must be larger than a threshold, *minsup.* [2]
2. The appearance of errors is considered. For example, if {NEW=23} ==> {RESULT=30} appears 99 times, and other cases also exist, such as {NEW=23} ==> {RESULT=77}, but it only appears once. We say the former has confidence 99%; the latter 1%[4]. When a database contains errors, we consider a rule value reduct (decision rule) as long as its confidence is above a specified minimal confidence, *minconf.*

The rest of this paper will present algorithms to search decision rules that satisfy the above conditions.

3 Algorithm 1: generic SQL implementation

We start from a database-oriented solution, i.e. using SQL queries for database mining. In the following part, we assume the database to be mined is table T, the set of condition attributes of table T, $COND$, include $c1$, $c2$, ... cm, and decision attributes, $DECS$, include $d1$, $d2$,..., dn.

Step 1: search for relative reducts

For a subset S of C to be checked, assuming it is { s1, s2,..., si }, at first, we will group the table T by S, and get a temporary table tmp1. Table tmp1 will contain all different groups by S. The SQL is as follows.

CREATE TABLE tmp1 as
SELECT s1, s2,..., si, count(*) AS c1
FROM T

GROUP BY s1, s2,..., si;

Secondly, we group the table T by both S and decision attributes, and get temporary table tmp2. Table tmp2 will contain all different groups for $S+DECS$ The SQL is as follows.

CREATE TABLE tmp2 as

SELECT s1, s2,..., si , count(*) AS c2

FROM T

GROUP BY s1, s2,..., si , d1, d2,...,dn;

Next, w e pick up inconsistent rules, which should ha ve different count(*) values in table tmp1 and tmp2; and the sum of the count(*) values of these inconsistent rules in table tmp1 is the total n umber of inconsistent rules. The SQL is as follows.

SELECT SUM(tmp1.c1) AS exCount

FROM tmp1, tmp2

WHERE tmp1.s1 = tmp2.s1

...

AND tmp1.sj = tmp2.sj

AND tmp1.c1 = tmp2.c2

If exCount <= E, the $maximum inconsistent rules$ introduced in section 3, S is a relative reduct.

Here, w e will use Table 1 to explain the above operations. After w e group Table 1 by CASE, the table tmp1 and tmp2 we get are as follows.

Table tmp1	
CASE	Count(*) AS c1
2	2
6	1
20	2
60	2

Table tmp2	
CASE	Count(*) AS c2
2	2
6	1
20	2
60	1
60	1

We will pick up the following row from tmp1,

CASE	Count(*)
60	2

This shows that inconsistent rules are those rules with {CASE = 60}, and there are totally tw o inconsistent rules. Thus the exCount is 2. If E is equal to 2, {CASE} is a relative reduct.

Step 2: search for value reducts

T o searc h for value reducts from a relativ e reduct, all of its subsets are chec ked. The operations will be explained using an example. For T able1, after the relative reduct {TEST, LOW, NEW} is obtained, here how to get value reducts related with {NEW} is shown as follows. It is assumed that, $minsup = 3$, and $minconf = 70\%$.

A t first, we group table T by {NEW, RESULT}, and record the counts. SQL is as follows,

CREATE TABLE tmpA as
SELECT NEW, RESULT, count(*)as countA
FROM T
GROUP by NEW, RESULT

Assume the table tmpA we get is as follows (a little different from the original one).

T able tmpA		
NEW	RESULT	countA
11	1	2
11	3	8
10	1	1
12	10	3
23	30	4

Next, we group the table T only by {NEW}, and also record the count. SQL is as follo ws,

CREATE TABLE tmpA as
SELECT NEW, count(*)as countB
FROM T
GROUP by NEW

w e get the table tmpB.

Next, we will pick up value reducts, which are some tuples of table tmpA. These tuples have iden tical values in the subset {NEW} between tmpA and tmpB, countA > minsup, and confidence > minconf. For example, in table tmpA, we will pick up the following tuples,

The SQL we use in this step is as follows,

SELECT tmpA.NEW, tmpA.RESULT, tmpA.countA, tmpA.countA / tmpB.countB

Table tmpB	
NEW	countB
11	10
10	1
12	3
23	4

NEW	RESULT	support	confidence
11	3	8	80%
12	10	3	100%
23	30	4	100%

FROM tmpA, tmpB
where tmpA.NEW=tmpB.NEW
and tmpA.countA >= minsup
and tmpA.countA / tmpB.countB >= minconf;

Step3: removing superset from value reduct table.

The last step for mining decision rules is to remove those supersets from value reduct table. The SQL we use is as follows. Here, we use a special function provided by ORACLE RDBMS, NVL(x, y), which returns x if x is not null, otherwise y.

DELETE FROM valReductTb t1
WHERE ROWID > (
SELECT min(rowid)
FROM valReductTb t2
WHERE t1.decs=t2.decs
and ((t2.conds is null and t1.conds is null)
or NVL(t1.conds, t2.conds)=NVL(t2.conds, t1.conds))
)

Analysis

Mining decision rules need to loop all subsets of condition attributes. Therefore, SQL queries for data mining are very long and complicated because it does not provide a for or **do/while** repetition structure like C. This also forces us to use dynamic SQL. On the other hand, SQL data mining has bad performance. When checking whether {NEW} is a relative reduct, once $exCount > E$, further checking the rest of rows should be stopped immediately. How ever, using SQL, the checking is always executed to all rows. If assuming N is the number of rows, the running time for searching relative reducts is calculated as follows.

1. For two "group by" operations, the time complexity is $O(2NlgN)$.
2. For a "table join" operation, the average running time is $O(NlgN)$.

Thus, the total running time is $O(3NlgN)$ for one subset. For a decision table, usually most of COND subsets are not relative reducts, so SQL method wastes a lot of time on non-reduct COND subsets.

Searching value reducts uses the similar methods: tw o "group by" operations, then "table join". Thus the running time for searching value reducts for eac h COND subset is also $O(3NlgN)$.

4   Algorithm 2: a faster algorithm using sorting

Before w e start chec king a subset S is a relativ e reduct or not, w e sort the table by S using an nlog(n) sorting algorithm. Now, starting from the first row, through comparing tw o neighboring rows we can find inconsistent rules easily. A counter exCount is set to record the number of inconsistent rules. For the kth and (k+1)th row, if they have the same values in attributes { s1, s2, ..sp } but different in decision attributes, they are inconsistent rules; exCount is increased by one. If $exCount > E$, { s1, s2, ..sp } is not a relative reduct, and checking is stopped immediately;

Assuming N is the number of rows of the table to be mined, this algorithm giv es us running time O(NlgN+E) in the best case and O(NlgN+N) in the worst case during checking relative reducts. Since most of COND subsets are not relativ e reducts, the worst case does not happen a lot, and actual running time for this algorithm is much faster than the worst case.

This sorting method can also be used on searching value reducts. How ever, as we show before, searching value reducts requires chec king every ro w, no matter in the best case or w orst case. Thus the total running time is O(NlgN + N). Comparing it with the first algorithm, this does not improve a lot.

5   Algorithm 3: a much faster algorithm using tree structure.

F or algorithm 2, the running time for sorting always exists. Sometimes, this sorting is not necessary. For example, if the first five rows are inconsistent rules, then exCount = 5. If the maximum exceptions allowed, E, is 4, we immediately kno w that{NEW} is not a relative reduct, then stop. In this case, sorting is not necessary at all.

In this section, w e study ho w to avoid sorting process and get better performance. We will use a binary-search-tree data structure. Each node of the tree represents the information of an entity. Assume S is the subset of condition attributes to be check ed. The node of this binary tree has the form:

struct node { values of S,

total number of rules,

list of v alues of decision attributes,

},

Before checking relative reducts, we initialize exCount = 0. When an entity is check ed, a node is built, the tree is searc hed, and the node is inserted in to the tree according to the values of S. During this process, there will be following cases happening:

1. if no node in the tree has identical v alues of S to the new node, the new node is inserted into the tree;
2. if one node in the tree has identical values of S to the new node, in its decision attributes' value list, as long as we find a different value from the new node's values, the total number of rules for this node will be computed as part of exCount. If exCount is larger than E, S is not a relative reduct. We stop immediately.

This algorithm does not require sorting the table. F or the best case, the running time is $O(E)$. For thew orst case, the running time is $O(NlgN)$ when the tree is balanced, but $O(N^2)$ when the tree is linear. In a real-world decision table, there are not many relative reducts, so the worst case happens few times. Comparing with algorithm 1 and 2, this algorithm greatly improves the performance of searching relative reducts.

This algorithm can also be used on searching value reducts. How ever, w e always need to build a tree using all entities so the running time is $O(NlgN)$ or $O(N^2)$. This tree-structure algorithm does not improv e a lot on the second step.

6   Conclusion

1. Rough set methodology is a good tool for mining decision rules from traditional RDBMS, it can be adapted to different noise levels.
2. T raditionalSQL can be used for database mining, but is costly; some adjustments to query optimizers deem necessary.
3. An algorithm using sorting method gives a little better performance.
4. A much faster algorithm using binary-search-tree data structure is developed, which av oids the complete sorting process. The loop stops as soon as it finds sufficient inconsistent tuples.

References

1. Z. P awlak. Rough Sets: Theoretical Aspects of Reasoning about Data, Kluwer Academic, Dordrech t (1991)
2. T. Y. Lin, "Rough Set Theory in Very Large Databases," Symposium on Modeling, Analysis and Simulation, IMACS Multi Conference (Computational Engineering in Systems Applications), Lille, France, July 9-12, 1996, Vol. 2 of 2, 936-941.
3. T. Y. Lin. "An Overview of Rough Set Theory from the Point of View of Relational Databases," Bulletin of International Rough Set Society, vol. 1, no. 1, March 1997.
4. R. Agraw al, R. Srikan t. "F ast Algorithms for Mining Association Rules," in Proceeding of 20th VLDB Conference San Tiago, Chile, 1994.

On Efficient Construction of Decision Trees From Large Databases

Nguyen Hung Son

Institute of Mathematics, Warsaw University
Banacha 2, 02–097, Warsaw, Poland
email: son@mimuw.edu.pl

Abstract. The main task in decision tree construction algorithms is to find the "best partition" of the set of objects. In this paper, we investigate the problem of optimal binary partition of continuous attribute for large data sets stored in *relational databases*. The critical for time complexity of algorithms solving this problem is the number of simple SQL queries necessary to construct such partitions. The straightforward approach to optimal partition selection needs at least $O(N)$ queries, where N is the number of pre-assumed partitions of the searching space. We show some properties of optimization measures related to discernibility between objects, that allow to construct the partition very close to optimal using only $O(\log N)$ simple queries.

Key words: Data Mining, Rough set, Decision tree.

1 Introduction

The philosophy of "rough set" data analysis methods is based on *handling of discernibility between objects* (see [9, 13]). In recent years, one can find a number of applications of rough set theory in Machine Learning, Data Mining or KDD. One of the main tasks in data mining is the classification problem. The most two popular approaches to classification problems are "decision tree" and "decision rule set". Most "rough set" methods are dealing with classification problem by extracting a set of decision rules (see [14, 13, 10]). We have shown in previous papers that the well known discernibility property in rough set theory can be used to build decision tree with high accuracy from data.

The main step in methods of decision tree construction is to find optimal partitions of the set of objects. The problem of searching for optimal partitions of real value attributes, defined by so called cuts, has been studied by many authors (see e.g. [1–3, 11]), where optimization criteria are defined by e.g. height of the obtained decision tree or the number of cuts. In general, all those problems are hard from computational point of view. Hence numerous heuristics have been investigated to develop approximate solutions of these problems. One of major tasks of these heuristics is to define some approximate measures estimating the quality of extracted cuts. In rough set and Boolean reasoning based methods, the quality is defined by the number of pairs of objects discerned by the partition (called *discernibility measure*).

W. Ziarko and Y. Yao (Eds.): RSCTC 2000, LNAI 2005, pp. 354–361, 2001.
© Springer-Verlag Berlin Heidelberg 2001

We consider the problem of searching for optimal partition of real value attributes assuming that the large data table is represented in relational data base. The straightforward approach to optimal partition selection needs at least $O(N)$ simple queries, where N is the number of pre-assumed partitions of the searching space. In this case, even the linear complexity is not acceptable because of the time for one step (see [5, 8]). We assume that the answer time for such queries does not depend on the interval length. We show some properties of considered optimization measures allowing to reduce the size of searching space. Moreover, we prove that using only $O(\log N)$ simple queries, one can construct the partition very close to optimal. We have showing that the main part of the formula estimating the quality of the best cut for independent variables from [8] is the same in case of "fully" dependent variables. Comparing [8], we also extend the algorithm of searching for the best cut by adding the global searching strategy.

2   Basic notions

An *information system* [9] is a pair $\mathbb{A} = (U, A)$, where U is a non-empty, finite set called the *universe* and A is a non-empty finite set of *attributes (or features)*, i.e. $a : U \to V_a$ for $a \in A$, where V_a is called *the value set of a*. Elements of U are called *objects or records*. Two objects $x, y \in U$ are said to be discernible by attributes from A if there exists an attribute $a \in A$ such that $a(x) \neq a(y)$. Any information system of the form $\mathbb{A} = (U, A \cup \{dec\})$ is called *decision table* where $dec \notin A$ is called *decision attribute*. Without loss of generality we assume that $V_{dec} = \{1, \ldots, d\}$. Then the set $DEC_k = \{x \in U : dec(x) = k\}$ will be called the kth *decision class of* $\mathbb{A}$ for $1 \leq k \leq d$. Any pair (a, c), where a is an attribute and c is a real value, is called *a cut*. We say that *"the cut (a, c) discerns a pair of objects x, y"* if either $a(x) < c \leq a(y)$ or $a(y) < c \leq a(x)$.

   The decision tree for a given decision table is (in simplest case) a binary directed tree with *test functions* (i.e. boolean functions defined on the information vectors of objects) labelled in internal nodes and decision values labelled in leaves. In this paper, we consider decision trees using cuts as test functions. Every cut (a, c) is associated with test function $f_{(a,c)}$ such that for any object $u \in U$ the value of $f_{(a,c)}(u)$ is equal to 1 (true) if and only if $a(u) > c$. The typical algorithm for decision tree induction can be described as follows:

1. For a given set of objects U, select a cut (a, c_{Best}) of high quality among all possible cuts and all attributes;
2. Induce a partition U_1, U_2 of U by (a, c_{Best}) ;
3. Recursively apply Step 1 to both sets U_1, U_2 of objects until some stopping condition is satisfied.

   Developing some decision tree induction methods [3, 11] and some supervised discretization methods [2, 6], we should often solve the following problem: *"for a given real value attribute a and set of candidate cuts $\{c_1, \ldots, c_N\}$, find a cut (a, c_i) belonging to the set of optimal cuts with high probability."*.

Definition 1 *The d-tuple of integers $\langle x_1, ..., x_d \rangle$ is called class distribution of the set of objects $X \subset U$ iff $x_k = card(X \cap DEC_k)$ for $k \in \{1, ..., d\}$. If the set of objects X is defined by $X = \{u \in U : p \le a(u) < q\}$ for some $p, q \in \mathbb{R}$ then the class distribution of X can be called* **the class distribution in** $[p; q)$.

Any cut $c \in \mathbf{C}_a$ splits the domain $V_a = (l_a, r_a)$ of the attribute a into two intervals: $I_L = (l_a, c)$; $I_R = (c, r_a)$. We will use the following notation:

- U_{L_j}, U_{R_j} – the sets of objects from j^{th} class in I_L and I_R. Let $U_L = \bigcup_j U_{L_j}$ and $U_R = \bigcup_j U_{R_j}$ where $j \in \{1, ..., d\}$;
- $\langle L_1, .., L_d \rangle$ and $\langle R_1, .., R_d \rangle$ – class distributions in U_L and U_R. Let $L = \sum_{j=1}^{d} L_j$ and $R = \sum_{j=1}^{d} R_j$
- $C_j = L_j + R_j$ – number of objects in the jth class;
- $n = \sum_{i=1}^{d} C_j = L + R$ – the total number of objects;

Usually, for a fixed attribute a and the set of all relevant cuts $\mathbf{C}_a = \{c_1, ..., c_N\}$ on a, we use some *measure (or quality functions)* $F : \{c_1, ..., c_N\} \to \mathbb{R}$ to estimate the quality of cuts. For a given measure F, the *straightforward algorithm* should compute the values of F for all cuts: $F(c_1), .., F(c_N)$. The cut c_{Best} which maximizes or minimizes the value of function F is selected as the result of searching process. The most popular measures for decision tree induction are *"Entropy Function"* and *"Gini's index"* [4, 1, 11]. In this paper we consider the *discernibility measure* as a quality function. Intuitively, energy of the set of objects $X \subset U$ can be defined by the number of pairs of objects from X to be discerned called $conflict(X)$. Let $\langle N_1, ..., N_d \rangle$ be a class distribution of X, then $conflict(X)$ can be computed by $conflict(X) = \sum_{i<j} N_i N_j$. The cut c which divides the set of objects U into U_1, and U_2 is evaluated by $W(c) = conflict(U) - conflict(U_1) - conflict(U_2)$ i.e. the more is number of pairs of objects discerned by the cut (a, c), the larger is chance that c can be chosen to the optimal set of cut. Hence, in the decision tree induction algorithms based on Rough Set and Boolean reasoning approach, the quality of a given cut c is defined by

$$W(c) = \sum_{i \ne j}^{d} L_i R_j = \sum_{i=1}^{d} L_i \sum_{i=1}^{d} R_i - \sum_{i=1}^{d} L_i R_i \qquad (1)$$

This algorithm is called Maximal-Discernibility heuristics or *the MD-heuristics*. The high accuracy of decision trees constructed by using MD-heuristic and their comparison with Entropy-based decision methods has been reported in [7].

For given set of candidate cuts $\mathbf{C}_a = \{c_1, .., c_N\}$ on a, by median of the kth decision class (denoted by $Median(k)$) we mean the cut $c \in \mathbf{C}_a$ minimizing the value $|L_k - R_k|$. Let $c_{min} = \min_i\{Median(i)\}$ and $c_{max} = \max_i\{Median(i)\}$ we have shown (in [8]) the technique for irrelevant cut eliminating called *"Tail cuts can be eliminated"* as follows.

Theorem 1 *The quality function $W : \{c_1, .., c_N\} \to \mathbb{N}$ defined over the set of cuts is increasing in $\{c_1, ..., c_{min}\}$ and decreasing in $\{c_{max}, ..., c_N\}$.*

This property is interesting because it implies that the best cut c_{Best} can be found in the interval $\{c_{min}, ..., c_{max}\}$ using only $O(d \log N)$ queries to determine the medians of decision classes (by applying Binary Search Algorithm) and to eliminate all tail cuts. Let us also observe that if all decision classes have similar medians then almost all cuts can be eliminated.

Example We consider a data table consisting of 12000 records. Objects are classified into 3 decision classes with the distribution $\langle 5000, 5600, 1400 \rangle$, respectively. One real value attribute has been selected and $N = 500$ cuts on its domain has generated class distributions as shown in Figure 1.

The medians of three decision classes are c_{166}, c_{414} and c_{189}, respectively. The median of every decision class has been determined by *binary search algorithm* using $\log N = 9$ simple queries. Applying Theorem 1 we conclude that it is enough to consider only cuts from $\{c_{166}, ..., c_{414}\}$. In this way 251 cuts have been eliminated by using 27 simple queries only.

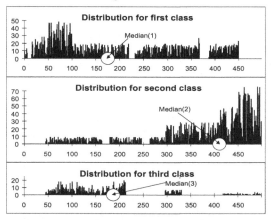

Fig. 1. Distributions for decision classes 1, 2, 3.

3   Divide and Conquer Strategy

The main idea is to apply the *"divide and conquer"* strategy to determine the best cut $c_{Best} \in \{c_1, ..., c_n\}$ with respect to a given quality function.

First we divide the set of possible cuts into k intervals (e.g. $k = 2, 3, ..$). Then we choose the interval to which the best cut may belong with the highest probability. We will use some approximating measures to predict the interval which probably contains the best cut with respect to discernibility measure. This process is repeated until the considered interval consists of one cut. Then the best cut can be chosen between all visited cuts.

The problem arises how to define the measure evaluating the quality of the interval $[c_L; c_R]$ having class distributions: $\langle L_1, ..., L_d \rangle$ in $(-\infty; c_L)$; $\langle M_1, ..., M_d \rangle$ in $[c_L; c_R]$; and $\langle R_1, ..., R_d \rangle$ in $(c_R; \infty)$. This measure should estimate the quality of the best cut among those belonging to the interval $[c_L; c_R]$. In next Section we present some theoretical considering about the quality of the best cut in $[c_L; c_R]$. These results will be used to construct the relevant measure to estimate the quality of the whole interval. We consider two specific probabilistic models for distribution of objects in the interval $[c_L; c_R]$.

Independency model: Let us consider an arbitrary cut c lying between c_L and c_R and let us assume that $\langle x_1, x_2, ..., x_d \rangle$ is a class distribution of the interval $[c_L; c]$. In this model we assume that $x_1, x_2, ..., x_d$ are independent random variables with uniform distribution over sets $\{0, ..., M_1\}, ..., \{0, ..., M_d\}$, respectively. One can observe that for all $i \in \{1, .., d\}$ $E(x_i) = \frac{M_i}{2}$ and $D^2(x_i) = \frac{M_i(M_i+2)}{12}$. We have shown in [8] the following theorem:

Theorem 2 *The mean $E(W(c))$ of quality $W(c)$ for any cut $c \in [c_L; c_R]$ satisfies*

$$E(W(c)) = \frac{W(c_L) + W(c_R) + conflict([c_L; c_R])}{2} \tag{2}$$

and the standard deviation of $W(c)$ is equal to

$$D^2(W(c)) = \sum_{i=1}^{n} \left[\frac{M_i(M_i + 2)}{12} \left(\sum_{j \neq i} (R_j - L_j) \right)^2 \right] \tag{3}$$

where $conflict([c_L; c_R]) = \sum_{i<j} M_i M_j$.

Full dependent model: In this model, we assume that the values $x_1, ..., x_d$ are proportional to $M_1, ..., M_d$, i.e.

$$\frac{x_1}{M_1} \simeq \frac{x_2}{M_2} \simeq ... \simeq \frac{x_d}{M_d}$$

In this model we have the following theorem:

Theorem 3 *In full independent model, quality of the best cut in interval $[c_R; c_L]$ is equal to*

$$W(c_{Best}) = \frac{W(c_L) + W(c_R) + conflict([c_L; c_R])}{2} + \frac{[W(c_R) - W(c_L)]^2}{8 \cdot conflict([c_L; c_R])} \tag{4}$$

if $|W(c_R) - W(c_L)| < 2 \cdot conflict([c_L; c_R])$. Otherwise it is evaluated by

$$\max\{W(c_L), W(c_R)\}.$$

3.1   Evaluation measures

These are two extreme cases of independent and "fully" dependent random variables of object distribution between decision classes. For real–life data one can expect that the variables are "partially" dependent. Hence we base our heuristic on hypothesis that the derived formula for the quality of the best cut in $[c_L; c_R]$

$$Eval([c_L; c_R]) = \frac{W(c_L) + W(c_R) + conflict([c_L; c_R])}{2} + \Delta \tag{5}$$

where the value of Δ is defined by:

$$\Delta = \frac{[W(c_R) - W(c_L)]^2}{8 \cdot conflict([c_L; c_R])} \quad \text{(in the dependent model)}$$

$$\Delta = \alpha \cdot \sqrt{D^2(W(c))} \quad \text{for some } \alpha \in [0; 1]; \quad \text{(in the independent model)}$$

The choice of Δ and the value of parameter α from $[0; 1]$ can be tuned in learning process or are given by expert.

3.2 Local and Global Search

We present two strategies of searching for the best cut using formula 5 called *local* and *global search*. In local search algorithm, first we discover the best cuts on every attribute separately. Next, we compare all locally best cuts to find out the globally best one. The details of local algorithm can be described as follows:

ALGORITHM: **Searching for semi-optimal cut**
PARAMETERS: $k \in \mathbb{N}$ **and** $\alpha \in [0; 1]$.
INPUT: **attribute** a; **the set of candidate cuts** $\mathbf{C}_a = \{c_1, .., c_N\}$ **on** a;
OUTPUT: **The optimal cut** $c \in \mathbf{C}_a$

begin
  $Left \leftarrow \min$; $Right \leftarrow \max$;   {see Theorem 1}
  while $(Left < Right)$
    1.**Divide** $[Left; Right]$ **into** k **intervals with equal length by** $(k + 1)$
    boundary points i.e.

$$p_i = Left + i * \frac{Right - Left}{k};$$

    for $i = 0, .., k$.
    2.**For** $i = 1, .., k$ **compute** $Eval([c_{p_{i-1}}; c_{p_i}], \alpha)$ **using Formula (5). Let**
    $[p_{j-1}; p_j]$ **be the interval with maximal value of** $Eval(.)$;
    3.$Left \leftarrow p_{j-1}$; $Right \leftarrow p_j$;
  endwhile;
  Return the cut c_{Left};
end

One can see that to determine the value $Eval([c_L; c_R])$ we need only $O(d)$ simple SQL queries of the form:SELECT COUNT FROM ... WHERE attribute BETWEEN c_L AND c_R. Hence the number of queries necessary for running our algorithm is of order $O(dk \log_k N)$. In practice we set $k = 3$ because the function $f(k) = dk \log_k N$ over positive integers is taking minimum for $k = 3$. For $k > 2$, instead choosing the best interval $[p_{i-1}; p_i]$, one can select the best union $[p_{i-m}; p_i]$ of m consecutive intervals in every step for a predefined parameter $m < k$. The modified algorithm needs more – but still $O(\log N)$ – simple questions only.

The global strategy is searching for the best cut over all attributes. At the beginning, the best cut can belong to every attribute, hence for each attribute we keep the interval in which the best cut can be found (see Theorem 1), i.e. we have a collection of all potential intervals

$$\mathbf{Interval\_Lists} = \{(a_1, l_1, r_1), (a_2, l_2, r_2), ..., (a_k, l_k, r_k)\}$$

Next we iteratively run the following procedure

- remove the interval $I = (a, c_L, c_R)$ having highest probability of containing the best cut (using Formula 5);
- divide interval I into smaller ones $I = I_1 \cup I_2 ... \cup I_k$;
- insert $I_1, I_2, ..., I_k$ to **Interval_Lists**.

This iterative step can be continued until we have one–element interval or the time limit of searching algorithm is exhausted. This strategy can be simply implemented using priority queue to store the set of all intervals, where priority of intervals is defined by Formula 5.

3.3   Example

In Figure 2 we show the graph of $W(c_i)$ for $i \in \{166, ..., 414\}$ and we illustrated the outcome of application of our algorithm to the reduce set of cuts for $k = 2$ and $\Delta = 0$.

First the cut c_{290} is chosen and it is necessary to determine to which of the intervals $[c_{166}, c_{290}]$ and $[c_{290}, c_{414}]$ the best cut belongs. The values of function $Eval$ on these intervals is computed:

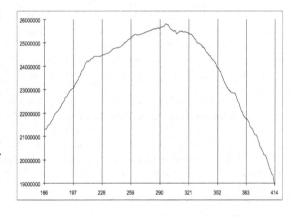

Fig. 2. Graph of $W(c_i)$ for $i \in \{166, .., 414\}$.

$$Eval([c_{166}, c_{290}]) = 23927102, \qquad Eval([c_{290}, c_{414}]) = 24374685.$$

Hence, the best cut is predicted to belong to $[c_{290}, c_{414}]$ and the search process is reduced to the interval $[c_{290}, c_{414}]$. The above procedure is repeated recursively until the selected interval consists of single cut only. For our example, the best cut c_{296} has been successfully selected by our algorithm. In general, the cut selected by the algorithm is not necessarily the best. However, numerous experiments on different large data sets shown that the cut c^* returned by the algorithm is close to the best cut c_{Best} (i.e. $\frac{W(c^*)}{W(c_{Best})} \cdot 100\%$ is about 99.9%).

4   Conclusions

The problem of optimal binary partition of continuous attribute domain for large data sets stored in *relational data bases* has been investigated. We reduced the number of simple queries from $O(N)$ to $O(\log N)$ to construct the partition very close to the optimal one. We plan to extend these results for other measures.

Acknowledgement: This paper was supported by KBN grant 8T11C02519 and Wallenberg Foundation – WITAS project.

References

1. Chmielewski, M. R., Grzymala-Busse, J. W.: Global discretization of attributes as preprocessing for machine learning. In. T.Y. Lin, A.M. Wildberger (eds.). Soft Computing. Rough Sets, Fuzzy Logic Neural Networks, Uncertainty Management, Knowledge Discovery, Simulation Councils, Inc., San Diego, CA 294–297
2. Dougherty J., Kohavi R., Sahami M.: Supervised and unsupervised discretization of continuous features. In. Proceedings of the Twelfth International Conference on Machine Learning, Morgan Kaufmann, San Francisco, CA
3. Fayyad, U. M., Irani, K.B.: On the handling of continuous-valued attributes in decision tree generation. Machine Learning **8**, 87–102
4. Fayyad, U. M., Irani, K.B.: The attribute selection problem in decision tree generation. In. Proc. of AAAI-92, San Jose, CA. MIT Press
5. J. E. Gehrke, R. Ramakrishnan, and V. Ganti. RAINFOREST - A Framework for Fast Decision Tree Construction of Large Datasets. In Proc. of the 24th International Conference on Very Large Data Bases, New York, New York, 1998.
6. Nguyen, H. Son: Discretization Methods in Data Mining. In L. Polkowski, A. Skowron (Eds.): *Rough Sets in Knowledge Discovery* **1**, Springer Physica-Verlag, Heidelberg, 451–482.
7. H.S. Nguyen and S.H. Nguyen. From Optimal Hyperplanes to Optimal Deciison Trees, *Fundamenta Informaticae* **34**No 1–2, (1998) 145–174.
8. Nguyen, H. Son: Efficient SQL-Querying Method for Data Mining in Large Data Bases. Proc. of Sixteenth International Joint Conference on Artificial Intelligence, IJCAI-99, Morgan Kaufmann Publishers, Stockholm, Sweden, pp. 806-811.
9. Pawlak Z.: *Rough sets: Theoretical aspects of reasoning about data*, Kluwer Dordrecht.
10. Polkowski, L., Skowron, A. (Eds.): *Rough Sets in Knowledge Discovery* **Vol. 1,2**, Springer Physica-Verlag, Heidelberg.
11. Quinlan, J. R. *C4.5. Programs for machine learning.* Morgan Kaufmann, San Mateo CA.
12. Skowron, A., Rauszer, C.: The discernibility matrices and functions in information systems. In. R. Słowiński (ed.). Intelligent Decision Support – Handbook of Applications and Advances of the Rough Sets Theory, Kluwer Academic Publishers, Dordrecht 311–362
13. J. Komorowski, Z. Pawlak, L. Polkowski and A. Skowron,(1998). Rough sets: A tutorial. In: S.K. Pal and A. Skowron (eds.), Rough - fuzzy hybridization: A new trend in decision making, Springer-Verlag, Singapore, pp. 3-98.
14. Ziarko, W.: Rough set as a methodology in Data Mining. In Polkowski, L., Skowron, A. (Eds.): *Rough Sets in Knowledge Discovery* **Vol. 1,2**, Springer Physica-Verlag, Heidelberg, pp. 554–576.

An Approach to Statistical Extention of Rough Set Rule Induction

Shusaku Tsumoto

Department of Medicine Informatics, Shimane Medical University, School of Medicine,
89-1 Enya-cho Izumo City, Shimane 693-8501 Japan
E-mail: tsumoto@computer.org

Abstract. This paper introduces a new approach to induced rules for
quantitative evaluation, which can be viewed as a statistical extention
of rough set methods. For this extension, chi-square distribution and
F-distribution play an important role in statistical evaluation.

1 Introduction

Rough set based rule induction methods have been applied to knowledge discovery in databases[1, 2, 6, 7]. The empirical results obtained show that they are very
powerful and that some important knowledge has been extracted from datasets.
However, quantitative evaluation of induced rules are based not on statistical
evidence but on rather naive indices, such as conditional probabilities and functions of conditional probabilities.

In this paper, we introduce a new approach to induced rules for quantitative
evaluation, which can be viewed as a statistical extention of rough set methods.
For this extension, chi-square distribution and F-distribution play an important
role in statistical evaluation.

The paper is organized as follows: Section 2 discusses the characteristics of
contingency tables. Section 3 shows the definitions of statistical measures used
for contingency tables and their assumptions. Section 4 presents an approach
to statistical evaluation a rough set model and an illustrative example. Finally,
Section 5 concludes this paper.

2 From Information Systems to Contingency Tables

2.1 Accuracy and Coverage

In the subsequent sections, we adopt the following notations, which is introduced
in [5]. Let U denote a nonempty, finite set called the universe and A denote
a nonempty, finite set of attributes, i.e., $a : U \rightarrow V_a$ for $a \in A$, where V_a
is called the domain of a, respectively. Then, a decision table is defined as an
information system, $A = (U, A \cup \{d\})$. The atomic formulas over $B \subseteq A \cup \{d\}$
and V are expressions of the form $[a = v]$, called descriptors over B, where $a \in B$
and $v \in V_a$. The set $F(B, V)$ of formulas over B is the least set containing all

W. Ziarko and Y. Yao (Eds.): RSCTC 2000, LNAI 2005, pp. 362–369, 2001.

atomic formulas over B and closed with respect to disjunction, conjunction and negation. For each $f \in F(B,V)$, f_A denote the meaning of f in A, i.e., the set of all objects in U with property f, defined inductively as follows.

1. If f is of the form $[a = v]$ then, $f_A = \{s \in U | a(s) = v\}$
2. $(f \wedge g)_A = f_A \cap g_A$; $(f \vee g)_A = f_A \vee g_A$; $(\neg f)_A = U - f_a$

By the use of this framework, classification accuracy and coverage, or true positive rate is defined as follows.

Definition 1.
Let R and D denote a formula in $F(B,V)$ and a set of objects which belong to a decision d. Classification accuracy and coverage(true positive rate) for $R \to d$ is defined as:

$$\alpha_R(D) = \frac{|R_A \cap D|}{|R_A|}(= P(D|R)), \text{ and } \kappa_R(D) = \frac{|R_A \cap D|}{|D|}(= P(R|D)),$$

where $|A|$ denotes the cardinality of a set A, $\alpha_R(D)$ denotes a classification accuracy of R as to classification of D, and $\kappa_R(D)$ denotes a coverage, or a true positive rate of R to D, respectively.

It is notable that these two measures are equal to conditional probabilities: accuracy is a probability of D under the condition of R, coverage is one of R under the condition of D.

2.2   Contingency Tables

From the viewpoint of information systems, contingency tables summarizes the relation between attributes with respect to frequencies. These viewpoints have already been discussed in [8, 9]. However, in this study, we focus on more statistical interpretation of this table. Let R_1 and R_2 denote a formula in $F(B,V)$. A contingency tables is a table of a set of the meaning of the following formulas: $|[R_1 = 0]_A|, |[R_1 = 1]_A|, |[R_2 = 0]_A|, |[R_1 = 1]_A|, |[R_1 = 0 \wedge R_2 = 0]_A|, |[R_1 = 0 \wedge R_2 = 1]_A|, |[R_1 = 1 \wedge R_2 = 0]_A|, |[R_1 = 1 \wedge R_2 = 1]_A|, |[R_1 = 0 \vee R_1 = 1]_A|(= |U|)$. This table is arranged into the form shown in Table 1. From this table, accuracy and coverage for $[R_1 = 0] \to [R_2 = 0]$ are

Table 1. Two way Contingency Table

	$R_1 = 0$	$R_1 = 1$	
$R_2 = 0$	a	b	$a + b$
$R_2 = 1$	c	d	$c + d$
	$a + c$	$b + d$	$a + b + c + d$
			$(= |U| = N)$

Table 2. A Small Dataset

a	b	c	d	e
0	0	0	0	1
1	0	1	1	1
0	1	1	1	0
1	1	1	1	1
0	0	1	0	0

Table 3. Corresponding Contingency Table

	b=0	b=1	
e=0	1	1	2
e=1	2	1	3
	3	2	5

defined as:

$$\alpha_{[R_1=0]}([R_2=0]) = \frac{|[R_1=0 \wedge R_2=0]_A|}{|[R_1=0]_A|} = \frac{a}{a+c}, and$$

$$\kappa_{[R_1=0]}([R_2=0]) = \frac{|[R_1=0 \wedge R_2=0]_A|}{|[R_2=0]_A|} = \frac{a}{a+b}.$$

For example, let us consider an information table shown in Table 2. When we examine the relationship between b and e via a contingency table, first we count the frequencies of four elementary relations, called *marginal distributions*: $[b = 0]$, $[b = 1]$, $[e = 0]$, and $[e = 1]$. Then, we count the frequencies of four kinds of conjunction: $[b = 0] \wedge [e = 0]$, $[b = 0] \wedge [e = 1]$, $[b = 1] \wedge [e = 0]$, and $[b = 1] \wedge [e = 1]$. Then, we obtain the following contingency table (Table 3). From this table, accuracy and coverage for $[b = 0] \rightarrow [e = 0]$ are obtained as $1/(1 + 2) = 1/3$ and $1/(1 + 1) = 1/2$.

3   Chi-square Test

The chi-square test is based on the following theorem[4].

Theorem 1. *When a contingency table shown in Table 4 is given, the test statistic:*

$$\chi^2 = \sum_{i,j=1}^{n,m} \frac{(x_{ij} - a_i b_j/N)^2}{a_i b_j/N} \tag{1}$$

follows chi-square distribute with the freedom of $(n-1)(m-1)$.

In the case of binary attributes shown in Table 1, this test statistic can be transformed into the following simple formula and it follows the chi-square destruction

Table 4. Contingency Table

	A_1	A_2	$\cdots$	A_n	Sum
B_1	x_{11}	x_{21}	$\cdots$	x_{n1}	b_1
B_2	x_{21}	x_{22}	$\cdots$	x_{n2}	b_2
$\cdots$	$\cdots$	$\cdots$	$\cdots$	$\cdots$	$\cdots$
B_m	x_{m1}	x_{m2}	$\cdots$	x_{m2}	b_m
Sum	a_1	a_2	$\cdots$	a_n	Sum

with the freedom of one.

$$\chi^2 = \frac{N(x_{11}x_{22} - x_{12}x_{21})^2}{a_1 a_2 b_1 b_2} = \frac{N(ad - bc)^2}{(a + c)(b + d)(a + b)(c + d)} \tag{2}$$

One of the core ideas of chi-square test is that the test statistic measures the square of difference between the real value and the expected value of one column. In the example shown in Table 4, $(x_{11} - a_1 b_1/N)^2$ measures the difference between x_{11} and the expected value of this column $a_1 b_1/N$ where b_1/N is a marginal distribution of B_1.

Another core idea is that $a_1 b_1/N$ is equivalent to the variance of marginal distributions if they follow multinomial distributions. [1] Thus, chi-square test statistic is equivalent to total sum of the ratio of the square distance s^2 to the corresponding variance σ^2. Actually, the theorem above comes from more general theorem as a corollary if a given multinomial distribution converges into a normal distribution.

Theorem 2. *If $x_1, x_2, \cdots, x_n$ are randomly selected from the population following a normal distribution $N(m, \sigma^2)$, the formula*

$$y = \frac{1}{\sigma^2} \sum_{i=1}^{n} (x_i - m)^2$$

follows the χ^2 distribution with the freedom of $(n - 1)$.

In the subsequent sections, we assume all the assumptions discussed above.

4   Towards Statistical Extension of Rough Sets

4.1   Rough Set Approximations and Contingency Tables

The important ideas in rough sets is that real-world concepts can be captured by two approximations: lower and upper approximations[3]. Although these ideas are deterministic, they can be extended into naive probabilistic models if we set up precision as shown in Ziarko's variable precision rough set model(VPRS)[10].

---

[1] If the probabilities p and q come from the multinomial distribution, Npq is equal to variance.

Table 5. Contingency Table for Lower Approximation

	$R_1 = 0$	$R_1 = 1$	
$R_2 = 0$	a	b	$a + b$
$R_2 = 1$	0	d	d
	a	$b + d$	$a + b + d$

Table 6. Contingency Table for Upper Approximation

	$R_1 = 0$	$R_1 = 1$	
$R_2 = 0$	a	0	a
$R_2 = 1$	c	d	$c + d$
	$a + c$	d	$a + c + d$

From the ideas of Ziarko's VPRS, Tsumoto shows that lower and upper approximations of a target concept correspond to sets of examples which satisfy the following conditions[7]:

Lower Approximation of D: $\cup\{R_A | \alpha_R(D) = 1.0\}$,
Upper Approximation of D: $\cup\{R_A | \kappa_R(D) = 1.0\}$,

where R is a disjunctive or conjunctive formula. Thus, if we assume that all the attributes are binary, we can construct contingency tables corresponding to these two approximations as shown in the following subsubsections.

Lower approximation. From the definition of accuracy shown in Section 3, the contingency table for lower approximation is obtained if c is set to 0. That is, the following contingency table corresponds to the lower approximation of $R_2 = 0$. In this case, the test statistic is simplified into:

$$\chi^2 = \frac{N(ad)^2}{a(b + d)(a + b)d} \tag{3}$$

Upper approximation. From the definition of coverage shown in Section 3, the contingency table for lower approximation is obtained if c is set to 0. That is, the following contingency table corresponds to the lower approximation of $R_2 = 0$. In this case, the test statistic is simplified into:

$$\chi^2 = \frac{N(ad)^2}{(a + c)da(c + d)} \tag{4}$$

4.2   Measuring Distance from Two Approximations

As discussed in Section 3, the core idea of χ^2 test is to measure the distance from the ideal marginal distributions. In the above subsections, the equations 3 and

Table 7. Two Tables for χ^2-test

	$R_1 = 0$	$R_1 = 1$	
$R_2 = 0$	x_{11}	x_{12}	b_0
$R_2 = 1$	x_{21}	x_{22}	b_1
	a_0	a_1	N

	$R_1 = 0$	$R_1 = 1$	
$R_2 = 0$	$a_0 b_0 / N$	$a_1 b_0 / N$	b_0
$R_2 = 1$	$a_0 b_1 / N$	$a_1 b_1 / N$	b_1
	a_0	a_1	N

4 measure the distance between lower and upper approximations and marginal distributions, respectively. In statistical analysis, if all columns (or row) match with marginal distributions, then we can say that these a variable R_1 have no correlation to the other one R_2. For the evaluation of this assertion, the χ^2-test statistic is used. From this statistic, we obtain the corresponding p-value that measures the probability if no correlation between R_1 and R_2 is assumed.

On the other hand, from the viewpoint of information tables, the test statistic can be viewed as the distance between the existing table and the table with no correlation (Table 7). Thus, intuitively we can conclude that the χ^2 square test measures a similarity(quasi-distance) between two tables.

From the statistical point of view, this χ^2 test statistic represents statistical information about information tables. Thus, if comparison between these test statistics is allowed, this test statistic can be used to measure a similarity between two tables. For this purpose, we can use the following theorem[4].

Theorem 3. *Let $x_1, x_2, \cdots, x_n$ and $y_1, y_2, \cdots, y_m$ be randomly selected from two populations following a normal distribution $N(\bar{x}, \sigma^2)$ and a normal distribution $N(\bar{y}, \sigma^2)$ (with the same value of variances), respectively. the test statistic*

$$y = \frac{\sum_{i=1}^{n}(x_i - \bar{x})^2}{\sum_{i=1}^{m}(y_i - \bar{x})^2}$$

follows the F-distribution with the freedom of $(n - 1, m - 1)$.

The important assumption of this theorem is that the variances of two samples are equal. In the case of two-way contingency table, this assumption is translated into that which four marginal distributions of one table is equivalent to that of the other table. Therefore,

Corollary 1. *If the four marginal distributions of an information table T_1 is equal to those of the other table T_2, then the test-statistic*

$$f(T_1, T_2) = \frac{\chi^2(T_1)}{\chi^2(T_2)}$$

follows F-distribution with the freedom of $(m - 1, n - 1)$, where $m - 1$ and $n - 1$ are the freedom of $\chi^2(T_1)$ and $\chi^2(T_2)$.

In this way, if we assume that the marginal distribution of an information table is the same as that of another table, we can compare these two tables. Thus, if we prepare the sample tables for lower and upper approximation, we can discuss whether the contingency table of an information system is statistically different from that of a sample table of lower or upper approximation.

4.3   Example

Let us consider an example shown in Table 3.[2] The test statistic for this table is equal to:

$$\chi^2 = \frac{5 * (1 - 2)^2}{3 * 2 * 2 * 3} = \frac{5}{36} = 0.14 \tag{5}$$

The p-value of this test statistic is equal to 0.709. Thus, the probability that these two attributes have no correlation is equal to 0.709. Especially, this test statistic measures a similarity between Table 3 and Table 8. From the above

Table 8. Compared Contingency Table

	b=0	b=1	
e=0	1.2	0.8	2
e=1	1.8	1.2	3
	3	2	5

tables, let us calculate the similarity between Table 3 and a table for upper approximation shown in Table 9. The test statistic for Table 9 is equal to:

Table 9. Table Corresponding to Upper Approximation

	b=0	b=1	
e=0	2	0	2
e=1	1	2	3
	3	2	5

$$\chi^2 = \frac{5 * (4 - 0)^2}{3 * 2 * 2 * 3} = \frac{80}{36} = 2.22, \tag{6}$$

and p-value is 0.136. From the equations 5 and 6, a similarity between these two tables is:

$$f = \frac{2.22}{0.14} = 16,$$

---

[2] Please note that this example is for illustration. For real-word statistical analysis, the values for each cell and the sample size should be much larger because all the statistical theorem describes the asymptotic characteristics.

which can be tested by F-distribution with the freedom of $(1,1)$. Since the p-value of f is equal to 0.156, the probability that these two tables are different from each other is equal to 0.156.

Thus, if we set up the precision(critical value) to be 0.05 as in statistical analysis, the null hypothesis that these two tables are similar to each other will be accepted. In other words, from the viewpoint of conservative (or strict) criteria, the conclusions in Table 3 is similar to those in Table 9 (upper approximation). Thus, it is weakly concluded that the meaning of $[e = 0]$ can be regarded as upper approximation of $[b = 0]$.

5   Conclusion

In this paper, we introduce a new approach to induced rules for quantitative evaluation, which can be viewed as a statistical extention of rough set methods. For this extension, chi-square distribution and F-distribution play an important role in statistical evaluation. Chi-square test statistic measures statistical information about an information table and F-test statistic is used to measure the difference between two tables. This paper is a preliminary study on a statistical evaluation of information tables, and the discussions are very intuitive, not mathematically rigor. Also, for simplicity of discussion, we assume that all conditional attributes and decision attributes in information tables are binary. More formal analysis will appear in the future work.

References

1. Polkowski, L. and Skowron, A.(Eds.) *Rough Sets and Knowledge Discovery 1*, Physica Verlag, Heidelberg, 1998.
2. Polkowski, L. and Skowron, A.(Eds.) *Rough Sets and Knowledge Discovery 2*, Physica Verlag, Heidelberg, 1998.
3. Pawlak, Z., *Rough Sets.* Kluwer Academic Publishers, Dordrecht, 1991.
4. Rao, C.R. *Linear Statistical Inference and Its Applications, 2nd Edition*, John Wiley & Sons, New York, 1973.
5. Skowron, A. and Grzymala-Busse, J. From rough set theory to evidence theory. In: Yager, R., Fedrizzi, M. and Kacprzyk, J.(eds.) *Advances in the Dempster-Shafer Theory of Evidence*, pp.193-236, John Wiley & Sons, New York, 1994.
6. Tsumoto, S. Extraction of Experts' Decision Rules from Clinical Databases using Rough Set Model Journal of Intelligent Data Analysis, 2(3), 1998.
7. Tsumoto, S., Knowledge discovery in clinical databases and evaluation of discovered knowledge in outpatient clinic. *Information Sciences*, **124**, 125-137, 2000.
8. Yao, Y.Y. and Wong, S.K.M., A decision theoretic framework for approximating concepts, *International Journal of Man-machine Studies*, **37**, 793-809, 1992.
9. Yao, Y.Y. and Zhong, N., An analysis of quantitative measures associated with rules, N. Zhong and L. Zhou (Eds.), *Methodologies for Knowledge Discovery and Data Mining, Proceedings of the Third Pacific-Asia Conference on Knowledge Discovery and Data Mining*, LNAI **1574**, Springer, Berlin, pp. 479-488, 1999.
10. Ziarko, W., Variable Precision Rough Set Model. *Journal of Computer and System Sciences*, 46, 39-59, 1993.

The Inconsistency in Rough Set Based Rule Generation [1]

Guoyin Wang and Feng Liu

Institute of Computer Science and Technology
Chongqing University of Posts and Telecommunications
Chongqing, 400065, P. R. China
e-mail: wanggy@cqupt.edu.cn

Abstract. As the amount of information in the world is steadily increasing, there is a growing demand for tools for analyzing the information. The problem of data mining is investigated in this paper. It is very important and useful to generate decision rules and reason under inconsistency. Propositional default rules are generated in this paper. Based on analysis of inconsistency, Skowron's default rule generation algorithm is improved. A corresponding reasoning method with a rule-choosing stratagem of lower frequency first under inconsistency is also developed. A suitable decision can be generated for any yet unseen object including one with unknown attribute values and one that is even inconsistent (conflicting) with objects of the training decision table. The rule-choosing stratagem is shown to be valid by our experiments.

1 Introduction

As the amount of information in the world is steadily increasing, there is a growing demand for tools for analyzing the information, finding patterns in terms of implicit dependencies in data. Realizing that much of the collected data will not be handled or even seen by human beings, data mining technology will be of increasing importance in the future. Although simple statistical techniques for data analysis were developed long ago, advanced techniques for intelligent data analysis are not yet mature. As a result, there is a growing gap between data generation and data understanding.

Rough sets have been introduced as a tool to process inexact, uncertain or vague knowledge in AI, like for example knowledge based systems in medicine, natural language processing, decision systems, approximate reasoning [1], [2], [3], [4]. Some rough set based methods and algorithms were developed to generate rules from a decision table without any conflicting objects in the last years [5], [6], [7], [8]. In these cases, definite rules may be generated. Unfortunately, there are lots of inconsistencies, or uncertainties in real life. It is needed to be able to reason also under inconsistency. Different experts may disagree on the classification of one particular object, in which case it is desirable to assign different trust to the respective conclusions. Also, if objects are classified inconsistently, we want still to generate

---

[1] This paper is partially supported by Foundation for University Key Teacher by the Ministry of Education, National Science Foundation of China and Application Science Foundation of Chongqing.

W. Ziarko and Y. Yao (Eds.): RSCTC 2000, LNAI 2005, pp. 370–377, 2001.

rules that can reflect the normal situation. The inconsistency problem may be caused by factors such as insufficiency of condition attributes, errors in measuring process and mistakes in recording process, etc. Some researchers have done some work in this filed trying to generate default rules from an inconsistent decision table [9], [10].

Skowron proposed a rough set framework that is able to generate and reason about classes for which no unique decision can be made [10]. His basic idea is to create indeterminacy in information systems, and generate rules covering the majority of the cases. Skowron considered the generation of indeterminacy through selecting projections over the condition attributes, allowing certain attributes to be excluded from consideration. The rules that result generally have at least two advantages as compared to deterministic rules: they are always simpler in structure, and though not entirely correct, they will in many cases prove to be better when handling yet unseen cases, being less susceptible to noise. Unfortunately, Skowron did not study the inconsistency problem thoroughly, so there are still some conflicts in his method. If a yet unseen case that is inconsistent with a case in the training decision table occurs, conflict results will be generated by different default rules extracted from the training decision table. In this case, the rules can not work, and no decision can be made.

We will examine both explicit inconsistency and implicit inconsistency in an information system thoroughly in this paper. A new default rule extracting algorithm and its corresponding reasoning algorithm will be developed based on Skowron's default rule extracting algorithm. A new rule-choosing stratagem under inconsistency will be presented. In section 2, we will briefly introduce Skowron's default rule extracting algorithm and analyze its shortcomings. In section 3, we will examine the explicit and implicit inconsistency in a decision table thoroughly. An improved default rule extracting algorithm and its corresponding reasoning algorithm with a new rule-choosing stratagem of lower frequency first will be developed in section 4. In section 5, we will illustrate the validity of our methods through some experiments. At last, we will conclude our work in section 6.

2   Skowron's Default Rule Extracting Algorithm

Skowron developed an algorithm to extract default rules from a decision table even if it contains some inconsistent cases [10].

Algorithm 1: Skowron's default rule extracting algorithm.

Input: A training decision table $A^* = (U, A^*)$, where $A^* = (C^*, \{D\})$, U is a finite and nonempty set called the universe that contains all cases (samples, or objects), A^* is a finite, nonempty set of attributes, C^* is a finite, nonempty set of condition attributes, and D is the decision attribute.

Output: default rules.

Step 1: Calculate the indiscernibility relation $U/IND(C^*)$. If a class $E_{(k,C^*)}$ $(E_{(k,C^*)} \in U/IND(C^*), k=1,\ldots,|U/IND(C^*)|)$, of cases in $U/IND(C^*)$ can be classified into a decision class X_j with a membership that is greater than some threshold, then a corresponding default rule can be extracted. That is,

If $\mu_{C^*}(E_{(k,C^*)}, X_j) = |E_{(k,C^*)} \cap X_j| / |E_{(k,C^*)}| \geq \mu_{jr}$, then the following default rule can be generated:

R: $\mathrm{Des}(E_{(k,C^*)},C^*) \to \mathrm{Des}(X_j,D) \mid |E_{(k,C^*)} \cap X_j|/|E_{(k,C^*)}|$,

where, $|E_{(k,C^*)} \cap X_j|/|E_{(k,C^*)}|$ is the certainty factor (CF) of rule $\mathrm{Des}(E_{(k,C^*)},C^*) \to \mathrm{Des}(X_j,D)$.

Step 2: Put the decision table $\mathbf{A}^*$ into a decision table set ψ, that is, $\psi=\{\mathbf{A}^*\}$.

Step 3: If ψ is empty ($\psi=\phi$), then stop; else, take a decision table $\mathbf{A}=(U,A)$ out of ψ ($\psi=\psi-\{\mathbf{A}\}$) and calculate its core attributes set $\mathrm{Core}(C)$, where, $A=(C,\{D\})$. We can select projections $C_{Pr}=C-C_{Cut}$ ($r=1,\ldots,\mathrm{Card}(\mathrm{Core}(C))$) over the attributes. The projections are selected such that new indeterminacy results. The following 5 sub-steps should be done to these projections.

1. If $C_{Pr}=\phi$, then do nothing to the projection; else, do the following 4 steps to it.

2. Insert the projection decision table $\mathbf{A'}=(U,A')$ into ψ, that is, $\psi=\psi\cup\{\mathbf{A'}\}$, where, $A'=(C_{Pr},\{D\})$.

3. Calculate the indiscernibility relation $U/\mathrm{IND}(C_{Pr})$.

4. If a class $E_{(k,CPr)}$ ($E_{(k,CPr)} \in U/\mathrm{IND}(C^*)$, $k=1,\ldots,|U/\mathrm{IND}(C_{Pr})|$) in $U/\mathrm{IND}(C_{Pr})$ of cases can be classified into a decision class X_j with a membership that is greater than some threshold, then a corresponding default rule can be extracted. That is,

If $\mu_{CPr}(E_{(k,CPr)},X_j)=|E_{(k,CPr)} \cap X_j|/|E_{(k,CPr)}| \geq \mu_{jr}$, then the following default rule can be generated:

R': $\mathrm{Des}(E_{(k,CPr)}, C_{Pr}) \to \mathrm{Des}(X_j,D) \mid |E_{(k,CPr)} \cap X_j|/|E_{(k,CPr)}|$.

5. Facts are constructed that may potentially block the application of a default rule.

If there is an E_i, $E_i \in U/\mathrm{IND}(C) \wedge E_i \subseteq E_{(k,CPr)} \wedge E_i \cap X_j=\phi$, then the following fact can be generated:

F': $\mathrm{Des}(E_i,C_{Cut}) \to \neg(R')$,

where, $\neg$ is a logical NOT operator.

Step 4: go to step 3.

To make it clearer, let's look at a simple training decision table. The information system $\mathbf{A}=(U,A)$, displayed in table 1, resulted from having observed a total of one hundred objects (the universe U) that were classified according to condition attributes $C=\{a,b,c\}$. The decision attribute is d. The partition of the universe induced by the condition attributes contains n=5 classes, namely E_1 through E_5. The class E_5 is shown split into two disjoint sets of objects, $E_5=E_{5,1} \cup E_{5,2}$, reflecting the different decisions, d=3 (for $E_{5,1}$) and d=4 (for $E_{5,2}$). Hence, the system is inconsistent with respect to the objects in class E_5.

The discernibility matrix $M_D(C)=\{m_D(i,j)\}_{n*n}$ (over the condition attributes $C=\{a,b,c\}$) of the decision system is given in table 2. Its core attribute set is $\{a,c\}$. If the threshold value (μ_{jr}) is set to be 0.55, we can obtain the following rules:

Table 1. Example Information System

	a	b	c	d
E_1	1	2	3	1 (50x)
E_2	1	2	1	2 (5,x)
E_3	2	2	3	2 (30x)
E_4	2	3	3	2 (10x)
$E_{5,1}$	3	5	1	3 (4x)
$E_{5,2}$	3	5	1	4 (1x)

Table 2. Discernibility Matrix of the Decision System

	E_1	E_2	E_3	E_4	E_5
E_1		c	a	ab	abc	ac
E_2	c			ab	c(a∨b)
E_3	a			abc	a
E_4	ab			abc	a∨b
E_5	abc	ab	abc	abc	a∨b

R_1: $a_1c_3{\to}d_1$ | 1.0, R_2: $a_1c_1{\to}d_2$ | 1.0, R_3: $b_2c_1{\to}d_2$ | 1.0,

R_4: $a_2{\to}d_2$ | 1.0, R_5: $b_3{\to}d_2$ | 1.0, R_6: $a_3{\to}d_3$ | 0.8,

R_7: $b_5{\to}d_3$ | 0.8,

where, a rule $a_ib_jc_k{\to}d_p$ | μ means that if the values of the condition attributes a, b and c are separately i, j and k, then the decision of this object should be p with a certainty factor μ.

According to algorithm 1, the following rules and facts can also be generated from the projections (C-{a}, C-{c}, C-{a,b}, and C-{a,c}) of table 1:

R_8 (C-{a}): $b_2c_3{\to}d_1$ | 0.62, R_9 (C-{c}): $a_1{\to}d_1$ | 0.91,

R_{10} (C-{a,b}): $c_3{\to}d_1$ | 0.56, R_{11} (C-{a,c}): $b_2{\to}d_1$ | 0.59,

F_1 (C-{a}): $a_2{\to}{\neg}R_8$, F_2 (C-{c}): $c_1{\to}{\neg}R_9$,

F_3 (C-{a,b}): $b_3{\to}{\neg}R_{10}$, F_4 (C-{a,c}): $a_2{\to}{\neg}R_{11}$, $c_1{\to}{\neg}R_{11}$.

One can find that the premise parts of default rules are short. They can reason in absence on knowledge. Assume now that a new object is observed, for which the value of the attribute a is 1, whereas the values for all the other attributes of the object are unknown. The definite rules, those rules which certainty factor is 1, do not sanction and conclude in this case, we may however apply the default rule R_9 to conclude (by assumption) that the decision in this case should be 1 with a certainty factor 0.91. If, later, further knowledge is made available, the assumption (and therefore the conclusion) may have to be retracted. Unfortunately, Skowron's algorithm is not complete. For instance, if the object to be classified is $a_1b_3c_3$, then the decision should be 1 (CF=1.0) according to rule R_1 while it should be 2 (CF=1.0) according to rule R_5. We can still not get the decision. Again, if the case to be classified is $a_1b_5c_2$, then the decision should be 1 (CF=0.91) according to rule R_9 while 3 (CF=0.8) according to rule R_7. How should the decision be derived in this case? Obviously, this problem is caused by conflicts between rules. Skowron's analysis for the inconsistency in a decision table is not complete.

3   On the Inconsistency of an Information System

Through examining a decision table, we find that there may be 3 kinds of inconsistent information in a decision table from which default rules need to be extracted.

1. There are some inconsistent objects in the training decision table. This kind of inconsistency may be caused in the following 3 cases:

• The condition attributes is insufficient to describe objects. Some additional attributes will be needed to distinguish objects.

• There may be errors in measuring process of the values for attributes and mistakes in recording process.

• Some inconsistency may be generated in the preprocessing process of the original data. For example, in the discretizing process, some continuous values were converted into discrete ones. Some objects that can be distinguished from each other according to their original continuous attribute values before may become indistinguishable.

2. Inconsistency generated through selecting projections over the condition attributes.

3. There may exist some objects outside of the training decision table that are inconsistent with objects in the training decision table. There is usually only a part of objects of the universe in the training decision table. So, we don't know whether there is inconsistency in the universe even if the training decision table is consistent. Rules can generate inconsistent decisions for a yet unseen object even if they are all consistent in the training decision table.

The first kind of inconsistency is explicit and can be discovered directly from the training decision table. The 2nd kind of inconsistency is generated by the rule generation algorithm and thus can be discovered in the rule generation process. Unfortunately, the 3rd kind of inconsistency is implicit and therefore unpredictable in the process of rule generation. We don't know whether there is any 3rd kind of inconsistency until it happens. Skowron considered only the former 2 inconsistent cases in his algorithm. The conflict problems stated in section 2 were caused by the 3rd kind of inconsistency. Rules generated from a limited training decision table that is a subset of the universe can only be consistent in the training decision table, and may be inconsistent in the universe. For example, rule R_1 and R_5 are consistent in table 1. However, if there is an object $a_1b_3c_3$ in the universe, these two rules will be conflict. That is, they are inconsistent in the universe.

4   Rule Generation under Inconsistency and its Corresponding Reasoning Method

First, we modify Skowron's algorithm. Rules are written in the following new style:

R: $Des(E_{(k,C)}, C) \rightarrow Des(X_j, D) \mid (\mid E_{(k,C)} \cap X_j \mid, \mid E_{(k,C)} \mid)$.

That is, the certainty factor is not recorded directly in a rule. Two parameters, the number of objects of the intersection of the equivalence class $E_{(k,C)}$ and the decision class X_j and the number of objects of the equivalence class $E_{(k,C)}$, are recorded as its parameters instead. Then, the following rules can be extracted from table 1.

R_1: $a_1c_3 \rightarrow d_1 \mid (50,50)$, R_2: $a_1c_1 \rightarrow d_2 \mid (5,5)$, R_3: $b_2c_1 \rightarrow d_2 \mid (5,5)$,

R_4: $a_2 \rightarrow d_2 \mid (40,40)$, R_5: $b_3 \rightarrow d_2 \mid (10,10)$, R_6: $a_3 \rightarrow d_3 \mid (4,5)$,

R_7: $b_5 \rightarrow d_3 \mid (4,5)$, R_8: $b_2c_3 \rightarrow d_1 \mid (50,80)$, R_9: $a_1 \rightarrow d_1 \mid (50,55)$,

R_{10}: $c_3 \rightarrow d_1 \mid (50,90)$, R_{11}: $b_2 \rightarrow d_1 \mid (50,85)$.

The fact set remains unchanged.

One can find that each rule has not only its certainty factor ($\mid E_{(k,C)} \cap X_j \mid / \mid E_{(k,C)} \mid$) information, but also the information of the frequency ($\mid E_{(k,C)} \mid$) of the objects described by the premise of the rule occurring in the training decision table. $\mid E_{(k,C)} \mid$ is called the frequency of the rule. The 3rd kind of inconsistency can be processed using this information.

Algorithm 2: Reasoning method under inconsistency

Input: A rule set Ω, $\Omega = \{R_i \mid i=1,...,n\}$, that can be matched by an object to be classified. Where, the parameters of rule R_i is (α_i, β_i), its conclusion is γ_i, and $\gamma_i \neq \gamma_j (i \neq j)$. If there are more than one rules (e.g. m rules) which map the object to a same conclusion, then the rule $R_i \mid \alpha_i / \beta_i^2 = Max\{\alpha_i / \beta_i^2 \mid i=1,...,m\}$ is selected to be the representative of these m rules. If there are several rules (e.g. k rules) in these m rules having the maximum α_i/β_i^2 at the same time, then rule $R_i|\alpha_i/\beta_i=Max\{\alpha_i/\beta_i|i=1,...,k\}$ should be selected.

Output: The decision for the object with its certainty factor.

Step 1: Select the decision (γ) for the object:

$\gamma=\gamma_i$, $\alpha_i/\beta_i^2=Max\{\alpha_i/\beta_i^2|i=1,...,n\}$.

If there are several decisions (e.g. k rules) have the maximum α_i/β_i^2, then decision $\gamma_i|\alpha_i/\beta_i=Max\{\alpha_i/\beta_i|i=1,...,k\}$ should be selected.

Step 2: Calculate the certainty factor (CF) for the decision.

$CF=Min\{\alpha_i/\beta_i|i=1,...,n\}$.

That is, the certainty factor of the final decision is the minimum one of the certainty factors of all matched rules. Logical AND operation is adopted here.

Our basic idea to select a suitable decision for a yet unseen object is that the classes containing fewer objects of a decision table may represent some special cases in the universe, thus, rules generated from these classes should have more priority in the reasoning process. This rule-choosing stratagem is called lower frequency first, that is, the rule with the lowest frequency has the greatest priority. Let's take a look at the characteristics of this reasoning method.

Suppose there are two inconsistent rules that can match the object to be classified.

1. If $\alpha_1/\beta_1=\alpha_2/\beta_2$, then $\gamma=\gamma_i$, $\beta_i=Min\{\beta_i|i=1,2\}$;
2. If $\beta_1=\beta_2$, then $\gamma=\gamma_i$, $\alpha_i=Max\{\alpha_i|i=1,2\}$;
3. If $\beta_1>\beta_2$, then

- If $\alpha_2=\beta_2$, then $\gamma=\gamma_2$;
- If $\alpha_1=\beta_1$, we might suppose $\beta_2=\beta_1-a$, $\alpha_2=\beta_2-b$, then

$$\gamma = \begin{cases} \gamma_1 , & a \leq b \\ \gamma_1 , & (a > b) \wedge (a^2/(a-b) > \beta_1) \\ \gamma_2 , & (a > b) \wedge (a^2/(a-b) < \beta_1) \\ \gamma_i , & (a > b) \wedge (a^2/(a-b) = \beta_1) \wedge (\beta_i = Min\{\beta_i \mid i = 1,2\}) . \end{cases}$$

Using this reasoning method and rule-choosing stratagem, a suitable decision can be generated for any object according to default rules generated from a training decision table. The two objects, $a_1b_3c_3$ and $a_1b_5c_2$, can not be classified by Skowron's method, can be classified into suitable classes using this method now. The object $a_1b_3c_3$ can be matched with Rule R_1 and R_5, its final decision is d_2 with a certainty factor 1.0. The object $a_1b_5c_2$ can be matched with Rule R_9 and R_7, its final decision is d_3 with a certainty factor 0.8.

5   Simulation Result

To test the validity of the rule-choosing stratagem of this paper, we compare it with the rule-choosing stratagem of higher frequency first. In the stratagem of higher frequency first, we can reason in the following way.

Suppose there are two inconsistent rules (R_1 and R_2) that can match the object to be classified.

If $\alpha_1/\beta_1=\alpha_2/\beta_2$, then $\gamma=\gamma_i$, $\beta_i=Max\{\beta_i|i=1,2\}$;

Else If $\beta_1=\beta_2$, then $\gamma=\gamma_i$, $\alpha_i=Max\{\alpha_i|i=1,2\}$;

Else   If $\beta_1 > \beta_2$, then
If $\alpha_1/\beta_1 > \alpha_2/\beta_2$, then $\gamma = \gamma_1$;
Else $\gamma = \gamma_i$, $\alpha_i^2/\beta_i = \text{Max}\{\alpha_i^2/\beta_i | i=1,2\}$;

5 data sets are used in our simulation experiments. Each data set is randomly divided into 2 equal parts, one (training set) is used for rule generation. The modified Skowron's default rule generation algorithm is used. The experiments result is shown in table 3.

Table 3. Experiments Result

Data set (S)	|S|	C(S)	|TS|	C(TS)	VS	C(R)	LFF		HFF	
							RR	CCR	RR	CCR
Protein Localization Sites	336	10	168	4	TS	168	99%	99%	67%	67%
					S	336	82%	82%	68%	68%
Glass Identification Database	214	23	107	3	TS	107	99%	99%	44%	44%
					S	214	85%	85%	47%	47%
BUPA liver disorders	345	181	173	83	TS	173	86%	86%	71%	71%
					S	345	74%	74%	66%	66%
Hayes-Roth & Hayes-Roth(1997) Database	160	33	80	11	TS	79	95%	95%	64%	64%
					S	159	88%	87%	58%	58%
Iris Plants Database	150	82	75	41	TS	75	83%	83%	75%	75%
					S	90	81%	81%	71%	71%
Postoperative Patient Data	90	11	45	4	TS	45	96%	96%	67%	67%
					S	90	81%	81%	71%	71%

In Table 3, S is the data set, |S| is the cardinality of set S, C(S) is the number of conflicting samples in S, TS is the training set, |TS| is the cardinality of TS, C(TS) is the number of conflicting samples in TS, VS is the set used to test the recognition rate, C(R) is the number of test samples which can be matched with more than one rules, RR is the correct recognition rate, CCR is the rate of choosing a correct rule. LFF is the stratagem of lower frequency first, while HFF is the stratagem of higher frequency first. There are some conflicting rules for every sample to be recognized when the default rules generated by Skowron's algorithm are used. Thus, RR=CCR for every testing. In order to get decision tables with many conflicting samples, Navi algorithm and Semi-Navi algorithm in Rosetta are used for discretization.

From the experiments result of Table 3, one can find that the stratagem of lower frequency first is much better than the stratagem of higher frequency first when a training set itself is used to test the generated rules. And, the mis-recognized samples in the stratagem of lower frequency first are all conflicting samples in the training set. Thus, the rules are a good representation of the information of the training set if the stratagem of lower frequency first is used. Otherwise, the rules can not represent the information of the training set if the stratagem of higher frequency first is used. We can also find from Table 3 the recognition rate of the stratagem of lower frequency first is also much higher than the stratagem of higher frequency first when the whole data set is used in the recognition test. Moreover, the recognition rates of 3 data sets are higher than the recognition rate of their training set. This is unreasonable.

6   Conclusion

We studied the problem of extracting default rules under inconsistency in this paper. Based on Skowron's default rule generation algorithm, we examined the inconsistency that may happen in an information system thoroughly. We developed a new default rule representing style, rule generation algorithm from an inconsistent decision table and its corresponding reasoning method with the rule-choosing stratagem of lower frequency first under inconsistency. The default rules generated have strong ability to match new objects to be processed. A suitable decision can be generated for any yet unseen object including one with unknown attribute values and one that is even inconsistent with objects of the training decision table.

References

1. Pawlak, Z.: Rough Sets. Int. J. of Computer and Information Sciences. 5 (1982) 341-356
2. Pawlak, Z.: Rough Sets, Theoretical Aspects of Reasoning about Data. Dordrecht Kluwer, (1991)
3. Pawlak, Z., Grzymala-Busse, J., Slowinski, R., Ziarko, W.: Rough Sets. Communications of the ACM. 11 (1995) 89-95
4. Pawlak, Z., Vagueness-a Rough Set View. In: Mycielski, J., Rozenberg, G., Salomaa, A. (eds.): Structures in Logic and Computer Science. Springer-Verlag (1997) 106-117
5. Ziarko, W., Cercone, N., Hu, X.: Rule Discovery from Databases with Decision Matrices. In: Proc. of 9th Int. Symp. on Foundations of Intelligent Systems. (1996) 653-662
6. Hu, X., Cercone, N.: Mining Knowledge Rules from Databases: A Rough Set Approach. In: Proc. of 12th Int. Conf. on Data Engineering, (1996) 96-105
7. Hu, X., Cercone, N.: Learning Maximal Generalized Decision Rules Via Discretization, Generalization and Rough Set Feature Selection. In: Proc. of Int. Conf. on Tools with Artificial Intelligence, (1997) 548-556
8. Chang, L.Y., Wang, G.Y., Wu, Y.: An Approach for Attribute Reduction and Rule Generation Based on Rough Set Theory. J. of Software. 11 (1999) 1206-1211
9. Shan, N., Hamilton, H.J., Cercone, N.: Induction of Classification Rules from Imperfect Data. In: Proc. of 9th Int. Symp. on Foundations of Intelligent Systems, (1996) 118-127
10.Mollestad, T., Skowron, A.: A Rough Set Framework for Data Mining of Propositional Default Rules. In: Proc. of 9th Int. Symp. on Foundations of Intelligent Systems, (1996) 448-457
11.Wang, G.Y., Wu, Y., Liu, F.: Generating Rules and Reasoning under Inconsistencies. In: 2000 IEEE Int. Conf. on Industrial Electronics, Control and Instrumentation, accepted and to appear.

A Comparison of Several Approaches to Missing Attribute Values in Data Mining

Jerzy W. Grzymala-Busse[1] and Ming Hu[2]

[1] Department of Electrical Engineering and Computer Science
University of Kansas
Lawrence, KS 66045, U.S.A.
E-mail: Jerzy@eecs.ukans.edu
http://lightning.eecs.ukans.edu/index.html
[2] JP Morgan
New York, NY 10260, U.S.A.
E-mail: Hu_ming@jpmorgan.com

Abstract: In the paper nine different approaches to missing attribute values are presented and compared. Ten input data files were used to investigate the performance of the nine methods to deal with missing attribute values. For testing both naive classification and new classification techniques of LERS (Learning from Examples based on Rough Sets) were used. The quality criterion was the average error rate achieved by ten-fold cross-validation. Using the Wilcoxon matched-pairs signed rank test, we conclude that the C4.5 approach and the method of ignoring examples with missing attribute values are the best methods among all nine approaches; the most common attribute-value method is the worst method among all nine approaches; while some methods do not differ from other methods significantly. The method of assigning to the missing attribute value all possible values of the attribute and the method of assigning to the missing attribute value all possible values of the attribute restricted to the same concept are excellent approaches based on our limited experimental results. However we do not have enough evidence to support the claim that these approaches are superior.

Key words: Data mining, knowledge discovery in databases, machine learning, learning from examples, attribute missing values.

1 Introduction

One of the main tools of data mining is rule induction from raw data represented by a database. Real-life data are frequently imperfect: erroneous, incomplete, uncertain and vague. In the reported research we investigated one of the forms of data incompleteness: missing attribute values.

We assume that the format of input data files is in the form of a table, which is called a *decision table*. In this table, each column represents one *attribute*, which

W. Ziarko and Y. Yao (Eds.): RSCTC 2000, LNAI 2005, pp. 378–385, 2001.

represents some feature of the examples, and each row represents an *example* by all its attribute values. The *domain* of each attribute may be either symbolic or numerical. We assume that all the attributes of input data are symbolic. Numerical attributes, after discretization, become symbolic as well. For each example, there is a *decision value* associated with it. The set of all examples with the same decision value is called a *concept*. Members of the concept are called *positive* examples, while all other examples are called *negative* examples.

The table is *inconsistent* if there exist two examples with all attribute values identical, but belonging to different concepts. For inconsistent data tables, we can induce rules which are called *certain* and *possible* [5].

2 Description of Investigated Approaches to Missing Attribute Values

We used the following nine approaches to missing attribute values:

1. Most Common Attribute Value. It is one of the simplest methods to deal with missing attribute values. The CN2 algorithm [3] uses this idea. The value of the attribute that occurs most often is selected to be the value for all the unknown values of the attribute.

2. Concept Most Common Attribute Value. The most common attribute value method does not pay any attention to the relationship between attributes and a decision. The concept most common attribute value method is a restriction of the first method to the concept, i.e., to all examples with the same value of the decision as an example with missing attribute vale [9]. This time the value of the attribute, which occurs the most common within the concept is selected to be the value for all the unknown values of the attribute. This method is also called maximum relative frequency method, or maximum conditional probability method (given concept).

3. C4.5. This method is based on entropy and splitting the example with missing attribute values to all concepts [12].

4. Method of Assigning All Possible Values of the Attribute. In this method, an example with a missing attribute value is replaced by a set of new examples, in which the missing attribute value is replaced by all possible values of the attribute [4]. If we have some examples with more than one unknown attribute value, we will do our substitution for one attribute first, and then do the substitution for the next attribute, etc., until all unknown attribute values are replaced by new known attribute values.

5. Method of Assigning All Possible Values of the Attribute Restricted to the Given Concept. The method of assigning all possible values of the attribute is not related with a concept. This method is a restriction of the method of assigning all possible values of the attribute to the concept, indicated by an example with a missing attribute value.

6. Method of Ignoring Examples with Unknown Attribute Values. This method is the simplest: just ignore the examples which have at least one unknown attribute value, and then use the rest of the table as input to the successive learning process.

7. Event-Covering Method. This method, described in [2] and [14], is also a probabilistic approach to fill in the unknown attribute values. By event-covering we mean covering or selecting a subset of statistically interdependent events in the outcome space of variable-pairs, disregarding whether or not the variables are statistically independent [14].

8. A Special LEM2 Algorithm. A special version of LEM2 that works for unknown attribute values omits the examples with unknown attribute values when building the block for that attribute [6]. Then, a set of rules is induced by using the original LEM2 method.

9. Method of Treating Missing Attribute Values as Special Values. In this method, we deal with the unknown attribute values using a totally different approach: rather than trying to find some known attribute value as its value, we treat "unknown" itself as a new value for the attributes that contain missing values and treat it in the same way as other values.

3   Classification

Frequently rules induced from raw data are used for classification of unseen, testing data. In the simplest form of classification, if more than one concept was indicated by rules for a given example, the classification of the example was counted as an error. Likewise, if an example was not completely classified by any of rules, it was considered an error. This classification scheme is said to be *naive* LERS classification scheme.

The new classification system of LERS is a modification of the *bucket brigade algorithm* [1, 7]. The decision to which concept an example belongs is made on the basis of three factors: strength, specificity, and support. They are defined as follows: *Strength* is the total number of examples correctly classified by the rule during training. *Specificity* is the total number of attribute-value pairs on the left-hand side of the rule. The matching rules with a larger number of attribute-value pairs are considered more specific. The third factor, *support*, is defined as the sum of scores of all matching rules from the concept. The concept C for which the support, i.e., the following expression

$$\sum_{\text{matching rules } R \text{ describing } C} \text{Strength}(R) * \text{Specificity}(R)$$

is the largest is a winner and the example is classified as being a member of C.

If an example is not completely matched by any rule, some classification systems use *partial matching*. System AQ15, during partial matching, uses the probabilistic sum of all measures of fit for rules [10]. Another approach to partial matching is presented in [13]. Holland *et al.* [8] do not consider partial matching as a viable alternative of complete matching and rely on a default hierarchy instead. In the new classification system of LERS, if complete matching is impossible, all partially matching rules are identified. These are rules with at least one attribute-value pair matching the corresponding attribute-value pair of an example.

For any partially matching rule R, the additional factor, called *Matching factor* (R), is computed. Matching_factor is defined as the ratio of the number of matched attribute-value pairs of a rule with an example to the total number of attribute-value pairs of the rule. In partial matching, the concept C for which the following expression is the largest

$$\sum_{\text{partially matching rules } R \text{ describing } C} \text{Matching\_factor}(R) * \text{Strength }(R) * \text{Specificity}(R)$$

is the winner and the example is classified as being a member of C.

Rules induced by a new version of LERS are preceded by three numbers: specificity, strength, and the total number of training examples matching the left-hand side of the rule.

4 Experiments

Table 1 describes input data files, in terms of the number of examples, the number of concepts, and the number of attributes that describe the examples, that were used for our experiments. All ten data files were taken from real world where unknown attribute values frequently occur.

Table 1. Description of data files

Name of Data Files	No. of Examples	No. of Attributes	No. of Concepts
Breast cancer	286	9	2
Echocardiogram	74	13	2
Hdynet	1218	73	2
Hepatitis	155	19	2
House	435	16	2
Im85	201	25	86
New-o	213	30	2
Primary tumor	339	17	21
Soybean	307	35	19
Tokt	6608	67	2

The *breast cancer* data set was obtained from the University Medical Center, Institute of Oncology, Ljubljana, Yugoslavia, due to donations from M. Zwitter and M. Soklic. Breast cancer is one of three data sets provided by the Oncology Institute that has repeatedly appeared in the machine learning literature. There are nine out of 286 examples containing unknown attribute values.

The *echocardiogram* data set is donated by Steven Salzberg, and this data has been used several times to predict the survival of a patient. There are a total of 132 missing values among all the attribute values.

The *hdynet* data set, which comes from real life, presents the premature birth described by 73 attributes. There were 814 out of 1218 examples containing unknown attribute values.

The *hepatitis* data set was donated by G. Gong, Carnegie-Mellon University, via Bojan Cestnik of Jozef Stefan Institute. There were 75 out of 155 examples that contain unknown attribute values in this data set.

Table 2. Error rates of input data sets by using LERS new classification

Data file	Methods								
	1	2	3	4	5	6	7	8	9
Breast	34.62	34.62	31.5	28.52	31.88	29.24	34.97	33.92	32.52
Echo	6.76	6.76	5.4	—	—	6.56	6.76	6.76	6.76
Hdynet	29.15	31.53	22.6	—	—	28.41	28.82	27.91	28.41
Hepatitis	24.52	13.55	19.4	—	—	18.75	16.77	18.71	19.35
House	5.06	5.29	4.6	—	—	4.74	4.83	5.75	6.44
Im85	96.02	96.02	100	—	96.02	94.34	96.02	96.02	96.02
New-o	5.16	4.23	6.5	—	—	4.9	4.69	4.23	3.76
Primary	66.67	62.83	62.0	41.57	47.03	66.67	64.9	69.03	67.55
Soybean	15.96	18.24	13.4	—	4.1	15.41	19.87	17.26	16.94
Tokt	31.57	31.57	26.7	32.75	32.75	32.88	32.16	33.2	32.16

Table 3. Error rates of input data sets by using LERS naive classification

Data file	Methods							
	1	2	4	5	6	7	8	9
Breast	49.30	52.1	46.98	47.32	48.38	52.8	52.1	47.55
Echo	27.03	25.68	—	—	31.15	29.73	33.78	22.97
Hdynet	67.49	69.62	—	—	65.27	69.21	56.98	61.33
Hepatitis	38.06	28.39	—	—	32.5	37.42	41.29	34.84
House	10.11	7.13	—	—	9.05	10.57	12.87	11.72
Im85	97.01	97.01	—	97.01	94.34	97.01	97.01	97.01
New-o	11.74	11.74	—	—	11.19	11.27	10.33	10.33
Primary	83.19	77.29	53.16	60.09	81.82	80.53	82.1	79.94
Soybean	25.41	22.48	—	4.86	24.06	24.10	21.82	22.15
Tokt	63.62	63.62	62.82	62.82	64.15	63.36	63.62	63.89

The *house* data set, which has 203 examples that contain unknown attribute values, consists of votes of 435 congressmen in 1984 on 16 key-issues (yes or no).

The *im85* data set is from a 1985 Automobile Imports Database, and it consists of three types of entities: a) the specification of an auto in terms of various characteristics, b) its assigned insurance risk rating, and c) its normalized losses in use as compared to other cars.

The *new-o* data set is another set of breast cancer data that uses different attributes from the breast cancer data set. In this approach, there are 30 attributes to describe the examples. There were a total of 213 examples, and 70 of them have at least one unknown attribute value.

The *primary-tumor* data set was obtained from the University Medical Center, Institute of Oncology, Ljubljana, Yugoslavia. The data set primary-tumor has 21 concepts and 17 attributes, and 207 out of 339 examples contain at least one missing value.

For the *soybean* data set, R. S. Michalski used this data set in the context of developing an expert system for soybean disease diagnosis. There are 19 classes, but, only the first 15 classes have been used in prior work. And, the last four classes have very few examples and there are 41 examples that contain unknown attribute values.

The *tokt* data set, which is the largest data file in this experiment, came from the practical data about premature birth, which is similar to the hdynet data set. Among 6619 examples in this data set, only 11 examples contain unknown attribute values.

In our experiments, we required that no decision value is unknown. If some unknown decision values existed in the input data files, the input data files were pre-processed to remove them.

Our experiments were conducted as follows. All of the nine methods from Section 2 were applied to all the ten data sets. Both original data sets and our new data sets, except for C4.5 method, were sampled into ten pairs of training and testing data. Then the sampled files were used as input to LEM2 single local covering [5] to generate classification rules, except the special LEM2 method, where rules were induced directly from the data file with missing attribute values. Other data mining systems based on rough set theory are described in [11]. We used *ten-fold cross validation* for the simple and extended classification methods. The performance of different methods was compared by calculating the average error rate. Here, we did a slight modification using *leaving-one-out* for the data set echocardiogram since it has less than 100 examples.

In Tables 2 and 3, the error rates that were not available, because of the limited system memory, are indicated by '–'.

5   Conclusions

Our main objective was comparison of the methods to deal with missing attribute values. Results of our experiments are presented in Table 2 and Table 3. In order to rank those methods in a reasonable way we used the Wilcoxon matched-pairs signed rank test [7].

The very first observation is that the extended (LERS) classification is always better than the simple classification method.

Results of the Wilcoxon matched-pairs signed rank test are: using LERS new classification method, C4.5 (method 3) is better than method 1 with a significance level 0.005. Also, method 6 is better than method 1, LEM2 (method 8) and method 9 with significance level 0.1. Differences in performance for other combinations of methods are statistically insignificant. Similarly, for LERS naive classification, results of the Wilcoxon matched-pairs signed rank test are: method 2 is better than method 7 with significance level 0.1, method 9 is better than methods 1 and 7, in both cases with the significance level 0.05, and, finally, method 6 performs better than method 1 with significance level 0.05. Differences in performance for other combinations of methods are statistically insignificant.

For methods that do not differ from each other significantly with respect to the Wilcoxon matched-pairs signed rank test, we estimated their relative performance by the number of test cases that have smaller error rate. If one method performs better than the other in more than 50% of the test cases, we—heuristically—conclude that it performs better than the other one. For example, in Table 2, since the C4.5 approach gives a smaller error rate than method 6 in 6 out of 10 test cases, we can conclude that using LERS new classification, the C4.5 approach performs better than method 6. Based on this heuristic evaluation principle, among all the indistinguishable methods except for method 4 and method 5, we observe that using LERS new classification, the C4.5 approach performs better than any other method; method 6 performs better than any other method except for the C4.5 approach; and method 1 performs worse than any other method. When using the LERS naive classification, method 9 performs better than any other method; method 2 performs better than any other methods except for method 9; and method 1 performs worse than any other method.

We do not have enough experimental results for method 4 and method 5. But from our available results, they perform very well. These methods are promising candidates for the best-performance methods. However, it is risky for us to conclude that they are the best methods among all nine methods because we do not have enough test files to support this conjecture statistically, using the Wilcoxon matched-pairs signed rank tests. Using both new and naive classification of LERS, the error rate of method 4 is smaller than that of any other method in more than 50% of the applicable test cases; method 5 has a smaller error rate than any other methods, except method 4, in more than 50% of the applicable test cases. The approaches of method 4 and method 5 are similar. By substituting missing value by all possible values of an attribute in our substitution, we can get as much information as possible, but the size of the resulting table may increase exponentially, thus we cannot get the results for some of our data sets because of insufficient system memory.

References

[1]   Booker, L. B., Goldberg, D. E., and Holland, J. F.: Classifier systems and genetic algorithms. In *Machine Learning. Paradigms and Methods*. Carbonell, J. G. (ed.), The MIT Press, Cambridge MA (1990) 235–282.

[2]   Chiu, D. K. and Wong A. K. C.:   Synthesizing knowledge: A cluster analysis approach using event-covering.   IEEE Trans. Syst., Man, and Cybern. **SMC-16** (1986), 251–259.

[3]   Clark, P.   Niblett, T.: The CN2 induction algorithm.   Machine Learning **3** (1989) 261–283.

[4]   Grzymala-Busse, J. W.:   On the unknown attribute values in learning from examples. Proc. of the ISMIS-91, 6th International Symposium on Methodologies for Intelligent Systems, Charlotte, North Carolina, October 16–19, 1991, Lecture Notes in Artificial Intelligence, vol. 542.   Springer-Verlag, Berlin Heidelberg New York (1991) 368–377.

[5]   Grzymala-Busse, J. W.:   LERS—A System for Learning from Examples Based on Rough Sets.   In: Slowinski, R. (ed.): *Intelligent Decision Support.   Handbook of Applications and Advances of the Rough Sets Theory.*   Kluwer Academic Publishers, Boston MA (1992) 3–18.

[6]   Grzymala-Busse, J. W. and Wang A. Y.:   Modified algorithms LEM1 and LEM2 for rule induction from data with missing attribute values.   Proc. of the Fifth International Workshop on Rough Sets and Soft Computing (RSSC'97) at the Third Joint Conference on Information Sciences (JCIS'97), Research Triangle Park, NC, March 2–5, 1997, 69–72.

[7]   Hamburg, M.:   *Statistical Analysis for Decision Making.*   Harcourt Brace Jovanovich, Inc., New York NY (1983) 546–550, 721.

[8]   Holland, J. H., Holyoak K. J., and Nisbett, R. E.:   *Induction. Processes of Inference, Learning, and Discovery.*   The MIT Press, Cambridge MA (1986).

[9]   Knonenko, I., Bratko, and I. Roskar, E.:   Experiments in automatic learning of medical diagnostic rules.   Technical Report, Jozef Stefan Institute, Lljubljana, Yugoslavia, 1984.

[10]   Michalski, R. S., Mozetic, I., Hong, J. and Lavrac, N.:   The AQ15 inductive learning system: An overview and experiments.   Department of Computer Science, University of Illinois, Rep. UIUCDCD-R-86-1260, 1986.

[11]   Polkowski, L. and Skowron, A. (eds.):   *Rough Sets in Knowledge Discovery*, 2, *Applications, Case Studies and Software Systems*,   Appendix 2: Software Systems. Physica Verlag, Heidelberg New York (1998) 551–601.

[12]   Quinlan, J. R.:   *C4.5: Programs for Machine Learning.*   Morgan Kaufmann Publishers, San Mateo CA (1993).

[13]   Stefanowski, J.:   On rough set based approaches to induction of decision rules. In Polkowski L., Skowron A. (eds.) *Rough Sets in Data Mining and Knowledge Discovery.*   Physica Verlag, Heidelberg New York (1998) 500–529.

[14]   Wong, K. C. and Chiu, K. Y.:   Synthesizing statistical knowledge for incomplete mixed-mode data.   IEEE Transactions on Pattern Analysis and Machine Intelligence **9** (1987) 796–805.

Discovering Motiv Based Association Rules in a Set of DNA Sequences

Hoang Kiem, Do Phuc

University of Natural Sciences, HCMC
Faculty of Information Technology
227 Nguyen Van Cu St, District 5, HCM city, Vietnam
hkiem@htco.com.vn

Abstract.

The research of similarity between DNA sequences is an important problem in Bio-Informatics. In the traditional approach, the dynamic programming based pair-wise alignment is used for measuring the similarity between two sequences. This method does not work well in a large data set. In this paper, we consider motif like the phrase of document and use text mining techniques for finding the frequent motifs, maximal frequent motifs, motif based association rules in a group of genes.

1. Introduction

In text mining, the phrases of document play an important role in developing the text mining algorithms [1]. We consider motif - sub sequences that occur relatively often in a set of DNA sequences- as a phrase of document and develop the algorithms for discovering the motif based association rules. The study of motif has been considered in [2,5], these techniques do not work well with a large data set of genes which are popular in the Internet. In this paper, we propose algorithms for finding the motifs, motif based association rules based on the idea of association rule mining. Based on the association rule discovery algorithms [3,7], we develop algorithms for discovering the motif, the motif based association rules. We also tested our proposed algorithms and present the experiment results from the data of 106 DNA promoter sequences of the UCI repository of machine learning database. The paper is organized as follows 1) Introduction 2) The problem of frequent motif discovery 3) The problem of discovering the relationship among motifs and classification rules 4) Conclusions and future works.

2. The Problem of Frequent Motif Discovery

Let $A = \{$"A", "C", "T", "G"$\}$ be the set of bases forming DNA sequences, each base is a nucleotide [6]. Each DNA sequence is considered as a text string $s_1,s_2, ...s_n$ where $s_k \in A$, $k=1,...,n$. We denote $|s|$ as the length of sequence s. Let s be a DNA

W. Ziarko and Y. Yao (Eds.): RSCTC 2000, LNAI 2005, pp. 386–390, 2001.

sequence, P(s) be the set of sub sequences (sub-string) of s. We define in S an order relation "<" as: $\forall s_a, s_b \in P(s)$, $s_a < s_b \Leftrightarrow s_a$ is a sub sequence of s_b. Given $s_m \in P(s)$, s_m is called a maximal element of s_a if there is no $s_b \in P(s)$ other than s_m such that $s_m < s_b$.

2.1. Frequent Motifs and Maximal Frequent Motifs in a Set of DNA Sequences

Let S be a set of m DNA sequences, $P(s_i)$ be the set of all sub sequences of s_i and PU be the union of all $P(s_i)$ for i=1,...,m. Let $N(s_t)$ be the number of DNA sequences containing s_t. Given $s_t \in PU$ and a threshold $\tau \in [0,1]$, s_t is called a frequent motif if $N(s_t)/ m >= \tau$. We denote $F(S,\tau) = \{ s_t \in PU \mid N(s_t)/m >= \tau \}$.

It is easy to hold that if $s_a \in F(S, \tau)$ and $s_t \in P(s_a) \Rightarrow s_t \in F(S, \tau)$.

Let S be a set of m DNA sequences and $F(S, \tau)$ be the set of all the frequent motifs with threshold τ. Given $a \in F(S,\tau)$, a is called a maximal frequent motif of S, if and only if i) $a \in F(S, \tau)$ and ii) $\sim\exists b \in F(S,\tau)$, $a \neq b$, $a < b$.

2.2. A Proposed Algorithm for Finding the Frequent Motifs

Given a set S of m DNA sequences and a threshold $\tau \in [0,1]$, find all the frequent motifs.

Example 1: Given $S = \{s_1, s_2, s_3, s_4 \}$ containing four DNA sequences as follows:
s_1 = 'ACGTAAAAGTCACACGTAGCCCCACGTACAGT'
s_2 = 'CGCGTCGAAGTCGACCGTAAAAGTCACACAGT'
s_3 = 'GGTCGATGCACGTAAAATCAGTCGCACACAGT'
s_4 =' ACGTAAAAGTAGCTACCCGTACGTCACACAGT'

With threshold τ=1.0, some frequent motifs are as follows:
TCA, GTC, CACA, ACAGT, CGTAAAA.

We develop a proposed algorithm for finding frequent motifs. Let $L(S,k,\tau)$ be a set of all frequent motifs with the threshold τ and k is the length (number of bases) of these motifs. The proposed algorithm is as follows:

> Answer = $\emptyset$
> Generate $L(S,1, \tau)$ from $\{$ "A", "C", "T", "G"$\}$
> For (k=2; $L(S,1,\tau) <> \{\}$; k++) do begin
>   Generate $L(S,k, \tau)$ from $L(S,k-1,\tau)$
>   Answer = $\cup_k L(S,k, \tau)$
> end
> Return Answer

a) Generate $L(S,1,\tau)$

The one letter motifs are possible "A", "C", "T", "G", so we need to check each of them, if it satisfies the definition of motif then save it into $L(S,1, \tau)$.

b) Generate $L(S,k, \tau)$ from $L(S,k-1, \tau)$ and $L(S,1,\tau)$

It is easy to hold that $s_a \in F(S, \tau)$ and $s_t \in P(s_a) \Rightarrow s_t \in F(S, \tau)$, we employ this proposition to generate $L(S, k, \tau)$ from $L(S,k-1, \tau)$ and $L(S,1,\tau)$.

The proposed algorithm is summarized as follows:

Create a matrix which row and column are L $(S,1,\tau)$.

L$(S,k,\tau) = \varnothing$

For (each $s_y \in$ L$(S,k-1, \tau)$) do

For (each $s_x \in$ L$(S,1,\tau)$) do

begin

$s_t = s_y + s_x$ // string concatenation

If ($N(s_t)/m \geq \tau$) and |st| == k) then SaveFreqMotif (s_t, L(S,k,τ))

end;

Answer = L(S,k,τ)

Return Answer

SaveFreqMotif(s_t,L(S,k,τ)) is the function for saving the frequent motif s_t into L(S,k,τ).

In the data set of 106 DNA promoter sequences which is divide into two classes: promoter class and non- promoter class. With threshold T=0.3, we discover 97 frequent motifs as follows:

A; C; T; G; AA; AC; AT; AG; CA; CC; CT; CG; TA; TC; TT; TG; GA; GC; GT; GG; AAA; AAC; AAT; AAG; ACA; ACC; ACT; ACG; ATA; ATC; ATT; ATG; AGA; AGC; AGT; AGG; CAA; CAC; CAT;CAG; CCA; CCT; CCG; CTA; CTC; CTT; CTG; CGA; CGC;CGT; CGG; TAA; TAC; TAT; TAG; TCA; TCC; TCT; TCG; TTA;TTC; TTT; TTG; TGA; TGC; TGT; TGG; GAA; GAC; GAT; GAG; GCA; GCC; GCT; GCG; GTA; GTC; GTT; GTG; GGA; GGC; GGT; AACT; AATG; ATGC; CAAT; CTTT; CTTG; TAAC; TACT; TACG; TTTT; TTGA; TTGT; GCAT; GCCT; GCTT and 58 maximal frequent motifs as follows:

AAA; AAG; ACA; ACC; ATA; ATC; ATT; AGA; AGC; AGT; AGG; CAC; CAG; CCA; CCG; CTA; CTC; CTG; CGA; CGC; CGT; CGG; TAT; TAG; TCA; TCC; TCT; TCG; TTA; TTC; TGG; GAA; GAC; GAT; GAG; GCG; GTA; GTC; GTT; GTG; GGA; GGC; GGT; AACT; AATG; ATGC; CAAT; CTTT; CTTG; TAAC; TACT; TACG; TTTT; TTGA; TTGT; GCAT; GCCT; GCTT

3. The Problem of Discovering the Relationship among Motifs and Classification Rules

Given an finite set O of m objects and a finite set D_T of n descriptors, let R be the binary relation from O to D_T. Binary relation R is represented by a matrix B_T. B_T is called the information matrix. Let b_{ij} (i=1,...,m and j=1,...,n) be the element of matrix B_T, $b_{ij}=1$ if $(o_i,d_j) \in$ R or object o_i has descriptor d_j , otherwise $b_{ij}=0$. Given O and D_T, let $P(D_T)$ be a power set of D_T and $P(O)$ be a power set of O.

We define functions ρ and λ ρ: $P(D_T) \rightarrow P(O)$ and λ : $P(O) \rightarrow P(D_T)$ as follows:

• Given $S \subseteq D_T$ then $\rho(S) = \{o \in O / \forall d \in S, (o,d) \in$ R $\}$
• Given $X \subseteq O$ then $\lambda(X) = \{d \in D_T / \forall o \in X, (o,d) \in$ R $\}$

3.1. Large Descriptor Ssets and Maximal Large Descriptor Sets

Given an information matrix B_T of O, D_T and a threshold MINSUP$\in[0,1]$, Let $S\in P(D_T)$, S is called a large descriptor set S of B_T if S satisfies the condition:

$$Card(\ \rho(S))/Card(O)>=MINSUP$$

where Card is the cardinality of set. Let L be the set of all large descriptor sets, given $s_M \in L$, s_M is called a maximal large descriptor set if and only if $\sim \exists\ s_L \in L$ and $s_M \subset s_L$. It is easy to hold that $S_A \in L$ and $S_B \subset S_A. \Rightarrow S_B \in L$.

3.2. Association Rules and Confidence Factor

Given O, D_T, B_T and a threshold MINSUP, let S be a large descriptor set of B, and $S=S_A \cup S_B$ and $S_A \cap S_B=\varnothing$. An association rule is a mapping from S_A to S_B and is denoted as $S_A \rightarrow S_B$. The confidence factor (CF) of this rule is calculated by:

$$Card(\ \rho(S_A) \cap \rho(S_B)\ )\ /\ Card(\rho(S_A)).$$

The CF shows the confidence of the possibility of occurrence of S_B if S_A is given.

In normal, given a threshold MINCONF$\in[0,1]$ and MINSUP $\in[0,1]$, find the association rules which have support greater than MINSUP and CF greater than threshold MINCONF. This problem was solved by many algorithms in [3,7].

3.3. Discovering the Relationships among Motifs

A large descriptor set is really a significant combination of frequent motifs. We employ the set of discovered frequent motifs D' as a part of descriptor set D_T and O as a set of DNA sequences. From O and $D_T=D' \cup \{promoter+, promoter-\}$, we create an information matrix and employ the algorithms in [3,7] for discovering large descriptor sets, maximal large descriptor sets and the association rules.

With the data set is 106 DNA promoters and MINSUP=30%, we discover 97 large descriptor sets and 58 maximal large descriptor sets. Some maximal large descriptor sets are listed as follows:

{ AAG,ACA} support = 0.39 ; {ACC, CCA} support =0.34 ;
{AGA,ATC} support = 0.32 ; {TCT,TTA,TTC} support = 0.37;

Some typical association rules discovered from 106 DNA promoter sequences are listed as follows:

a) For promoter class

{ATA , ATT}　　⇒ Promoter + 　confidence = 　0.80; support = 0.30
{ATA, TTA}　　⇒ Promoter + 　confidence = 　0.78; support = 0.30

b)For non promoter class

{AGA, CTC}　　⇒ Promoter - 　confidence = 　0.75; support = 0.31
{AGA, GAC}　　⇒ Promoter - 　confidence = 　0.72; support = 0.31

We plan to employ the above association rules as a mean for classification rules [6]. With rule {ATA , ATT} ⇒ Promoter + confidence = 0.80 support = 0.30 means that if DNA sequence contains motif ATA and motif ATT then there are 80% this sequence belonging to the promoter class.

3.4. Negation of the Information Matrix and Negative Association Rules

Given a binary relation R from O to $D_T=\{d_1,...,d_n\}$. Let $D_N=\{\sim d_1,...,\sim d_n\}$, we define a binary relation R' from O to D_N. Let B_N be the information matrix of R' and B_T be the information matrix of R then $B_N = \sim B_T$. This matrix is called the negation of the information matrix B_T. We employed the same methods with B_N for discovering the large descriptor sets, maximal large descriptor sets and association rules. This kind of association rules is called negative association rules [3].

With the MINSUP=0.3 and MINCONF=0.7, we discover the following negative association rules from 106 DNA promoters, some of them are as follows:

$\{\sim AAA, \sim CTTG, \sim TTTT\}$ ⇒ ~ Promoter +, confidence = 0.86, support = 0.30
$\{\sim CAAT, \sim TACG\}$ ⇒ ~ Promoter -, confidence = 0.71, support = 0.32

We employ the motif based association rules and the negative association rules for developing the gene classification problem as we did in text classification [4].

4. Conclusions and Future Works

From the view of association rule mining techniques, we have developed the algorithms for finding the frequent motifs in a set of DNA sequences. The algorithms for discovering motifs, significant combinations of motifs, the association rules discovery algorithms are proposed and tested on the data set of 106 DNA promoters of UCI. The experimental results encourage us to use the frequent motif as features for DNA biological sequences for gene identification problems.

References
[1] A Chouchoulas and Q. Shen, A Rough Set based approach to text classification, In the Proceedings of RFDGRC'99 international conference, Yamaguchi-UBE, Japan, 1999.
[2] Anders Krogh: An introduction to Hidden Markov Models for Biological Sequences, Computer Methods in Molecular Biology, Elservier, 1998
[3] Hoang Kiem, Do Phuc: Discovering the binary and fuzzy association rules from database: In the proceedings of the AFSS2000 international conference, Tsukuba, Japan, 2000
[4] Hoang Kiem, Do Phuc: On the Extension of lower approximation in rough set theory for classification problem in data mining , the WCC2000 conference , Beijing, August 2000 (to be accepted for presentation).
[5] Timothy L. Bailey: Discovering motifs in DNA and protein sequence: the approximate common sub-string problem: Ph D dissertation, Univ California, San Diego, USA, 1995
[6] Robert Giegerich and David Wheeler : Pair wise Sequence Alignment, 1996 website: http://www.techfak.uni-bielefeld.de/bcd/Curric/PrwAli/prwali.html
[7] R. Agrawal , R. Srikant, Fast Algorithm for Mining Association Rules in large database, Research report RJ, IBM Almaden Research Center, San Jose, CA ,1994

Inducing Theory for the Rule Set

Marzena **Kryszkiewicz**

Institute of Computer Science, Warsaw University of Technology
Nowowiejska **15/19**, 00-665 Warsaw, Poland
mkr@ii.pw.edu.pl

Abstract. An important data mining problem is to restrict the number of association rules to those that are novel, interesting, useful. However, there are situations when a user is not allowed to access the database and can deal only with the rules provided by somebody else. The number of rules can be limited e.g. for security reasons or the rules are of low quality. Still, the user hopes to find new interesting relationships. In this paper we propose how to induce as much knowledge as possible from the provided set of rules. The algorithms for inducing theory as well as for computing maximal covering rules for the theory are provided. In addition, we show how to test the consistency of rules and how to extract a consistent subset of rules.

1 Introduction

The problem of discovery of strong association rules was introduced in [1] for sales transaction database. The association rules identify sets of items that are purchased together with other sets of items. An important data mining problem is to restrict the number of association rules to those that are novel, interesting, useful. However, there are situations when a user is not allowed to access the database and can deal only with the rules provided by somebody else. The number of rules can be limited e.g. for security reasons or the rules are of low quality. Still, the user hopes to find new interesting relationships. The user may be even willing to induce as much knowledge as possible from the provided set of rules. We addressed this problem in [5]. We offered there how to use the cover and extension operators in order to augment the original knowledge. The cover operator does not require any information on statistical importance (support) of rules and produces at least as good rules as original ones; the extension operator requires information on support of original rules. The newly induced rules can be of higher quality than the original one [5]. Additionally, it was shown in [5] how to compute the least set of rules called maximal covering rules that represents all rules that can be induced by the two operators from the original rule set. It was shown that in general, maximal covering rules do not constitute a subset of the original rules set. Some induced rules can be maximal covering rules as well.

In this paper we propose another approach to inducing maximal knowledge from the given rule set. The new approach utilizes information both on supports and confidences of original rules. The algorithms for inducing theory as well as for computing maximal covering rules for the theory are provided. In addition, we show how to test the consistency of the rule set. A simple method of extracting a consistent subset of rules is shown.

W. Ziarko and Y. Yao (Eds.): RSCTC 2000, LNAI 2005, pp. 391–398, 2001.

2 Association Rules, Rule Cover, and Maximal Covering Rules

Let $I = \{i_1, i_2, \ldots, i_m\}$ be a set of distinct literals, called *items*. In general, any set of items is called an *itemset*. The itemset consisting of k items will be called k-*itemset*. Let D be a set of transactions, where each transaction T is a subset of I. An *association rule* is an expression of the form $X \Rightarrow Y$, where $0 \neq X, Y \subset I$ and $X \cap Y = 0$. *Support* of an itemset X is denoted by $sup(X)$ and defined as the number (or the percentage) of transactions in D that contain X. *Support* of the association rule $X \Rightarrow Y$ is denoted by $sup(X \Rightarrow Y)$ and defined as $sup(X \cup Y)$. *Confidence* of $X \Rightarrow Y$ is denoted by $conf(X \Rightarrow Y)$ and defined as $sup(X \cup Y) / sup(x)$. The problem of mining association rules is to generate all rules that have sufficient support and confidence. In the sequel, the set of all association rules whose support is greater than s and confidence is greater than c will be denoted by $AR(s,c)$. If s and c are understood, then $AR(s,c)$ will be denoted by AR.

A notion of a *cover operator* was introduced in [3] for deriving a set of association rules from a given association rule without accessing a database. The cover C of the rule $X \Rightarrow Y, Y \neq 0$, was defined as follows:

$$C(X \Rightarrow Y) = \{X \cup Z \Rightarrow V \mid Z, V \subseteq Y \wedge Z \cap V = 0 \text{ and } V \neq \varnothing\}.$$

Each rule in $C(X \Rightarrow Y)$ consists of a subset of items occurring in the rule $X \Rightarrow Y$. The antecedent of any rule r covered by $X \Rightarrow Y$ contains X and perhaps some items from Y, whereas r's consequent is a non-empty subset of the remaining items in Y. The following properties of the cover operator will be used further in the paper:

Property 1 [3]. Let $r: (X \Rightarrow Y)$ and $r': (X \Rightarrow Y)$ be association rules.

$$r' \in C(r) \text{ iff } X' \cup Y' \subseteq X \cup Y \text{ and } X' \supseteq X.$$

Next property states that every rule in the cover of another rule has support and confidence at least as good as those of the covering rule.

Property 2 [3]. Let r and r' be association rules.

$$\text{If } r' \in C(r), \text{ then } sup(r') \geq sup(r) \text{ and } conf(r') \geq conf(r).$$

It follows from Property 2 that if r belongs to $AR(s,c)$, then every rule r' in $C(r)$ also belongs to $AR(s,c)$. The number of different rules in the cover of the association rule $X \Rightarrow Y$ is equal to $3" \cdot 2^m$, where $m = |Y|$ (see [3]).

Example 1. Let $T_1 = \{A,B,C,D,E\}$, $T_2 = \{A,B,C,D,E,F\}$, $T_3 = \{A,B,C,D,E,H,I\}$, $T_4 = \{A,B,E\}$ and $T_5 = \{B,C,D,E,H,I\}$ are the only transactions in the database D. Let $r: (B \Rightarrow DE)$. Then, $C(r) = \{B \Rightarrow DE, B \Rightarrow D, B \Rightarrow E, BD \Rightarrow E, BE \Rightarrow D\}$. The support of r is equal to 4 and its confidence is equal to 80%. The support and confidence of all other rules in C(r) are not less than the support and confidence of r.□

The knowledge one can induce from the rule set R by means of the cover operator is the union of the covers of all rules in R. The covers of different rules can overlap. It was shown in [5] that the intersection of covers of a set of rules $r_1, r_2, \ldots, r_n$ is equal to 0 or is a cover of the rule: $X_1 \cup X_2 \cup \ldots \cup X_n \Rightarrow Y_1 \cap Y_2 \cap \ldots \cap Y_n$, where X_i denotes the antecedent and Y_i denotes the consequent of respective r_i. In particular, $C(r') \cap C(r) = C(r')$ for $r' \in C(r)$, so r' is less *covering* than r. The union of covers of the rule set R can be quite large, hence it is useful to keep only a subset of rules that are most representative and such that the union of their covers is equal to the union of covers of all rules in R. This is the idea behind the notion of *maximal covering rules* for R introduced in [5]. The *maximal covering rules* (*MCR*) for R were defined as follows:

$$MCR(R) = \{r \in R | \neg \exists r' \in R, r' \neq r \wedge r \in C(r')\}.$$

Whatever can be induced from R by the cover operator will be also induced from its subset $MCR(R)$ (see [5] for the algorithm for computing $MCR(R)$).

3 Consistency of Rules

3.1 Testing Consistency of Rules

Having been provided with the set of rules, it is reasonable to check if it is consistent i.e. if the are no somehow contradicting rules. Obviously, even if we know the method of checking the consistency of the rule set we cannot be sure if we were not cheated by the rules' provider who was smart enough to deform the rules by adding/removing items or changing rules supports and confidences in such a way that the modified set of rules is still consistent. In the sequel of this section, we will concentrate on checking the consistency of the delivered set of rules.

Let R be a set of rules the supports and confidences of which are known. Then the supports of itemsets of these rules as well as the supports of itemsets of the antecedents of these rules are also known. The support of the antecedent of a rule $r \in R$ is equal to $sup(r) / conf(r)$. Applying this simple observation, we introduce the notion of *known itemsets* for R denoted by $KIS(R)$ and defined as follows:

$$KIS(R) = \{X \cup Y | X \Rightarrow Y \in R\} \cup \{X | X \Rightarrow Y \in R\}.$$

Clearly, support of $Z \in KIS(R)$ is *determined in R uniquely* provided there is no pair of different rules $X \Rightarrow Y, X' \Rightarrow Y'$ in R such that:
- $Z = X \cup Y = X' \cup Y'$ and $sup(X \Rightarrow Y) \neq sup(X' \Rightarrow Y')$ or
- $Z = X \cup Y = X'$ and $sup(X \Rightarrow Y) \neq sup(X')$ or
- $Z = X = X'$ and $sup(X) \neq sup(X')$.

We define the set of rules as *inconsistent* iff some of the condition below is met:
C1. There is a rule in R or its antecedent the support of which is greater than 100% or is not greater than 0;
C2. There is $X \in KIS(R)$ the support of which is not determined in R uniquely;
C3. There are $X,Y \in KIS(R)$ such that their supports are determined in R uniquely and $sup(X) < sup(Y)$ for $X \subset Y$.

Example 2. Let us consider two rules discovered from some hospital database: $r_1:\{X\} \Rightarrow \{U,M\}$ (sup.=10%, conf.=90%), $r_2:\{X,U\} \Rightarrow \{O\}$ (sup.=90%, conf.=10%), where X stands for (*medical treatment* = X), U for (*result* = *Unsuccessful*), M for (*marital status* = *Married*) and O for (*age* = *Old*). Hence, $KIS(R) = \{\{X,U,M\}, \{X\}, \{X,U,O\}, \{X,U\}\}$. The rules r_1 and r_2 determine supports of itemsets in $KIS(R)$ uniquely: $sup(\{X,U,M\}) = 10\%$, $sup(\{X\}) = 10\% / 90\% \approx 11\%$, $sup(\{X,U,O\}) = 90\%$, $sup(\{X,U\}) = 90\% / 10\% = 900\%$. Nevertheless, the set of rules is inconsistent since: $sup(\{X,U\}) > 100\%$, $sup(\{X\}) < sup(\{X,U\})$, $sup(\{X\}) < sup(\{X,U,O\})$.   □

3.2 Extracting Consistent Rule Set

The problem of extracting a consistent subset of rules from an inconsistent rule set reminds the non-monotonic logics problem of computing a model for an inconsistent theory. There may be several models that are more or less suited to the reality. Different models can overlap partially. The *ExtractConsistentRules* algorithm, we present in this section, finds a consistent subset of rules by eliminating rules that violate Conditions C1-C3.

```
Algorithm. ExtractConsistentRules(set of rules R);
forall rules r in R do
  r.antecedent.sup = r.sup / r.conf;
R' = {r∈R| 0 < r.sup ≤ 100% and 0 < r.antecedent.sup ≤ 100%};   // (C1)
F = KIS(R');
forall itemsets f in F do {
  f.supList = {sup(X⇒Y)|X⇒Y∈R' and X∪Y=f} ∪ {sup(X)|X⇒Y∈R' and X=f};
  if |f.supList| > 1 then {                                      // (C2)
    R' = R' \ {X⇒Y∈R'| X∪Y=f or X=f};
    remove f from F; }
  else f.visited = false; };
while F ≠ ∅ do {
  f = a maximal itemset in F;
  if f.visited = false then {
    V = all subsets of f in F with support lower than f.sup;
    if V ≠ ∅ then {                                             // (C3)
      R' = R' \ {X⇒Y∈R'| X∪Y=f or X=f};
      forall itemsets v in V do
        if v.visited = false then {
          v.visited = true;
          R' = R' \ {X⇒Y∈R'| X∪Y=v or X=v}; }; }; };
  remove f from F; };
return R';
```

At first, all rules whose support is incorrect are removed as well as rules whose antecedents have incorrect support value (Condition C1). The remaining rules are assigned to R'. Known itemsets F are derived from R'. Next, for each itemset in F, it is created a list of support values based on the information on rules in R'. The itemsets with non-unique support values are removed from F altogether with the rules built from these itemsets or having antecedents equal to these itemsets (Condition C2). The remaining itemsets in F are initially marked as not visited. For each itemset $f \in F$, if not visited, it is checked whether there are subsets V having lower support. If so, then all rules corresponding to f or its subsets in V are removed (Condition C3). In the algorithm, f is chosen arbitrarily from among currently maximal itemsets. This ensures f cannot invalidate other itemsets, so it is not kept in F after its evaluation. On the other hand, the itemsets in V are not deleted since they may happen to invalidate also other supersets (unless they are currently maximal itemsets). Nevertheless, the itemsets in V do not relate to any rule in R' or its antecedent any longer. In order to avoid unnecessary evaluations, the itemsets in V are marked as visited.

In the case of Example 2, the removal of the rule r_2, the antecedent of which has support greater than 100%, is sufficient to obtain the consistent knowledge.

Example 3. Let us consider the following rule set $R = \{r_1, r_2, r_3, r_4, r_5\}$, where: r_1: $\{A\} \Rightarrow \{B,C\}$ (sup.=35%, conf.=7/12), r_2: $\{B\} \Rightarrow \{C\}$ (sup.=30%, conf.=3/4), r_3: $\{A,B\} \Rightarrow \{C,D\}$ (sup.=30%, conf.=3/4), r_4: $\{B,C\} \Rightarrow \{E\}$ (sup.=25%, conf.=5/6), r_5: $\{A\} \Rightarrow \{E\}$ (sup.=40%,conf.=1/2)}.

Now, we list the contents of $KIS(R)$ extended by the information on support(s) for each itemset: $KIS(R) = \{\{A\}$ (60%, 80%), $\{B\}$ (40%), $\{A,B\}$ (40%), $\{A,E\}$ (40%), $\{B,C\}$ (30%), $\{A,B,C\}$ (35%), $\{B,C,E\}$ (25%), $\{A,B,C,D\}$ (30%)$\}$. All rules in R and their antecedents have acceptable values of support. However, there are two support values associated with the itemset $\{A\}$: the value 60% was computed as the support of the antecedent of the rule r_1, the value 80% was computed as the support of the antecedent of the rule r_5. Since, we do not know which value is correct we decide to remove the itemset $\{A\}$ altogether with the rules that were built from it (here none) and the rules whose antecedent is equal to $\{A\}$ (here r_1 and r_5). Let us also note that the itemset $\{A,B,C\}$ (35%) has greater support than its subset $\{B,C\}$ (30%). Hence, the rules built from $\{A,B,C\}$ (here r_1) or $\{B,C\}$ (here r_2) and the rules whose antecedents are equal to one of the two itemsets (here r_4) will be removed (unless they were removed earlier as in the case of r_1). The final consistent set of rules is equal to $\{r_3\}$. □

4 Inducing Theory

In the sequel, we assume R is a consistent set of rules whose supports and confidences are known. In this section we propose how to induce as much knowledge as possible from R (i.e. *theory* for R). In order to augment the initial knowledge R we are going to use the information on supports of itemsets which is available in R. We note that for any itemsets X,Y,Z such that $Y \subseteq X \subseteq Z$ the following holds: $sup(Y) \geq sup(X) \geq sup(Z)$. Hence, the support of any (unknown) itemset X can be estimated if there are Y, Z in $KIS(R)$ such that $Y \subseteq X \subseteq Z$. Applying the information on supports of itemsets in $KIS(R)$, the support of X can be assessed as follows:

$$\min\{sup(Y)|\ Y \in KIS(R) \wedge Y \subseteq X\} \geq sup(X) \geq \max\{sup(Z)|\ Z \in KIS(R) \wedge X \subseteq Z\}.$$

Now we introduce the notion of *derivable itemsets* for R. *Derivable itemsets* for R will be denoted by $DIS(R)$ and will be defined as follows:

$$DIS(R) = \{X|\ \exists Y,Z \in KIS(R),\ Y \subseteq X \subseteq Z \}.$$

Obviously, $DIS(R) \supseteq KIS(R)$. Let *pessimistic support* (*pSup*) and *optimistic support* (*oSup*) of an itemset $X \in DIS(R)$ wrt. R be defined as follows:

$$pSup(X,R) = \max\{sup(Z)|\ Z \in KIS(R) \wedge X \subseteq Z\},$$

$$oSup(X,R) = \min\{sup(Y)|\ Y \in KIS(R) \wedge Y \subseteq X\}.$$

Then the real support of $X \in DIS(R)$ belongs to $[pSup(X,R),\ oSup(X,R)]$. Clearly, if $X \in KIS(R)$, then $sup(X)=pSup(X,R)=oSup(X,R)$.

Property 3. Let $X,Y \in DIS(R)$ and $X \subset Y$. Then:
- $pSup(X,R) \geq pSup(Y,R)$,
- $oSup(X,R) \geq oSup(Y,R)$.

Knowing $DIS(R)$ one can induce (approximate) rules $X \Rightarrow Y$ provided $X \cup Y \in DIS(R)$ and $X \in DIS(R)$. The *pessimistic confidence* (*pConf*) of induced rules is defined as follows:

$$pConf(X \Rightarrow Y,R) = pSup(X \cup Y,R)\ /\ oSup(X,R).$$

Property 4. Let $X,Y \in DIS(R)$ and $V \subset X$. Then:

$$pConf(X \Rightarrow Y, R) \geq pConf(X/V \Rightarrow Y \cup V, R).$$

Now we will introduce the notion of *theory* for the rule set R formally. *Theory* for R will be denoted by T and defined as follows:

$$T(R) = \{X \Rightarrow Y | X \cup Y \in DIS(R) \text{ and } X \in DIS(R)\}.$$

It is guaranteed for every rule $r \in T(R)$ that its support is at least as good as $pSup(r,R)$ and its confidence is at least as good as $pConf(r,R)$. Sometimes, we may be interested only in deriving the rules from R whose (pessimistic) support is greater than s and whose (pessimistic) confidence is greater than c. All rules derivable from R that satisfy these conditions will be denoted by $T(R,s,c)$. In particular, one may be interested in discovery of $T(R,s,c)$, where $s = \min\{sup(r) | r \in R\}$ and $c = \min\{conf(r) | r \in R\}$. In what follows, we offer the *GenTheory* algorithm which computes $T(R,s,c)$. In particular, for $s=0$ and $c=0$ the result will be equal to $T(R)$. The algorithm is a modification of the *AprioriGenRules* algorithm [2].

```
Algorithm. GenTheory(set of rules R, min. sup. s, min. conf. c);
D = {Z∈DIS(R) | pSup(Z)>s};
forall k-itemsets Z∈D, k ≥ 2, do {;
  H₁ = {{Y}| Y∈Z};                          // 1-item consequents
  for (i = 1; (Hᵢ ≠ ∅) and (i < k); i++) do {
    forall itemsets Y∈Hᵢ do {
      X = Z\Y;
      if X ∈ D then {
        pConf = pSup(Z) / oSup(X);
        if pConf > c then
          print the rule X⇒Y with confidence = pConf and support = pSup(Z);
      else
        delete Y from Hᵢ; }; };
    Hᵢ₊₁ = AprioriGen(Hᵢ); }; };           // k-item consequents are generated
```

The *GenTheory* algorithm assigns to D the itemsets from $DIS(R)$ whose pessimistic support is greater than s. From each k-itemset in Z, $k \geq 2$, there are created candidate rules of the length k. At first, there are considered the candidate rules with single item consequents and derivable antecedents (i.e. belonging to D). If the pessimistic confidence of a candidate rule is greater than c, then the candidate rule belongs to $T(R,s,c)$. Next, the *AprioriGen* function (see [2] for details) generates candidate rules with 2-item consequents from the 1-item consequents of the discovered rules. (The consequents Y of candidate rules $X \Rightarrow Y$ that turned out not to have sufficient confidence are not taken into account since any rule created from $Z = X \cup Y$ with consequent containing Y, e.g. $X/V \Rightarrow Y \cup V$, will not have sufficient pessimistic confidence either). In general, each i-th iteration looks as follows:

- Evaluate candidate rules with i-item consequents and derivable antecedents;
- If a candidate rule has sufficient confidence print it out; otherwise remove its consequent from the set of i-item consequents;
- Generate $(i+1)$-item consequents from the remaining i-item consequents.

5 Maximal Covering Rules for Theory

In this section we consider generation of maximal covering rules for the theory for R (i.e. $MCR(T(R))$). Let us start with the property of rules generated from $DIS(R)$:

Property 6. Let $Z \in DIS(R)$, $Z' \in KIS(R)$, $Z' \supseteq Z \supseteq X$, and $pSup(Z,R) = sup(Z')$.

- $pSup(X \Rightarrow Z/X) = sup(X \Rightarrow Z'/X)$,
- $pConf(X \Rightarrow Z/X) = pConf(X \Rightarrow Z'/X)$,
- If $X \Rightarrow Z/X \in T(R,s,c)$, then $X \Rightarrow Z'/X \in T(R,s,c)$,
- $X \Rightarrow Z/X \in C(X \Rightarrow Z'/X)$.

Observations:

O1. It follows from the definition of $DIS(R)$ and Property 6 that for every rule in $T(R,s,c)$ built from an itemset in $DIS(R) \backslash KIS(R)$ there is a covering rule in $T(R,s,c)$ built from an itemset in $KIS(R)$. This implies that no rule built from an itemset in $DIS(R) \backslash KIS(R)$ is a maximal covering rule. Hence, generation of candidate maximal covering rules can be restricted to generating rules from $KIS(R)$.

O2. Property 6 implies that no rule $X \Rightarrow Z/X$ built from $Z \in KIS(R)$ is maximal covering if there is a proper superset $Z' \in KIS(R)$ of Z having the same (pessimistic) support as Z.

The two observations were used in the *FastGenMaxCoveringRules* algorithm that computes $MCR(T(R))$. Our algorithm is a modification of the *FastGenAllRepresentatives*, we proposed in [4].

```
Algorithm. FastGenMaxCoveringRules(set of rules R, min. sup. s, min. conf. c);
 K = {Z∈KIS(R)| pSup(Z)>s};
 D = {Z∈DIS(R)|∃Y∈K, Y⊃Z};
 forall k-itemsets Z ∈ K, k ≥ 2, do {
  maxSup = max({sup(Z')| Z⊂Z'∈K} ∪ {0});
  if Z.sup ≠ maxSup then {
   A₁ = {{X}| X∈Z};                         // create 1-item antecedents
   for (i = 1; (Aᵢ ≠ ∅) and (i < k); i++) do {// loop
    forall itemsets X ∈ Aᵢ ∩ D do {
     pConf = sup(Z) / oSup(X);
     /* Is X ⇒ Z\X an association rule? */
     if pConf > c then {
     /*Isn't any longer assoc. rule X⇒Z'\X that covers X⇒Z\X?*/
      if (maxSup / oSup(X) ≤ c) then
       print the rule X⇒Z\X with support = sup(Z) and confidence = pConf;
       /* Antecedents of association rules are not extended */
       Aᵢ = Aᵢ \ {X}; }; };
    Aᵢ₊₁ = AprioriGen(Aᵢ); }; }; };         // compute (i+1)-item antecedents
```

The *FastGenMaxCoveringRules* starts with computing subsets K and D that consist from the itemsets in $KIS(R)$ and $DIS(R)$, respectively, which have sufficient support. The algorithm computes maximal covering rules from each k-itemset, $k \geq 2$, in K. Let Z be a considered itemset in K. Only k-rules are generated from Z. First, $maxSup$ is determined as a maximum from the supports of the itemsets in K that are proper supersets of Z. If there is no proper superset of Z in K, then $maxSup=0$. If $sup(Z)$ is the same as $maxSup$, then no maximal covering rule can be generated from Z because there is some proper superset of Z with support equal to $sup(Z)$ (Observation O2). Otherwise, 1-item antecedents of candidate rules are created. The *loop* starts. In general, the i-th iteration of *loop* looks as follows:

Each candidate $X \Rightarrow Z/X$, where $X \subset Z$ belongs to derivable i-itemsets in A_i, is considered. The candidate rule with sufficient pessimistic confidence that does not belong to the cover of a rule created from a proper superset of Z is maximal covering.

When all maximal covering rules with i-item antecedents are found then the $(i+1)$-item antecedents are created by the *AprioriGen* function from the i-item antecedents of rules whose pessimistic confidence is not greater than the threshold c. This fact can be justified as follows: Let $X \Rightarrow Z\backslash X$ be a rule with sufficient pessimistic confidence. Then a rule $X' \Rightarrow Z\backslash X'$, where $X' \supset X$, belongs to $C(X \Rightarrow Z\backslash X)$. So, $X' \Rightarrow Z\backslash X'$ is not maximal covering. Hence, it does not make sense to create candidate rules with antecedents being supersets of antecedents of rules with sufficient pessimistic confidence since we know apriori that they are not maximal covering.

The algorithm ends if the set of antecedents of candidate rules is empty.

6 Related Work

It was shown in [5] how to apply the cover operator C and so called *extension operator E* in order to induce as much as possible from the original rule set R. The *extension operator for* the rule set R (denoted by $E(r,R)$) was defined as follows:

$$E(r: X \Rightarrow Y, R) = \{X \Rightarrow (X' \cup Y') \setminus X | \exists r': X' \Rightarrow Y', X \cup Y \subseteq X' \cup Y' \wedge sup(r) = sup(r')\}.$$

Let us stress that the cover operator does not need any information on support of rules so that to generate rules, which are not weaker than the original rules in R. The extension operator requires the knowledge on supports of rules, but does not apply the information on supports of the antecedents of rules. Therefore, we deduce that $T(R)$, computed with the use of the knowledge on supports of rules and their antecedents produces a superset of the rules one would obtain by applying both the cover and extension operators to R (possibly many times - as long as no new knowledge can be generated [5]).

7 Conclusions

It was shown in the paper how to induce as much knowledge as possible from the given rule set. The algorithms for inducing theory as well as for deriving maximal covering rules for theory were offered. Unlike in [5], the whole theory for the rule set can be computed at once. It was shown how to test the consistency of rules and how to extract consistent subset of rules.

References

1. Agrawal, R., Imielinski, T., Swami, A.: Mining Associations Rules between Sets of Items in Large Databases. In: Proc. of the ACM SIGMOD Conference on Management of Data. Washington, D.C. (1993) 207-216
2. Agrawal, R., Mannila, H., Srikant, R., Toivonen, H., Verkamo, A.I.: Fast Discovery of Association Rules. In: Fayyad, U.M., Piatetsky-Shapiro, G., Smyth, P., Uthurusamy, R. (eds.): Advances in Knowledge Discovery and Data Mining. AAAI, Menlo Park, California (1996) 307-328
3. Kryszkiewicz, M.: Representative Association Rules. In: Proc. of PAKDD '98. Melbourne, Australia. LNAI 1394. Springer-Verlag (1998) 198-209
4. Kryszkiewicz, M.: Fast Discovery of Representative Association Rules. In: Proc. of RSCTC '98. Warsaw, Poland. LNAI 1424. Springer-Verlag (1998) 214-221
5. Kryszkiewicz, M.: Mining with Cover and Extension Operators. In: Proc. of PKDD'00. Lyon, France. To appear as a Springer-Verlag LNAI volume

Some Basics of a Self Organizing Map (SOM) Extended Model for Information Discovery in a Digital Library Context

Jean-Charles Lamirel[1], Jacques Ducloy[2], and Hager Kammoun[1]

[1]Loria, Campus Scientifique, BP 229,
54506 Vandoeuvre Cedex, France
{lamirel, kammoun}@loria.fr
[2]Inist, 4 allée du Parc de Brabois
54500 Vandoeuvre, France
Ducloy@loria.fr

Abstract. This paper presents the MicroNOMAD Discovering tool. The tool combines both topographic structures, textual and iconographic interaction, as well as viewpoint exchanges by the mean of a major extension of the classical Kohonen SOM model. It may be used for various discovering tasks in a multimedia Digital Library context. The tool basic principles are firstly described. Finally, a tool experimentation that has been carried out on the multimedia database associated to the BIBAN "Art Nouveau" server is presented.

1 Introduction

Most of the Digital Library architectures are derived from Information Retrieval models and are designed to help end-users in retrieving information «they already know but they have lost a link to». Indeed, such architectures give no ways to user for exploring the knowledge of a corpus in order to answer questions like this: «what is the most important feature in this topic? what's new? etc.». The first aim of our approach is then to build up Information Discovering Systems rather than Information Retrieval Systems. The second aim is to show that exploiting the existing interaction between texts and images in a multimedia Digital Library context may well facilitate the contents interpretation. To achieve this, our MicroNOMAD tool core model strongly derives from the multimap topographic model, which has been successfully tested on textual data in the framework of the NOMAD IR System [4]. This latter model, which may be considered as an extension of the basic Kohonen's topographic map model, enables the user to browse through a documentary database by means of an advanced topographic interface. To take benefit of the discovering and browsing properties of the NOMAD multimap model in a multimedia context, we have mainly based our adaptation of the original model to the MicroNOMAD approach on a parallel implementation of a thematic mapping and of an image mapping on the same maps.

In a first part we will explain the basics of our new Discovering tool. In a second part we will conclude with experiments on this tool.

W. Ziarko and Y. Yao (Eds.): RSCTC 2000, LNAI 2005, pp. 399–403, 2001
© Springer-Verlag Berlin Heidelberg 2001

2   The MicroNOMAD Adaptive Model

The MicroNOMAD basic image classification process is based on the Kohonen topographic map model [3]. In this model a data classification is seen as a mapping on a 2D neuron grid in which neurons establish predefined neighborhood relation. After the classification process, each neuron of the map will then play the role of a data class representative. The main advantages of the Kohonen map model, as compared to other classification models, are its natural robustness and its very good illustrative power. Indeed, it has been successfully applied for several classification tasks [5][6] [7]. In our own case, each topographic map is initially built up by unsupervised competitive learning carried out on the whole multimedia database. This learning takes place through the profile vectors extracted from the image descriptions, which describe the characteristics of these images in the viewpoint associated to the map. For each neuron n of a map M, the basic competitive learning function has the following classical form:

$$W_n^{t+1} = W_n^t + \alpha(t)k(t)(W_{n*}^t - P_n^t) . \tag{1}$$

The topological properties associated with the Kohonen maps make it then possible to project the original images (i.e. data) onto a map, so that their proximity on the map corresponds as closely as possible to their proximity in the viewpoint[1] associated to said map. Once associated to a neuron, an image could be considered as a member of the class described by this neuron.

After the preliminary learning phase, each map is organized through analysis of the main components of the neuron profiles, so as to be legible for the user. A first phase of this analysis consists in defining class names that can optimally represent the class contents when the map is displayed to the user. The second phase of the analysis consist in dividing the map into coherent logical areas or neurons groups [4][5]. Each area, which can be regarded as a macro-class of synthesis, yields a very reliable information on the relative importance of the different themes described by the map.

The communication between Kohonen maps that has been first introduced in the NOMAD IR model [4] represents a major amelioration of the basic Kohonen model. In MicroNOMAD, this communication is based on the use of the images that have been projected onto the maps as intermediaries neurons or activity transmitters between maps. The communication process between maps can be divided in two successive steps: original activity setting on source maps (1) and activity transmission to target maps (2).

The original activity may be directly set up by the user on the neuron or on the logical areas of a source map. This protocol can be interpreted as the user's choices to highlight (positively or negatively) different themes representing his centers of interest relatively to the viewpoint associated to the source map. The original activity may also be indirectly set up by the projection of an user's query on the neurons of a source map. The effect of this process will then be to highlight the themes that are more or less related to that query.

The activity transmission can be considered as a process of evaluation of the semantic correlations existing between themes of a source viewpoint (source map)

---

[1] The "viewpoint" notion is an original notion that has been firstly introduced in the NOMAD IR system for playing the role of semantic context of retrieval [4].

and themes belonging to several other viewpoints (target maps). It can be also interpreted as a general form of unsupervised inference. The activity of a class i of a target map T derived from the activity of a source map S is computed by the formula:

$$A_i^T = f_{n \in i}(g(A_n)), A_n = g(A_{j_n}^S) . \tag{2}$$

Two modes (corresponding to the above mentioned f function) are used for the computation of the transmitted activity. In the "possibilistic" mode, each class of the target maps inherited of the activity transmitted by its most activated associated data. This approach could help the user to detect weak semantic correlation (weak signals) existing between themes belonging to different viewpoints. In the probabilistic mode, each class inherited of the average activity transmitted by its associated data, whether activated or not. As opposed to the possibilistic computation, the probabilistic computation yields a more reliable measure of the semantic correlations and may be then used to differentiate between strong and weak matching.

3   Experimentation

A first experiment was carried out with the MicroNOMAD Discovering tool on the multimedia database "Art Nouveau" managed by the BIBAN server [2]. This database contains approximately 300 images related to the various artistic works of the Art Nouveau School. It covers several domains, such as architecture, painting and sculpture. The images have associated bibliographic description containing optionally title, indexer keywords and author information. We have decided to use 3 different viewpoints (profiles) in our experiment:
1. The "Indexer keywords" viewpoint. Its is represented by the keywords set used by the indexer in the keyword description field of the images.
2. The "Title keywords" viewpoint. Its associated keywords set is automatically build up through a basic keywords extraction (use of a stop word list and plural to singular conversion) of image titles. After the keywords extraction a new "Title keywords" field is added to the image description.
3. The "Authors" viewpoint. It is represented by the set of authors cited in the image descriptions.
The first step of the experiment consists in transforming the image description associated to the chosen viewpoints in profile vectors. For that step, we have also chosen to apply a classical Log-Normalization step [9], in order to reduce the influence of the most widespread words of the profiles. The second step is the original classifications building. It has been implemented through the classical Kohonen SOMPACK algorithm [7]. The results, which consist in three different classifications associated with the three different viewpoints are then "dressed" and converted to XML format. For the sake of portability, the core of the MicroNOMAD Discovering tool has been developed as a Java application. Its entries are the XML classification files produced in the preceding step and it implements the class naming strategies, the maps division into logical areas, and the above described intermap communication process.

Original multiple viewpoints classification approach have directly produced very interesting results proving again the relevance of such an approach which aims at reducing the noise which is inevitably generated in an overall classification approach whereas flexibility and granularity of the analyses are increased. As an example, in our experiment we found that a "Title keywords" classification can highlight information that is very complementary to the one highlighted by an "Indexer keywords" classification.

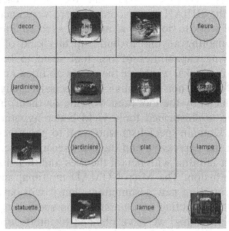

Fig. 1. Partial view of the "Title keywords" map

Maps also represent an useful tool for the indexation specialists. They help them in estimating the quality of the indexation of a database. Thanks to the classification method, strong indexation incoherences can be easily found out on the map: such incoherences are obvious if themes that specialists judged of equal weight in a domain appear with strongly different surface areas on a map.

After experimentation with several users, the opportunity to have simultaneously images and coherently organized textual information on the same support (map) seems to be definitely of great utility. Classification results interpretation are really made easier by the presence of images, as well as text represent a good help in the choice of reliable browsing points in the multimedia database.

According to user's opinion, the intermap communication process appears to be a very interesting and original feature of the model. It provides the system with a new capability that may be called a dynamic and flexible browsing behavior. As opposed to classical browsing mechanism, like hypertext links, the browsing effect could then be directly tied to the user's information and explanation needs. Moreover the number and the type of viewpoints (i.e. concurrent or complementary) that can possibly be simultaneously used are not limited by the model.

These last properties could lead us to consider our approach as a good basis for building intelligent multimedia discovering systems, especially for the ones that are strongly tied to image interpretation. Let us mention that the model is now tested for two important applications:

1. Interactive browsing through museum database and intelligent setting up of exhibitions in the framework of the technical collection of the French scientific museum "Musée de la Villette".
2. Management of multiple classifications of butterflies (color, shape, ...) in the Taiwanese NSC Digital museum of butterflies [2].

4   Conclusion

The MicroNOMAD discovering tool development represents obviously an important step for providing an iconographic interface to Digital Library Server with a high level of interactivity. We have said that the first reactions we received in demonstrating it in the BIBAN server context were very encouraging. Nevertheless, we have still a lot of work to do if we want to put such an interface on the Internet or to produce a tool enabling anyone to build up this kind of application. The basic browsing and querying capabilities of our tool seem to be well-suited to overall browsing and querying tasks, whatever the user's abilities may be. Nevertheless, the relative difficulty for the non-specialists of precisely analyzing the classification results that are produced by the tool (and working on them) is the real challenge. As shown in this paper, sophisticated tools yield better hypotheses but they are more difficult to validate. Domain specialists who want to get effective results by a deeper exploitation of both the expressive and the discovering power of the MicroNOMAD tool, but who are not familiar with neural theories and their background behavior are also a source of confusion. The MicroNOMAD multimap core model will be very useful to them in proposing new assumptions, but it will have to be interfaced with very simple tools, thus enabling non classification specialists to check the proposed assumptions. To achieve this, we have planned to interface our model with such a validation tool based on Gallois lattice and logical inference [8].

References

1. Ducloy J. : DILIB, une plate-forme XML pour la génération de serveurs WWW et la veille scientifique et technique. In : Le Micro Bulletin Thématique, No. 3, april 99.
2. Hong J. & al. : A digital museum of Taiwanese butterflies. In : Proceedings ACM/DL 00, San Antonio, Texas, june 2000.
3. Kohonen T. : Self-Organisation and Associative Memory. Springer Verlag, New York, USA, 1984.
4. Lamirel J.C. and Crehange M.: Application of a symbolico-connectionist approach for the design of a highly interactive documentary database interrogation system with on-line learning capabilities. In : Proceedings ACM/CIKM 94, Gaitherburg, Maryland, november 94.
5. Lin X., Soergel D. and Marchionini G. : A Self-Organizing Semantic Map for Information Retrieval. In : Proceedings of SIGIR, Chicago, USA, 1991.
6. Martinetz T. M. and Schulten K. J. : Topology Representing Networks. In : Neural Networks, 7 (3) : 507–522, 1994.
7. SOM papers : http://www.cis.hut.fi/nnrc/refs/
8. Simon A. and Napoli A. : Building Viewpoints in an Object-Based Representation System for Knowledge Discovery in Databases. In : Proceedings IRI-99.
9. W. J. Wilbur and L. Coffee. : The Effectiveness of Document Neighboring in Search Enhancement. In : Information Processing and Management, 30 (2) : 253–266, 1994.

Handling Data with Plural Sources in Databases

Michinori Nakata[1], Tetsuya Murai[2], and Germano Resconi[3]

[1] Department of Information Science, Chiba-Keizai College,
4-3-30 Todoroki-cho, Inage-ku, Chiba 263-0021, Japan
nakata@chiba-kc.ac.jp
[2] Division of Systems and Information Engineering,
Graduate School of Engineering, Hokkaido University,
Kita 13, Nishi 8, Kita-ku, Sapporo 060-8628, Japan
murahiko@main.eng.hokudai.ac.jp
[3] Dipartimento di Matematica, Universita Cattolica,
Via Trieste 17, Brescia, Italy
resconi@numerica.it

Abstract. Data with plural sources are handled in an extended relational model under a modal logic approach. A source of information
corresponds to a possible world and relationships between sources are
expressed by an accessibility relation. An attribute value in a database
relation is the triplet that consists of a set of possible worlds, an accessibility relation and a set of value assignment functions. A tuple in a
relation consists of a tuple value and a membership attribute value to
which the tuple value belongs to the relation. The degree, not always
equal to 1 or 0, comes from different values being obtained from plural
sources. Furthermore, the degree is expressed by two approximate values;
namely, one is a degree in the lower limit that means necessity; the other
is a degree in the upper limit that means possibility. This comes from
sources having some relationships with others. Simple queries containing
elementary formulas and logical operators are shown in the extended relational model.

Keywords: Plural sources, Data modeling, Extended relational databases,
Modal logic.

1 Introduction

Imperfection pervades the real world. Environments without imperfection are
exceptional for the real world[9]. We realize some aspects of the real world in
database systems by using information obtained from the real world. Thus, we
cannot obtain practically useful databases in various fields without considering
imperfect information.

Thus far, many investigations have been carried out to make databases truly
realistic. Following Lipski's pioneering works[5], several extended versions of relational models have been proposed to deal with a kind of imperfect information[2,
9]. These extended relational models are constructed under the premise that all

W. Ziarko and Y. Yao (Eds.): RSCTC 2000, LNAI 2005, pp. 404–411, 2001.

data for an entity are obtained from the only one source of information. As a matter of fact, we usually encounter plural sources for an entity in environments accompanied with estimates. For example, when more than one analyst publishs their report for estimating a growth rate of a company, investors have plural sources of the estimated growth rate for the company. In addition, an analyst may have some relationships with others. Our objective is to show a fundamental framework handling data with plural sources of information in databases.

The present paper is organized as follows: in section 2 data with plural sources of information are expressed on the basis of modal logic approach; section 3 is assigned to describing an extended relational model and simple select operators; the last section is concluding remarks.

2   Modal Logic and Sources of Information

2.1   Modal Logic

Modal logic is an extension of classical propositional logic. Its language consists of the set of atomic propositions or propositional variables, logical connectives $\neg, \wedge, \vee, \rightarrow, \leftrightarrow$, modal operators of necessity $\square$ and possibility $\diamond$, and supporting symbols $(,), \{, \}, \ldots$. An atomic proposition is a formula. The meaning of a formula is its truth value in a given context. Various contexts are usually expressed in terms of modal logic. A model M of modal logic is the triple $\langle \mathcal{W}, \mathcal{R}, \mathcal{V} \rangle$ where $\mathcal{W}$, $\mathcal{R}$, and $\mathcal{V}$ denote a set of possible worlds w_i, a binary relation on $\mathcal{W}$, and a set of value assignment functions ν_i, respectively. One for each world in $\mathcal{W}$, by which truth(t) or falsity(f), is assigned to each atomic proposition. Value assignment functions are inductively extended to all formulas in the usual way. The only interesting cases are:

$\nu_i(\square p) = t$ iff for all $w_j \in \mathcal{W}$ $\langle w_i, w_j \rangle \in \mathcal{R}$ implies $\nu_j(p) = t$,
and
$\nu_i(\diamond p) = t$ iff there is some $w_j \in \mathcal{W}$ such that $\langle w_i, w_j \rangle \in \mathcal{R}$ and $\nu_j(p) = t$.

A relation $\mathcal{R}$ is usually called an accessibility relation; we say that a world w_j is accessible from a world w_i when $\langle w_i, w_j \rangle \in \mathcal{R}$. It is convenient to represent the relation $\mathcal{R}$ by an $n \times n$ matrix $\mathcal{R} = [r_{ij}]$ when n is the number of possible worlds, where

$$r_{ij} = \begin{cases} 1 \; if \; \langle w_i, w_j \rangle \in \mathcal{R}, \\ 0 \; otherwise, \end{cases}$$

and to define for each world $w_i \in \mathcal{W}$ and each formula p

$$T(\nu_i(p)) = \begin{cases} 1 \; if \; \nu_i(p) = t, \\ 0 \; otherwise. \end{cases}$$

2.2   Plural Sources of Information

When an en tit y has plural sources of information, those sources do not alw ays have the same contents. By data obtained from a source we only ha ve the possibility that the data are correct. Thus, each source of information for an entity corresponds to a possible world and in the possible world it is true that the entity is characterized by the information. When an en tit y has plural sources of information with the same con ten ts, this is equivalent to that the entity has the only source of information.

A possible world is characterized by that a proposition is true. To model this, we use the following proposition:

$$a_x : \text{"A value is a given element } x\text{,"}$$

where x is an element in a given universe of discourse $\mathcal{X}$. For each world $w_i \in \mathcal{W}$, it is assumed that $\nu_i(e_{\{x\}}) = $ t for one and only one $x \in \mathcal{X}$. Thus, in multiple worlds of some modal logic, a proposition a_x for some x is valued differently in different worlds. In other words, the proposition a_x is true for a world w_i, but false for another world w_j.

Example 1

Let an entity have u and v as an attribute value for possible worlds w_1 and w_2, respectively. Then, propositions $e_{\{u\}}$ and $e_{\{v\}}$ are true for the possible worlds w_1 and w_2, respectively. This means that v alue assignmen t functions are $\nu_1(e_{\{u\}}) = $ t and $\nu_2(e_{\{v\}}) = $ t for the possible worlds w_1 and w_2, respectively.

Sources of information usually ha v e some relationships with others. This is expressed by an accessibility relation that is a binary relation. For the sake of simplicity, let two sources of information exist for an attribute of an entity. This is expressed by a set of possible worlds $\mathcal{W}(= \{w_1, w_2\})$. When the t w o sources have no relationship with each other, an accessibility relation $\mathcal{R}$ is:

$$\mathcal{R} = \begin{pmatrix} 1 & 0 \\ 0 & 1 \end{pmatrix}.$$

The possible worlds are not accessible to each other; namely, the two sources are isolated.

When t wo sources have relationships with each other,

$$\mathcal{R} = \begin{pmatrix} 1 & 1 \\ 1 & 1 \end{pmatrix}.$$

This is an equivalence relation. Possible worlds are accessible to each other. This means that each possible world accepts that a proposition holds in the other possible world.

When the first possible w orld accepts that a proposition holds in the second possible world, but the second possible world does not so,

$$\mathcal{R} = \begin{pmatrix} 1 & 1 \\ 0 & 1 \end{pmatrix}.$$

The accessibility relation is asymmetry. This often appears in our daily life; for example, A accepts B, but B does not A.

Example 2
An entity has five sources of information for a value of an attribute A_1 and obtained values of the attribute are a, b, c, c, and d, respectively. The first source is isolated. The second and the third sources are accessible to each other. The fourth source is accessible to the fifth, but the fifth source is not to the fourth. Then, an accessibility relation $\mathcal{R}$ is

$$
\mathcal{R} = \begin{pmatrix} 1 & & & & 0 \\ & 1 & 1 & & \\ & 1 & 1 & & \\ & & & 1 & 1 \\ 0 & & & 0 & 1 \end{pmatrix}.
$$

A set $\mathcal{V}$ of value assignment functions consists of five value assignment valuation functions; namely, $\nu_1(e_{\{a\}}) = \mathsf{t}$, $\nu_2(e_{\{b\}}) = \mathsf{t}$, $\nu_3(e_{\{c\}}) = \mathsf{t}$, $\nu_4(e_{\{c\}}) = \mathsf{t}$, $\nu_5(e_{\{d\}}) = \mathsf{t}$.

3 Extended Relational Model

We develop an extended relational model where plural sources of information exist. First, we address the framework of the extended relational model, and then describe select operators that are most frequently used in query processing. The other operators consisting of relational algebra will be addressed in another paper.

3.1 Framework

Definition
A extended relational scheme R consists of a set of conventional attributes $\mathcal{A} = \{A_1, A_2, \ldots, A_m\}$ and a membership attribute μ; namely,

$$
R = \mathcal{A} \cup \{\mu\} = \{A_1, A_2, \ldots, A_m, \mu\},
$$

where m is the number of conventional attributes.

Definition
The value $t[A_i]$ of an attribute A_i in a tuple t is represented by:

$$
\langle \mathcal{W}, \mathcal{R}, \mathcal{V} \rangle_{t[A_i]} = \langle \mathcal{W}_{t[A_i]}, \mathcal{R}_{t[A_i]}, \mathcal{V}_{t[A_i]} \rangle,
$$

where $\mathcal{W}_{t[A_i]}$, $\mathcal{R}_{t[A_i]}$, and $\mathcal{V}_{t[A_i]}$ for $t[A_i]$ denote, respectively, a set of possible worlds, an accessibility relation, and a set of value assignment functions.

In addition to the above extension, we introduce a membership attribute, as is done in data models handling a kind of imperfect information [1, 3, 4, 6].

Definition
Each tuple value $t[\mathcal{A}]$ in a database relation r has its membership value $t[\mu]$ that is expressed by $(t[\mu_\square], t[\mu_\diamond])$, where $t[\mu_\square]$ (resp. $t[\mu_\diamond]$) is a degree in necessity(resp. in possibility) that the tuple value $t[\mathcal{A}]$ belongs to r; namely,

$$r = \{\langle t[\mathcal{A}], (t[\mu_\square], t[\mu_\diamond])\rangle \mid t[\mu_\diamond] > 0\}.$$

As is addressed in the definition, a relation is an extended set having tuple values as elements.

A degree that a tuple value belongs to a relation comes from to what extent the tuple value is compatible with imposed restrictions on the relation. At the level of tuple values restrictions imposed on a relation specify a set of values that each tuple can take. When an attribute value in a tuple contains data obtained from plural sources of information, the number of possible worlds that is related with the tuple is plural and generally different possible worlds are associated with different values. A tuple value associated with a possible world is contained in a set of tuple values specified by restrictions, but another value with a different possible world does not so. Thus, A value of membership attribute is not always equal to 0 or 1. Moreover, some possible worlds are associated with others by an accessibility relation. This leads to that the value of membership attribute is not obtained in a exact single value, but a pair of approximate values; namely, the lower limit and the upper limit that means necessity and possibility, respectively.

The membership values are calculated, but not given in the present framework. Suppose that $t[\mathcal{A}]$ is expressed by $\langle \mathcal{W}_{t[\mathcal{A}]}, \mathcal{R}_{t[\mathcal{A}]}, \mathcal{V}_{t[\mathcal{A}]}\rangle$ and $|\mathcal{W}_{t[\mathcal{A}]}| = n$. Compatibility degrees of the tuple value $t[\mathcal{A}]$ with restrictions C in necessity and in possibility are:

$$Nec(C|t[\mathcal{A}]) = T(\square(C|t[\mathcal{A}]))/n, \ Pos(C|t[\mathcal{A}]) = T(\diamond(C|t[\mathcal{A}]))/n,$$

where $T(\square(C|t[\mathcal{A}]))$ (resp. $T(\diamond(C|t[\mathcal{A}]))$) is the number of worlds in which the tuple value $t[\mathcal{A}]$ is compatible with the restrictions C in necessity(resp. in possibility):

$$T(\square(C|t[\mathcal{A}])) = \sum_i T(\nu_i(\square(C|t[\mathcal{A}]))), \ T(\diamond(C|t[\mathcal{A}])) = \sum_i T(\nu_i(\diamond(C|t[\mathcal{A}]))),$$

where $T(\nu_i(\square(C|t[\mathcal{A}])))$ (resp. $T(\nu_i(\diamond(C|t[\mathcal{A}]))))$ is equal to one when the tuple value $t[\mathcal{A}]$ in a possible world w_i is compatible with the restrictions C in necessity (resp. in possibility). For each possible world $w_i \in \mathcal{W}_{t[\mathcal{A}]}$,

$$\nu_i(\square(C|t[\mathcal{A}])) = t \equiv \forall w_j \in \mathcal{W}_{t[\mathcal{A}]} \ \langle w_i, w_j\rangle \in \mathcal{R}_{t[\mathcal{A}]} \Rightarrow \nu_j(C|t[\mathcal{A}]) = t.$$

This formula means that for each possible world $w_i \in \mathcal{W}_{t[\mathcal{A}]}$, $\nu_i(\square(C|t[\mathcal{A}])) = t$ if and only if $\nu_j(C|t[\mathcal{A}]) = t$ in all possible worlds w_j that are accessible from the possible world w_i. Similarly,

$$\nu_i(\Diamond(C|t[\mathcal{A}])) = \mathrm{t} \equiv \exists w_j \in \mathcal{W}_{t[\mathcal{A}]} \; \langle w_i, w_j \rangle \in \mathcal{R}_{t[\mathcal{A}]} \land \nu_j(C|t[\mathcal{A}]) = \mathrm{t}.$$

This formula means that for each possible world $w_i \in \mathcal{W}_{t[\mathcal{A}]}$, $\nu_i(\Diamond(C|t[\mathcal{A}])) = \mathrm{t}$ if and only if $\nu_j(C|t[\mathcal{A}]) = \mathrm{t}$ in at least one possible world accessible from the possible world w_i. Thus, $Nec(C|t[\mathcal{A}])$ (resp. $Pos(C|t[\mathcal{A}])$) is the number ratio of possible worlds that are compatible with the restrictions C in necessity (resp. in possibility).

Example 3

$\langle \mathcal{W}, \mathcal{R}, \mathcal{V} \rangle_{A_1}$ for an attribute value A_1 of an entity is already obtained as is shown in example 2. We suppose that the only restriction C is $A_1 = \{b, c\}$ that is imposed on the relation to which the entity belongs. We obtain a membership attribute value $(2/5, 3/5)$ from $Nec(C|t[\mathcal{A}]) = 2/5$ and $Pos(C|t[\mathcal{A}]) = 3/5$ by using the above formulas.

Tuples with low values of membership attributes may appear in database relations. We use a pair of values $(t[\mu_{\Box,r}], t[\mu_{\Diamond,r}])$ as a membership attribute value for a tuple t in a relation r. These values are degrees in necessity and in possibility that the tuple value $t[\mathcal{A}]$ belongs to the relation. From the degrees we obtain degrees $(t[\mu_{\Box,\bar{r}}], t[\mu_{\Diamond,\bar{r}}])$ that $t[\mathcal{A}]$ does not belong to r; namely,

$$t[\mu_{\Box,\bar{r}}] = 1 - t[\mu_{\Diamond,r}], \; t[\mu_{\Diamond,\bar{r}}] = 1 - t[\mu_{\Box,r}].$$

We set the following criterion for accepting a tuple in a relation:

The degrees in necessity (resp. in possibility) to which a tuple value $t[\mathcal{A}]$ belongs to a relation r are greater than or equal to the degree to which the tuple value does not so.

Namely, $t[\mu_{\Box,r}] \geq t[\mu_{\Box,\bar{r}}]$, which is equivalent to $t[\mu_{\Diamond,r}] \geq t[\mu_{\Diamond,\bar{r}}]$, for an accepted tuple t in a relation r. This means that $t[\mu_{\Box,r}] + t[\mu_{\Diamond,r}] \geq 1$. The tuple that does not satisfy this criterion should be discarded by users.

3.2 Query Evaluations

We address how to extract desired information from our extended relational database. A query is performed by using select operations in relational databases, where it is evaluated to what degree a data value is compatible with a select condition $\mathcal{F}$. $\mathcal{F}$ consists of elementary formulas and logical operators $and(\land)$, $or(\lor)$, and $not(\neg)$. A typical elementary formula is "A_k is m" where m is a predicate.

The compatibility degree of a tuple value $t[\mathcal{A}]$ with a select condition $\mathcal{F}$ is expressed by a pair of degrees $Nec(\mathcal{F}|t[\mathcal{A}])$ and $Pos(\mathcal{F}|t[\mathcal{A}])$ that denote to what extent the tuple value is compatible with $\mathcal{F}$ in necessity and in possibility, respectively. Suppose that $t[\mathcal{A}]$ is expressed by $\langle \mathcal{W}_{t[\mathcal{A}]}, \mathcal{R}_{t[\mathcal{A}]}, \mathcal{V}_{t[\mathcal{A}]} \rangle$ and $|\mathcal{W}_{t[\mathcal{A}]}| = n$.

$$Nec(\mathcal{F}|t[\mathcal{A}]) = T(\Box(\mathcal{F}|t[\mathcal{A}]))/n, \ Pos(\mathcal{F}|t[\mathcal{A}]) = T(\Diamond(\mathcal{F}|t[\mathcal{A}]))/n,$$

where $T(\Box(\mathcal{F}|t[\mathcal{A}]))$ and $T(\Diamond(\mathcal{F}|t[\mathcal{A}]))$ are evaluated by the same method as addressed in the previous subsection. For a negative select condition $\neg\mathcal{F}$,

$$Nec(\neg\mathcal{F}|t[\mathcal{A}]) = 1 - Pos(\mathcal{F}|t[\mathcal{A}]), \ Pos(\neg\mathcal{F}|t[\mathcal{A}]) = 1 - Nec(\mathcal{F}|t[\mathcal{A}]).$$

The results of select operations are database relations having the same structure as the original ones. A membership attribute value for a tuple t in a derived relation is

$$(Nec(\mathcal{F}|t[\mathcal{A}]) \times t[\mu_\Box], Pos(\mathcal{F}|t[\mathcal{A}]) \times t[\mu_\Diamond]),$$

where $(t[\mu_\Box], t[\mu_\Diamond])$ is a membership attribute value in the original relation and $\times$ is arithmetic product, because the compatibility degree is the number ratio of possible worlds that are compatible with $\mathcal{F}$.

Queries can be classified into atomic ones and compound ones. Atomic queries are ones in which their select condition is expressed by an elementary formula or its negation. We show an example for an atomic query.

Example 4

The entity addressed in example 3 has another attribute A_2. Values of the attribute that are obtained from four sources are 30 that comes from the first to the third and 20 that does from the fourth, respectively. The first and the second sources have nothing to do with the others. The third and the fourth sources accept each other. Suppose a select condition $\mathcal{F}$: A_2 is $\{30, 40\}$; namely, A_2 is equal to 30 or 40. We calculate a compatibility degree of the tuple value $t[\mathcal{A}]$ with the select condition. The first and the second possible worlds are certainly compatible with the select condition. All possible worlds are possibly compatible with the condition. So, we get $Nec(\mathcal{F}|t[\mathcal{A}]) = 2/4$ and $Pos(\mathcal{F}|t[\mathcal{A}]) = 4/4$. Considering the membership attribute value $(2/5, 3/5)$ in the original relation, we obtain a membership attribute value $(1/5, 3/5)$ from $(2/4 \times 2/5, 4/4 \times 3/5)$.

Compound queries have a select condition $\mathcal{F}$ containing logical operators $and(\wedge)$ and/or $or(\vee)$. We suppose that the select condition $\mathcal{F}$ is composed of elementary formulas f_1 and f_2 in a compound query. If f_1 and f_2 is noninteractive to each other; in other words, they do not contain common attributes at all, the compound query can be calculated by evaluations of two atomic queries; namely,

$$Nec(f_1 \wedge f_2|t[\mathcal{A}]) = Nec(f_1|t[\mathcal{A}]) \times Nec(f_2|t[\mathcal{A}]),$$
$$Pos(f_1 \wedge f_2|t[\mathcal{A}]) \leq Pos(f_1|t[\mathcal{A}]) \times Pos(f_2|t[\mathcal{A}]),$$
$$Nec(f_1 \vee f_2|t[\mathcal{A}]) \geq Nec(f_1|t[\mathcal{A}]) + Nec(f_2|t[\mathcal{A}]) - Nec(f_1|t[\mathcal{A}]) \times Nec(f_2|t[\mathcal{A}]).$$
$$Pos(f_1 \vee f_2|t[\mathcal{A}]) = Pos(f_1|t[\mathcal{A}]) + Pos(f_2|t[\mathcal{A}]) - Pos(f_1|t[\mathcal{A}]) \times Pos(f_2|t[\mathcal{A}]),$$

where the equality holds when f_1 and f_2 are noninteractive; namely, f_1 and f_2 do not contain common attributes.

When f_1 and f_2 are interactive, suppose that f_1 is A_i is S_1 and f_2 is A_i is S_2. We calculate the compatibility degree $Pos(f_1 \wedge f_2|t[\mathcal{A}])$ in possibility by resetting such that $\mathcal{F}$ is A_i is $S_1 \cap S_2$ in the case of $\mathcal{F} = f_1 \wedge f_2$ and the compatibility degree $Nec(f_1 \vee f_2|t[\mathcal{A}])$ in necessity by resetting such that $\mathcal{F}$ is A_i is $S_1 \cup S_2$ in the case of $\mathcal{F} = f_1 \vee f_2$.

4　Concluding Remarks

We have developed relational databases with plural sources of information under a modal logic approach. Each source is expressed by a possible world where a proposition holds. Furthermore, the relationships between sources are considered by using an accessibility relation. An attribute value is the triplet that consists of a set of possible worlds, an accessibility relation, and a set of value assignment functions. In the approach all pieces of information obtained for an entity are kept without fusing. This leads to that update of information can be easily made. In addition, our approach can be extended into databases with imperfect infomation[7]. Thus, our approach gives significant bases to handling data with plural sources of information.

References

1. Baldwin, J. F. [1983] A Fuzzy Relational Inference Language for Expert Systems, *in Proceedings of the 13th IEEE Inter. Symp. on Multiple-Valued Logic* (Kyoto, Japan), pp. 416-421.
2. Dyreson C. E. [1997]A Bibliography on Uncertainty Management in Information Systems. in *Uncertainty Management in Information Systems: From Needs to Solutions*, Eds., A. Motro and P. Smets, Kluwer Academic Publishers, pp. 413-458.
3. Lee, S. K. [1992]Imprecise and Uncertain Information in Databases: An Evidential Approach. in Proceedings of the 8th International Conference on Data Engineering, IEEE 1992, pp. 614-621.
4. Li, D. and Liu, D. [1990] A Fuzzy Prolog Database System, Research Studies Press, 1990.
5. Lipski, W. [1979]On Semantics Issues Connected with Incomplete Information Databases. ACM Transactions on Database Systems, **4**:3, 262-296.
6. Nakata, M. [1996] Unacceptable Components in Fuzzy Relational Databases, International Journal of Intelligent Systems, **11**:9, 633-647.
7. Nakata, M. [2000] Imperfect Information with Plural Sources in Databases, in preparation.
8. Nakata, M., Resconi, G., and Murai, T. [1997] Handling Imperfection in Databases: A Modal Logic Approach, Database and Expert Systems Applications(A. Hameurlain and A. M. Tjoa., eds.), Lecture Notes in Computer Science 1308, Springer 1997, pp. 613-622.
9. Parsons, S. [1996] Current Approaches to Handling Imperfect Information in Data and Knowledge Bases, IEEE Transactions on Knowledge and Data Engineering, **8**:3, 353-372. Hierarchical Uncertainty Metatheory Based upon Modal Logic, International Journal of General Systems, **21**, 23-50.

Research Topics Discovery from WWW by Keywords Association Rules

David Ramamonjisoa[1], Einoshin Suzuki[2], and Issam Hamid[1]

[1] Faculty of Software and Information Science, Iwate Prefectural University (IPU)
152-52 Sugo, Takizawa, Iwate, 020-0193, Japan
{david, issam}@soft.iwate-pu.ac.jp
[2] Electrical and Computer Engineering, Yokohama National University,
79-5 Tokiwadai, Hodogaya, Yokohama 240-8501, Japan
suzuki@dnj.ynu.ac.jp

Abstract. In this paper, we present agents based tool to discover new research topics from the information available on the World Wide Web (WWW). Agents are using KAROKA (Keywords Association Rules Optimizer Knobots Advisers). KAROKA is a model of discovery in text database used in WWW. The WWW sources are converted to a highly structured collection of text. Then, KAROKA tries to extract association rules, regularities and useful information in the collection of text. KAROKA techniques are described such as information retrieval similarity metrics for text, generation and pruning of keywords combination, and summary proposal of discovered information.

1 Introduction

When a user explores a new domain, attempting to summarize the essence of an area previously unknown to the user, it is called *information and knowledge discovery* [1].

Information Discovery Agent (Web Mining Agent) is an important kind of information seeking agent trying to realize the previously mentioned tasks for text documents with images or sounds on the Web.

Existing Information Discovery agents are currently unable to produce themselves rational analyses of retrieved information on the World Wide Web because the Web is not structured information sources. These agents are processing user-queries and returning high quality results to the user satisfying the users preferences without bringing new ideas or unexpected interesting discovery. Intelligent Information Discovery Agents should have some mechanisms to distinguish the irrelevant data and unexpected interesting results.

Current systems are focusing on users relevance feedback, preferences as described to supervised-learning systems. Some systems are using the event detection and tracking described as unsupervised-clustering algorithms to search high quality information and extract the maximum knowledge. These methods do not bring novelty, utility, and understandability to the results [5].

The domain we focus on is the discovery of Research Topics on "Data Mining and Knowledge Discovery and Their Applications." We try to determine "What

W. Ziarko and Y. Yao (Eds.): RSCTC 2000, LNAI 2005, pp. 412–419, 2001.
© Springer-Verlag Berlin Heidelberg 2001

are possible and promising research topics according to the information on the Web in this research domain."

We introduce *KAROKA (Keywords Association Rules Optimizer Knobots Advisers)*, a personalized system that pro-actively tries to discover information from various distributed sources and presents it to the user in the form of a digest. KAROKA is using a tool similar to "AltaVista Discovery" [2], Karnak [3] or CiteSeer [4] for the exploration of World Wide Web.

In section 2, we describe the related works to our research. The KAROKA model is explained in section 3. Section 4 details the example of KAROKA use and experimental results in *"data mining trends and forecasts."* Section 5 summarizes the paper and describes our future work.

2   Related Works

We describe here the status of the current research on information retrieval and information discovery agents which may have impact on our own research.

2.1   Information Retrieval Agents

Although our work is different from information retrieval, we are using search engine to our system. We briefly describe here a survey.

Search engines have been built in the past that focus on indexing the WWW, thus providing an easier way of finding documents. A common problem of WWW indexing engines is that they frequently return documents that are irrelevant to what the user is interested in. Sometimes this is caused by a poor choice of keywords by the user and sometimes by the poor indexing of the documents by the engine. Personalization of the system can form more exact queries.

Another problem of search engines is the processing of information available in the Web. For any given query, there are often simply too many relevant documents for the user to cope with the result efficiently. This work is beyond the search engine processing.

In the survey described recently by Mladenic [6], personalized agents are developed with machine learning (text-learning) or data-mining techniques. Text-learning is applied on collected information to help users browsing the Web or ameliorating their searches.

Information retrieval utilizes the weighted keyword vector as a general method of document representation. Text-learning systems are using feature selection and classification methods to customize to individual users for the personalization.

The intelligent agents in information retrieval are now able to find and filter important and useful information according to their user preferences [6] [7].

Some intelligent agents are capable to compare and advise similarity of documents in order to refine as much as possible search or recommendation web pages [8] [9].

2.2   Agents for information and Knowledge Discovery

We mean Discovery Agents the autonomous programs which process discovery of information from the Web. We do not use and generalize the term for binary and image database. Our area of discovery is the text or keywords.

The process of information discovery could be greatly assisted if there was a way of making it personalized. For instance a personalized agent could form more exact queries.

Current Information Discovery Agents specializing in WWW visit directly the site of interest and analyze the document to find if it has changed and how much, by using the database where the URLs are stored. New and unknown information sources will be progressively detected and added to the database.

For example, "Altavista Discovery" [2] or "Karnak" [3] has a purpose to personalize the information at the user level. "CiteSeer" [4] is focusing its discovery based on group of users or clustering.

The Ideal Discovery Agents are able to navigate, read, summarize, and again surf the collection of text to form an abstraction of what new research or domain is all about.

The methods used in discovery agents are same in data mining. Mining functions are based on statistics, classification, correlation-detection, factorial analysis methods, and graphical representation. For examples, text clustering is used to create thematic overview of text collections and co-citation analysis are developed to find general topics within a collection.

Our methods are related to "Text Data Mining" described in [10]. These include the exploration strategies and hypothetical sequence of operations.

Until now, discovery agents do not suggest a *possible new research topic* yet. As the goal of our research is to discover new research topics from the web, we will describe in the following our methods and experiments.

3   KAROKA Model and Architecture

KAROKA objectives are to design agents that can process user queries in domain specific research areas, collect WWW sources relevant to the queries, extract keywords and rules from the retrieved sources, infer to determine possible new research topics in the domain, and present results as *'list of possible new research topics'* in the domain. KAROKA (see figure 1) uses keywords extracted from technical and project presentation articles available in the World-Wide-Web documents.

KAROKA does not search the WWW itself but instead launches multiple agents that utilize existing indexing engines and perform a "meta-search" in order to collect and discover information that is broadly of interest to the user. Then the system further analyzes the retrieved documents in building keywords database and structured tree such as indexes to the research topics domain.

It uses information discovery agents to monitor frequently changing information resources and update the database.

According to the research topic area, it generates and combines randomly keywords of research sub-topics. Strategies and constraints for the new topics

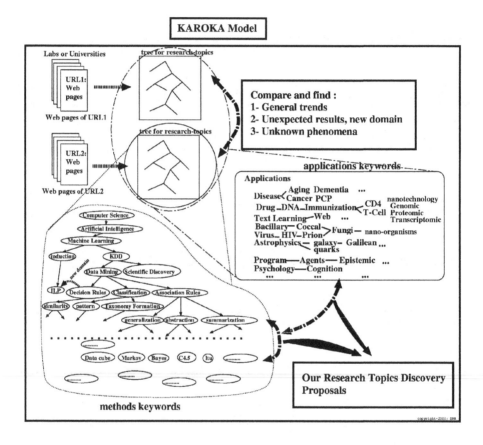

Fig. 1. KAROKA model and analysis : User focuses to the research topics in found web pages by a search engine according to a query in a specific domain (in this figure, the domain is in *Computer Science/Artificial Intellicenge/Machine Learning/Knowledge Discovery and Data mining*). Documents are retrieved from *URL* and are structured to research topics tree. *Research topics* are composed of one or several keywords. Keywords are classified into *method keywords* or *application keywords*. According to our strategies and algorithms, new research topics are derived

selection are based on classification techniques, association rules, and verification on the WWW.

3.1   Preliminary Preparation

Some research topics are already indexed and categorized hierarchically in the search engines via the Web. Those research topics are general and common. For example, the index of the research topics on Data Mining and Knowledge Discovery has many sub-topics.

```
Computer Science>Artificial Intelligence>Machine Learning
Machine Learning>Knowledge Discovery>...
                Data Mining>Application>...
                    Classification>...
                    Feature Selection>...
Classification>Decision Rules - Winnow - TFIDF - Naive Bayes - ...
...
```

This preliminary preparation can be realized with an user agent. The following tasks should be done first.

User uses search engine to find general 'research topics' to the research domain he/she is interested.

User selects some URLs from the search results to start the discovery of new research topics.

For each URL pages, there is a list of contents. User focuses to the content *research topics or research areas'* if it exists.

User retrieves all documents in found section and stores in his computer. If the document is in HTML format, then document is transformed to structured text, the headers are treated as a special type of keyword.

3.2   Keywords Extraction

Each retrieved document for every URL is processed one by one. Document can have different format such as text ascii, PDF or postscript. With a conversion tools, the document is always converted into ascii text before the keyword extraction.

One obvious methods to extract keywords is to find the keywords as the authors defined in the document. Some documents such as articles or technical papers contain explicitly the keywords. These explicit keywords are treated with priority.

Documents without explicit keywords are processed with the document representation (bag of words) [11]. In defining rules through a training, we can extract the important words in the bag of words. The rules concern to eliminate the words with low frequency and high weight. These words are too common. In HTML documents, the 'header' keyword and 'body' keywords are weighted differently. The 'header' of some URL may contain the 'Research topics'.

3.3   Knobots Adviser

The Knobots advisers are autonomous programs to collect results of queries with Web search engines.

Each generated keyword by the association rules optimizer module is checked in the Web. The first criteria for the relevance is the occurrence and the rank of the related URL given by the search engine such as first 10 or 100 matches.

The knobot then checks the URL to identify the quality of the page and collects possible technical papers or reports.

3.4   Keywords Association Rules Optimizer

The collected keywords are processed to discover new topics.

This module classifies the keywords as "method keywords" and "application keywords." The "method keywords" are related to the research topics and their sub-topics generally already known in the research community.

The "application keywords" are related to the other domains which the methods are applied.

First keywords selection: Each *method keyword* is checked with the Web for the occurrence, then they are sorted.

The keywords with high frequency (occurrence) are eliminated. They are too common for the topics.

Combining keywords: Remaining keywords from the first selection are combined two-by-two to obtain the trends of the research within these research topic keywords.

We then check with the knobot adviser module the relevance of these generated combination of keywords in the Web.

Second keywords selection: According to the relevance of the keywords in the Web, a second selection is necessary to eliminate again the high frequency keywords. The same method as in first selection is used here.

At this stage, we observed the existence of research topics classes. The high frequency keywords class is belong to the known research topics. Low frequency keywords class may be new research topics or irrelevant keywords.

Second keywords combination: We applied the first selection method to the *application keywords.*

We added the *application keywords* to two combined keywords result of the previous second keywords selection method. We then generated trends of research topics based on three keywords as two method keywords and one application keyword. A final check with the Web is realized to get the new research topics proposals.

4   Experiments

Our URLs starting points are Laboratories and Universities in Europe, Asia or North America with english pages (for examples, ATT, IBM, Microsoft Research, CMU, UCI, ORST, ...). We focused on the research topics *Data mining and Knowledge Discovery* to each URL.

KAROKA is then launched to extract research topics keywords and try to propose the promising new research topics.

We describe here different values of the experiment.

200 extracted explicit keywords from articles and technical papers on *"data mining and knowledge discovery."* and 600 keywords from HTML documents are retained.

Classification of keywords as "methods keywords" and "application keywords", elimination of common keywords using the term frequency, selection of keywords using the knobots gave the results as 110 method keywords and 50 application keywords.

Combination of methods keywords two-by-two to find new topics, internet ranking according to the occurrence and elimination gave 316 research topics.

Combination of the methods keywords with application keywords and selection returned 13450 research topics.

Top 200 of these research topics are proposed as final results. These are research topics resulted from the KAROKA experiment.

```
::::::::::::::::::::::::  cognitive impairment ,8330 expectation-maximization ,1220
methodkeyword results    combining classifiers,190  experimental comparisons ,427
::::::::::::::::::::::::  concept learning ,3490      explanation-based learning ,1769
attribute focusing ,37   datacube ,1467             feature selection ,4200
back propagation ,5070   deductive learning ,170    indicator variables ,1107
bayesian networks ,4750  density estimation ,4369   inductive learning ,4353
bayesian statistics ,3460dependence rules ,45        instance-based learning ,829
c4.5,3810                duplicate elimination ,598 lattice traversal,9
categorical data,6277    ensemble learning ,160     >>>>>>>>>> cut here>>>>>>>>
closure properties ,1424 exemplar-based learning ,122
```

```
::::::::::::::
research topics results (application keyword+two method keywords)
::::::::::::::
>>>>>>>>>> cut here>>>>>>>>>>>>>>>>>
astrophysics   and lazy learning  and tree-structured classifiers       20
hiv    and deductive learning  and exemplar-based learning              20
hiv    and optimal classification and rule consistency                 20
mutations   and concept learning  and maximal hypergraph cliques       20
pharmacy   and lattice traversal and model uncertainty                 20
astrophysics   and dependence rules  and multi-strategy learning       20
episodes   and lattice traversal and statistical learning theory       20
mutations   and experimental comparisons  and tree-structured classifiers  20
episodes   and first order decision trees  and rule-based systems      20
mutations   and lattice traversal and model uncertainty                20
episodes   and duplicate elimination  and lattice traversal            20
pharmacy   and density estimation  and vcdimension                     20
```

```
astrophysics    and attribute focusing  and lazy learning                    20
hiv    and categorical data and first order decision trees                   20
query    and first order decision trees  and rule learning                   20
pharmacy    and perceptron  and vcdimension                                  20
pharmacy    and lattice traversal and rule induction                         20
query    and curse-of-dimensionality  and vcdimension                        20
episodes    and vcdimension  and winnow                                      20
astrophysics    and lattice traversal and rule induction                     20
drug resistance    and lazy learning  and montecarlo methods                 19
dementia    and lattice traversal and statistical reasoning                  19
>>>>>>>>>> cut here>>>>>>>>>>>>>>>>
```

Our first experimental results show the new research topics. We are sure that these topics are new and rare in the WWW. User should conclude to their utility and understandability.

5   Conclusions and Future Works

In this paper, we presented a model for research topics discovery from the Information World Wide Web. The model is based on KAROKA system. KAROKA is a personalized tool using keywords association rules and knobots. With KAROKA, we have partially automated the discovery.

Our experiment results show the KAROKA system applied to discover new research topics on *Data Mining and Knowledge Discovery.*

At the stage of the KAROKA program, the user must interpret the result given by KAROKA as a support for his/her research topics finding.

In the future, we are refining the KAROKA program to be more precise and flexible for any language.

References

1. Crimmins, F. et al. : "TetraFusion: Information Discovery on the Internet" IEEE Intelligent Systems Journal, pp.55-62, July-August, 1999.
2. http://discovery.altavista.com
3. http://www.Karnak.com
4. http://www.researchindex.com
5. Pazzani, M. : "Trends and Controversies: Knowledge discovery from data ?" IEEE Intelligent Systems Journal, pp.10-13, March-April, 2000.
6. Mladenic, D. and Stefan, J. : "Text-learning and Related Intelligent Agents: A Survey" IEEE Intelligent Systems Journal, pp.44-54, July-August, 1999.
7. Levy, A.Y. and Weld, D.S. : "Intelligent Internet systems" Artificial Intelligence Journal, vol. 118, numbers 1-2, pp.1-14, April, 2000.
8. Cohen, W.W. and Fan W. : "Web-Collaborative Filtering: Recommending Music by Crawling The Web" in The 9th International WWW Conference, May, 2000.
9. Good, N. et al. : "Combining collaborative filtering with personal agents for better recommendations" in the Proceedings of AAAI-99, pp.439-446, 1999.
10. Hearst, M. : "Untangling Text Data Mining," in the Proceedings of ACL'99, June, 1999.
11. Salton, G. : "Automatic Text Processing: The Transformation, Analysis, and Retrieval of Information by Computer;", Addison-Wesley, 1989.

Temporal Templates and Analysis of Time Related Data

Piotr Synak

Polish-Japanese Institute of Information Technology
Koszykowa 86, 02-008 Warsaw, Poland
email: synak@pjwstk.waw.pl

Abstract. In the paper we investigate the problem of analysis of time related information systems. We introduce notion of temporal templates, i.e. homogeneous patterns occurring in some periods. We show how to generate temporal templates and time dependencies among them. We also consider decision rules describing dependencies between some temporal features of objects. Finally, we present how temporal templates can be used to discover behaviour of such temporal features.

Keywords: temporal template, temporal feature extraction, rough sets

1 Introduction

The intelligent analysis of data sets describing real life problems becomes a very important topic of current research in computer science. Different kind of data sets, as well as different types of problems they describe, cause that there is no universal methodology nor algorithm to solve these problems. For example, analysis of a given data set (information system) may be completely different, if we define a time order on a set of objects described by this data set, because the problem may be redefined to include time dependencies. Also, the expectation of an analyst may be different for the same data set, according to the situation. Let us consider a decision problem described by an information system (decision table), where objects are ordered according to time. In one situation the analyst may want to extract typical decision rules, e.g. "*if a = v and c = w then decision = d*", but another time, information about how the change of given condition (attribute) influences change of decision, e.g. "*if Δa=high positive and Δc=high negative then $\Delta decision$ = neutral*". Much more general problem is to find, for a given property of condition ($\Delta a = positive$ is an example of property *positive change* of condition a), temporal dependencies giving the idea about periods of occurrences of this property, as well as temporal dependencies between different properties. For example, we can discover, that, if properties "$\Delta a = positive$" and "Δc=negative" appear together in the same time and last for some long period, it means, that after a certain time another set of properties will appear together (e.g. properties "$\Delta b = neutral$" and "$\Delta c = positive$").

In the paper we introduce the notion of temporal template, i.e. homogeneous pattern occurring in time. We show how to extract several different temporal templates and how to discover dependencies among them. We also present a method of generation of more general decision rules, which contain information about, e.g changes of values in some period. We show how temporal templates can be used to track temporal behaviour of attribute properties.

W. Ziarko and Y. Yao (Eds.): RSCTC 2000, LNAI 2005, pp. 420–427, 2001.

2   Temporal templates

First, let us define the way we represent data sets. The basic notion is an *information system* [8] which is defined as a pair $\mathbf{A} = (U, A)$, where U is a non-empty, finite set called the *universe* and A is a non-empty, finite set of *attributes*. Each $a \in A$ corresponds to function $a : U \to V_a$, where V_a is called the *value set* of a. Elements of U are called *objects*, *situations* or *rows*, interpreted as, e.g., cases, states, patients, observations.

A special case of information system is a *decision table* $\mathbf{A} = (U, A \cup \{d\})$, where $d \notin A$ is a distinguished attribute called *decision*. The elements of A are called *conditional attributes (conditions)*.

Any expression of the form $(a \in V)$, where $a \in A \cup \{d\}, V \subseteq V_a$ we call a *descriptor*. A descriptor $(a \in V)$ is a *conditional descriptor* iff $a \in A$, or *decision descriptor* iff $a \in \{d\}$.

We say that an object $x \in U$ *matches* a descriptor $(a \in V)$ iff $a(x) \in V$.

The notion of templates was intensively studied in literature (see e.g. [1], [5], [7]). For a given information system $\mathbf{A} = (U, A)$ by *generalized template* we mean a set of descriptors

$$T = \{(a \in V) : V \subseteq V_a\} \tag{1}$$

such, that, if $(a \in V) \in T$ and $(b \in W) \in T$ then we have $a \neq b$. An object $x \in U$ *matches* a generalized template T, if it matches all descriptors of T. A special case of generalized templates are templates with one-value descriptors, i.e. of form $(a = v)$. Templates can be understood as patterns which determine homogeneous subsets of information system

$$T(\mathbf{A}) = \{x \in U : \forall_{(a \in V) \in T} \ a(x) \in V\} \tag{2}$$

In many practical problems a very important role plays time domain. In this case, the set of objects is ordered according to time $\mathbf{A} = (\{x_1, x_2, ..., x_n\}, A)$, and we say, that t is the time of occurrence of object $x_t, 1 \leq t \leq n$. From now on by information system we mean one in a sense presented above. In such systems one can consider templates that occur in some period. This kind of templates we call *temporal templates* and define as

$$\mathbf{T} = (T, t_s, t_e), \ 1 \leq t_s \leq t_e \leq n \tag{3}$$

We say, that two temporal templates $\mathbf{T_1} = (T_1, t_s, t_e)$ and $\mathbf{T_2} = (T_2, t_s', t_e')$ are *equal* if $T_1 = T_2$.

Now, let us define two important properties of temporal templates. By *width* of temporal template $\mathbf{T} = (T, t_s, t_e)$ we mean $width(\mathbf{T}) = t_e - t_s + 1$, i.e. the length of the period which T occurs in. Please notice, that not all objects $x_{t_s}, x_{t_s+1}, ..., x_{t_e}$ have to match T and intuitively the more objects match T the better. Thus, by *support* of $\mathbf{T}$ we understand $supp(\mathbf{T}) = card(T(\mathbf{A_T}))$, where $\mathbf{A_T} = (\{x_{t_s}, ..., x_{t_e}\}, A)$ is the information system determined by $\mathbf{T}$ and $T(\mathbf{A_T})$ is set of objects of this system that match T. The *quality* of temporal template is a function that maximizes width, support as well as number and precision of descriptors.

In Figure 1 we show two examples of temporal templates $\mathbf{T_1} = (\{(a \in \{u\}), (c \in \{v\})\}, 2, 8)$ and $\mathbf{T_2} = (\{(b \in \{x\}), (d \in \{y\})\}, 10, 13)$.

$\mathbf{A}$	a	b	c	d	e
x_1	·	·	·	·	·
x_2	u	·	v	·	·
x_3	u	·	v	·	·
x_4	·	·	·	·	·
x_5	u	·	v	·	·
x_6	u	·	v	·	·
x_7	·	·	·	·	·
x_8	u	·	v	·	·
x_9	·	·	·	·	·
x_{10}	·	x	·	y	·
x_{11}	·	x	·	y	·
x_{12}	·	·	·	y	·
x_{13}	·	x	·	y	·
x_{14}	·	x	·	·	·
x_{15}	·	·	·	·	·

T_1 spans rows x_1 to x_9; T_2 spans rows x_{10} to x_{13}.

Fig. 1. An example of temporal templates.

3   Temporal templates generation

In this section we present an algorithm that generates several temporal templates which are disjoint according to time. The main idea is based on scanning the information system with some time window and generation of best generalized template within this window. There is a chance, that after a light shift of a time window, the previously found template is also the best one in a new window. By shifting the window we may discover the beginning (x_{t_s}) and the end (x_{t_e}) of the area where a given template is optimal or close to the optimal one. Shifting the time window through the whole information system generates a set of temporal templates.

Let $\mathbf{A} = (\{x_1, x_2, ..., x_n\}, A)$ be an information system. By *time window* on $\mathbf{A}$ of size s in point t we understand an information system $win_{\mathbf{A}}^s(t) = (\{x_t, x_{t+1}, ..., x_{t+s-1}\}, A)$, where $1 \leq t$ and $t + s - 1 \leq n$. In the process of temporal rules generation both size s of a window and number of used windows are parameters that have to be tuned according to the type of data. Below we present details of the algorithm.

Algorithm: Temporal templates generation

Input: Information system $\mathbf{A}$, size of time window $size$, length of the shift of time window $step$, quality threshold τ

Output: Set of temporal templates

1. $i := 1, T = NULL, t_s = 1$
2. **while** $i < n - size$ **begin**
3. $best = FindBestTemplate(win_{\mathbf{A}}^{size}(i))$
4.   **if** $best \neq T$ **then begin**
5. $t_e = i$
6.   **if** $Quality((T, t_s, t_e)) \geq \tau$ **then**
7.   **output** (T, t_s, t_e)
8. $t_s = i$
9. $T = best$
10.   **end if**
11. $i := i + step$
12. **end while**

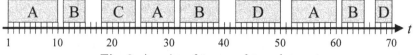

Fig. 2. A series of temporal templates.

At each run of the loop the algorithm is shifting the time window. The task of subroutine $FindBestTemplate(\mathbf{A})$ is to return the optimal (or semi-optimal) template for $\mathbf{A}$. Because the problem of optimal template generation is NP-hard (see e.g. [5]) some approximation algorithm to be used in this procedure should be considered. In [5], [7] there can be found several very fast heuristics for templates generation.

The input parameter τ is used to filter out temporal templates of low quality. The definition of template quality may be different and it depends on the formulation of a problem.

4   Discovery of dependencies between temporal templates

Generation of series of temporal templates defines a very interesting problem related to discovery of dependencies among them. Several templates may occur many times and it can happen, that one template is always (or almost always) followed by another one. Generally, one set of templates may imply occurrence of other templates and extraction of such information may be very useful. Let us consider the situation, that we are observing occurrence of some temporal template in current data. On the basis of template dependencies we predict occurrence of another template in the near future. In this section we present a simple method that allows to discover such template dependencies.

Let us observe, that the process of temporal templates generation results with a series of templates. On time axis each template has its time of occurrence and width. In Figure 2 we have an example series of four temporal templates $\mathbf{A}$, $\mathbf{B}$, $\mathbf{C}$, $\mathbf{D}$. From such a series we want to extract three kinds of dependencies. First, the dependencies between templates, e.g. decision rules of form *"if template in time t-3 is* $\mathbf{B}$ *and template in time t-1 is* $\mathbf{A}$ *then template in time t is* $\mathbf{D}$*"*. Second, if we have such a rule then on the basis of widths of templates $\mathbf{B}$ and $\mathbf{A}$ we want to predict the width of $\mathbf{D}$. Third kind of information, we need, is the time of occurrence (t_s) of template $\mathbf{D}$.

Now, let us focus on the problem of decision rules generation, that map dependencies between templates. Templates can be treated as events and the problem may be reformulated in terms of frequent episodes detection in time series of events. This problem is investigated in e.g. [4], however, the difference is, that here events are not points on time axis (as in [4]), but intervals of different length. Thus, we propose another method which takes advantage of rough set theory.

The idea is based on construction of a decision table from time series of templates and further computation of decision rules. The number of condition attributes n is a parameter of this method and it reflects how many steps in past we want to consider. The objects of this table we construct from series of templates - one object is a consecutive sequence of $n + 1$ template labels. For example, if our template series is one presented in Figure 2, i.e. **A, B, C, A, B, D, A, B, D** and $n = 2$ we obtain the decision table as in Table 1.

	T_{t-2}	T_{t-1}	T_t
x_1	**A**	**B**	**C**
x_2	**B**	**C**	**A**
x_3	**C**	**A**	**B**
x_4	**A**	**B**	**D**
x_5	**B**	**D**	**A**
x_6	**D**	**A**	**B**
x_7	**A**	**B**	**D**

Table 1. Decision table constructed from a sequence of temporal templates

From this table we can generate decision rules using, e.g. rough set methods (see [2], [3], [8]). In our example we obtain, among other, the following rules:

if $T_{t-1} = $ **A then** $T_t = $ **B**
if $T_{t-2} = $ **B then** $T_t = $ **A**
if $T_{t-2} = $ **A** and $T_{t-1} = $ **B then** $T_t = $ **D**

Once we have a set of decision rules we can compute, for each rule, how widths of predecessor templates of the rule determine width of successor template. For a given rule, we construct a new decision table with condition attributes responding to widths of predecessor templates of the rule and decision attribute from width of successor template. As condition attributes we also consider length of gaps on time axis between consecutive templates. The set of objects we create on the basis of all objects of input decision table matching the rule.

Suppose, we consider decision rule *if* $T_{t-2} = $ **A** *and* $T_{t-1} = $ **B** *then* $T_t = $ **D**. There are two objects x_4, x_7 matching this rule. We check widths and gaps between templates represented by these objects and create a decision table as in Table 2a. It is obvious, that before further processing of this table, it should be scaled to contain more general values (Table 2b). From such a table we can compute decision rules expressing widths of templates.

	A	Gap_{AB}	B	D
x_1	7	1	8	9
x_2	9	1	5	4

(a)

	A	Gap_{AB}	B	D
x_1	high	small	high	high
x_2	high	small	medium	medium

(b)

	A	Gap_{AB}	B	Gap_{BD}
x_1	7	1	8	3
x_2	9	1	5	2

(c)

	A	Gap_{AB}	B	Gap_{BD}
x_1	high	small	high	small
x_2	high	small	medium	small

(d)

Table 2. Decision tables describing gaps between and widths of templates generated for sample decision rule.

If we already have the information about which template is going to appear next and what is its expected width, what we still need to know is the estimated time of its appearance. This information we can generate in an analogous way as in case of width. The difference is, that when constructing decision table, we take as decision attribute not the template, which is in the successor of the rule, but the length of the gap between this template and last predecessor template (see Table 2cd). Finally, we compute rules expressing time of template occurrence.

5   Temporal features and decision rules

In this section we investigate the problem of decision rules generation that contain new features describing behaviour of objects in time. We assume, that the input data are numerical, so we can say about the degree of change of attribute value. An example of such a rule is "*if change of attribute a is positive and change of attribute c is negative then change of decision is negative*". Decision rules of this kind are very useful for analysis of time related information systems. We also show, how to compute templates built from descriptors of form $(\Delta a = positive)$, that can be further used to more general association rules generation.

In our method, first, we construct an information system which is a result of scanning of input data with a time window of some size. The size of a window is the parameter and can be tuned up according to type of data. When looking at the data through a time window we observe a history of changes of attribute values within some period (which is related to the size of the window). On the basis of this history we can construct new features describing behaviour of attributes in time, e.g. characteristics of plots of values, information about trends of changes. In our method let us focus on one example of such a temporal feature, which is the change of value within a time window.

Let $\mathbf{A} = (\{x_1, x_2, ..., x_n\}, \{a_1, ..., a_m\})$ be an information system and let us consider time windows of size s on $\mathbf{A}$. We construct a new information system $\mathbf{A}_s = (\{y_1, y_2, ..., y_{n-s+1}\}, \{\Delta a_1, ..., \Delta a_m\})$ in the following way:

$$\Delta a_i(y_j) = a_i(x_{j+s-1}) - a_i(x_j). \qquad (4)$$

In Table 3a we have a sample information system, which after scanning with a time window of size $s = 3$ is as one in Table 3b.

A	a_1	a_2	a_3	d
x_1	1.1	2.5	2.1	0.8
x_2	2.0	3.0	1.8	1.0
x_3	2.3	2.6	0.6	0.5
x_4	1.0	1.8	0.7	2.5
x_5	1.8	1.7	1.2	1.6
x_6	1.2	1.9	2.0	1.7
x_7	0.5	1.5	0.2	0.4
x_8	0.7	1.7	1.5	0.9
x_9	1.0	2.5	1.4	1.9

(a)

$\mathbf{A}_3''$	Δa_1	a_1	Δa_2	a_2	Δa_3	a_3	Δd
y_1''	positive	low	neutral	med.	negative	high	neutral
y_2''	negative	high	negative	high	negative	high	positive
y_3''	neutral	high	negative	high	positive	low	positive
y_4''	neutral	low	neutral	low	positive	low	negative
y_5''	negative	high	neutral	low	negative	med.	negative
y_6''	neutral	med.	neutral	low	neutral	high	negative
y_7''	neutral	low	positive	low	positive	low	positive

(d)

$\mathbf{A}_3$	Δa_1	Δa_2	Δa_3	Δd
y_1	1.2	0.1	-1.5	-0.3
y_2	-1.0	-1.2	-1.1	1.5
y_3	-0.5	-0.9	0.6	1.1
y_4	0.2	0.1	1.3	-0.8
y_5	-1.3	-0.2	-1.0	-1.2
y_6	-0.5	-0.2	-0.5	-0.8
y_7	0.5	1.0	1.2	1.5

(b)

$\mathbf{A}_3'$	Δa_1	Δa_2	Δa_3	Δd
y_1'	positive	neutral	negative	neutral
y_2'	negative	negative	negative	positive
y_3'	neutral	negative	positive	positive
y_4'	neutral	neutral	positive	negative
y_5'	negative	neutral	negative	negative
y_6'	neutral	neutral	neutral	negative
y_7'	neutral	positive	positive	positive

(c)

Table 3. (a) Sample information system, (b) after scanning with a time window of size 3, (c) scaled, (d) scaled and containing new attributes (levels).

The obtained information system should be scaled next, so the results of analysis could be more general. In our example we use three-value scale, which describes the degree of change - *negative, neutral* or *positive* (see Table 3c). Obtained information system is a base for our further analysis.

First, let us consider templates computed for such system (see e.g. [7], [5]). They are built from descriptors of form $(\Delta a = v)$ (or $(\Delta a \in \{v_1, v_2, ...\})$ in more general case) and long templates with large support may contain very useful knowledge about higher-order dependencies between attributes. These templates can be used to generation of approximate association rules (see [6]) built from higher-order descriptors. An example of templates for Table 3c is

$T_1 = \{(\Delta a_1 = neutral), (\Delta a_3 = positive)\}$
$T_2 = \{(\Delta a_2 = neutral), (\Delta d = negative)\}$

Now, suppose the decision attribute is defined and we consider a decision table $\mathbf{A} = (\{x_1, x_2, ..., x_n\}, \{a_1, ..., a_m\} \cup \{d\})$ describing some decision problem. Using the method described above we can generate a decision table built from information about attribute changes (including decision), which we can compute decision rules for. Obtained rules contain knowledge about how value changes of conditional attributes determine change of decision values.

For example, if we consider the information system presented in Table 3a as decision table, with last attribute being a decision, after processing, we obtain Table 3c. From this table we can extract, e.g., the following rules:

if $\Delta a_2 = negative$ **then** $\Delta d = positive$
if $\Delta a_1 = neutral$ **and** $\Delta a_2 = neutral$ **then** $\Delta d = negative$

Another extension of this method is to include, in the final decision table, more conditional attributes describing, e.g. levels of attribute values. It can happen, that the degree of change of decision attribute depends not only on degrees of conditional attributes changes, but also on the levels of values. In Table 3d we have an example of such attributes which are scaled into three classes - *low, medium, high*. From this table we extract, among other, the following rules:

if $\Delta a_2 = neutral$ **and** $a_2 = low$ **then** $\Delta d = negative$
if $\Delta a_2 = negative$ **and** $a_2 = high$ **then** $\Delta d = positive$

The first rule, for example, can be interpreted as "*if value of a_2 isn't changing much (in some period) when it is low, then the value of decision is decreasing*".

6   Behaviour of temporal features in time

In the previous section we showed how new features, that express behaviour of attributes in a time window, can be extracted from an information system. There can be several features considered - as an example we took a difference between value at the end and the beginning of time window. Now, let us investigate the problem of temporal behaviour discovery of a group of features.

Suppose, we consider a feature "a_1 grows fast" discovered from time related information system. There can be considered several different problems related to this feature. The basic one is to discover periods that this feature holds in. Besides, we would like to know what are other properties (e.g. "a_3 behaves stably") that hold at the same time. Another question is what are the symptoms that this feature is about to finish and what features are going to appear next.

We believe, that temporal templates generation is a tool which helps to answer above questions. First, we analyze the information system using time window method and construct new system built from temporal features. Then, we compute temporal templates and generate dependencies among them. One can notice, that temporal template, for so processed information system, contains information about a set of features that hold at the same period, as well as about beginning and end time of this period. Analysis of dependencies between temporal templates gives the idea about what new set of features may appear after a current one.

7   Summary

We claim that the notion of temporal templates can be very useful for analysis of time related information systems. Investigation of dependencies between temporal templates gives the idea how the knowledge hidden in data is changing in time. Very important topic is the extraction of new features from data, describing temporal properties of data. Finally, temporal templates can be used to check how these features behave in time.

Acknowledgments

This work was supported by the grants of Polish National Committee for Scientific Research (KBN) No. 8T11C02417 and 8T11C02519.

References

1. Agrawal, R., Mannila, H., Srikant, R., Toivonen, H., Verkamo, I.: Fast Discovery of Association Rules, *Proceedings of the Advances in Knowledge Discovery and Data Mining*. AAAI Press/The MIT Press, CA (1996) 307–328.
2. Bazan, J.: Dynamic reducts and statistical inference, *Proceedings of Information Processing and Management of Uncertainty on Knowledge Based Systems (IPMU-96)*, July 1-5, Granada, Spain, Universidad de Granada, vol. III, (1996) 1147–1152.
3. Bazan, J., Skowron, A. and Synak, P.: Discovery of Decision Rules from Experimental Data, *Proceedings of the Third International Workshop on Rough Sets and Soft Computing*. San Jose, California (1994) 526–533.
4. Mannila, H., Toivonen, H.,Verkamo, A.,I.: Discovery of frequent episodes in event sequences. *Report C-1997-15*, University of Helsinki, Department of Computer Science, Finland (1997).
5. Nguyen S.H.: Regularity Analysis And Its Applications In Data Mining. PhD Dissertation, Warsaw University, Poland (2000).
6. Nguyen, H.S., Nguyen, S.H.: Rough Sets and Association Rule Generation. Fundamenta Informaticae **40/4** (2000) pp. 383–405.
7. Nguyen, S.H., Skowron, A., Synak, P.: Discovery of data pattern with applications to decomposition and classification problems. In L. Polkowski, A. Skowron (eds.):*Rough Sets in Knowledge Discovery* **2**. Physica-Verlag, Heidelberg (1998) 55–97.
8. Pawlak, Z., Skowron, A.: A rough set approach for decision rules generation, *ICS Research Report 23/93*, Warsaw University of Technology, *Proceedings of the IJCAI'93 Workshop W12: The Management of Uncertainty in AI*, France (1993).
9. Ziarko, W., Shan, N.: An incremental learning algorithm for constructing decision rules, *Proceedings of the International Workshop on Rough Sets and Knowledge Discovery*. Banff. (1993) 335–346.

Constraint Based Incremental Learning of Classification Rules

Arkadiusz Wojna

Institute of Informatics, Warsaw University
ul. Banacha 2, 02-097 Warsaw, Poland
http://www.mimuw.edu.pl/~awojna
wojna@mimuw.edu.pl

Abstract. We present a modification of a simple incremental procedure maintaining the set of all current reduct rules. It reduces searching to the part of the rule space limited by a dynamic monotonic constraint. Efficiency problems and their solutions for the class of coverage based constraints are discussed and an illustrative example is provided.
Keywords: rough sets, machine learning, incremental learning, decision algorithms.

1 Introduction

In recent years rough sets were intensively studied as a method for approximative concept synthesis from data tables. Many data sources have dynamic character and their size is still increasing. In order to maintain the validity of knowledge extracted from dynamically changing data one should develop incremental learning strategies.

Incremental learning has been already widely studied in machine learning and for the exhaustive overview of these methods the reader is referred e.g. to [1] and [3]. This paper examines the problematics on the ground of rough sets introduced by Pawlak [5]. Different incremental algorithms maintaining reducts were proposed e.g. [4], [7] and experimental results comparing nonincremental and incremental methods for reduct generation may be found in [9].

The subject of the paper is an incremental method maintaining a set of reduct rules. Shan and Ziarko [7] described an algorithm generating all reduct rules. This paper presents its more practical version based on the notion of dynamic monotonic constraint that reduced the size of the rule space to be searched. The idea of searching for rules satisfing user requirements has been already used in nonincremental approach e.g. [2], [8]. Different properties and accelerating methods of the proposed solution are described and an experimental example that demonstrates potential advantages of the constraint based approach is provided.

2 Classification Rules and Constraints

We denote a finite set of binary attributes by A and a finite set of decisions by V. The domain of all objects is defined by $U = \{0, 1\}^A$. The input of an incremental

W. Ziarko and Y. Yao (Eds.): RSCTC 2000, LNAI 2005, pp. 428–435, 2001.

algorithm is a finite sequence of pairs (u_i, d_i) called a *sample*, where $u_i \in U$ is an object and $d_i \in V$ is a decision for u_i. The notion of a sample corresponds to the notion of a decision table [5] in nonincremental approach. For a given sample s we denote the set of all examples from s with a decision d by $Class_s(d)$.

A *classification rule* is an implication $\alpha \Rightarrow d$ where α is a conjunction of literals of attributes from A and $d \in V$. The support of a sample s for a conjunction α is defined by $[\alpha]_s = \{(u, d) \in s : u \text{ satisfies } \alpha\}$ and for a rule $\alpha \Rightarrow d$ is defined by $[\alpha \Rightarrow d]_s = [\alpha]_s \cap Class_s(d)$.

A rule $\alpha \Rightarrow d$ is *certain* for a sample s if for each pair (u_i, d_i) in s such that u_i satisfies α the decisions are equal $d_i = d$. A certain rule $\alpha \Rightarrow d$ is a *reduct rule* if α is a minimal conjunction in the sense of literal set inclusion among all conjunctions occurring on the lefthand side of a certain rule with the same decision d. The set of all reduct rules with the decision d for a sample s is denoted by $RedRul_s(d)$. We use two measures for rules: *confidence* and *coverage* [2], [5], [6], [8]:

$$confidence_s(\alpha \Rightarrow d) = \begin{cases} 0 & \text{if } [\alpha]_s = \emptyset \\ \frac{||[\alpha \Rightarrow d]_s||}{||[\alpha]_s||} & \text{if } [\alpha]_s \neq \emptyset \end{cases}$$

$$coverage_s(\alpha \Rightarrow d) = \frac{||[\alpha \Rightarrow d]_s||}{||Class_s(d)||}$$

Usually the set of all reduct rules is very large and only a small subset, that can be described by a monotonic constraint, is relevant. A monotonic constraint is a set of rules C such that if a rule $\alpha \Rightarrow d$ belongs to C then for each $B \subseteq Literals(\alpha)$ the rule $\bigwedge B \Rightarrow d$ also belongs to C. We restrict the space of reduct rules to bounded by C: $RedRul_s^C(d) = C \cap RedRul_s(d)$. Throughout the paper, somewhat informally, we denote the description of a monotonic constraint and the set of rules defined by the monotonic constraint with the same symbol C.

In the next sections we focus our attention on two types of a monotonic constraint: the first one $RedRul_s^{coverage > \theta}(d)$ bases on a fixed coverage threshold $\theta \in [0, 1]$:

$$\{r \in RedRul_s(d) : coverage_s(r) > \theta\}$$

and the other one $RedRul_s^{best-k}(d)$ includes always the set of exactly k best reduct rules:

$$\{r \in RedRul_s(d) : ||\{r' \in RedRul_s(d) : coverage_s(r') > coverage_s(r)\}|| < k\}$$

3   Incremental Constraint Based Algorithm

The algorithm [7] computing all reduct rules starts with the set of the most general rules one for each decision class and after each new example is added it extends each rule that is inconsistent with the example by adding the literals excluding the example.

Since the space of rules is usually too large for searching for all reduct rules, we propose a modified version of the incremental algorithm using a dynamic monotonic constraint that may change after each new example is added. The algorithm limits the set of maintained reduct rules to rules satisfying the constraint. Let C denote the considered monotonic constraint. During computation the algorithm always maintains the following sets: s — the set of training examples, $Rules(d)$ — the set of reduct rules with the decision d, $CCand(d)$ — the set of candidates for reduct rules with the decision d satisfying the constraint C and $nonCCand(d)$ — the set of candidates for reduct rules with the decision d not satisfying C.

Like in [7] the algorithm starts with the set of the most general rules one for each decision class and for each new example it executes procedure *learn*. The difference is that the constraint based algorithm extends candidates only from the sets $CCand$ set leaving the sets $nonCCand$ unchanged:

Algorithm 1 *learn(u,d)*
$s := s + (u, d);$
update the constraint C;
for each $d' \in V$ do
step 1:
  move all rules $r \in Rules(d')$ such that $r \notin C$ to $nonCCand(d')$;
  move all rules $r \in Rules(d')$ inconsistent with (u,d) to $CCand(d')$;
  move all certain rules $r \in nonCCand(d')$ such that $r \in C$ to $Rules(d')$;
  move all rules $r \in nonCCand(d')$ such that $r \in C$ to $CCand(d')$;
step 2:
  while $CCand(d') \neq \emptyset$ do
    remove an arbitrary rule $\alpha \Rightarrow d'$ from $CCand(d')$;
    find an example (u'', d'') inconsistent with the rule $\alpha \Rightarrow d'$;
    for each attribute $a \in A \setminus Attributes(\alpha)$ do
    $l :=$ literal for a which excludes u'';
    if $\alpha \wedge l \Rightarrow d'$ is not subsumed
        by another rule from $Rules(d') \cup CCand(d') \cup nonCCand(d')$ then
        if $\alpha \wedge l \Rightarrow d' \notin C$ then $nonCCand(d') := nonCCand(d') \cup \{\alpha \wedge l \Rightarrow d'\}$
        else if $\alpha \wedge l \Rightarrow d'$ is certain then $Rules(d') := Rules(d') \cup \{\alpha \wedge l \Rightarrow d'\}$
        else $CCand(d') := CCand(d') \cup \{\alpha \wedge l \Rightarrow d'\};$

At the beginning of the procedure $learn(u, d)$ the sets $Rules(d')$ are assumed to contain all reduct rules satisfying the constraint C and $nonCCand(d')$ are assumed to contain all generated up to now rules not satisfying C, both according to a sample s before adding a new example (u, d). The sets $CCand(d')$ should be empty.

In the step 1 the procedure moves rules according to changes in the sample s and the constraint C: reduct rules for a previous sample may be inconsistent with a new example (u, d) and the modified constraint may both include new candidate and reduct rules and exclude previously covered reduct rules. Time needed for this step may vary significantly in dependence on a used constraint.

For the constraint $coverage > 0$ migration only for rules that cover a new object u is possible, for constraints with positive coverage threshold other rules with the decision d can migrate and for $best - k$ constraints checking constraint satisfiability becomes much more complex. In the last case a good solution is to assume the ranking based on the current set of reduct rules and do the step 2 correcting the ranking every time when a new reduct rule is found.

In the step 2 the procedure extends all candidates satisfing C. Candidates that were previously in $nonCCand(d')$ may be inconsistent with any example in the sample s, not always with the last one (d, u). Therefore the procedure must search the sample s for an inconsistent example. In order to avoid searching the whole sample for each candidate the procedure may assign to each extended rule $\alpha \wedge l \Rightarrow d'$ the position in s where an inconsistent example for the previous rule $\alpha \Rightarrow d'$ was found and continue searching from this place. After an inconsistent example (u'', d'') is found, the candidate is extended with all literals excluding u''. The next time consuming operation is subsumption checking. If an extension is not subsumed by another rule it is directed to the appropriate set, otherwise it is removed.

Theorem 1. *At the end of the procedure learn the union $\bigcup_{d \in V} Rules(d)$ is always equal to the set of all reduct rules satisfying the constraint C for the sample s.*

4   Improving Efficiency

One of the properties of the algorithm presented in the previous section is that it never reduces rules. Generating more and more new rules without any reduction prolongs checking for subsumptions and leads to the lack of memory. In order to avoid the problem the following solution may be used. Every time after a rule $\alpha \Rightarrow d'$ is added to the set $nonCCand(d')$ it is also reduced as much as it is possible:

Algorithm 2 *reduce($\alpha \Rightarrow d'$)*
reduce the rule $\alpha \Rightarrow d'$ to $\beta \Rightarrow d'$
    where β is any minimal conjunction subsuming α such that $\beta \Rightarrow d' \notin C$;

The presented improvement applies to constraints that have "shrinking" property what means that new examples may lead to excluding a rule from a constraint. An example of a "shrinking" constraint is $coverage > \theta$ for any $\theta > 0$, whereas the constraint $coverage > 0$ does not have this property.

However, this modification brings another undesirable phenomenon affecting efficiency namely "shimmering" of rules what means that a single rule may be generated and reduced many times while the constraint is changing dynamically and repeated computation of rule parameters significantly slowers the performance. We present two methods to deal with this problem.

The first one consists in maintaining two buffers: $BufExt$ saves rules for which the extending operation was already performed and $BufRed$ saves rules that were reduced. The buffers are usually too limited for keeping all rules that appeared in the process of learning. Therefore a certain measure is applied to estimate which rules are the most probable to be reused in the near future. For coverage based constraints coverage is a good measure for it. The following procedure *saveExtended* is executed each time when a rule is extended:

Algorithm 3 *saveExtended(r)*
if the buffer $BufExt$ is not full then add r to BufExt;
else if $coverage_s(r) < \max_{r' \in BufExt} coverage_s(r')$
    then replace a rule with the maximal coverage in $BufExt$ with r ;

The analogical procedure *saveReduced* is executed when a rule is reduced:

Algorithm 4 *saveReduced(r)*
if the buffer $BufRed$ is not full then add r to BufRed;
else if $coverage_s(r) > \min_{r' \in BufRed} coverage_s(r')$
    then replace a rule with the minimal coverage in $BufRed$ with r ;

When the procedure *learn* needs to compute parameters for a new generated or reduced rule first it checks whether the rule is still available in the corresponding buffer.

Another solution that reduces "shimmering" is grouping examples. Instead of learning each new example separately first the algorithm gets a large group of examples and then starts learning rules. The learning process for a group of examples may last much longer than for a single example. However, notice that the procedure *learn* may be easily split into two parts: the first one corrects the contents of the maintained sets and the parameters of rules according to the sample including a new group and the next one generates new rules. The first part is always short hence the second one is critical for time performance. Therefore a good assumption for the second part is to be ready to stop learning and classify a new object with a current set of rules every time when the classification procedure is called. It requires from the algorithm to use rules with confidence less than 1 for classification. In this proposition a strategy of choosing rules for extension is important. The higher confidence a rule saves after updating by a new group of examples the more reliable it is for the classification procedure. Therefore a good strategy is to start extending with a rule having confidence nearest to 1 and move towards rules with lower confidence. In this way more reliable rules are adapted to a new group first. The latter solution provides also a good background for distributed computation.

The presented algorithm may be also adapted to the case when it is given a very large set of examples s at once. Like in the incremental algorithm it executes the procedure *learn* for successive examples in s. Because of the size the computation for the whole sample would last very long and would block classification procedure calls. To avoid it the learning procedure is always stopped when the classification procedure is called and waits until the classification is completed. Classification uses a current set of computed rules. Many of them may be still inconsistent with a number of examples, therefore before classification the algorithm needs to calculate qualitative parameters of rules: confidence and coverage, according to the whole sample s. It imposes the additional condition that a used classifier accepts rules with confidence less than 1.

Computing parameters for a set of rules consumes much less time than generating this set but computing them every time when the classification procedure is called is usually still too expensive for a large set of rules and a large set of objects. In order to avoid the problem the algorithm may perform the following operations. For a particular object to classify it may compute parameters only for rules covering the object. Once computed parameters for a rule may be preserved as long as the rule is held in the corresponding union $Rules(d) \cup Cand(d)$. Independently of classification procedure calls the learning procedure may stop at regular intervals and compute parameters for rules generated since the previous stop. The choice of appropriate data structures may significantly accelerate computation of parameters for rules and objects.

In case when all methods of improving efficiency fail, the exhaustive search may be replaced immediately by any heuristic search.

5   Illustrative Example

We present experimental results for the data set Income (13 attributes, 30162 training cases, 15060 testing cases) from the repository at University of California, Irvine (http://kdd.ics.uci.edu). In preprocessing discretization was used and 32 binary attributes were chosen by greedy heuristic algorithm optimizing discernibility.

The learning procedure was executed for groups of examples. We used the incremental constraint based algorithm with the modification that rules were extended not in all possible directions but only with rules that have the best confidence and the best coverage if there are ties in confidence for at least one covered object.

For each coverage threshold 0, 0.05 and 0.2 we performed series of computation in the following way. First the procedure *learn* was executed for the first 1/8 part of the training set and the testing set was classified. In each next step the number of examples equal to the number of examples received in all previous steps was added and learned and the next test of the testing set was performed. In this way the size of successive groups of examples grew exponentially. Each test object was classified with the decision of the best covering rule in the union $\bigcup_{d \in V} Rules(d) \cup nonCCand(d)$ according to the confidence and in case of ties to the coverage.

The results are presented on the graphs below. Left side graphs present the classification error, time and number of rules obtained in three series of incremental learning with different constraints: *coverage* > 0.2 (light line with boxes), *coverage* > 0.05 (medium dark line with circles) and *coverage* > 0 (dark line with diamonds). Right side graphs present the final results of incremental (medium dark line with crosses) and nonincremental (dark line with circles) learning for different coverage based constraints.

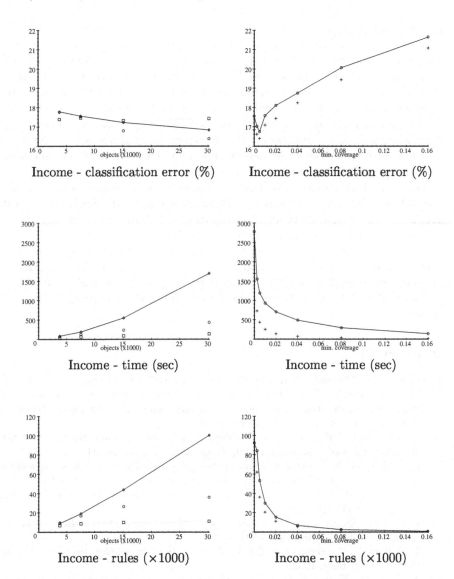

In the presented example the application of stronger constraints brought a significant reduction of used memory and time and very small deterioration of accuracy or even improvement for low coverage thresholds. The results show also that accuracy obtained with the small part of the training set used in a learning process is not significantly lower than for the whole training set and finally incremental learning reached better results than nonincremental. Similar properties of test results on other data sets (Shuttle, Letter) indicate that the combination of the incremental approach and a coverage based constraint may be an effective tool for learning concepts from both dynamic and large data sets.

6   Conclusions

We have shown how rough set methods can be adapted to dynamically changing data. We proposed a method based on a special type of monotonic constraints that allowed us to reduce searching in the space of rules without substantial changes in the classification quality. The presented method may be adapted to large data sets especially when one implements it using cluster of computers. The experimental example indicates that the incremental approach may preserve all advantages of nonincremental methods and add new ones like reduction in used time and memory and continuous improvement.

The following related problems are the subject for future study: methods for coding arbitrary attributes by binary ones e.g. by discretization or value grouping and efficient methods for computing confidence and coverage for large rule sets because this is the most time consuming operation.

Acknowledgement The author is grateful to professor Andrzej Skowron for useful remarks on this presentation. This work was supported by Research Program of European Union, CRIT 2 Esprit Project No. 20288 and grants 8 T11C 025 19 and 8 T11C 009 19 from the Polish National Committee for Scientific Research.

References

1. P. Langley, *Elements of machine learning*, The MIT Press, 1996
2. M. Kryszkiewicz, H. Rybiński, *Knowledge discovery from large databases using rough sets*, in: Proceedings of the 6th European Congress on Intelligent Techniques and Soft Computing, Aachen, Germany, Vol. 1, 85-89.
3. R. Michalski, *A theory and methodology of inductive learning*, Machine Learning: An Artificial Intelligence Approach, Tioga, 1983, Vol. 1, 83-134.
4. M. Orłowska, M. Orłowski, *Maintenance of knowledge in dynamic information systems*, in: R. Słowiński (editor), Intelligent decision support - handbook of applications and advances of the rough sets theory, Kluwer Academic Publishers, Dordrecht, 1992, 315-330.
5. Z. Pawlak, *Rough sets - theoretical aspects of reasoning about data*, Kluwer Academic Publishers, Dordrecht, 1991.
6. L. Polkowski, A. Skowron (editors), *Rough sets in knowledge discovery*, Physica-Verlag, Heidelberg, 1998.
7. N. Shan, W. Ziarko, *Data-based acquisition and incremental modification of classification rules*, Computational Intelligence, 11(2), 1995, 357-370.
8. J. Stefanowski, *On rough set based approaches to induction of decision rules*, in: L. Polkowski, A. Skowron (editors), Rough sets in knowledge discovery 1, Physica-Verlag, Heidelberg, 1998, 500-529.
9. R. Susmaga, *Experiments in incremental computation of reducts*, in: L. Polkowski, A. Skowron (editors), Rough sets in knowledge discovery 1, Physica-Verlag, Heidelberg, 1998, 530-553.

A Hybrid Model for Rule Discovery in Data

Ning Zhong[1], Juzhen Dong[1], Chunnian Liu[2], and Setsuo Ohsuga[3]

[1] Dept. of Information Engineering, Maebashi Institute of Technology
[2] School of Computer Science, Beijing Polytechnic University
[3] Dept. of Information and Computer Science, Waseda University

Abstract. This paper presents a hybrid model for rule discovery in real world data with uncertainty and incompleteness. The hybrid model is created by introducing an appropriate relationship between deductive reasoning and stochastic process, and extending the relationship so as to include abduction. Furthermore, a Generalization Distribution Table (GDT), which is a variant of transition matrix in stochastic process, is defined. Thus, the typical methods of symbolic reasoning such as deduction, induction, and abduction, as well as the methods based on soft computing techniques such as rough sets, fuzzy sets, and granular computing can be cooperatively used by taking the GDT and/or the transition matrix in stochastic process as mediums. Ways for implementation of the hybrid model are also discussed.

1 Introduction

In order to deal with the complexity of real world, we argue that an ideal rule discovery system should have such features as:

- The use of background knowledge can be selected according to whether background knowledge exists or not.
  That is, on the one hand, background knowledge can be used flexibly in the discovery process; on the other hand, if no background knowledge is available, it can also work.
- Imperfect data can be handled effectively, and the accuracy affected by imperfect data can be explicitly represented in the strength of the rule.
- Biases can be flexibly selected and adjusted for constraint and search control.
- Data change can be processed easily.
  Since the data in most databases are ever changed (e.g., the data are often added, deleted, or updated), a good method for real applications has to handle data change conveniently.
- The discovery process can be performed in a distributed cooperative mode.

It is clear that no method can contain all of the above performances. A unique method is to combine several techniques together to construct a hybrid approach. We argue that the hybrid approach is an important way to deal with real world problems. Here "hybrid" means the way of combining many advantages of existing methods, and avoiding their disadvantages or weaknesses when

W. Ziarko and Y. Yao (Eds.): RSCTC 2000, LNAI 2005, pp. 436–444, 2001.

the existing methods are used separately. There are ongoing efforts to integrate logic (including non-classical logic), artificial neural networks, probabilistic and statistical reasoning, fuzzy set theory, rough set theory, genetic algorithm and other methodologies in the soft computing paradigm [1, 14, 9].

In this paper, a hybrid model is proposed for discovering *if-then* rules in data in the environment with uncertainty and incompleteness. The central of the hybrid model is the Generalization Distribution Table (GDT) that is a variant of transition matrix in stochastic process. W e will also discuss wa ys for implementation of the hybrid model.

2   A Hybrid In telligen Model

2.1   An Overview

In general, hybrid models involve a variety of different types of processes and representations in both learning and performance. Hence, multiple mechanisms interact in a complex way in most models.

Figure 1 shows a hybrid intelligent model for discovering *if-then* rules in data, which is created by introducing an appropriate relationship between deductive reasoning and stochastic process [9], and extending the relationship so as to include abduction [14]. Then a Generalization Distribution Table (GDT), which is a variant of transition matrix (TM) in stochastic process, is defined [14, 15]. Thus, the typical methods of symbolic reasoning such as deduction, induction, and abduction, as well as the methods based on soft computing techniques such as rough sets, fuzzy sets, and granular computing can be cooperatively used by taking the GDT and/or the transition matrix in stochastic process as mediums.

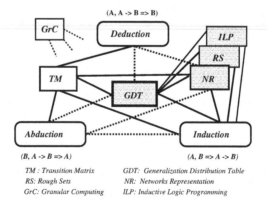

Fig. 1. A hybrid model for knowledge discovery

The shadow parts in Figure 1 are the major parts of our current study. The central idea of our methodology is to use the GDT as a hypothesis search space for generalization, in which the probabilistic relationships between concepts and instances over discrete domains are represented. By using the GDT as a probabilistic search space, (1) unseen instances can be considered in the rule discovery process and the uncertainty of a rule, including its ability to predict unseen instances, can be explicitly represented in the strength of the rule [15]; (2) biases can be flexibly selected for search control and background knowledge can be used as a bias to control the creation of a GDT and the rule discovery process [17].

Based on the GDT, we have developed or are developing three hybrid systems. The first one is called *GDT-NR*, which is based on a network representation of the GDT; the second one is called *GDT-RS*, which is based on the combination of GDT and Rough Set theory; the third one is called *GDT-RS-ILP*, which is based on the combination of GDT, Rough Set theory, and Inductive Logic Programming, for extending GDT-RS for relation learning. Furthermore, Granular Computing (GrC) can be used as a preprocessing step to change granules of individual objects for dealing with continuous values and imperfect data [12, 13].

In Figure 1, we further distinguish two kinds of lines: the solid lines and the dotted lines. The relationships denoted by the solid lines will be described in this paper, but the ones denoted by the dotted lines will not.

2.2   Deductive Reasoning vs. Stochastic Process

Deductive reasoning can be analyzed ultimately into the repeated application of the strong syllogism:

$$\frac{\text{If A is true, then B is true}}{\text{Hence, B is true.}}$$
$$\text{A is true}$$

That is, $(A \wedge (A \rightarrow B) \Rightarrow B)$ in short, where, A and B are logic formulae. Let us consider two predicates F and G, and let d be a finite set. For simplicity, we assume that F and G are single place predicates. They give descriptions on an object in d. Or, in other words, F (the definition for G is the same as the one for F from now) classifies all elements in the set d into two classes:

$\{x \in d |\ F(x) \text{ is true (or x satisfies F)}\}$, and
$\{x \in d |\ F(x) \text{ is false (or x does not satisfy F)}\}$.

In the following, $F(x)$ and $\overline{F}(x)$ mean "$F(x)$ is true (or x satisfies F)" and "$F(x)$ is false (or x does not satisfy F)" respectively for $x \in d$. Thus, one of the most useful forms of $A \rightarrow B$ can be denoted by multi-layer logic [8, 9] into

$[\forall X/d](F(X) \rightarrow G(X))$.

That is, the multi-layer logic formula is read "for any X belonging to d, if $F(X)$ is true then $G(X)$ is also true". Notice that since the set d can be looked upon as an ordered set, F is represented by a sequence of n binary digits. Thus, the ith binary digit is 1 or 0 for $F(a_i)$ or $\overline{F}(a_i)$ corresponding to the ith element a_i of the set d. Based on the preparation, several basic concepts are described first for creating an appropriate relationship between deductive reasoning and stochastic process.

1. Expansion function.

   If the domain set d of formula F is finite, $d = \{a_1, a_2, \ldots, a_n\}$, then the multi-layer logic formulae $[\forall X/d]F(X)$ and $[\exists X/d]F(X)$ can be expanded as

   $$F(a_1) \wedge F(a_2) \wedge \ldots \wedge F(a_n) \text{ (or } \wedge_{a_i \in d} F(a_i) \text{ in short) and}$$
   $$F(a_1) \vee F(a_2) \vee \ldots \vee F(a_n) \text{ (or } \vee_{a_i \in d} F(a_i) \text{ in short), respectively.}$$

   These are called *expansion functions* of multi-layer logic formulae.

   An expansion function is used to extract from a set the elements that possess specified properties. Furthermore the multi-layer logic formulae

   $$[\forall X\#/d](F(X) \rightarrow G(X)) \text{ and}$$
   $$[\exists X\#/d](F(X) \rightarrow G(X))$$

   can be expanded as

   $$\wedge_{a_i \in d}(F(a_i) \rightarrow G(a_i)) \text{ and } \vee_{a_i \in d}(F(a_i) \rightarrow G(a_i)), \text{ respectively.}$$

2. States of d with respect to F

   Let $d = \{a_1, a_2, \ldots, a_n\}$ be a finite set. The *states* of d with respect to F are defined as the conjunctions of either $F(a_i)$ or $\overline{F}(a_i)$ for every element a_i in d. That is, the states of d corresponding to F are

   $S_1(d, F) : \overline{F}(a_1) \wedge \overline{F}(a_2) \wedge \ldots \wedge \overline{F}(a_{n-1}) \wedge \overline{F}(a_n)$,
   $S_2(d, F) : \overline{F}(a_1) \wedge \overline{F}(a_2) \wedge \ldots \wedge \overline{F}(a_{n-1}) \wedge F(a_n)$,
   ..., and
   $S_{2^n}(d, F) : F(a_1) \wedge F(a_2) \wedge \ldots \wedge F(a_{n-1}) \wedge F(a_n)$.

   Let $Prior\_S(d, F) = \{S_1(d, F), \ldots, S_{2^n}(d, F)\}$. Each $S_i(d, F)$ in $Prior\_S(d, F)$ is called a prior state of d with respect to F.

   For example, if $d = \{a_1, a_2, a_3\}$, its possible prior and posterior states are

   $Prior\_S(d, F) = \{\overline{F}(a_1) \wedge \overline{F}(a_2) \wedge \overline{F}(a_3), \overline{F}(a_1) \wedge \overline{F}(a_2) \wedge F(a_3), \ldots,$
   $$F(a_1) \wedge F(a_2) \wedge F(a_3)\},$$
   $Posterior\_S([\forall X/d]F(X)) = \{F(a_1) \wedge F(a_2) \wedge F(a_3)\}$,
   $Posterior\_S([\exists X/d]F(X)) = \{ \overline{F}(a_1)\overline{F} \wedge (a_2) \wedge F(a_3), \ldots,$
   $$F(a_1) \wedge F(a_2) \wedge F(a_3) \}.$$

   Using binary digit 1 and 0 instead of $F(a_i)$ and $\overline{F}(a_i)$, the above states can be expressed as follows:

   $Prior\_S(d, F) = \{000, 001, 010, \ldots, 111\}$,
   $Posterior\_S([\forall X/d]F(X)) = \{111\}$,
   $Posterior\_S([\exists X/d]F(X)) = \{001, 010, \ldots, 111\}$.

3. Probability vector of state occurring.

   A probability vector with respect to $[\forall X/d]F(X)$ is
   $$P^{[\forall X/d]F(X)} = (0, 0, \ldots, 1), \text{ and}$$

   a probability vector with respect to $[\exists X/d]F(X)$ is
   $$P^{[\exists X/d]F(X)} = (0, \alpha, \ldots, \alpha),$$

   where, $\sum \alpha = 1$.

Based on the basic concepts stated above, our purpose is to create an equivalent relationship between deductive reasoning and stochastic process [9]. In other words, $(F \wedge (F \rightarrow G) \Rightarrow G)$ is explained by an equivalent stochastic process $P^{[\forall X/d]F(X)}T = P^{[\forall X/d]G(X)}$, where T is a transition matrix that is equivalent to $(F \rightarrow G)$. In order to create the equivalent relationship, the following three conditions must be satisfied for creating T :

 - The elements t_{ij} of T are the probability $p(S_j(d,G)|S_i(d,F))$.
 - $p(S_j(d,G)|S_i(d,F))$ must satisfy the truth table of implicative relation.
 - $\sum_{j=1}^{2^n} t_{ij} = 1$.

Here $S_i(d,F)$ denotes the ith prior state in $Prior\_S(d,F)$ and $S_j(d,G)$ denotes the jth prior state in $Prior\_S(d,G)$. That is, since
$$[\forall X/d](F(X) \rightarrow G(X)) \text{ (or } [\exists X/d](F(X) \rightarrow G(X)))$$
is equivalent to
$$\wedge_{a_i \in d}(F(a_i) \rightarrow G(a_i)) \text{ (or } \vee_{a_i \in d}(F(a_i) \rightarrow G(a_i)))$$
(i.e., by using the expansion function stated above). According to the truth table of implicative relation, if the value of $F(a_i)$ is known, to satisfy $F(a_i) \rightarrow G(a_i)$, it must follow that $\overline{F}(a_i) \vee G(a_i) = 1$. In other words, the creation of T must satisfy the condition:

if $F(a_i)$ is true, $G(a_i)$ is true, otherwise any value of $G(a_i)$ is correct.

Table 1 shows an example of transition matrix corresponding to $[\forall X/d](F(X) \rightarrow G(X))$ and $d = \{a_1, a_2, a_3\}$. In Table 1, the states in the left column denote respectively

000: $\overline{F}(a_1) \wedge \overline{F}(a_2) \wedge \overline{F}(a_3)$
001: $\overline{F}(a_1) \wedge \overline{F}(a_2) \wedge F(a_3)$
......
111: $F(a_1) \wedge F(a_2) \wedge F(a_3)$,

and the states in the top row denote respectively

000: $\overline{G}(a_1) \wedge \overline{G}(a_2) \wedge \overline{G}(a_3)$
001: $\overline{G}(a_1) \wedge \overline{G}(a_2) \wedge G(a_3)$
......
111: $G(a_1) \wedge G(a_2) \wedge G(a_3)$,

and the elements t_{ij} of T denoted in the transition matrix are the probability distribution corresponding to $[\forall X/d](F(X) \rightarrow G(X))$ and the elements of T not displayed are all zero. Furthermore, since any background knowledge is not used to create the probability distribution shown in Table 1, the probabilities of states occurring are equiprobable. For example, if the states of F is $\{010\}$, to satisfy $F(X) \rightarrow G(X)$, the possible states of G are $\{010, 011, 110, 111\}$, and the probability of each of them is 1/4.

Table 1. A transition matrix equivalent to $[\forall\, X/d](F(X) \to G(X))$

	000	001	010	011	100	101	110	111
000	1/8	1/8	1/8	1/8	1/8	1/8	1/8	1/8
001		1/4		1/4		1/4		1/4
010			1/4	1/4			1/4	1/4
011				1/2				1/2
100					1/4	1/4	1/4	1/4
101						1/2		1/2
110							1/2	1/2
111								1

2.3   Hypothesis Generation Based on the Transition Matrix

Our purpose is to create a hybrid model, as shown in Figure 1, for rule discovery in data with uncertainty and incompleteness. For the purpose, we would like to discuss here a kind of weaker reasoning, or call weaker syllogisms:

$$\frac{\begin{array}{c} \text{If A is true, then B is true} \\ \text{B is true} \end{array}}{\text{Hence, A becomes more plausible.}}$$

That is, $(B \wedge (A \to B) \mapsto A)$ in short. The evidence does not prove that A is true, but verification of one of its consequences does give us more confidence in A. This is a kind of plausible reasoning for hypothesis generation, which is called "abduction". In other words, from the observed fact B and known rule $A \to B$, A can be guessed. That is, according to the transition matrix shown in Table 1, from each element $x \in d$ such that $G(x)$ is $true$ and the rule $[\forall\, X/d](F(X) \to G(X))$, $F(X)$ can be guessed. Thus, an appropriate relationship between deductive reasoning and abductive reasoning is created by using the transition matrix as a medium as shown in Figure 1.

2.4   Generalization Distribution Table (GDT)

The central idea of our methodology is to use a variant of transition matrix, called a *Generalization Distribution Table (GDT)*, as a hypothesis search space for generalization, in which the probabilistic relationships between concepts and instances over discrete domains are represented [14]. Thus, the representation of the original transition matrix introduced in Section 2.2 must be modified appropriately and some concepts must be described for our purpose.

A GDT is defined as consisting of three components. The first one is *possible instances*, which are all possible combinations of attribute values in a database. They are denoted in the top row of a GDT. The second one is *possible generalizations* for instances, which are all possible cases of generalization for all possible instances. They are denoted in the left column of a GDT. "$*$", which specifies a wild card, denotes the generalization for instances. For example, the generalization $*b_0c_0$ means the attribute a is unimportant for describing a concept.

The third component of the GDT is *probabilistic relationships* between the possible instances and the possible generalizations, which are represented in the elements G_{ij} of a GDT. They are the probabilistic distribution for describing the strength of the relationship between every possible instance and every possible generalization. The prior distribution is equiprobable, if any prior background knowledge is not used. Thus, it is defined by the Eq. (1), and $\sum_{j} G_{ij} = 1$:

$$G_{ij} = p(PI_j|PG_i) = \begin{cases} \dfrac{1}{N_{PG_i}} & \text{if } PI_j \in PG_i \\ 0 & \text{otherwise} \end{cases} \tag{1}$$

where PI_j is the jth possible instance, PG_i is the ith possible generalization, and N_{PG_i} is the number of the possible instances satisfying the ith possible generalization, that is,

$$N_{PG_i} = \prod_{k \in \{l| \ PG[l]=*\}} n_k \tag{2}$$

where $PG_i[l]$ is the value of the kth attribute in the possible generalization PG_i, $PG[l] = *$ means that PG_i doesn't contain attribute l.

Thus, in our approach, the basic process of hypothesis generation is to generalize the instances observed in a database by searching and revising the GDT. Here, two kinds of attributes need to be distinguished: *condition* attributes and *decision* attributes (sometimes called class attributes) in a database. Condition attributes as possible instances are used to create the GDT, but the decision attributes are not. The decision attributes are normally used to decide which concept (class) should be described in a rule. Usually a single decision attribute is all that are required.

3   W ays of Implemen tation

We have tried several ways for implementing some aspects of the hybrid model stated in the above section. One possible way is to use the transition matrix in stochastic process as a medium for implementing a hybrid system [9]. Let us assume that some causal relation seems to exist betw een observations. Let the observations be classified into finite classes and represented by a state set. The scheme of transition process stated in Section 2.2 is used as the framework to represent the causal relation. Through learning in this framework a tendency of the transition between input and output is learned. If the transition matrix reveals the complete or approximate equiv alence with logical inference, the logical expression can be discovered. Furthermore, the transition process can be represented by connectionist network for solving the space complexity.

Another way is to use the GDT as a medium for generalization and dealing with uncertainty and incompleteness. A GDT can be represented by a variant of connectionist networks (GDT-NR for short), and rules can be discovered by learning on the network representation of the GDT [16]. Furthermore, the GDT is combined with the rough set methodology (GDT-RS for short) [15]. By using GDT-RS, a minimal set of rules with larger strengths can be acquired from databases with noisy, incomplete data. The strength of a rule represents the uncertainty of the rule, which is influenced by both unseen instances and noises. Two algorithms have been developed for implementing the GDT-RS [2].

GDT-NR and GDT-RS stated above are two hybrid systems belonging to *attribute-value learning*, which is a main stream in inductive learning and data mining communities up to date. Another type of inductive learning is *relation learning* or called Inductive Logic Programming (ILP) [3, 6].

ILP is a relatively new method in machine learning. ILP is concerned with learning from examples within the framework of predicate logic. ILP is relevant to data mining, and compared with the attribute-value learning methods, it possesses the following advantages:

- ILP can learn knowledge which is more expressive than that by the attribute-value learning methods, because the former is in predicate logic while the latter is usually in propositional logic.
- ILP can utilize background knowledge more naturally and effectively, because in ILP the examples, the background knowledge, as well as the learned knowledge are all expressed within the same logic framework.

However, when applying ILP to large real-world applications, we can identify some weak points compared with the attribute-value learning methods, such as:

- It is more difficult to handle numbers (especially continuous values) prevailing in real-world databases, because predicate logic lacks effective means for this.
- The theory, techniques and experiences are much less mature for ILP to deal with imperfect data (uncertainty, incompleteness, vagueness, impreciseness, etc. in examples, background knowledge as well as the learned rules) than in the traditional attribute-value learning methods (see [3, 11], for instance).

The discretization of continuous valued attributes as a pre-processing step is a solution for the first problem mentioned in the above [7]. Another way is to use Constraint Inductive Logic Programming (CILP), an integration of ILP and CLP (Constraint Logic Programming) [4].

For the second problem, a solution is to combine GDT (also GDT-RS) with ILP, that is, GDT-ILP and GDT-RS-ILP to deal with some kinds of imperfect data which occur in large real-world applications [5]. The GDT-RS-ILP system has been developing on the way.

4   Conclusion

In this paper, a hybrid model for rule discovery in real world data with uncertainty and incompleteness was presented. The central idea of our methodology is to use the GDT as a hypothesis search space for generalization, in which the probabilistic relationships between concepts and instances over discrete domains are represented. By using the GDT as a probabilistic search space, (1) unseen instances can be considered in the rule discovery process and the uncertainty of a rule, including its ability to predict unseen instances, can be explicitly represented in the strength of the rule; (2) biases can be flexibly selected for search control, and background knowledge can be used as a bias to control the creation of a GDT and the rule discovery process. Several hybrid discovery systems, which are based on the hybrid model, have been developed/developing.

The ultimate aim of the research project is to create an *agent-oriented* and *knowledge-oriented* hybrid intelligent model and system for knowledge discovery and data mining in an evolutionary, distributed cooperative mode. That is, the work that we are doing takes but one step toward this model and system.

References

1. Banerjee, M., Mitra, S., and Pal, S.K. "Rough Fuzzy MLP: Knowledge Encoding and Classification", *IEEE Tran. Neural Networks*, Vol.9, No.6 (1998) 1203-1216.
2. Dong, J.Z., Zhong, N., and Ohsuga, S. "Probabilistic Rough Induction: The GDT-RS Methodology and Algorithms", Z.W. Ras and A. Skowron (eds.) *Foundations of Intelligent Systems*. LNAI 1609, Springer (1999) 621-629.
3. Lavrac, N., Dzeroski, S., and Bratko, I. "Handling Imperfect Data in Inductive Logic Programming", L. de Raedt (ed.) *Advances in Inductive Logic Programming*, IOS Press (1996) 48-64.
4. Liu, C., Zhong, N., and Ohsuga, S. "Constraint ILP and its Application to KDD", *Proc. of IJCAI-97 Workshop on Frontiers of ILP* (1997) 103-104.
5. Liu, C. and Zhong, N. "Rough Problem Settings for Inductive Logic Programming", Zhong, N., Skowron, A., and Ohsuga, S. (eds.) *New Directions in Rough Sets, Data Mining, and Granular-Soft Computing*, LNAI 1711, Springer-Verlag (1999) 168-177.
6. Muggleton, S. "Inductive Logic Programming", New Generation Computing, Vol. 8, No 4 (1991) 295-317.
7. Nguyen S.H. and Nguyen, H.S. "Quantizationof Real Value Attributes for Control Problems", *Proc. Forth European Congress on Intelligent Techniques and Soft Computing EUFIT'96*, Aachen, Germany (1996) 188-191.
8. Ohsuga, S. and Yamauchi, H. "Multi-Layer Logic - A Predicate Logic Including Data Structure as Knowledge Representation Language", *New Generation Computing*, Vol.3, No.4 (1985) 403-439.
9. Ohsuga, S. "Symbol Processing by Non-Symbol Processor", *Proc. 4th Pacific Rim International Conference on Artificial Intelligence (PRICAI'96)* (1996) 193-205.
10. Pawlak, Z. *Rough Sets, Theoretical Aspects of Reasoning about Data*, Kluwer Academic Publishers (1991).
11. Yao, Y.Y. and Zhong, N. "An Analysis of Quantitative Measures Associated with Rules", *Proc. PAKDD'99*, Springer-Verlag LNAI 1574, 479-488, 1999.
12. Yao, Y.Y. and Zhong, N. "Potential Applications of Granular Computing in Knowledge Discovery and Data Mining", *Proc. The 5th.International Conference on Information Systems Analysis and Synthesis (IASA'99)*, edited in the invited session on Intelligent Data Mining and Knowledge Discovery (1999) 573-580.
13. Yao, Y.Y. Granular Computing: Basic Issues and Possible Solutions, *Proc. JCIS 2000*, invited session on Granular Computing and Data Mining, Vol.1, 186-189, 2000.
14. Zhong, N. and Ohsuga, S. "Using Generalization Distribution Tables as a Hypotheses Search Space for Generalization", *Proc. 4th International Workshop on Rough Sets, Fuzzy Sets, and Machine Discovery (RSFD-96)* (1996) 396-403.
15. Zhong, N., Dong, J.Z., and Ohsuga, S. "Data Mining: A Probabilistic Rough Set Approach", L. Polkowski and A. Skowron (eds.) *Rough Sets in Knowledge Discovery*, Vol.2, Physica-Verlag (1998) 127-146.
16. Zhong, N., Dong, J.Z., Fujitsu, S., and Ohsuga, S. "Soft Techniques to Rule Discovery in Data", *Transactions of Information Praessing Society of Jap an*, Vol.39, No.9 (1998) 2581-2592.
17. Zhong, N., Dong, J.Z., and Ohsuga, S. "Using Background Knowledge as a Bias to Control the Rule Discovery Process", *Proc. PKDD'2000* (2000) (in press).

Rough Sets in Approximate Spatial Reasoning

Thomas Bittner and John G. Stell

[1] Centre de recherche en geomatique, Laval University, Quebec, Canada.
Thomas.Bittner@scg.ulaval.ca
[2] Department of Computer Science, Keele University, UK. john@cs.keele.ac.uk

Abstract. In spatial reasoning the qualitative description of relations between spatial regions is of practical importance and has been widely studied. Examples of such relations are that two regions may meet only at their boundaries or that one region is a proper part of another. This paper shows how systems of relations between regions can be extended from precisely known regions to approximate ones. One way of approximating regions with respect to a partition of the plane is that provided by rough set theory for approximating subsets of a set. Relations between regions approximated in this way can be described by an extension of the RCC5 system of relations for precise regions. Two techniques for extending RCC5 are presented, and the equivalence between them is proved. A more elaborate approximation technique for regions (boundary sensitive approximation) takes account of some of the topological structure of regions. Using this technique, an extension to the RCC8 system of spatial relations is presented.

Keywords: qualitative spatial reasoning, approximate regions.

1 Introduction

Rough set theory [Paw91] provides a way of approximating subsets of a set when the set is equipped with a partition or equivalence relation. Given a set X with a partition $\{a_i \mid i \in \mathcal{I}\}$, an arbitrary subset $b \subseteq X$ can be approximated by a function $\varphi_b : \mathcal{I} \to \{\mathsf{fo}, \mathsf{po}, \mathsf{no}\}$. The value of $\varphi_b(i)$ is defined to be fo if $a_i \subseteq b$, it is no if $a_i \cap b = \varnothing$, and otherwise the value is po. The three values fo, po, and no stand respectively for 'full overlap', 'partial overlap' and 'no overlap'; they measure the extent to which b overlaps the elements of the partition of X.

In spatial reasoning it is often necessary to approximate not subsets of an arbitrary set, but parts of a set with topological or geometric structure. For example the set X above might be replaced by a regular closed subset of the plane, and we might want to approximate regular closed subsets of X. This approximation might be with respect to a partition of X where the cells (elements of the partition) might overlap on their boundaries, but not their interiors. Because of the additional topological structure, it is possible to make a more detailed classification of overlaps between subsets and cells in the partition. An account of how this can be done was given in our earlier paper [BS98]. This is, however, only one of the directions in which the basic rough sets approach to approximation can be generalized to spatial approximation.

W. Ziarko and Y. Yao (Eds.): RSCTC 2000, LNAI 2005, pp. 445–453, 2001.

Our concern in the present paper is relationships between spatial regions when these regions have been given approximate descriptions. The study of relationships between spatial regions is of practical importance in Geographic Information Systems (GIS), and has resulted in many papers [EF91,RCC92,SP92]. Examples of relationships might be that two regions meet only at their boundaries or that one region is a proper part of another. While such relationships have been widely studied, the topic of relationships between approximate regions has received little attention.

The structure of the paper is as follows. In section 2 we set out the particular type of approximate regions we use in the main part of the paper. In section 3 we discuss one particular scheme of relationships between regions, known as the RCC5, and in section 4 we show how this can be generalized to approximate regions. In section 5 we briefly consider how our work can be extended to deal with more detailed boundary-sensitive approximations and the RCC8 system of relationships between regions. Finally in section 6 we present conclusions and suggest directions for further work.

2   Approximating Regions

Spatial regions can be described by specifying how they relate to a frame of reference. In the case of two-dimensional regions, the frame of reference could be a partition of the plane into cells which may share boundaries but which do not overlap. A region can then be described by giving the relationship between the region and each cell.

2.1   Boundary   Insensitive Approximation

Approximation functions. Suppose a space R of detailed or precise regions. By imposing a partition, G, on R we can approximate elements of R by elements of Ω_3^G. That is, we approximate regions in R by functions from G to the set $\Omega_3 = \{\text{fo}, \text{po}, \text{no}\}$. The function which assigns to each region $r \in R$ its approximation will be denoted $\alpha_3 : R \to \Omega_3^G$. The value of $(\alpha_3 r)g$ is fo if r covers all the of the cell g, it is po if r covers some but not all of the interior of g, and it is no if there is no overlap between r and g. We call the elements of Ω_3^G the boundary insensitive approximations of regions $r \in R$ with respect to the underlying regional partition G.

Each approximate region $X \in \Omega_3^G$ stands for a set of precise regions, i.e. all those precise regions having the approximation X. This set which will be denoted $[\![X]\!]$ provides a semantics for approximate regions: $[\![X]\!] = \{r \in R \mid \alpha_3 r = X\}$.

Operations on approximation functions. The domain of regions is equipped with a meet operation interpreted as the intersection of regions. In the domain of approximation functions the meet operation between regions is approximated by pairs of greatest minimal, $\wedge_{min}$, and least maximal, $\wedge_{max}$, meet operations on approximation mappings [BS98].

Consider the operations $\wedge_{min}$ and $\wedge_{max}$ on the set $\Omega_3 = \{\text{fo}, \text{po}, \text{no}\}$ that are defined as follows.

$\wedge_{min}$	no	po	fo
no	no	no	no
po	no	no	po
fo	no	po	fo

$\wedge_{max}$	no	po	fo
no	no	no	no
po	no	po	po
fo	no	po	fo

These operations extend to elements of Ω_3^G (i.e. the set of functions from G to Ω_3) by

$$(X \wedge_{min} Y)g = (Xg) \wedge_{min} (Yg)$$

and similarly for $\wedge_{max}$. This definition of the operations on Ω_3^G is equivalent to the construction for operations given by Bittner and Stell [BS98, page 108].

2.2 Boundary Sensitive Approximation

We can further refine the approximation of regions R with respect to the partition G by taking boundary segments shared by neighboring partition cells into account. That is, we approximate regions in R by functions from $G \times G$ to the set $\Omega_5 = \{\text{fo}, \text{fbo}, \text{pbo}, \text{nbo}, \text{no}\}$. The function which assigns to each region $r \in R$ its boundary sensitive approximation will be denoted $\alpha_5 : R \to \Omega_5^{G \times G}$. The value of $(\alpha_5 r)(g_i, g_j)$ is fo if r covers all of the cell g_i, it is fbo if r covers all of the boundary segment, (g_i, g_j), shared by the cell g_i and g_j and some but not all of the interior of g_i, it is pbo if r covers some but not all of the boundary segment (g_i, g_j) and some but not all of the interior of g_i, it is nbo if r does not intersect with boundary segment (g_i, g_j) and some but not all of the interior of g_i, and it is no if there is no overlap between r and g_i.

Let $bs = (g_i \cap g_j)$ be the boundary segment shared by the cell g_i and g_j. We define boundary sensitive approximation, α_5, in terms of pairs of approximation functions, α_3, as follows [BS98] (left table below):

$(\alpha_5 r)(g_i, g_j)$	$(\alpha_3 r)bs$ = fo	$(\alpha_3 r)bs$ = po	$(\alpha_3 r)bs$ = no
$(\alpha_3 r)g_i = $ fo	fo	-	-
$(\alpha_3 r)g_i = $ po	fbo	pbo	nbo
$(\alpha_3 r)g_i = $ no	-	-	no

$\wedge_{max}$	no	nbo	pbo	fbo	fo
no	no	no	no	no	no
nbo	no	nbo	nbo	nbo	nbo
pbo	no	nbo	pbo	pbo	pbo
fbo	no	nbo	pbo	fbo	fbo
fo	no	nbo	pbo	fbo	fo

Each approximate region $X \in \Omega_3^{G \times G}$ stands for a set of precise regions, i.e. all those precise regions having the approximation X: $[\![X]\!] = \{r \in R \mid \alpha_5 r = X\}$.

We define the operation $\wedge_{max}$ on the set $\Omega_5 = \{\text{fo}, \text{fbo}, \text{pbo}, \text{nbo}, \text{no}\}$ as shown in the right table above. This operation extends to elements of $\Omega_5^{G \times G}$ (i.e. the set of functions from $G \times G$ to Ω_5) by $(X \wedge_{max} Y)(g_i, g_j) = (X(g_i, g_j)) \wedge_{max} (Y(g_i, g_j))$. The definition of the operation $\wedge_{min}$ is similar but slightly more complicated. The details can be found in [BS98].

3   RCC5 Relations

Qualitative spatial reasoning (QSR) is a well-established subfield of AI. It is concerned with the representation and processing of knowledge of space and activities which depend on space. However, the representations used for this are qualitative, rather than the quantitative ones of conventional coordinate geometry. One of the most widely studied formal approaches to QSR is the Region-Connection Calculus (RCC) [CBGG97]. This system provides an axiomatization of space in which regions themselves are primitives, rather than being constructed from more primitive sets of points. An important aspect of the body of work on RCC is the treatment of relations between regions. For example two regions could be overlapping, or perhaps only touch at their boundaries. There are two principal schemes of relations between RCC regions: five boundary insensitive relations known as RCC5, and eight boundary sensitive relations known as RCC8.

In this paper we propose a specific style of defining RCC relations. This style allows to define RCC relations exclusively based on constraints regarding the outcome of the meet operation between (one and two dimensional) regions. Furthermore this style of definitions allows us to obtain a partial ordering with minimal and maximal element on the relations defined. Both aspects are critical for the generalization of these relations to the approximation case.

Given two regions x and y the RCC5 relation between them can be determined by considering the triple of boolean values:

$$(x \wedge y \neq \bot, \ x \wedge y = x, \ x \wedge y = y).$$

The correspondence between such triples, the RCC5 classification, and possible geometric interpretations are given below.

$x \wedge y \neq \bot$	$x \wedge y = x$	$x \wedge y = y$	RCC5
F	F	F	DR
T	F	F	PO
T	T	F	PP
T	F	T	PPi
T	T	T	EQ

The set of triples is partially ordered by setting $(a_1, a_2, a_3) \leq (b_1, b_2, b_3)$ iff $a_i \leq b_i$ for $i = 1, 2, 3$, where the Boolean values are ordered by $F < T$. This is the same ordering induced by the RCC5 conceptual graph [GC94]. But note that the conceptual graph has PO and EQ as neighbors which is not the case in the Hasse diagram for the partially ordered set. The ordering is indicated by the arrows in the figure above. We refer to this as the RCC5 *lattice* to distinguish it from the conceptual neighborhood graph.

4   Semantic and Syntactic Generalizations of RCC5

The original formulation of RCC dealt with ideal regions which did not suffer from imperfections such as vagueness, indeterminacy or limited resolution. However, these are factors which affect spatial data in practical examples, and which are significant in applications such as geographic information systems (GIS)[BF95]. The issue of vagueness and indeterminacy has been tackled in the work of [CG96]. The topic of the present paper is not vagueness or indeterminacy in the widest sense, but rather the special case where spatial data is approximated by being given a limited resolution description.

4.1   Syntactic and Semantic Generalizations

There are two approaches we can take to generalizing the RCC5 classification from precise regions to approximate ones. These two may be called the semantic and the syntactic. The syntactic has many variants.

Semantic generalization. We can define the RCC5 relationship between approximate regions X and Y to be the set of relationships which occur between any pair of precise regions representing X and Y. That is, we can define

$$\mathcal{SEM}(X,Y) = \{RCC5(x,y) \mid x \in [\![X]\!] \text{ and } y \in [\![Y]\!]\}.$$

Syntactic generalization. We can take a formal definition of RCC5 in the precise case which uses operations on R and generalize this to work with approximate regions by replacing the operations on R by analogous ones for Ω^{G1}.

4.2   Syntactic Generalization

The above formulation of the RCC5 relations can be extended to approximate regions. One way to do this is to replace the operation $\wedge$ with an appropriate operation for approximate regions. If X and Y are approximate regions (i.e. functions from G to Ω_3) we can consider the two triples of Boolean values:

$$(X \wedge_{min} Y \neq \bot, \; X \wedge_{min} Y = X, \; X \wedge_{min} Y = Y),$$
$$(X \wedge_{max} Y \neq \bot, \; X \wedge_{max} Y = X, \; X \wedge_{max} Y = Y).$$

In the context of approximate regions, the bottom element, $\bot$, is the function from G to Ω_3 which takes the value no for every element of G. Each of the above triples provides an RCC5 relation, so the relation between X and Y can be measured by a pair of RCC5 relations. These relations will be denoted by $R_{min}(X,Y)$ and $R_{max}(X,Y)$.

Theorem 1 *The pairs $(R_{min}(X,Y), R_{max}(X,Y))$ which can occur are all pairs (a,b) where $a \leq b$ with the exception of* (PP, EQ) *and* (PPi, EQ).

Proof   First we show that $R_{min}(X,Y) \leq R_{max}(X,Y)$. Suppose that $R_{min}(X,Y) = (a_1, a_2, a_3)$ and that $R_{max}(X,Y) = (b_1, b_2, b_3)$. We have to show that $a_i \leq b_i$ for $i = 1, 2, 3$. Taking the first component, if $X \wedge_{min} Y \neq \bot$ then for each g such that $Xg \wedge_{min} Yg \neq$ no, we also have, by examining the tables for $\wedge_{min}$ and $\wedge_{max}$, that $Xg \wedge_{max} Yg \neq$ no. Hence $X \wedge_{max} Y \neq \bot$. Taking the second component, if $X \wedge_{min} Y = X$ then $X \wedge_{max} Y = X$ because from $Xg \wedge_{min} Yg = Xg$ it follows that $Xg \wedge_{max} Yg = Xg$. This can be seen from the tables for $\wedge_{min}$ and $\wedge_{max}$ by considering each of the three possible values for Xg. The case of the third component follows from the second since $\wedge_{min}$ and $\wedge_{max}$ are commutative.

Finally we have to show that the pairs (PP, EQ) and (PPi, EQ) cannot occur. If $R_{max}(X,Y) = $ EQ, then $X = Y$ so $X \wedge_{min} Y = X$ must take the same value as $X \wedge_{min} Y = Y$. Thus the only triples which are possible for $R_{min}(X,Y)$ are those where the second and third components are equal. This rules out the possibility that $R_{min}(X,Y)$ is PP or PPi.   ☐

---

[1] This technique has many variants since there are many different ways in which the RCC5 can be formally defined in the precise case, and some of these can be generalized in different ways to the approximate case. The fact that several different generalizations can arise from the same formula is because some of the operations in R (such as $\wedge$ and $\vee$) have themselves more than one generalization to operations on Ω^G.

4.3   Correspondence of Semantic and Syntactic Generalization

Let the syntactic generalization of RCC5 defined by

$$\mathcal{SYN}(X,Y) = (R_{min}(X,Y), R_{max}(X,Y)),$$

where R_{min} and R_{max} are as defined above.

Theorem 2 *For any approximate regions X and Y, the two ways of measuring the relationship of X to Y are equivalent in the sense that*

$$\mathcal{SEM}(X,Y) = \{\rho \in RCC5 \mid R_{min}(X,Y) \leq \rho \leq R_{max}(X,Y)\},$$

where RCC5 is the set $\{\mathsf{EQ}, \mathsf{PP}, \mathsf{PPi}, \mathsf{PO}, \mathsf{DR}\}$, and $\leq$ is the ordering in the RCC5 lattice.

The proof of this theorem depends on assumptions about the set of precise regions R. We assume that R is a model of the RCC axioms so that we are approximating continuous space, and not approximating a space of already approximated regions.

Proof   There are three things to demonstrate. Firstly that for all $x \in [\![X]\!]$, and $y \in [\![Y]\!]$, that $R_{min}(X,Y) \leq RCC5(x,y)$. Secondly, for all x and y as before, that $RCC5(x,y) \leq R_{max}(X,Y)$, and thirdly that if ρ is any RCC5 relation such that $R_{min}(X,Y) \leq \rho \leq R_{max}(X,Y)$ then there exist particular x and y which stand in the relation ρ to each other. To prove the first of these it is necessary to consider each of the three components $X \wedge_{min} Y \neq \bot$, $X \wedge_{min} Y = X$ and $X \wedge_{min} Y = Y$ in turn. If $X \wedge_{min} Y \neq \bot$ is true, we have to show for all x and y that $x \wedge y \neq \bot$ is also true. From $X \wedge_{min} Y \neq \bot$ it follows that there is at least one cell g where one of X and Y fully overlaps g, and the other at least partially overlaps g. Hence there are interpretations of X and Y having non-empty intersection. If $X \wedge_{min} Y = X$ is true then for all cells g we have $Xg = \mathsf{no}$ or $Yg = \mathsf{fo}$. In each case every interpretation must satisfy $x \wedge y = x$. Note that this depends on the fact that the combination $Xg = \mathsf{po} = Yg$ cannot occur. The case of the final component $X \wedge_{min} Y = Y$ is similar. Thus we have demonstrated for all $x \in [\![X]\!]$ and $y \in [\![Y]\!]$ that $R_{min}(X,Y) \leq RCC5(x,y)$. The task of showing that $RCC5(x,y) \leq R_{max}(X,Y)$ is accomplished by a similar analysis. Finally, we have to show that for each RCC5 relation, ρ, where $R_{min}(X,Y) \leq \rho \leq R_{max}(X,Y)$, there are $x \in [\![X]\!]$ and $y \in [\![Y]\!]$ such that the relation of x to y is ρ. This is done by considering the various possibilities for $R_{min}(X,Y)$ and $R_{max}(X,Y)$. We will only consider one of the cases here, but the others are similar. If $R_{min}(X,Y) = \mathsf{PO}$ and $R_{max}(X,Y) = \mathsf{EQ}$, then for each cell g, the values of Xg and Yg are equal and there must be some cells where this value is po and some cells where the value is fo. Precise regions $x \in [\![X]\!]$ and $y \in [\![Y]\!]$ can be constructed by selecting sub-regions of each cell g say x_g and y_g, and defining x and y to be the unions of these sets of sub-regions. In this particular case, there is sufficient freedom with those cells where $Xg = Yg = \mathsf{po}$ to be able to select x_g and y_g so that the relation of x to y can be any ρ where $\mathsf{PO} \leq \rho \leq \mathsf{EQ}$.

☐

5   Generalizing RCC8 Relations

5.1   RCC8 Relations

RCC8 relations take the topological distinction between interior and boundary into account. In order to describe RCC8 relations we define the relationship between x and y by using a triple, but where the three entries may take one of three truth values rather than the two Boolean ones. The scheme has the form

$$(x \wedge y \not\approx \bot, x \wedge y \approx x, x \wedge y \approx y)$$

where

$$x \wedge y \not\approx \bot = \begin{cases} \mathsf{T} & \text{if the interiors of } x \text{ and } y \text{ overlap,} \\ & i.e., x \wedge y \neq \bot \\ \mathsf{M} & \text{if only the boundaries } x \text{ and } y \text{ overlap,} \\ & i.e., x \wedge y = \bot \text{ and } \delta x \wedge \delta y \neq \bot \\ \mathsf{F} & \text{if there is no overlap between } x \text{ and } y, \\ & i.e., x \wedge y = \bot \text{ and } \delta x \wedge \delta y = \bot \end{cases}$$

and where

$$x \wedge y \approx x = \begin{cases} \mathsf{T} & \text{if } x \text{ is contained in } y \text{ and the boundaries are either disjoint or identical} \\ & i.e., x \wedge y = x \text{ and } (\delta x \wedge \delta y = \bot \text{ or } \delta x \wedge \delta y = \delta x) \\ \mathsf{M} & \text{if } x \text{ is contained in } y \text{ and the boundaries are not disjoint and not identical,} \\ & i.e., x \wedge y = x \text{ and } \delta x \wedge \delta y \neq \bot \text{ and } \delta x \wedge \delta y \neq \delta x \\ \mathsf{F} & \text{if } x \text{ is not contained within } y, i.e., x \wedge y \neq x \end{cases}$$

and similarly for $x \wedge y \approx y$.

The correspondence between such triples, the RCC8 classification, and possible geometric interpretations are given below.

$x \wedge y \not\approx \bot$	$x \wedge y \approx x$	$x \wedge y \approx y$	RCC8
F	F	F	DC
M	F	F	EC
T	F	F	PO
T	M	F	TPP
T	T	F	NTPP
T	F	M	TPPi
T	F	T	NTPPi
T	T	T	EQ

DC(x,y)   EC(x,y)   PO(x,y)   TPP(x,y)   NTPP(x,y)   EQ(x,y)

The RCC5 relation DR refines to DC and EC and the RCC5 relation PP(i) refines to TPP(i) and NTPP(i). We define F < M < T and call the corresponding Hasse diagram RCC8 *lattice* (figure above) to distinguish it from the conceptual neighborhood graph.

5.2   Syntactic Generalization of RCC8

Let X and Y be boundary sensitive approximations of regions x and y. The generalized scheme has the form

$$((X \wedge_{min} Y \not\approx \bot, X \wedge_{min} Y \approx X, X \wedge_{min} Y \approx Y),$$
$$(X \wedge_{max} Y \not\approx \bot, X \wedge_{max} Y \approx X, X \wedge_{max} Y \approx Y))$$

where

$$X \wedge_{min} Y \not\approx \perp = \begin{cases} \text{T} & X \wedge_{min} Y \neq \perp \\ \text{M} & X \wedge_{min} Y = \perp \text{ and } \delta X \wedge_{min} \delta Y \neq \perp \\ \text{F} & X \wedge_{min} Y = \perp \text{ and } \delta X \wedge_{min} \delta Y = \perp \end{cases}$$

and where

$$X \wedge_{min} Y \approx X = \begin{cases} \text{T} & X \wedge_{min} Y = X \text{ and } (\delta X \wedge_{min} \delta Y = \perp \text{ or } X \wedge_{min} Y = Y) \\ \text{M} & X \wedge_{min} Y = X \text{ and } \delta X \wedge_{min} \delta Y \neq \perp \text{ and } X \wedge_{min} Y \neq Y \\ \text{F} & X \wedge_{min} Y \neq X \end{cases}$$

and similarly for $X \wedge_{min} Y \approx Y$, $X \wedge_{max} Y \not\approx \perp$, $X \wedge_{max} Y \approx X$, and $X \wedge_{max} Y \approx Y$. In this context the bottom element, $\perp$, is the function from $G \times G$ to Ω_5 which takes the value no for every element of $G \times G$. The formula $\delta X \wedge_{min} \delta Y \neq \perp$ is true if we can derive from boundary sensitive approximations X and Y that for all $x \in [\![X]\!]$ and $y \in [\![Y]\!]$ the least relation that can hold between x and y involves boundary intersection[2]. Correspondingly, $\delta X \wedge_{max} \delta Y \neq \perp$ is true if the largest relation that can hold between $x \in [\![X]\!]$ and $y \in [\![Y]\!]$ involves boundary intersection.

Each of the above triples defines a RCC8 relation, so the relation between X and Y can be measured by a pair of RCC8 relations. These relations will be denoted by $R^8_{min}(X, Y)$ and $R^8_{max}(X, Y)$. [BS00] show the correspondence between this syntactic generalization and the semantic generalization corresponding to the RCC5 case.

6   Conclusions

In this paper we discussed approximations of spatial regions with respect to an underlying regional partition. We used approximations based on approximation functions and discussed the close relationship to rough sets. We defined pairs of greatest minimal and least maximal meet operations on approximation functions that constrain the possible outcome of the meet operation between the approximated regions themselves. The meet operations on approximation mappings provide the basis for approximate qualitative spatial reasoning that was proposed in this paper.

Approximate qualitative spatial reasoning is based on: (1) Jointly exhaustive and pair-wise disjoint sets of qualitative relations between spatial regions, which are defined in terms of the meet operation of the underlying Boolean algebra structure of the domain of regions. As a set these relations must form a lattice with bottom and top element. (2) Approximations of regions with respect to a regional partition of the underlying space. Semantically, an approximation corresponds to the set of regions it approximates. (3) Pairs of meet operations on those approximations, which approximate the meet operation on exact regions.

Based on those 'ingredients' syntactic and semantic generalizations of jointly exhaustive and pair-wise disjoint relations between regions were defined. Generalized relations hold between approximations of regions rather than between (exact) regions themselves. Syntactic generalization is based on replacing the meet operation defining

---

[2] For details see [BS00].

relations between regions by its greatest minimal and least maximal counterparts on approximations. Semantically, syntactic generalizations yield upper and lower bounds (within the underlying lattice structure) on relations that can hold between the corresponding approximated regions.

There is considerable scope for further work building on the results in this paper. We have assumed so far that the regions being approximated are precisely known regions in a continuous space. However, there are practical examples where approximate regions are themselves approximated. This can occur when spatial data is required at several levels of detail, and the less detailed representations are approximations of the more detailed ones. One direction for future investigation is to extend the techniques in this paper to the case where the regions being approximated are discrete, rather than continuous. This could make use of the algebraic approach to qualitative discrete space presented in [Ste00]. Another direction of ongoing research is to apply techniques presented in this paper to the temporal domain [Bit00].

References

[BF95]   Peter Burrough and Andrew U. Frank, editors. *Geographic Objects with Indeterminate Boundaries*. GISDATA Series II. Taylor and Francis, London, 1995.

[Bit00]   T. Bittner. Approximate temporal reasoning. In *Workshop proceedings of the Seventeenth National Conference on Artificial Intelligence, AAAI 2000*, 2000.

[BS98]   T. Bittner and J. G. Stell. A boundary-sensitive approach to qualitative location. *Annals of Mathematics and Artificial Intelligence*, 24:93–114, 1998.

[BS00]   T. Bittner and J. Stell. Approximate qualitative spatial reasoning. Technical report, Department of Computing and Information Science, Queen's University, 2000.

[CBGG97]   A.G. Cohn, B. Bennett, J. Goodday, and N. Gotts. Qualitative spatial representation and reasoning with the region connection calculus. *geoinformatica*, 1(3):1–44, 1997.

[CG96]   A.G. Cohn and N.M. Gotts. The 'egg-yolk' representation of regions with indeterminate boundaries. In P. Burrough and A.U. Frank, editors, *Geographic Objects with Indeterminate Boundaries*, GISDATA Series II. Taylor and Francis, London, 1996.

[EF91]   Max J. Egenhofer and Robert D. Franzosa. Point-set topological spatial relations. *International Journal of Geographical Information Systems*, 5(2):161–174, 1991.

[GC94]   J.M. Goodday and A.G. Cohn. Conceptual neighborhoods in temporal and spatial reasoning. In *ECAI-94 Spatial and Temporal Reasoning Workshop*, 1994.

[Paw91]   Zdzis aw Pawlak. *Rough sets : theoretical aspects of reasoning about data*. Theory and decision library. Series D, System theory, knowledge engineering, and problem solving ; v. 9. Kluwer Academic Publishers, Dordrecht ; Boston, 1991.

[RCC92]   D. A. Randell, Z. Cui, and A. G. Cohn. A spatial logic based on regions and connection. In *3rd Int. Conference on Knowledge Representation and Reasoning*. Boston, 1992.

[SP92]   T.R. Smith and K. K. Park. Algebraic approach to spatial reasoning. *Int. J. Geographical Information Systems*, 6(3):177 – 192, 1992.

[Ste00]   J. G. Stell. The representation of discrete multi-resolution spatial knowledge. In A. G. Cohn, F. Giunchiglia, and B. Selman, editors, *Principles of Knowledge Representation and Reasoning: Proceedings of the Seventh International Conference (KR2000)*, pages 38–49. Morgan-Kaufmann, 2000.

A Study of Conditional Independence Change in Learning Probabilistic Network

Tao Lin[1] and Yichao Huang[2]

University of Regina, Regina SK S4S 0A2, CANADA
{taolin, huangy}@cs.uregina.ca

Abstract. This paper discusses the change for the conditional independence set in learning *Probabilistic Network* based on markov property.
They are generalized into several cases for all of possible changes. We
show that these changes are sound and complete. Any structure learning
methods for the *Decomposable Markov Network* and *Bayesian Network*
will fall into these cases. This study indicates which kind of domain model
can be learned and which can not. It suggests that prior knowledge about
the problem domain decides the basic frame for the future learning.

1 Introduction

A probabilistic network (PN) combines a qualitative graphic structure which
encodes dependencies of domain variables, with a quantitative probability distribution which specify the firmness of these dependencies [1], [3]. As many
effective probabilistic inference techniques have been developed and the probabilistic network has been applied in many artificial intelligence domains, many
researchers turn their attention to automatically obtain a model $\mathcal{M}$ of PN from
data [2], [4], [8].

Counting from a complete graph with n variables, there are various models
with $1, 2, ...m, m \leq n(n-1)/2$ links missing. We call all of these models*searching
space* for *structure learning* [9]. The *structure learning* is to find an efficient way
to pick up the one in *searching space* which incorporates CIs of $\mathcal{M}$ the most.
Starting from a PN with certain graphical structure which characterizes an
original CIs set, there exist some "false" CIs which are superfluous or missing
actual one of $\mathcal{M}$, links are picked up to be deleted or added on account of
different approaches (constraint-based or score-based). When a certain criterion
is satisfied, the "pick-up" action comes to stop and the final CIs set is obtained
which is intuitively looked as the representation of $\mathcal{M}$ [4], [5]. [6], [7] A more
extensive review of literature for learning PN can be found in [8],

However, not all of CIs set can be incorporated into DMN or BN. [1] indicates that a class of domain models cannot be learnded by a certain procedure.
To find which kind of CIs can be learned and which can not, normally we have
to use *d-seperation* to make the new CIs set campared with the original one [3]
In this paper, we discuss all of possible branchings coming from an original CIs
set correspoding to DMN or BN respectively and generalize into serveral cases.

W. Ziarko and Y. Yao (Eds.): RSCTC 2000, LNAI 2005, pp. 454–461, 2001.
© Springer-Verlag Berlin Heidelberg 2001

It is proved that these cases are sound and complete. Any *structure learning* methods for DMN or BN will fall into this category.

We start from any probabilistic structure, which could be looked as the prior knowledge about the problem. Links will be added or deleted in order to match the data. As soon as the original structure is fixed, we argue that all of the possible changes are also fixed correspondingly, which means the prior knowledge actually already draws the frame for the future refreshment of the domain knowledge.

2   The Cases for Decomposable Markov Network

In this section, we describe a set of rules on how CI is changed after one link is deleted each time from the undirected graphical structure of DMN. It is also proved that these rules are sound and complete. A *Decomposable Markov Network* is a pair (G,P) where G is a chordal graph over n variables and P is an associated JPD which can be factorized as

$$P(v_1,\ldots,v_n) = (\prod_{i=1}^{m} p(C_i))/(\prod_{i=1}^{m} p(S_i)) \qquad (1)$$

where $p(C_i)$ are probability functions defined on the *maximal cliques* $C_1,\ldots,C_m$ of G, $p(S_i)$ are probability functions defined on the separator of C_i as $S_i = C_i \cap (C_1 \cup \ldots \cup C_{i-1})$ [2].

After a link (x,y) is deleted from a DMN, there should be corresponding changes taken place in the CIs set as well. All kinds of possible changes are generalized as follows. Such cases could apply concurrently if their conditions are satisfied at the same time. We use "$\Rightarrow$" to denote the CIs set change after delete one link from the original graph.

Proposition 1. *Given two cliques C_1, C_2 in the graph structure G of a DMN. Link $(x,y) \in C_1 \cup C_2$. Suppose this link is deleted from the G, following rules are applied to justify the change of* CI:

R1: Suppose nodes $x,y \in C_1$, this clique will break into two smallers:

$$\emptyset \Rightarrow I(C_1 - \{x\}, C_1 - \{xy\}, C_1 - \{y\})$$

R2: Suppose node $y \in C_1$ but $y \notin C_1 \cap C_2$, node $x \in C_1 \cap C_2$, $y \in C_1$:

$$I(C_1, C_1 \cap C_2, C_2) \Rightarrow I(C_1 - \{x\}, C_1 \cap C_2 - \{x\}, C_2)$$
$$\wedge\ I(C_1 - \{y\}, C_1 \cap C_2, C_2)\ \wedge\ I(C_1 - \{x\}, C_1 - \{x,y\}, C_1 - \{y\}$$

R3: Suppose nodes $x,y \notin C_1 \cap C_2$, but $x,y \in C_1$

$$I(C_1, C_1 \cap C_2, C_2) \Rightarrow I(C_1 - \{x\}, C_1 \cap C_2, C_2)\ \wedge\ I(C_1 - \{y\}, C_1 \cap C_2, C_2)$$

R4: Suppose nodes $x,y \in C_1 \cap C_2$, $C_1 \nsubseteq C_2$, the DMN *could become non-chordal.*
R5: Suppose $C_1 \cap C_2 = \emptyset$, node $x \in C_1$, node $y \in C_2$

$$I(C_1, \{x\}, \{x,y\}) \wedge I(C_2, \{y\}, \{x,y\}) \Rightarrow I(C_1, \emptyset \cap C_2)$$

Proof: After a link (x, y) is deleted from a complete graph, node x could connect all of other nodes except y, variable y could connect all of other nodes except x as well. Other variable nodes still connect to each other. Therefore, there are two cliques after deletion: one includes variable x and all of other nodes except y; the other includes variable y and all of other nodes except x. Their separator is all of variables set except x, y, thus, we may get $I(C - \{x\}, C - \{x, y\}, C - \{y\})$ (R1 applies).

Suppose $y \in C_1$ but $y \notin C_1 \cap C_2$. According to the proof of R1, after (x, y) is deleted, C_1 will break into two cliques $C_1 - \{x\}$ and $C_1 - \{y\}$. Since $x \in C_1 \cap C_2$, we get $(C_1 - \{x\}) \cap C_2 = C_1 \cap C_2 - \{x\}$, $(C_1 - \{y\}) \cap C_2 = C_1 \cap C_2$. Therefore, $I(C_1 - \{x\}, C_1 \cap C_2 - \{x\}, C_2)$ and $I(C_1 - \{y\}, C_1 \cap C_2, C_2)$ hold, respectively (R2 applies).

According to the proof of R1, after (x, y) is deleted, C_1 will break into two cliques $C_1 - \{x\}$ and $C_1 - \{y\}$. Since $x, y \notin C_1 \cap C_2$, we get $(C_1 - \{x\}) \cap C_2 = C_1 \cap C_2$, $(C_1 - \{y\}) \cap C_2 = C_1 \cap C_2$. Therefore, $I(C - \{x\}, C - \{x, y\}, C - \{y\})$, $I(C_1 - x, C_1 \cap C_2, C_2)$, $I(C_1 - y, C_1 \cap C_2, C_2)$ hold, respectively (R3 applies).

Suppose $C_1 \nsubseteq C_2$, there must be at least one variable in one cliques but not in the other, thus, $u \in C_1$, $v \in C_2$, while $u \notin C_2$, $v \notin C_1$, which means they are not connected with each other, thus, (u, v) doesn't exist. If $x, y \in C_1 \cap C_2$, these two variables u, v must connect to x, y, respectively. Therefore, $u - x - y - v$ construct into a circle with length 4. There are only two chords (x, y) and (u, v) in the graph, since (u, v) doesn't exist, after we delete (x, y), this graph would be non-chordal (R4 applies).

According to given condition, (x, y) is the only link connect C_1 with C_2. After deletion, cliques C_1 and C_2 will not connected any more. (R5 applies)

Proposition 2. *Rules R1-R5 are complete for all possible CIs change by deleting one link each time in the graphical structure of a DMN.*

Proof: By induction, there is no other possibility to choose a link except above. Therefore, no matter which link we choose to delete in the graphical structure of M_k, there are always some of rules R1-R5 could apply to the corresponding CIs set change.

Example 1. Given a complete graph with 4 variables, figure 1 indicates how to apply rules above to find the change of CIs set correspondingly.

3   The Cases for Ba y esian Netowrk

If we might have to consider the directionality in a directed graphical structure, it is much more complicated to study the CIs set change in a Bayesian Network(BN) than that of DMN. However, Pearl [3] showed that it is the directionality that makes BN a richer language in expression of dependencies. Some CIs can be expressed by a BN but not a DMN. As BN and DMN are so closely related, study on one of both will benefit the study on the other as well.

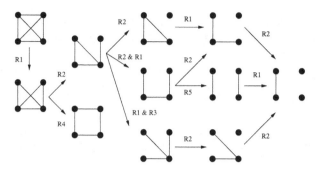

Fig. 1. Applying rules for CIs set change to a graph with 4 variables at different topological structure.

A *Bayesian network* is a pair (D,P) where D is a directed acyclic graph (DAG) over n variables and P is an associated JPD which can be factorized as

$$P(v_1, \ldots, v_n) = \prod_{i=1}^{n} p(v_i|pa(v_i)) \qquad (2)$$

where $pa(v_i)$ is the parents set of node v_i in D [2].

3.1 Local directed Mark ov property

For the DAG D in a BN, it is proved that if $P(v_1, \ldots, v_n)$ admits factorization on account of D, P must obey the local directed Marko v propert y(DL) [2], thus, any variable v is conditionally independent of its non-descendants, $nd(v)$ namely, given its parents $pa(v)$ in the DAG, thus, $I(v, pa(v), nd(v))$.

This property paves the way on how CIs set changes correspond to the arrow variation in the BN. It is discussed in this section how the DL property is changed among t wo nodes, v_i and v_j, their corresponding ancestors, A_i and A_j and corresponding descendants, D_i and D_j after the arrow direction between these two nodes is changed or the arrow is deleted. W e take "$v_i \rightarrow v_j$" to denote an arrow from v_i to v_j. If there is a path from v_i to v_j, it is denoted as $v_i \longmapsto v_j$. Assuming it is clear enough in the con text, we also use $A_i \longmapsto v_i$ to denote there is a path from a node in A_i to v_i and $A_i \longmapsto A_j$ to denote there is path from a node in A_i to a node in A_j.

Lemma 1. *For those nodes without path to or from v_i and v_j in a DAG, the arrow variation of $v_i \rightarrow v_j$ (delete or change arrow direction) will not change their DL properties.*

Proof: Assume v_k is one of these nodes, for a node v_a in $A_i \cup A_j$, there is no path from v_k to it because $v_a \longmapsto v_i$. It follows that $A_i \cup A_j \subseteq nd(v_k)$.

There is no path from the node $v_d \in D_i \cup D_j$ to v_k, otherwise $v_i \longmapsto v_k$ or $v_j \longmapsto v_k$. If $v_k \longmapsto v_d$, then $nd(v_d) \subseteq nd(v_k)$. After arrow variation, if there is a path from $v' \in de(v_d)$ to a node $v'' \in nd(v_d)$, it must come to node v'' via v_i or v_j. It is impossible because $v' \longmapsto v_i(v_j) \longmapsto v_d \longmapsto v'$ forms a cyclic which will conflict with the definition of BN.

Lemma 2. *After the direction of an arrow $v_i \to v_j$ is changed in BN, a path forms a cycle if and only if $v_i \longmapsto A_j$. And we call this path as* weak path.

Proof: It is easy to verify $v_i \longmapsto A_j$ is a weak path. The remaining part of proof is to show there exists only this kind of weak path.

If $v_i \longmapsto A_j$, there is a node v with $v_i \longmapsto v, v \longmapsto A_j$, which means $v \in D_i$, $D_i \longmapsto A_j$. Furthermore, since $A_j \longmapsto v_j$ is trivial, $D_i \longmapsto v_j$ holds.

If $D_i \longmapsto v_j$, there must be a node $v \in D_i$ with $v \longmapsto v_j$. Therefore $v \in A_j$, $D_i \longmapsto A_j$ holds. Since $v_i \longmapsto D_i$ is trivial, if $D_i \longmapsto A_j$, $v_i \longmapsto A_j$ is trivial.

If $D_i \longmapsto A_j$, $v_i \to D_i$ is trivial, $v_i \to A_j$ exists. Since $A_j \longmapsto v_j$ is trivial, $D_i \longmapsto v_j$ holds.

There is no other possibility to form a weak path related to $v_i \to v_j$.

Lemma 3. *For v_i, v_j, their ancestors set A_i, A_j and their descendants set D_i, D_j, the only possible kinds of path are: $A_i \longmapsto A_j$, $A_j \longmapsto A_i$, $A_i \longmapsto D_j$, $A_j \longmapsto D_i$, $D_i \longmapsto D_j$, $D_j \longmapsto D_i$.*

The possible paths are shown in the Figure 2. Since the discussion of possible paths for v_j is exactly the same as that of v_i after arrow direction is changed, the proposition for them will apply symmetrically for that of v_j and v_j itself.

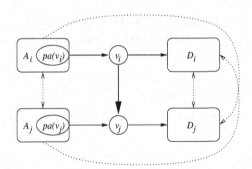

Fig. 2. Possible paths from v_i, A_i and D_i to other components which is shown in dash arrow.

3.2   Arrow direction change

After arrow direction is changed from $v_i \to v_j$ to $v_i \leftarrow v_j$, the non-descendant sets of v_i, $v_{a_i} \in A_i$ and $v_{d_i} \in D_i$ could be changed correspondingly. In this subsection, we discuss how these non-descendant set change take impact on the related CI changes.

Proposition 3. *After arrow direction is changed from $v_i \rightarrow v_j$ to $v_i \leftarrow v_j$, if there is a node $v_{d_i} \in D_i$ with $v_{d_i} \longmapsto D_j$, then*

$$I(v_i, pa(v_i), ((A_i \cup A_j) - pa(v_i))) \Rightarrow I(v_i, pa(v_i) \cup v_j, ((A_i \cup A_j \cup D'_j) - pa(v_i))); \quad (3)$$

where $D'_j = D_j - (D_j \cap nd(v_{d_i}))$; Otherwise

$$I(v_i, pa(v_i), ((A_i \cup A_j) - pa(v_i))) \Rightarrow I(v_i, pa(v_i) \cup v_j, ((A_i \cup A_j \cup D_j) - pa(v_i))). \quad (4)$$

Proposition 4. *After arrow direction is changed from $v_i \rightarrow v_j$ to $v_i \leftarrow v_j$, for a node $v_{a_i} \in A_i$, if there is a path $A_i \longmapsto A_j$, then its corresponding CI will not change; or else if there is a node $v_{d_i} \in D_i$ with $v_{d_i} \longmapsto D_j$, then*

$$I(v_{a_i}, pa(v_{a_i}), (A'_i \cup A_j) - pa(v_{a_i})) \Rightarrow I(v_{a_i}, pa(v_{a_i}), (A'_i \cup A_j \cup v_j \cup D'_j) - pa(v_{a_i})); \quad (5)$$

where $A'_i = A_i - de(v_{a_i})$, $D'_j = D_j - (D_j \cap nd(v_{d_i}))$; Otherwise

$$I(v_i, pa(v_i), (A'_i \cup A_j) - pa(v_{a_i})) \Rightarrow I(v_i, pa(v_i) \cup v_j, (A'_i \cup A_j \cup v_j \cup D_j) - pa(v_{a_i})). \quad (6)$$

Proposition 5. *After arrow direction is changed from $v_i \rightarrow v_j$ to $v_i \leftarrow v_j$, for a node $v_{d_i} \in D_i$, there is no any change on its corresponding CI.*

Proposition 3 and 4 provide rules for the corresponding nodes in the DAG. It should be noticed that another node could also characterizes its own DL property in terms of the same CI.

3.3   Arrow deleted

After arrow $v_i \rightarrow v_j$ is deleted from the DAG of a BN, the non-descendant sets of v_i, $v_{a_i} \in A_i$ and $v_{d_i} \in D_i$ could also be changed correspondingly. We discuss these related CIs change in this subsection.

Proposition 6. *If there is a weak path related to $v_i \rightarrow v_j$, then deleting this arrow $v_i \rightarrow v_j$ will not change any CI.*

Proposition 7. *After arrow $v_i \rightarrow v_j$ is deleted, if there is a node $v_{d_i} \in D_i$ with $v_{d_i} \longmapsto D_j$, then*

$$I(v_i, pa(v_i), A_i \cup A_j) \Rightarrow I(v_i, pa(v_i), A_i \cup A_j \cup v_j \cup D'_j) \quad (7)$$

where $D'_j = D_j - de(v_{d_i})$; Otherwise

$$I(v_i, pa(v_i), A_i \cup A_j) \Rightarrow I(v_i, pa(v_i), A_i \cup A_j \cup v_j \cup D_j). \quad (8)$$

Proposition 8. *After arrow $v_i \rightarrow v_j$ is deleted from the graphical structure of a BN, for the node $v_{a_i} \in A_i$, if there is path $A_i \longmapsto A_j$, there is no change for corresponding CI of node v_{a_i}, else if there is a node $v_{d_i} \in D_i$ with $v_{d_i} \longmapsto D_j$, then*

$$I(v_{a_i}, pa(v_{a_i}), A'_i \cup A_j) \Rightarrow I(v_{a_i}, pa(v_{a_i}), A'_i \cup A_j \cup v_j \cup D'_j) \quad (9)$$

where $D'_j = D_j - de(v_{d_i})$, $A'_i = A_i - de(v_{a_i})$; Otherwise

$$I(v_{a_i}, pa(v_{a_i}), A'_i \cup A_j) \Rightarrow I(v_{a_i}, pa(v_{a_i}), A'_i \cup A_j \cup v_j \cup D_j). \quad (10)$$

Proposition 9. *After arrow $v_i \to v_j$ is deleted from the graphical structure of a BN, there is no any change on its corresponding CI for the node $v_{d_i} \in D_i$.*

3.4   Soundness and Completeness

Proposition 10. *Proposition 3-6 are sound and complete for the corresponding CIs set changes of a given BN after an arrow on its graphical structure is deleted or changed.*

Proof: Soundness is easy because it is followed directly from the analysis of DL property change on each kind of possible path which is shown in Lemma 3. Here we will prove completeness by induction,

Suppose therer are only 2 links in the DAG of a given BN. Figure 3 lists all of the possible topological structures they may have:

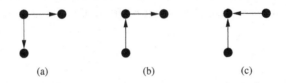

(a)   (b)   (c)

Fig. 3. All of possible topological structures for two links in a DAG

For figure 3(a), suppose we pick up arrow $x \to y$ and change its direction, the corresponding CIs change is: $I(x, \emptyset, \emptyset) \Rightarrow I(x, y, \emptyset)$ and $I(y, x, z) \Rightarrow I(y, \emptyset, \emptyset)$.

There is still $I(y, x, z)$ for node z before and after the arrow direction change. Therefore, after arrow direction is changed, the CIs set does not change. The same result is followed by figure 3(b) and (c).

For all of these figures, as soon as one arrow is deleted, A_i, A_j, D_i and D_j will be empty. From the equation(7), (8), (9) and (10), there is no CI exists.

Assuming there are k links in the DAG, each time when an arrow is picked up to change the arrow direction or be deleted, it follows the Proposition 3-6.

Given a DAG with $k + 1$ links, we may only constrain our consideration on a chosen link's nodes, their ancestors set and descendant set according to lemma 1. If arrow direction is to be changed, we may have to choose the link without *weak path* related to it. Lemma 3 lists all kinds of possibility to choose. In this way, proposition 3- 5 apply. If arrow is to be deleted, suppose there is weak path related to this arrow, then nothing is changed in the CIs set according to proposition 6. If there is no weak path related to this arrow, then proposition 7 to 9 will apply to the corresponding CIs change. After this arrow is deleted, there are only k links left in the DAG, assumption (2) will apply.

4   Conclusion

In this paper, we discussed how the CIs set is changed corresponding to link variation in the DMN and arrow variation in the BN. It is proved that these changes are sound and complete in the DMN or BN, respectively. We argue that any *structure learning* aimed to obtain a CIs set in terms of DMN or BN, then our proposition should apply. It is also indicted clearly to the inside mechanism of *structure learning* step by step. The question which kind of CIs set could be learned and which can not be is also answered. In other point of view, it suggests that the prior knowledge actually decides the scope of problem domain we can learned from the data in the future. In this way we pave the way to the feasibility study of *structure learning* algorithm.

Acknowledgemen t

We are gratefull to S. K. M. W ong posing the problem of CI change in the PN structure. The authors have also benefited from discussions with him.

References

1. Y.Xiang, S.K.M.W ong and N.Cercone: A 'Microscopic' Study of Minimum Entropy Search in Learning Decomposable Markov Networks. Machine Learning, Vol.26, No.1, (1997) 65–92
2. Robert G. Cowell, A. Philip Dawid, Steffen L. Lauritzen, David J. Spiegelhater: Probabilistic Networks and Expert Systems. Springer-Verlag, Berlin Heidelberg New York (1999)
3. J.Pearl: Probailistic Reasoning in Intelligent Systems: Nerworks of Plausible Inference. Morgan Kaufmann (1988).
4. D.Heckerman, D.Geiger, and D.M.Chickering. Learning Bayesian Network: the combination of knowledge and statistical data. *Machine Learning*, 20:197-243, 1995.
5. S.K.M.W ong and Y.Xiang: Construction of a Markov network from data for probailistic inference. In Proc. 3rd Inter. Workshop on Rough Sets and Soft Computing, San Jose, (1994) 562–569.
6. P.Spirtes and C.Glymour: An algorithm for fast recovery of sparse causal graphs. Social Science Computer Review, 9(1), (1991) 62–73.
7. R.M.Fung and S.L.Crawford: Constructor:a system for the induction of probabilistic models. In Proc. of AAAI, MIT Press, (1990) 762–769.
8. W.Buntine: A guide to the literature on learning probabilistic networks from data. IEEE Trans. on Knowledge and data engineering, 8(2), (1996) 195–210.
9. D.M.Chickering: Learning equivalence classes of Bayesian network structures. In Proc. 12th conf. on Uncertainty in Artificial Intelligence, Morgan Kaufmann, (1996) 150–157.
10. A.P.Dawid and S.L.Lauritzen: Hyper Markov laws in the statistical analysis of decomposable graphical models. Annals of Statistics, 21(3), (1993) 1272–1317.

λ-Level Rough Equality Relation and the Inference of Rough Paramodulation

Qing Liu

Department of Computer Science, NanChang University
NanChang, JiangXi 330029, P.R. China
Email: qliu@263.net

Abstract. In the paper, it defines a λ-level rough equality relation in rough logic and establishes a rough sets on real-valued information systems with it. We obtain some related properties and relative rough paramodulation inference rules. They will be used in the approximate reasoning and deductive resolutions of rough logic. Finally, it proves that λ-level rough paramodulant reasoning is sound.

1 Introduction

Pawlak introduced the rough sets via equivalence relation[1]. It provides a basic theory of handling uncertainty knowledge, hence it is used in studying the partial truth values in logic. The theory have been an inspiration for logical investigation and computer scientists. They try to establish a logical systems of approximate reasoning, so that it is used to handle incomplete informations and to deduce partial truth values. Partial truth values are usually researched and interpreted through employing information table, hence information system is used as studying start point of rough logic. Equality objects are same description with respect to attribute sets, thus to lead to approximate definition for the sets of objects with indiscernibility relation. We define the equality of two objects employing the λ-level rough equality of their attribute values under given accuracy, we call it a λ-level (or λ-degree) rough equality relation, denoted by $=_{\lambda R}$. We may derive several of its properties and relative inference rules of λ-level rough palamodulation. It is illustrated in the successful applications of approximate reasoning and deductive resolution with real examples.

Pawlak defined the rough equality of two sets in his works[1], namely the set X is equal to the set Y if $R_*(X) = R_*(Y) \wedge R^*(X) = R^*(Y)$, written by $X \approx_R Y$, where $X, Y \subseteq U$ is any subset on universe U of objects, R is an equivalence relation, R_* and R^* is the lower and upper approximate sets of X respectively. Banerjee and Chakraborty introduced a new binary conjunctive $\approx$, it is defined by two modal operators L (necessary) and M (possible $\Diamond$) in the S_5 of modal logic. Namely, the formula α is rough equal to formula β iff $(L\alpha \leftrightarrow L\beta) \wedge (M\alpha \leftrightarrow M\beta)$. It points out the rough equality of two formulas in modal logic. In the fact, the rough equality relation conjunctive $\approx$ is added into the S_5 of modal logic, we will derive several properties about the rough

W. Ziarko and Y. Yao (Eds.): RSCTC 2000, LNAI 2005, pp. 462–469, 2001.
© Springer-Verlag Berlin Heidelberg 2001

equality relation conjunctive $\approx$. Hence we bear in mind the idea of defining a λ-level rough equality relation conjunctive $=_{\lambda R}$ in the rough logic. we may also obtain the related several properties in rough logic from it. And to have relative λ-level rough paramodulation inference rules.

2 λ-level Rough Equality Relation

The comparison of arbitrary two real numbers avoids always to use equality relation sign $=$, because absolute equal phenomenon is always rarely seen. Hence, we may use the comparison between absolute value of their subtraction to arbitrary small positive number ϵ if it is needful. Such as, $x - y$ might be written $abs(x-y) < \epsilon$, where abs is an abbreviation of the word $absolute$, ϵ is an enough small positive number. Thus we introduce a λ-level rough equality relation conjunctive $=_{\lambda R}$ to be useful and necessary.

Definition 1. Let $K = (U, A, V, f)$ be a knowledge representation system[5], where

(1). U is an universe of discourse objects;

(2). A is a set of attributes ;

(3). $V = \bigcup_{a \in A} V_a$ is the union of sets of attribute values for each attribute $a \in A$, here the $a(x)$ of each object $x \in U$ with respect to attribute $a \in A$ is a real number. Namely, The knowledge representation systems we give are an information table based on real values used as attribute values;

(4). $f : A \times U \to V$ is a mapping, it is interpreted as: for each attribute $a \in A$ and object $x \in U$, the feature of x with respect to attribute a or the character of x possessing attribute a is transformed into a value of quantities, namely a real number $a(x) = v_a \in [0, 1]$. x don't have the feature of attribute a entirely if $v_a = 0$; x has full feature of the attribute a if $v_a = 1$; x has v_a degree feature of the attribute a if $v_a > 0$.

For example, the attribute a is the symptom of temperature for a patient, thus the temperature is transformed into the value of quantities, $a(x) = v_a \in [0, 1]$ is a real number. $v_a = 1$ if temperature ≥ 40 degree c, which represents high symptom of temperature for the patient; $v_a = 0$ if 36 degree $c \leq$ temperature < 37 degree c, which represents formal temperature for the patient; $v_a > 0$ if 40 degree $c >$ temperature ≥ 37 degree c, which represents v_a degree fervor symptom for the patient.

For a given knowledge representation system, due to the set of attribute values is of real numbers, in order to conveniently using equality relation reasoning on set of real numbers, we define a λ-level rough equality relation $=_{\lambda R}$. Which describes the equal degree of attribute values for two objects x_1 and x_2 with respect to attribute a. If the equality degree greater or equal to the value λ, then two objects are thought to be indiscernibility, we call it λ-level rough equality. Its formal definition is described as follow:

Definition 2. Let $S = (U, A, V, f)$ be an information system, for $\forall x_1, \forall x_2 \in U$ and $a \in A$,

$$x_1 =_{\lambda R} x_2 \quad iff \quad |f(a, x_1) - f(a, x_2)| < \epsilon,$$

where ϵ is a given approximate accuracy, we call x_1 λ-level rough equality to x_2.

Obviously, ϵ and equal degree or equal grade fact λ is the following relation:

$$\epsilon = 1 - \lambda.$$

It follows that they are complementary.

For example, let $S = (P, A, V, f)$ represent a system of Chinese Tradition Medicine, where P is the set of the patients, A is the set of the symptoms, V is the set of real values of attributes. It is represented as following table.

$P \backslash A$	a	b	c	d
p_1	0.25	0.03	0.01	0.95
p_2	0.25	0.03	0.01	0.9

We may obtain that p_1 and p_2 is 0.4-level rough equality with respect to attribute d from the table.

Property 1. Let $S = (U, A, V, f)$ be an information system, $u_{\lambda RI}$ is an assignment symbol to formulas, then which satisfies:

(1). $u_{\lambda RI}(x =_{\lambda R} x) = 1$, for all $x \in U$;

(2). $u_{\lambda RI}(x =_{\lambda R} x') =_{\lambda R} u_{\lambda RI}(x' =_{\lambda R} x)$,for all $x, x' \in U$;

(3). $u_{\lambda RI}(x =_{\lambda R} x'') \geq max(u_{\lambda RI}(x =_{\lambda R} x'), u_{\lambda RI}(x' =_{\lambda R} x''))$, for all $x, x', x'' \in U$.

In usual case, $=_{\lambda R}$ relation is the λ-level indiscernibility[5], we call also it λ-level equivalence[2], hence for $\forall \alpha \leq \beta$, it also satisfies:

$$=_{\beta R} \subseteq =_{\alpha R}.$$

Thus we may define the lower approximation and upper approximation of $X \subseteq U$ with respect to the λ-level rough equality relation $=_{\lambda R}$.

Definition 3. Let $A = (U, =_{\lambda R})$ be an approximate space, where U is a nonempty finite universe of objects, $=_{\lambda R}$ is a binary λ-level rough equality relation on U. The lower and upper approximate sets of $X \subseteq U$ from it are defined as follows:

$$=_{\lambda R*}(X) = \{x \in U : [x]_{=_{\lambda R}} \subseteq X\},$$

and

$$=_{\lambda R}^{*}(X) = \{x \in U : [x]_{=_{\lambda R}} \cap X \neq \Phi\},$$

respectively.

Nakamura defined a rough grade modalities[5], he combined the common feature of three models of rough logic, fuzzy logic and modal logic. Stepaniuk defined two structures of approximate first-order logic[6], namely $\sigma_m = (IND, R_1, \cdots, R_m, =)$ and $\tau_m = (R_1, \cdots, R_m)$. The former has an equal relation symbol $=$, hence it may give several similar results to classical first-order logic with equal conjunctive $=$. Lin and Liu defined a first-order rough logic via the neighborhood topological interior and closure used as two operators[9]. Yao and Liu defined α-degree true in generalized decision logic[11], that is,

$v(\phi) = \alpha \in [0,1]$, we say that the formula ϕ is α-degree true. we may move the λ-level rough equality $=_{\lambda R}$ into the logical systems[3,4,7,9,11] and to allow us to create an axiomatic systems of λ-level rough equality relation $=_{\lambda R}$. We may still inference to use the systems of relative approximate reasoning with the λ-level rough equality $=_{\lambda R}$. Liu defined an operator ξ in rough logic[4,7]. It is used in the rough logic. we put it in front of the formulas, which is called a degree of the formulas taking true. Therefore, we will move λ-level rough equality relation conjunctive $=_{\lambda R}$ into the rough logic, obviously we may obtain some properties and rough paramodulation inference rules about λ-level rough equality $=_{\lambda R}$

3   The Significance of λ-level Rough Equality Relation in Rough Logic

Let $\Sigma = \{p, q, \cdots, \phi, \psi, \cdots, \alpha, \beta, \cdots, x, y, \cdots\}$ be a set of the language symbols[4,5,6,7,12], they are the propositional variables, well-formed formulas, terms and individual variable respectively. I is an interpretation function for the formulas, $u_{\lambda RI}$ is an assignment function to the formulas in the interpretation I.

For $\forall \phi \in \Sigma, u_{\lambda RI}(\phi) =_{\lambda R} \xi \in [0,1]$. Let α, β, γ be the term in the $\Sigma, \lambda \in [0,1]$ is a given degree or λ-level rough equality relation $=_{\lambda R}$.

Property 2

(1). $u_{\lambda RI}(\alpha =_{\lambda R} \alpha) \geq \lambda$;

(2). If $u_{\lambda RI}(\alpha =_{\lambda R} \beta) \geq \lambda$, then $u_{\lambda RI}(\beta =_{\lambda R} \alpha) \geq \lambda$;

(3). If $u_{\lambda RI}(\alpha =_{\lambda R} \beta) \geq \lambda \wedge u_{\lambda RI}(\beta =_{\lambda R} \gamma) \geq \lambda$, then $u_{\lambda RI}(\alpha =_{\lambda R} \gamma) \geq u_{\lambda RI}(\alpha =_{\lambda R} \beta \wedge \beta =_{\lambda R} \gamma) \geq \lambda$;

(4). If $u_{\lambda RI}(\alpha =_{\lambda R} \beta) \geq \lambda$, then $u_{\lambda RI}(p(\cdots \alpha \cdots)) =_{\lambda R} u_{\lambda RI}(p(\cdots \beta \cdots))$.

Property 3 Let $A = (U, =_{\lambda R})$ be an approximate space,$\lambda \geq 0.5, \xi \in [0,1], \xi \geq \lambda$, x,y,z are individual variables and $p(\cdots)$ is an atom in the Σ, thus we have

(1). $u_{\lambda RI}(\xi(x =_{\lambda R} x)) = \xi$;

(2). $u_{\lambda RI}(\xi(x =_{\lambda R} y) \vee (1 - \xi)(y =_{\lambda R} x)) \geq max(u_{\lambda RI}(\xi(x =_{\lambda R} y), u_{\lambda RI}((1 - \xi)(y =_{\lambda R} x))) \geq \xi$;

(3). $u_{\lambda RI}((1 - \xi)(x =_{\lambda R} y) \vee (1 - \xi)(y =_{\lambda R} z) \vee \xi(x =_{\lambda R} z)) \geq max(u_{\lambda RI}((1 - \xi)(x =_{\lambda R} y)), u_{\lambda RI}((1 - \xi)(y =_{\lambda R} z)), u_{\lambda RI}(\xi(x =_{\lambda R} z))) \geq \xi$;

(4). If $u_{\lambda RI}(x_i =_{\lambda R} x_0) \rightarrow u_{\lambda RI}(p(\cdots x_i \cdots)) =_{\lambda R} u_{\lambda RI}(p(\cdots x_0 \cdots))$, then $u_{\lambda RI}((1 - \xi)(x_i =_{\lambda R} x_0) \vee (1 - \xi)(p(\cdots x_i \cdots)) \vee \xi(p(\cdots x_0 \cdots))) \geq max(u_{\lambda RI}((1 - \xi)(x_i =_{\lambda} x_0)), u_{\lambda RI}((1 - \xi)p(\cdots x_i \cdots)), u_{\lambda RI}(\xi p(\cdots x_0 \cdots))) \geq \xi$;

(5). If $u_{\lambda RI}(x_i =_{\lambda R} x_0) \rightarrow u_{\lambda RI}(f(\cdots x_i \cdots) =_{\lambda R} f(\cdots x_0 \cdots))$,then $u_{\lambda RI}((1 - \xi)(x_i =_{\lambda R} x_0) \vee \xi(f(\cdots x_i \cdots) =_{\lambda R} f(\cdots x_0 \cdots)))) \geq max(u_{\lambda RI}((1 - \xi)(x_i =_{\lambda R} x_0)), u_{\lambda RI}(\xi(f(\cdots x_i \cdots) =_{\lambda R} f(\cdots x_0 \cdots)))) \geq \xi$.

To sum up, we may have the axiomatic set of rough logic with λ-level rough equality relation $=_{\lambda R}$

$A_1 :|\sim \xi(p =_{\lambda R} p)$;

$A_2 :|\sim (1 - \xi)(p =_{\lambda R} q) \vee \xi(q =_{\lambda R} p)$;

$A_3 :|\sim (1 - \xi)(p =_{\lambda R} q) \vee (1 - \xi)(q =_{\lambda R} \gamma) \vee \xi(p =_{\lambda R} \gamma)$;

$A_4 :|\sim (1 - \xi)(\alpha_j =_{\lambda R} \alpha_0) \vee (1 - \xi)p(\cdots \alpha_j \cdots) \vee \xi p(\cdots \alpha_0 \cdots)$;

$A_5 :|\sim (1 - 1)(\alpha_j =_{\lambda R} \alpha_0)) \vee \xi(f(\cdots \alpha_j \cdots) =_{\lambda R} f(\cdots \alpha_0 \cdots))$.

where $j = 1, \cdots, n$, f is a function term, α_j and α_0 are a general term,$|\sim$ is the axiom or theorem symbol in rough logic, it denotes the difference from $\vdash$ in the classical two valued logic.

4 λ-Level Rough Paramodulation and Reasoning

Equality relation is a most basic and most often application in mathematics. The important character of equality relation is that it may be used in the λ-level rough equality substitution in rough logic. The λ-level rough equality relation $=_{\lambda R}$ defined in the paper is introduced in the operator rough logic[4,7], we may have the λ-level rough equality of two formulas in the logic.

Definition 4 Let ϕ and ψ be the two formulas in the operator rough logic[4,7], λ-level rough equality of ϕ and ψ is defined as follows:

$$\phi =_{\lambda R} \psi \ \ iff \ \ u_{\lambda RI}(\phi) =_{\lambda R} u_{\lambda RI}(\psi) \ \ iff \ \ |u_{\lambda RI}(\phi) - u_{\lambda RI}(\psi)| < \epsilon.$$

Similarly, it can also obtain the relationship between λ and ϵ is

$$\lambda = 1 - \epsilon.$$

Definition 5 Let C_1 and C_2 be the clauses of no common variable[4,7]. $\lambda \geq 0.5$,the structures of C_1 and C_2 are as follow:
$C_1 : \lambda_1 L[t] \vee C_1'$, for $\lambda_1 > \lambda \vee \lambda_1 < 1 - \lambda$;
$C_2 : \lambda_2 (r =_{\lambda R} s) \vee C_2'$, for $\lambda_2 \geq \lambda$,
where $\lambda_1 L[t]$ denotes to contain the rough literal[4,7] of term t and C_1' and C_2' are still clauses. If t and r have most general unifier σ,then we call

$$\lambda^* L^\sigma[s^\sigma] \cup C_1'^{\sigma]} \cup C_2'\sigma$$

a binary λ-level rough paramodulation of C_1 and C_2, written by $P_{\lambda R}(C_1, C_2)$, where $\lambda_1 L[t]$ and $\lambda_2 (r =_{\lambda R} s)$ are λ-level rough paramoculation literal, where λ^* is defined as follows:

$$\lambda^* = (\lambda_1 + \lambda_2)/2, \ \ for \ \ \lambda_1 > \lambda,$$

or

$$\lambda^* = (1 + \lambda_1 - \lambda_2)/2, \ \ for \ \ \lambda_1 < 1 - \lambda,$$

and $L^\sigma[s^\sigma]$ denotes the result obtained by replacing t^σ of one single occurring in L^σ by s^σ.

Example 1. consider the following clauses,

$$C_1 : P(a) \vee Q(b),$$
$$C_2 : a =_{\lambda R} b \vee R(b),$$

where L is a literal, C_1' is $Q(b)$, $P(a)$ may also be written $L[a]$; C_2 contains an equality literal $a =_{\lambda R} b$. Thus, $L[a]$ will be paramodulated as $L[b]$, it may also be written as $P(b)$. Hence, λ-level rough paramodulation of C_1 and C_2 is a new clause

$$C : P(b) \vee Q(b) \vee R(b)$$

Example 2, Consider following the clauses

$$C_1 : P(g(f(x))) \vee Q(x),$$
$$C_2 : f(g(b)) = \lambda Ra \vee R(g(c)),$$

where L is $P(g(f(x)))$, C_1' is $Q(x)$, r is $f(g(b))$, s is a; C_2' is $R(g(c))$. L contains the term $f(x)$ that can be unified to r. Now, let t be $f(x)$, a most general unifier of t and r is $\sigma = \{g(b), x\}$, hence, $L^\sigma[t^\sigma]$ is $P(g(f(g(b))))$, $L^\sigma[t^\sigma]$ is $P(g(a))$. Since $C_1'^\sigma$ is $Q(g(b))$ and $C_2'^\sigma$ is $R(g(c))$. Thus, we obtain a binary λ-level rough paramodulation of C_1 and C_2 is

$$C : P(g(a)) \vee Q(g(b)) \vee R(g(c))$$

In the example, two paramodulation literals are $P(g(f(x)))$ and $f(g(b)) =_{\lambda R} a$.
Definition 6. λ-level rough paramodulation of C_1 and C_2 is one of the following binary λ-level rough paramodulation:
(1) A binary λ-level rough paramodulation of C_1 and C_2;
(2) A binary λ-level rough paramodulation of C_1 and a fact of C_2;
(3) A binary λ-level rough paramodulation of a fact of C_1 and C_2;
(4) A binary $\lambda - lavel$ rough paramodulation of a fact of C_1 and a fact of C_2.
Theorem Let C_1 and C_2 be two clauses, $\lambda \geq 0.5$, $P_{\lambda R}(C_1, C_2)$ is a binary λ-level rough paramodulation of C_1 and C_2, thus we have

$$C_1 \wedge C_2 \rightarrow P_{\lambda R}(C_1, C_2).$$

This theorem shows that given two clauses are λ-level rough true[11], so is the result $P_{\lambda R}(C_1, C_2)$ obtained by λ-level rough paramodulation of C_1 and C_2).
Proof: No less general, let C_1 is $\lambda_1 L[t] \vee C_1'$ and $u_{\lambda RI}(C_1) > \lambda$; C_2 is $\lambda_2(r =_{\lambda R} s) \vee C_2'$ and $u_{\lambda RI}(C_2) > \lambda$, where $\lambda_2 > \lambda$. A mgu of t and r is σ, $u_{\lambda RI}$ is an assignment function to the formulas in interpretation of the rough logic[3,4,5,6,7,11]. we need only to prove $u_{\lambda RI}(C_1 \wedge C_2) > \lambda$ and $u_{\lambda RI}(P_{\lambda R}(C_1, C_2)) > \lambda$ respectively. $u_{\lambda RI}(C_1 \wedge C_2) > \lambda$ is obvious. Thus, we discuss only following two cases:
(1) $\lambda_1 > \lambda$.   If $_{\lambda RI}(L^\sigma[t^\sigma]) < 0.5$, $u_{\lambda RI}(\phi) < 0.5$, it means that ϕ is false[3,4,7,11], due to $t^\sigma =_{\lambda R} r^\sigma$ and $r^\sigma =_{\lambda R} s^\sigma$, so is $u_{\lambda RI}(L^\sigma[s^\sigma]) < 0.5$ and $u_{\lambda RI}(C_1') > \lambda$.

$$u_{\lambda RI}(P_{\lambda R}(C_1, C_2)) = u_{\lambda RI}(\lambda^* L^\sigma[s^\sigma] \vee C_1'^\sigma \vee C_2'^\sigma) > \lambda.$$

If $u_{\lambda RI}(L^\sigma[t^\sigma]) > 0.5$, $u_{\lambda RI}(\phi) > 0.5$, it means that ϕ is true[3,4,7,11], de to $t^\sigma =_{\lambda R} r^\sigma$ and $r^\sigma =_{\lambda R} s^\sigma$, so is $u_{\lambda RI}(L^\sigma[s^\sigma]) > 0.5$ and $\lambda^* > \lambda$ by definition of λ^* the above. Hence,

$$u_{\lambda RI}(P_{\lambda R}(C_1, C_2)) = u_{\lambda RI}(\lambda^* L^\sigma[s^\sigma] \vee C_1'^\sigma \vee C_2'^\sigma) > \lambda.$$

(2) $\lambda_1 < \lambda$.   Similar, we may also have $u_{\lambda RI}(P_{\lambda R}(C_1, C_2)) > \lambda$. Therefore,

$$C_1 \wedge C_2 \rightarrow P_{\lambda R}(C_1, C_2)$$

The proof is finished.

5   Conclusion

λ-level rough equality relation is used in rough logic, it will partition a grade for the approximate concept, namely it gives an approximate precision in approximate theory. This rough equality relation conjunctive $=_{\lambda R}$ is introduced in rough logic, we may have an axiomatic system with λ-level equality $=_{\lambda R}$. It can be used in the rough reasoning, we introduce λ-level rough paramodulation and its inference methods. In fact, it is a substitution by λ-level rough equality or we call it λ-level rough paramodulation reasoning.

The further works in the paper will be to study the combining strategies between λ-level rough paramodulation and rough resolution[4,7,8]. It is possible to generate some new resolution strategies and reasoning methods, which will raise the speed and efficiency of resolution algorithm running.

Acknowledgement

This study is supported by National Nature Science Fund #69773001 and JiangXi Province Nature Science Fund #9911027 in China.

References

1. Z. Pawlak, Rough Sets, Theoretical Aspects of Reasoning about Data, *Dordrecht: Kluwer* (1991).
2. M. Banerjee and M. K. Chakraborty, Rough Algebra, *Ics Research Report, Institute of Computer Science, Warsaw University of Technology,*47/93.
3. Q. Liu, The Resolution for Rough Propositional Logic with Lower (L) and Upper (H) Approximate Operators,*LNAI 1711, Springer,*11,(1999), 352-356.
4. Q. Liu, Operator Rough Logic and Its Resolution Principles, *Journal of Computer,*5, (1998), 476-480,(in Chinese).
5. A. Nakamura, Graded Modalities in Rough Logic, *Rough Sets in Knowledge Discovery I, by L. Polkaowski and A. Skowron, Phsica-verlag,Heidelberg,* (1998).
6. J. Stepaniuk, Rough Relation and Logics, *Rough Sets in Knowledge Discovery I, by L. Polkaowski and A. Skowron, Physica-Verlag, Heidelberg,*(1998).
7. Q. Liu, The OI-Resolution of Operator Rough Logic, *LNAI 1424, Springer,* **6,** (1998), 432-435.
8. C. L. Chang and R. C. T. Lee, Symbolic Logic and Mechanical Theorem Proving, *Academic Press., INC,*1973, 163-180.
9. T. Y. Lin and Q. Liu, First-order Rough Logic I: Approximate Reasoning via Rough Sets,*Fundamenta Informaticae, Vol.27, No.2,3,8* (1996),137-154.

10. E. Orlowska, A logic of Indiscernibility Relation, *Lecture Notes on Computer Science*, **208,** (1985), 172-186.

11. Y. Y. Yao and Q. Liu, A Generalized Decision Logic in Interval-Set-Valued Information Tables, *LNAI 1711, Springer,***11,**(1999),285-293.

12. Z. Pawlak, Rough Logic, *Bulletin of the Polish Academy of Sciences Techniques,*(1987), 253-258.

13. X. H. Liu, Fuzzy Logic and Fuzzy Reasoning,*Publisher of JiLin University,* **6,** (1989),(in Chinese).

14. Q. Liu and Q. Wang, Rough Number Based on Rough Sets and Logic Values of λ Operator, *Journal of Software,* **Sup.,** (1996),455-461), (in Chinese).

15. Q. Liu, The Course of Artificial Intelligence, *High College Join Publisher in JiangXi,* **9,**(1992), (in Chinese).

16. Q. Liu, Accuracy Operator Rough Logic and Its Resolution Reasoning, *In: Proc. RSFD'96, Inc. Conf. The University of Tokyo, Japan,* (1996), 55-59.

17. T. Y. Lin, Q. Liu and Y. Y. Yao, Logic Systems for Approximate Reasoning: Via Rough Sets and Topology,*LNAI 869,* 1994, 65-74.

18. R. C. T. Lee, Fuzzy Logic and the Resolution Principles, *Journal of the Association for Computing Machinery, Vol.19, No. 1,***1,** (1972), 109-119.

19. R. C. T. Lee and C. L. Chang, Some Properties of Fuzzy Logic, *Information Control,* **5,** (1971).

Annotated Semantics for Defeasible Deontic Reasoning

Kazumi Nakamatsu[1], Jair Minoro Abe[2] and Atsuyuki Suzuki[3]

[1]Himeji Institute of Technology, Shinzaike HIMEJI 670-0092 JAPAN
nakamatu@hept.himeji-tech.ac.jp
[2]Faculdade SENAC de Ciencias Exatas e Tecnologia, Rua Galvao Bueno, 430
01506-000 Sao Paulo SP-BRAZIL
jmabe@lsi.usp.br
[3]Shizuoka University, Johoku 3-5-1, HAMAMATSU 432-8011 JAPAN
suzuki@cs.inf.shizuoka.ac.jp

Abstract. In this paper, we propose an annotated logic program called an EVALPSN (Extended Vector Annotated Logic Program with Strong Negation) to formulate the semantics for a defeasible deontic reasoning proposed by D.Nute. We propose a translation from defeasible deontic theory into EVALPSN and show that the stable model of EVALPSN provides an annotated semantics for D.Nute's defeasible deontic logic. The annotated semantics can provide a theoretical base for an automated defeasible deontic reasoning system.

keywords : annotated logic, defeasible deontic logic, extended vector annotated logic program with strong negation, stable model

1 Introduction

Various non-monotonic reasonings are used in intelligent systems and the treatment of inconsistency is becoming important. Moreover, it is indispensable to deal with more than two kinds of non-monotonic reasoning having different semantics in such intelligent systems. If we consider computer implementation of such complex intelligent systems. we need a theoretical framework for dealing with some non-monotonic reasonings and inconsistency uniformly. We take annotated logic programming as a theoretical framework. Actually we have provided some annotated logic program based semantics for some non-monotonic reasonings such as Reiter's default reasoning, Dressler's non-monotonic ATMS, etc. in ALPSN(Annotated Logic Program with Strong Negation) [7, 8] and Billington's defeasible reasoning [10, 1] in VALPSN(Vector Annotated Logic Program with Strong Negation), which is a new version of ALPSN.

In this paper, we provide a theoretical framework based on EVALPSN (Extended VALPSN) for dealing with D.Nute's defeasible deontic reasoning. We propose a translation from defeasible deontic theory into EVALPSN. By the translation, the derivability of the defeasible deontic logic can be translated into the satisfiability of EVALPSN stable models, and the derivability can be computed by the computation of the corresponding EVALPSN stable models. Therefore the stable models of EVALPSN can be an annotated semantics for the

W. Ziarko and Y. Yao (Eds.): RSCTC 2000, LNAI 2005, pp. 470–478, 2001.

defeasible deontic logic and the annotated semantics can provide a theoretical foundation for an automated defeasible deontic reasoning system based on the translation and the computation of EVALPSN stable models.

Defeasible logics are well-known formalizations of defeasible reasoning, however, some of them do not have appropriate semantics. In [9], we have shown that Billington's defeasible theory [1] can be translated into V ALPSN and VALPSN can deal with defeasible reasoning. On the other hand, deontic logics are formalizations of normative reasoning and have been developed as a special case of modal logic. Moreover, they are also focused on as tools of modeling legal argument [12] and some applications to computer science have been realized [6, 5]. D.Nute introduced a defeasible deontic logic based on both defeasible logic and deontic logic, which can deal with both defeasible and normative reasoning [11].

2   VALPSN and EVALPSN

We have formally defined VALPSN and its stable model semantics in [9]. After reviewing VALPSN briefly, we describe the extended parts of EVALPSN.

Generally, a truth value called an annotation is explicitly attached to each atom of annotated logic. For example, let p be an atom, μ an annotation, then $p:\mu$ is called an annotated atom. A partially ordered relation is defined on the set of annotations which constitutes a complete lattice structure. An annotation in VALPSN is 2 dimensional vector such that its components are non-negative integers. We assume the complete lattice of vector annotations as follows :

$$\mathcal{T}_v = \{(x,y)|\ 0 \leq x \leq m, 0 \leq y \leq m, x, y \text{ and } m \text{ are in tegers}\}.$$

The ordering of $\mathcal{T}_v$ is denoted in the usual fashion by a symbol $\preceq_v$. Let $\mathbf{v_1} = (x_1, y_1)$ and $\mathbf{v_2} = (x_2, y_2)$.

$$\mathbf{v_1} \preceq_v \mathbf{v_2} \quad \text{iff} \quad x_1 \leq x_2 \text{ and } y_1 \leq y_2.$$

In a v ector annotated literal $p:(i, j)$, the first component i of the vector annotation (i, j) indicates the degree of positive information to support the literal p and the second one j indicates the degree of negative information. For example, a vector annotated literal $p:(3, 2)$ can be informally interpreted that p is known to be true of strength 3 and false of strength 2.

Annotated logics have two kinds of negations, an epistemic negation($\neg$) and an ontological negation($\sim$). The epistemic negation follo w edb y an annotated atom is a mapping betw een annotations and the ontological negation is a strong negation which appears in classical logics. The epistemic negation of vector annotated logic is defined as the following exc hange betw een the components of v ector annotations,

$$\neg(p : (i, j)) = p : \neg(i, j) = p : (j, i).$$

Therefore, the epistemic negation follow ed b y annotated atomic formulas can be eliminated by the above syntactic operation. On the other hand, the epistemic one followd by a non-atom is interpreted as strong negation [2].

Definition 1(Strong Negation, $\sim$)   Let A be an arbitrary formula in VALPSN.

$$\sim A =_{def} A \to (\neg(A \to A) \wedge (A \to A)).$$

Definition 2 (V ALPSN)   If $L_0, \cdots, L_n$ are v ector annotated literals,

$$L_1 \wedge \cdots \wedge L_i \wedge \sim L_{i+1} \wedge \cdots \wedge \sim L_n \to L_0$$

is called a *vector annotated logic program clause with strong negation*(VALPSN clause). A VALPSN is a finite set of VALPSN clauses.

We assume that all interpretations of a V ALPSN P ha ve a Herbrand base B_P(the set of all v ariable-free atoms) under consideration as their domain of interpretation. A Herbrand interpretation can be considered to be a mapping I : $B_P \to \mathcal{T}_v$. Usually, I is denoted by the set $\{p : \sqcup \mathbf{v_i} | I \models p : \mathbf{v_1} \wedge \cdots \wedge p : \mathbf{v_n}\}$, where $\sqcup \mathbf{v_i}$ is the least upper bound of $\{\mathbf{v_1}, \ldots, \mathbf{v_n}\}$. The ordering $\preceq_v$ is extended to interpretations in the natural way. Let I_1 and I_2 be any interpretations, and A be an atom.

$$I_1 \preceq_I I_2 =_{def} (\forall A \in B_P)(I_1(A) \preceq_v I_2(A)).$$

In order to provide the stable model semantics for VALPSN, we define a mapping T_P betw een Herbrand interpretations associated with every VALP(Vector Annotated Logic Program without strong negation) P.

$$T_P(I)(A) = \sqcup\{\mathbf{v} \mid B_1 \wedge \cdots \wedge B_m \to A : \mathbf{v} \text{ is a ground instance of}$$
$$\text{a VALP clause in } P \text{ and } I \models B_1 \wedge \cdots \wedge B_m \},$$

where the notation $\sqcup$ denotes the least upper bound. We define a special interpretation Δ which assigns the truth value $(0, 0)$ to all members of B_P. Then, the upw ard iteration $T_P \uparrow \lambda$ of the mapping T_P is defined as :

$$T_P \uparrow 0 = \Delta$$
$$T_P \uparrow \lambda = \sqcup_{\alpha < \lambda} T_P(T_P \uparrow \alpha) \text{ for an y ordinals } \alpha, \lambda.$$

Now we describe the Gelfond-Lifschitz transformation [3] for VALPSN. Let I be any interpretation and P a VALPSN. P^I, the *Gelfond-Lifschitz transformation* of the VALPSN P with respect to I, is a VALP obtained from P by deleting

1) each clause that has a strongly negated vector annotated literal $\sim (C : \mathbf{v})$ in its body with $I \models (C : \mathbf{v})$, and

2) all strongly negated vector annotated literals in the bodies of the remaining V ALPSN clauses.

Since P^I con tains no strong negation, P^I has a unique least model that is given by $T_{P^I} \uparrow \omega$ [4]

Definition 3(Stable Model of VALPSN) If I is a Herbrand interpretation of a V ALPSN P, I is called a *stable model* of P *iff* $I = T_{P^I} \uparrow \omega$.

The main difference between VALPSN and EVALPSN is annotation and its lattice structure. An annotation of EVALPSN has a form of $[(i, j), \mu]$ such that the first component (i, j) is a 2-dimentional vector as same as that of VALPSN and the second one μ is a kind of indices which expresses fact(α), obligation(β), and so on. An annotation of EVALPSN is called an extended vector annotation. We assume the complete lattice $\mathcal{T}$ of extended vector annotations as follo ws : $\mathcal{T} = \mathcal{T}_v \times \mathcal{T}_d$, where

$$\mathcal{T}_v = \{(x, y)|\ 0 \le x \le 3,\ 0 \le y \le 3\} \quad \text{and} \quad \mathcal{T}_d = \{\perp, \alpha, \beta, \gamma, *_1, *_2, *_3, \top\},$$

where x and y are integers. The ordering of $\mathcal{T}_d$ is denoted by a symbol $\preceq_d$ and described by the Hasse's diagram(cube) in **Fig.1**. The intuitiv e meanings of the members of $\mathcal{T}_d$ are : $\perp$ (unknown), α (fact), β (obligation), γ (not obligation), $*_1$ (both fact and obligation), $*_2$ (both obligation and not obligation), $*_3$ (both fact and not obligation) and $\top$ (inconsistent). The diagram $\mathcal{T}_d$ shows that $\mathcal{T}_d$ is a trilattice in which the direction of a line $\gamma\beta$ indicates *de ontic truth*, the direction of a line $\perp *_2$ indicates *deontic knowledge* and the direction of a line $\perp\alpha$ indicates *actuality*. The ordering of $\mathcal{T}$ is denoted by a symbol $\preceq$ and defined as follows : let $[(i_1, j_1), \mu_1]$ and $[(i_2, j_2), \mu_2]$ be extended vector annotations,

$$[(i_1, j_1), \mu_1] \preceq [(i_2, j_2), \mu_2] \quad \textit{iff} \quad (i_1, j_1) \preceq_v (i_2, j_2) \text{ and } \mu_1 \preceq_d \mu_2.$$

We provide an intuitive interpretation for some members of $\mathcal{T}$. For example, an extended vector annotatedliteral $p : [(3, 0), \alpha]$ can be informally interpreted as "it is known that it is a fact that p is true of strength 3", and $q : [(0, 2), \beta]$ can be also informally interpreted as "it is known that it is obligatory that q is false of strength 2.

There are tw o kinds of epistemic negation,$\neg_1$ and $\neg_2$, in the extended vector annotated logic, which are regarded as mappings over $\mathcal{T}_v$ and $\mathcal{T}_d$, respectively.
Definition 4(Epistemic Negation of EVALPSN)

$$\forall \mu \in \mathcal{T}_d \qquad \neg_1([(i, j), \mu]) = [(j, i), \mu],$$
$$\neg_2([(i, j), \perp]) = [(i, j), \perp],\ \neg_2([(i, j), \alpha]) = [(i, j), \alpha],$$
$$\neg_2([(i, j), \beta]) = [(i, j), \beta],\ \neg_2([(i, j), \gamma]) = [(i, j), \gamma],$$
$$\neg_2([(i, j), *_1]) = [(i, j), *_3],\ \neg_2([(i, j), *_2]) = [(i, j), *_2],$$
$$\neg_2([(i, j), *_3]) = [(i, j), *_1],\ \neg_2([(i, j), \top]) = [(i, j), \top].$$

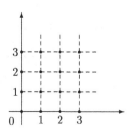 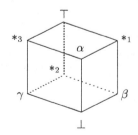

Fig. 1.   Lattice $\mathcal{T}_v$ and   Lattice $\mathcal{T}_d$

We can eliminate syntactically the epistemic negation followed by annotated atom based on the definition. We can also define the strong(ontological) negation ($\sim_s$) in EVALPSN by the epistemic negations as well as **Definition 1**. The formal interpretations of the epistemic negations and the strong negation can be defined as well as the case of ALPSN [7, 2].

Definition 5(EVALPSN) If $L_0, \cdots, L_n$ are extended vector annotated literals,

$$L_1 \wedge \cdots \wedge L_i \wedge \sim_s L_{i+1} \wedge \cdots \wedge \sim_s L_n \rightarrow L_0$$

is called a *extended vector annotated logic program clause with strong negation*(EVALPSN clause). An EVALPSN is a finite set of EVALPSN clauses.

3   Defeasible Deontic Logic

We introduce D.Nute's defeasible deontic logic in [11]. A literal is an y atomic formula or its negation. If ϕ is a literal, then $\bigcirc\phi$ and $\sim \bigcirc\phi$ are deontic formulas. All and only literals and deontic formulas are formulas. If ϕ is a literal, ϕ and $\sim \phi$ are called the complements of each other, and $\bigcirc\phi$ and $\sim \bigcirc\phi$ are called the complements eac h other. $\neg\phi$ denotes the complement of an y formula ϕ, positiv e or negatie. Rules are expressions distinct from formulas. Rules are constructed by using three primitive symbols : $\rightarrow$, $\Rightarrow$, and $\rightsquigarrow$. If $A \cup \{\phi\}$ is a set of formulas, $A \rightarrow \phi$ is a *strict rule*, $A \Rightarrow \phi$ is a *defeasible rule*, and $A \rightsquigarrow \phi$ is an *undercutting defeater.* In each case, we call A the *antecedent* of the rule and ϕ the *consequent* of the rule. If $A = \{\psi\}$, we denote $A \rightarrow \phi$ as $\psi \rightarrow \phi$, and similary for defeasible rules and defeaters. Antecedents for strict rules and defeaters must be non-empty, and antecedents for defeasible rules may be empty. We call a rule of form $\emptyset \Rightarrow \phi$ a *presumption* and represent it more simply as $\Rightarrow \phi$. All rules are read as 'if-then' statements. We read $A \Rightarrow \phi$ as "If A, then evidently(normally, typically, presumably) ϕ", we read $\Rightarrow \phi$ as "Presumably, ϕ", and we read $A \rightsquigarrow \phi$ as "If A, it might be that ϕ". The role of defeater is only to interfere with the process of drawing an inference from a defeasible rule. Defeaters never support inferences directly although they may support inferences indirectly by undercutting potential defeaters.

Definition 6 (Defeasible Theory)   A *defeasible theory* is a quadraple $\langle F, R, C, \prec \rangle$ such that F is a set of formulas(intuitively F can be regarded as a set of facts), R is a set of rules, C is a set of finite sets of formulas such that for every formula ϕ, either $\{\phi\} \in C$ or $\{\neg\phi\} \in C$, or $\{\phi, \neg\phi\} \in C$, and $\prec$ is an acyclic binary relation on the non-strict rules in R. The members of C are called *conflict sets.*

The ultimate purpose of the set of conflict sets is to determine sets of competing rules. We introduce some notations. There are four kinds of defeasible consequence relation : strict derivability($\vdash$), strict refutability($\dashv$), defeasible derivability($\vDash$), and defeasible refutability($\sim$). In Nute[11], firstly he presented a defeasible logic which includes deontic operators in its language, then in order to provide a defeasible deontic logic he extended the proof theory for the original defeasible logic to include some inference conditions, deontic inheritance, deontic detachment, and so on. We introduce only some of the inference conditions.

Definition 7(DSD-proof) A **DSD**-proof in a defeasible theory $\langle F, R, \mathcal{C}, \prec \rangle$ is a sequence σ of defeasible assertions such that for each $k \leq l(\sigma)$, one of inference conditions [**M**$^+$], [**DM**$^+$], [**DM**$^-$], [**E**$^+$], [**SS**$_\mathbf{D}^+$], [**DSS**$^+$], [**DSD**$^+$], [**DDD**$^+$] and [**DSD**$-$] holds. A defeasible theory having **DSD**-proof is called a *defeasible deontic theory*.

[**M**$^+$] $\sigma_k = T \vdash \phi$ and either $\phi \in F$ or there is $A \rightarrow \phi \in R$ such that $T \vdash A$ succeeds at σ_k.

[**DM**$^+$] $\sigma_k = T \vdash \bigcirc\phi$ and there is $A \rightarrow \phi \in R$ or $A \rightarrow \bigcirc\phi \in R$ such that $T \vdash \bigcirc A$ succeeds at σ_k.

[**DM**$^-$] $\sigma_k = T \dashv \phi$,

1. $\phi \notin F$,
2. for each $A \rightarrow \phi \in R$, $T \dashv A$ succeeds at σ_k, and
3. if $\phi = \bigcirc\psi$, then for each $A \rightarrow \psi \in A$ or $A \rightarrow \bigcirc\psi \in R$, $T \dashv \bigcirc A$ succeeds at σ_k.

[**E**$^+$] $\sigma_k = T \mathrel{\mid\!\sim} \phi$ and $T \vdash \phi$ succeeds at σ_k.

[**SS**$_\mathbf{D}^+$] $\sigma_k = T \mathrel{\mid\!\sim} \phi$ and there is $A \rightarrow \phi \in R$ such that

1. $T \mathrel{\mid\!\sim} A$ succeeds at σ_k,
2. for each $\phi \asymp_T C_\phi$ there is $\psi \in C_\phi$ such that $T \dashv \psi$ succeeds at σ_k, and
3. for each $\phi \asymp_T C_\phi$ and $\mathcal{C}$ d-covering C_ϕ in T such that every rule in C_R is strict, either
   (a) there is a literal $\psi \in C_\phi$ and $B \rightarrow \psi \in C_R$ such that $T \sim |B$ succeeds at σ_k,
   (b) there is $\bigcirc\psi \in C_\phi$ and $B \rightarrow \psi \in C_R$ such that $T \mathrel{\rightarrowtail} \bigcirc B$ succeeds at σ_k, or
   (c) there is $\bigcirc\psi \in C_\phi$ and $B \rightarrow \bigcirc\psi \in C_R$ such that both $T \sim |B$ and $T \mathrel{\rightarrowtail} \bigcirc B$ succeed at σ_k.

[**DSS**$^+$], [**DSD**$^+$], [**DDD**$^+$] and [**DSD**$-$] can be found in [11].

4   From Defeasible Deontic Theory into EVALPSN

We assume the following correspondence as a basis of a translation from a defeasible deontic theory into an EVALPSN.

[**Assumption**] Let $T = \langle F, R, \mathcal{C}, \prec \rangle$ be a defeasible deontic theory, I be the stable model of an EVALPSN which is the translation of T, and ϕ be a literal.

$T \vdash \phi$	iff	$I \models \phi : [(3,0), \alpha]$,	$T \dashv \phi$	iff	$I \not\models \phi : [(3,0), \alpha]$,
$T \vdash \bigcirc\phi$	iff	$I \models \phi : [(3,0), \beta]$,	$T \dashv \bigcirc\phi$	iff	$I \not\models \phi : [(3,0), \beta]$,
$T \vdash\sim \bigcirc\phi$	iff	$I \models \phi : [(3,0), \gamma]$,	$T \dashv\sim \bigcirc\phi$	iff	$I \not\models \phi : [(3,0), \gamma]$,
$T \mathrel{\mid\!\sim} \phi$	iff	$I \models \phi : [(2,0), \alpha]$,	$T \sim |\phi$	iff	$I \not\models \phi : [(2,0), \alpha]$,
$T \mathrel{\mid\!\sim} \bigcirc\phi$	iff	$I \models \phi : [(2,0), \beta]$,	$T \mathrel{\rightarrowtail} \bigcirc\phi$	iff	$I \not\models \phi : [(2,0), \beta]$,
$T \mathrel{\mid\!\sim}\sim \bigcirc\phi$	iff	$I \models \phi : [(2,0), \gamma]$,	$T \sim |\sim \bigcirc\phi$	iff	$I \not\models \phi : [(2,0), \gamma]$.

It is considered that the negation $\sim$ followed by a literal is translated to the epistemic negation $\neg_1$ in EVALPSN and the negation $\sim$ followed by a deontic operator $\bigcirc$ is translated into the epistemic negation $\neg_2$ as well. The intuitiv e meanings of some translated EVALPSN clauses are as follows :

$\phi : [(3,0),\alpha]$ ϕ is strictly derivable as a fact,

$\phi : [(2,0),\beta]$ $\bigcirc\phi$ is defeasibly derivable,

$\phi : [(1,0),\gamma]$ $\sim\bigcirc\phi$ is not defeasibly derivable,

$\phi : [(0,0),\alpha]$ ϕ is unknown as a fact.

$I \models \phi : [(1,0),\gamma]$ expresses that the antecedent of a rule in which the consequent is $\sim \bigcirc\phi$ is deriv ablebut the consequent is not deriv ablebecause the rule is defeated b y other conflicting rules. The vector annotation $(1,0)$ indicates not only the refutation of ϕ but also a defeated evidence of defeasible reasoning and we will utilize such information for inductive defeasible deontic reasoning in the future.

Based on the assumption, the translation rule is given for facts, obligations, and four kinds of rules, $A \rightarrow \phi$, $A \rightarrow \bigcirc\phi$, $A \Rightarrow \phi$ and $A \Rightarrow \bigcirc\phi$. Since defeaters $A \rightsquigarrow \phi$ or $A \rightsquigarrow \bigcirc\phi$ cannot be used to derive the consequents ϕ or $\bigcirc\phi$, they do not need to be translated. We describe the translation only for facts and strict rules.

[T ranslationRule] Let $T = \langle F, R, \mathcal{C}, \prec \rangle$ be a defeasible deontic theory, and ϕ and ψ be literals.

[1] for $\phi, \bigcirc\phi, \sim\bigcirc\phi \in F$, they are translated into

$$\phi : [(3,0),\alpha], \phi : [(3,0),\beta] \text{ and } \phi : [(3,0),\gamma], \text{ respectively.}$$

Let $A = \{a_1, \bigcirc a_2\}$ and $B = \{b_1, \bigcirc b_2\}$ for simplicity.

[2] for $r = A \rightarrow \phi \in R$. From **[M$^+$]**. Suppose $\exists A \rightarrow \phi \in R$. Since $T \vdash A$ implies $T \vdash \phi$, the rule r is translated into

$$a_1 : [(3,0),\alpha] \wedge a_2 : [(3,0),\beta] \rightarrow \phi : [(3,0),\alpha].$$

F rom **[DM$^+$]**. Suppose $\exists A \rightarrow \phi \in R$. Since $T \vdash \bigcirc A$ implies $T \vdash \bigcirc\phi$, the rule r is translated into

$$a_1 : [(3,0),\beta] \wedge a_2 : [(3,0),\beta] \rightarrow \phi : [(3,0),\beta].$$

F rom **[SS$^+_{\mathbf{D}}$]**. Suppose $\exists \psi \in C_\phi$ and $\exists B \rightarrow \psi \in C_R$. From 3.(a), since $T \hspace{1mm}\vdash\hspace{-2mm}\sim A$, $T \dashv \psi$ and $T \sim |B$ imply $T \hspace{1mm}\vdash\hspace{-2mm}\sim \phi$, the rule r is translated into

$$a_1 : [(2,0),\alpha] \wedge a_2 : [(2,0),\beta] \wedge \sim_s \psi : [(3,0),\alpha] \wedge \sim_s b_1 : [(2,0),\alpha] \rightarrow \phi : [(2,0),\alpha]$$
$$a_1 : [(2,0),\alpha] \wedge a_2 : [(2,0),\beta] \wedge \sim_s \psi : [(3,0),\alpha] \wedge \sim_s b_2 : [(2,0),\beta] \rightarrow \phi : [(2,0),\alpha].$$

Suppose $\exists \bigcirc \psi \in C_\phi$ and $\exists B \rightarrow \psi \in C_R$. From 3.(b), since $T \hspace{1mm}\vdash\hspace{-2mm}\sim A$, $T \dashv \psi$ and $T \hspace{1mm}\dashv\hspace{-1mm}\sim \bigcirc B$ imply $T \hspace{1mm}\vdash\hspace{-2mm}\sim \phi$, the rule r is translated into

$$a_1 : [(2,0),\alpha] \wedge a_2 : [(2,0),\beta] \wedge \sim_s \psi : [(3,0),\alpha] \wedge \sim_s b_1 : [(2,0),\beta] \rightarrow \phi : [(2,0),\alpha]$$
$$a_1 : [(2,0),\alpha] \wedge a_2 : [(2,0),\beta] \wedge \sim_s \psi : [(3,0),\alpha] \wedge \sim_s b_2 : [(2,0),\beta] \rightarrow \phi : [(2,0),\alpha].$$

Suppose $\exists\, \bigcirc \psi \in C_\phi$ and $\exists B \to \bigcirc\psi \in C_R$. From 3.(b), since $T \mathrel{\vdash\!\!\!\sim} A$, $T \dashv \psi$, $T \sim |B$ and $T \mathrel{\sim\!\!\!\vdash} \bigcirc B$ imply $T \mathrel{\vdash\!\!\!\sim} \phi$, the rule r is translated into

$a_1 : [(2,0),\alpha] \wedge a_2 : [(2,0),\beta] \wedge \sim_s \psi : [(3,0),\alpha] \wedge \sim_s b_1 : [(2,0),\alpha] \wedge \sim_s b_1 : [(2,0),\beta]$
$\to \phi : [(2,0),\alpha]$

$a_1 : [(2,0),\alpha] \wedge a_2 : [(2,0),\beta] \wedge \sim_s \psi : [(3,0),\alpha] \wedge \sim_s b_1 : [(2,0),\alpha] \wedge \sim_s b_2 : [(2,0),\beta]$
$\to \phi : [(2,0),\alpha]$

$a_1 : [(2,0),\alpha] \wedge a_2 : [(2,0),\beta] \wedge \sim_s \psi : [(3,0),\alpha] \wedge \sim_s b_1 : [(2,0),\beta] \wedge \sim_s b_2 : [(2,0),\beta]$
$\to \phi : [(2,0),\alpha]$

$a_1 : [(2,0),\alpha] \wedge a_2 : [(2,0),\beta] \wedge \sim_s \psi : [(3,0),\alpha] \wedge \sim_s b_2 : [(2,0),\beta] \to \phi : [(2,0),\alpha].$

We take an example of defeasible deontic theories from [11] called *Chisholm Example*. Then we translate it into an EVALPSN and compute the stable model of the EVALPSN.

Example (*Chisholm Example*) We consider the following situation.

1. Jones ought to visit his mother.
2. If Jones visits his mother, he ought to call her and tell her he is coming.
3. It ought to bethat if Jones does not visit his mother, he does not call her and tell her he is coming.
4. In fact, Jones does not visit his mother.

If w e formalize 1-4 as a defeasible deontic theory, it includes :

$$1.\ \bigcirc v \qquad 2.\ v \Rightarrow \bigcirc c \qquad 3.\ \sim v \Rightarrow \bigcirc \sim c \qquad 4.\ \sim v$$

The above formulas $1 - 4$ are translated into EVALPSN clauses according to the translation rules. $\bigcirc v$ and $\sim v$ are translated into $v : [(3,0),\beta]$ and $v : [(0,3),\alpha]$, respectively. The defeasible rule $v \Rightarrow \bigcirc c$ is translated into

$$v : [(2,0),\alpha] \wedge \sim_s c : [(0,3),\beta] \wedge \sim_s v : [(0,2),\alpha] \to c : [(2,0),\beta]$$
$$v : [(2,0),\beta] \wedge \sim_s c : [(0,3),\beta] \wedge \sim_s v : [(0,2),\alpha] \to c : [(2,0),\beta]$$
$$v : [(2,0),\beta] \wedge \sim_s c : [(0,3),\alpha] \wedge \sim_s v : [(0,2),\beta] \to c : [(2,0),\beta]$$
$$v : [(2,0),\alpha] \wedge \sim_s c : [(0,3),\beta] \wedge v : [(0,2),\alpha] \to c : [(1,0),\beta]$$
$$v : [(2,0),\beta] \wedge \sim_s c : [(0,3),\beta] \wedge v : [(0,2),\alpha] \to c : [(1,0),\beta]$$
$$v : [(2,0),\beta] \wedge \sim_s c : [(0,3),\alpha] \wedge v : [(0,2),\alpha] \to c : [(1,0),\beta]$$
$$v : [(2,0),\beta] \wedge \sim_s c : [(0,3),\alpha] \wedge v : [(0,2),\beta] \to c : [(1,0),\beta]$$

The defeasible rule $\sim v \Rightarrow \bigcirc \sim c$ is also translated into

$$v : [(0,2),\alpha] \wedge \sim_s c : [(3,0),\beta] \wedge \sim_s v : [(2,0),\alpha] \to c : [(0,2),\beta]$$
$$v : [(0,2),\beta] \wedge \sim_s c : [(3,0),\beta] \wedge \sim_s v : [(2,0),\alpha] \to c : [(0,2),\beta]$$
$$v : [(0,2),\beta] \wedge \sim_s c : [(3,0),\alpha] \wedge \sim_s v : [(2,0),\beta] \to c : [(0,2),\beta]$$
$$v : [(0,2),\alpha] \wedge \sim_s c : [(3,0),\beta] \wedge v : [(2,0),\alpha] \to c : [(0,1),\beta]$$
$$v : [(0,2),\beta] \wedge \sim_s c : [(3,0),\beta] \wedge v : [(2,0),\alpha] \to c : [(0,1),\beta]$$
$$v : [(0,2),\beta] \wedge \sim_s c : [(3,0),\alpha] \wedge v : [(2,0),\alpha] \to c : [(0,1),\beta]$$
$$v : [(0,2),\beta] \wedge \sim_s c : [(3,0),\alpha] \wedge v : [(2,0),\beta] \to c : [(0,1),\beta]$$

Then the EVALPSN has only one stable model

$$\{ \, v\!:\![(3,0),\beta], \; v\!:\![(0,3),\alpha], \; c\!:\![(0,2),\beta], \; c\!:\![(1,0),\beta] \, \},$$

which says that $T \vdash \bigcirc v$, $T \vdash \sim v$, $T \mathrel{\vdash\!\!\!\!-} \bigcirc \sim c$ and $T \mathrel{\rightthreetimes} \bigcirc c$.

5   Conclusion

In this paper, w e have provided an annotated semantics for Nute's defeasible deontic logic. That is one of theoretical frameworks for intelligen t reasoning systems based on annotated logic programming. What we have done in this paper includes to provide a theoretical framework for automated reasoning systems of defeasible deontic logic. Since inference conditions in the defeasible deontic logic are too complicated, ha ving an automated reasoning system for the defeasible deontic logic is quite useful and conv enient. Actually, w e have implemented an EVALPSN based automated reasoning system for the defeasible deontic logic.

References

1. Billington,D. : Conflicting Literals and Defeasible Logic. Proc. 2nd Australian Work-shop on Commonsense Reasoning, (1997) 1–15
2. Da Costa,N.C.A., Subrahmanian,V.S., and Vago,C. : The Paraconsistent Logics $P\mathcal{T}$. Zeitsc hrift für MLGM **37** (1989) 139–148
3. Gelfond,M. and Lifschitz, V. : The Stable Model Semantics for Logic Programming. Proc. 5th Int'l Conf. and Symp. on Logic Programming (1989) 1070–1080
4. Llo yd,J.W. : *F oundations of Logic Programming*(2nd edition). Springer (1987)
5. Meyer,C.J. and Wieringa,J.R. : *Deontic Logic in Computer Science*. John Wiley & Sons (1993)
6. McNamara,P. and Prakken,H.(eds.) : *Norms, Logics and Information Systems*. New Studies in Deontic Logic and Computer Science, Fron tiers in Artificial Intelligence and Applications Vol.49 IOS Press (1999)
7. Nakamatsu,K. and Suzuki, A. : Annotated Semantics for Default Reasoning. Proc. 3rd Pacific Rim International Conference on Artificial Intelligence (1994) 180–186
8. Nakamatsu,K. and Suzuki,A. : A Nonmonotonic ATMS Based on Annotated Logic Programs. Agents and Multi-Agents Systems, LNAI **1441** Springer (1998) 79–93
9. Nakamatsu,K.and Abe, J.M. : Reasonings Based on Vector Annotated Logic Pro-grams. Computational Intelligence for Modelling, Control & Automation, Concurrent Systems Engineering Series **55**. IOS Press (1999) 396–403
10. Nute,D. : Basic Defeasible Logics. In tensional Logics for Programming. Oxford University Press (1992) 125–154
11. Nute,D. : Apparent Obligation. Defeasible Deontic Logic. Kluw er Academic Pub-lisher (1997) 287–316
12. Prakken,H. : *L ogical T ools for Modelling L egal A r gument* A Study of Defeasible Reasoning in Law. Law and Philosophy Library Vol.32 Kuw er Academic (1997)

Spatial Reasoning via Rough Sets

Lech Polkowski

Polish–Japanese Institute of Information Technology
Koszykowa 86, 02-008 Warsaw, Poland
and
Department of Mathematics and Information Sciences
Warsaw University of Technology
Pl. Politechniki 1,00-650 Warsaw, Poland
e-mail: polkow@pjwstk.waw.pl

Abstract. Rough set reasoning may be based on the notion of a part to a degree as proposed in rough mereology. Mereological theories form also a foundation for spatial reasoning. Here we show how to base spatial reasoning on rough–set notions.
Keywords: rough sets, mereology, rough mereology, connection, spatial reasoning

1   Introduction

For expressing relations among entities, computer science has two basic langua-ges: the language of set theory, based on the opposition element–set, where enti-ties are considered as consisting of distinct points, and languages of mereology, based on the opposition part–whole, for discussing entities continuous in their nature.[1] Mereological theories active nowadays go back to ideas of S. Leśniewski and A. N. Whitehead; mereological theory of Leśniewski is based on the notion of a part and the notion of a (collective) class cf. [5]. Mereological ideas of Whi-tehead were formulated as Calculus of Individuals and were expressed in terms of connection cf. [1]. [2] In [13] a new paradigm for approximate reasoning, rough mereology, has been introduced. Rough mereology is based on the notion of a part to a degree and thus falls in the province of part–based mereologies.[3] In this paper we introduce rough mereology in the ontological universe of Leśniewski [5] (cf.[8], [15], [16]). We define Čech quasi–topology [6] and we apply it in a study of connections.[4] We introduce some notions of connection viz. the limit connection C_T and graded connections C_α and we study their properties. In

---

[1] For instance, Spatial Reasoning relies extensively on mereological theories of part cf. [11]

[2] Mereology based on connection gave rise to spatial calculi based on topological no-tions derived therefrom (mereotopology) [3], [11].

[3] Rough mereology inherits the general idea of approximations by means of member-ship functions from rough set theory [12].

[4] A quasi–topology was introduced in the connection model of mereology [1] under additional assumptions of regularity.

W. Ziarko and Y. Yao (Eds.): RSCTC 2000, LNAI 2005, pp. 479–486, 2001.

particular, we demonstrate that they induce the same notion of an element as the original mereology. We also discuss the case of distributed reasoning systems in which connections in potentially infinite information systems may be studied and external connections may be effected in a simple way. Our results show that rough mereology offers an inference mechanism based on connection applicable e.g. to spatial reasoning. [5] The reader will find all notions and bibliography in [14]

2   Preliminaries

We introduce consecutively basic notions of ontology, mereology and rough mereology.

2.1   Ontology

We adopt Ontology introduced by Leśniewski [5] based on the primitive notion of a copula *is* denoted ε whose meaning is expressed by

The Ontology Axiom $X\varepsilon Y \iff \exists Z. Z\varepsilon X \wedge \forall U, W.(U\varepsilon X \wedge W\varepsilon X \implies U\varepsilon W) \wedge \forall Z.(Z\varepsilon X \implies Z\varepsilon Y)$.[6]

We have thus a mechanism to discuss and understand statements like X *is* Y without resorting to intuition. The reader will observe that X *is* Y (formally written down as $X\varepsilon Y$) is analogous to the formula $X \in Y$ if we know that X is a singleton, say $X = \{a\}$ for some a. In ontological setting we do not need to know this: it is encoded in the ontological axiom.

2.2   Mereology

Mereology is a theory of collective classes i.e. individual objects representing distributive classes (names). We adopt the notion of a part as the primitive notion of mereology cf. [5], [10]. We assume that the ontological copula ε is given and that the Ontology Axiom holds. A predicate *pt* of *part* satisfies the following conditions.

(ML1) $X\varepsilon pt(Y) \implies X\varepsilon X \wedge Y\varepsilon Y$ (X, Y are individual entities).

(ML2) $X\varepsilon pt(Y) \wedge Y\varepsilon pt(W) \implies X\varepsilon pt(W)$ (*pt* is transitive).

(ML3) $\forall X.(\neg(Z\varepsilon pt(Z)))$ (*pt* is non–reflexive).

On the basis of the notion of a part, we define the notion of an *element* as a predicate *el* (originally called an *ingredient*).

---

[5] Rough set ideas have already been applied explicitly in Spatial Reasoning e.g. in fields of multi-resolution data management, epistemology of rough location cf. [14] and in "egg-yolk" representation of vague regions [4].

[6] The meaning of $X\varepsilon Y$ can be made clear now: shortly, X is an individual (i.e. any $U, W\varepsilon X$ are such that $U\varepsilon W$ (and vice versa)) and this individual is Y. In particular, $X\varepsilon X$ holds iff X is an individual object.

Definition 2.1 $X \varepsilon el(Y) \iff X \varepsilon pt(Y) \vee X = Y.$[7]

A fundamentally important component of Mereology is the *class functor, Kl*, intended to make distributive classes (names) into collective classes (individuals).[8]

Definition 2.2

$$X \varepsilon Kl(Y) \iff \exists Z. Z \varepsilon Y \wedge \forall Z.(Z \varepsilon Y \implies Z \varepsilon el(X)) \wedge \forall Z.(Z \varepsilon el(X) \implies$$

$$\exists U, W.(U \varepsilon Y \wedge W \varepsilon el(U) \wedge W \varepsilon el(Z))).$$[9]

We now recall mereology based on the notion of a Connection.

2.3 Connection

This approach cf. [1] is based on the functor C of *being connected*.[10]

Definition 2.3

(C1) $X \varepsilon C(Y) \iff X \varepsilon X \wedge Y \varepsilon Y$ (X, Y are individuals).

(C2) $X \varepsilon C(X)$ (reflexivity).

(C3) $X \varepsilon C(Y) \implies Y \varepsilon C(X)$ (symmetry).

(C4) $[\forall Z.(Z \varepsilon C(X) \iff Z \varepsilon C(Y))] \implies (X = Y)$ (extensionality).[11]

From C, other functors are derived; we recall basic for topological issues functors (TP), (NTP).

Definition 2.4

(O) $X \varepsilon O(Y) \iff \exists Z(Z \varepsilon el_C(X) \wedge Z \varepsilon el_C(Y))$ (X and Y overlap).

(EC) $X \varepsilon EC(Y) \iff X \varepsilon C(Y) \wedge \neg(X \varepsilon O(Y))$ (X is externally connected to Y).

(TP) $X \varepsilon TP(Y) \iff X \varepsilon pt_C(Y) \wedge \exists Z(Z \varepsilon EC(X) \wedge Z \varepsilon EC(Y))$ (X is a tangential part of Y).

(NTP) $X \varepsilon NTP(Y) \iff X \varepsilon pt_C(Y) \wedge \neg(X \varepsilon TP(Y))$ (X is a non–tangential part of Y).[12]

2.4 Rough Mereology

Rough mereology is based on the predicate of being a *part to a degree* rendered here as a family μ_r called *a rough inclusion* where $r \in [0, 1]$. The formula $X \varepsilon \mu_r(Y)$ reads X *is a part of* Y *to degree at least* r. We assume an ontology of ε and a mereology inducing a functor el. Basic postulates of rough mereology are as follows.

---

[7] We recall that X, Y are *external to each other*, in symbols $X \varepsilon ext(Y)$ when there is no Z with $Z \varepsilon el(X)$ and $Z \varepsilon el(Y)$.

[8] Mereology we adopt here is thus the classical (maximal) mereology cf. [11].

[9] Thus, $Kl(Y)$ is an individual containing as elements all individuals in Y and such that each of its elements has an element in common with an individual in Y; notice obvious analogies to the union of a family of sets in set theory.

[10] For the uniformity of exposition sake, we will formulate all essentials of this theory in the ontological language applied above.

[11] Notice that some schemes dispense with extensionality cf. [3], [11].

[12] Connection allows for topological notions cf. [1], [11].

(RM0) $\exists r(X\varepsilon\mu_r(Y)) \implies X\varepsilon X \wedge Y\varepsilon Y$ (X and Y are individuals).
(RM1) $X\varepsilon\mu_1(Y) \iff X\varepsilon el(Y)$ (being a part to degree 1 is equivalent to being an element).
(RM2) $X\varepsilon\mu_1(Y) \implies \forall Z.(Z\varepsilon\mu_r(X) \implies Z\varepsilon\mu_r(Y))$ (monotonicity in the object position).
(RM3) $X = Y \wedge X\varepsilon\mu_r(Z) \implies Y\varepsilon\mu_r(Z))$ (the identity is a μ–congruence).
(RM4) $X\varepsilon\mu_r(Y) \wedge s \le r \implies X\varepsilon\mu_s(Y)$ (the meaning *a part to a degree at least* r).[13]

3   Mereotopology

We now are concerned with topological structures arising in a mereological universe endowed with a rough inclusion μ_r.[14]. For a different approach where connection may be derived from the axiomatized notion of a boundary see [16]. We show that in this framework one defines Čech quasi–topologies. [15]

3.1   Mereotopology : Čech topologies

Here, we induce a Čech quasi–topology in any rough mereological universe. [16]
Definition 3.1 (i) We introduce the name $M_r X$ for the property expressed by μ_r with respect to X i.e. $Z\varepsilon M_r X \iff Z\varepsilon\mu_r(X)$; (ii) We let $Z\varepsilon Kl_r(X) \iff Z\varepsilon Kl(M_r X)$.

Thus $Kl_r(X)$ is the class of objects having the property $\mu_r(X)$. We define a functor int.
Definition 3.2 We define a name $I(X)$ by letting

$$Z\varepsilon I(X) \iff \exists s < 1.(Kl_s(Z)\varepsilon el(X))$$

and we let $int(X) = Kl(I(X))$.
Then we have the following properties of int cf. [14].

Proposition 1. *(i)* $int(X)\varepsilon el(X)$; *(ii)* $X\varepsilon el(Y) \implies int(X)\varepsilon el(int(Y))$.

Properties (i)-(ii) witness that int introduces a *Čech quasi–topology*. We denote it by the symbol τ_μ.

---

[13] It follows that μ_1 coincides with the given el establishing a link between rough mereology and mereology while predicates μ_r with $r < 1$ diffuse el to a hierarchy of a part in various degrees. The reader may use as an archetypical rough inclusion the Łukasiewicz measure $\mu(X,Y) = \frac{card(X \cap Y)}{card(X)}$, X, Y being non–empty finite sets in a universe U.

[14] As recalled in Section 2.3 topological structures may be defined within the connection framework via the notion of a non–tangential proper part. The predicate of connection allows also for some calculi of topological character based directly on regions e.g. $RCC--calculus$ [2]

[15] A quasi–topology is a topology without the null element (the empty set).

[16] Recall that a Čech topology [6] is a closure structure in which the closure operator cl satisfies the following (i) $cl\emptyset = \emptyset$ (ii) $X \subseteq clX$ (iii) $X \subseteq Y \implies clX \subseteq clY$, so the associated (by duality $intX = U - cl(U - X)$) Čech interior operator int should only satisfy the following: $int\emptyset = \emptyset$; $intX \subseteq X$; $X \subseteq Y \implies intX \subseteq intY$.

4   Connections from Rough Inclusions

In this section we investigate some methods for inducing connections from rough inclusions.[17]

4.1   Limit Connection

We define a functor C_T as follows.

Definition 4.1 $X \varepsilon C_T(Y) \iff \neg(\exists r, s < 1.ext(Kl_r(X), Kl_s(Y)))$.[18] Clearly, (C1-C3) hold with C_T irrespective of μ. For (C4), consult [14].

4.2   From Graded Connections to Connections

We begin with a definition of an individual $Bd_r X$.

Definition 4.2 $Bd_r X \varepsilon Kl(\mu_r^+(X))$ where $Z \varepsilon \mu_r^+(X) \iff Z \varepsilon \mu_r(X) \wedge \neg(\exists s > r.Z \varepsilon \mu_s(X))$.

We introduce a *graded (r, s)–connection* $C(r, s)$ $(r, s < 1)$ via

Definition 4.3 $X \varepsilon C(r, s)(Y) \iff \exists W.W \varepsilon el(Bd_r X) \wedge W \varepsilon el(Bd_s Y)$.

We have then

Proposition 2. *(i)* $X \varepsilon C(1, 1)(X)$; *(ii)* $X \varepsilon C(r, s)(Y) \implies Y \varepsilon C(s, r)(X)$.

Concerning the property (C4), we adopt here a new approach. It is valid from theoretical point of view to assume that we may have "infinitesimal" parts i.e. objects as "small" as desired.[19]

Infinitesimal parts model We adopt a new axiom of infinitesimal parts (IP) $\neg(X \varepsilon el(Y)) \implies \forall r > 0.\exists Z \varepsilon el(X), s < r.Z \varepsilon \mu_s^+(Y)$.

Our rendering of the property (C4) under (IP) is as follows:

Proposition 3.

$$\neg(X \varepsilon el(Y))$$
$$\implies \forall r > 0.\exists Z, s \geq r.Z \varepsilon C(1, 1)(X) \wedge Z \varepsilon C(1, s)(Y).$$

Connections from Graded Connections Our notion of a connection will depend on a threshold, α, set according to the needs of the context of reasoning. Given $0 < \alpha < 1$, we define a functor C_α as follows.

Definition 4.4 $X \varepsilon C_\alpha(Y) \iff \exists r, s \geq \alpha.X \varepsilon C(r, s)(Y)$.

Then the functor C_α has all the properties of a connection:

Proposition 4. *(IP) For any* α : *(i)* $X \varepsilon C_\alpha(X)$; *(ii)* $X \varepsilon C_\alpha(Y) \implies Y \varepsilon C_\alpha(X)$; *(iii)* $X \neq Y \implies \exists Z.(Z \varepsilon C_\alpha(X) \wedge \neg(Z \varepsilon C_\alpha(Y)) \vee Z \varepsilon C_\alpha(Y) \wedge \neg(Z \varepsilon C_\alpha(X)))$.

---

[17] See Section 2.3 for basic notions related to mereological theories based on the notion of a connection.

[18] Thus, X and Y are connected in the limit sense whenever they cannot be separated by means of their open neighborhoods.

[19] Cf. an analogous assumption in mereology based on connection [11]).

5   Examples

In this section, we will give some examples related to notions presented in the preceding sections.

5.1   Concerning C_α: case of points

Our universe will be selected from a quad–tree in the Euclidean plane formed by squares $[k + \frac{i}{2^s}, k + \frac{i+1}{2^s}] \times [l + \frac{j}{2^s}, l + \frac{j+1}{2^s}]$ where $k, l \in \mathbf{Z}, i, j = 0, 1, ..., 2^s - 1$ and $s = 1, 2, ...$.[20]

The choice of points will depend on the level of granularity of knowledge we assume.[21] We assume that our sensor system perceives each square X as the square X' whose each side length is that of X plus 2α where $\alpha = 2^{-s}$ for some $s > 1$ (we then express uncertainty of location applying 'hazing" of objects cf. [18]). We restrict ourselves to squares with side length at least 4α (as smaller squares may be localized with a too high uncertainty). We let for simplicity $4\alpha = 1$ so points are squares of the form $[k, k+1] \times [l, l+1]$, $k, l \in \mathbf{Z}$. We define μ_r by letting

$$X \varepsilon \mu_r(Y) \Leftrightarrow \frac{\lambda(X' \cap Y')}{\lambda(X')} \geq r$$

where X', Y' are enlargements of X, Y defined above and λ is the area (Lebesgue) measure in the two-dimensional plane. We may check straightforwardly that

Proposition 5. *Functors* μ_r *satisfy (RM1)-(RM5).*

Applying our notion of the connection C_α defined in Section 4, we find that two adjacent squares (e.g. $X = [0, 1] \times [0, 1]$, $Y = [0, 1] \times [1, 2]$) are connected in degree 0.3(3) (i.e. $X \varepsilon C_{0.3(3)}(Y)$) while two squares having one vertex in common (e.g. $X = [0, 1] \times [0, 1]$ and $Y = [1, 2] \times [1, 2]$) are connected in degree 0.1(1) (i.e. $X \varepsilon C_{0.1(1)}(Y)$). Pairs of disjoint squares are connected in degree 0 at most. Observe that in this case C_α with $\alpha > 0$ is a connection even if (IP) does not hold.

5.2   Connections in Distributed Systems

We refer to a model for approximate synthesis in a distributed system M_A proposed in [13] .[22] Reasoning in M_A goes by means of standards and rough inclusions

---

[20] $\mathbf{Z}$ denotes the set of integers.

[21] A point is an object X with the property that $Y \varepsilon el(X) \Longrightarrow Y = X$.

[22] We recall briefly its main ingredients. Consider a distributed (multi-agent) system $M_A = \{Ag, Link, Inv\}$ where Ag is a set of agents, $Link$ is a finite list of words over Ag and Inv is a set of inventory objects. Each t in $Link$ is a word $ag_1 ag_2 ... ag_k ag$ meaning that ag is the parent node and $ag_1 ag_2 ... ag_k$ are children nodes in an elementary team t; both parties are related by means of the operation o_t which makes from a tuple

μ_{ag} at any ag. Instrumental in the reasoning process are *rough connectives* $f_{\sigma,t}$ where $t \in Link$ and $\sigma = (st_1, ..., st_k, st)$ is admissible. They propagate rough inclusion values from children nodes to the parent node according to the formula $\forall i.x_i \varepsilon \mu_{ag_i,r_i}(st_i) \Longrightarrow o_t(x_1, .., x_k) \varepsilon \mu_{ag,f(r_1,..,r_k)}(st)$.

We assume that rough connectives f are cofinal i.e. for any $r < 1$ there are $u_1, ..., u_k < 1$ with $f(u_1, ..., u_k) \geq r$ (this assumption clearly implies some form of potential infinity in data tables). Then C_T is preserved in the sense

Proposition 6. $\forall i.x_i \varepsilon C_{T,\top}(st_i) \Longrightarrow o_t(x_1, .., x_k) \varepsilon C_{T,\top}(st).$

Similarly, one has a minimax formula for C_α.

Proposition 7. $\forall i.x_i \varepsilon C_{\alpha_i}(y_i) \Longrightarrow o_t(x_1, .., x_k) \varepsilon C_{F(\alpha_1,..,\alpha_k)}(o_t(y_1, ..., y_k))$
$$where\ F(\alpha_1, .., \alpha_k) = inf_{i=1}^k \{f(r_1, .., r_k), f(s_1, ..., s_k) : r_i, s_i \geq$$
$\alpha_i\}.$[23]

Case of C_T Consider a system M_A where $Ag = \{ag_0^0\} \cup \bigcup_{i,j=1}^\infty \{ag_i^j\}$. *Link* consists of words of the form $ag_{i_1}^{j_1} ag_{i_2}^{j_2} ... ag_{i_k}^{j_k} ag_i^j$ with $i_1 \leq i_2 \leq .... \leq i_k < i$ i.e. any agent ag_i^j of the level i may form a local team with agents of lower levels $i_1, i_2, ..., i_k$. For any agent $ag = ag_i^j$ of the level i, the universe U_{ag} consists of points i.e. squares of side length 2^{-i} and of classes of squares sent by children agents in local teams with the head ag. Thus, any agent has potentially infinite universe. In particular, the agent ag_0^0 has in its universe all points of side length 1 as well as all classes (unions) of finite collections of squares of smaller size sent by lower level agents. In this context, one may define a connection C_T. We define the rough inclusion μ_{ag} at any agent ag by the formula: $X \in \mu_{ag,r}(Y) \Leftrightarrow \frac{\lambda(X \cap Y)}{\lambda(X)} \geq r$ thus el is $\subseteq$.[24] Then one checks in a direct way that two squares whose union is connected topologically (e.g. $X = [0,1] \times [0,1], Y = [1,2] \times [0,1]$) satisfy the formula $X \in C_T(Y)$ while for X, Y with $Kl(X,Y) = X \cup Y$ not connected topologically we have $\neg(X \in C_T(Y))$.[25],[26]

Acknowledgement *This work has been prepared under the grant no 8T11C 02417 from the State Committee for Scientific Research (KBN) of the Republic of Poland.*

$(x_1, ..., x_k)$ of objects, resp. at $ag_1, .., ag_k$ the object $o_t(x_1, .., x_k)$ at ag; the tuple $(x_1, ..., x_k, o_t(x_1, .., x_k))$ is *admissible*. Leaf agents $Leaf$ are those ag which are not any parent node. They operate on objects from Inv. Each agent ag is equipped with an information system $\mathbf{A_{ag}} = (U_{ag}, A_{ag})$ and a rough inclusion $\mu_a g$ on U_{ag}; a set $St_{ag} \subseteq U_{ag}$ of *standard objects* is also defined for any ag.

[23] These results form a basis for distributed connection calculi to be explored in a future research.

[24] I.e. we apply the formula from Section 5.1 but without hazing.

[25] Thus, in distributed environments it is possible to define naturally an external connection EC (i.e. some objects may connect to each other without overlapping) even with objects simple from the topological point of view.

[26] Observe that (IP) holds in this case.

References

1. B. L. Clarke, A calculus of individuals based on connection, Notre Dame Journal of Formal Logic 22(2), 1981, pp.204-218.
2. A. G. Cohn, Calculi for qualitative spatial reasoning, in: J. Calmet, J. A. Campbell, J. Pfalzgraf (eds.), Artificial Intelligence and Symbolic Mathematical Computation, Lecture Notes in Computer Science, vol. 1138, Springer Verlag, Berlin, pp. 124–143.
3. A. G. Cohn, A. C. Varzi, Connections relations in mereotopology, in: H. Prade (ed.), Proceedings ECAI'98. 13th European Conference on Artificial Intelligence, Wiley and Sons, Chichester, 1998, pp. 150–154.
4. A. G. Cohn, N. M. Gotts, The "egg-yolk" representation of regions with indeterminate boundaries, in: P. Burrough, A. M. Frank (eds.), Proceedings GISDATA Specialist Meeting on Spatial Objects with Undetermined Boundaries, Fr. Taylor, 1996, pp.171–187.
5. J. Srzednicki, S. J. Surma, D. Barnett, V. F. Rickey (eds.), Collected Works of Stanisław Leśniewski, Kluwer, Dordrecht, 1992.
6. E. Čech, Topological Spaces, Academia, Praha, 1966.
7. C. Freksa, D. M. Mark (eds.), Spatial Information Theory. Cognitive and Computational Foundations of Geographic Information Science, Lecture Notes in Computer Science,vol. 1661, Springer Verlag, Berlin, 1999.
8. N. Guarino, The ontological level, in: R. Casati, B. Smith, G. White (eds.), Philosophy and the Cognitive Sciences, Hoelder-Pichler-Tempsky, Vienna, 1994.
9. R. Kruse, J. Gebhardt, F. Klawonn, Foundations of Fuzzy Systems, John Wiley & Sons, Chichester, 1984.
10. S. Leśniewski On the foundations of mathematics, Topoi 2, 1982, pp. 7–52.
11. C. Masolo, L. Vieu, Atomicity vs. infinite divisibility of space, in: [7], pp. 235–250.
12. Z. Pawlak, Rough Sets: Theoretical Aspects of Reasoning about Data, Kluwer, Dordrecht, 1992.
13. L. Polkowski and A. Skowron, Rough mereology: a new paradigm for approximate reasoning, International Journal of Approximate Reasoning 15(4), 1997, pp. 333–365.
14. L. Polkowski, On connection synthesis via rough mereology, Fundamenta Informaticae, to appear.
15. B. Smith, Logic and formal ontology, in: J. N. Mohanty, W. McKenna (eds.), Husserl's Phenomenology: A Textbook, Lanham: University Press of America, 1989, pp. 29–67.
16. B. Smith, Boundaries: an essay in mereotopology, in: L. Hahn (ed.), The Philosophy of Roderick Chisholm, Library of Living Philosophers, La Salle: Open Court, 1997, pp. 534–561.
17. J. G. Stell, Granulation for graphs, in: [7], pp. 416–432.
18. T. Topaloglou, First-order theories of approximate space, in Working Notes of the AAAI Workshop on Spatial and Temporal Reasoning, Seattle, 1994, pp. 47–53.
19. M. F. Worboys, Imprecision in finite resolution spatial data, Geoinformatica, 1998, pp. 257–279.

Decomposition of Boolean Relations and Functions in Logic Synthesis and Data Analysis

P. Sapiecha, H. Selvaraj*, and M. Pleban

Institute of Telecommunication,
Warsaw University of Technology,
Nowowiejska 15/19, 00-665 Warsaw, Poland,
e-mail: sapiecha@tele.pw.edu.pl

*University of Nevada, Las Vegas,
4505, Maryland Parkway, Las Vegas, NV 89154-4026, USA

Abstract. This paper shows that the problem of decomposing a finite function $f(A, B)$ into the form $h(g(A), B)$, where g is a Boolean function, can be resolved in polynomial time, with respect to the size of the problem. It is also shown that omission of the characteristic of the g function can significantly complicate the problem. Such a general problem belongs to the NP-hard class of problems. The work shows how the problem of decomposition of a finite function can be reduced to the problem of coloring the vertices of a graph. It is also shown that the problem of decomposition of relations can be reduced to coloring the vertices of their hypergraphs. In order to prove the validity of the theorems, combinatory properties of Helly are used.

1   Introduction

Decomposition methods of finite functions have wide application in many areas of computer science. One such area is artificial intelligence, in particular, machine learning, logic synthesis and image processing. In machine learning and image processing decomposition methods can be used in the descavering process of some hidden properties of the data [13], [7]. These methods can also be used in the process of compressing decision rules [10], [8]. In logic synthesis, decomposition methods are very effective in FPGA/PLA based digital circuit design [5], [10], [6]. Considering the wide variety of applications of functional decomposition, it becomes necessary to ponder over the computational complexity of certain decomposition problems and the effectiveness of the algorithms used. Apart from these analytical issues, there are other questions like: how general are the existing algorithms? and can they be further generalized? This work deals with decomposition of relations, which is generalization of the problem of functional decomposition. This problem was briefly discussed in the paper [6]. In order to justify the answers to the questions posed above, we have to use combinatorics and Helly's properties.

W. Ziarko and Y. Yao (Eds.): RSCTC 2000, LNAI 2005, pp. 487–494, 2001.

2   Basic Principles

Graph $G = (V, E)$ is a pair of sets V and E, where V is the given set, and E is the binary relation defined on V. Members of set V are called *vertices* and members of set E are called *edges*. A *stable set*, or *independent set*, is a subset of vertices with the propert that no two vertices in the stable set are adjacent. Complete and correct *coloring of vertices* of a graph G is understood as completely defined function $c : V \rightarrow Colors$, such that, if $\{v, w\} \in E$, then $c(v) \neq c(w)$. A graph is said to be *k-colorable* if the set of colors has cardinality k. The least natural number k for which the graph is k-colorable is called *chromatic number* of graph G and is denoted as $\chi(G) = k$. So, any independent set of vertices can be monochromatic. The problem of computing $\chi(G)$ for any given graph G is $NP-$ hard [4].

Hypergraph $H = (X, F)$ is an extension of the idea of graph. Edges of a hypergraph H can be connected to any number of vertices X, that is, $F \subseteq P(X)$. A subset of vertices T of a hypergraph H, which is connected to all the edges of the hypergraph is called a *transversal* set of vertices. The strength of the smallest transversal (with the least number of members) of a hypergraph H is denoted by $\tau(H)$. The problem of computing $\tau(H)$ for a given hypergraph H is NP-hard [4]. Moreover, by an *independent set* of vertices of hypergraph H we mean a set of vertices no two of which are connected by an edge of H and by the *coloring of hypergraph* we mean any function that is not monochromatic on the vertices of any edge.

Let $Card(X)$ denote the cardinality of set X. It is said that hypergraph $H = (X, F)$ has *Helly's property* [1], if every subset of set F is a *star*:

$$\forall J \subseteq \{1, 2, \ldots, Card(F)\} \forall i, j \in J : E_i \cap E_j \neq \emptyset \Rightarrow \bigcap_{k \in \{1, \ldots, n\}} E_k \neq \emptyset$$

Let $H = (X, F)$ be a hypergraph. Graph $G_H = (V, E)$ is called the *representative graph* of hypergraph H, if $V = F$ and $(f_1, f_2) \in E \Leftrightarrow f_1 \cap f_2 \neq \emptyset$.

Theorem 1. *Let H be a hypergraph with Helly's property. If $co-G = (V, V^2 \backslash E)$ denotes the co-graph of any graph $G = (V, E)$, then the sets that generate an independent set of vertices in $co - G_H$ is a star.*

Proof. Refer [1].

Theorem 2. *Let H be a hypergraph with Helly's property. This leads to the following equation:*

$$\tau(H) = \chi(co - G_H).$$

Proof. Refer [9].

Properties of Helly are generalized as *k-Helly properties*. Hypergraph $H = (X, F)$ is said to have *k-Helly property*, if every subset of set F is *k-star*:

$$\forall J \subseteq \{1, 2, \ldots, Card(F)\} \forall I \subseteq J : Card(I) \leq k \,\&\, \bigcap_{k \in I} E_k \neq \emptyset \Rightarrow \bigcap_{k \in \{1, \ldots, n\}} E_k \neq \emptyset$$

Therefore, Helly's property is the same as 2-Helly property. The following fact can be derived straight from the definitions.

Theorem 3. *If hypergraph* $H = (X, F)$ *is k-Helly and* $k \leq h$*, then the hypergraph is h-Helly as well.*

Properties of k-Helly were stated for the first time in combinatory geometry and are closely related to the classic definition named after Helly [1].

Theorem 4. (Helly) *Finite convex sets in* w R^{n-1} *have properties of n-Helly.*

This leads to a simple combinatory corollary:

Theorem 5. *If a given hypergraph* $H = (X, F)$ *satisfies* $X = [n]$*, then it is n-Helly.*

Let $H = (X, F)$ be a hypergraph. Hypergraph $H_k = (V, E)$ is called k-*representative hypergraph* of hypergraph H, if $V = F$ and

$$I \in E \Leftrightarrow (I \subseteq \{1, \ldots, Card(F)\} \,\&\, Card(I) \leq k \,\&\, \bigcap_{k \in I} E_k \neq \emptyset)$$

Therefore, $H_2 = G_H$. From Theorems 1.1 i 1.2 we can derive the following fact:

Theorem 6. *Let hypergraph* H *be k-Helly. The sets generating an independent set of vertices in the co-representative hypergraph,* $co - H_k$ *is k-star.*

Theorem 7. *Let hypergraph* H *be k-Helly. Then the following equation is true:*

$$\tau(H) = \chi(co - H_k).$$

3   Decomposition of Finite Functions

Basic concepts on decomposition of partially defined finite functions are presented in this section. Let $F : [m]^n \rightarrow [m]^k$ ($F = \{f_i\}_{i \in [k]}$ be a set of finite functions, where $f_i : [m]^n \rightarrow [m]$). Let A *(bound set)* and B *(free set)* be disjoint partitions of the set of variables $Var(F)$. The *decomposition chart* of the function F consists of a two dimensional matrix with indexed columns using the values of the bound set variables and indexed rows using the values of the free set variables. Elements of the matrix m_{ij} are the values assumed by the function F for the vectors constructed from i-th row and i-th column. *The column multiplicity of the matrix is denoted by* $v(B|A)$.

Theorem 8. *Let* $F : [m]^n \rightarrow [m]^k$*,* $F = \{f_i\}_{i \in [k]}$ *be a group of partial finite functions, where* $f_i : [m]^n \rightarrow [m]$*. Let* A *and* B *be a pair of disjoint subsets of variables* $Var(F)$*.*

$$F(A, B) = H(G(A), B) \qquad \Leftrightarrow v(B|A) \leq d^j,$$

where: $G = \{g_k(A)\}_{k=1,\ldots,j}$ *   for* $g_k : [m]^{Card(A)} \rightarrow [d]$ *   and*

$\qquad \quad H = \{h_l(G(A), B)\}_{l=1,\ldots,n}$*,   for* $h_l : [u]^{j+Card(B)} \rightarrow [m]$*,* $u = \max\{d, m\}$*.*

Proof. Refer [8].

This theorem suggests that the fundamental problem of decomposing a *partial* finite function is nothing but finding an expansion of the function for which $v(B|A)$ has the least value.

> **Problem: k – decompose partial finite functions**
> **for a given pair of disjoint sets of input variables**
> *Given:* finite group of finite functions F,
>   the partition of the set of input variables into a pair of subsets A and B, and k∈N,
> *Problem:* Is there an expansion for the function F, for which $v(B|A) = k$?

The problem of finding an expansion of the function for which $v(B|A)$ has the least value is nothing but a graph coloring problem [12].

Theorem 9. *Problem k-decomposition of partially defined finite functions leads to the problem of k-coloring the graph (in PTIME).*

Proof outline. It can be observed that the columns of the decomposition matrix can be treated as an interval in the lattice $K = ([m]^n, \leq_n)$, where n and m are natural numbers, and $\leq_n$ is a lexicographical order. Our aim is to verify if we can "paste" the columns so that there remains only k columns. This problem is equivalent to answering if there exists a k - element transversal hypergraph for the given section. Therefore, the question is equivalent (refer Theorem 1.2., an interval in the lattice has 2-Helly's property) to asking if there exists a k - color representative graph (denoted by $G_{F(A,B)}$) for the hypergraph. So, we need to construct the representative to constructing the incidence graph. The time needed is in the order of $O(c^2 r)$, where c and r are the number of columns and rows of the decomposition matrix.

On the basis of the above explanation, the following heuristic is proposed for decomposing finite functions.

Decomposition heuristic

> *Input:* finite group of finite functions F and
>   a partition of the input variables into a pair of subsets A and B;
> *Output:* decomposition F(A,B) = H(G(A),B), minimizing the value $v(B|A)$;
>   1. find the graph $G_{F(A,B)}$ of incompatible columns in the decomposition matrix;
>   2. find the minimum coloring for the graph $G_{F(A,B)}$;
>   3. find the decomposition H(G(A),B) for the function F from the results obtained in step 2.

Theorem 10. *The problem of k - decomposition for $k = 1, 2$ for partially defined Boolean functions (and also for partial finite functions) is in PTIME.*

Proof. (Sketch) The theorem is derived from Theorem 2.2 and from the fact that the question of finding if a given graph is two-colorable can be resolved in polynomial time.

Theorem 11. *The problem of* k *- decomposition of partial finite functions for* $k \geq 3$ *is* NP *- complete.*

Proof. (Sketch) The proof is based on two observations. First, the problem of k - decomposition of partial finite functions belongs to NP class. Secondly, the problem of k - coloring of graphs is reduced (in PTIME) to the problem of k - decomposition of partial finite functions. The reduction procedure is given in Algorithm 1. The procedure generates the decomposition table for the given graph G, so that $\chi(G) = v(B|A)$.

Algorithm 1. Symbol '-' denotes *don't-care value*

```
procedure Reduction;
begin
    for i := 1 to n do
    begin
        for k := 1 to (i - 1) do v[k, i] := "-";
        v[i, i] := "1";
        for k := (i + 1) to n do
            if { vᵢ, vₖ } ∈ E then    v[k, i]  = "0"
                                else  v[k, i] := "-";
    end;
end;
```

Example 1. This example demonstrates the reduction algorithm for the graph G in figure. The algorithm results in:

Table 1.

1	2	3	4	5
1	-	-	0	0
-	1	-	-	0
-	-	1	0	-
-	-	-	1	0

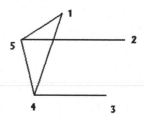

Indeed, $\chi(G) = v(B|A) = 3$.

4 Decomposition of Finite Relations

This section generalizes the problem of decomposition of partial finite functions into the problem of decomposition of finite relations. Finite relation is a function expressed as $r : X^n \rightarrow P(X)$. As the images of a finite relation belong to a certain set X, they *need not possess the property of* (eg.: $\{1, 2\}$, $\{1, 3\}$, $\{2, 3\}$). This is unlike the images of a group of partial functions, which were intervals in the lattice $K = ([m]^n, \leq_n)$. Therefore, we have to apply more subtle tools to decompose relations. These factors were not considered in the earlier work [6].

Retaining the definitions given in section three and changing the discussion from functions to relations, we can state the problem of decomposing a relation as follows:

Problem: k – decomposition of finite relation
for a given pair of disjoint sets of input variables

Given: *finite relation r, a partition of the input variables into a pair of subsets A and B, and k∈N,*

Problem: Is there an expansion for the relation r, for which $v(B|A) = k$?

Theorem 12. *Let* $r : X^n \to P(X)$ *be a relation. The problem of decomposing relation r can be reduced to the problem of coloring the hypergraph.*

Proof. From Theorem 1.7, on the basis of our understanding of decomposition table for relation r, we have to construct k - representative hypergraph for $k = Card(X)$. Algorithm 2 reduces the problem of decomposing a relation into the problem of coloring the hypergraph. If the decomposition table consists of r-rows and c-columns, the complexity of the algorithm is approximated to $O(cr^{k+1})$. In the algorithm, symbols $X|_A$ denote the projection of set X on the members of set A.

Algorithm 2.

```
procedure ConstructH_r(A,B);
begin
  k := Card(Y);
  X_r(AB) := X|_A;  F_r(AB) := ∅;
  for i := 2 to k do
    begin
      V := sucessive i-element subsets of the set X_r(AB) in lexicographical order;

      if(V is a solitory set of the hypergraph H_r(AB)) and ( ∃ w∈X|_B  ⋂  r(v,w) = ∅ )
                                                                    v∈V
      then F_r(AB) := F_r(AB) ∪ {V};
    end;
end;
```

5   Computer Test Results

The presented algorithms were implemented as a package of computer programs in the Institute of Telecommunication at Warsaw University of Technology. This section presents the results obtained.

Table 2 presents the results for logic synthesis problems of decomposing digital circuits to be implemented using $PLAs$. Silicon area is measured using the formula $S = P(4n_v + 2n_b + m)$, where P - number of terms, n_v - number of 4 - valued inputs, n_b - number of 2 - valued inputs, and m - number of outputs. Character 'b' means decomposition in binary system and 'm' means decomposition in multiple valued system.

Table 3 presents results for decomposition of information systems specified using "if_then_else_" type rules. Therefore, silicon area is measured using the formula $S = p \sum q_i$, where p - number of rules, and q_i - number of bits needed to represent the i-th attribute.

Table 2.

Function	Original silicon area	Silicon area after decomposition (b)	Silicon area after decomposition (m)
Rd84	5120	1280	1408
Rd73	2178	776	1142
Alu2	26624	24399	25124
Misex1	5888	1984	1984
Sao	24576	8192	8192

Table 3.

Example	Silicon area before decomposition	Silicon area after decomposition
Monks1te	4752	487
Monks2te	4752	507
Monks1tr	1364	933
Monks2tr	1859	1095
Monks3tr	1342	929
House	3944	1049
Nurse	1365	475

Additionally, table 4 contains the computational results of decomposition algorithms based on different approximation coloring heuristics (namely: the simple sequential algorithm, the largest-first sequential algorithm, the smallest-last sequential algorithm, the maximum independent set algorithm) applied to digital circuits to be implemented using PLA matrixes.

Table 4.

Example	Original size	After SSA	After LFSA	After SLSA	After MISA
Bbtas	210	157	171	162	162
Ex7	306	288	288	288	208
9sym	1615	530	541	515	543
Tra11	340	335	335	335	335
opus	588	544	544	549	576
Sqrt8	760	386	386	386	386
Adr4	1575	593	574	593	620
clip	2691	1414	1393	1337	1421
Z4	1062	377	366	366	424
Sao2	1392	810	820	792	774
root	1197	879	899	873	864

It is evident that the application of decomposition methods has resulted in significant reduction of silicon area. It has also been proved that the generalized decomposition methods for multiple valued functions compare well with the classical binary decomposition methods.

6   Conclusion

This work has shown that the problem of decomposing a finite function $f(A, B)$ into $h(g(A), B)$, where g is a Boolean function, can be resolved in polynomial time with respect to the size of the problem. It has been shown that omission of the characteristic of the g function can significantly complicate the problem. Such generalized problems belong to NP-hard class of problems. The work presented the reduction of the problem of decomposition of a finite function into the problem of coloring the vertices of a graph. It was also shown that the problem of decomposition of a relation could be reduced to coloring the vertices of a hypergraph. Presented algorithms were implemented in a packet of computer programs at the Warsaw University of Technology. Two important conclusions were drawn from the experiments conducted with the packet of software. First of all, the experiments show that the application of decomposition methods leads to reduction in silicon area for examples taken from logic synthesis and also for the multiple valued examples taken from information systems. Secondly, use of the generalized methods - methods to decompose multiple valued functions, instead of decomposition methods for Boolean functions, did not have negative influence on the effective performance of the presented system. In effect, a more tolerant software has been developed that can analyze a wider spectrum of problems.

References

1. Berge C.: *Hypergraphs, Combinatorics of Finite Sets*, North-Holland, 1989.
2. Boros E., Guvich V., Hammer P. L., Ibaraki T., Kogan A.: *Decompositions of Partially Defined Boolean Functions*, DIMACS-TR-94-9, 1994.
3. Garey M.R., Johnson D.S.: *Computers and Intractability: a guide to the theory of NP-Completeness*, W.H. Freeman and Company, San Framncisco, 1979.
4. Łuba T.: *Decomposition of Multiple-Valued Functions*, 25 ISMVL, US, 1995.
5. Łuba T., Selvaraj H.: *A General Approach to Boolean Function Decomposition and its Applications in FPGA-based Synthesis*, VLSI Design. Special Issue on Decompositions in VLSI Design, vol 3, Nos. 3-4, 289-300, 1995.
6. Perkowski M., Marek-Sadowska M., Jóźwiak L., Łuba T., Grygiel S., Nowicka M., Malivi R., Wang Z., Zhang J. S.: *Decomposition of Multiple-Valued Relations*, Proc. of International Symposium AA on MVL,1997.
7. Ross T., Noviskey, Taylor T., Gadd D.: *Pattern Theory: An Engineering Paradigm For Algorith Design*, Wright Laboratory, Wrigh-Patterson Air Force Base, 1990.
8. Sapiecha P., Perkowski M., Łuba T.: *Decomposition of Information Systems Based on Graph Coloring Heuristics*, Proc. CESA'96 IMACS Multiconference, France, 1996.
9. Sapiecha P.: *Algorytmy syntezy funkcji i relacji boolowskich w aspekcie metod reprezentacji i kompresji danych*, Ph.D. Thesis (in Polish), PW, 1998.
10. Selvaraj H., Łuba T., Nowicka M., Bignall B.: *Multiple-Valued Decomposition and its Applications in Data Compression and Technology Mapping*, International Conference on Computational Intelligence and Multimedia Applications, Australia, 1997.
11. Selvaraj H., Nowicka M., Łuba T.: *Non-Disjoint Decomposition Strategy in Decomposition-Based Algorithms and Tools.* Proc. International Conferece on Computational Intelligence and Multimedia Applications. Eds.: Selvaraj H. and Verma B., World Scientific, Singapore 1998.
12. Wan W., Perkowski M. A.: *A New Approach to the Decomposition of Incompletely Specified Multi-Output Functions Based on Graph Coloring and Local Transformations and Its Applications to FPGA Mapping*, Proc. European Design Automation Conference, 1992.
13. Zupan B., Bohanec M., Bratko I., Demšar.: *Machine Learning by Function Decomposition*, ICML, 1997.

Diagnostic Reasoning from the viewpoint of Rough Sets

Shusaku Tsumoto

Department of Medicine Informatics, Shimane Medical University, School of Medicine,
89-1 Enya-cho Izumo City, Shimane 693-8501 Japan
E-mail: tsumoto@computer.org

Abstract. In existing studies, diagnostic reasoning has been modeled as if-then rules in the literature. However, closer examinations suggests that medical diagnostic reasoning should consist of multiple strategies, in which one of the most important characteristics is that domain experts change the granularity of rules in a flexible way. First, medical experts use the coarsest information granules (as rules) to select the foci. For example, if the headache of a patient comes from vascular pain, we do not have to examine the possibility of muscle pain. Next, medical experts switches the finer granules to select the candidates. After several steps, they reach the final diagnosis by using the finest granules for this diagnostic reasoning. In this way, the coarseness or fineness of information granules play a crucial role in the reasoning steps. In this paper, we focus on the characteristics of this medical reasoning from the viewpoint of granular computing and formulate the strategy of switching the information granules. Furthermore, using the proposed model, we introduce an algorithm which induces if-then rules with a given level of granularity.

1 Introduction

One of the most important problems in developing expert systems is knowledge acquisition from experts[2]. In order to automate this problem, many inductive learning methods, such as induction of decision trees[1, 10], induction of decision list[3] rule induction methods[4–7, 10, 11] and rough set theory[8, 13, 17, 18], are introduced and applied to extract knowledge from databases, and the results show that these methods are appropriate.

However, it has been pointed out that conventional rule induction methods cannot extract rules, which plausibly represent experts' decision processes[13, 14]: the description length of induced rules is too short, compared with the experts' rules (Those results are shown in Appendix B). For example, rule induction methods, including AQ15[7] and PRIMEROSE[13], induce the following common rule for muscle contraction headache from databases on differential diagnosis of headache[14]:

```
[location=whole] & [Jolt Headache=no] & [Tenderness of M1=yes]
  => muscle contraction headache.
```

W. Ziarko and Y. Yao (Eds.): RSCTC 2000, LNAI 2005, pp. 495–502, 2001.
© Springer-Verlag Berlin Heidelberg 2001

This rule is shorter than the following rule given by medical experts.

```
[Jolt Headache=no]
& [Tenderness of M1=yes] & [Tenderness of B1=no]
& [Tenderness of C1=no]
  => muscle contraction headache,
```

where [Tenderness of B1=no] and [Tenderness of C1=no] are added.

These results suggest that conventional rule induction methods do not reflect a mechanism of knowledge acquisition of medical experts.

In this paper, we focus on the characteristics of this medical reasoning from the viewpoint of granular computing and formulate the strategy of switching the information granules. Furthermore, using the proposed model, we introduce an algorithm which induces if-then rules with a given level of granularity.

The paper is organized as follows: Section 2 discusses the characteristics of medical reasoning. Section 3 shows formalization of this diagnostic reasoning from the viewpoint of information granulation. Section 4 presents a formal model of medical differential diagnosis and rule induction algorithm for this model. Finally, Section 5 concludes this paper.

2   Medical Reasoning

As shown in Section 1, rules acquired from medical experts are much longer than those induced from databases the decision attributes of which are given by the same experts. This is because rule induction methods generally search for shorter rules, compared with decision tree induction. In the latter cases, the induced trees are sometimes too deep and in order for the trees to be learningful, pruning and examination by experts are required. One of the main reasons why rules are short and decision trees are sometimes long is that these patterns are generated only by one criteria, such as high accuracy or high information gain. The comparative study in this section suggests that experts should acquire rules not only by one criteria but by the usage of several measures.

Those characteristics of medical experts' rules are fully examined not by comparing between those rules for the same class, but by comparing experts' rules with those for another class. For example, a classification rule for muscle contraction headache is given by: [1]

```
[Jolt Headache=no]
& ([Tenderness of M0=yes]   or [Tenderness of M1=yes]
    or [Tenderness of M2=yes])
& [Tenderness of B1=no]   & [Tenderness of B2=no]
& [Tenderness of B3=no]   & [Tenderness of C1=no]
```

---

[1] Readers may say that these two rules are too long to satisfy. However, these attribute-value pairs are required for accurate diagnosis although they are redundant and some of them are related to others. This redundancy is one of the characteristics of medical reasoning.

```
& [Tenderness of C2=no]   & [Tenderness of C3=no]
& [Tenderness of C4=no]
 => muscle contraction headache
```

This rule is very similar to the following classification rule for disease of cervical spine:

```
[Jolt Headache=no]
& ([Tenderness of M0=yes]  or [Tenderness of M1=yes]
   or [Tenderness of M2=yes])
& ([Tenderness of B1=yes]     or [Tenderness of B2=yes]
   or [Tenderness of B3=yes] or [Tenderness of C1=yes]
   or [Tenderness of C2=yes] or [Tenderness of C3=yes]
   or [Tenderness of C4=yes])
 => disease of cervical spine
```

The differences between these two rules are attribute-value pairs, from tenderness of B1 to C4. Thus, these two rules can be simplified into the following form:

$$a_1 \& A_2 \& \neg A_3 \rightarrow \textit{muscle contraction headache}$$
$$a_1 \& A_2 \& A_3 \rightarrow \textit{disease of cervical spine}$$

The first two terms and the third one represent different reasoning. The first and second term a_1 and A_2 are used to differentiate muscle contraction headache and disease of cervical spine from other diseases. The third term A_3 is used to make a differential diagnosis between these two diseases. Thus, medical experts firstly selects several diagnostic candidates, which are very similar to each other, from many diseases and then make a final diagnosis from those candidates.

3   Formalization of Medical Reasoning

3.1   Accuracy and Coverage

In the subsequent sections, we adopt the following notations, which is introduced in [12].

Let U denote a nonempty, finite set called the universe and A denote a nonempty, finite set of attributes, i.e., $a : U \rightarrow V_a$ for $a \in A$, where V_a is called the domain of a, respectively.Then, a decision table is defined as an information system, $A = (U, A \cup \{d\})$.

The atomic formulas over $B \subseteq A \cup \{d\}$ and V are expressions of the form $[a = v]$, called descriptors over B, where $a \in B$ and $v \in V_a$. The set $F(B, V)$ of formulas over B is the least set containing all atomic formulas over B and closed with respect to disjunction, conjunction and negation.

For each $f \in F(B, V)$, f_A denote the meaning of f in A, i.e., the set of all objects in U with property f, defined inductively as follows.

1. If f is of the form $[a = v]$ then, $f_A = \{s \in U | a(s) = v\}$

2. $(f \wedge g)_A = f_A \cap g_A$; $(f \vee g)_A = f_A \vee g_A$; $(\neg f)_A = U - f_a$

By the use of this framework, classification accuracy and coverage, or true positive rate is defined as follows.

Definition 1 (Accuracy and Coverage).
Let R and D denote a formula in $F(B, V)$ and a set of objects which belong to a decision d. Classification accuracy and coverage(true positive rate) for $R \to d$ is defined as:

$$\alpha_R(D) = \frac{|R_A \cap D|}{|R_A|} (= P(D|R)), \quad and$$

$$\kappa_R(D) = \frac{|R_A \cap D|}{|D|} (= P(R|D)),$$

where $|A|$ denotes the cardinality of a set A, $\alpha_R(D)$ denotes a classification accuracy of R as to classification of D, and $\kappa_R(D)$ denotes a coverage, or a true positive rate of R to D, respectively.

It is notable that these two measures are equal to conditional probabilities: accuracy is a probability of D under the condition of R, coverage is one of R under the condition of D. It is also notable that $\alpha_R(D)$ measures the degree of the sufficiency of a proposition, $R \to D$, and that $\kappa_R(D)$ measures the degree of its necessity.[2]

For example, if $\alpha_R(D)$ is equal to 1.0, then $R \to D$ is true. On the other hand, if $\kappa_R(D)$ is equal to 1.0, then $D \to R$ is true. Thus, if both measures are 1.0, then $R \leftrightarrow D$.

Also, Pawlak recently reports a Bayesian relation between accuracy and coverage[9]:

$$\alpha_R(D)P(D) = P(R|D)P(D) = P(R, D)$$
$$= P(R)P(D|R) = \kappa_R(D)P(R)$$

This relation also suggests that *a priori* and *a posteriori* probabilities should be easily and automatically calculated from database.

3.2   Definition of Characterization Set

In order to model these three reasoning types, a statistical measure, coverage $\kappa_R(D)$ plays an important role in modeling, which is a conditional probability of a condition (R) under the decision D ($P(R|D)$).

Let us define a characterization set of D, denoted by $L(D)$ as a set, each element of which is an elementary attribute-value pair R with coverage being larger than a given threshold, δ_κ. That is,

$$L_{\delta_\kappa}(D) = \{[a_i = v_j] | \kappa_{[a_i = v_j]}(D) \geq \delta_\kappa\}.$$

---

[2] These characteristics are from formal definition of accuracy and coverage. In this paper, these measures are important not only from the viewpoint of propositional logic, but also from that of modelling medical experts' reasoning, as shown later.

Then, according to the descriptions in Section 2, three models of reasoning about complications will be defined as below:

1. Independent type: $L_{\delta_\kappa}(D_i) \cap L_{\delta_\kappa}(D_j) = \phi$,
2. Boundary type: $L_{\delta_\kappa}(D_i) \cap L_{\delta_\kappa}(D_j) \neq \phi$, and
3. Subcatgory type: $L_{\delta_\kappa}(D_i) \subseteq L_{\delta_\kappa}(D_j)$.

All three definitions correspond to the negative region, boundary region, and positive region[8], respectively, if a set of the whole elementary attribute-value pairs will be taken as the universe of discourse. Thus, reasoning about complications are closely related with the fundamental concept of rough set theory and approximate reasoning[16].

3.3   Characterization as Exclusive Rules

Characteristics of characterization set depends on the value of δ_κ. If the threshold is set to 1.0, then a characterization set is equivalent to a set of attributes in exclusive rules[13]. That is, the meaning of each attribute-value pair in $L_{1.0}(D)$ covers all the examples of D. Thus, in other words, some examples which do not satisfy any pairs in $L_{1.0}(D)$ will not belong to a class D.

Construction of rules based on $L_{1.0}$ are discussed in Subsection 4.4, which can also be found in [14, 15]. The differences between these two papers are the following: in the former paper, independent type and subcategory type for $L_{1.0}$ are focused on to represent diagnostic rules and applied to discovery of decision rules in medical databases. On the other hand, in the latter paper, a boundary type for $L_{1.0}$ is focused on and applied to discovery of plausible rules.

3.4   Characterization in Diagnostic Reasoning

Let us return to the example in Section 2. The two rules are represented as:

$$a_1 \& A_2 \& \neg A_3 \rightarrow muscle\ contraction\ headache$$
$$a_1 \& A_2 \& A_3 \rightarrow disease\ of\ cervical\ spine$$

From the viewpoint of characterization set, a_1 and members of A_2 should be included in the characterization sets of both classes. On the other hand, members of A_3 are included only in that of muscle contraction headache.[3] That is,

$$a_1 \in L_{1.0}(m.c.h.), A_2 \subset L_{1.0}(m.c.h.),$$
$$a_1 \in L_{1.0}(d.c.s.), A_2 \subset L_{1.0}(d.c.s.),$$
$$A_3 \subset L_{1.0}(m.c.h), A_3 \not\subset L_{1.0}(d.c.s.),$$

---

[3] For simplicity, the thresholdδ_κ is set to 1.0 in the following discussion because $\delta_\kappa = 1.0$ corresponds to exclusive reasoning(characterization)[13, 14]. However, it is easy to show that this condition is not actually necessary for the discussion. It is only needed for interpretation from the medical side.

where m.c.h. and d.c.s. denote muscle contraction headache and disease of cervical spine, respectively. These facts are summarized into:

$$\{a_1, A_2\} \subseteq L_{1.0}(m.c.h.) \cap L_{1.0}(d.c.s),$$
$$A_3 \subseteq L_{1.0}(m.c.h.) - L_{1.0}(d.c.s).$$

Thus, the relation between characterization sets of m.c.h. and d.c.h. is boundary type, and the difference set between these two characterization sets are important to discriminate between these two diseases. From the above discussion, it is easy to see that there are two types of attribute-value pairs: the first one describe the characteristics shared by these two diseases and the second one describe the discrimination between them. In this way, discrimination can be viewed as the use of attribute-value pairs belonging to the difference set between decision attributes (target classes). That is,

Definition 2 (Discriminating Descriptors).
Let D_1 and D_2 denote decision attributes and let $\Delta(D_1, D_2) = L_{1.0}(D_1) - L_{1.0}(D_2)$. Formulae in $\Delta(D_1, D_2)$ are called discriminating descriptors. On the other hand, formulae in $L_{1.0}(D_1) \cap L_{1.0}(D_2)$ are called indistinguishable descriptors.

To find rules discriminating between D_1 and D_2 is equivalent to find discriminating descriptors of $L_{1.0}(D_1)$ and $L_{1.0}(D_2)$.

For the above example shown in Section 2, $\Delta(m.c.h, d.c.s)$ is equal to A_3. It is notable that the domain for the characterization sets, in other words, selection of attributes as domain is very important to classify a type of relations between characterization sets. If we change the domain for the characterization sets, then they will have a different view for each decision attribute. For example, if we select $\{a_1, A_2\}$ as a domain, we cannot distinguish $L_{1.0}(m.c.h.)$ and $L_{1.0}(d.c.s)$ (subcategory type). On the other hand, if we select A_3 as a domain, we can distinguish between these two sets. For evaluation of the nature of characterization, we can define a index for discrimination power.

Definition 3 (Discrimination Power). *Let D_1 and D_2 denote two decision attributes and let a set of all descriptors denote $\mathcal{D}$. Discriminant power of $\mathcal{D}$ for D_1 and D_2 is defined as:*

$$\frac{|\Delta(D_1, D_2)|}{|\mathcal{D}|}.$$

4   Rule Induction based on Medical Diagnosis

4.1   Modelling Medical Diagnosis

As shown in the above subsection, if a set of discriminating descriptors($\Delta(D_1, D_2)$) is equal to the selected domain(D_s), then corresponding characterization sets are independent. If $\Delta(D_1, D_2)$ is empty, then $L_{1.0}(D_1)$ and $L_{1.0}(D_2)$ are of the sub-category type.

In the case of subcategory type, we can group these decision attributes into more generalized attributes. For example, if we take a_1, A_2 a D_s, then m.c.h. and d.c.s. can be grouped into one decision attribute, say m.c.h._and_d.c.s. Thus, medical differential diagnosis process can be modeled as follows.

1. Detect Subcategory and Independent type (to group several d_i), For each subcategory case, make a generalized decision attribute(D_i).
2. From observations of a case, check whether this case belongs to a group D_i.
3. Then, discrimination between D_i and D_j will be applied.

4.2   Rule Induction Method

Based on the above medical diagnosis model, we obtain the following algorithm for rule induction with grouping.

1. Generate Characterization Sets: $L_{1.0}(d_i)$
2. Detect Subcategory and Independent type
3. Grouping Decision Attributes (g_i) of Subcategory type: make partition of grouped attributes.
4. Apply Rule Induction to g_i: $C_i \rightarrow g_j$, which discriminate between grouped attributes
5. Apply Rule Induction within g_i: $gi \wedge C_j \rightarrow d_k$
6. Integrate Rules: $(C_i \rightarrow g_j) + (g_i \wedge C_j \rightarrow d_k)$

This algorithm was first introduced in [14] and extended into probabilistic case($\kappa <$ 1.0). In both papers, these algorithms were evaluated on three medical databases, the experimental results of which show that these algorithms generate rules more similar to medical experts' rules than the conventional rule induction methods.

5   Conclusions

In existing studies, diagnostic reasoning has been modeled as if-then rules in the literature. However, closer examinations suggests that medical diagnostic reasoning should consist of multiple strategies, in which one of the most important characteristics is that domain experts change the granularity of rules in a flexible way. In this paper, we focus on the characteristics of this medical reasoning from the viewpoint of granular computing and formulate the strategy of switching the information granules. Furthermore, using the proposed model, we introduce an algorithm which induces if-then rules with a given level of granularity. This paper is a preliminary study on application of granular computing method to medical diagnosis. Further formal studies will be shown in the near future.

References

1. Breiman, L., Freidman, J., Olshen, R., and Stone, C., *Classification And Regression Trees*, Wadsworth International Group, Belmont, 1984.
2. Buchnan, B. G. and Shortliffe, E. H., *Rule-Based Expert Systems*, Addison-Wesley, New York, 1984.
3. Clark, P. and Niblett, T., The CN2 Induction Algorithm. *Machine Learning*, 3, 261-283, 1989.
4. Langley, P. *Elements of Machine Learning*, Morgan Kaufmann, CA, 1996.
5. Mannila, H., Toivonen, H., Verkamo, A.I., Efficient Algorithms for Discovering Association Rules, in *Proceedings of the AAAI Workshop on Knowledge Discovery in Databases (KDD-94)*, pp.181-192, AAAI press, Menlo Park, 1994.
6. Michalski, R. S., A Theory and Methodology of Machine Learning. *Machine Learning - An Artificial Intelligence Approach*. (Michalski, R.S., Carbonell, J.G. and Mitchell, T.M., eds.) Morgan Kaufmann, Palo Alto, 1983.
7. Michalski, R. S., Mozetic, I., Hong, J., and Lavrac, N., The Multi-Purpose Incremental Learning System AQ15 and its Testing Application to Three Medical Domains, in *Proceedings of the fifth National Conference on Artificial Intelligence*, 1041-1045, AAAI Press, Menlo Park, 1986.
8. Pawlak, Z., *Rough Sets*. Kluwer Academic Publishers, Dordrecht, 1991.
9. Pawlak, Z. Rough Sets and Decision Analysis, *Fifth IIASA workshop on Decision Analysis and Support*, Laxenburg, 1998.
10. Quinlan, J.R., *C4.5 - Programs for Machine Learning*, Morgan Kaufmann, Palo Alto, 1993.
11. *Readings in Machine Learning*, (Shavlik, J. W. and Dietterich, T.G., eds.) Morgan Kaufmann, Palo Alto, 1990.
12. Skowron, A. and Grzymala-Busse, J. From rough set theory to evidence theory. In: Yager, R., Fedrizzi, M. and Kacprzyk, J.(eds.) *Advances in the Dempster-Shafer Theory of Evidence*, pp.193-236, John Wiley & Sons, New York, 1994.
13. Tsumoto, S. Automated Induction of Medical Expert System Rules from Clinical Databases based on Rough Set Theory *Information Sciences* **112**, 67-84, 1998.
14. Tsumoto, S. Extraction of Experts' Decision Rules from Clinical Databases using Rough Set Model Journal of Intelligent Data Analysis, 2(3), 1998.
15. Tsumoto, S., Automated Discovery of Plausible Rules based on Rough Sets and Rough Inclusion, *Proceedings of PAKDD'99*, LNAI, Springer-Verlag, 1999.
16. Yao, Y.Y. and Wong, S.K.M., A decision theoretic framework for approximating concepts, *International Journal of Man-machine Studies*, **37**, 793-809, 1992.
17. Ziarko, W., The Discovery, Analysis, and Representation of Data Dependencies in Databases. in: *Knowledge Discovery in Databases*, (Shapiro, G. P. and Frawley, W. J., eds.), AAAI press, Menlo Park, pp.195-209, 1991.
18. Ziarko, W., Variable Precision Rough Set Model. *Journal of Computer and System Sciences*. 46, 39-59, 1993.

Rough Set Approach to CBR

Jan Wierzbicki

Polish-Japanese Institute of Information Technology
Koszykowa Str. 86
02- 008 Warsaw, Poland
e-mail:jwierzbi@oeiizk.waw.pl

Abstract. We discuss how Case Based Reasoning (CBR) (see e.g. [1],
[4]) philosophy of adaptation of some known situations to new similar
ones can be realized in rough set framework [5] for complex hierarchical
objects.

We discuss how various problems can be represented by means of complex objects described by hierarchical attributes, and how to use similarity between them for predicting the relevant algorithms corresponding to
these objects. The complex object attributes are of different types: basic
attributes related to problem definition (e.g. features of object parts),
attributes reflecting some additional characteristic of problem (e.g. features of more complex objects inferred from properties of their parts and
their relations), and attributes representing algorithm structures (e.g.
order and/or properties of operations used to solve the given problem).
We show how to define these particular attributes sets, and how to recognize the similarity of objects in order to transform algorithms corresponding to these objects to a new algorithm relevant for the new
incompletely defined object [1,4].

Object similarity is defined on several levels; basic attribute recognition
level, characteristic attribute recognition level and algorithm operation
recognition level. Dependencies between attributes are used to link different levels. These dependencies can be extracted from data tables specifying the links.

We discuss how to classify new objects, and how to synthetize algorithm
for such new object, on the basis of algorithms corresponding to similar
objects. The main problem is the generation of rules enabling to create
operation sequences for a new algorithm. These rules are generated using
rough set approach [5].

1   Problem Representation As an Complex Object

We discuss methods of solving various problems, which can be specified, by
means of some hierarchical attributes. Examples of such problems are simple
mathematical tasks or finding the way out in the maze.

The problem-case is represented as a complex object (constructed hierarchically) defined by some attributes and its information signature (attribute value
vector). Information signature of any object $O \in U$ is defined by attribute value
vector, i.e., $Inf_A(O) = \{(a, a(O)) : a \in A\}$, where $a(O)$ is the value of attribute
a, on the object O [5].

W. Ziarko and Y. Yao (Eds.): RSCTC 2000, LNAI 2005, pp. 503–510, 2001.

Objects described by the same attributes are represented in information system [5]. A complex object can be described as an uncomposed one - using some general attributes (e.g. specifying general object type - e.g. geometrical figure or maze, its main components - e.g. two triangles, square maze) (Table 1a), or it can be described by some more detailed attributes after decomposition it up to some level, (Table 1b). Values of attributes are binary ones $(0, 1)$: 1 - means the attribute value is present (e.g. crossings in the maze are perpendicular) or its real value (which can be taken from the content of task) is known (e.g. square edge length is known), 0 - means the attribute value is not present or its real value is unknown.

Some of complex object attributes values point out to its subobjects (specified in other information systems) and specify how they are related (Table 1b).

Among object attributes we distinguish attributes, called algorithm (operations) attributes, with values informing if the operation is enabled. Such enabled attributes fire the corresponding operations transforming the object, and let to solve a given problem or to make some decisions.

Table 1a

object	g_1	...	g_n	op_1	...	op_n
O_1	0		0	1		0
O_2	0		1	1		1
O_3	1		1	0		1
...						
O_n	0		1	0		1

Table 1b - (objects of type A&B)

object	a_1	...	a_n	b_1	...	b_m	r_1	...	r_i	c_1	...	c_k	op_1	...	op_l
O_1'	1		0	1		1	1		1	0		0	1		0
O_2'	0		1	1		1	1		1	0		0	1		0
O_3'	1		1	0		1	0		0	0		1	1		0
...															
O_n'	0		0	1		0	0		1	1		1	1		1

$(g_1, .., g_n)$ are general attributes describing the object, $(op_1, .., op_m)$ algorithm attributes (operations) transforming the object.

In Table 1b several subsets of the complex object attribute set are distinguished:
(i) problem definition attributes - basic attributes - $(a_1, .., a_n)$, $(b_1, .., b_m)$,
(ii) relational attributes - declaring relations between component objects of the complex object - $(r_1, .., r_i)$,
(iii) problem additional characteristics attributes - $(c_1, .., c_k)$,
(iv) algorithm operations transforming object - the given problem - $(op_1, .., op_l)$.

By solving the problem we mean finding for a given object all values of attributes corresponding to operations to be executed to solve the problem - e.g. the way out in the maze is found (decision - answer for question if we are in the point of maze exit, is true), or to find value vector of some other missing object attributes, e.g. finding area of geometrical figure.

We solve the given problem by computing step by step algorithm operations for which its value vector is 1 (true) in an order specified by hierarchical structure of tables. This let us to find some required values of object attributes. To execute the particular operation, we need some other object attributes values. Some of them are from the main information system, but some of them must be taken from other information systems for some properly chosen subobjects (Fig.1).

We know attribute values for a, b, A, D, but we are looking for attribute values of c, d.

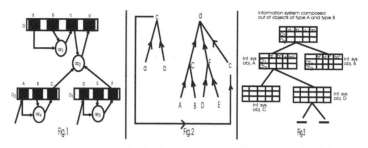

Algorithm operations: $op_1(a, b) = c$; $op_2(C, F, c) = d$; $op_3(D, E) = F$; $op_4(A, B) = C$, can be represented by dependencies [5]: $ab \to c$; $CFc \to d$; $DE \to F$; $AB \to C$, which in turn represent graphically a dependency (Fig.2) $ABDEab \to d$.

One may notice, that to execute the main object algorithm operation op_2 we need attribute values of c, C, F, which are obtained from operations op_1, op_3, op_4. Operation op_1 is the main object algorithm operation, while operations op_3, op_4 are subobjects algorithms operations.

Each complex object (from information system) representing the problem - case, is represented in a hierarchical system (Fig.3).

The components of such system are information systems [5] related in a specific order. Component objects are complex objects (defined hierarchically) or simple object (defined by primary attributes - primary concepts of domain in which the problem is included). Basic attributes are taken from the problem definition (content); other specify (map) subobjects - component objects (from other information systems - Table 2, Table 3). They are formulated (by the expert) in the way which let to specify some attributes and its value vectors from the sets of basic, relational and characteristic attributes for the subobject [2]. Thus we may obtain particular subobject (subobjects) of the complex object.

Table 2 - (objects of type A)

object	a'_1	...	a'_n	b'_1	...	b'_m	r'_1	...	r'_i	c'_1	...	c'_k	op'_1	...	op'_l
O'_1	0		0	1		1	1		1	1		1	1		1
O'_2	0		1	0		0	0		0	0		0	0		1
O'_3	1		1	1		1	1		1	0		0	0		0
...															
O'_n	0		1	0		1	0		1	0		1	0		1

Table 3 - (objects of type B)

object	a''_1	...	a''_n	b''_1	...	b''_m	r''_1	...	r''_i	c''_1	...	c''_k	op''_1	...	op''_l
O'_1	1		0	1		1	1		1	1		1	1		1
O'_2	0		0	0		0	1		1	0		0	1		1
O'_3	0		1	0		1	1		1	1		0	0		1
...															
O'_n	0		1	0		1	0		1	0		1	O		1

Relational attributes declare relations between component objects from which the complex object is constructed. Characteristics attributes are defined by an expert or system rules, and declare an extra description of the complex object, which let the system to input the proper object transformation rules. Algorithm operations are attributes transforming the object in order to find the missing value vector of some attribute [2].

From each information system one may derive some rules defining algorithm attributes (operation) using the basic, relational and characteristic object attributes. The rules are of the form $\alpha \to op(O)$ for some $op \in OP$, where α is a conjunction of descriptors (a, v) for some $a \in A$ and $v \in V_a$. On the right hand side of the rule more than one operation attribute can appear.

To execute for main complex object an algorithm operation we may need values obtained by subobject algorithm operations. To specify such situation we introduce special information table (Table 4.), consisting of necessary additional information from lower levels of hierarchical structure.

Table 4.

op'_1	...	op'_n	op''_1	...	op''_m	w_1	...	w_i	op_1	...	Op_k
1		0	0		1	1		1	1		0
0		1	1		1	0		1	1		1
...											
0		0	0		1	1		0	1		1

$(op'_1, .., op'_n)$ and $(op''_1, .., op''_m)$ operations for subobject algorithm and $(op_1, .., op_k)$ are operations of main object algorithm; and $(w_1, .., w_i)$ some additional attributes specifying hierarchical structure of main algorithm operations.

Example 1. Let us consider a simple geometrical figure (or maze) (Fig.4). This figure represents a hierarchical system. We may decompose such a figure to component objects (Fig.5). Such object decomposition can be done in different ways (Fig.5a). The most difficult problem is to define (find) the proper decomposition method for the case. Component object may represent a complex object itself, and can be decomposed to its component objects - complex or simple (Fig.6). Component objects of a complex object are related by relation defined by relational attributes - e.g. describing how such component objects are placed towards themselves.

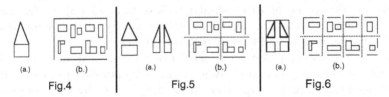

(a.)   (b.)   (a.)   (b.)   (a.)   (b.)

Fig.4   Fig.5   Fig.6

Each component object is labeled by algorithm transforming it - e.g. which let to obtain some of its attributes from others attributes or components (Fig.7).

Each component object can be related to another component object of a different complex object. Such relation let us to transform this object to the other one.

2   Similarity (Closeness) of Objects

One of the main problem to be solved is to construct relevant similarity measures between complex objects. The similarity of objects is relevant, if on the basis of it we may specify a proper set of algorithm attribute value (operations) for the new object, on the basis of known objects at some level. To do this complex objects must be decomposed up to some level. We compare complex objects (from the same or different information systems) by checking consistency of attributes and its value vectors. We start to check similarity of such objects by comparison attributes and its value vectors for the main complex object and next, by comparison attributes and its value vectors for its component objects. The number of concordant attributes and their values for the main complex objects and its component objects define the level of similarity, and thus some relations on the basis of which one may define similarity.

Let us consider Example 1. Complex objects are decomposed (Fig.4-Fig.6) in order to find the common set of object attributes, which can be measured. If such decomposition is proper, we may define on the basis of concordant attributes and their value vectors relations of similarity degree of objects. We measure similarity on several levels. First it is the similarity of objects from main information system (Fig.4), next it is the similarity of subobjects (from subinformations systems) (Fig.5, Fig.6). Similarity of main objects is more general, while similarity definition for subobjects is more detailed. If the main complex objects are not similar in a sufficient degree on high level attributes, we may try to define their similarity in a more detailed way, by taking into account their subobjects and similarity between them.

Objects from different information systems are described by different attributes and their value vectors, that is why we may define two types of similarity relation:

- consistency of attributes, and consistency of their value vectors,
- consistency of attributes, and inconsistency of their value vectors.

Objects from the same information system are described by the same attributes but with different their value vectors. For such objects there is one relation type - consistency or inconsistency of their value vectors.

One object can occur to be similar to several other objects by using the different similarity relation, depending on a quantity of consistent attributes and their value vectors and level of such consistency.

For any given object O it is extracted from hierarchical information system a set $\mathbf{O}$ of objects similar to O. Algorithms corresponding to objects from $\mathbf{O}$ should determine a proper algorithm for O. This is done by learning procedure. However, some simple heuristics can be also used based on similarity of objects (Fig. 8) what will be discussed later.

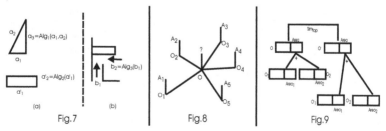

Fig.7   Fig.8   Fig.9

$\mathbf{O} = \{O, O_1, .., O_5\}$ is the set of objects similar to O and $A_1, .., A_5$ are algorithms transforming objects $O_1, .., O_5$.

Example 2. If on the top level $O\,Sim_{top}\,O'$ (where Sim_{top} - similarity on top level) and the value of attribute e.g. Area for O has been computed by decomposition of O into O_1 and O_2 and by performing the addition of $Area_1(O_1)$ and $Area_2(O_2)$, then we try to find first a decomposition of O' into O'_1 and O'_2 and next to compute its area, using the same operation of addition (Fig. 9). In some cases, for O'_1, O'_2 we search on corresponding levels of decomposition for similar objects and we follow the decomposition procedure for them.

3   New Object and Its Decomposition

Let us consider a new complex object. Now, the main problem is how to construct hierarchical structure of component objects for the new object by taking into account its similarity to the others objects and relations between extracted similar objects on the basis of their attributes. The new object is matched against complex objects represented in our knowledge base. The main goal of general strategy is to isolate possible component objects, by using similarity to others component objects and relations between them. In this way objects similar to a given new object in a satisfactory degree are extracted. On the basis of the extracted complex objects the proper operations of algorithms are selected: Attributes of extracted object and rules for the information system to which new object is assigned, point out operations of algorithm transforming such new object.

Here one can find the problem that not all values of attributes of the new object are known. Let us consider the rule $\{(a, a(O)), (b, b(O)), (c, c(O))\} \rightarrow op(O)$. For a new object attributes can be known only to $(a, a(O))$ and $(b, b(O))$. There is a problem if the system should start operation $op(O)$ or not. We may try to solve such problem by taking into account some attributes or rules for the subinformation systems for component objects. Considering subinformation systems one may find the missing attribute $(c, c(O))$ which let to start the operation $op(O)$, or just to skip this operation, and execute successfully the following operation of the main object algorithm. In some of the problems some missing attributes can be skipped, e.g. going trough the maze - we may try to find another way, but in some problems the missing attribute can not be skipped easily, e.g. in mathematical tasks - we have to find particular attribute value vector in order to solve the whole problem. That is why for some objects we have to declare very precise algorithm, with all attributes defined precisely (e.g. mathematical tasks), but for some objects we may declare more general algorithm with more general attributes (e.g. maze problem) which can be modified during its execution.

Example 3. We are looking for similar object to the new one. This step in CBR cycle [1] is called Retrieve. Let us consider a new complex object - new geometrical figure or maze. To assign the new proper algorithm for such object we have to decompose it up to some level. First we try to find the most similar main information system for the new object, taking into account general problem definition attributes - basic attributes and general relational attributes - declaring general relations between component objects of the complex object. If some algorithm operations - transforming the new object are known, e.g. we have some experience in going trough the new maze - we passed some its corridors successfully, we take them into account as well, while the most similar information system for the new object is assigned. Next we try to specify the most similar object or objects from the chosen information system for new object, taking into account basic attributes, detailed relational attributes, problem additional characteristics attributes and some known algorithm operation - attributes. The object or objects with maximum number of attributes and their values vectors consistent with attributes and their values vectors for the new object is chosen. If the most similar object or objects for the new object are extracted, we specify component objects (by its attributes) of the complex object. To do this we consider information systems for the component objects. In this way the most similar complex object (objects) is chosen for the new object. Known attributes for the new object and rules (obtained from information system to which the new object is assigned) defining object algorithm attributes let us to obtain some algorithm attributes (operations) for the new object (Fig. 8, 9).

It can happen that some chosen operations for the new object will not return the expected attribute values. In such case these "wrong" algorithm operations must be corrected and new missing operations specified. These steps in CBR cycle are called Reuse and Revise.

Let us consider the new algorithm Alg_n (defined by the rules) for the new object. $Alg_n = (op_{n1}, op_{n2}, .., op_{nm})$. We start to perform the operations $op_{n1}, ..,$ op_{nm}. To execute some of the operations we may need values got from previous operation or operations. If such needed value is missing the next operation can not be executed. In such situation we try to find the missing value (e.g. the edge length of some geometrical figures). To do this we consider another information system (sub-information system) (Table 2 or 3) for the component object of the complex object (we perform algorithm for the component object from which we try to get the missing value). If the missing value is obtained, we perform next operations of the main algorithm, if not we must modify some operations of the main algorithm.

For a given object with a strict structure, e.g. some mathematical tasks, to execute corresponding to this object algorithm it is necessary to perform all operations, and that is why we need all attributes value vectors. If some values are missing and if we can't obtain it from sub-information systems, we can't execute the operation. For some another objects, e.g. maze, if some attributes value vectors needed to execute the algorithm operation are missing, we may try to skip such operation, and execute another one. To do so, we must sometimes return to some already executed operations, and next perform some other operations. For example, if we can't pass the chosen corridor in the maze, we must go back to the corridors crossing and choose another corridor. In this way we correct the wrong algorithm operations.

4   Conclusions

We have presented the main idea on which a software system for problem solving is under the development.

We have discussed how to represent the various problems by means of complex objects represented by some hierarchical attributes, and how to use similarity between them for predicting the relevant algorithms corresponding to these objects.

The most difficult problem is the proper decomposition of the new complex object. On the basis of object attributes and their value vector we may predict similarity of the new object to the known ones. The level of such similarity specifies chances for developing the proper algorithm for the new object. Here we may notice three categories of the new objects: those which are similar to the known objects in a satisfactory degree, partial satisfactory degree and unsatisfactory degree. Any object similar in a satisfactory degree to the known objects allows to construct a correct algorithm. For an object similar in a partial satisfactory degree there is only a chance to construct a correct algorithm. Finally, for any objects similar in an unsatisfactory degree it is not possible to construct a correct algorithm.

We have distinguished at least two types of objects, those which are specified by some precise attributes - e.g. in case of mathematical tasks, and those which are specified by less precise attributes - e.g. in case of maze problems. For these two types different types of algorithms must be created.

We have outlined methods of retrieving similar objects for the new case, and reusing known algorithms for new objects using ideas of CBR cycle [1].

Acknowledgment. The author is due to thank Professor Andrzej Skowron for formulating the subject and for his numerous discussions and helpful critical remarks throughout the investigation.

References

1. A.Aamodt & E.Plaza (1994). *Case Based Reasoning: Foundational Issues, Methodological Variations, and System Approaches.* AI Communications.
2. K.Hammond (1986). *CHEF: A model of case-based planing.* In Proc. of AAAI-86, Cambridge MA: AAAI Press/MIT Press.
3. J.Kolodner (1993). *Case Based Reasoning.* Morgan Kaufmann.
4. J.Wierzbicki (1998) - *"CBR for complex objects represented in hierarchical information systems"*, (procedings First International Conference - Rough Sets and Current Trends in Computing, Warsaw 1998, Springer-Verlag Berlin Heidelberg.
5. Z.Pawlak (1991). *Rough Sets. Theoretical Aspects of Reasoning about Data.* Kluwer Academic Publishers, Boston, London, Dordrecht.
6. L.Polkowski, A.Skowron, J.Komorowski (1996). *Approximate case-based reasoning: A rough mereological approach.* In: H.D.Barkhard, M.Lenz (eds.), Fourth German Workshop on Case-Based Reasoning. System Development and Evaluation, Informatik Berichte 55, Humboldt University, Berlin, 144-151.

Rough Sets for Uncertainty Reasoning

S.K.M. Wong[1] and C.J. Butz[2]

[1] Department of Computer Science, University of Regina,
Regina, Canada, S4S 0A2, wong@cs.uregina.ca
[2] School of Information Technology & Engineering, University of Ottawa,
Ottawa, Canada, K1N 6N5, butz@site.uottawa.ca

Abstract. Rough sets have traditionally been applied to decision (classification) problems. We suggest that rough sets are even better suited for reasoning. It has already been shown that rough sets can be applied for reasoning about knowledge. In this preliminary paper, we show how rough sets provide a convenient framework for uncertainty reasoning. This discussion not only presents a new topic for future research, but further demonstrates the flexibility of rough sets.

1 Introduction

The theory of rough sets [4, 5, 9, 10] generalizes traditional set theory by allowing a concept to be described approximately by a lower and upper bound. Although rough sets have been extensively studied, most of these investigations demonstrated the usefulness of rough sets in decision (classification) problems. Wong [8] first demonstrated that rough sets can also be applied for reasoning about knowledge. This observation was also made later by Salonen and Nurmi [6].

In this preliminary paper, we extend the work in [6, 8] by demonstrating that rough sets can also be applied for *uncertainty* management. In [6, 8], rough sets are used as a framework to represent formulas such as "player 1 *knows* ϕ". By incorporating probability, we can now represent sentences such as "the probability of ϕ, according to player 1, is at least α", where ϕ is a formula and α is a real number in $[0, 1]$. Thereby, not only does this discussion present a new topic for future research, but it further demonstrates the flexibility of rough sets.

The remainder of this paper is organized as follows. Kripke semantics for modal logic are given in Section 2. The key relationships between rough sets and the Kripke semantics for modal logic are stated in Section 3. In Section 4, probability is incorporated into the logical framework. In Section 5, we demonstrate that rough sets are also a useful framework for uncertainty reasoning. The conclusion is given in Section 6.

2 Kripke semantics for Modal Logic

Consider an ordered pair $< W, R >$ consisting of a nonempty set W of possible worlds and a binary relation R on W. Let Q denote the set of sentence letters

W. Ziarko and Y. Yao (Eds.): RSCTC 2000, LNAI 2005, pp. 511–518, 2001.

(primitive propositions). An evaluator function f:

$$f : W \times Q \rightarrow \{\top, \bot\},$$

assigns a truth-value, $\top$ or $\bot$, to each ordered pair (w, q), where $w \in W$ is a *possible world*, and $q \in Q$ is a sentence letter. We call the triple $M = < W, R, f >$ a *model* (structure), and R a *possibility* (accessibility) relation.

The function of an evaluator is to determine which primitive proposition q is to be *true* at which world w in a model. We write $(M, w) \models q$, if $f(w, q) = \top$. We can now define what it means for a proposition (formula) to be true at a given world in a model by assuming that $\models$ has been defined for all its subformulas of φ. That is, for all propositions φ and ψ,

$$(M, w) \models \varphi \wedge \psi \text{ iff } (M, w) \models \varphi \text{ and } (M, w) \models \psi,$$
$$(M, w) \models -\varphi \text{ iff } (M, w) \not\models \varphi,$$

and

$$(M, w) \models \square\varphi \text{ iff } (M, x) \models \varphi, \text{ for all } x \text{ such that } (w, x) \in R.$$

The above definition enables us to infer inductively the truth-value, i.e., $(M, s) \models \varphi$, of all other propositions from those of the primitive propositions. We say "φ is true at (M, s)" or "φ holds at (M, s)" or "(M, s) satisfies φ", if $(M, s) \models \varphi$.

In order to establish a connection with rough set theory, we review the notion of an *incidence* mapping [1], denoted by I. To every proposition φ, we can assign a set of worlds $I(\varphi)$ defined by:

$$I(\varphi) = \{w \in W | (M, w) \models \varphi\}.$$

This function is used in establishing the relationship between a Kripke structure and an Auman structure in the recent work of Fagin et al. [2]. The important point of this discussion is that the incidence mapping I provides a set-theoretic interpretation of Kripke semantics.

3   Rough Sets versus Kripke semantics

The original motive of rough sets [5] was to characterize a particular *concept* (represented by a subset of a finite universe W of interest) based on the information (knowledge) on hand. This *knowledge* is represented by a binary relation R on W. Rough sets can be viewed as an extension of ordinary sets, in which a set $A \subseteq W$ is described by a pair $(\underline{A}, \overline{A})$ of subsets of W. Note that $\underline{A}$ and $\overline{A}$ are not necessarily distinct. For our exposition here, we may assume that R is an equivalence relation. In this case, rough sets are defined by the following *knowledge operator* K: for all $A \subseteq W$

$$\underline{A} = K(A) = \{w \in W \mid [w]_R \subseteq A\},$$

and

$$\overline{A} = -K(-A) = \{w \in W \mid [w]_R \cap A \neq \emptyset\},$$

where $[w]_R$ denotes the equivalence class of R containing the elements $w \in W$. In the theory of rough sets, we call $\underline{A}$ the lower approximation and $\overline{A}$ the upper approximation of A.

It was shown [7] that the Kripke semantic model is equivalent to the characterization of modal propositions by a rough-set model. That is, each proposition $\varphi \in L$ can be represented by a subset of possible worlds and the modal operator $\Box$ by the knowledge operator K defined above. The key relationships between the Kripke semantic model and the rough-set model are summarized as follows:

(i) $(M, w) \models \varphi$ iff $w \in I(\varphi)$,

(ii) $(M, w) \models \Box\varphi$ iff $w \in K(I(\varphi))$.

The above results enable us to adopt rough sets for reasoning about knowledge instead of using the framework based on modal logic as suggested by Fagin et al. [2].

We conclude this section with an example [2] to illustrate how the rough-set model is used in reasoning. Consider a deck of cards consisting of three cards labeled X, Y and Z. Assume there are two players (agents), i.e., $G = \{1, 2\}$. Players 1 and 2 each gets one of these cards. The third card is left face down. We describe a possible world by the cards held by each player. Clearly, there are six possible worlds, i.e., $W = \{(X, Y), (X, Z), (Y, X), (Y, Z), (Z, X), (Z, Y)\}$ $= \{w_1, w_2, w_3, w_4, w_5, w_6\}$. For example, $w_2 = (X, Z)$ says that player 1 holds card X and player 2 holds card Z. The third card Y is face down. We can easily construct the two partitions π_1 and π_2 of W, which respectively represent the knowledge of the two players. For example, $w_1 = (X, Y)$ and $w_2 = (X, Z)$ belong to the same block of π_1 because in a world such as $w_1 = (X, Y)$, player 1 considers two worlds possible, namely $w_1 = (X, Y)$ itself and $w_2 = (X, Z)$. That is, when player 1 holds card X, he considers it possible that player 2 holds card Y or card Z. Similarly, in a world $w_1 = (X, Y)$, player 2 considers the two worlds $w_1 = (X, Y)$ and $w_6 = (Z, Y)$ possible, i.e., w_1 and w_6 belong to the same block of π_2. Based on this analysis, one can easily verify that:

$$\pi_1 = \{[w_1, w_2]_{1X}, \ [w_3, w_4]_{1Y}, \ [w_5, w_6]_{1Z}\},$$
$$\pi_2 = \{[w_3, w_5]_{2X}, \ [w_1, w_6]_{2Y}, \ [w_2, w_4]_{2Z}\}.$$

It is understood that in both worlds w_1 and w_2 of the block $[w_1, w_2]_{1X}$ in π_1, player 1 holds card X; in both worlds w_1 and w_6 of the block $[w_1, w_6]_{2Y}$, player 2 holds card Y, and so on. The corresponding equivalence relations R_1 and R_2 can be directly inferred from π_1 and π_2. In this example, we have six primitive propositions: $1X$ denotes the statement "player 1 holds card X", $1Y$ denotes the statement "player 1 holds card Y", ..., and $2Z$ denotes the statement "player

2 holds card Z". Each of these propositions is represented by a set of possible worlds. By the definition of the mapping I, we obtain:

$$I(1X) = \{w_1, w_2\}, \quad I(1Y) = \{w_3, w_4\}, \quad I(1Z) = \{w_5, w_6\},$$

$$I(2X) = \{w_3, w_5\}, \quad I(2Y) = \{w_1, w_6\}, \quad I(2Z) = \{w_2, w_4\}.$$

Using these primitive representations, the representations of more complex propositions can be easily derived from properties $(i1) - (i5)$. For example,

$$\begin{aligned}
I(1X \wedge 2Y) &= I(1X) \cap I(2Y) \\
&= \{w_1, w_2\} \cap \{w_1, w_6\} = \{w_1\},
\end{aligned}$$

$$\begin{aligned}
I(2Y \vee 2Z) &= I(2Y) \cup I(2Z) \\
&= \{w_1, w_6\} \cup \{w_2, w_4\} = \{w_1, w_2, w_4, w_6\}.
\end{aligned}$$

More interesting is the following expression which indicates that if player 1 holds card X, then he *knows* that player 2 holds card Y or card Z:

$$\begin{aligned}
I(\Box_1(2Y \vee 2Z)) &= K_1(I(2Y \vee 2Z)) \\
&= \{w \mid [w]_{\pi_1 \cap \pi_2} \subseteq I(2Y \vee 2Z)\} \\
&= \{w \mid [w]_{\pi_1 \cap \pi_2} \subseteq \{w_1, w_2, w_4, w_6\}\} \\
&= \{w_1, w_2\}.
\end{aligned}$$

4   Incorporating Probability

The discussion here draws from that given by Halpern [3]. The language is extended to allow formulas of the form $P_i(\phi) \geq \alpha$, $P_i(\phi) \leq \alpha$, and $P_i(\phi) = \alpha$, where ϕ is a formula and α is a real number in the interval $[0,1]$. A formula such as $P_i(\phi) \geq \alpha$ can be read "the probability of ϕ, according to player i, is at least α".

To give semantics to such formulas, we augment the Kripke structure with a probability distribution. Assuming there is only one agent, a *simple probability structure* M is a tuple (W, p, π), where p is a discrete probability distribution on W. The distribution p maps worlds in W to real numbers in $[0,1]$ such that $\sum_{w \in W} p(w) = 1.0$. We extend p to subsets A of W by $p(A) = \sum_{w \in A} p(w)$. We can now define satisfiability in simple probability structures: the only interesting case comes in dealing with formulas such as $P_i(\phi) \geq \alpha$. Such a formula is true, if:

$$(M, w) \models P(\phi) \geq \alpha \text{ if } p(\{w | (M, w) \models \phi\}) \geq \alpha.$$

That is, if the set of worlds where ϕ is true has probability at least α. The treatment of $P_i(\phi) \leq \alpha$, and $P_i(\phi) = \alpha$ is analogous.

Simple probability structures implicitly assume that an agent's (player's) probability distribution is independent of the state (world). We can generalize simple probability structures with *probabilistic Kripke structures* by having p depend on the world and allowing different agents to have different probability distributions.

A *probabilistic Kripke structure* M is a tuple $(W, p_1, \ldots, p_n, \pi)$, where for each agent i and world w, we take $p_i(w)$ to be a discrete probability distribution, denoted $p_{i,w}$, over W. To evaluate the truth of a statement such as $P_i(\phi) \geq \alpha$ at world w we use the distribution $p_{i,w}$:

$$(M, w) \models P_i(\phi) \geq \alpha \text{ if } p_{i,w}(\{w | (M, w) \models \phi\}) \geq \alpha.$$

We now combine reasoning about knowledge with reasoning about probability. A *Kripke structure for knowledge and probability* is a tuple $(W, \mathcal{K}_1, \ldots, \mathcal{K}_n, p_1, \ldots, p_n, \pi)$. This structure can give semantics to a language with both knowledge and probability operators. A natural assumption in this case is that, in world w, agent i only assigns probability to those worlds $\mathcal{K}_i(w)$ that he considers possible. (However, in some cases this may not be appropriate [3].)

We use the following example from [3] to illustrate a logical approach to reasoning about uncertainty. Alice has two coins, one of which is fair while the other is biased. The fair coin has equal likelyhood of landing heads and tails, while the biased coin is twice as likely to land heads as to land tails. Alice chooses one of the coins (assume she can tell them apart by their weight and feel) and is about to toss it. Bob is not given any indication as to which coin Alice chose.

There are four possible worlds:

$$W = \{w_1 = (F, H), \ w_2 = (F, T), \ w_3 = (B, H), \ w_4 = (B, T)\}.$$

The world $w_1 = (F, H)$ says that the fair coin is chosen and it lands heads. We can easily construct two partitions π_{Alice} and π_{Bob} of W, which represent the respective knowledge of Alice and Bob:

$$\pi_{Alice} = \{[w_1, w_2], [w_3, w_4]\},$$
$$\pi_{Bob} = \{[w_1, w_2, w_3, w_4]\}.$$

The corresponding equivalence relations R_{Alice} and R_{Bob} can be directly inferred from π_{Alice} and π_{Bob}. In this example, we consider the following four propositions: f - Alice chooses the fair coin; b - Alice chooses the biased coin; h - The coin will land heads; t - The coin will land tails.

We first define a probability distribution $p_{Alice,w}$, according to Alice, for each of the worlds $w \in W$. In world $w_1 = (H, T)$, $p_{Alice,w_1}(w_1) = 1/2$, $p_{Alice,w_1}(w_2) = 1/2$, $p_{Alice,w_1}(w_3) = 0.0$, $p_{Alice,w_1}(w_4) = 0.0$. For world $w_3 = (B, T)$, $p_{Alice,w_3}(w_1) = 0.0$, $p_{Alice,w_3}(w_2) = 0.0$, $p_{Alice,w_3}(w_3) = 2/3$, $p_{Alice,w_3}(w_4) = 1/3$. These definitions are illustrated in Figure 1.

It can be verified that $p_{Alice,w_2} = p_{Alice,w_1}$ and $p_{Alice,w_4} = p_{Alice,w_3}$. Moreover, Bob's probability distributions are the same as Alice's, namely,

$$p_{Bob,w_i} = p_{Alice,w_1}, \quad i = 1, 2, 3, 4.$$

	Coin	Lands	p_{Alice,w_1}	p_{Alice,w_3}
w_1	fair	heads	1/2	0
w_2	fair	tails	1/2	0
w_3	biased	heads	0	2/3
w_4	biased	tails	0	1/3

Fig. 1. A knowledge system for Alice.

The truth evaluation function π maps $\pi(h, w_1) = true$, $\pi(h, w_2) = true$, $\pi(h, w_3) = false$, $\pi(h, w_4) = false$. Thus, $I(h) = \{w_1, w_2\}$.

It can now be shown that

$$(M, w_1) \models P_{Alice}(h) = 1/2,$$

since $p_{Alice,w_1}(\{w_1, w_3\}) = 1/2$. Similarly,

$$(M, w_2) \models P_{Alice}(h) = 1/2.$$

This means that Alice *knows* the probability of heads is 1/2 in world w_1:

$$(M, w_1) \models \Box_{Alice}(P_{Alice}(h) = 1/2),$$

since $(M, w_1) \models P_{Alice}(h) = 1/2$, $(M, w_2) \models P_{Alice}(h) = 1/2$, and $[w_1, w_2]$ is an equivalence class in R_{Alice}.

The same is not true for Bob. Note that

$$(M, w_1) \models P_{Bob}(h) = 1/2,$$

since $p_{Bob,w_1}(\{w_1, w_3\}) = 1/2$. However,

$$(M, w_3) \not\models P_{Bob}(h) = 1/2,$$

since $p_{Bob,w_3}(\{w_1, w_3\}) = 2/3$. Therefore,

$$(M, w_1) \not\models \Box_{Bob}(P_{Bob}(h) = 1/2),$$

since for instance $(w_1, w_3) \in R_{Bob}$. This says that Bob does *not* know that the probability of heads is 1/2 in world w_1.

5   Rough Sets for Uncertain t y Reasoning

Recall that each proposition is represented by a set of possible worlds. The proposition $P_{Alice}(h) = 1/2$, for instance, is represented by

$$I(P_{Alice}(h) = 1/2) = \{w_1, w_2\}.$$

Similarly, proposition $P_{Bob}(h) = 1/2$ is represented by

$$I(P_{Bob}(h) = 1/2) = \{w_1, w_2\}.$$

Recall the following results obtained in a previous section using a logical framework:

$$(M, w_1) \models \Box_{Alice}(P_{Alice}(h) = 1/2),$$
$$(M, w_2) \models \Box_{Alice}(P_{Alice}(h) = 1/2).$$

This knowledge can be expressed using the following proposition in rough sets:

$$K_{Alice}(P_{Alice}(h) = 1/2).$$

By definition, this proposition is represented by the following worlds:

$$\begin{aligned}
K_{Alice}(I(P_{Alice}(h) = 1/2)) &= K_{Alice}(\{w_1, w_2\}) \\
&= \{w \in W \mid [w]_{Alice} \subset \{w_1, w_2\}\} \\
&= \{w_1, w_2\}. \tag{1}
\end{aligned}$$

This result is consistent with our earlier result that:

$$I(\Box_{Alice}(P_{Alice}(h) = 1/2)) = \{w_1, w_2\}.$$

Even though Bob using the same probability distributions, he is still uncertain as to when the fair coin is used:

$$(M, w_1) \not\models \Box_{Bob}(P_{Bob}(h) = 1/2).$$

The same knowledge (or lack there of) can be expressed using rough sets as:

$$A = I(P_{Bob}(h) = 1/2) \quad = \quad \{w_1, w_2\}.$$

However,

$$\underline{A} = K(A) \quad = \quad \{w \mid [w]_{Bob} \subset A\} \quad = \quad \emptyset,$$

since

$$[w_1]_{Bob} \quad = \quad \{w_1, w_2, w_3, w_4\} \quad = \quad [w_2]_{Bob} \quad = \quad [w_3]_{Bob} \quad = \quad [w_4]_{Bob}.$$

Finally, let us determine when Alice *knows* that the coin is fair and also *knows* that the probability of heads is $1/2$. This sentence is represented in rough sets as:

$$K_{Alice}(f) \wedge K_{Alice}(P_{Alice}(h) = 1/2).$$

Now

$$I(K_{Alice}(f)) = \{w_1, w_2\}.$$

By Equation (1),

$$K_{Alice}(I(_{Alice}(h) = 1/2)) = \{w_1, w_2\}.$$

By the definition of the incidence mapping:

$$\begin{aligned}
&I(K_{Alice}(f) \wedge K_{Alice}(P_{Alice}(h) = 1/2)) \\
&= I(K_{Alice}(f)) \cap K_{Alice}(I(_{Alice}(h) = 1/2)) \\
&= \{w_1, w_2\} \cap \{w_1, w_2\} \\
&= \{w_1, w_2\}.
\end{aligned}$$

6   Conclusion

Rough sets have primarily been applied to classification problems. Recently, it has been shown that rough sets can also be applied to reasoning about knowledge [6, 8]. In this preliminary paper, we have added probability. This allows us to represent formulas such as "the probability of ϕ, according to player 1, is at least α", where ϕ is a formula and α is a real number in $[0, 1]$. Thus, the only extension to the work in [8] is to allow *formulas* involving probability.

On the other hand, our original objective was to introduce a probability *operator P* in the same spirit as the knowledge *operator K* in [8]. Unfortunately, while P behaves nicely with K, P does not always interact nicely with itself. W e are currently working to resolve these problems.

References

1. Bundy, A., Incidence calculus: a mechanism for probability reasoning. *International Journal of Automated Reasoning*, 1, 263-283, 1985.
2. Fagin, R., Halpern, J.Y., Moses, Y. and Vardi, M.Y., *Reasoning about knowledge*. MIT Press, Cambridge, Mass., 1996.
3. Halpern, J.: A logical approach to reasoning about uncertainty: a tutorial. in: Arrazola, X., Korta, K., and Pelletier, F.J., (Eds.), Discourse, Interaction, and Communication. Kluwer, 1997.
4. Pawlak, Z., W ong, S.K.M. and Ziarko, W., Probabilistic v ersus deterministic approach. *International Journal of Man-Machine Studies*, **29**, 81-95, 1988.
5. Pawlak, Z., *Rough sets - Theoretical Aspects of Reasoning about Data.* Kluwer Academic Publishers, 1991.
6. Salonen, H. and Nurmi, H.: A note on rough sets and common knowledge events. European Journal of Operational Research, 112, 692–695, 1999.
7. W ong, S.K.M., W ang, L.S. and Bollmann-Sdorra, P., On qualitative measures of ignorance. *International Journal of Intelligent Systems*, **11**, 27-47, 1996.
8. Wong, S.K.M.: A rough-set model for reasoning about knowledge. In *Rough Sets in Knowledge Discovery*, L. Polkowski and A. Skowron (Eds.), Physica-Verlag, 276-285, 1998.
9. Yao, Y.Y., Li, X., Lin, T.Y. and Liu, Q., Representation and classification of rough set models. In *the 3rd International Workshop on Rough Sets and Soft Computing*, 630-637, 1994.
10. Yao, Y.Y. and Lin, T.Y., Generalization of Rough Sets using Modal Logics, Intelligent Automation and Soft Computing, Vol. 2, No. 2, 103-120, 1996.

Feature Subset Selection by Neuro-rough Hybridization

Basabi Chakraborty

Faculty of Software and Information Science
Iwate Prefectural University
152-52 Aza Sugo, Takizawamura, Iwate 020-0193, Japan
email: basabi@soft.iwate-pu.ac.jp

Abstract. Feature subset selection is of prime importance in pattern classification, machine learning and data mining applications. Though statistical techniques are well developed and mathematically sound, they are inappropriate for dealing real world cognitive problems containing imprecise and ambiguous information. Soft computing tools like artificial neural network, genetic algorithm fuzzy logic, rough set theory and their integration in developing hybrid algorithms for handling real life problems are recently found to be the most effective. In this work a neuro-rough hybrid algorithm has been proposed in which rough set concepts are used for finding an initial subset of efficient features followed by a neural stage to find out the ultimate best feature subset. The reduction of original feature set results in a smaller structure and quicker learning of the neural stage and as a whole the hybrid algorithm seems to provide better performance than any algorithm from individual paradigm as is evident from the simulation results.

1 Introduction

Selection of a good subset of available features is not only the prime concern of pattern classification problems but also plays an important role in the fields of machine learning, knowledge discovery and data mining. Irrelevant and redundant features generally affect the performance of mostly all common machine learning or pattern classification algorithms. A good choice of an useful feature subset from a vast set of features helps in devising a compact and efficient learning algorithm for pattern classification or machine learning as well as results in better understanding and interpretation of data in knowledge discovery and data mining problems.

The problem of feature selection got an immemnse attention from statistical community from long back. Significant contributions [1] from statisticians have come in the field of pattern recognition, ranging from techniques that find optimal feature subset to suboptimal or near optimal solutions. Most of the statistical approaches are based on some assumption about probability distribution of the data set which, in practice, rarely follows the ideal one. Presently artificial neural networks (ANN) are becoming popular for analysis of vast data sets [2], [3]. The neural approach is specially efficient when the only source of available information is provided by the training data. They are known to be capable of extracting information from raw data and generalizes well.

W. Ziarko and Y. Yao (Eds.): RSCTC 2000, LNAI 2005, pp. 519–526, 2001.

Generally ANN's consider a fixed topology of layers of neurons interconnected by links in a predefined manner. Connection weights are usually initialized by small random values. The main drawbacks of a neural network learning system is that it is time consuming, specially when the number of input is large and though it is effective in presence of noise, the proper choice of network architecture till remains an unsolved problem. Recently fuzzy set theory and rough set theory [4] are widely used as the tools for knowledge extraction from large databases with imprecise and uncertain information. These soft computing tools have been proved to provide adaptivity and fault tolerance in modern intelligent systems. Hybrid systems [5], [6] have been developed by integrating the merits of different paradigms to handle real life problems more efficiently.

Motivated by the improved performance with hybridization, in this work a neuro-rough hybrid algorithm has been proposed to solve the problem of feature subset selection. The theory of rough set has been used in the first stage to have a rough idea about useful features and their subsets from the raw data. In the second stage a neural network has been used with the reduced feature set for finding out the ultimate best subset of features. The reduction of original feature set results in a smaller structure and quicker learning of the neural stage and as a whole the hybrid algorithm seems to provide better performance than any algorithm from individual paradigm. The algorithm has been simulated on two different data sets and it has been found that the present algorithm considerably reduces the time required for finding the best feature subset compared to our previous work reported in [7] where only neural network has been used for solving the problem. In the next section a brief introduction to rough set preliminaries and its use to feature subset selection problem has been discussed.

2   Rough Set Theory and Feature Subset selection

This section describes the basic concepts of rough set theory and how it can be used to have an initial idea of useful feature subset. The detail concepts of rough set theory can be found in [8].

2.1   Rough Set Preliminaries

According to *rough set* theory an *information system* is a four-tuple $S = (U, Q, V, f)$ where

U, a non-empty finite set, represents the universe of objects,

Q, a non-empty finite set, represents the set of attributes or features,

V, a non-empty finite set, represents the set of possible attribute or feature values and f is the information function which given an object and a feature, maps it to a value, i,e. $f : U \times Q \rightarrow V$.

An information system is represented by an attribute-value table in which rows are labeled by objects of the universe and columns by the attributes.

An *indiscernibility relation* is an equivalence relation with respect to a set of attributes (features) which partitions the universe of objects into a number of classes in such a manner that the member of same classes are indiscernible while the member of different classes are distinguishable with respect to the particular set of attributes.

Let P be a subset of Q, that is, P is a subset of features. The P-*indiscernibility relation*, denoted by $IND(P)$, is defined as,

$$IND(P) = \{(x,y) \in U^2 \text{ for every feature } a \in P, \ f(x,a) = f(y,a)\}$$

Then $IND(P) = \bigcap_{a \in P} IND(a)$

For any concept (class or label) X where $X \subseteq U$ and for any subset of features P, $P \subseteq Q$ the P-*lower*($\underline{P}$) and the P-*upper* approximation ($\overline{P}$) of X are defined as follows:

$$\underline{P}(X) = \cup\{Y \in U/IND(P) : Y \subseteq X\}$$
$$\overline{P}(X) = \cup\{Y \in U/IND(P) : Y \cap X \neq \phi\}$$

The boundary region for the concept X with respect to the subset of features P is defined as:

$$BND_P(X) = \overline{P}(X) - \underline{P}(X).$$

$POS_P(X) = \underline{P}(X)$ and $NEG_P(X) = U - \overline{P}(X)$ are known as P-positive region of X and P-negative region of X.

If $BND_P(X) = \phi$ then X is *definable or classifiable* using P. Otherwise the class X is a *rough set* with respect to the feature subset P.

Let $F = \{X_1, X_2, \ldots, X_n\}, X_i \in U$ be a classification of U and let $P \subseteq Q$, then $\underline{P}F = \{\underline{P}X_1, \underline{P}X_2, \ldots, \underline{P}X_n\}$ and $\overline{P}F = \{\overline{P}X_1, \overline{P}X_2, \ldots, \overline{P}X_n\}$ denote the P-lower and P-upper approximation of classification F (family of classes).

Inexactness of a rough set is due to the existence of boundary region. The greater the boundary region, the lower the accuracy of the set. Measures for accuracy or approximation of a rough set are defined below.

Approximation Measures Two measures to describe inexactness of approximate classifications have been defined in rough set theory as follows:

The *accuracy of approximation* expresses the possible correct decisions when classifying objects using attribute subset P and is defined as

$$\alpha_P(F) = \frac{\sum card\underline{P}X_i}{\sum card\overline{P}X_i}$$

The *quality of approximation* expresses the percentage of objects which can be correctly classified employing the attribute subset P. and is defined as

$$\gamma_P(F) = \frac{\sum card\underline{P}X_i}{cardU}$$

Reduct and Core of Attributes These two are fundamental concepts in rough set theory in connection with the knowledge reduction. A *reduct* denotes the essential part of the knowledge while *core* is the most important part. The P-reduct of A is the minimal subset of A which provides the same classification of objects as the whole set A.

An attribute or feature $a \in P$ is superfluous or redundant in P if $IND(P) = IND(P - \{a\})$; otherwise the attribute a is indispensible in P. P is an independent set of features if there does not exist a strict subset P' of P such that $IND(P) = IND(P')$.

A subset R ($R \subseteq P$) is a *reduct* of P if it is independent and $IND(R) = IND(P)$. Each reduct has the property that a feature cannot be removed from it without changing the indiscernibility relation. Many reducts for a given set of features P may exist.

The set of features belonging to the intersection of all reducts of P is called *core* of P (P-core). In fact P-core is the union of all indispensible features in P.

Thus $core(P) = \bigcap_{R \in Reduct(P)} R$

The indispensable features, reducts and core can be similarly defined relative to the output class or label known as relative reducts or relative core. In this paper we will use the term reduct and core to mean relative reduct and relative core with respect to the output labels or classes.

2.2   Feature Subset Selection with Rough Set

The basic idea of the first step of the two-stage neuro-rough hybrid feature subset selection algorithm proposed in this paper following the concepts of rough set theory are represented by the following steps.

1. The multidimensional data (each dimension representing individual attribute or feature) of known classification i,e the training data is expressed as the attribute value table with rows as the objects or the instances and columns as the attribute values and the corresponding class label. As continious feature value is difficult to handle by the proposed algorithm the feature values are discretized to some predefined levels. Let the number of attributes be Q and the number of classes be n in the data set where the number of instances are N. The objective is to select the subset of attributes P in such a way that minimizes $BND_P(X_i)$ for all i corresponding to the classification $F = \{X_1, X_2, \ldots, X_n\}, X_i \in U$.

2. Now the indiscernibility relation induced by any feature $(a \in Q)$ or a subset of features P from the feature set Q and the corresponding classification (partition) is examined. The feature subset P for which the accuracy measures $\alpha_P(F)$ and $\gamma_P(F)$ of the classification $F = \{X_1, X_2, \ldots, X_n\}, X_i \in U$ are highest, is selected as an approximation for the possible good feature subset.

3. In the next step it is examined whether the selected feature subset P has any relative reduct $P_1, P_2 \ldots$, relative to the classification F, considering the attribute-value table or the discernibility matrix.

4. The reduct of P obtained in the previous step (if multiple reduct exists, reduct containing the fewer number of features) is considered in the next step for presenting as inputs to neural network for finding out the ultimate best feature subset.

3   Neural Network for Best Feature Subset Selection

A fractal neural network model, a modified version of feedforward multilayer perceptron with statistically fractal connection structure, used earlier and reported in [7] has been used here. The proposed model and the feature subset selection algorithm by using it is presented in short in the next subsections.

3.1   Fractal neural network model

The fractal neural network model is a modified version of feedforward multilayer neural network in which upper layer neurons are connected to the lower layer neurons with a probability following an inverse power law which generates a sparse network with statistically fractal connection structure. However the final hidden layer is fully connected to the output layer. Each layer is an array of neurons in one or two dimension depending on the type of input to be processed. The probability that ith processing element in the kth layer receives connection from the jth processing element of the previous layer, defined by CP_{ijk} follows the law

$$CP_{ijk} = Ar_{ijk}^{D_k - d} \tag{1}$$
$$i = 1, 2 \ldots n_k$$
$$j = 1, 2 \ldots n_{k-1}$$
$$0 \leq D_k \leq d$$

where r_{ijk} is the Euclidean distance between ith processing element in the kth layer (considering one dimensional layers) and jth processing element of the previous layer defined as

$$r_{ijk} = ||Q_{ik} - Q_{j(k-1)}||, \; r_{ijk} \geq 1 \tag{2}$$

d denotes dimension of the array of neurons in kth layer. A represents a constant, D_k represents the fractal dimension (similarity dimension) of the synaptic connection distribution of k the layer. Q_{ik}, and $Q_{j(k-1)}$ denotes the spatial position of the ith processing element in the kth layer and jth processing element of the previous layer defined by

$$Q_{ik} = [\lceil n_{k-1}(2i - 1)/2n_k \rceil, k] \text{ for } i = 1, 2, \ldots, n_k \tag{3}$$

where n_{k-1} and n_k represents the number of neurons in the $(k-1)$th and kth layers respectively.

To implement such a sparse neural network, for each i, j, k, a uniform random number ρ on the interval [0,1] has to be generated and the connectivity C_{ijk} of the link from the ith processing element in the kth layer to the jth processing element of the previous layer is to be assigned as

$$C_{ijk} = 1, \text{ if } CP_{ijk} \geq \rho \tag{4}$$
$$= 0, \text{ Otherwise}$$

The operation of the network is similar to the operation of any multilayer feedforward backpropagation network. The connection structure of the network allows low probability of long range connection links and high probability of short range connection links.

3.2   Feature Subset Selection Algorithm

A simple algorithm for selecting best feature subset by the proposed fractal neural network has been presented below. The network is trained for optimum efficiency determined by the highest classification rate for the problem at hand by suitable set up of the different parameters using the feature subset selected in the first step as rough set reduct. The features are then removed one by one, selection for their removal is done by examining the change in classification rate.

Depending on the problem and the required classification accuracy, the final subset of features has to be determined. The actual steps of the algorithm are as follows.

1. For a selected feature subset of P features, a fractal neural network with the input layer of P neurons and the output layer of n neurons (for a n class problem) is set up. The number of hidden layers, the number of neurons in each hidden layer and the fractal dimension of the synaptic connections are chosen heuristically by trial. The connection structure is set up according to Eq. 1 and Eq. 4 with the proper selection of the value of d and A.
2. The network is trained several times for selecting the optimum number of hidden layers, the number of neurons in each hidden layer and the optimum value of the fractal dimension for which the classification rate for the test samples is the highest. This optimum net configuration is retained for later steps.
3. The fractal network set up in the previous step is retrained with the subset of input features one less from the initial subset of P features. The classification rate with the one less input is calculated with the same test samples.
4. All the inputs are removed one by one and the whole procedure of the previous step is repeated. The inputs are ranked according to the classification rate of the network without that particular input. The highest classification rate obtained, corresponds to the most irrelevant input.
5. After removal of the most irrelevant input selected in the previous step, step 3 and step 4 is repeated for removal of the next irrelevant feature.
6. The process is stopped when any one of the following stopping criteria is met.
    (a) The total number of features attains a pre-assigned limit.
    (b) The classification score falls below a preassigned limit.

4   Simulation and Results

The proposed algorithm has been simulated by two data sets, Sonar data set used for underwater target recognition [9] and Iris data set [10], commonly used to test pattern recognition problems.

4.1   Simulation with IRIS data

This data set contains three classes each with 50 sample vectors. Each sample has four feature vectors (F_1, F_2 , F_3 & F_4). As the number of features in this case are small, the feature set has been extended and twelve features have been generated from the primary four features and all togther sixteen features are considered as the feature set for our experiment. The generated features according to the increasing order of feature number are $(F_1, F_2, (F_3, F_4, F_1 - F_2, F_1 - F_3, F_1 - F_4, F_2 - F_3, F_2 - F_4, F_3 - F_4, F_1/F_2, F_1/F_3, F_1/F_4, F_2/F_3, F_2/F_4, F_3/F_4,)$.

Following *rough set theoretic* concepts, the initial approximation of the best feature subset has come out to be the following subset

$(3, 4, 8, 9, 11, 12, 5, 6, 14)$

In the second stage a the fractal network with 9 neurons in the input layer and 3 neurons in the output layer has been set up. One hidden layer with different number of neurons $(2, 4, 6)$ and different values of fractal dimension ranging from 0.85 to 0.98 has been used for experiment. The initial weight values are selected randomly from 0.1 to -0.1. The network with one hidden layer of 4 neurons and fractal dimension 0.9 has been chosen for the optimum network for feature selection. The final best feature subset came out to be $(3, 4, 8, 9)$.

Table 1 represents the comparison in terms of time and recognition score of the present algorithm and the algorithm presented in [7] in which only fractal neural network have been used as the tool feature subset selection. The table shows that the hybrid neuro-rough algorithm performs better than only neural algorithm in terms of time though the recognition rate and the ultimate selected feature subset are same.

No. of features in subset	Average recognition score		Time taken	
	for neural algorithm	for neuro-rough algorithm	for neural algorithm	for neuro-rough algorithm
9	97.2%	96.8%	1.38 hrs	0.25 hrs
4	98.2%	98.2%	1.72 hrs	0.58 hrs

Table 1. Comparison of Neural and Neuro-Rough algorithm for IRIS data set

4.2 Simulation with SONAR data

This data set is produced from taking 60 sample points per signal (making 60 features) from power spectral envelope of sonar returns from two types of targets. These samples were normalized to take on values between 0.0 and 1.0, details can be found in [9].

For this data set, initial approximation of the efficient feature subset following rough set theoretic algorithm has come out to be a set of 15 features. In the second stage the fractal network with 15 neurons in the input layer and 2 neurons in the output layer has been used. The connection structure has been set up according to Eq. 1 and Eq. 4 with the values of A and d taken as 1 as before. The number of neurons in the hidden layer is varied between 4 to 10 for experiment. The value of the fractal dimension has also been varied (from 0.8 to 0.95) to find out the optimum connection structure of the fractal network in this particular problem. Table 2 represents the values of time taken and the recognition score of the present algorithm and the neural only algorithm for Sonar data set. The table shows that the hybrid neuro-rough algorithm has better performance in the case of Sonar data also than only neural algorithm in terms of time while the recognition rate and the number of features in the ultimate selected feature subset are more or less same.

5   Conclusion

Feature subset selection is very important in pattern classification, machine learning or data mining problems. Most of the collected real data set contains redundant or irrelevant information. While statistical techniques to the problem

No. of	Average recognition score		Time taken	
features in subset	for neural algorithm	for neuro-rough algorithm	for neural algorithm	for neuro-rough algorithm
15	90.2%	93.8%	7.12hrs	0.57 hrs
5	98.2%	98.2%	7.48 hrs	1.67 hrs

Table 2. Comparison of Neural and Neuro-Rough algorithm for SONAR data set

of the best feature subset selection are well known and mathematically strong, they are computationally unattractive specially in case of real world large data set problems. Artificial neural networks are nowadays becoming popular tools in pattern classification.

In this work a hybrid two stage feature subset selection has been proposed to lessen time and computational burden. In the first stage rough set theoretic concepts are applied to extract information from the raw data set to find out approximate set of efficient features. In the second stage a fractal neural network model has been used to find out the ultimate best feature subset. As the number of features are reduced in the first stage the time taken for finding out the best feature subset is comparatively less than the neural only approach to feature subset selection problem. The simulation results also reflect the benefit of hybridization. Though extensive simulations by different data sets, specially from real world applications, are yet to be done, the proposed hybridization clearly shows a way for quick algorithm for solution of the feature subset selection problem.

References

1. P. A. Devijver and J. Kittler,"Pattern Recognition: A Statistical Approach", Prentice Hall International, 1982.
2. C. M. Bishop," Neural Networks for Pattern Recognition", Oxford University Press, 1995.
3. R. L. Kennedy et al, "Solving Data Mining Problems Through Pattern Recognition", Prentice Hall International, 1997.
4. J. W. Grzymala-Buss, " Knowledge acquisition under uncertainty- A rough set approach", Journal of Intelligent and Robotics Systems, Vol 1, No. 1, pp. 3–16, 1988.
5. R. Hashemi et al., " A fusion of rough sets, modified rough sets and genetic algorithms for hybrid diagonostic systems.", Rough Sets and Data Mining Analysis for Imprecise Data, Kluwer Academic Publishers, 1997.
6. M. Banerjee et al., " Rough fuzzy MLP: Knowledge encoding and Classification, IEEE Trans. on NN, Vol 9, No. 6, pp 1203–1215, 1998.
7. B. Chakraborty and Y. Sawada,"Fractal Neural Network Feature Selector for Automatic Pattern recognition System", IEICE Trans. on Fundamentals of Electronics, Communications and Computer Sciences, Vol E82-A, No. 9, pp. 1845–1850, September, 1999.
8. Z. Pawlak, "Rough sets, Theoretical Aspects of Reasoning About Data", Kluwer Academic Publishers, 1991.
9. P. R. Gorman and T. J. Sejnowski, "Analysis of Hidden Units in a Layered Network Trained to Classify Sonar Targets", Neural Networks, vol 1, pp . 75–89, 1988.
10. J. C. Bezdek, "Pattern Recognition with Fuzzy Objective Functions", Plenum Press, NY, 1981.

Accurate Estimation of Motion Vectors using Active Block Matching

Seok-Woo Jang, Kyu-Jung Kim and Hyung-Il Choi

Soongsil University, 1-1, Sangdo-5 Dong, Dong-Jak Ku, Seoul, Korea
hic@computing.soongsil.ac.kr

Abstract. We propose an active block matching algorithm for motion estimation. The proposed algorithm dynamically determines the search area and the matching metric. We exploit the constraint of small velocity changes of a block along the time to determine the origin of the search area. The range of the search area is adjusted according to the motion coherency of spatially neighboring blocks. Our matching metric includes multiple features. The degree of overall match is computed as the weighted sum of matches of individual features. We adjust the weights depending on the distinctiveness of features in a block, so that we may discriminate features according to the characteristics of an involved block. The experimental results show that the proposed algorithm can yield very accurate block motion vectors.

1 Introduction

The technique of block motion estimation is currently favored by many researchers in the field. The process of block matching is to find a candidate block, within a search area in the previous frame, that is most similar to the current block in the present frame, according to a predetermined criterion [1,2]. In this paper, we propose an active search algorithm for a candidate block. The search origin for each block is adjusted by a motion vector of the block in the previous frame to make use of the constraint of small velocity changes of a block along the time. We also adjust the range of the search area according to the motion coherency of spatially neighboring blocks. A smaller search area will be assigned to a block having more coherent motion in its neighboring blocks.

Most block matching algorithms just consider the difference of color or gray intensities of corresponding blocks when they compute the degree of match [3,4,5]. This criterion of match may be acceptable for the case of video coding, since the primary concern of coding is to reduce the redundancy between successive frames. However, when we need an accurate estimation of block motion vectors as in video conference, it may cause the problem. To resolve such a situation, we involve multiple features in a matching metric. The degree of overall match is computed as the weighted sum of matches of individual features. We adjust the weights depending on the distinctiveness of features in a block, so that we may discriminate features according to the characteristics of an involved block.

W. Ziarko and Y. Yao (Eds.): RSCTC 2000, LNAI 2005, pp. 527–531, 2001.

2   Adaptive Setting of Search Area

The blocks to be examined in the previous frame are within a search area whose origin and size are determined by exploiting the motion vectors of blocks in the previous frame. We denote a search area as in (1).

$$SA = (p(x,y), s(x,y)) = ((x,y)(xs, ys)) \tag{1}$$

where $p(x,y)$ denotes the origin of a search area and $s(x,y)$ denotes its size. In most image sequences, motions are smooth and slow-varying. Discontinuity of motion vectors only occurs at the boundary of objects moving in different directions [3]. Since the moving object often covers several blocks, motion vectors between adjacent blocks are highly correlated. In our proposed approach, we utilize these characteristics. The blocks of the previous frame already have motion vectors, since their corresponding blocks in the second previous frame have been identified. We presume that the motion vector of a block is likely to be similar to the motion vectors of its neighboring blocks. We also presume that the motion of a block does not change rapidly along a relatively small time interval. We therefore use, as the origin of a search area, the location (i.e., block) in the previous frame which points to the current block by its motion vector.

$$p_{B_i^{cur}}(x,y) = p_{B_{j(i)}^{prev}}(x,y) + MV(B_{j(i)}^{prev}) \tag{2}$$

In (2), $B_{j(i)}^{prev}$ denotes the block-j in the previous frame whose motion vector, $MV(B_{j(i)}^{prev})$, points to the current block-i in the present frame, B_i^{cur}. This equation depicts how the search origin of B_i^{cur} is computed.

Typically, only a small number of blocks have a large displacement in most image sequences. Therefore it is not efficient to fix the search range for each block. We take advantage of the inter-block motion correlation to adaptively determine the size of a search area. The size of a search area is allowed to vary within its maximum range of $s_{max}(x,y)$ and its minimum range of $s_{min}(x,y)$ as in (3).

$$s_{B_i^{cur}}(x,y) = s_{min}(x,y) + (1 - CF(MV(B_{j(i)}^{prev}))) \cdot (s_{max}(x,y) - s_{min}(x,y)) \tag{3}$$

In (3), $CF(MV(B_{j(i)}^{prev}))$ is a certainty factor that reflects the reliability of the motion vector $MV(B_{j(i)}^{prev})$. It is designed to have a value between 0 and 1, so that the size of range is adjusted depending on the reliability of the related motion vector. This strategy is based on the assumption of slow-varying motion. To determine the reliability of a motion vector, we utilize the smoothness constraint of motion. We represent the motion coherency of spatially neighboring blocks in a form of a certainty factor.

$$CF(MV(B_{j(i)}^{prev})) = \frac{K_1}{1 + K_2 \cdot VD(B_{j(i)}^{prev})} \tag{4}$$

$$VD(B_{j(i)}^{prve}) = ||MV(B_{j(i)}^{prev}) - \mu||^2 \cdot \sigma^2$$

$$\mu = mean\ of\ MV(B_{j*}^{prev}),\ j^* \in neighborhood\ of\ j(i)$$

$$\sigma^2 = variance\ of\ MV(B_{j*}^{prve}),\ j^* \in neighborhood\ of\ j(i)$$

In (4), $VD(B^{prev}_{j(i)})$ denotes the variance-compensated distance of $MV(B^{prev}_{j(i)})$ from the mean of its neighboring motion vectors. $VD(B^{prev}_{j(i)})$ becomes small when $MV(B^{prev}_{j(i)})$ is close to the mean and the variance is small. In other words, if the motion coherency of spatially neighboring blocks is high and also the motion vector under consideration is close to the mean motion vector of the neighboring blocks, then $VD(B^{prev}_{j(i)})$ becomes small and $CF(MV(B^{prev}_{j(i)}))$ gets large. A large certainty factor indicates that $MV(B^{prev}_{j(i)})$ is highly reliable and the size of a search area can be reduced.

3   Adaptive Setting of Matching Metric

Given a block of size N × N, the block motion estimation looks for the best matching block within a search area. One can consider various criteria as a measure of the match between two blocks [6]. We claim that the intensity difference between two blocks may not provide an accurate estimation of block motion, since it does not consider the internal structure. We suggest to involve various types of multiple features in a matching metric.

When multiple features are used in a matching metric, one has to take into consideration the following two issues. The first is the issue of normalizing the scale of features. The second issue is how to properly weigh the features according to their importance. We normalize each feature by dividing it with the highest value that it can have, so that the similarity of an individual feature between two blocks ranges from 0 to 1. For example, at each search point, the displaced block similarity (DBS) according to the k-th feature, f_k, is computed as in (5). In (5), the index $(n; i, j)$ denotes the block at (i, j) in the present frame, the index $(n - 1; i+x, j+y)$ denotes a candidate block at $(i+x, j+y)$ within a search area in the previous frame, and the displacement (x, y) denotes the corresponding disparity between two blocks.

$$DBS(f_k; i, j; x, y) = \tag{5}$$
$$1 - \frac{1}{N^2} \sum_{u=0}^{N-1} \sum_{v=0}^{N-1} \left| \frac{f_k(n; i+u, j+v) - f_k(n-1; i+x+u, j+y+v)}{f_{k\max}} \right|$$

The overall displaced block similarity $(ODBS)$ is then formed as the weighted sum of the similarities of individual features as in (6). The candidate block that maximizes (6) is selected as the best matched block and the corresponding displacement (x, y) becomes the motion vector of the block $(n; i, j)$.

$$ODBS(i, j; x, y) = \frac{1}{N} \sum_{k=1}^{N} w_k \cdot DBS(f_k; i, j; x, y) \tag{6}$$

To determine the weights w_k, we use the entropy value of the corresponding feature f_k in the search area under consideration. We compute weights of features as the normalized entropies as in (7), so that they have values from 0 to 1 and the sum of them becomes 1. In (7), H_{f_k} denotes the entropy of the corresponding feature and $P(\cdot)$ denotes the probability density of the feature value which is to be evaluated in a given search area.

$$w_k = \frac{H_{f_k}}{\sum_{k=1}^{N} H_{f_k}} \tag{7}$$

$$H_{f_k} = -\sum P(f_k(n; i, j)) \cdot \ln P(f_k(n; i, j))$$

4   Experimental Results and Discussions

In this section, we evaluate the performance of the proposed active block matching algorithm in terms of the accuracy of resulting motion vectors. As for features of each block, we used three different types; brightness, gradient, and laplacian.

Fig. 1 shows two adjacent frames in sequence of test images. In this sequence, frames are captured with such camera operations as the rotation by two degrees per frame in a clockwise direction, translation by two pixels per frame in a southeast direction, and zooming by 1.05 magnification per frame. Fig. 2 depicts motion vectors for the images of Fig. 1. Ideally, the motion vectors should diverge out in a form of spiral whose origin is a couple of pixels off to a southeast direction. We can clearly see that our approach outperforms others.

In this paper, we have presented an active block matching algorithm for motion estimation. Our algorithm dynamically determines the search area and the matching metric. Experimental results show that our algorithm outperforms other algorithms in terms of accuracy of the estimated motion vectors, though our algorithm requires some computational overhead.

Acknowledgement

This work was partially supported by the KOSEF through the AITrc and BK21 program (E-0075)

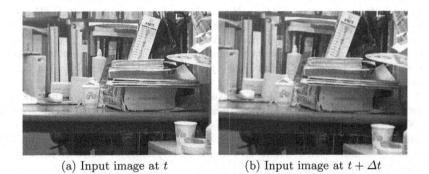

(a) Input image at t (b) Input image at $t + \Delta t$

Fig. 1. Test images with multiple camera operations

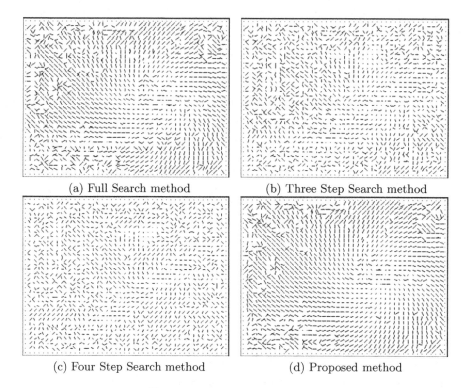

(a) Full Search method   (b) Three Step Search method

(c) Four Step Search method   (d) Proposed method

Fig. 2. Estimated motion vectors

References

1. T. Koga, K. Iinuma, and T. Ishiguro: Motion Compensated Interframe Coding for Video Conferencing. Proc. NTC81, New Orleans, LA, November (1981) G5.3.1-5.3.5
2. R. Srinivasan and K. R. Rao: Predictive Coding based on Efficient Motion Estimation. IEEE Trans. on Communications, Vol. COM-33, September (1985) 1011-1015
3. B. Liu and A. Zaccarin: New Fast Algorithms for the Estimation of Block Motion Vectors. IEEE Trans. on Circuits and Systems for Video Technology, Vol. 3, No. 2 (1994) 438-441
4. R. Li, B. Zeng, and M. L. Liou: A New Three-step Search Algorithm for Block Motion Estimation. IEEE Trans. on Circuits and Systems for Video Technology, Vol. 4, No. 4 (1994) 438-441
5. Lai-Man Po and Wing-Chung Ma: A Novel Four-Step Algorithm for Fast Block Motion Estimation. IEEE Trans. on Circuits and Systems for Video Technology, Vol. 6, No. 3 (1996) 313-317
6. Ramesh Jain, Rangachar Kasturi, and Brian G. Schunck: Machine Vision. McGraw-Hill, (1995)

Anytime Algorithm for Feature Selection

Mark Last[1], Abraham Kandel[1], Oded Maimon[2], Eugene Eberbach[3]

[1]Department of Computer Science and Engineering, University of South Florida, 4202 E. Fowler Avenue, ENB 118, Tampa, FL 33620 USA
{mlast,kandel}@csee.usf.edu
[2]Department of Industrial Engineering, Tel-Aviv University, Tel-Aviv 69978, Israel
maimon@eng.tau.ac.il
[3]Jodrey School of Computer Science, Acadia University, Wolfville, NS B0P 1X0, Canada
eugene.eberbach@acadiau.ca

Abstract. Feature selection is used to improve performance of learning algorithms by finding a minimal subset of relevant features. Since the process of feature selection is computationally intensive, a trade-off between the quality of the selected subset and the computation time is required. In this paper, we are presenting a novel, anytime algorithm for feature selection, which gradually improves the quality of results by increasing the computation time. The algorithm is interruptible, i.e., it can be stopped at any time and provide a partial subset of selected features. The quality of results is monitored by a new measure: fuzzy information gain. The algorithm performance is evaluated on several benchmark datasets.

Keywords: feature selection, anytime algorithms, information-theoretic network, fuzzy information gain

1 Introduction

Large number of potential features constitutes a seriously obstacle to efficiency of most learning algorithms. Such popular methods as k-nearest neighbors, C4.5, and backpropagation do not scale well in the presence of many features. Moreover, some algorithms may be confused by irrelevant or noisy attributes and construct poor classifiers. A successful choice of features provided to a classifier can increase its accuracy, save the computation time, and simplify its results.

In practical applications, like data mining, there is no better solution than using the knowledge of a domain expert, who can identify manually all relevant predictors of a given variable. However, in many learning problems, such an expert is not available, and we have to use *automated methods* of feature selection that choose an optimal subset of features according to a given criterion. A detailed overview of feature selection methods is presented by Liu and Motoda (1998).

Since classification accuracy is an important objective of learning algorithms, the most straightforward method (called the *wrapper model*) is to evaluate each subset of

W. Ziarko and Y. Yao (Eds.): RSCTC 2000, LNAI 2005, pp. 532–539, 2001.

features by running a classifier and measuring its validation accuracy. Obviously, this approach requires a considerable computation effort. Another approach (the *filter model*) uses indirect performance measures (like information, distance, consistency, etc.). The filter algorithms are computationally cheaper, but they are evaluating features in a random order, which makes their intermediate results hardly useful.

Whether the wrapper model or the filter model is applied to a set of features, the user may have to stop the execution of the algorithm, because there is no more time left for continuing the computation. Moreover, the time constraints may be unknown in advance and they can vary from seconds in real-time learning systems to hours or days in large-scale knowledge discovery projects. In both cases, we may be interested to find a good, but not necessarily the optimal, set of features as quickly as possible. However, as appears from (Liu and Motoda, 1998), the existing methods of feature selection do not consider the trade-off between time and performance.

Anytime algorithms (e.g., Dean and Boddy, 1988, Horvitz, 1987, Russell and Wefald, 1991, Zilberstein, 1996) offer such a trade-off between the solution quality and the computational requirements of the search process. The approach is known under a variety of names, including flexible computation, resource bounded computation, just-in time computing, imprecise computation, design-to-time scheduling, or decision-theoretic metareasoning. All these methods attempt to find the best answer possible given operational constraints. A formal model for anytime algorithms is provided by $-calculus (Eberbach, 2000), which is a higher-order polyadic process algebra with a utility (cost) allowing to capture bounded optimization and metareasoning typical for distributed interactive AI systems.

In section 2, we are describing the information-theoretic connectionist method of feature selection, initially introduced by us in (Maimon, Kandel, and Last, 1999) and (Last and Maimon, 1999). This paper shows for the first time that the method is much faster than the wrapper techniques and it can be implemented as an anytime algorithm, when the computation time is limited. Section 3 reports initial experiments that study the performance of the information-theoretic method and suggests possible enhancements using $-calculus. Finally, in section 4 we summarize the benefits and the limitations of our approach and discuss some directions for future research in the field of resource-bounded feature selection.

2   Information-Theoretic Method of Feature Selection

Our method selects features by constructing an *information-theoretic connectionist network*, which represents interactions between the predicting (*input*) attributes and the classification (*target*) attributes. The minimum set of input attributes is chosen by the algorithm from a set of *candidate input* attributes. The network construction procedure is outlined in sub-section 2.1. The theoretical properties of the algorithm in the context of anytime computation are evaluated in sub-section 2.2.

2.1   Network Construction Algorithm

An information-theoretic network is constructed for each target attribute separately. It consists of the root node, a changeable number of hidden layers (one layer for each input attribute), and a target layer. Each hidden (target) layer consists of nodes representing different values of an input (target) attribute. The network differs from the structure of a standard decision tree (see Quinlan, 1986 and 1993) in two aspects: it is restricted to the same input attribute at all the nodes of each hidden layer and it has interconnections between the terminal (unsplitted) nodes and the final nodes, representing the values of the target attribute.

The network construction algorithm starts with a single-node network representing an empty set of input attributes. A node is splitted if it provides a statistically significant decrease in the conditional entropy of the target attribute (based on a pre-defined significance level). A new input attribute is selected to maximize the total significant decrease in the conditional entropy. The nodes of a new hidden layer are defined for a Cartesian product of splitted nodes of the previous hidden layer and values of the new input attribute.   If there is no candidate input attribute significantly decreasing the conditional entropy of the target attribute, then the construction stops. Detailed descriptions of the algorithm steps are provided in (Maimon, Kandel, and Last, 1999) and (Last and Maimon, 1999).

An example of an information-theoretic connectionist network, which has three hidden layers (related to three selected attributes), is shown in Fig. 1.   The performance of the algorithm is evaluated in Section 3 below.

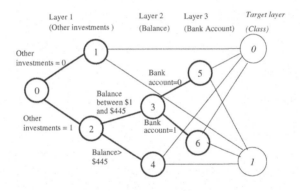

Fig. 1. Information-Theoretic Network: Credit Dataset

2.2   Anytime Properties of the Information-Theoretic Algorithm

According to Zilberstein (1996), the desired properties of anytime algorithms include the following: *measurable solution quality*, which can be easily determined at run time, *monotonicity* (quality is a non-decreasing function of time), *consistency* of the quality w.r.t. computation time and input quality, diminishing returns of the quality over time, interruptibility of the algorithm (from here comes the term *any time*), and preemptability with minimal overhead.   Thus, measuring the quality of the intermediate results is the key concept of anytime algorithms.

To represent the automated perception of the network quality, we will use here a new measure, called *fuzzy information gain*, which is defined as follows:

$$FGAIN = \frac{2}{1+e^{out}}, \quad out = \frac{\beta\alpha H(A_i/I_i)}{MI(A_i;I_i)} \tag{1}$$

Where

$H(A_i/I_i)$ - estimated conditional entropy of the target attribute A_i, given the set of input attributes I_i

$MI(A_i;I_i)$ - estimated mutual information between the target attribute A_i and the set of input attributes I_i

α - significance level, used by the algorithm

β - scaling factor, representing the perceived utility ratio between the significance level and the estimated mutual information. The meaning of different values of β is demonstrated in Fig. 2. The shape of *FGAIN (MI)* varies from a step function for low values of β (about 1) to almost a linear function, when β becomes much higher (about 500). Thus, β can be used to represent the level of user-specific quality requirements. A general fuzzy-theoretic approach to automating the human perception of data is described in (Last and Kandel, 1999).

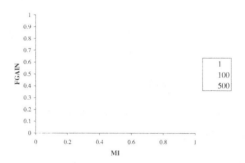

Fig. 2. Fuzzy Information Gain as a function of *MI*, for three different values of β.

Interpretation. FGAIN is defined above as a continuous monotonic function of three parameters: α, $H(A_i/I_i)$, and $MI(A_i;I_i)$. It is non-increasing in the significance level α, because lower α means higher confidence and, consequently, higher quality. In the ideal case $\alpha = 0$, which implies that *FGAIN* is equal to one. *FGAIN* is also non-increasing in the conditional entropy $H(A_i/I_i)$, because lower conditional entropy represents lower uncertainty of the target attribute, given the values of the input attributes. If the target attribute is known perfectly $(H(A_i/I_i) = 0)$, FGAIN obtains the highest value (one). On the other hand, *FGAIN* is non-decreasing in the mutual information $MI(A_i;I_i)$ that represents the decrease in the uncertainty of the target. When $MI(A_i;I_i)$ becomes very close to zero, FGAIN becomes exponentially small.

Now we need to verify that our method of feature selection has the desired properties of anytime algorithms, as defined by Zilberstein (1996). The conformity with each property is checked below.

- *Measurable quality*. According to equation (1), the Fuzzy Information Gain can be calculated directly from the values of conditional entropy and mutual information after each iteration of the algorithm.
- *Recognizable quality*. In (Last and Maimon, 1999), we have shown that the mutual information can be calculated incrementally by adding the conditional mutual information of each step to the mutual information at the previous step. This makes the determination of *FGAIN* very fast.
- *Monotonicity*. A new attribute is added to the set of input attributes only if it causes an increase in the mutual information. This means that the mutual information is a non-decreasing function of run time. Since one can easily verify, that the Fuzzy Information Gain is a monotonic non-decreasing function of *MI*, the monotonicity of the quality is guaranteed.
- *Consistency*. The theoretical run time of the algorithm has been shown by us in (Last and Maimon, 1999) to be quadratic-logarithmic in the number of records and quadratic polynomial in the number of initial candidate input attributes. In the next section, we are going to analyze experimentally the performance profile of the algorithm on datasets of varying size and quality.
- *Diminishing returns*. This property is very important for algorithm's practical usefulness: it means that after a small part of the running session, the results are expected to be sufficiently close to the results at completion time. We could prove this property mathematically, if we could show that the mutual information is a concave function of the number of input attributes. Though the last proposition is not true in a general case, it is possible to conclude from Fano's inequality (see Cover, 1991) that the mutual information is *bounded* by a function, which behaves this way. This conclusion is confirmed by the results of the next section.
- *Interruptibility*. The algorithm can be stopped at any time and provide the current list of selected attributes. Each iteration forms, what is called, a *contract anytime algorithm*, i.e. the corrections of *FGAIN* are available only after termination of an iteration.
- *Preemptability*. Since the algorithm maintains the training data, the list of input attributes, and the structure of the information-theoretic network, it can be easily resumed after an interrupt. If the suspension is expected to be long, all the relevant information may be stored in files on a hard disk.

3   Experimental Results

According to Zilberstein (1996), the performance profile (PP) of an anytime algorithm denotes the expected output quality as a function of the execution time t. To study the performance profile of the information-theoretic method for feature selection, we have applied it to several benchmark datasets, available from the UCI Machine Learning Repository (Blake and Merz, 1998). Rather than measuring the absolute execution time of the algorithm on every dataset, we have normalized it with respect to the *completion time*, which is the minimal time, when the expected quality is maximal (Zilberstein, 1993). Obviously, this relative time is almost independent of the hardware platform, used for running the algorithm.

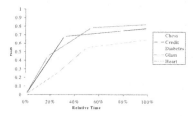

Fig. 3. Performance profile of the information-theoretic algorithm

We have used seven datasets for our analysis (see Table 1), but in two datasets (Breast and Iris), the run time was too short to be detectable by the computer system (Pentium II 400 MHZ). Thus, we are presenting in Fig. 3 performance profiles for five datasets only. Two important observations can be made from this chart. First, we can see that *FGAIN* is a non-decreasing function of execution time. The second observation is about the diminishing returns: except for the Chess dataset, the performance profiles are *concave* functions of time. We have explained the theoretical background of this result in sub-section 2.2 above.

The number of selected features in each dataset and the absolute execution times are shown in Table 1. The size of the datasets varies between 150 and 3,196 cases. The total number of candidate input attributes is up to 36, including nominal and continuous features. On average, less than 30% of the attributes have been selected by the algorithm, when it was run to its termination. The completion time starts with undetectable (less than 0.1 sec.) and goes up to 1.65 sec. for the Diabetes dataset, which has 768 records and 8 continuous attributes. These times are significantly lower than the execution times of a wrapper selector, which may vary between 16sec and several minutes for data sets of similar size (see Liu and Motoda, 1998).

Another question is how useful are the selected features for the classification task? The selected features can be considered useful, if a classifier's accuracy remains at approximately the same level. To verify this assumption, we have partitioned each dataset into training and validation records, keeping the standard 2/3 : 1/3 ratio (Liu and Motoda, 1998). The C4.5 algorithm (Quinlan, 1993) has been trained on each dataset two times: before and after feature selection. The error rate of both models has been measured on the same validation set. The minimum and the maximum error rates have been calculated for a 95% confidence interval. As one can see from Table 2, the error rate of C4.5 after feature selection is not significantly different from its error rate with all the available features. Moreover, it tends to be slightly lower after applying the feature selection algorithm. One exception is the Chess dataset, where the error rate has increased beyond the upper bound of the confidence interval. Due to the feature selection procedure, the stability of the error rate is accompanied, in most datasets, by a considerable reduction in the size of the decision tree model (measured by the number of tree nodes).

The novelty of our approach is that it allows capturing the trade-off between the solution quality and the time saved and/or complexity of classification represented by the number of input attributes. This can be crucial for classification algorithms working with a large number of input attributes, or with real time constraints. Alternative quality measures and costs of meta-reasoning can be studied in the process algebra framework provided by $-calculus (Eberbach, 2000) which formalizes anytime algorithms. For example, in terms of $-calculus expressing the tradeoff between the quality solution and the time, can be thought as a new measure $FGAIN_{tot}=(1\text{-}t)\ FGAIN$, where t is a normalized execution time (assuming that the execution time is bounded), or alternatively *out* in *FGAIN* can be modified.

Table 1. Feature selection: summary of results

Dataset	Data Size	Classes	Continuous	Nominal	Total Attributes	Selected Attributes	Completion Time (sec.)
Breast	699	2	9	0	9	3	-
Chess	3196	2	0	36	36	9	0.28
Credit	690	2	6	8	14	3	1.04
Diabetes	768	2	8	0	8	4	1.65
Glass	214	6	9	0	9	3	0.61
Heart	297	2	6	7	13	3	0.22
Iris	150	3	4	0	4	1	-
Mean	**859**	**3**	**6**	**7**	**13.3**	**3.6**	**0.76**

Table 2. Error rate and tree size of C4.5 before and after feature selection

Dataset	Validation Items	Before F.S. Tree Size	Error Rate	Min.	Max.	After F.S. Tree Size	Error Rate
Breast	204	29	5.4%	2.3%	8.5%	19	4.9%
Chess	1025	45	1.3%	0.6%	2.0%	29	3.0%
Credit	242	26	14.5%	10.0%	18.9%	3	14.0%
Diabetes	236	63	28.4%	22.6%	34.1%	23	23.3%
Glass	71	39	36.6%	25.4%	47.8%	39	33.8%
Heart	93	33	19.4%	11.3%	27.4%	16	24.7%
Iris	49	9	0.0%	0.0%	9.5%	5	2.0%

4   Summary

In this paper, we have presented a novel algorithm for feature selection, which can be interrupted at any time and provide us with a partial set of selected features. The quality of the algorithm results is evaluated by a new measure, the fuzzy information gain, which represents the user perception of the model quality. The performance profile of the algorithm has been shown to be a non-decreasing and mostly concave function of execution time. The quality of the final output has been confirmed by applying a data mining algorithm (C4.5) to a set of selected features.

Topics for future research include consideration of alternative quality measures, predicting expected quality for a given run time (and vice versa), and integrating anytime feature selection with real-time learning systems.

Acknowledgment

This work was partially supported by the USF Center for Software Testing under grant no. 2108-004-00.

References

1. C.L. Blake and C.J. Merz , UCI Repository of machine learning databases, , 1998.
2. T. M. Cover, Elements of Information Theory, Wiley, New York, 1991.
3. T. Dean and M. Boddy, An Analysis of Time-Dependent Planning, Proc. AAAI-88, pp.49-54, AAAI, 1988.
4. E. Eberbach, Expressiveness of $-Calculus: What Matters?, Proc. The Ninth Intern. Symp. on Intelligent Information Systems IIS'2000, Springer-Verlag, Bystra, Poland, June 2000.
5. E. Eberbach, Expressing Evolutionary Computation, Genetic Programming, Artificial Life, Autonomous Agents and DNA-Based Computing in $-Calculus – Revised Version, Proc. Congress on Evolutionary Computation CEC'2000, San Diego, CA, July 2000.
6. E.J. Horvitz, Reasoning about Beliefs and Actions under Computational Resource Constraints, Proc. of the 1987 Workshop on Uncertainty in AI, Seattle, Washington, 1987.
7. M. Last and O. Maimon, An Information-Theoretic Approach to Data Mining, Submitted to Publication, 1999.
8. M. Last and A. Kandel, Automated Perceptions in Data Mining, Proc. 1999 IEEE International Fuzzy Systems Conference, Seoul, Korea, pp. 190-197, 1999.
9. H. Liu and H. Motoda, Feature Selection for Knowledge Discovery and Data Mining, Kluwer, Boston, 1998.
10. O. Maimon, A. Kandel, and M. Last, Information-Theoretic Fuzzy Approach to Knowledge Discovery in Databases. In Advances in Soft Computing - Engineering Design and Manufacturing, R. Roy, T. Furuhashi and P.K. Chawdhry, Eds. Springer-Verlag, London, 1999.
11. S. Russell and E. Wefald, Do the Right Thing: Studies in Limited Rationality, The MIT Press, 1991.
12. J.R. Quinlan, Induction of Decision Trees, Machine Learning, vol. 1, no. 1, pp. 81-106, 1986.
13. J. R. Quinlan, C4.5: Programs for Machine Learning, Morgan Kaufmann, San Mateo, CA, 1993.
14. S. Zilberstein, Operational Rationality through Compilation of Anytime Algorithms, Ph.D. Dissertation, University of California at Berkeley , 1993.
15. S. Zilberstein, Using Anytime Algorithms in Intelligent Systems, AI Magazine, vol. 17, no. 3, pp. 73-83, 1996.

Reconstruction in an Original Video Sequence of the Lost Region Occupied by Caption Areas

Byung Tae Chun[1], Younglae Bae[1] and Tai-Yun Kim[2]

[1] Image Processing Department, Computer and Software Technology Lab.
ETRI (Electronics and Telecommunications Research Institute)
161 Kajong-Dong, Yusong-Gu, Taejon, 305-350, Korea
E-mail :chunbt@etri.re.kr
[2] Dept. of Computer Science and Engineering, Korea University,
Anam-dong, Seongbuk-ku, Seoul 136-701, Korea

Abstract. Conventional researches on image restoration have focused on restoring blurred images to sharp images using frequency filtering or video coding for transferring images.

In this paper, we proposes a method for recovering original images using camera motion and video information such as caption regions and scene changes. The method decides the direction of recovery using the caption information and scene change information. According to direction of recovery, a rough estimate of the direction and position of the original image is obtained using calculated motion vector from camera motion. Because the camera motion dose not reflects local object motion, some distortion can happen in the recovered image. To solve this problem, block matching algorithm that is applied in units of caption character components on the obtained recovery positions. Experimental results show that the case of images having little motions is well recovered. We see that the case of images having motion in complex background is also recovered.

1 Introduction

Captions are frequently inserted into broadcast images or video images to aid the understanding of audience. For such images already broadcast, it is sometimes necessary to remove the captions and recover the original images. When the number of images requiring such recovery is small, manual processing is possible, but as the number grows it would be very difficult to do it manually. Therefore, a method for recovering original image data for the caption areas is needed. Research on image restoration has focused on restoring blurred images to sharp images using frequency filtering [1] or video coding for transferring images [2] . Other research include recovery of cultural heritage using interpolation [3]. The method used is based on lines and therefore are not suitable for caption areas with larger sizes. Restoration methods using *BMA(Block Matching Algorithm)* [4] are done by simple comparison with previous frames such that errors can propagate.

W. Ziarko and Y. Yao (Eds.): RSCTC 2000, LNAI 2005, pp. 540–544, 2001.

2   A prior-information extraction for reconstructing original images

To reconstruct the lost region occupied by caption areas, we extract a prior-information in videos such as the information for caption area, cut detection and camera motion.

The caption information consists of the start frame and the end frame of caption and character components of extracted caption areas. The caption extraction method [5] we use is based on graph-theoretic clustering . We can decide the direction and the starting point for recovering using the information for the start frame and the end frame. And the extracted character components are used as basic units for recovering.

The information for cut detection is used to decide the direction and the ending point for recovery. We use the method [6] for detecting cuts in video. This method proposes a new algorithm of detecting cuts using motion vector(Mv). The motion vector(Mv) consists of magnitude(M) and direction(D).

The information for camera motion is used to decide the location of the reference original image. We use a method based on the method   [7] to extract information for camera motion in video. The camera motions are classified into *Pr(Panning-right)*, *Pl(Panning-left)*, *Tu(Tilting-up)*, *Td(Tilting-down)*, and *Zm(zooming)*.

3   Reconstruction of the lost region occupied by caption areas

Because the frame just before the caption appeared or the frame just after the caption disappeared has the original image, we find a position of the start frame(St-fr_i) and the end frame(Ed-fr_i), and use the frames as the basis for recovery.

The direction of recovery is to decide which direction to proceed for recovery starting from the start frame or the end frame. We decide the end points of recovery in relation to the information for cut detection. There are three cases.

In first case, there is no scene change, the order for recovery is from the start frame(St-fr_i) to the middle frame of caption and from the end frame(Ed-fr_i) to the middle frame.

In second case, if there is a scene change, the order is from the start frame(St-fr_i) to the scene change frame and from the end frame(Ed-fr_i) to the scene change frame.

In third case, if there are more than two scene changes, the direction of recovering is from the start frame(St-fr_i) to the scene change(REd_1) in the forward direction, and from the end frame(Ed-fr_i) to the scene change(REd_2) in the reverse direction. In this paper, we don't process frames between scene changes. Because the frames between the first and second scene change have no original image for reference and the recovered character region is too big, the traditional method or our method is not able to recover the original image. Therefore, these frames should be processed by another method.

We recover the original image for caption area by using extracted video information(caption information and scene change information) and camera motion information. The character components are extracted from extracted caption areas as shown in Fig 1. They are used as the basic units for recovery. The camera motion information gives each frame general information for camera motion and motion vector information which are the motion direction and magnitude as shown in Fig 2. Here *No* stands for no camera operation.

Fig. 1. Extracting character components in extracted caption region

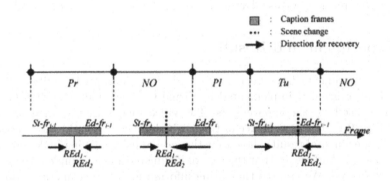

Fig. 2. Camera motion and direction information for recovery

A method for original image recovery for caption area is that firstly we find where the character component in start frame(St-fr_i) is located in the ogrinal image. The caption areas are replaced by founded orignal areas. We recover current frame(St-fr_i) for all character components by above processing step. The recovered caption area is used for recovering the next frame(St-fr_i+1). If the recovery is finished from the start frame to REd_1, then the recovery is done from the start frame(Ed-fr_i) to REd_2 in reverse direction.

However how can we find where character components are located in the original image ? We may find it using camera motion information. But because we determined the camera motion from the whole images, we are not able to reflect local object movements, which is a cause of distortion in recovery. To solve this problem, we use *BMA*. Block matching is very popular and used for motion compensation. Our matching criterion is *MAD (Minimum Absolute Difference)* criterion. The *MAD* is defined as in formula(1).

Block matching compares the contour pixels of character component with the original image which has no caption. And then we find the position of minimum color distance with original image. If we find the position for minimum color distance in the original image, the color of the recovered caption region is taken from the position.

$$MAD(d_1, d_2) = \sum_{(x,y) \in R} | F(x, y, t) - F(x + d_1, y + d_2, t - 1) | \qquad (1)$$

(Here F is a frame in video sequence, R specifies the contour pixels of character component for which the translation vector has to be calculated. $|d1|,|d2|$ < the search range).

When there is a camera motion, *d1*, *d2* values are determined using the motion vector($MV(D,M)$) obtained from each frame. We consider some deviation of motion vectors and adjust the $d1(= d1 \pm 1)$, $d2(=d2 \pm 1)$ values. When there is no camera motion, we use default values of $d1(= 16)$, $d2(=16)$.

4   Experiment and Results

The experiments have been performed on a Pentium PC with 550 MHz CPU. The program is implemented in Visual C^{++} Ver.6.0. The MPEG1 data have been used for the experiments, The video image used are a movie as in Fig. 3. Fig. 3(a) show a sequence of the original image. Fig. 3(b) show recovered image. Experimental results show that the case of image having little motions is well recovered. And we see that the case of having motion in complex background is also recovered. We can see that using information about camera motion and video information gives more accurate recovery. In case the movement of objects is sudden or large, we see some distortion in recovering original image.

5   Conclusion

As a result of experiment, we know that the stationary image and image having little motion is well recovered. But the images having a lot of movement in complex background show some distortion. Therefore the following should be researched more in the future. Firstly, more sophisticated recovery method is needed for processing images with large and complex motions such as action movies. Secondly, for recovering in case when more than two scene changes occur in caption region, we will need a method using panorama technology to recovery it. For dissolve captions aimed at smooth feeling to audiences, we need a method of interpolation by frame interval control.

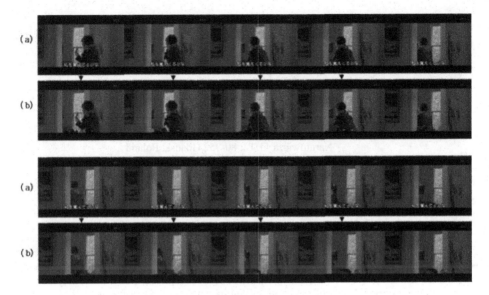

Fig. 3. A sequence of reconstructed original images in movie video

References

1. Rafael C. Gonzalez and Richard E. Woods, Digital Image Processing, Addison-Wesley Publishing Company, 1992.
2. Hui Liu, "Ordered Kohonen vector quantization for very low rate interframe video coding", Proc. of SPIE, Vol.2419, pp.71-80, San Jose, 1995.
3. W.K.Oh and S.Y.Gi,: Restoration of Josahdang wall painting using archaeological image processing. Technical Report E21131, KIST/SERI, May, Korea, 1991.
4. Byung Tae Chun, Younglae Bae and Tai-Yun Kim, "A method for original image recovery for caption area in video," IEEE Inter. Conference on SMC'99, Vol.II, pp.930-935, Japan, 10. 1999.
5. Byung Tae Chun, Younglae Bae, Tai-Yun Kim, Gil-Rok Oh," Digital video caption segmentation using graph-theoretic clustering," ICAPRDT'99, pp.406-410, India, 12. 1999.
6. Ok-Bae Chang, Myung-Sup Yang and jae-Hyun Lee, "Segmentation of gradual scene transitions using motion vector", IS and T/SPIE, Vol. 3422, pp.187-198, July. 1998.
7. F. Idris, S. Panchanathan, "Detection of Camera Operation in Compressed Video Sequences," SPIE Proceedings : Storage and Retrieval for Imgage and Video Database V, Vol.3022, pp.493-505, 1997.

Determining Influence of Visual Cues on the Perception of Surround Sound Using Soft Computing

Andrzej Czyzewski, Bozena Kostek, Piotr Odya, Slawomir Zielinski

Technical University of Gdansk, Sound Engineering Department,
Narutowicza 11/12; 80-952 Gdansk, Poland

Abstract. Contemporary digital video, film or multimedia presentations are often accompanied by the surround sound. Techniques and standards involved in digital video processing are much more developed than concepts underlying creating recording and mixing of the multichannel sound. The main challenge in the sound processing in the multichannel system is to create an appropriate basis for the relating multimodal context of visual and sound domains. Therefore, one of the purposes of experiments is to study in which way and how the surround sound interferes or is associated with the visual context. This kind of study was hitherto carried out when two-channel sound technique was associated with a stereo TV. However, there is not much study done yet that associates surround sound and digital video presented at the TV screen. The main issue in such experiments is the analysis of the influence of visual cues on perception of the surround sound. This problem will be solved with the application of fuzzy logic to the processing of subjective test results.

1   Introduction

There are many scientific reports showing that human perception of sound is affected by image and vice versa. For example, Stratton in his experiments carried out at the end of 19th century proved that visual cues can influence directional perception of sound. This conclusion was confirmed by Klemm [ ], Held [ ] and others. Gardner experimentally demonstrated how the image can affect the perceived distance between the sound source and the listener [ ]. The phenomenon of interference between the audio and video stimuli was reported also by Thomas, Witkin, Wapner and Leventhal [ ]. Very important experiments demonstrating interaction between audio and video in stereo TV were made by Brook, Danilenko, Strasser [ ] and Wladyka [ ]. However, still there is no clear answer to the question how the video influences the localization of virtual sound sources in multichannel surround systems (e.g. DTS). Therefore, there is a need of systematic research in this area, especially as sound and video engineers seek such information in order to optimize the surround sound. The results of this kind of research may improve production of movie soundtracks, recording of music events and live transmissions, thus the resulting surround sound may seem more natural to the listener. The experiments are based on the subjective testing of a group of people, so-called experts, listening to the sound with- and without vision. The obtained results are

W. Ziarko and Y. Yao (Eds.): RSCTC 2000, LNAI 2005, pp. 545–552, 2001.
© Springer-Verlag Berlin Heidelberg 2001

processed in order to find some hidden relations underlying the influence of video to the perception of audio, particularly with regard to the influence of video to the directivity of localization of sound sources in the surrounding acoustical space. Some soft computing methods could be used to the processing of subjective test results, bringing better results of the analysis than statistical methods, particularly if the number of tests and involved experts are reasonably small. An approach to such an application is formulated in the paper. The proposed method of analysis of subjective opinion scores could be also used in other domains than audio-video perception investigation (public opinion analysis etc.).

2   Experimental Background

Results of such experiments may show in which cases and in what way the video can affect the localization of virtual sound sources. In most cases the video "attracts" the attention of the listener and, as a consequence, he or she localizes the sound closer to the screen center. Therefore, this effect can be called audio-visual proximity effect.

In the experiments two rooms are used: auditory room and control room (Fig. 1), which are acoustically separated. A window between these two rooms allows for projection of video from the control room to the auditory room. A view of the auditory room is presented in Fig. 2. The place for the listener is positioned in the so-called "sweet spot" (the best place for listening).

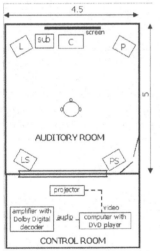

Fig. 1. Setup used during experiments   **Fig. 2.** Auditory room fragment

During tests AC-3 (Dolby Digital) audio encoded and MPEG2 video encoded files were used. Sound files were prepared in the Samplitude 2496 application and then exported to the AC-3 encoder. The following equipment was used during the tests: computer with built in DVD player, amplifier with Dolby Digital decoder, video projector, screen (dimensions: 3x2 m), loudspeakers.

Preliminary Listening Tests

In the preliminary experiments the arrangement of loudspeakers was as follows: four loudspeakers were aligned along the left-hand side of the screen (Fig. 3). In this case, the first loudspeaker was placed at the edge of the room, whereas the fourth one was positioned under the screen. This arrangement of loudspeakers allowed for how the visual object can affect the angle of the subjectively perceived sound source.

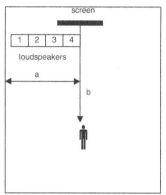

Fig. 3. Arrangement of loudspeakers during the tests

The experiment scenario was as follows. In the first phase of experiment white noise was presented from the loudspeakers with random order. The expert's task was to determine from which loudspeaker sound was heard. Then, in the second phase of experiment a blinking object was displayed in the center of the screen with synchronously generated white noise. In the center of the circle a one-digit number was displayed. Each time the circle was displayed the number was changed. The reason of that was the need of drawing the attention of the listener to the picture. Obtained results show that image proximity effect is speaker dependent, however most experts' results clearly demonstrate the mentioned effect. The most prominent data showing this effect is shown in Fig. 4. The shift in the direction to the centrally located loudspeaker is clearly visible.

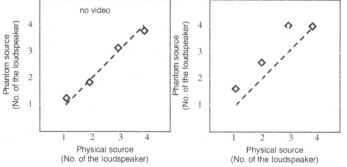

Fig. 4. Comparison of answers of the expert S.Z. for two types of experiments: without video / with video

3. Fuzzy Logic Processing of Subjective Results of Surround Sound Directivity Testing

The subjective tests presented below aimed at finding a relation between precise surround directivity angles and semantic descriptors of horizontal plane directions. It is hardly to expect an expert to be exact in localizing phantom sources in the surround stereophonic base and to provide precise values of angles. On the other hand, it seems quite natural that an expert will localize a sound using such directional descriptors as: *left, left-front, front, right-front, right, rear-right, rear, rear-left*. Thus, first series of the experiment should consist in mapping these descriptors to angles as in Fig. 5.

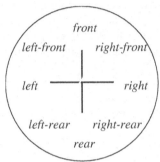

Fig. 5. Questionnaire form used in the first stage of experiments

In this step of the investigations, sound samples recorded in the anechoic chamber should be presented to the group of experts. Experts, while listening to sounds excerpts, are instructed to rate their judgements of the performance using such descriptions as introduced above. In order to obtain statistically validated results various sound excerpts should be presented to the sufficiently large number of experts during experiments. This procedure is based on the concept of the Fuzzy Quantization Method (FQM) applied to acoustical domain [ ], [ ]. Since the experimenter knows to what angle a given sound was assigned, thus this stage of experiments will result in mapping semantic descriptors received from experts to particular angles describing the horizontal plane.

In order to simplify this phase of tests, localization sphere should be divided into 5° steps. Fig. 6 shows exemplary mapping of the front membership function. All other membership function should be estimated in a similar way (see Fig. 7).

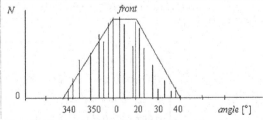

Fig. 6. Experts' votes for the front membership function, N - number of experts voting for particular values of localization (variable: *angle*)

As shown in Fig. 6 and 7, distribution of the observed instances may suggest a typical trapezoidal shape of a membership function. In the next step of the analysis, such membership functions should be identified with the use of some statistical methods. This can be done by several techniques. The most common technique is linear approximation, where the original data range is transformed to the interval [0,1]. Thus, triangular or trapezoidal membership functions may be used in this case. In the linear regression method, one assigns minimum and maximum attribute values. Assuming that the distribution of parameters provides a triangular membership function for the estimated parameter, the maximum value may thus be assigned as the average value of the obtained results. This may, however, cause a loss of information and bad convergence. The second technique uses bell shaped functions. The initial values of parameters can be derived from the statistics of the input data. Further, the polynomial approximation of data, either ordinary or Chebyshev, may be used. This technique is justified by a sufficiently large number of results or by increasing the order of polynomials; however, the latter direction may lead to a weak generalization of results. Another approach to defining the shape of the membership function involves the use of the probability density function. The last mentioned technique was discussed in the given context more thoroughly in the literature [ ].

Fig. 7. Directivity membership functions on the horizontal plane

Intuitively, it seems appropriate to build the initial membership function by using the probability density function and by assuming that the parameter distribution is trapezoidal or triangular. The estimation of the observed relationships is given by the function shown in Fig. 8.

The f_1 membership function from Fig. 8 is defined by a set of parameters: A, a, b, c, d and is determined as follows:

$$f_1(x, A, a, b, c, d) = \begin{cases} 0 & \text{if } x < a \text{ or } x > d \\ A(x-a)/(b-a) & \text{if } a \le x \le b \\ A & \text{if } b < x < c \\ -A(x-d)/(d-c) & \text{if } c \le x \le d \end{cases} \quad (1)$$

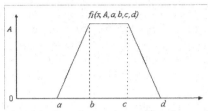

Fig. 8. Trapezoidal membership function estimated by the probability density function

The equation describing the *m*th moment of the probability density for the function $f_1(x,A,a, b,c,d)$ is calculated as follows:

$$m_x = \int_{-\infty}^{+\infty} x^x f_1(x)dx \qquad (2)$$

The estimate of the *m*th moment of the probability density function from the test (assuming that all observation instances fall into the interval *j*, where: $j=1,2...k$) is calculated according to the formula:

$$\hat{m}_x = \sum_{j=1}^{k} x^x \cdot P(x = x_j) \qquad (3)$$

where: $P(x=x_j)$ represents the probability that the attribute value of instance *x* falls into the interval *j*.

Next, the subsequent statistical moments of order from 0 to 4 for this function should be calculated. Then, by substituting the observed values into Eq. (3), the consecutive values of m_n are calculated. From this, the set of 5 linear equations with 5 unknown variables *A,a,b,c,d* should be determined. After numerically solving this set of equations, the final task of the analysis will be validation of the observed results using Pearson's χ^2 test with *k*-1 degrees of freedom [ ].

Using the above outlined statistical method, a set of fuzzy membership functions for the studied subjective sound directivity can be estimated.

4   Inter-Modal Testing Phase

In order to proceed with testing the inter-modal relation between sound localization and video images, another questionnaire should be used. This time, experts' task would be assigning the crisp angle value to the incoming sound excerpt while watching a TV screen. Having previously estimated membership functions, it would be then possible to check whether the observation of video images can change sound localization and if yes then to what degree. This can be done by performing a fuzzification process. The data representing the actual listening tests would then pass trough the fuzzification operation in which degrees of membership should be assigned for each crisp input value.

The process of fuzzification is illustrated in Fig. 9. The pointers visible in this figure refer to the degrees of membership for the precise value of localization angle 335°. Thus, this value belongs, respectively, to the *left-rear* fuzzy set with the degree of 0, to the *left-front* fuzzy set with the degree of 0.65 and to the *front* set with the degree of 0.35. The same procedure should be applied to every sound-image instance.

Consequently, the process of fuzzy inference can be started, allowing to find winning rules. The (exemplary) fuzzy rules have the following form:

1. if *front* AND *FRONT* than *no_shift*
2. if *left_front* AND *LEFT_FRONT* than *no_shift*
3. if *left_front* AND *FRONT* than *slight_shift*
4. if *left* AND *LEFT* than *no_shift*
5. if *left* AND *LEFT_FRONT* than *slight_*

where small italic labels denote current directivity indices and the capital italic labels denote the directivity of the same sound played back during the previous tests (in the absence of vision).

It was assumed that the presence of vision is causing the shifting of sound localization to the front direction only (not to opposite directions in relation to the frontal one) and there is no possibility for phantom sources to migrate from the left to the right hemisphere and vice versa. These assumptions have been justified in practice. The rules applying to the right: front lateral and rear directions are similar to above ones. The AND function present in the rules is the *"fuzzy and"* [ ], thus it chooses the smaller value from among these which provide arguments of this logical function. The consequences: *no_shift*; *slight_shift*; *medium_shift*; *strong_shift* are also fuzzy notions, so if it is necessary to change them to the concrete (crisp) angle values, a defuzzification process should be performed basing on the output prototype membership functions.

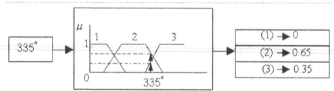

Fig. 9. Fuzzification process of localization angle: (1) - *left-rear*, (2) - *left-front*, and (3) - *front membership functions*

All rules are evaluated once the fuzzy inference is executed and finally the strongest rule is selected as the winning one. These are standard procedures related to fuzzy logic processing of data [ ]. The winning rule demonstrates the existence and the intensity of the phantom sound source shifting due to the presence of vision. Since the fuzzy rules are readable and understandable for human operators of the system, thus this application provides a very robust method for studying complex phenomena related to influence of vision coming from frontal TV screen to the subjective localization of sound sources in surround space. The mentioned defuzzification procedure [ ] allows also for mapping of fuzzy descriptor to the crisp angle measure every time when it is necessary to estimate such a value.

5 Conclusions

Audio and video interact with each other. Mechanisms of such interaction are currently investigated in two domains: perceptual and aesthetic ones employing fuzzy logic in the process of analysis of tested subjects' answers. The results of such experiments could yield the recommendations to sound engineers producing surround movie sound tracks, digital video and multimedia.

6 Acknowledgements

The research is sponsored by the Committee for Scientific Research, Warsaw, Poland. Grant No. 8 T11D 00218.

7 References

[1] O. Klemm, "Untersuchungen uber die Lokalisation von Schallreizen III: Uber den Anteil des beidohrigen Horens", Arch. ges. Psychol. 38, pp. 71-114, 1918.
[2] R. Held, "Shifts in binaural localization after prolonged exposures to atypical combination of stimuli", Amer. J. Psychol., vol. 68, pp. 526-548, 1955.
[3] M.B. Gardner, "Proximity Image Effect in Sound Localization", J.A.S.A, vol. 43, pp. 163, 1968.
[4] H.A. Witkin, S. Wapner, T. Leventhal, "Sound Localization with Conflicting Visual and Auditory Cues", J. Exp. Psych., vol. 43, pp. 58-67, 1952.
[5] M. Brook, L. Danilenko, W. Strasser, "Wie bewertet der Zuschauer das stereofone Fernsehes?", 13 Tonemeistertagung; Internationaler Kongres, pp. 367-377, 21-24 Nov. 1984.
[6] M. Wladyka, "Examination of Subjective Localization of Two Sound Sources in Stereo Television Picture", Master Degree Thesis, Sound Eng. Dept., Technical Univ. of Gdansk, Poland, 1987. (in Polish)
[7] B. Kostek, "Soft Computing in Acoustics, Applications of Neural Networks, Fuzzy Logic and Rough Sets to Musical Acoustics, Studies in Fuzziness and Soft Computing, Physica Verlag", Heilderberg, New York 1999.
[8] B. Kostek, "Assessment of Concert Hall Acoustics Using Rough Set and Fuzzy Set Approach", chapter in "Rough-Fuzzy Hybridization: A New Trend in Decision-Making", Pal S.K., Skowron A. (Eds.), pp. 381-396, Springer-Verlag, Singapore 1999.
[9] B. Kosko, "Fuzzy Engineering", Prentice-Hall Intern. Ed., New Yersey, 1997.

Application of normalized decision measures to the new case classification

Dominik Ślęzak[1,2], Jakub Wróblewski[1,2]

[1] Polish-Japanese Institute of Information Technology
Koszykowa 86, 02-008 Warsaw, Poland
[2] Institute of Mathematics, Warsaw University
Banacha 2, 02-097 Warsaw, Poland

Abstract. The optimization of rough set based classification models with respect to parameterized balance between a model's complexity and confidence is discussed. For this purpose, the notion of a parameterized approximate inconsistent decision reduct is used. Experimental extraction of considered models from real life data is described.

1 Introduction

While reasoning about a domain specified by our needs, we usually base on the information gathered by the analysis of a sample of objects. The rough set theory ([3]) assumes that a universe of known objects is the only source of knowledge, which can be applied to construct models of reasoning about new cases. Reasoning can be stated, e.g., as a classification problem, concerning prediction of a decision attribute under information provided over conditional attributes. For this purpose, one stores data within decision tables, where each training case drops into one of predefined decision classes.

Classification of new objects is performed by analogy, e.g., by the usage of "if...then..." decision rules calculated over the universe of a given table. Theoretical studies related to the *Minimum Description Length Principle (MDLP)* (cf. [5]), as well as practical experiences, lead to the same conclusion: *Optimal rule-based classification models should be extracted from data by tuning up a parameterized tradeoff between the overall confidence and complexity of the decision rule collections.*

Confidence of a rule-based model can be interpreted as the expected chance of correct classification of new cases. To express such a chance numerically, we need to set up the model of representing inexact *conditions→decision* dependencies. Then, we are able to evaluate the degree of decision information provided by each particular subset of conditional attributes, and to express the dynamics of this degree under the attribute reduction. In the same way, one can interpret the complexity of a given collection as opposite to the expected chance of recognizing new cases by its decision rules. It leads to a rough set based version of MDLP, related to the fundamental concept of searching for approximate decision

W. Ziarko and Y. Yao (Eds.): RSCTC 2000, LNAI 2005, pp. 553–560, 2001.
© Springer-Verlag Berlin Heidelberg 2001

reducts: *Given a rule-based decision model, any simplification which approxima-*
tely preserves the expected chance of correct classification should be performed to
increase the expected chance of the new case recognition.

The above principle can be regarded as the starting point for the design of
the process of the rule-based classification model optimization. In the paper, we
discuss exemplary methodology of setting up the foregoing items, concerning the
adjustment of thresholds, voting measures, etc.. Accordingly, Section 2 includes
the basics of rough set based classification techniques (cf. [4], [6]). Sections 3 and
4 outline exemplary extensions of these methods by introducing the notion of an
approximate reduct based on a *normalized decision function* (cf. [7]). In Sections
5 and 6 we present the main contribution – the classification algorithm based on
the family of parameterized decision functions. Section 7 contains experimental
verification of the performance of the proposed classification framework.

2   Decision rules and reducts

In the rough set theory sample of data takes the form of an information system
$\mathbb{A} = (U, A)$, where each attribute $a \in A$ is a function $a : U \to V_a$ into the set of all
possible values on a. Reasoning about data can be stated as, e.g., a classification
problem, where a distinguished decision is to be predicted under information over
conditional attributes. In this case, we consider a triple $\mathbb{A} = (U, A, d)$, called a
decision table, where, for the decision attribute $d \notin A$, values $v_d \in V_d$ correspond
to mutually disjoint decision classes of objects.

Definition 1. *Let $\mathbb{A} = (U, A, d)$, where $A = \langle a_1, \ldots, a_{|A|} \rangle$, be given. For any*
$B \subseteq A$, $B = \langle a_{i_1}, \ldots, a_{i_{|B|}} \rangle$, the B-information function over U is defined by

$$\overrightarrow{Inf}_B(u) = \langle a_{i_1}(u), \ldots, a_{i_{|B|}}(u) \rangle \tag{1}$$

The B-indiscernibility relation is the equivalence relation defined by

$$IND_{\mathbb{A}}(B) = \{(u, u') \in U \times U : \overrightarrow{Inf}_B(u) = \overrightarrow{Inf}_B(u')\} \tag{2}$$

Each $u \in U$ induces a B-indiscernibility class of the form

$$[u]_B = \{u' \in U : (u, u') \in IND_{\mathbb{A}}(B)\} \tag{3}$$

which can be identified with vector $\overrightarrow{Inf}_B(u)$.

Indiscernibility enables us to express global dependencies as follows:

Definition 2. *Given $\mathbb{A} = (U, A, d)$, we say that $B \subseteq A$ defines d in $\mathbb{A}$ iff*

$$IND_{\mathbb{A}}(B) \subseteq IND_{\mathbb{A}}(\{d\}) \tag{4}$$

or, equivalently, iff for any $u \in U$ the following u-oriented rule is valid in $\mathbb{A}$:

$$\bigwedge_{a \in B}(a = a(u)) \Rightarrow (d = d(u)) \tag{5}$$

We say that $B \subseteq A$ is a decision reduct iff it defines d and none of its proper
subsets does it.

Given a collection of subsets $\mathcal{B} \subseteq \mathcal{P}(A)$ which define d, we can classify any new case $u_{new} \notin U$ by using the bunch of decision rules of the form (5). The only requirement is that u_{new} must be comparable with U with respect to at least one (several) of $B \in \mathcal{B}$. To improve such understood recognition of new cases, we base on (approximate) decision reducts of possibly low complexity, expressible in various terms (cf. [2], [8]).

3   Normalized decision functions

In a consistent decision table $\mathbb{A} = (U, A, d)$ – where each indiscernibility class of $IND_{\mathbb{A}}(A)$ has one decision value – decision rules lead to deterministic classification within the universe U. In case of inconsistent decision tables (non-deterministic dependencies among attributes), we should specify the way of dealing with uncertainty.

Definition 3. *Let* $\mathbb{A} = (U, A, d)$, *linear ordering* $V_d = \langle v_1, \ldots, v_r \rangle$, $r = |V_d|$, *and* $B \subseteq A$ *be given. By a* B**-rough membership distribution** *we call the function* $\overrightarrow{\mu}_{d/B} : U \to \triangle_{r-1}$ *defined by*[1]

$$\overrightarrow{\mu}_{d/B}(u) = \langle \mu_{d=1/B}(u), \ldots, \mu_{d=r/B}(u) \rangle \tag{6}$$

where, for $k = 1, \ldots, r$, $\mu_{d=k/B}(u) = |\{u' \in [u]_B : d(u') = v_k\}| / |[u]_B|$ *is the rough membership function (cf. [4]) labeling* $u \in U$ *with the degree of hitting the k-th decision class with its B-indiscernibility class* $[u]_B$.

Distributions of the form (6) seem to express the most accurate knowledge about dependencies of the decision on conditions (cf. [4], [7], [9]). Thus, it should be possible to model various B-based reasoning strategies as functions acting over $\overrightarrow{\mu}_{d/B}$ by "forgetting" a part of frequency information, which is redundant with respect to a given approach.

Definition 4. *([7]) Let* $\mathbb{A} = (U, A, d)$, $B \subseteq A$ *and* $\phi : \triangle_{r-1} \to \triangle_{r-1}$, $r = |V_d|$, *be given. We say that* ϕ *is a* **normalized decision function** *(ND-function) iff it satisfies the following, logical and monotonic consistency assumptions:*

$$\forall_k (s[k] = 0 \Rightarrow \phi(s)[k] = 0) \quad \wedge \quad \forall_{k,l} (s[k] \le s[l] \Rightarrow \phi(s)[k] \le \phi(s)[l]) \tag{7}$$

Function $\overrightarrow{\phi}_{d/B}(u) = \phi(\overrightarrow{\mu}_{d/B}(u))$ *is called a normalized* $\phi_{d/B}$*-decision function.*

According to (7), a positive weight cannot be attached to a non-supported event and the relative chances provided by the reasoning strategy cannot contradict those derived directly from an information source.

---

[1] For any $r \in \mathbb{N}$, $\triangle_{r-1}$ denotes the $(r-1)$-dimensional simplex of real valued vectors $s = \langle s[1], \ldots, s[r] \rangle$ with non-negative coordinates, such that $\sum_{k=1}^{r} s[k] = 1$.

Example 1. Consider ND-functions $\partial, m : \triangle_{r-1} \to \triangle_{r-1}$ defined by

$$\partial(s)[k] = \left\{ \begin{array}{ccc} |\{l : s[l] > 0\}|^{-1} & for & s[k] > 0 \\[2mm] (\ |\{l : s[l] = \max(s)\}|^{-1} & for & s[k] = \max(s)\ ) \\[2mm] 0 & otherwise \end{array} \right. \tag{8}$$

where $\max(s) = \max_k s[k]$. One can see that by combining $\overrightarrow{\mu}_{d/B}(u)$, $u \in U$, with ∂ and m we obtain the uniform distributions spanned over the subsets: (1) $\partial_{d/B}(u) \subseteq V_d$ induced by the generalized decision (cf. [6]); (2) $m_{d/B}(u) \subseteq V_d$ of decision values taking the maximum over the coordinates of $\overrightarrow{\mu}_{d/B}(u)$.

4   Normalized decision measures

In real life applications, the search for attributes which approximately preserve ϕ-decision distributions seems to be promising. We are likely to understand an approximate ϕ-decision reduct as a minimal subset of conditions, which almost preserves information about decision in terms of a given ND-function.

Definition 5. *Let* $\mathbb{A} = (U, A, d)$ *and* $\phi \in \triangle_{r-1} \to \triangle_{r-1}$, $r = |V_d|$, *be given. The* **normalized ϕ-decision measure** $E_{\phi/\mathbb{A}} : \mathcal{P}(A) \to [0,1]$ *is defined by*[2]

$$E_{\phi/\mathbb{A}}(B) = \tfrac{1}{|U|} \sum_{u \in U} \langle \overrightarrow{\mu}_{d/B}(u) | \overrightarrow{\phi}_{d/B}(u) \rangle \tag{9}$$

The value of (9) equals to the average probability that objects $u \in U$ will be correctly classified by a random $\overrightarrow{\phi}_{d/B}(u)$-weighted choice among decision classes ([7]). Thus, ϕ-decision measures enable us to evaluate subsets numerically with respect to their capabilities of ϕ-defining decision.

Definition 6. *Let* $\mathbb{A} = (U, A, d)$, $\varepsilon \in [0,1)$ *and* $\phi \in \triangle_{r-1} \to \triangle_{r-1}$, $r = |V_d|$, *be given. We say that subset* $B \subseteq A$ ε**-approximately ϕ-defines** d *iff*

$$E_{\phi/\mathbb{A}}(B) \geq (1 - \varepsilon) E_{\phi/\mathbb{A}}(A) \tag{10}$$

We say that $B \subseteq A$ *is an* ε**-approximate ϕ-decision reduct** *iff it ϕ-defines* d ε*-approximately and none of its proper subsets does it.*

Two parameters can be tuned up while searching for optimal conditions for classification: (1) ND-function ϕ responsible for the way of understanding inexact *conditions→decision* dependencies, and (2) the degree $\varepsilon \in [0,1)$ up to which we are likely to neglect the decrease of ϕ-decision information provided by smaller subsets $B \subseteq A$ with respect to the whole of A. In case of the first parameter, it is easier to handle a numeric factor responsible for adjusting a specific function:

---

[2] By "$\langle \cdot | \cdot \rangle$" we mean the inner product of two distribution vectors.

Definition 7. *Let* $\mathbb{A} = (U, A, d)$ *be given. For any* $x \in (0, +\infty)$, *we define the* **normalized** x**-decision function** *by putting, for any* $s \in \triangle_{r-1}$, $r = |V_d|$,

$$x(s)[k] = (s[k])^x / \sum_{l=1}^{r}(s[l])^x \tag{11}$$

Proposition 1. *All* x-*decision functions satisfy* (7). *For any* $s \in \triangle_{r-1}$,

$$\lim_{x \to 0^+} x(s) = \partial(s) \quad \wedge \quad \lim_{x \to +\infty} x(s) = m(s) \tag{12}$$

For any $\mathbb{A} = (U, A, d)$, $B \subseteq A$, $0 < x_1 < x_2 < +\infty$, *we have*

$$E_{\partial/\mathbb{A}}(B) \leq E_{x_1/\mathbb{A}}(B) \leq E_{x_2/\mathbb{A}}(B) \leq E_{m/\mathbb{A}}(B) \tag{13}$$

where equalities hold iff $E_{\partial/\mathbb{A}}(B) = E_{m/\mathbb{A}}(B)$.

One can see that the obtained x-parameterized family covers densely enough all possible ways of performance of ND-functions over training data.

5   Optimization of approximate reducts

In Section 1 we suggested to relate the overall confidence of a rule-based decision model to the expected chance of correct classification of new cases. Above, we argued that the quantities of normalized decision measures can be interpreted in that way. Analogously, let us now propose an exemplary measure of the expected chance of recognizing new cases by decision rules generated by a given $B \subseteq A$.

Definition 8. *Let* $\mathbb{A} = (U, A, d)$ *be given. The* **normalized coverage measure** $cov_{\mathbb{A}} : \mathcal{P}(A) \to [0, 1]$ *is defined by*

$$cov_{\mathbb{A}}(B) = \frac{1}{|U|} \sum_{u \in U} \mu_B(u) \tag{14}$$

where $\mu_B(u) = |[u]_B|/|U|$ *is the frequency of occurrence of vector* $\overrightarrow{Inf}_B(u)$ *in* $\mathbb{A}$.

One should realize that this is just one of possibilities of estimating the recognition probability (cf. [8]). Still, we would like to proceed with this measure, because it turns out to be flexible enough with respect to applications.

   The exemplary procedure presented below searches for an ε-approximate x-decision reduct $B \subseteq A$ with the highest possible coverage $cov_{\mathbb{A}}(B)$, by following a randomly generated permutation $\sigma \in \Sigma_{|A|}$ over conditional attributes[3,4]. First, starting with $B = \{a_{\sigma(1)}\}$, we add the foregoing attributes until B begins to x-define d in ε-approximate way. The second part reflects the optimization principle formulated at the beginning of this paper: *We try to reduce a model until it ε-approximately preserves x-decision information, to increase $cov_{\mathbb{A}}$ as the measure of predicted average chance of the new case recognition.*

---

[3] We denote by Σ_n the set of all n-element permutations, i.e. "1–1" functions $\sigma : \{1, \ldots, n\} \to \{1, \ldots, n\}$. We use $\sigma \in \Sigma_{|A|}$ to re-order $A = \langle a_1, \ldots, a_{|A|} \rangle$.

[4] We base the search for $cov_{\mathbb{A}}$-maximal approximate decision reducts on random heuristics, because this optimization problem is NP-hard (cf. [7]).

Algorithm: Approximate reducts generation
Input: Decision table $\mathbf{A} = (U, A, d)$, permutation σ of A, $\varepsilon \in [0,1)$, $x \in (0, +\infty)$.
Output: ε-approximate x-decision reduct.
1. $B := \{a_{\sigma(1)}\}$; $cov := cov_{\mathbf{A}}(B)$; $app := E_{x/\mathbf{A}}(B)$; $i := 2$
2. **while** $app < (1 - \varepsilon)E_{x/\mathbf{A}}(A)$ **begin**
3. $B := B \cup \{a_{\sigma(i)}\}$
4. $(cov, app) = Test(B, x)$
5. $i := i + 1$
6. **end while**
7. $maxcov := 0$
8. **do**
9. $stop := 1$
10.   **for each** $a_j \in B$
11. $cov := cov_{\mathbf{A}}(B \setminus \{a_j\})$
12. $app := E_{x/\mathbf{A}}(B \setminus \{a_j\})$
13.   **if** $app < (1 - \varepsilon)E_{x/\mathbf{A}}(A)$ **and** $cov > maxcov$ **then begin**
14. $maxcov := cov$
15. $maxj := j$
16. $stop := 0$
17.   **end if**
18.   **end for**
19.   **if** $stop = 0$ **then** $B := B \setminus \{a_{maxj}\}$
20. **while** $stop = 0$
21. **return** B

6   Optimization of the approximate reduct collections

A rule-based decision model should correspond to more than one subset of conditions. Thus, we construct systems composed of collections of *classifying agents* based on different ε-approximate x-decision reducts, obtained as $cov_{\mathbf{A}}$-optimal while following a number of randomly generated permutations. Since it is not known how to adjust the best configuration of $\varepsilon \in [0,1)$ and $x \in (0, +\infty)$, we consider collections of (ε, x)-parameterized agents initialized randomly, to simulate a kind of the adaptation process searching through the space of $[0,1) \times (0, +\infty)$.

Given $\mathcal{B} \subseteq \mathcal{P}(A) \times (0, +\infty)$ as the collection of obtained parameterized reducts, one needs also to specify the way of voting between particular agents. In general, negotiations concerning prediction of the decision value for a given u lead to the choice of $v_k \in V_d$ with a maximal value of a voting measure, calculated from u-oriented x-decision rules induced by particular elements of $\mathcal{B}$. Below, one can find examples of such voting measures:

$$VOTE1(\mathcal{B}) = \sum\nolimits_{(B,x)\in\mathcal{B}: \mu_B(u)>0} \mu_B(u) x_{d=k/B}(u)$$

$$VOTE2(\mathcal{B}) = \sum\nolimits_{(B,x)\in\mathcal{B}: \mu_B(u)>0} \mu_{d=k/B}(u) x_{d=k/B}(u) \qquad (15)$$

$$VOTE3(\mathcal{B}) = \sum\nolimits_{(B,x)\in\mathcal{B}: \mu_B(u)>0} x_{d=k/B}(u)$$

Another problem concerns the fact that although all agents can be used to classify new objects, a subset of them often performs much better (cf. Fig. 1). We apply a specific genetic algorithm to search for optimal sub-collections of agents. In particular, it results with an indirect optimization process concerned with the ranges of (ε, x)-parameters.

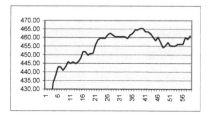 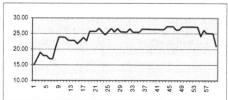

Fig. 1. Classification quality (vertical axis) and the number of agents in a team (horizontal axis) – examples obtained for "DNA splices" and "primary tumor" data sets.

7   Experimental results

The methodology described in the paper was implemented and tested on several data sets obtained from [1]. Results presented in Table 1 concern two of them: (1) "DNA splices" – 2000 objects, 20 symbolic attributes, 3 decision classes; (2) "Primary tumor" – 339 objects, 17 symbolic attributes, 22 decision classes.

Voting	Approx	Mode	Result	Voting	Approx	Mode	Result
1	–	–	82.06	1	–	–	43.01
2	–	–	86.96	2	–	–	**43.29**
3	–	–	**93.75**	3	–	–	42.44
3	0.5	–	68.74	2	0.5	–	43.28
3	0.4	–	87.05	2	0.4	–	**44.08**
3	0.3	–	93.39	2	0.3	–	43.48
3	0.2	–	**94.97**	2	0.2	–	40.80
3	0.1	–	93.14	2	0.1	–	39.53
3	0.2	exp(−2)	94.27	2	0.4	exp(−2)	44.46
3	0.2	exp(−1)	94.68	2	0.4	exp(−1)	44.67
3	0.2	exp(0)	95.01	2	0.4	exp(0)	**44.70**
3	0.2	exp(1)	94.93	2	0.4	exp(1)	42.85
3	0.2	exp(2)	**95.10**	2	0.4	exp(2)	42.79

Table 1. Experimental results for "DNA splices" and "primary tumor" data sets, obtained by voting among optimized collections of ε-approximate x-decision reducts: **(1)** The choice of a measure from (15) corresponds to the *Voting* column; **(2)** Quantities of ε and x are chosen randomly from small intervals around values in the *Approx* and *Mode* columns, where symbol "–" means the uniform random choice from a wider interval; **(3)** Average percent of tested objects classified correctly for particular settings is presented in the *Result* column. Cross-validation (CV-5) was used in case of "primary tumor" data.

Experiments presented in Table 1 were performed with various settings:

1. Optimal voting measure was selected by setting other parameters randomly;
2. For the best voting method, several values of $\varepsilon \in [0, 0.5]$ were tested;
3. For the best voting method and ε-thresholds selected from the small interval around the best value found previously, different values of parameter $x \in [\exp(-2), \exp(2)]$ were tested.

It is worth noting that the best results obtained in our experiments are close to the best results ever found.

8   Conclusions

Parameterized tradeoff between model complexity and its accuracy was discussed. To handle it in a flexible way, the notion of a parametrized ε-approximate x-decision reduct was used. Main issues concerning implementation of the classification algorithm based on described methodology were outlined.

Experiments performed on two "benchmark" data sets show that our technique is relatively fast and very efficient. It is worth noting that best results for these sets were obtained using significantly different voting and (x, ε)-settings. It suggests us to consider the adaptive mechanisms of tuning up these parameters in the nearest future.

Acknowledgements

This work was supported by the grants of Polish National Committee for Scientific Research (KBN) No. $8T11C02319$ and $8T11C02419$. "Primary tumor" was obtained from the University Medical Centre, Institute of Oncology, Ljubljana, Yugoslavia. Thanks go to M. Zwitter and M. Soklic for its providing.

References

1. Michie D., Spiegelhalter D.J., Taylor C.C. (eds.): Machine Learning, Neural and Statistical Classification. Ellis Horwood Limited (1994). Data avaliable at: http://www.ics.uci.edu/~mlearn/MLRepository.html.
2. Nguyen, S.H., Skowron, A., Synak, P.: Discovery of data patterns with applications to decomposition and classification problems. In: L. Polkowski, A. Skowron (eds.), Rough Sets in Knowledge Discovery, Physica Verlag, Heidelberg (1998) pp. 55–97.
3. Pawlak, Z.: Rough sets – Theoretical aspects of reasoning about data. Kluwer Academic Publishers, Dordrecht (1991).
4. Pawlak, Z., Skowron, A.: Rough membership functions. In: R.R. Yaeger, M. Fedrizzi, and J. Kacprzyk (eds.), Advances in the Dempster Shafer Theory of Evidence, John Wiley & Sons, Inc., New York, Chichester, Brisbane, Toronto, Singapore (1994) pp. 251–271.
5. Rissanen, J.: Modeling by the shortest data description. Authomatica, **14** (1978) pp. 465–471.
6. Skowron, A., Rauszer, C.: The discernibility matrices and functions in information systems. In: R. Słowiński (ed.), Intelligent Decision Support. Handbook of Applications and Advances of the Rough Set Theory, Kluwer Academic Publishers, Dordrecht (1992) pp. 311–362.
7. Ślęzak, D.: Normalized decision functions and measures for inconsistent decision tables analysis. To appear in Fundamenta Informaticae (2000).
8. Wróblewski J.: Genetic algorithms in decomposition and classification problem. In: L. Polkowski, A. Skowron (eds.), Rough Sets in Knowledge Discovery, Physica Verlag, Heidelberg (1998) pp. 471–487.
9. Ziarko, W.: Decision Making with Probabilistic Decision Tables. In: N. Zhong, A. Skowron and S. Ohsuga (eds.), Proc. of the Seventh International Workshop RSFDGrC'99, Yamaguchi, Japan, LNAI **1711** (1999) pp. 463–471.

An Application of Rough Sets and Haar Wavelets to Face Recognition

R. Swiniarski

Department of Mathematical and Computer Sciences,
San Diego State University, San Diego, CA 92182 U.S.A.
rswiniar@sciences.sdsu.edu

Abstract. The paper presents an application of data mining methods for face recognition. The proposed methods are based on wavelets, principal components analysis, rough sets and neural networks. The features from the face images have been extracted based on the Haar wavelets followed by the principal component analysis (PCA), and rough sets processing. We have applied the rough sets methods for selection of facial features based on the minimum concept description paradigm. The recognition of facial images, for the reduced features, has been carried on using error backpropagation neural network.
Keywords: rough sets, face recognition, feature selection, wavelets

1   Introduction

The selection of the best feature sets representing recognized objects is an important step in the classifier design. The feature selection process is generally goal, data, and classifier type dependent [1],[2]. We present an application of feedforward error backpropagation neural network for face images classification using Haar wavelets for feature extraction, and Principal Component Analysis followed by rough sets method for feature reduction and selection.
We emphasize rough sets methodology [1],[2],[3] and a minimum concept description paradigm, for selection of the final face feature vector.
The paper begins with the brief description of Haar wavelets transform for extraction of facial features. In the following sections we shortly present the principal component analysis, and the rough sets theory as a foundation of methods for feature selection/reduction. Finally, the description of numerical experiments of face recognition, using the neural network classifier and the presented methods of feature extraction, projection and selection, concludes the paper.

2   The 2D Haar Wavelets Transform

The wavelet transform is a method of approximating a given function $f(t) \in L^2(\mathbf{R})$ using other function $\psi(t)$ (a wavelet function) representing a scalable approximation curve localized on definite time (or space) interval. Two-dimensional

W. Ziarko and Y. Yao (Eds.): RSCTC 2000, LNAI 2005, pp. 561–568, 2001.

parametrization, with a dilation parameter a and and a translation parameter b, yields a family of continuous wavelets

$$\psi_{a,b}(t) = |a|^{-\frac{1}{2}} \psi \left(\frac{t-b}{a} \right), \quad a, b \in \mathbf{R}, \ a \geq 0 \tag{1}$$

where a and b may vary over $\mathbf{R}$. For a given function $f(t)$ the continuos wavelet transform is defined as

$$WT\,f(a,b) = |a|^{-\frac{1}{2}} \int_{-\infty}^{\infty} f(t)\psi \left(\frac{t-b}{a} \right) dt \tag{2}$$

The discrete wavelets are obtained, by sampling parameters a and b as $a = a_0^j$ and $b = kab_0 = ka_0^j b$ (where $j, k \in \mathbf{Z}$, $i, j = \overline{+} 1, \overline{+} 2, \cdots$)

$$\psi_{j,k}(t) = |a_0|^{-\frac{1}{2}} \psi(a_0^{-j} t - kb_0), \quad j, k \in \mathbf{Z} \tag{3}$$

where $\psi_{j,k}(t)$ constitutes basis for $L^2(\mathbf{R})$.

The discrete wavelet transform is defined for sampled parameters by the equation

$$DWT\,f(j,k) = a^{-\frac{1}{2}} \int_{-\infty}^{\infty} f(t)\psi(a_0^{-j} t - kb_0)\,dt \tag{4}$$

For $a_0 = 2$ ($a = 2^j$) and $b_0 = 1$ ($b = 2^j k$), functions $\psi_{j,k}(t)$ form an orthogonal wavelet base for $L^2(\mathbf{R})$: $\psi_{j,k}(t) = 2^{-\frac{1}{2}} \psi(2^{-j} t - k)$. Multi-resolution analysis of a function $f(t)$ can be realized using dilated and translated scaling function $\phi_{j,k} = |a|^{-\frac{1}{2}} \phi(a^{-1} - k)$, $a = 2$. For discrete parameters a and b, a discrete wavelets transformation decomposes a function, determined by N discrete samples $\{f(1), f(2), \cdots, f(N)\}$, into an expansion of two function: a scaling function $\phi(t)$ and a wavelet function $\psi(t)$. The basis set for a scaling function (non-normalized) are

$$\phi_{L,k}(t) = \phi(2^L t - k), \quad k = 1, 2, \cdots, K_L, \quad K_L = N2^{-L} \tag{5}$$

where L is an expansion level, and for the wavelet function

$$\psi_{j,k}(t) = \psi(2^j t - k), \quad j = 1, 2, \cdots, L; \quad k = 1, 2, \cdots, K; \quad K = N2^{-j} \tag{6}$$

where the level of expansion L satisfies: $0 < L \leq log_2(N)$.

An L-level discrete wavelets transform of function $f(t)$ described by N samples contains:

1. a set of parameters $\{a_{L,k}\}$ defined by the inner products of $f(t)$ with $N2^{-j}$ translations of scaling function $\phi(t)$ at L different widths

$$\{a_{L,k}\} = \{< f(t), \phi_{L,k}(t) >; \quad k = 1, 2, \cdots, K_L, \quad K_L = N2^{-L}\} \tag{7}$$

2. a set of parameters $\{b_{j,k}\}$ defined by the inner products of $f(t)$ with $N2^{-L}$ translations of wavelet function $\psi_{j,k}(t)$ at a single width

$$\{b_{j,k}(t)\} = \{< f(t), \psi_{j,k}(2^j t - k) >, j = 1, 2, \cdots, L; k = 1, 2, \cdots, K; K = N2^{-j}\}$$

The one of the simplest wavelets, the Haar wavelet is defined as follows

$$\psi(t) = \begin{cases} 1, & if & t \in [0, 0.5) \\ -1, if & t \in [0.5, 1) \\ 0, & otherwise \end{cases} \tag{8}$$

The dilations and translations of the Haar wavelet function form an orthogonal wavelet base for $L^2(\mathbf{R})$. The mother Haar wavelets is defined as

$$\psi_{j,k}(t) = \psi(2^j t - k), \quad j, k \in \mathbf{Z} \tag{9}$$

The Haar scaling function $\phi(t)$ is the unit-width function $\phi(t)$

$$\phi(t) = \begin{cases} 1, if & 0 \leq t \leq 1, \\ 0, otherwise \end{cases} \tag{10}$$

For function $f(t) \in L^2(\mathbf{R})$, the discrete wavelet expansion of $f(t)$ is represented as

$$f(t) = a_{0,0}\phi_{0,0}(t) + \sum_{j=1}^{d} \sum_{k=0}^{2^{j-1}} b_{j,k}\psi_{j,k}(t) \tag{11}$$

where $\phi_{0,0}(t)$ is a scale function on interval $[0, 1)$, $\psi_{j,k}(t)$ is the set of wavelets with different resolution.

The two-dimensional wavelet transform can be realized by successive applying the one-dimensional wavelet transform to data in every dimension. The two-variable wavelet function can be defined as a product of two one-dimensional mother (generating) wavelets $\psi(t_1)$ and $\psi(t_2)$

$$\psi(t_1, t_2) = \psi(t_1)\psi(t_2) \tag{12}$$

With dilation and translation parameters the two-dimensional wavelets function is defined as

$$\psi_{(a_1,a_2),(b_1,b_2)}(t_1, t_2) = \psi_{(a_1,b_1)}(t_1)\psi_{(a_2,b_2)}(t_2) \tag{13}$$

where $\psi_{a,b}(t) = |a|^{-\frac{1}{2}}\psi(\frac{t-b}{a})$. Assuming that $a_1 = a_2$ the two-dimensional wavelet expansion of two-variable function $f(t_1, t_2)$ can be expressed as

$$WT f((a), (b_1, b_2)) = |a|^{-1} \int_{-\infty}^{\infty} \int_{-\infty}^{\infty} f(t_1, t_2)\psi_{(a),(b_1,b-2)}(t_1, t_2) \, dt_1 \, dt_2 \tag{14}$$

For a two-dimensional grid of $2^n \times 2^n$ values (for example an image pixels) and for discrete parameters $a_1 = 2^{n-j_1}$, $a_2 = 2^{n-j_2}$, $b_1 = 2^{n-j_1}k_1$, $b_2 = 2^{n-j_2}k_2$, with integer values for j_1, j_2, k_1 and k_2 the 2D discrete wavelet function can be defined

$$\psi_{j_1,j_2,k_1,k_2}(t_1, t_2) = 2^{\frac{(j_1+j_2)}{2}-n}\psi(2^{j_1-n}t_1 - k_1)\psi(2^{j_2-n}t_2 - k_2) \quad (15)$$

where j_1, j_2, k_1 and k_2 are the dilation and the translation coefficients for each variable. Additionally, defining the scaling function

$$\phi_{j_1,j_2,k_1,k_2}(t_1, t_2) = 2^{\frac{j_1+j_2}{2}-n}\phi(2^{j_1-n}t_1 - k_1) \; \phi(2^{j_2-n}t_2 - k_2) \quad (16)$$

allows to define a complete basis to reconstruct a discrete function $f(t_1, t_2)$ (for example a discrete image):

$$\phi_{0,0,0,0}(t_1, t_2) = 2^{-n}\phi(2^{-n}t_1)\phi(2^{-n}t_2) \quad (17)$$

$$\gamma^H_{0,j_2,0,k_2}(t_1, t_2) = 2^{\frac{j_2}{2}-n}\phi(2^{-n}t_1)\psi(2^{j_2-n}t_2 - k_2)\gamma^V_{j_1,0,k_1,0}(t_1, t_2)$$

$$= 2^{\frac{j_1}{2}-n}\psi(2^{j_1-n}t_1 - k_1)\phi(2^{-n}t_2)$$

The 2D discrete wavelets coefficients are defined as

$$2DWT \; w_{j_1,j_2,k_1,k_2} = \int\int f(t_1, t_2)\lambda_{j_1,j_2,k_1,k_2} \; dt_1 \; dt_2 \quad (18)$$

where $\lambda_{j_1,j_2,k_1,k_2}$ denotes any of previously defined orthonormal bases. These coefficient can be formed as the Haar wavelets coefficient matrix $\mathbf{P}$.

2.1   Pattern forming based on the Haar wavelet transform

An image pattern can be formed with coefficients of one or few levels of 2D discrete wavelet transform of an image, represented by the rectangular array of gray level pixels. We have used the first level of 2D discrete Haar wavelet transform of centered $r \times r = 2^6 \times 2^6$ ($r = 2^{n_i} = 64$, $n_i = 6$) subimage of a face.

The $r \times r$ element original first level Haar wavelets matrix $\mathbf{P}$ can be used to form a $n_{Haar} = r \times r$ element Haar pattern as a concatenation of the matrix rows $\mathbf{x}_{Haar} = [\mathbf{p}_1, \mathbf{p}_2, \cdots, \mathbf{p}_r]^T \in \mathbf{R}^{n_{Haar}}$. where $\mathbf{p}_i$ is a ith row of the first level Haar wavelets transform coefficient matrix $\mathbf{P}$.

Despite of the expressive power of the Haar wavelets transformation it is difficult to say arbitrarily how powerful the Haar wavelets-based features could be for a classification of face images.

Experiments have shown that the Haar pattern can be heuristically reduced by removing r_r trailing element of each row of the original 2D Haar wavelets matrix $\mathbf{P}$. The heuristically reduced Haar wavelets-based pattern is formed with reduced rows of Haar wavelets coefficient matrix $\mathbf{P}_r = [\mathbf{p}_{r,1}, \mathbf{p}_{r,2}, \cdots, \mathbf{p}_{r,r}]^T$ where $\mathbf{p}_{r,i}$ is the reduced by r_r columns ith row of the Haar wavelets matrix $\mathbf{P}$. The reduction number r_r is chosen by assuring that $r - r_r$ leading elements in all rows of the matrix $\mathbf{P}_r$ have values below heuristically selected threshold ϵ_{Haar}. This can result in $n_{Haar,r} = r \times (r - r_r)$ element reduced Haar wavelets patterns $\mathbf{x}_{Haar,r}$. In the next sections we discuss techniques of finding reduced set of face image features.

3   Principal Component Analysis

We have applied Principal Component Analysis (PCA) [1] for the orthonormal projection (and reduction) of reduced Haar wavelets patterns $\mathbf{x}_{Haar,r}$ of facial images. Let us assume that a limited size sample of N random $n_{Haar,r}$-*dimensional patterns* $\mathbf{x}_{Haar,r} \in \mathbf{R}^{n_{Haar,r}}$ representing extracted features by the Haar decomposition of face image matrices has been gathered as an unlabeled training data set $T_{Haar,r} = \{\mathbf{x}^1_{Haar,r}, \mathbf{x}^2_{Haar,r}, \cdots, \mathbf{x}^N_{Haar,r}\}$ represented as a $N \times n$ data pattern matrix $\mathbf{X} = \left[\mathbf{x}^1_{Haar,r}, \mathbf{x}^2_{Haar,r}, \cdots, \mathbf{x}^N_{Haar,r}\right]^T$. The optimal linear transformation of reduced Haar patterns $\mathbf{y} = \hat{\mathbf{W}}\mathbf{x}_{Haar,r}$, into $m = n_{Haar,r,pca,r}$ element reduced PCA pattern $\mathbf{y} = \mathbf{x}_{Haar,r,pca,r}$, is provided using the $m \times n$ optimal Karhunen-Loéve transformation matrix $\hat{\mathbf{W}} = \mathbf{W}_{KLT} = \left[\mathbf{e}^1, \mathbf{e}^2, \cdots, \mathbf{e}^m\right]^T$ composed with m rows being the first m orthonormal eigenvectors (corresponding to eigenvalues arranged in decreasing order) of the original data $\mathbf{X}$ covariance matrix $\mathbf{R}_X$. The optimal matrix $\hat{\mathbf{W}}$ transforms the original $n_{Haar,r}$-*dimensional* patterns $\mathbf{x}_{Haar,r}$ into $m = n_{Haar,r,pca,r}$ -*dimensional* $(m \leq n_{Haar,r})$ reduced PCA feature patterns $\mathbf{y} = \mathbf{x}_{Haar,r,pca,r}$; $\mathbf{Y} = (\hat{\mathbf{W}}\mathbf{X}^T)^T = \mathbf{X}\hat{\mathbf{W}}^T$, minimizing the mean least square reconstruction error. The open question remains, which principal components to select as the best for a given processing goal. We discuss in the next section an application of rough sets for feature selection of reduced PCA patterns.

4   Rough Sets

The rough sets theory has been developed by Professor Pawlak [1] for knowledge discovery in databases and experimental data sets. Let us consider an *information system* given in the form of the decision table

$$DT = <U, \ C \cup D, \ V, \ f> \tag{19}$$

where U is the *universe*, afinite set of N objects $\{x_1, x_2, ..., x_N\}$, $Q = C \cup D$ is a finite set of *attributes*, C is a set of *condition* attributes, D is a set of *decision* attributes, $V = \bigcup_{q \in C \cup D} V_q$, where V_q is the set of *domain (value)* of attribute $q \in Q$, $f : U \times (C \cup D) \rightarrow V$ - is a total *decision function* (information function, decision rule in DT) such that $f(x, q) \in V_q$ for every $q \in Q$ and $x \in V$. For a given subset of attributes $A \subseteq Q$ the $IND(A)$ (denoted by $\tilde{A}$)

$$IND(A) = \{(x, \ y) \in U : \ for \ all \ a \in A, \ f(x, \ a) = f(y, \ a)\} \tag{20}$$

is an *equivalence relation* on universe U (called an *indiscernibility relation*).

For a given information system S a given subset of attributes $A \subseteq Q$ determines the approximation space $AS = (U, IND(A))$ in S. For a given $A \subseteq Q$ and $X \subseteq U$ (a concept X), the *A-lower approximation* $\underline{A}X$ of set X in AS and the *A-upper approximation* $\bar{A}X$ of set X in AS are defined as follows:

$$\underline{A}X = \{x \in U : [x]_A \subseteq X\} = \bigcup\{Y \in A^* : Y \subseteq X\} \qquad (21)$$

$$\bar{A}X = \{x \in U : [x]_A \cap X \neq \emptyset\} = \bigcup\{Y \in A^* : Y \cap X \neq \emptyset\} \qquad (22)$$

A *reduct* is the essential part of an information system (related to a subset of attributes) which can discern all objects discernible by the original information system. A *core* is a common part of all reducts. Given an information system S condition and decision attributes $Q = C \cup D$, for a given set of condition attributes $A \subset C$ we can define a positive region A $POS_A(D)$ in the relation $IND(D)$, as

$$POS_A(D) = \bigcup\{\underline{A}X | X \in IND(D)\} \qquad (23)$$

The positive region $POS_A(D)$ contains all objects in U which can be classified without error (ideally) into distinct classes defined by $IND(D)$ based only on information in the relation $IND(A)$.

For an information system S and a subset of attributes $A \subseteq Q$ an attribute $a \in A$ is called *dispensable* in the set A if $IND(A) = IND(A - \{a\})$ (it means that indiscernibility relations generated by sets A and $A - \{a\}$ are identical). The set of all indispensable attributes in the set $A \subseteq Q$ is called a *core* of A in S. and it is denoted by $CORE(A)$.

We have applied an idea of rough sets reduct to the proposed technique of feature selection-reduction of face images and corresponding data sets reduction.

5   Rough sets for feature reduction/selection

The PCA does not guarantee that selected first principal components, as a feature vector, will be adequate for classification. One of possibilities for selecting features from principal components is to apply rough sets theory [1], [2]. Specifically, defined in rough sets computation of a reduct can be used for selection some of principal components being a reduct. these principal components will describe all concepts in a data set.

The rough sets method is used for selection of a reduct from the discretized reduced PCA patterns. The final pattern is formed from real-valued reduced PCA patterns based on the selected reduct. We have applied the following algorithm for feature selection:

Algorithm: Feature extraction/selection using PCA and rough sets

Given: A $N - case$ data set T containing $n - dimensional$ $(n = n_{Haar,r,pca,r})$ patterns $\mathbf{x} = \mathbf{x}_{Haar,r,pca,r}$, with real-valued attributes, labeled by l associated classes
$\{(\mathbf{x}^1, c^1_{target}), (\mathbf{x}^2, c^2_{target}), \cdots, (\mathbf{x}^N, c^N_{target})\}$.

1. Isolate from the original class labeled data set T, a pattern part as $N \times n$ data pattern matrix $\mathbf{X}$.

2. Compute the covariance matrix $\mathbf{R}_x$ for $\mathbf{X}$.
3. Compute for the matrix $\mathbf{R}_x$ the eigenvalues and corresponding eigenvectors, and arrange them in descending order.
4. Select the reduced dimension $m \leq n$ of a feature vector in principal components space using defined selection method.
5. Compute the optimal $m \times n$ transform matrix $\mathbf{W}_{KLT}$.
6. Transform original patterns from $\mathbf{X}$ into $m - dimensional$ feature vectors in the principal component space by the formula $\mathbf{Y} = \mathbf{X}\mathbf{W}_{KL}$ for a whole set of patterns.
7. Discretize the patterns in $\mathbf{Y}$ with resulting matrix $\mathbf{Y}_d$.
8. Compose the decision table DT_m constituted with the patterns from the matrix $\mathbf{Y}_d$ with the corresponding classes from the original data set T.
9. Compute a selected reduct from the decision table DT_m treated as a selected set of features $X_{feature,reduct}$ describing all concepts in DT_m.
10. Compose the final (reduced) real-valued attribute decision table $DT_{f,r}$ containing these columns from the projected discrete matrix $\mathbf{Y}_d$ which are correspond to the selected feature set $X_{feature,reduct}$. Label patterns by corresponding classes from the original data set T.

The results of discussed method of feature extraction/selection depend on a data set type and three designer decisions: a) selection of dimension $m \leq n_{Haar,r}$ in the principal component space, b) discretization method applied, and selection of a reduct.

6   Numerical experiments - face recognition

As a demonstration of role of rough sets methods for feature selection/reduction we have carried out numerical experiments of recognition of 13 classes of facial images, with 30 different instances for each class. Each gray scale face image was of the dimension 112×92 pixels. Given original face image set, the first level Haar wavelets transform was applied to centered 64×64 pixel sub-windows of an original face image. The resulting 64×64 Haar wavelets coefficient matrix has been used to form an original $64 \times 64 = 4096$ Haar wavelets feature pattern. The Haar wavelets patterns have been heuristically reduced to the size 2048. Then, according to the proposed method, we have applied Principal Components Analysis (PCA) for feature projection/reduction, followed by the heuristic reduction of projected PCA patterns to the length of 200. In the final processing step we have applied and rough sets method for the final feature selection/reduction based on reduced PCA patterns. The discretized training set was used to find the minimal 6-element reduct [1]. This reduct was used to form the final pattern. The entire image data set was divided into training, testing sets: 70% of these sub-images were used for the training set, 15% for the validation set, and the final 15% for the test set.

The training, validation, and the test sets (decision tables) with real-value pattern attributes were reduced according to the selected reduct. Classification of face images have been performed by the single hidden layer, 80 neuron, error back propagation neural network, with the resulting accuracy 99.24% for the test set.

7   Conclusion

The sequence of data mining steps, including application of 2D Haar wavelets for feature extraction, PCA, and rough sets for projection and feature selection, has showed a potential for designing of neural network classifiers for face images. Rough sets methods have showed ability to reduce significantly a pattern dimensionality.

References

1. Cios, K. Pedrycz, W.. and R. Swiniarski. (1998). Data Mining Methods for Knowledge Discovery. Kluwer Acad. Publ., Boston.
2. Pawlak, Z. (1991). Rough sets, Theoretical aspects of reasoning about data, Kluwer, Dordrecht 1991.
3. Skowron, A. (1990).The rough sets theory and evidence theory. Fundamenta Informaticae vol.13 pp.245-262.
4. R. Swiniarski. (1999). " Rough Sets and Principal Component Analysis and Their Applications in Data Model Building and Classification". In S. K. Pal and A. Skowron (Eds.), "Rough Fuzzy Hybridization : New Trend in Decision Making", Springer Verlag, Singapore, January 1999.
5. Yao, Y.Y., Wong, S.K.M., and Lin, T.Y., (1997), "A review of rough set models", in: T.Y. Lin, N. Cercone (eds.), *Rough Sets and Data Mining. Analysis for Imprecise Data*, Kluwer Academic Publishers, Boston, London, Dordrecht, 47–75.
6. Ziarko, W., (1993), "Variable Precision Rough Set Model", *Journal of Computer and System Sciences*, 40, 39–59.

Detecting Spots for NASA Space Programs Using Rough Sets[1]

Zbigniew M. Wojcik

Lockheed Martin Space Mission Systems & Services
Science, Engineering, Analysis and Test.
2400 NASA Road 1, Houston, TX 77058
wojcik@texas.net

Abstract. This new method of detecting spots is based on the general concept of the rough sets. The lower approximation collects the set of objects which are assigned to a class in question without any doubt, and the upper approximation is composed of the lower approximation and the objects which are classified to the class with some uncertainty. In the initial step all objects detected are assigned to the class of spot-like objects (i.e. spot candidates). Subsequent refinements tend to extract spots with higher and higher degree of certainty, based on the lower approximation. Learning system is defined based on the rough sets making the learning phase automatic (by exploitation of the lower approximation) refining the set of candidates.

1 Introduction

Space operations require one spacecraft be mated with another spacecraft on an orbit. Fig,1 shows two spacecraft, one near another. Several disks on the spacecraft are used to define their relative locations (see Fig.1). Centers of the disks were computed by an on-board computer and then used for the mating operation. Difficulties are with clutter (incl. objects reminding spots), reflections, obscurations and shadows. The first step in the processing is usually to detect the spot and then to work on the small image tile containing the spot (e.g., of the dimensions of 32 by 32 pixels) to precisely compute the spot location. Fig.2 is an example of a 32 by 32 tile with a spot. You can see a real shadow and clutter on it. The computer processing can admit several spot-like objects detected in addition to real spots which can be further rejected by using previous knowledge about the relative distances between the real spots and knowledge about expected approximate relative locations of the two spacecrafts at any time. The task is to reduce the number of the extra spot-like objects detected. This paper presents Wojcik's methods to solve this task.

The possibility of computing the position of disk with resolution much better than one pixel spacing was discovered by Z. Wojcik and published in 1976 [2]. The resolution achieved by Wojcik was 0.05 pixel spacing. The significance of using the disk-shaped spot is an increased resolution of the camera image by about 20 times in the horizontal direction and 20 times in the vertical direction.

---

[1] The work was under NASA contract No. NAS9-19100

W. Ziarko and Y. Yao (Eds.): RSCTC 2000, LNAI 2005, pp. 569-576, 2001.
© Springer-Verlag Berlin Heidelberg 2001

2 Rough Sets Model for Detection of Spots in Space Image

Our rough sets [1] incorporates the Universe (or data) and an equivalence relation R.
The equivalence class describes the class of the objects looked for.

Fig.1. The input image with two spacecraft. Black spots are used to define accurate positions
of the crafts with respect to each other when mating in space

Connected edges of regions detected by applying subsequent thresholds to a gray-
level image are used. Figs. 3 – 4 show examples of such edges of a gray-level image
(Fig. 2) with a spot. The brightest pixels in Figs. 3 – 4 represent such edges. The
regions were detected by using several subsequent thresholds. The set of geometrical
centers x_e, y_e of each connected edge e of each region detected at any threshold is used
as the Universe for the rough sets analysis. The initial equivalence relation R_d is the
operation collecting the set of geometrical centers x,y located within the distance d
from the center x_e, y_e. The equivalence class $[x_e, y_e]/R_d$ extracted by R_d is the set of
geometrical centers within the distance d from x_e, y_e:

$$[x_e, y_e]/R_d = \{ \ x,y \in U: \ |x-x_e| < d \ \wedge \ |y-y_e| < d \ \} \ \} \tag{1}$$

That is, the centroids of the connected edge features lie close together. For
instance, the equivalence class of a spot-like object is made by each geometrical
center of the connected circular edges located about the image center in Figs. 3–4,
because they are within the distance d=3 from the connected circular edge shown in
Fig. 4.

The cardinal number (count of the elements) of the equivalence class must exceed
an assumed threshold cThr (e.g. cThr=3):

$$\mathrm{Card}([x_e, y_e]/R_d) \ > \mathrm{cThr} \tag{2}$$

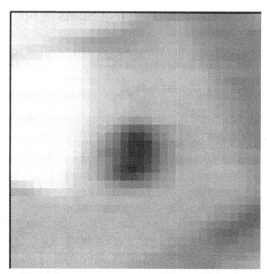

Fig.2. The input tile   (size of 32 by 32 pixels)

That is, if Eq.(2) is not satisfied, then we do not consider the candidate equivalence class (Eq.(1)) of connected edges to be a valid spot candidate (compare also Figs 3-4). The upper approximation US of the set of spots S in an input image is given by the set of the geometrical centers satisfying Eq.(2):

$$US = \{ [x_e, y_e] \in U: \ Card([x_e, y_e]/R_d) > cThr \} \tag{3}$$

The final equivalence relation R detecting the true spots (i.e., the spots and nothing but the spots) is unknown. We define the lower approximation LS of the true spots S in an input image detected with certainty:

$$LS = \{ [x_e, y_e] \in U: \ [x_e, y_e]/R \subseteq S \} \tag{4}$$

The lower approximation is defined for some threshold lTc higher than cThr:

$$LS = \{ [x_e, y_e] \in U: \ Card([x_e, y_e]/R_d) > lThr \} \tag{5}$$

the set of true spots, which is larger than LS and smaller than US, is approached by eliminating from US all objects which are not the spots with the aid of additional features F (relations on the US). We hope to define the relation R_d and the features F so that:

$$R = R_d \bigcap F \tag{6}$$

Each spot-like object detected by R_d and subjected to detection of any feature $f \subseteq F$ may not be eliminated from the set of spots, and still must make a part of the US:

$$US_f = \{ [x_e, y_e] \in U : [x_e, y_e]/(R_d \cap f) \cap S \neq \varnothing \} \subseteq US \qquad (7.a)$$

$$\bigcup_{f \subset F} US_f = US \qquad (7.b)$$

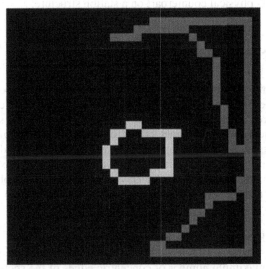

Fig. 3. Input image tile (Fig.2) subjected to a threshold. All non-black pixels are below the threshold selected. Then the regions below the threshold were shrunk by one pixel, and finally the edges of the shrunken objects were detected. One connected edge followed (gray level turned darker)

where $\varnothing$ is the empty set. Equations (7.a,b) say, that the set of edges of geometrical centers near x_e, y_e extracted by R_d and additionally refined by a proper relation (feature) f must still be the member of the class of spot-likes objects US for as long as this set of edges represents a true spot. Each feature f refines the equivalence class (Eq.(1)) by simply adding an additional constraint f to it in the form of the term $[x_e, y_e]/(R_d \cap f)$. This constraint, however, may not eliminate the candidate from the set of spots if the candidate is the true spot, what is represented by the condition: $[x_e, y_e]/(R_d \cap f) \cap S \neq \varnothing$. The set of the candidates $[x_e, y_e]/(R_d \cap f) \cap S \neq \varnothing$ refined by the feature f is then still a subset of the upper approximation US of the true spots S.

We define features F so that the application of all of them makes the true spots S:

$$\bigcap_{f \subset F} US_f = S \supseteq LS \qquad (8)$$

An example of the feature f is the ratio of the gray level at the center of the spot candidate to the gray level around the spot candidate (below an assumed threshold to remain the member of US).   Example of another feature f is the number of zero-crossings (transitions of the gradient sign from positive to negative) around the spot candidate (below an assumed threshold to remain the member of US).

Although it is not demonstrated here, the result US of this spot detection can be further fed into a "constellation sieve" or matching with a knowledge database of anticipated spot constellations for the final refinement.   That process associates a detected spot with its physical counterpart on a station structure.

3 Fast Rough Sets Learning System

Very effective is a rough sets system, which: 1. finds of the true spots (i.e. the spot-like objects with the highest number of concentric edges), then 2. learns about their features and measures of the features, and finally 3. rejects from the set of candidates all the spot-like objects whose parameters are too far from the parameters of the true spots.   This learning is from examples of true spots provided by the lower approximation given by Eq.(5). For instance, given the upper approximation (i.e., all the spot candidates), the average diameter Dav of the spots (i.e., of objects LS_C in the upper range of the number of concentric edges) is measured in the first step of the refinement:

$$LS_C=[x_e,y_e]/R_C=\{ [x_e,y_e]\in US: MaxC\text{-}Card([x_e,y_e]/R_d) <cT) \} \tag{8}$$

where MaxC is the maximum number of concentric edges of the spot-like objects US, Card($[x_e,y_e]/R_d$) is the number of concentric edges of the spot-like object defined by the equivalence class $[x_e,y_e]/R_d$, cT is the range in the number of concentric edges around the MaxC making the lower approximation. and LS_C is assumed to be the essential part of the true spots because of the highest number of concentric edges in each object.

After the first stage of learning about Dav in the set LS_C, all those spot-like objects are rejected from US whose diameter D is different from Dav by more than an assumed threshold dT:

$$US_D=\{ [x_e,y_e]\in US: [x_e,y_e]/(R_d\cap f=\text{lD-Davl}<cT)\cap S\neq \varnothing \}\subseteq US \tag{9}$$

The set US_D approaching the true spots no longer has objects whose diameter is different from that of the true spots.

The advantage of using rough sets for the learning system is making the learning phase automatic (by exploitation of the lower approximation).   The lower approximation of the true spots provides automatically the best examples from which the system can learn the parameters of the true elements of the class. Lower and upper bounds of the parameters are then used to refine the upper approximation to detect the true class.

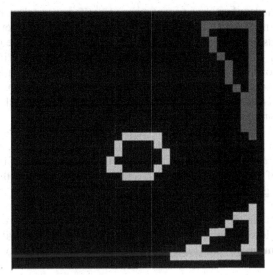

Fig. 4. Input image tile (Fig.2) subjected to a lower threshold. Regions below the threshold were shrunk by one pixel, and edges of the shrunken objects are represented by the brightest pixels. One connected edge followed (gray level turned darker)

4   Results

Selected steps in processing a gray-level image of a spot with a bright reflection are shown in Figs. 2-4.   The methodology uses multiple thresholds [3], based on which all binary objects obtained at all thresholds are processed to detect the equivalence class of edges.   Image shown in Fig. 3 was received by using the first simple thresholding producing the first binary representation, then shrinking all the objects below the threshold by one pixel, then detection of edges of the shrunk objects. The shrinking is by leaving only these pixels representing an object which do not have any background pixel in the direct neighborhood. The edge detection is by leaving only these pixels representing an object which do have a background pixel in the direct neighborhood [3]. This simple edge detector satisfies the underlying requirement for edge connectivity exploited in further processing [3].

The image presented in Fig.3 was received by tracing one connected edge [3]. Since our edges are always connected, we segment the binary image into objects by the edge following algorithm [3]. The algorithm turns to a different color each edge traced and moves to a neighbor edge pixel not traced yet [3]. It completes, if there are no more edge pixels in the direct neighborhood not traced yet. Note, that shrinking helps to separate binary objects from each other.

Image in Fig.4 was obtained by using a threshold lower than threshold applied to image in Fig.3.   Again, shrinking helps to separate binary objects from each other (compare three connected edge components represented by the brightest pixels on two binary objects).

Each connected edge component was traced separately, Fig.4 shows one connected edge traced and two more left for edge tracing. Note that objects for smaller thresholds become smaller, and if the object is circular, the geometrical center of its connected edge stays in relatively stable location in the image plane. The connected edge of large shrunken object shown in Fig.3, for example, was split into two smaller objects by subsequent smaller threshold (compare Fig.4), so its geometrical center happened to be unstable. This stability of circular connected edges is exploited in our algorithm.

As shown in Figs 2-4, subsequent lower thresholds, shrinking, edge detection and edge following result in connected edges of the circular object (spot) at relatively stable location of its geometrical center, and connected edges of the other objects of much less stable geometrical centers.

Fig.1 presents the input space image with spots located on the two spacecraft, and Fig.5 shows the lower approximations of the spots, all marked with the white cross. The spot at the image bottom (Fig.5) was not processed because it was too close to the image frame.

Fig.5. Image with two spacecraft. The true set of the spots detected by a computer program is marked with white crosses

5 Conclusion

The equivalence class is a powerful tool capable to detect a class or an object from data. The equivalence class of concentric connected edges defines a spot candidate in our application. Wojcik's method applies the concentricity (the equivalence relation) and uses additional features refining the true class (spots) from all candidates (e.g., from the upper approximation). High level of concentricity (high value of the cardinal number of the equivalence class) defines the true spots (class) but not all the set of the true class. High level of concentricity (high range of the cardinal number of the equivalence class) defines the lower approximation of the true spots (class). Wojcik's learning system defines the examples as the lower approximation, then learns parameters and features from the examples, computes their ranges, and based on them refines the candidates to achieve the true class of objects looked for. The rough sets provides a model for both the detection of the spots (class) and for the learning system (automatic learning the true examples and then learning from the examples).

Further acknowledgement: The author expresses thanks for Dr. Richard Juday (NASA at Houston) for valuable comments to this article and for the supervision of the project, and Dr. Mike Rollins for the collaboration.

References

[1] Z. Pawlak, Rough Sets, Theoretical Aspects of Reasoning about Data, Norwell, Massachusetts: Kluwer Academic Publishers, 1991.
[2] Z. M. Wojcik, "The Alignment of Graphic Images in Solid-State Technology", Micro-electronics and Reliability, Vol. 15, Pergamon Press, pp. 613-618, 1976.
[3] Z. M. Wojcik, "A Natural Approach in Image Processing and Pattern Recognition: Rotating Neighborhood, Self-Adapting Threshold, Segmentation and Shape Recognition", Pattern Recognition, Vol. 18, No.5, Pergamon Press, pp. 299-326, 1985.

Solving the Localization-Detection Trade-Off in Shadow Recognition Problem Using Rough Sets[1]

Zbigniew M. Wojcik

Lockheed Martin Space Mission Systems &
Services Science, Engineering, Analysis and Test.
2400 NASA Road 1, Houston, TX 77058
wojcik@texas.net

Abstract. This new method of detecting and compensating shadow is based on the general principle of the rough sets. Shadow recognition is constrained by the rough sets principle according to which the upper approximation of objects must contain non-empty lower approximation – the true class of objects in question. By imposing this constraint, the well known localization-detection trade-off is solved. In the first step the shadow is detected reliably by using a high threshold. Reliable classification (shadow detection) with the aid of a high threshold makes the lower approximation of shadow. Then, the upper approximation is constructed based on the lower approximation (reliably detected shadow) by using a low threshold. Directly using a low threshold would detect a lot of clutter and noise rather than shadow. Rough sets principle prevents this: each shadow candidate must contain the lower approximation. On the other hand, making the threshold high detects shadows reliably, but not accurately, for instance, shadow frequently begins at a low threshold. Rough sets solves this problem by tracking the upper approximation with the aid of a low threshold from the lower approximation.

1 Introduction

Spacecraft position during the mating operation on an orbit in space is determined based on centers of the disks attached to the craft. Shadow passing the disk image affects severely the results of computation of the disk center. Fig. 1.a presents a 32 by 32 pixel image tile with a disk and a shadow. The task is to neutralize the shadow presence so that the results of the spacecraft positioning are not affected by the shadow. The required accuracy is 0.05 pixel spacing, so even a small shadow gradient has an impact on the results. The discovery made by Z. Wojcik [4] indicates that the center of a disk image taken by a camera can be determined with the accuracy of 0.05 pixel spacing when shadow is not present.

For a shadow to be compensated it must be recognized first. But shadow is difficult to recognize. A contextual knowledge is needed about the scene objects. Image factorization can be used for shadow removal [2, 3]: shadow can be removed by appropriate weighting a factor associated with shadow gray level. Only somebody must tell which factor is the shadow. In addition, shadow edge is not represented by a

---

[1] The work was under NASA contract No. NAS9-19100

W. Ziarko and Y. Yao (Eds.): RSCTC 2000, LNAI 2005, pp. 577–583, 2001.
© Springer-Verlag Berlin Heidelberg 2001

single factor, and weighting a few factors may weight other image details to disappear.

This research is in a much better situation in which the spot context is very well defined. The spot is located on a uniform background, therefore, two significant gray level changes on two sides of the spot indicate the presence of a shadow. The two gray level changes on the two sides of the spot make an equivalence class. Shadow is detected if there is the equivalence class composed of two significant gray level changes on two sides of the spot. Other shadow pixels are attached to the equivalence class of the shadow detected, with some ambiguity, lying along the assumed shadow edge, making a shadow approximation. Ambiguity comes from using now a new threshold lower than the threshold used for shadow detection. Shadow upper approximation (detected at a lower threshold) is then compensated for based on the gray level changes of the equivalence class members.

Fig.1.a. Input image tile (32 by 32 pixels) with a spot and shadow passing through it

2   Rough Sets Model for Detection of Shadow around a Spot in Space Image

Our rough sets [1] incorporates the Universe (or data) U and an equivalence relation R. The equivalence relation describes the class of the objects looked for, in our case, the two gray level changes indicating the presence of a shadow. More sampling lines are allowed but are not used in this research.

The gray level changes are collected along the four sampling lines l, r, u, b around the spot: on the left, right, upper and bottom of the spot correspondingly. These four sampling lines around each spot make the Universe U. Four edges at the square inside the image shown in Fig. 1.b are examples of the four sampling lines.

Maximum ml, mr, mu, mb and minimum (negative values) nl, nr, nu, nb of the changes in the sampling lines l, r, u, b are computed.

The equivalence class $[ml, mr]/R_h$ or $[nl, nr]/R_h$ extracted by the equivalence relation R_h is the set collecting the two gray level changes, positive ml and mr or negative nl and nr, in the two sampling lines l and r, exceeding a threshold T:

$$[ml, mr]/R_h = \{ l,r \in U: ml > T \wedge mr > T \} \tag{1.a}$$
$$[nl, nr]/R_h = \{ l,r \in U: nl < -T \wedge nr < -T \} \tag{1.b}$$

R_h specified above detects horizontal shadows passing spots: with darker side on the left side on the spot image, and with darker side on the right side on the spot image respectively. The equivalence class $[mu, mb]/R_v$ or $[nu, nb]/R_v$ extracted by R_v is the set of two gray level changes in the two sampling lines u and b exceeding a threshold T:

$$[mu, mb]/R_v = \{ u,b \in U: mu > T \wedge mb > T \} \tag{1.c}$$
$$[nu, nb]/R_v = \{ u,b \in U: nu < -T \wedge nb < -T \} \tag{1.d}$$

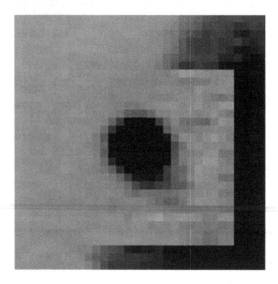

Fig. 1.b. The input image (Fig. 1.a) after shadow compensation in the spot area

R_v detects vertical shadows passing spots. The threshold T for the shadow gray level changes is relatively high so that there is the full certainty that the shadow is detected if it exists. The certainty of the classification to the class of shadows S satisfies the definition of the lower approximation of the rough sets. Thus, the lower approximation LS of a shadow S is:

$$LS = \{ l,r, u,b \in U: \quad [ml, mr]/R_h \subseteq S \cup [nl, nr]/R_h \subseteq S$$
$$\cup [mu, mb]/R_v \subseteq S \cup [nu, nb]/R_v \subseteq S \} \tag{2}$$

Once the shadow is detected reliably looking for high gray level changes in the places where they are not supposed to exist under a regular lighting without shadow, the upper approximation of the shadow is traced from the location of its detection by using a new threshold t lower than T.   Equivalence relations r_h and r_v now include thresholding with the aid of a small threshold t and the connectivity feature c with the lower approximation of shadow. The equivalence classes [ml , mr]/ r_h or [nl , nr]/ r_h of small intensity shadow are the same as Eqs (1.a,b,c,d) with the exception that t is used instead of T.   The upper approximation US of shadow finds accurately the begins and ends of shadows because it works at a low threshold t:

$$US=\{l,r,u,b\in U[ml,mr]/(r_h \cap c_h)\cap S \neq \varnothing \cup [nl, nr]/(r_h \cap c_h) \cap S \neq \varnothing$$
$$\cup \ [mu, mb]/(r_v \cap c_v) \cap S \neq \varnothing \cup [nu, nb]/(r_v \cap c_v) \cap S \neq \varnothing \} \qquad (3)$$

where c_h is the operation testing connectivity with the corresponding equivalence classes [ml, mr]/R_h or [nl, nr]/R_h detecting shadow at a high threshold, and c_v is the operation testing connectivity with the corresponding high threshold equivalence classes [mu, mb]/R_v or [nu, nb]/R_v.

Fig.2.a. Another input image with a real shadow passing the spot

Eq.(3) says, that shadow detection with the aid of relations r_h and r_v thresholding gray level changes by using a low threshold t must test a corresponding connectivity c_h or c_v with shadow detected at a high threshold to still represent shadows (i.e. to make a non-empty set of shadows S). The way of implementing Eq.(3) is to detect shadows reliably at a high threshold first and then to trace each high shadow gray level change to its ends by using a low threshold.

Fig. 2.b. The input image from Fig. 2.a after compensating shadow in the spot area

3 Overcoming the Localization-Detection Trade-Off by Using the Rough Sets Principle

Because of the overwhelming localization-detection trade-off, the direct thresholding either detects reliably with a low precision, or does not detect reliably (because the results contain in addition a lot of clutter and noise) with a high precision. As shown above in section 2, rough sets model splits the shadow recognition task into two phases. Phase one uses the lower approximation collecting shadows detected reliably by using a high threshold T. The detection of shadows is certain then on a uniform background. However, the localization of the beginnings and ends of the shadows is not accurate for a high threshold T. When lowering the threshold to detect the shadow more accurately on the level of the lower approximation, noise and clutter is detected even if shadow does not exist.   Keeping in mind that the lower approximation detects the shadows reliably, the upper approximation is used. The upper approximation involves the lower approximation. Starting from each lower approximation, continuity of each shadow reliably detected is traced by using a lower threshold. When starting from a low threshold it would not be clear whether shadow candidate detected contains the lower approximation of shadow, i.e. the true shadow. Rough sets imposes the constraint that the upper approximation can be constructed only when the lower approximation is not empty.

4   Fast Rough Sets Machine Learning

The lower approximation collects the set of true examples on which the system learns about the whole class. Thus, the lower approximation (Eq.(2)) firmly characterizes shadow detected between two respective sampling lines. The upper approximation, too, characterizes shadow located between two sampling lines by its threshold $t < T$. Thus, shadow is detected, measured and is certainly characterized at all locations between two sampling lines.

Shadow is compensated now between the two respective sampling lines by using the knowledge learned from the two firm examples delivered by the lower approximation. These lower approximation examples carry information on the pixels at which gray level change exceeds threshold T, what indicates the presence of a shadow, and the upper approximation provides all remaining pixels associated with the shadow detected. These upper approximation pixels represent shadow of a lower gray level change.

Characteristics of the upper approximation pixels are computed at the end of the learning phase. These characteristics learned provide specific gray level changes at all pixels making the upper approximation of shadow, based on which ratios are computed of the gray levels of the shadow upper approximation pixels to the gray levels of the pixels adjacent to the upper approximation (without shadow). The upper approximation pixels representing shadow (implied between the two sampling lines) are compensated, by multiplying the upper approximation pixels by the reciprocal of the corresponding ratios.

Wojcik's rough sets machine learning works by defining and collecting the lower approximation., then gathering (learning) characteristics and features of all elements (examples) of the lower approximation, and finally by applying these characteristics and features to the candidate elements to recognize or process the true class members. The candidate elements in this application are the pixels implied by the interpolation from the lower approximation of shadow detected at the two sampling lines. The characteristics learned are the ratios of the gray levels of the upper approximation pixels to the pixels adjacent to the upper approximation. The pixels processed (compensated) with the aid of the knowledge learned are the pixels implied by the interpolation, as making the shadow. In this application the upper approximation does not define the candidates of the true class, the upper approximation provides more accurate characteristics of shadow, what overcomes the well-known localization-detection trade-off.

5   Conclusion

Complexity of the localization-detection trade-off has been brought to computing the rough sets: finding the lower and upper approximations. The upper approximation can not be found reliably by using directly a low threshold because it is not known if the candidates detected contain the lower approximation. Therefore, Wojcik's method constructs the full rough sets representing shadows starting from the non-empty lower approximation detected reliably at a high threshold. Then, the upper approximation is traced at a low threshold from the lower approximation. The rough sets represents shadows reliably and accurately.

Wojcik's machine learning assumes that there is the lower approximation providing the true examples to find the set of features representing the class in question. Then these features are measured, and the characteristics collected are used as patterns to match the candidates. The candidates are finally refined by these characteristics. Finding true examples and the whole learning process can be automatic if the lower approximation is definable. Automation of the learning process is the advantage of this machine learning.

Further acknowledgement: The author expresses thanks for Dr. Richard Juday (NASA at Houston) for valuable comments to this article and for the supervision of the project, and Dr. Mike Rollins (Lockheed Martin) for the collaboration.

References

[1] Z. Pawlak, Rough Sets, Theoretical Aspects of Reasoning about Data, Norwell, Massachusetts: Kluwer Academic Publishers, 1991.
[2] J.R. Taylor, J.L. Johnson, "K-Factor Shadow Removal", SPIE Vol. 3715, pp. 328-334, 1999 (SPIE Confr., part on Optical Pattern Recognition, Orlando. Fl.)
[3] A. Toet, "Multiscale contrast enhancement with application to image fusion", Opt. Eng. Vol. 31, No. 5, pp. 1026-1031, 1992.
[4] Z. M. Wojcik, "The Alignment of Graphic Images in Solid-State Technology", Microelectronics and Reliability, Vol. 15, Pergamon Press, pp. 613-618, 1976.

A Modular Neural Network For Vague Classification

Gasser Auda[1,2] and Mohamed Kamel[2]

[1] Electrical and Computer Engineering Department,
Ryerson Polytechnic University, Toronto, Canada
gauda@ee.ryerson.ca http://www.ee.ryerson.ca/g̃auda
[2] Systems Design Engineering Departmen t
University of Waterloo, W aterloo, Canada
{gasser,mk amel}@w atfor.uwaterloo.ca

Abstract. A modular neural network classifier design is presented. The objective behind the design is to enhance the classification performance of conventional neural classifiers according to two criteria, namely, reducing the classification error, and allowing vague/boundary classification decisions. The proposed model uses an unsupervised network to decompose the classification task over a number of neural network modules. During learning, every module is trained using samples representing the other modules, and modules are trained in parallel. After the training phase, every module inhibits or enhances the responses of the other modules by "voting" for the existence of the input within their decision boundaries. If the result of the majority vote is a "tie", then the sample is classified as a vague class (or boundary) between the (two or more) classes that have the tie. The proposed classifier is tested using a two-dimensional illustrative benchmark classification problem. Results are showing an enhancement in the classification performance according to the above two criteria.

1 Introduction

The definition of rough sets [9] made several contributions to the field of classification, pattern recognition and knowledge discovery [5]. Defining "vague classes" is one of those contributions. Often, in real life systems, there are patterns/objects/attributes that cannot be naturally classified as belonging to any specific category. This is not because of a deficiency in the classification system. Even a human-expert w ould fail to classify such patterns. As an example, there w ere samples in the NCR's n umerals benchmark problem, w e previously used in [1], that can nev er be definitely classified as a digit "7" versus a digit "1", or a digit "8" versus a digit "9", and so on. It is more realistic, and more accurate, for the classification system to iden tify these patterns as "boundary" or "v ague" and, have this as the final classification decision for such samples.

W. Ziarko and Y. Yao (Eds.): RSCTC 2000, LNAI 2005, pp. 584–589, 2001.

The current generation of NN classifiers suffer from a major dra wback; their inability to cope with the increase of size and/or complexit y of the classification task [2]. This is often referred to as the scalability problem. Large networks tend to introduce high internal interference because of the strong coupling among their hidden-layer weights [7].

This paper introduces a modular neural net work structure that enhances the classification performance. The model uses a divide-and-conquer approach for decomposing the classification task in to a group of simpler sub-tasks. Each classification sub-task is handled by a simple, fast and efficient module. Then, sub-solutions are in tegrated via a m ulti-module decision-making strategy, which has the ability to classify a tested sample as a "vague class", or boundary, between two or more classes.

Section 2 describes the proposed model and the theoretical basis behind the different design aspects. Section 3 describes the preliminary experimental study that is carried out to pro ve the merit of the proposed model. Section 4 summarizes the paper's conclusions and outlines some future work.

2   The Proposed Modular Netw ork

First, an unsupervised network is used for task-decomposition, i.e. subgroups of classes are assigned to small modules rather than classifying all classes using one large non-modular net work. During learning, each module is trained to classify its o wn group of classes. What is more than this simple divide-and-conquer idea is that ev ery module is trained using samples represen ting the other modules (groups) as w ell. The structure, therefore, of a module's output la yer consists of "class outputs" (C_i) equal to the number of classes in the group, plus "group outputs" (O_{ij}) equal to the num ber of the "other" modules, i.e. n um ber-of-modules-1 (Fig. 1).

If the training sample is in one of the module's classes, its output bit is high $(C_i = 1.0)$ while all other outputs pointing to other classes in the group and other groups are low (0.0). Otherwise, the output of module i should point to the bit representing one other module j $(O_{ij} = 1.0)$, with the others are low (0.0).

During testing, every module inhibits or enhances the responses of the other modules by voting for the existence of the input within their decision boundaries. Group outputs (votes) approach 1.0 or 0.0 according to how near or far, respectively, the sample is from the corresponding class in the feature space. This "cooperation" in taking the decision takes place above the modules' output la yer in the voting block (Fig. 1). Multiple

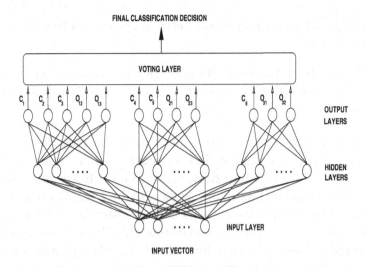

Fig. 1. An example of the proposed modular neural network.

neural network modules cooperating in taking a classification decision are modeled as m ultiple v oters electing one candidate in a single ballot election. All modules are considered "candidates" and "v oters." V oting "bids" are the different group outputs, and the highest for each module is considered the module's v ote. Pluralit y (Majorit y) v oting is the most common v oting sc heme in real-life collectiv e decision-making processes [8, 10]. Each voter votes for one alternative, and the alternative with the largest number of votes wins.

The advantage of this scheme, from the NN perspectiv e, is that it only uses the highest output value, which is the most probable output to be true, even if its value is way below "1.0". Note that the probability of correctness of a certain class is *not* proportional to the corresponding output value. Empirically, according to [1] and also [6], we noticed that the probability of having one of the lower outputs as the correct output is very low, unless the NN needs more training. Therefore, w e consider the lower outputs as "information noise", and rely only on the highest v alue for the module's v ote. F or a comparison of the differert voting schemes applied to NN-classifiers decision-making, refer to [3].

Therefore, the decision making process can be summarized as follows.

1. The *voting strategy* determines the group/module that is more likely to contain the tested sample within its decision boundaries.

2. If the result of the majority vote is a "tie", then the sample is classified

as a vague class (or boundary) between the classes (two or more) that have the tie.

3. If the result of the majority vote clearly defines a winner-module, the maxim um class-output *within* this module is taken as the final decision.

Without allo wing v ague/boundary classes, the only wato take a classification decision in case of a tie vote is the random choice between the winner classes. This will, naturally, cause a lot of classification errors. For example, the random class c hoice is lik ely to give a 50% error in case of a boundary between two classes. This is a drawback in all modular designs that use the majority vote as a m ulti-module decision-making strategy, for example, [3] and [4]. Therefore, the v ague classification decision will help reduce misclassifications. Another advantage of this decision mak ing process is that it mak es the classification system more realistic and practical, as outlined in Section 1.

3   Experiments

W e created a t wo-feature/attribute classification problem to use as a benchmark. A t w o-dimensional input space giv es an illustration of the shapes of the decision boundaries. Samples are randomly generated from Gaussian distributions around 20 random class means (Fig. 2). Each class consists of 200 samples, half for training and half for testing. A R T2 and Backpropagation (BP) schemes are used for the unsupervised and supervised networks, respectively.

The modular net w ork is compared to the non-modular net wrk (2 inputs and 20 outputs) classifying the same data. To guarantee a fair comparison, the BP learning parameters, the algorithm for terminating training, and the criteria for determining the n um ber of hidden nodes are unified across the modular and the non-modular neural net w orks. The unsupervised task-decomposition tec hnique clustered the data in to 12 groups, namely, 1-3-4, 2, 5, 6-7, 8, 9, 10, 11-12-14, 13, 15-19-20, 16-17, and 18. All modules are trained in parallel, and they all use an equal num ber of training samples per class, or group, output.

Table 1 summarizes the results. The proposed modular structure decreased the classification errors of the non-modular neural net work by 73.3% of the errors, i.e., from 72.4% to 92.65% correct classification rate. Given the accuracy of the voting scheme in iden tifying modules (groups),

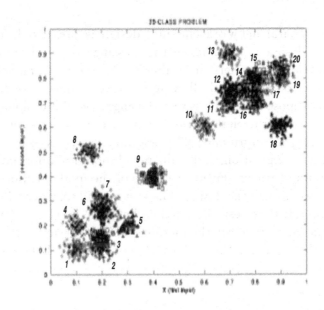

Fig. 2. The 20-class 2-dimensional classification data.

Network	# Miscl. samples	Correct cl. %
Non-modular BP	552	72.4%
Modular network	147	92.65%
Modular network (with vague classes)	98	95.1%

Table 1. A summary of the results.

and the efficiency of neural modules dealing with smaller and more homogeneous sub-tasks, decision boundaries are drawn much more accurately than the non-modular network that uses the same supervised learning scheme.

About 36% of the samples caused a tie between 2 or more of the voting modules. Without allowing vague/boundary decision, the classification outcome would depend on a random choice for the output class. This caused a percentage of errors that were eliminated by the vague-class decision. Hence, the percentage of certain/lower approximation classification increased to 95.1% for this benchmark.

4   Conclusion

A modular neural net w ork classifier design is presen ted. The objective behind the design is to enhance the classification performance of conventional neural classifiers according to two criteria, namely, reducing the classification error, and allowing vague/boundary classification decisions. The proposed classifier is tested using a two-dimensional illustrativ e benchmark classification problem. Results are sho wing an enhancemert in the classification performance. The new model's success is due to its m ulti-stage task-decomposition, and utilizing all modules' information through cooperation and voting. Future versions of the model will in tegrate more rough set concepts in the design. The system will select specific attributes for each module that best differen tiate bet w een its classes (similar to [9]). This is believed to increase the modules' abilit y for accurate classification and to further enhance the performance of the neural classifier.

References

1. G. Auda. *Cooperative modular neural network classifiers*. PhD thesis, University of W aterloo, 1996.
2. G. Auda, M. Kamel, and H. Raafat. A new neural network structure with cooperative modules. In *World Congr ess on Computational Intelligenc ę* volume 3, pages 1301–1306, Florida, USA, June 1994.
3. G. Auda, M. Kamel, and H. Raafat. V oting schemes for cooperative neural network classifiers. In *IEEE International Conference on Neural Networks, ICNN'95*, volume 3, pages 1240–1243, Perth, Australia, November 1995.
4. R. Battiti and A. Colla. Democracy in neural nets: Voting schemes for classification. *Neural Networks*, 7(4):691–707, 1994.
5. W. Ziark o (ed.). *Rough sets, fuzzy sets and knowledge discovery*. Springer-Verlag, 1993.
6. G. Goetsch. Maximization of mutual information in a context sensitive neural network. Technical Report CMU-CS-90-168, Carnegie Mellon Univ ersit y, Sept. 1990.
7. R. Jacobs, M. Jordan, and A. Barto. Task decomposition through competition in a modular con nectionist architecture: The what and where vision tasks. *Neural Computation*, 3:79–87, 1991.
8. H. Normi. *Comparing voting systems*. D. Reidel Publishing Company, 1987.
9. Z. Pawlak. *Rough sets: Theoretical aspects of reasoning about data*. Kluwer Academic Publishers, 1991.
10. P. Straffin. *Topics on the theory of voting*. The UMAP Expository Monograph Series, Birkhauser, 1980.

Evolutionary Parsing for a Probabilistic Context Free Grammar

L. Araujo

Dpto. Sistemas Informáticos y Programación. Universidad Complutense de Madrid. Spain. `lurdes@sip.ucm.es`

Abstract. Classic parsing methods are based on complete search techniques to find the different interpretations of a sentence. However, the size of the search space increases exponentially with the length of the sentence or text to be parsed, so that exhaustive search methods can fail to reach a solution in a reasonable time. Nevertheless, large problems can be solved approximately by some kind of stochastic techniques, which do not guarantee the optimum value, but allow adjusting the probability of error by increasing the number of points explored. Genetic Algorithms are among such techniques. This paper describes a probabilistic natural language parser based on a genetic algorithm. The algorithm works with a population of possible parsings for a given sentence and grammar, which represent the chromosomes. The algorithm produces successive generations of individuals, computing their "fitness" at each step and selecting the best of them when the termination condition is reached. The paper deals with the main issues arising in the algorithm: chromosome representation and evaluation, selection and replacement strategies, and design of genetic operators for crossover and mutation. The model has been implemented, and the results obtained for a number of sentences are presented.

keywords: Evolutionary programming, Parsing, Probabilistic Grammar

1 Introduction

Classic parsing methods are based on complete search techniques to find the different interpretations of a sentence. However, experiments on human parsing suggest that people do not perform a complete search of the grammar while parsing. On the contrary, human parsing seems to be closer to a heuristic process with some random component. This suggest exploring alternative search methods in order to improve the efficiency. Another central point when parsing is the need of selecting the "most" correct parsing from the multitude of possible parsings consistent with the grammar. In such a situation, some kind of disambiguation is required. Statistical parsing helps to tackle the previous questions, that is, avoids an exhaustive search and provides a way of dealing with disambiguation.

Stochastic grammars [1], obtained by supplementing the elements of algebraic grammars with probabilities, represent an important part of the statistical methods in computational linguistics. They have allowed important advances in

W. Ziarko and Y. Yao (Eds.): RSCTC 2000, LNAI 2005, pp. 590–597, 2001.

areas such as disambiguation and error correction. Another stochastic methods are genetic algorithms (GAs). They have been already applied to different issues of natural language processing. Davis and Dunning [3] use them for query translation in a multi-lingual information retrieval system. GAs have also been applied to the inference of context-free grammars [2]. Wyard [6] devised a genetic algorithm for the language of correctly balanced nested parentheses, while Smith and Witten [5] proposed a genetic algorithm for the induction of non recursive s-expressions.

This paper presents a stochastic parser based on a genetic algorithm which works with a population of possible parsings. The algorithm produces successive generations of individuals, computing their "fitness" at each step and selecting the best of them when the termination condition arises. Apart from the characteristic efficiency of these stochastic methods, the nature of the generation of solutions in a genetic algorithm brings the advantages of statistical approaches.

The rest of the paper proceeds as follows: Section 2 describes the evolutionary parser, presenting the main elements of the genetic algorithm; section 3 presents and discusses the experimental results, and section 4 draws the main conclusions of this work.

2   Evolutionary Parsing

The syntactic structure of a sentence is a necessary previous step to determine its meaning. Such structures assign a syntactic category (verb, noun, etc) to each word in the sentence and specify how these categories are clustered to form higher level categories (np, vp, etc) until building the whole sentence. The grammar specifies the permitted structures in a language. Context free grammars (GFGs), whose rules present a single symbol on the left-hand-side, are a sufficiently powerful formalism to describe most of the structure in natural language, while at the same time is sufficiently restricted as to allow efficient parsing.

Parsing according to a grammar amounts to assigning one or more structures to a given sentence of the language the grammar defines. If there are sentences with more than one structure, as in natural language, the grammar is ambiguous. Parsing can be sought as a search process that looks for correct structures for the input sentence. Besides, if we can establish some kind of preference between the set of correct structures, the process can be regarded as an optimization one. This suggests considering evolutionary programming techniques, which are acknowledged to be practical search and optimization methods [4].

Probabilistic grammars [1] offer a way to establish preferences between parsings. In a probabilistic CFG a weight is assigned to each rule in the grammar. The probability of each parsing is the product of the probabilities of all the rules used in the parsing. Probabilistic grammars not only offer a way to deal with issues such as ambiguity or ungrammaticality [1], but can also lead to an improvement in performance. Genetic algorithms and probabilistic grammars complement each other, for at least two reasons:

a) Large populations in a GA lead to a higher diversity at the expense of slowing down the convergence process, while higher percentages in the applications of genetic operators hasten the process but increase the selective pressure. The use of probabilistic grammars help to accelerate the convergence process. Although the selective pressure is increased for individuals composed of grammar rules of high probability, this will lead to better individuals for most sentences (since they will correspond to the most probable rules). Thus, in general there will not be a premature convergence to a wrong individual.

b) The nature of the GAs, which favours the exploration of new areas of the search space, helps to reach a correct result, even if the sentence to parse requires applying rules of low probability.

According to the previous considerations, a probabilistic GA has been designed, in which the parsings that compose the population correspond to a probabilistic CFG. When the algorithm finishes with correct parsings, the one for which the product of the probabilities of its genes is the largest is chosen. This is the answer of the algorithm for the most probable parsing of the sentence.

2.1   Chromosome Representation

Our system chromosomes represent parsings for the input sentence, corresponding to a fixed context-free grammar. The input sentence is given as a sequence of words with their set of categories attached to them (if they belong to several categories every of them is added). Nevertheless, this information could be easily obtained from a lexicon in a preprocessing step. Let us consider a simple example. The sentence "the man sings a song" will be given as *the(Det) man(Noun) sings(Verb) a(Det) song(Noun)*.

A chromosome is represented as a data structure containing the following information:

- Fitness of the chromosome.
- A list of *genes*, which represents the parsing of different sets of words in the sentence.
- The number of genes in the chromosome.
- The depth of the parsing tree.

Each gene represents the parsing of a consecutive set of words in the sentence. If this parsing involves no terminal symbols, the parsing of the subsequent partitions of the set of words is given in later genes. Accordingly, the information contained in a gene is the following:

- The sequence of words in the sentence to be analyzed by the gene. It is represented by two data: the position in the sentence of the first word in the sequence, and the number of words of the sequence.
- The rule of the grammar used to parse the words in the gene.
- If the right hand side of the rule contains no terminal symbols, the gene also stores the list of references to the genes corresponding to the parsing of these symbols.

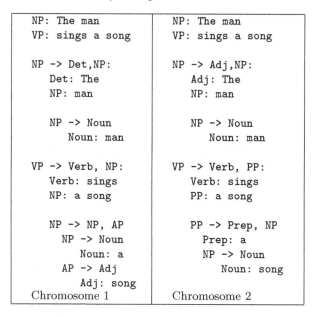

```
NP: The man                    NP: The man
VP: sings a song               VP: sings a song

NP -> Det,NP:                  NP -> Adj,NP:
     Det: The                       Adj: The
     NP: man                        NP: man

  NP -> Noun                      NP -> Noun
       Noun: man                       Noun: man

VP -> Verb, NP:               VP -> Verb, PP:
     Verb: sings                    Verb: sings
     NP: a song                     PP: a song

  NP -> NP, AP                    PP -> Prep, NP
     NP -> Noun                        Prep: a
          Noun: a                      NP -> Noun
     AP -> Adj                             Noun: song
          Adj: song
Chromosome 1                   Chromosome 2
```

Fig. 1. Possible chromosomes for the sentence *The man sings a song*. NP stands for nominal phrase, VP for verb phrase, *Det* for determiner, *Adj* for adjective, PP for prepositional phrase and AP for adjective phrase.

- The depth of the node corresponding to the gene in the parsing tree. It will be used in the evaluation function.

Figure 1 presents some possible chromosomes for the sentence of the example.

Initial Population The initial population consists of PS randomly generated (according to the probabilities of the different rules) individuals. The steps for the creation of chromosomes in the initial population are the following:

- The set of words in the sentence is randomly partitioned, making sure that there is at least one verb in the second part, which corresponds to the main VP.
- The set of words corresponding to the NP is parsed by randomly generating (consistently with the assigned probabilities) any of the possible NP rules. The same is done for generating the parsing of the VP with the VP rules. The process is improved by enforcing the application of those rules able to parse the right number of words of the gene.
- If the rules applied contain some non terminal symbol in its right hand side, the parsing process is applied to the set of words which are not yet assigned a category.
- The process continues until there are no terminal symbols left pending to be parsed.

2.2   Fitness: Chromosome Evaluation

Adaptation of individuals is revised after each new generation, testing the ability of every chromosome to parse the objective sentence. The evaluation of individuals is a crucial point in the evolutionary algorithms since the opportunities of an individual for survival depends on its fitness.

Fitness is computed as

$$\text{fitness} = \frac{\text{Number of coherent genes} - \sum_{i\in\text{incoherent genes}} \frac{\text{penalization}}{\text{depth}(i)}}{\text{Total number of genes}}$$

This formula is based on the relative number of *coherent* genes. A gene will be considered *coherent* if

a) it corresponds to a rule whose right hand side is only composed by terminal symbols, and they correspond to the categories of the words to be parsed by the rule.

b) it corresponds to a rule with non-terminal symbols in its right hand side and each of them is parsed by a coherent gene.

The formula takes into account the relative relevance of the genes: the higher in the parsing tree is the node corresponding to an incoherent gene, the worse is the parsing. Thus the fitness formula presents a penalization factor which decreases with the depth of the gene.

2.3   Genetic Operators

Chromosomes in the population of subsequent generations which did not appear in the previous one are created by means of two genetic operators: *crossover* and *mutation*. The crossover operator combines two parsings to generate a new one; mutation creates a new parsing by replacing a randomly selected gene in a previous chromosome. The rates of crossovers and mutations performed at each step are input parameters. The efficiency of parsing is very sensitive to them. At each generation a number of chromosomes equal to the number of offsprings is selected to be replaced. The selection is performed with respect to the relative fitness of the individuals: a chromosome with a worse than average fitness has higher chances to be selected for replacement. On the contrary, chromosomes adapted over the average have higher probability to be selected for reproduction.

Reproduction Crossover operator generates a new chromosome that is added to the population in the new generation. The part of one parent after a point randomly selected is exchanged with the corresponding part of the other parent to produce two offsprings, under the constraint that the genes exchanged correspond to the same type of parsing symbol (NP, VP, etc) in order to avoid wrong references of previous genes in the chromosome. Of course those exchanged which produce parsings inconsistent with the number of words in the sentence must be avoided. Therefore, the crossover operation performs the following steps:

- Select two parent chromosomes, C_1 and C_2.
- Randomly select a word from the input sentence.
- Identify the inner most gene to which the selected word corresponds in each parent chromosome.
- If the genes correspond to different sets of words, the next gene in the inner most order is selected. This process continues until the sequences of words whose parsings are to be exchanged are the same, or until the main NP or VP are reached.
- If the two selected genes parse the same sequence of words the exchange is performed.
- If the process to select genes lead to the main NP or VP, and the sequence of words do not match yet, the exchange can not be performed. In this case a new procedure is followed: in each parent one of the two halves is maintained while the other one is randomly generated to produce a parsing consistent with the number of words of the sentence. This produces four offsprings, out of which the best is selected.
- Finally, the offspring chromosome is added to the population.

Mutation Selection for mutation is done in inverse proportion to the fitness of a chromosome. Mutation operation changes the parsing of some randomly chosen sequence of words. The mutation operation performs the following steps:

- A gene is randomly chosen from the chromosome.
- The parsing of the selected gene, as well as every gene corresponding to its decomposition, are erased.
- A new parsing is generated for the selected gene.

3   Experimental Results

The algorithm has been implemented using C language and run on a Pentium II processor. In order to evaluate its performance we have considered the parsing of the sentences appearing in Table 1. The average length of the sentences is around 10 words. However, they present different complexities for the parsing, mainly the length and the number of subordinate phrases.

1	Jack(noun) regretted(verb) that(wh) he(pro) ate(verb) the(det) whole(adj) thing(noun)
2	The(det) man(noun) who(wh) gave(verb) Bill(noun) the(det) money(noun) drives(verb) a(det) big(adj) car(noun)
3	The(det) man(noun) who(wh) lives(verb) in(prep) the(det) red(adj) house(noun) saw(verb) the(det) thieves(noun) in(prep) the(det) bank(noun)

Table 1. Sentences used in the parsing experiments.

The results reported in this section have been obtained as the average of five runnings with different seeds. Results show that in most cases the correct parsing is reached in a small number of steps, less than 10 for populations of above 300 individuals.

Results obtained with a deterministic context free grammar are compared to the ones obtained using a probabilistic grammar. Figure 2 shows the results obtained for the sentences when using rates of crossover of 50% and one of mutation of 20%. Results clearly improve in all the cases by using the probabilistic grammar. The first observation is that while the deterministic CFG produces irregular convergence processes, the probabilistic one leads to highly regular processes as the population size grows. This indicates a higher robustness of the genetic algorithm. The difference between the results from the two kinds of grammar increases with the complexity of the sentence. Thus while the deterministic CFG leads to quick convergence for the sentence 1, the process is quite irregular for sentences 2 and 3. Another observation is that a threshold population size is required to achieve convergence.

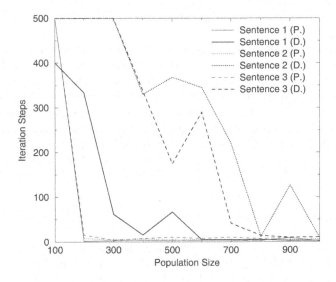

Fig. 2. Number of iteration required to reach the correct parsing with a probabilistic grammar (P) and a deterministic grammar (D).

The most relevant GA parameters have been studied (data not shown). It is clear that the population diversity and the selection pressure are related to the *population size*. If the population size is too small the genetic algorithm will converge too quickly to a bad result (all individuals correspond to similar incorrect parsings), but if it is too large the GA will take too long to converge. Results show that the behavior is quite different for each sentence: the higher the "sentence complexity", the larger the population size required to reach the correct parsing in a reasonable number of steps. The sentence complexity depends on its length and on the number of subordinate phrases it contains. Besides, as the population size increases, higher rates of crossover and mutation are required to increase the efficiency of the algorithm.

4   Conclusions

This paper presents a genetic algorithm that adapts a population of possible parsings for a given input sentence and a given grammar. Genetic algorithms allow a statistical treatment of the parsing process, providing at the same time the typical efficiency of stochastic methods.

Results from a number of tests indicate that the GA is a robust approach for parsing positive examples of natural language. A number of issues of the GA have been tackled, such as the design of the genetic operators and a study of the GA parameters. The tests indicate that the GA parameters need to be suitable for the input sentence complexity. The more complex the sentence (length and number of subordinate phrases), the larger the population size required to quickly reach a correct parsing.

Probabilistic grammars and genetic algorithms have been shown to complement each other. The use of a probabilist context free grammar instead of a deterministic one for the generation of the population of parsings in the algorithm has been investigated. Results obtained for these experiments show a clear improvement in the performance. For short sentences, though, greedy parser algorithm can be at least as fast. Nevertheless, the method proposed herein also allows dealing with problems such as ambiguity or ungrammaticality, and are expected to be advantageous for parsing long texts. Work along this line is currently in process.

References

1. E. Charniak. *Statistical Language Learning*. MIT press, 1993.
2. Paul Cohen and Ed Feigenbaum. Grammatical inference. In *HandBook of Artificial Intelligence*, volume 3, pages 494–511. Pitman Books Limited, 1984.
3. T. Dunning M. Davis. Query translation using evolutionary programming for multilingual information retrieval II. In *Proc. of the Fifth Annual Conf. on Evolutionary Programming*. Evolutionary Programming Society, 1996.
4. Z. Michalewicz. *Genetic algorithms + Data Structures = Evolution Programs*. Springer-Verlag, 2nd edition, 1994.
5. I.H. Witten T.C. Smith. A genetic algorithm for the induction of natural language grammars. In *Proc. IJCAI-95 Workshop on New Approaches to Learning Natural Language*, pages 17–24, Montreal, Canada, 1995.
6. P. Wyard. Context free grammar induction using genetic algorithms. In *Proc. of the 4th Int. Conf. on Genetic Algorithms*, pages 514–518, 1991.

The Application of Genetic Programming in Milk Yield Prediction for Dairy Cows

Chaochang Chiu[1], Jih-Tay Hsu[2], and Chih-Yung Lin[1]

[1] Department of Information Management, Yuan Ze University, Taiwan
imchiu@im.yzu.edu.tw

[2] Department of Animal Science, National Taiwan University, Taiwan
jthsu@ccms.ntu.edu.tw

Abstract. Milk yield forecasting can help dairy farmers to deal with the continuously changing condition all year round and to reduce the unnecessary overheads. Several variables (somatic cell count, pariety, day in milk, milk protein content, milk fat content, season) related to milk yield are collected as the parameters of the forecasting model. The use of an improved Genetic Programming (GP) technique with dynamic learning operators is proposed and achieved with acceptable prediction results.

Keywords: Genetic programming, dynamic mutation, milk yield prediction

1 Introduction

In Taiwan, milk consumption is getting more popular in the last few decades by the preaching from the government and the awareness of consumer right movement for better nutrition. Due to neglecting the importance of predicting the milk yield, dairy farmers usually could not deal well with the continuously changing condition and have high uncontrollable overheads of the production system. Many research works have tried to solve this problem by using traditional approaches such as regression and time series analysis. But they are restricted by the missing or incomplete data and may not generate sufficiently accurate results in the effort of milk yield prediction. According to previous studies (Dun, 1980; Wu, 1989; Tseng, 1992; Hu, 1994; Mo, 1996), it can be found that the nature of the milk yield-forecasting matter is complex, nonlinear, and continuous. Therefore the development of mathematical models by statistic methods for this effort may be difficult or complicated and lack of learning and adaptation capabilities in recognizing the behavior of data set.

Genetic Algorithms (GA) are heuristic-based search optimization techniques rooted on the principles of natural evolution (Holland, 1975). As an extension of GA paradigm, Genetic Programming (GP) is able to automatically construct computer programs by means of the Darwinian theory of natural selection (Koza, 1982). GP does not use an encoding of the problem into a finite alphabet string and does not require an assumption of any functional relationship between independent and dependent variables. This technique has been widely applied to classification,

W. Ziarko and Y. Yao (Eds.): RSCTC 2000, LNAI 2005, pp. 598–602, 2001.

forecasting, and model construction problems (Dworman et al., 1996; Lee et al., 1997; Jan, 1998). This research adopts an improved dynamic learning mechanism in GP that is able to construct a forecasting formula more efficiently (Chiu, 1999).

2 Basic Principles of GP

GP involves creating a mathematical or logical expression, in symbolic form, that provides a good, best, or perfect fitness between a given finite sampling of values of the independent variables and the associated values of the dependent variables. In other words, GP involves creating a model fitting a given sample of data. When the variables are real-valued, GP finds both the functional form and the numeric coefficients for the model. GP differs from conventional linear, quadratic, or polynomial regression, which merely finds the numeric coefficients for a function. GP expresses the model in a tree-type structure that allows expressing mathematical, logical, or functional structures within one expression tree. The tree structure follows the so-called "Polish notation" which puts the operator (the functional form) first and then the operands (the numeric coefficients). Thus, for example, the simple mathematical function $y = 5*x+3$ can be represented as $y = +(*(5,x), 3)$ or in tree structure shown in Figure 1. In applying GP to a problem, the user must define all the possible operator set and operand set used as nodes in a tree. The set of operators is to be used to generate the mathematical expression that attempts to fit the given finite sample of data. The set of operands (along with the set of operators) is the ingredient from which GP attempts to construct a model to solve, or approximately solve, the problem.

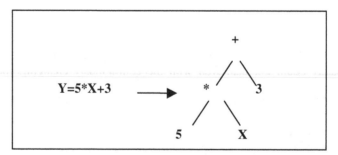

Fig. 1. A subroutine and an equivalent tree

Because of the tree-based representation, the three main genetic operators, selection, crossover and mutation, work differently from the way GA are constructed. In the selection operation, couples of parent trees are selected for reproduction on the basis of their fitness. The most usual selection mechanisms are Fitness Proportionate Reproduction, and Tournament Selection. In the crossover operation, two parent trees are sexually combined to form two new offspring. The parents are picked in random and parts of their trees are exchanged. Since different mutation operator may exhibit different impact on the evolution performance, this paper proposes the mixture approach that synergies the evolution advantages from multiple mutation operators (Chiu, 1999).

3   The Proposed Methodology

Basically the first step of mutation operation is to determine the mutation node. Depending upon the problem characteristics, the operation may apply to the following three types of mutation subjects single node, multiple nodes, or a sub-tree in programs under certain constraints of evolutionary consideration. Assume O1, O2, O3, are defined as the three mutation operators that apply to the three type of mutation subjects respectively. The basic principle of this approach is that, for the sake of encouraging the operator that is resulting better offspring fitness values during evolution, the mutation rate assigned to its corresponding operator is thereby strategically being increased. On the other hand, if the operator doesn't produce better results, its mutation rate is therefore decreased. Each mutation rate designated to its corresponding operator is dynamically re-allotted according to individual performance. Let r be the pre-defined mutation rate, and thus each operator is initially assigned with mutation rate r/3. A simpler form of the same crossover point in both program parents is adopted here. Right after mutation operation, each mutation operator is being evacuated and re-ranked according to individual performance. Among these three operators whose mutation rate of the best is increased by w% of r/3, the secondary is remained same of its present rate, while the worst one is decreased by w% of r/3. Though the adjustment processes, the mutation ratios are updated dynamically. The detailed algorithm can be found in (Chiu, 1999):

4   The Experiments and Results

Milk production data along with physiological data of cows were obtained from the Experimental Farm in College of Agriculture, National Taiwan University. According to previous studies, those factors including somatic cell count, pariety, day in milk, milk protein content, milk fat content, and month of the year are related to the milk yield (Mo, 1996; Wu, 1989). Thus data that consisting of the above factors had been collected from 30 from 30 Holstein cows that are of same species every month. Training data comprises 643 data records that are derived between August 1994 and July 1996; and testing data comprises 532 data records that are from September 1996 to September 1998. There were data missed in August, November 1996, April, May, June, July, August, September, November 1997. We pre-process the data of the month variable and classify 12 months into groups by clustering December, January, February, March as 1, April, May, June, October, November as 2, and July, August, September as 3 according to the climate characteristics of each month. Since GP randomly searches the model, the operator set, the operand set, and the parameters for the training process have to be determined in advance. The operator set includes +, -, *, ^, and / (prevent the division from zero denominators). The parameters for the training process are defined as follows: Population Size: *150;* Crossover Rate: *0.6;* Mutation Rate (r): *0.2;* Simplification of Solutions: *True;* Termination Criterion: *5000 generations.* The computer system used for this process was a single processor Intel Pentium III 450 MHz system with 128MB RAM, running on Microsoft 98.

In the training process, every created solution has a fitness value to evaluate the performance. In this study, the mean error for the best individual in every generation is used for model evaluation. The learning curve of the model is flat approximately between the 200th generation and the 750th generation; seeming that the model has already converged. But the model creates better solutions after the 750th generation. Although GP could not generate models with zero error, increasing the generation size may be a way to gain better model. The mean error of the last best generation is 4.278. Though the predicted results do not exactly coincide with the actual data, they conform to the entire trends of expected data. Also the comparative predicting performance between GP and regression method is depicted in Table 1. We will comment these results in Section 5.

Table 1. The comparative predicting performance between GP and Regression

Comparative Performance	Correlation Coefficient	Absolute Mean Error
GP	0.695	4.676
Regression	0.693	5.512

The derived formula

The model produced by the GP process for the milk yield is as follows:
$(46.9757563391006+((-0.00260077965386069*((X1^1)*(X3^1)))+$
$((-6.29595132173721*(X4^1))+((-0.0183952688341093*(X6^5))+$
$((-3.28103058073334E-5*(X3^2))+((-0.115769461379268*(X6^2))+$
$((-0.00731554801232839*(X3^1))+((-0.00618269438233685*(X4^4))+$
$((-0.0248432320556369*(X4^3))+((0.00025459980864606*(X1^1))+$
$((-9.12858920103003E-5*(X2^5))+((-0.00261453182147923*(X5^3))+$
$((0.0190437641263939*(X6^4))+((5.62962747051406E-7*((X3^2)*(X4^3)))+$
$(0.0481493551104286*(X5^1))))))))))))))))))$

where: X1: somatic cell count; X2: parietyl; X3: day in milk; X4: milk protein content
X5: milk fat content; X6: month of the year

5   Dsicussion and Conclusions

In this paper, we proposed a genetic programming-based approach to model the milk yield prediction problem. By introducing a dynamic mutation operator method in GP's learning model, acceptable results have been achieved with the best mean absolute errors of 4.278 for training process and of 4.676 for testing process. According to Table 1, both of the predicted errors are smaller than 5.512 obtained from the regression model. Furthermore, as to the correlation coefficient, GP exhibits better performance than regression model (0.695 vs. 0.693). It can be seen that the predicted milk yield derived from GP model approximately conforms to the actual data record. As to certain coefficient of E-20 accuracy from the complicated formula derived, it can be improved in the future research by reducing with less accuracy in order to produce simpler while better formulae. This is because so much time on very small mutations of coefficients can be saved and also is able to direct more of the search towards the structure of the function. It is undoubtedly that there might exist influencing factors other than those have been collected in this research (Holmes et al., 1987; Schmidt et al., 1988; Skidmore et al., 1996; Olori, 1997; Olsson et al. 1998). Efforts in accumulating more data as well as collecting data from other sources to improve prediction accuracy are undergoing in our dairy farms.

References

Chiu, C. "Dynamic Learning in Genetic Programming," the 1999 International Conference on Artificial Intelligence (IC-AI'99), June 28 - July 1, 1999, Monte Carlo Resort, Las Vegas, Nevada, USA, pp. 416-422.

Dun, S. L., Study of Lactation Curve of Holstein Lactating Cows in Taiwan, Master Thesis, National Chung Hsing University, 1980.

Dworman, G, Kimbrough, S O, and Laing, J K, "On Automated Discovery of Models using Genetic Programming: Bargaining in A Three-Agent Coalitions Game," Journal of Management Information Systems, Vol. 12, No. 3, pp.97-125, 1996.

Holland, John H, Adaptation in Natural and Artificial Systems, University of Michigan, 1975.

Holmes, C. W., Wilson, G. F., Mackenzie, D. D. S., Flux, D. S., Brookes, I. M., and Davey, A. W. F., Milk Production from Pasture, Butterworths Agriculture Books, Wellington, New Zealand, 1987.

Hu, G. C., Analysis and Investigation of Genetic Markers of Holstein Breeding Bulls, Master Thesis, National Chung Hsing University, 1994

Jan, J. F, "Genetic Programming for Classification of Remote Sensing Data," Taiwan Journal of Forest Science, 13(2), pp.109-118, 1998.

Koza, J. R., Genetic Programming: On the Programming of Computers by Means of Natural Selection, The MIT Press, 1982.

Lee, D G, Lee, B W, and Chang, S H, "Genetic Programming Model for Long-term Forecasting of Electric Power Demand," Electric Power Systems Research, Vol. 40, pp. 17-22, 1997.

Mo, J. R., The Relationship of Electrical Conductivity, Somatic Cell Count, Milk Production, and Udder Status in Holstein Cows, Master Thesis, National Taiwan University, 1996

Olori, V. E., Brotherstone, S., Hill, W. G., and McGuirk, B. J., Effect of gestation stage on milk yield and composition in Holstein Friesian Dairy Cattle, Livestock Production Science, 52:167-176, 1997.

Olsson, G., Emanuelson, M., and Wiktorsson, H., Effects of Different Nutritional Levels Prepartum on the Subsequent Performance of Dairy Cows, Livestock Production Science, 53:279-290, 1998.

Schmidt, G. H., Van Vleck, L. D., and Hutjens, M. F., Principles of Dairy Science, Prentice Hall, Englewood Cliffs, New Jersey, 1988.

Skidmore, A. L., Brand, A., and Sniffen, C. J., Monitoring Milk Production: Decision Making and Follow-up, In: Brand, A., Noordhuizen, J. P. T. M., and Schukken, Y. H. (Ed.) Herd Health and Production Management in Dairy Practice. Wageningen Pers, Wageningen, The Netherlands, 1996.

Tseng, C. U., Reproduction Characteristics and Milk Yield of Holstein Dairy Cows in Taiwan, Master Thesis, National Chung Hsing University, 1992.

Wu, L. S., Effect of Age, Nutrition and Environmental Temperature on the Adrenocortical Function of Holstein Cows, Ph.D. Thesis, National Taiwan University, 1989

Localization of Sound Sources by Means of Recurrent Neural Networks

Rafal Krolikowski, Andrzej Czyzewski, Bozena Kostek

Technical University of Gdansk, Sound Engineering Department, Narutowicza 11/12,
80-952 Gdansk, Poland
kid@sound.eti.pg.gda.pl

Abstract. The issue of localization of sound sources for videoconferencing is discussed in the paper. A new algorithm for estimating speaker locations, based on recurrent neural networks (RNN), is introduced and described. The scheme of experiments carried out in an acoustically adopted chamber, exploiting the engineered method is detailed.

1   Introduction

Localization of sound sources is a key issue in contemporary tele- and videoconferencing systems. Such a localization considerably influences the efficiency of the source acquisition, since it reduces influences of other sources on the chosen one, improves the signal-to-noise ratio and in result - efficiency of noise reduction- & dereverberation algorithms. Furthermore, due to sound source tracking it is feasible to change automatically video camera direction during a conference, which improves the technical support and organization of the conference.

Artificial neural networks are commonly applied in many areas of engineering and audio signal processing [6] [7], since they are capable to process uncertain information. They have already been applied to purposes of sound localization [8], however these attempts were based on feed-forward structures [28]. The feed-forward networks do not offer such feasibility as recurrent ones do, especially in the field of time series modeling [9] or mapping of a complex process dynamics [4] [10] [20] [21]. As is mentioned in par. 2, the localization theories explain human perception of directivity basing on temporal relationships. Therefore the question stands whether recurrent neural networks (RNN) can serve such a task as localization of sound sources.

In this work the focus is put on a general RNN proposed by Elman [10], despite that there is a number of other recurrent architectures. An important class among RNNs constitute so called NARX networks (Nonlinear Auto Regressive with eXogenous inputs) [17] [18]. They are reported to be robust, more straightforward to converge during the training than general RNNs [14], and to be equivalent to a certain extent to general RNNs [22] [23]. However due to their structure definition (only a single feedback loop between the output and input with a number delay taps in the loop) they seem to be unsuitable for the purposes of the sound localization.

W. Ziarko and Y. Yao (Eds.): RSCTC 2000, LNAI 2005, pp. 603–610, 2001.

Another issue concerns a choice of training algorithm. Although a number of various methods for training of RNN have been proposed so far [12] [20], including see ond-order methods [11], Conjugate Gradient Learning [4] and even genetic algorithms [20], the authors have focused on the standard approach introduced by Wiliams and Zipser [25].

The reason for such a choice of the structure and the training method is related to the planned systematical investigation of the various methods for sound source localization and their applications to real videoconferencing. From this point of view, this paper describes preliminary experiments and reflects some initial states of the long-term research.

2   Sound Localization - Problem Statement

Despite the great development of science in the field of human perception, issues related to sound localization are not thoroughly known and phenomena underlying thereof are still the subject of intense research [2] [13]. Actually, perception of sound directivity by the human binaural system is based on the following two principal entities [13]:

Interaural Level Difference (ILD): difference of intensities of waveforms in the left and in right ears.

Interaural Time Difference (ITD): difference of arrival times of relevant waveforms in the both ears, which is equivalent to a phase difference of the waveforms.

In the filed of digital signal processing, sound source localization can be performed by means of a microphone array which can be either linear or non-linear [16] [19]. Under the ideal conditions, the signal $x_i(t)$ received from the i-th microphone in the linear array and in the t-th moment of time can be described as follows:

$$x_i(t) = \alpha_i \cdot s(t - (i-1) \cdot \tau), \tag{1}$$

where: α_i - attenuation coefficient for the i-th microphone,

$s(t)$ - source signal,

τ - time delay between adjoining microphones,

and the estimation of the source location is a deterministic problem.

However, under real conditions there occurs various distortions and interfering signals such as: background noise and reverberated signals. Then, the signals received by a linear microphone array are expressed by the relationships:

$$\begin{cases} x_1(t) = \alpha_1 \cdot h_1(t) * s(t) + n_1(t) \\ x_2(t) = \alpha_2 \cdot h_2(t) * s(t-\tau) + n_2(t) \\ \quad \vdots \\ x_i(t) = \alpha_i \cdot h_i(t) * s(t-(i-1)\cdot\tau) + n_i(t) \\ \quad \vdots \end{cases}, \tag{2}$$

where: $h_i(t)$ - impulse response of the reverberant channel associated with the i-th
          microphone,
       $n_i(t)$ - ambient noise received by the i-th microphone.

These conditions make the problem of sound source localization more complex, and therefore a number of various methods have been proposed. Most of them are based on estimation of the time delay including cross-correlation techniques [3], adaptive filtration [5] or computation of relevant eigenvalue vectors and matrices [1]. In turn, in the case of tracking or localization of a number of sources, the Maximum Likelihood-based methods are exploited [27]. More details can be found in the abundant literature on the localization of acoustic sources for multimedia applications [15] [16] [19] [24] [26].

3   Neural System for Sound Localization

The proposed neural system for sound localization consists of L equally spaced microphones (forming a linear array) connected to a recurrent neural network which architecture is shown in Fig. 1. The received and discrete values of waveforms coming from all L microphones are normalized and fed into L input units of the network. In a given moment of time t, the received signals can be described by the L-dimensional measurement vector $u(t) = \begin{bmatrix} u_1(t) & \cdots & u_l(t) & \cdots & u_L(t) \end{bmatrix}^T$. The input data for the neural network in the moment t compose of a train of N such L-tuple vectors, which can be treated as the N-th dimensional measurement matrix $U(t) = \begin{bmatrix} u(t) & \cdots & u(t-n) & \cdots & u(t-N+1) \end{bmatrix}$.

3.1   Structure of the Recurrent Neural Network

The structure of the exploited neural network is presented in Fig. 1. The network is an extended version of the general recurrent neural network proposed by Elman [10], and is composed of:

– the *Input Layer*, consisting of $N \times L + 1$ units including a bias,
– the *Hidden Layer*, consisting of M neurons,

– the *Context Layer*, consisting of M units which outputs are delayed by a single cycle with regard to those in the hidden layer,
– the *Output Layer*, consisting of K neurons

The input vector (matrix) $x(t)$ consists of the measurement matrix $U(t)$ augmented by the fixed value -1, which represents signals received by the L microphones over the last N time units.

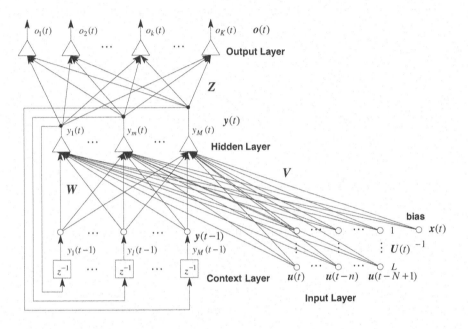

Fig. 1. Architecture of the extended globally recurrent neural network

3.2   Training of the Recurrent Neural Network

In general case, the training of such a network as presented in Fig. 1 is based on the relationships introduced by Wiliams and Zipser [25]. In order to simplify denotations, the *Input Layer* formed in the rectangular array (as in Fig. 1) can be "unfold", and hence can be referred to as a linear array of $N_1 = N \times L + 1$. In result, the input vector can be described as follows:

$$x(t) = \begin{bmatrix} u_1(t) & \cdots & u_L(t) & \cdots & u_1(t-N+1) & \cdots & u_L(t-N+1) & -1 \end{bmatrix}^T, \quad (3)$$

Assuming that the error measure in the t-th moment of discrete time is the mean-square error as below:

$$E(t) = \frac{1}{2} \cdot \sum_{k=1}^{K} [d_k(t) - o_k(t)]^2 , \tag{4}$$

where: $d_k(t)$ - the desired response of the k-th output neuron,

$o_k(t)$ - the current output of this neural unit,

it can be shown that the update expressions for particular weights and at given learning rate η are computed in the following way:

$$\Delta Z_{km}(t) = \eta \cdot [d_k(t) - o_k(t)] \cdot f_k'(netZ_k(t)) \cdot y_m(t) , \tag{5}$$

$$\Delta W_{ml}(t) = \eta \cdot \sum_{k=1}^{K} [d_k(t) - o_k(t)] \cdot f_k'(netZ_k(t)) \cdot \sum_{j=1}^{M} Z_{kj}(t) \cdot R_{ml}^{j}(t) , \tag{6}$$

$$\Delta V_{mn}(t) = \eta \cdot \sum_{k=1}^{K} [d_k(t) - o_k(t)] \cdot f_k'(netZ_k(t)) \cdot \sum_{j=1}^{M} Z_{kj}(t) \cdot S_{mn}^{j}(t) , \tag{7}$$

where: f_k' - the derivative of the activation function for the k-th neuron,

$y_m(t)$ - the output of the m-th hidden neural unit,

$k = 1,...,K$, $m = 1,...,M$, $l = 1,...,M$, $n = 1,...,N_1 = N \times L + 1$,

and the auxiliary terms $R_{mn}^{j}(t)$ and $S_{ml}^{j}(t)$ are defined as follows:

$$R_{ml}^{j}(t) = f_j'(netVW_j(t)) \cdot \left[\sum_{i=1}^{M} W_{ji}(t) \cdot R_{ml}^{i}(t-1) + \delta_{jm} \cdot y_l(t-1) \right], \tag{8}$$

$$S_{mn}^{j}(t) = f_j'(netVW_j(t)) \cdot \left[\sum_{i=1}^{M} W_{ji}(t) \cdot S_{mn}^{i}(t-1) + \delta_{jm} \cdot x_n(t) \right], \tag{9}$$

where δ_{jm} denotes the Kronecker's delta, and the weighted sums: $netZ_k(t)$ and $netVW_m(t)$ are calculated as below:

$$netVW_m(t) = \sum_{j=1}^{M} W_{mj}(t) \cdot y_j(t-1) + \sum_{j=1}^{N_1} V_{mj}(t) \cdot x_j(t) ; \quad netZ_k(t) = \sum_{j=1}^{M} Z_{kj}(t) \cdot y_j(t) \tag{5}$$

4   Experiments

The objectives of the planned experiments concern design and implementation of various methods for sound source localization, their mutual comparison and verifica tion under real conditions. Therefore the long-term experiments are divided into a number of surveys, including a number of recordings made in:

- acoustically chambers adopted (anechoic chamber-like conditions),
- different real conference rooms,
- including many speakers,
- with various distributions of participants

The objectives of the preliminary experiments concern design, implementation and tuning of relevant methods and the comparison of their efficiency with these of the standard ones. Therefore some simplifications of real conditions have been introduced.

4.1   Organization and Conditions of Experiments

The preliminary experiments have been carried out in a chamber acoustically adopted, i.e. all chairs and necessary stuff had been removed and curtains had been fixed at the doors, windows and other reflecting planes. The microphone array consisted of 4 electret microphones spaced by 5 cm, and was fixed 1.58 m from the floor and 1.25 m from the ceiling. Four mono tracks were recorded simultaneously using signals received from these microphones.

The recording parameters were as follows: 16bit/sample and the sampling frequency was equal to 8kHz and 48 kHz. There was one male speaker, distanced 1.5 m from the array. The speaker read a logatom list from the angles differing by $15°$. In result, 24 four-track recordings have been made, and every recording lasted approx. 50 s.

4.2   Surveys and Results of Experiments

For the preliminary experiments, the logatom lists were reduced and their number was limited to 7, which represented the sound directivity from $-45°$ to $+45°$ every $15°$ (7 classes). The structure of the RNN was as follows. The number of time units N ranged from 3 to 5, which for $L = 4$ microphones yielded $N_1 \in (13,17,21)$ input data. The input vectors formed sequences of various lengths (2, 3, 4 and 5). In turn, the number of hidden neurons was arbitrary set to 10 or 15, whereas the output layer consisted of 7 neurons. The activation function was set as the unipolar continuous function.

The training and testing vectors and sequences were different and selected randomly. The number of the vectors per class was equal to 100, which yielded totally 700 vectors for a training- and testing phase. In order to obtain statistically valid results, computations were repeated 10 times per a given survey.

The accuracy of the right direction detection ranged from 73 % up to 82 %. An interesting occurrence was observed: for too long input vectors the efficiency decreased. In the latter case it can be interpreted that the network began rather to approximate the waveforms themselves than their mutual relationships, which are essential for sound localization.

5   Conclusions

In this paper, the new algorithm, based on recurrent neural networks, for the estimation of sound source location has been proposed and described. Moreover, the scheme of experiments and some results of them have been included. The results of automatic discrimination of sound source direction obtained with the use of the implemented algorithm are promising. More experiments are planned which task is to compare obtained results with scores possible to get in the same acoustic conditions with some hitherto existing algorithms for the automatic sound source tracking.

References

1. Berdugo, B., Doron, M.A., Rosenhouse, J., Azhari, H.: On Direction Finding of an Emitting Source from Time Delays. J. of Acoustical Society of America 106 (1999) 3355–3363
2. Bodden, M.: Modeling Human Sound-Source Localization and the Cocktail-Party-Effect. Acta Acustica 1 (1993) 43–55
3. Brandstein, M.S.: A Pitch-Based Approach to Time-Delay Estimation of Reverberant Speech. Proc. of IEEE Workshop on Applications of Signal Processing to Audio and Acoustics, Mohonk, New Paltz, NY, USA (1997)
4. Chang, W.-F., Mak, M.W.: a Conjugate Gradient Learning Algorithm for Recurrent Neural Networks. Neurocomputing 24 (1999) 173–189
5. Chern, S.-J., Lin, S.-H.: An Adaptive Time Delay Estimation with Direct Computation Formula. J. of Acoustical Society of America 96 (1994) 811–820
6. Czyzewski, A., Krolikowski, R.: Application of Fuzzy Logic and Rough Sets to Audio Signal Enhancement. In: Pal, S.K., Skowron, A. (eds.): Rough Fuzzy Hybridization. A New Trend in Decision-Making. Springer-Verlag, Berlin Heidelberg New York (1999) 397–409
7. Czyzewski, A., Krolikowski, R.: Neuro-Rough Control of Masking Thresholds for Audio Signal Enhancement. Neurocomputing (to be appeared)
8. Datum, M.S., Palmieri, F., Moiseff, A.: An Artificial Neural Network for Sound Localization Using Binaural Cues. J. of Acoustical Society of America 100 (1996) 372–3383
9. Day, S.P., Davenport, M.R.: Continuous-Time Temporal Back-Propagation with Adaptable Time Delays. IEEE Trans. on Neural Networks (1993)
10. Elman, J.L.: Finding Structure in Time. Cognitive Science 14 (1990) 179–211
11. Goudreau, M.W., Giles, C.L., Chakradhar, S.T., Chen, D.: First-Order vs. Second-Order Single Layer Recurrent Neural Networks. IEEE Trans. on Neural Networks 5 (1994) 511–518
12. Goudreau, M.W., Giles, C.L.: Using Recurrent Neural Networks to Learn Structure of Interconnection Networks. Neural Networks 8 (1995) 793–820
13. Hartmann, W.M.: How We Localize Sound. Physics Today 11 (1999) 24–29

14. Horne, B.G., Giles, C.L.: An Experimental Comparison of Recurrent Neural Networks. In: Tesauro, G., Touretzky, D., Leen, T. (eds.): Neural Information Processing Systems, Vol. 7. MIT Press (1995) 697–705
15. Jacovitti, G., Scarano, G.: Discrete Time Techniques for Time Delay Estimation. IEEE Trans. on Signal Processing 41 (1993) 525–533
16. Khalil, F., Lullien, J.P., Gilloire, A.: Microphone Array for Sound Pickup in Teleconference Systems. J. of Audio Engineering Society 42 (1994) 691–700
17. Lin, T., Giles, C.L., Horne, B.G., Kung S.Y.: A Delay Damage Model Selection Algorithm for NARX Neural Networks. IEEE Trans. on Signal Processing, "Special Issue on Neural Networks", 11 (1997) 2719–2730
18. Lin, T., Horne, B.G., Tino P., Giles, C.L.: Learning Long-Term Dependencies in NARX Recurrent Neural Networks. IEEE Trans. on Neural Networks 7 (1996) 1329–1351
19. Mahieux, Y., le Tourneur, G., Saliou, A.: A Microphone Array for Multimedia Workstations. J. of Audio Engineering Society 44 (1996) 365–372
20. Mak, M.W., Ku, K.W., Lu, Y.L.: On the Improvement of the Real Time Recurrent Learning Algorithm for Recurrent Neural Networks. Neurocomputing 24 (1999) 13–36
21. Omlin, C.W., Giles, C.L.: Rule Revision with Recurrent Neural Networks. IEEE Trans. on Knowledge and Data Engineering 1 (1996) 183–196
22. Siegelmann, H.T., Horne, B.G., Giles, C.L.: Computational Capabilities of Recurrent NARX Neural Networks. IEEE Trans. on Systems, Man and Cybernetics - Part B: Cybernetics 2 (1997) 208–228
23. Sum, J.P.F., Kan, W.-K., Young, G.H.: A Note on the Equivalence of NARX and RNN. Neural Computing & Applications 8 (1999) 33–39
24. Wang, H., Chu, P.: Voice Source Localization for Automatic Camera Pointing System in Videoconferencing. Proc. of IEEE Workshop on Applications of Signal Processing to Audio and Acoustics, Mohonk, New Paltz, NY, USA (1997)
25. Wiliams, R.J., Zipser, D.: A Learning Algorithm for Continually Running Fully Recurrent Neural Netwroks. Neural Computation 1 (1989) 270–280
26. Zhang, M., Er, M.H.: An Alternative Algorithm for Estimating and Tracking Talker Location by Microphone Arrays. J. of Audio Engineering Society 44 (1996) 729–736
27. Ziskind, I., Wax, M.: Maximum Likelihood Localization of Multiple Sources by Alternating Projection. IEEE Trans. on Acoustics, Speech and Signal Processing 36 (1988) 1553–1560
28. Zurada, J.M.: Introduction to Artificial Neural Netwroks. West Publishing Company, St. Paul New York Los Angeles San Francisco (1992)

Acknowledgements

The research is sponsored by the Committee for Scientific Research, Warsaw, Poland. Grant No. 8 T11D 00218.

Towards Rough Neural Computing Based on Rough Membership Functions: Theory and Application

J.F. Peters[1], A. Skowron[2], L.Han[1], S.Ramanna[1]

[1] Electrical and Computer Engineering, University of Manitoba, Winnipeg, MB R3T 2N2
Canada
{jfpeters,liting,ramanna}@ee.umanitoba.ca
[2]Institute of Mathematics, Warsaw University, Banacha 2, 02-097 Warsaw, Poland

Abstract. This paper introduces a neural network architecture based on rough sets and rough membership functions. The neurons of such networks instantiate approximate reasoning in assessing knowledge gleaned from input data. Each neuron constructs upper and lower approximations as an aid to classifying inputs. Rough neuron output has various forms. In this paper, rough neuron output results from the application of a rough membership function. A brief introduction to the basic concepts underlying rough membership neural networks is given. An application of rough neural computing is briefly considered in classifying the waveforms of power system faults. Experimental results with rough neural classification of waveforms are also given.

1 Introduction

A form of rough neural computing based on based on rough sets, rough membership functions, and decision rules is introduced in this paper. Rough sets were introduced by Pawlak [1], and elaborated in [2]-[3]. Rough membership functions were introduced by Pawlak and Skowron [4]. Studies of neural networks in context of rough sets are extensive [5]-[12]. This paper considers the design and application of neural networks with two types of rough neurons: approximation neurons and decider neurons. The term *rough neuron* was introduced in 1996 [5]. In its original form, a rough neuron was defined relative to upper and lower bounds and inputs were assessed relative to boundary values. More recent work considers rough neural networks (rNNs) with neurons, which construct rough sets and output the degree of accuracy of an approximation [10]-[11], which is based on an earlier study [9]. The study of rough neurons is part of a growing number of papers on neural networks based on rough sets. Rough-fuzzy multilayer perceptrons (MLPs) in knowledge encoding and classification were introduced in [12]. Rough-fuzzy neural networks have recently been also used in classifying the waveforms of power system faults [10]. Purely rough membership function neural networks (rmfNNs) were introduced in [11] in the context of rough sets and the recent introduction of rough membership functions [4]. This paper considers the design of rough neural networks based on

W. Ziarko and Y. Yao (Eds.): RSCTC 2000, LNAI 2005, pp. 611–618, 2001.

rough membership functions, and hence this form of network is called a rough membership neural network (rmNN).   Preliminary computations in a rmNN are carried out with a layer of approximation neurons, which construct rough sets and where the output of each approximation neuron is computed with a rough membership function.   The values produced by a layer of approximation neurons are used to construct a condition vector.   Each new condition vector provides a stimulus for a decider neuron in the output layer of a rmNN.   A decider neuron enforces rules derived from decision tables based on rough set theory.   A decision table reflects our knowledge of the world at a given time.   This knowledge is represented by condition vectors and corresponding decisions.   Information granules in the form of rules are extracted from decision tables using rough set methods.   Discovery of decider neuron rules stems from an application of the rule derivation method given in [13]-[14].   This characterization of a decider neuron is based on the identification of information granules based on decision rules [15].   Each time a decider neuron is stimulated by a new condition vector constructed by the approximation neuron layer, it searches for the closest fit between each new condition vector and existing condition vectors extracted from a decision table.   Decider neurons are akin to what are known as logic neurons described in [16].

2   Rough Membership Functions

A brief introduction to the basic concepts underlying the construction of rough membership neural networks is given in this section.   A rough membership function (rm function) makes it possible to measure the degree that any specified object with given attribute values belongs to a given set X [4], [21].   A rm function μ_X^B is defined relative to a set of attributes $B \subseteq A$ in information system S = (U, A) and a given set of objects X.   The equivalence class $[x]_B$ induces a partition of the universe.   Let $B \subseteq A$, and let X be a set of observations of interest.   The degree of overlap between X and $[x]_B$ containing x can be quantified with the rough membership function:

$$\mu_X^B : U \to [0,1] \text{ defined by } \mu_X^B(x) = \frac{\left|[x]_B \cap X\right|}{\left|[x]_B\right|}$$

3   Design of Rough Neural Networks

Neural networks are collections of massively parallel computation units called neurons. A neuron is a processing element in a neural network.

3.1   Design of Rough Neurons

Typically, a neuron y maps its weighted inputs from $\mathbf{R}^n$ to [0, 1] [16].   Let T be a decision table (X, A, {d}) used to construct $\underline{B}X$, $\overline{B}X$, and let $X \subseteq Y$.   A selection of different types of neurons is given in Table 1: common neurons, approximation neurons and decider neurons.

Table 1. Selection of Different Types of Neurons

Common Neuron	Upper Approximation neuron
$y = f\left(\sum_{i=1}^{n} w_i x_i + \vartheta\right)$, where input x_i has connection (weight) w_i, which denotes a modifiable neural connection, and bias ϑ [16]	$y_x = f(\overline{B}X, \underline{B}X, X)$
	Lower approximation Neuron
	$y = f(\underline{B}X, X)$
	Decider Neuron
	$y_{rule} = \min(e_i, d_i([\mu_X^{A_1}(x)...\mu_X^{A_n}(x)]))$, with condition granule $[\mu_X^{A_1}(x)...\mu_X^{A_n}(x)]$

Let B, F, $[\,f\,]_B$ denote set of attributes, set of neuron inputs (stimuli), and equivalence class containing measurements derived from known objects, respectively. The basic computation steps performed by an approximation neuron are reflected in the flow graph in Fig. 1.

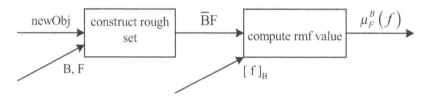

Fig. 1. Flow Graph for Basic Approximation Neuron Computation

An approximation neuron measures the degree of overlap of a set $[\,f\,]_B$ and $\overline{B}F$ representing certain as well as uncertain classifications of input signals.   A flow graph showing the basic computations performed by a decider neuron is given in Fig. 2.

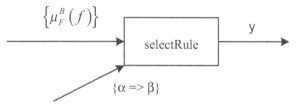

Fig. 2. Flow Graph for Decider Neuron

A decider neuron implements a selectRule algorithm.

Algorithm selectRule {
input set {α=>β}   //set of decision rules
input vector [c_{exp1}, c_{exp2}, ..., c_{expn}];   //condition vector input { $\mu_F^B(f)$ }
int chosenRule; //index used to identify decision rule
float[ ] sum; //stores sum of differences | $c_{exp\,j} - c_{ij}$ |
float bestMatch; //used to store value of best match
int vectorSize = 2, i = 1, j = 1;

$$bestMatch = \sum_{j=1}^{n} \left| c_{exp\,j} - c_{1,j} \right|; \text{ chosenRule = 1; } //\text{for vector } \alpha_1$$

while (vectorSize <= n) {

$$sum[vectorSize] = \sum_{j=1}^{n} \left| c_{exp\,j} - c_{ij} \right|;$$

if (sum[vectorSize] < bestMatch) { chosenRule = vectorSize }
vectorSize++; i++;   // use i to select i^{th} condition vector
j = 1;
}//while
return chosenRule;
} // **end** Algorithm selectRule

In Fig. 2, the set rmf = { $\mu_F^B(f)$ } consists of approximation neuron measurements in response to the stimulus provided a new object requiring classification. The elements of the set rmf are used by a decider neuron to construct an experimental condition vector α_{exp}. A second input to a decider neuron is the set R = {α=>β}. The elements of the set R are rules which have been derived from a decision table using rough set theory. After a decision rules has been selected, a decider neuron outputs min(e_i, d_i) where $d \in \{0,1\}$, and relative error $e_i = |c_{exp} - c_i|/c_i \in [0,1]$. In cases where $d = 0$, then $y_{rule} = min(e_i, d_i) = 0$, and the classification is unsuccessful. If $d = 1$, then $y_{rule} = min(e_i, d_i) = e_i$ indicates the relative error in a successful classification.

3.2 Rough Neural Network Example

By way of illustration, a rough neural network is constructed with two layers: input layer consisting of upper approximation neurons, and output layer with a single decider neuron (see Fig. 3). Using a sample of 61 fault files, a partial decision table has been constructed (see Table 2). Let v, i denote voltage, current, respectively. To complete the design of a decider neuron, rules are extracted from decision Table 2.

Table 2. Sample Power System Commutation Fault Decision Table

ac v error (a1)	phase i / i order (a2)	pole line v (a3)	6 pulse (a4)	phase i type (a5)	phase i ord (a6)	max phase i (a7)	d
0.0588	0	0.5	0	0	0	0	0
0.0588	0.06977	0.5	1	0.1875	0.05405	0.08571	1
0.0588	0.06977	0.5	1	0.1875	0.05405	0.08571	1

A sample of the rules derived from Table 2 using Rosetta [22] are as follows.

a1(0.058824) AND a5(0.187500) AND a7(0.000000) => d(0.00)
a1(0.058824) AND a5(0.187500) AND a7(0.085714) => d(1.00)

Rules like those given above are incorporated in a decider neuron repository (storage of rules associated with a decider neuron).   In the experiments described in this section, each approximation neuron is defined relative to a single attribute such as AC disturbance (see a1 in Fig. 3). The decider neuron in Fig. 3 implements the selectRule algorithm to produce its output.   The design of a particular decider neuron hinges on derivation of rules from a decision table reflecting our current knowledge of the world.

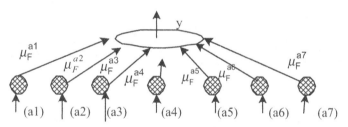

Fig. 3. Sample rough neural network

In the constructed networks, the weights are not primitive but they are functions of some other parameters like set of features (attributers).   The relationships between the weight values and these other parameters are expressed in the paper by the rough membership function. This function allows to measure a degree in which B-indiscernibility classes are included in a given set (in the considered example in the upper approximation of one of the decision class). Hence, the process of tuning weights in the network should be connected with tuning of parameters on which these weights depend.   In particular, in the considered example this can be related to searching for relevant feature set B of attributes.

3.3 Sample Verification

A comparison between the output from a rough neural network used to classify power system faults relative to 24 fault files and known classification of the sample fault data is given in Fig. 4.

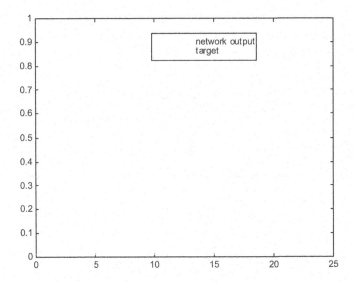

Fig. 4. Comparison of Rough Neural Network Output and Target Values

In all of the cases considered in Fig. 4, there is a close match between the target faults and the faults identified the neural network. Further, it should be observed that a total of eleven neural networks were used (one for each type of fault file) to generate the data used in Fig. 4, and carry out a complete classification of all fault files.

4 Concluding Remarks

Two basic types of rough neurons have been identified: approximation neurons and rule-based neurons.   The output of an approximation neuron is a rm function value, which indicates the degree of overlap between an approximation region and some other set of interest in a classification effort.   The output of rule-based neuron is a classification decision, which represents an assessment of the closeness of experimental data to a known feature of a feature space.   A sample application of these neurons in a power system fault classification system has been given. We consider the problem of learning schemes of information granule construction. These schemes transform input granules into output ones. It is necessary to tune parameters in these schemes to obtain the output granules of satisfactory quality from input granules.   One of the method of tuning these parameters can be based on finding function embedding these schemes into classical neural networks. Next known learning methods for neural networks can be applied. The weight values in such networks will reflect the inclusion degrees between granules. Hence the process of changing weights in neural networks should correspond to tuning degrees of granule inclusion. The paper presents an example of such situation.

References

1. Z. Pawlak, Rough sets, Int. J. of Computer and Information Sciences, Vol. 11, 1982, 341-356.
2. Z. Pawlak, Rough Sets: Theoretical Aspects of Reasoning About Data, Boston, MA, Kluwer Academic Publishers.
3. Z. Pawlak. Reasoning about data--A rough set persepective. Lecture Notes in Artificial Intelligence 1424, L. Polkowski and A. Skowron (Eds.). Berlin, Springer-Verlag, 1998, 25-34.
4. Z. Pawlak, A. Skowron, Rough membership functions. In: R. Yager, M. Fedrizzi, J. Kacprzyk (Eds.), Advances in the Dempster-Shafer Theory of Evidence, NY, John Wiley & Sons, 1994, 251-271.
5. P.J. Lingras, Rough neural networks. In: Proc. of the 6th Int. Conf. on Information Processing and Management of Uncertainty in Knowledge-based Systems (IPMU'96), Granada, Spain, 1996, 1445-1450.
6. H.S. Nguyen, M. Szczuka, D. Slezak, Neural networks design: Rough set approach to real-valued data. In: Proc. of PKDD'97, Trondheim, Norway. Lecture Notes in Artificial Intelligence 1263, Berlin, Springer-Verlag, 1997, 359-366.
7. P. Wojdyllo, Wavelets, rough sets and artificial neural networks in EEG analysis. In: Proc. of RSCTC'99, Lecture Notes in Artificial Intelligence 1424, Berlin, Springer-Verlag, 1998, 444-449.
8. M.S. Szczuka, Refining classifiers with neural networks, International Journal of Intelligent Systems [to appear].
9. L. Han, R. Menzies, J.F. Peters, L. Crowe, High voltage power fault-detection and analysis system: Design and implementation. Proc. CCECE99, 1253-1258.
10. L. Han, J.F. Peters, S. Ramanna, R. Zhai, Classifying faults in high voltage power systems: A rough-fuzzy neural computational approach. In: N. Zhong, A. Skowron, S. Ohsuga (Eds.), New Directions in Rough Sets, Data Mining, and Granular-Soft Computing, Lecture Notes in Artificial Intelligence 1711. Berlin: Springer, 1999, 47-54.
11. J.F. Peters, A. Skowron, Z. Surai, L. Han, S. Ramanna, Design of rough neurons: Rough set foundation and Petri net model, International Symposium on Methodologies for Intelligent Systems (ISMIS'2000) [submitted].
12. M. Banerjee, S. Mitra, S.K. Pal, Rough fuzzy MLP: Knowledge encoding and classification, IEEE Trans. Neural Networks, vol. 9, 1998, 1203-1216.
13. A. Skowron, C. Rauszer. The discernability matrices and functions in information systems. In: Intelligent Decision Support, Handbook of Applications and Advances of the Rough Sets Theory, Slowinski, R. (Ed.), Dordrecht, Kluwer Academic Publishers, 1992, 331-362.
14. A. Skowron and J. Stepaniuk. Decision rules based on discernability matrices and decision matrices. In: Proc. Third Int. Workshop on Rough Sets and Soft Computing, San Jose, California, 10-12 November 1994.
15. A. Skowron, J. Stepaniuk, Information granules: Towards foundations of granular computing, Internation Journal of Intelligent Systems [to appear].
16. W. Pedrycz, Computational Intelligence: An Introduction, Boca Raton, CRC Press, 1998.

17. H.S. Nguyen. Discretization of real-valued attributes: Boolean reasoning approach. Doctoral Thesis, Faculty of Mathematics, Computer Science and Mechanics, Warsaw University, 1997.

18. H.S. Nguyen. Rule induction from continuous data: New discretization concepts. Proc. of the Third Joint Conf. on Information Sciences, Raleigh, N.C., 1-5 March 1997.

19. A. Skowron, J. Stepaniuk, Information granules in distributed environment. In: N. Zhong, A. Skowron, S. Ohsuga (Eds.), New Directions in Rough Sets, Data Mining, and Granular-Soft Computing, Lecture Notes in Artificial Intelligence 1711. Berlin: Springer, 1999, 357-365.

20. A. Skowron, J. Stepaniuk. Constructive information granules. In: Proc. of the 15th IMACS World Congress on Scientific Computation, Modelling and Applied Mathematics, Berlin, Germany, 24-29 August 1997. Artificial Intelligence and Computer Science 4, 1997, 625-630.

21. J. Komorowski, Z. Pawlak, L. Polkowski, A. Skowron, Rough sets: A tutorial. In: S.K. Pal, A. Skowron (Eds.), Rough Fuzzy Hybridization: A New Trend in Decision-Making. Singapore: Springer-Verlag, 1999, 3-98.

22. Rosetta software system,

The Nature of Crossover Operator in Genetic Algorithms

Lihan Tian

Department of Computer Science, Jilin University
P.O.B. 209, 10-3 Jiefang South Lane, Changchun 130022, P. R. China

Abstract. Crossover is a main searching operator of genetic algorithms (GAs), which has distinguished GAs from many other algorithms. Through analyzing and imitating the implementation of crossover operator, this paper points out that crossover is intrinsically a heuristic mutation with reference. Its reference objective is just the other individual which is mated with the one which will be crossovered. On the basis of this conclusion this paper then explains and discusses the results obtained by other GA researchers through experiments.

1   Introduction

Genetic Algorithm (GA) was invented by John Holland in the 1960s, a professor of Michigan University. Holland's original goal was not to design algorithms to solve specific problems, but rather to formally study the phenomenon of adaptation as it occurs in nature and to develop ways in which the mechanisms of natural adaptation might be imported into computer systems[1]. However, after it was later developed by his students, colleagues and other researchers, Genetic Algorithm has been widely used in various problems as a robust search method, especially in optimum seeking[2,3].

As an important branch of Evolutionary Computation (EC), Genetic Algorithm (GA) is characterized by its current effectiveness, strong robustness, and simple implementation. It also has the advantage of not being restrained by certain restrictive factors of search space. Due to the advantages mentioned above, researchers have been showing increasing interest in GA. It has been applied successfully to many fields such as machine learning, engineering optimization, economy forecast, automatic programming, and so forth. Nowadays GA has become a very popular research subject in many branches of science[3--14].

Crossover operator is the main search operator of Genetic Algorithm, one of the most important features distinguishing it from other search algorithms. The effect of crossover operator in Genetic Algorithm is a disputed problem in GA field for a long time. Standard Genetic Algorithm and most other improved Genetic Algorithms adopt crossover operator as their main genetic operator[1-3]. Evolutionary Strategy only utilizes mutation operator in its early development and mutation operator is still its most important operator[9], although it later combines crossover operator. Evolutionary Programming only employs mutation operator and does not employ crossover operator at all.[10]

D. Fogel, one of the Evolutionary Programming advocates, declared that crossover is not superior to mutation[15]. Scheffer drew the conclusion by his

W. Ziarko and Y. Yao (Eds.): RSCTC 2000, LNAI 2005, pp. 619–623, 2001.

experiment that crossover is not always sufficient with only mutation [16]. Based on his theoretical analysis and experimental test on the model division and construction of mutation and crossover, Spears, an Evolutionary computation researcher in Artificial Intelligence Center of American Navy Laboratory, pointed out that mutation is more divisive than crossover and crossover is more constructive than mutation. One complements the other and both of them are indispensable[17].

So far, the study about crossover and mutation is only based on the macro effect of the two operators. This paper will simulate and analyze the performance process of crossover operator to discover the microcosmic nature of crossover operator.

2   The Nature of Crossover Operator

Crossover operator completes its performance by exchanging two selected parent genes, including point crossover and uniform crossover, to create two new individuals that respectively inherit some genes of their parents. The following will illustrate two-point crossover and uniform crossover with binary coding to simulate and analyze the crossover operator performance.

Suppose the two selected parents are:

$$\xi = \xi_1 \xi_2 ... \xi_L \qquad \xi \in \{0,1\}, \qquad i=1(1)L$$

$$\eta = \eta_1 \eta_2 .. \eta_L \qquad \eta \in \{0,1\}, \qquad i=1(1)L$$

where L is the length of chromosome bit string. According to ξ and η, we construct a new individual μ:

$$\mu = \mu_1 \mu_2 ... \mu_L \qquad \mu_i = \xi_i \text{ XOR } \eta_i, \quad i=1(1)L$$

Obviously the following equation should be true:

$$\sum_{i=1}^{L} \mu_i = H(\xi, \eta)$$

where $H(\xi,\eta)$ represents Hamming distance between ξ and η, and therefore μ is called Hamming distance of ξ and η.

2.1   Two-Point Crossover

With regard to two-point crossover, suppose the two selected crossover points are C_1 and C_2 ($C_1 < C_2$), and the new individuals generated by crossover operator are respectively:

$$\xi_T = \xi_1 .... \xi_{c_1-1} \eta_{c_1} ... \eta_{c_2} \xi_{c_2+1} ... \xi_L$$

$$\eta_T = \eta_1 .... \eta_{c_1-1} \xi_{c_1} ... \xi_{c_2} \eta_{c_2+1} ... \eta_L$$

Now use Hamming individual μ as reference, we mutate ξ. as follows:

$$\text{for } i=1 \text{ to } C_1-1$$
$$\xi'_T(i) = \xi_i;$$
$$\text{for } i=C_1 \text{ to } C_2$$
$$\text{if } (\mu_i == 0)$$
$$\text{then } \xi'_T(i) = \xi_I$$
$$\text{else } \xi'_T(i) = \sim \xi_i;$$
$$\text{for } i=C_2+1 \text{ to } L$$
$$\xi'_T(i) = \xi_i.$$

The nature of the above mutation is:
It remains the same gene value between 1 and C_1-1 and between C_2+1 and L. While between C_1 and C_2, if the gene value of μ is 0, it remains the loci, otherwise it inverses. That is, if the gene value of the corresponding loci is the same, it remains the gene value of the corresponding loci, otherwise it inverses. In other words, the transformation only exchanges the gene value between C_1 and C_2 of ξ and η. Therefore, the individual ξ'_T created by the above mutation is the same as ξ_T. Similarly, $\eta'_T = \eta_T$.

2.2   Uniform Crossover

With uniform crossover, suppose the created crossover template is $v = v_1 v_2 \ldots v_L$, the two new individuals generated by uniform crossover are:

$$\xi_U = \xi^U_1 \xi^U_2 \ldots \xi^U_L \quad \text{If } v_i=0 \quad \xi^U_i = \xi_i; \quad \text{If } v_i=1 \quad \xi^U_i = \eta_i \quad i=1(1)L$$

$$\eta_U = \eta^U_1 \eta^U_2 \ldots \eta^U_L \quad \text{If } v_i=0 \quad \eta^U_i = \eta_i; \quad \text{If } v_i=1 \quad \eta^U_i = \xi_i \quad i=1(1)L$$

Now use Hamming individual μ and uniform crossover v as reference, we mutate ξ as follows:

$$\text{for } i=1 \text{ to } L$$
$$\text{if } (v_i == 0)$$
$$\text{then } \xi'_U(i) = \xi_i$$
$$\text{else if } (\mu_i == 0)$$
$$\text{then } \xi'_U(i) = \xi_i$$
$$\text{else } \xi'_U(i) = \sim \xi_i$$

The nature of the above mutation is:
If the corresponding gene value of the created crossover template at a locus is 0, then the gene of the individual at the same locus remain the same as the original one. If the corresponding gene value of the created crossover template is 1, then inverse the gene. That is, , if $\mu_i=0$, then remain the gene value. If $\mu_i=1$, then inverse it. So the mutation has the same effect of crossover operator. Thus, $\xi'_U = \xi_U$. Similarly, we can obtain $\eta'_U = \eta_U$.

2.3   Analysis

With Hamming individual μ as reference objective, the two mutation processes respectively achieve two-point and uniform crossover operation of ξ and η and get the same result as normal crossover operator. In fact, any crossover operation defined on any alphabet can be made through corresponding mutation according to the media individual created by two parents. Therefore, we induce the following theorem: Crossover is intrinsically a heuristic mutation with reference, and the reference objective is just the other individual which is mated with the one which will be crossovered.

3   Discussion

The above conclusion may be used to explain the function and effect of crossover operator in evolutionary progress and help design more general operators with mutation and crossover effect at the same time.

• Because crossover operator does not alter the same genes of the two parents, and only changes different genes, "mutation" is defined in the different genes of two individuals and so it may produce certain heuristic effect. The inferior individual gets a certain part of superior individual gene, and may improve its fitness more probably. If the exchanged gene of superior individual is negative to the former individual because of gene epistasis, the fitness of superior individual is ameliorated to some extent. This explains the exploitation effect of crossover operator.

• Hamming distance between two offspring created by the two individuals with Hamming distance d and their parents is between 0 and d. According to the similar extent of two parents and difference of crossover points, crossover operation may generate more mutative effect than normal mutation (0≤mutation gene number≤d). This is why global search of crossover operator is superior.

• Because the nature of crossover operator is the heuristic mutation with reference, theoretically mutation can implement all what crossover operator can do. That is, it should be sufficient with only mutation. Because single aimless mutation lacks the referential heuristics of crossover operator, "it is not always sufficient with only mutation".

• Two parents with their Hamming distance d at most create Max{2d - 2, 0} by single crossover , at most d^2-d new individuals by two-point crossover, and at most create Max{2^d - 2,0} new individuals (the most individuals that crossover can generate). This is just the reason why two-point crossover is more constructive than single point crossover and less constructive than uniform crossover.

• Because crossover operator can not alter the same genes of two parents, the capacity of two individuals with Hamming distance d is 2^d, while the genetic space capacity of the whole problem is 2^L. Therefore, when the population diversity lacks, Genetic Algorithm performance, which only adopts crossover operator or search operator whose main operator is crossover operator, is greatly influenced.

• Because crossover operator only mutates the different genes of two parents, the individuals who are very similar should avoid crossover operation. For low fitness individuals usually lack referential heuristics, two low fitness individuals should avoid crossover operation too.

4   Conclusion

This paper simulates and analyzes the performance process of crossover operator and draws the conclusion that the nature of crossover operator is the heuristic mutation with reference.   Unlike normal mutation operator, the operator only mutates the different genes of two parents, it can achieve certain referential heuristics that normal operator can not do.   Genetic Algorithms and even Evolutionary Computation in nature just transform and select a group of feasible individuals.   The more heuristic transformation is, the more efficient algorithm search is.   It is not necessary to restrain Genetic Algorithms or Evolutionary Computation to the simulation of some principles or processes of evolutionism and biology genetics.   Research should focus on how to improve the heuristics of feasible transformation and rationality of selection, although it does not exclude the reference to the results from relative science.

5   Acknowledgments

The author gratefully acknowledges Fuhan Tian for his helpful advice and Huiling Li for her proof reading.

References

1.   John H. Holland, Aadaptation in Natural and Artificial Systems, MIT Press, 1992, Second edition.
2.   David E. Goldberg, Genetic Algorithms in search, Optimization, and Machine Learning, Addison-Wesley Publishing Company, 1989.
3.   Melanie Mitchell, An Introduction to Genetic Algorithms, MIT Press, 1996.
4.   Zbigniew Michalewicz, Genetic Algorithms + Data Structures = Evolution Programs, Springer, 1994, Second Extended Edition.
5.   Rechard K. Belew and Michael D. Vose ed., Foundations of Genetic Algorithms IV, Morgan Kaufamann Publishers, 1997.
6.   Lance Chambers ed., Practical Handbook of Genetic Algorithms, Volume I-II, CRC, 1995.
7.   Joachim Stender, Parallel Genetic Algorithms: Theory and Applications, IOS Press, 1993.
8.   Thomas Bäck ed., Proceedings of the Seventh International Conference on Genetic Algorithms, Morgan Kaufamann Publishers, 1997.
9.   Hans-Paul Schwefel, Evolution and Optimum seeking, John Wiley&Sons, 1994.
10.   Thomas Bäck ed., Evolutionary Algorithms in Theory and Practice: Evolution Strategies, Evolutionary Programming, Genetic Algorithms, Oxford University Press, 1996.
11.   Elie Sanchez etc. Ed., Genetic Algorithms and Fuzzy Logic Systems: Soft Computing Perspectives, World Scientific Publishing, 1997.
12.   K.F. Man etc., Genetic Algorithms for Control and Signal Processing (Advances in Industrial Control), Springer Verlag, 1997.
13.   G.Winter etc. Ed., Genetic Algorithms in Engineering and Computer Science, John Wiley and Sons, 1996
14.   T. Bäck,U. Hammel,H.-P. Schwefel, Evolutionary computation: comments on the history and current state, IEEE Tran. on Evolutionary Computation, 1997, 11, pp.3-17.
15.   D.B. Fogel, J.W. Atmar, Comparing genetic operators with Gaussian mutations in simulated evolutionary processes using linear systems, Biological Cybernetics, 1990, 63, pp111-114.
16.   J.D. Schaffer, L.J. Eshelman, On crossover as an evolutionarily viable strategy, Proceedings of the Fourth International Conference on Genetic Algorithms and Their Applications, Morgan Kaufmann, 1991.
17.   William M. Spears, The Role of Mutation and Recombination in Evolutionary, Ph.D. Thesis, George Mason University, 1998
18.   Mitchell A. Potter, The Design and Analysis of a Computational Model of Cooperative Coevolution, Ph.D. Thesis, George Mason University, 1997.

The Behavior of the Complex Integral Neural Network

Pan Yong[1], Shi Hongbao[1], Li Lei[2]

[1]Institute of Computing Technology, Tongji University(West Campus), Shanghai, 200331
[2]Suzhou Railway Teachers College, Suzhou, Jiangsu, 215009
ypan@shtdu.edu.cn

Abstract: In this paper, we introduce an Integral Neural Network based on complex domain. We describe the model of the neuron, analyze the behavior of the neuron, and indicate that in certain conditions it performs the calculation of Fourier Integration. Have studied on the neural network with hidden layers, we obtain the following facts: 1. This kind of structure can memorize a time variant function; 2. It calculates the convolution of input series and the function the neural network memorized; 3. This neural network structure also can calculate the correlation function; 4. In the case of many hidden layers, it can perform the Fourier Transform with many variants.

1. Introduction

As early in 1975, researchers had already extended the conventional LMS algorithm to the complex domain[5]. Until now, the papers published on complex domain neural network are not so much. The majority of those papers think of complex number as a pair of real numbers[9]. The studies in complex domain NN mainly focused on complex BP Neural Network and complex associative memory.

In the field of complex BP Neural Network, T. Nitta has already done many works[2]. His main results include the followings: It is not suitable to choose an holomorphic function as the activation function. The learning speed of the complex BP algorithm is faster than that of real-BP. The complex-BP network can learn the basic geometric transformation (rotation, similarity transformation and parallel displacement etc.) There are several works extends the complex domain neural networks to the associative memory[1][3][4]. The ways to study complex associative memory is similar to those to study real associative memory, except that the data range is extended to complex domain[6].

There also have some reports on Neural Networks for Time Series Processing. The basic idea of introducing the time variant function to neural network is based on the fact that the world is always changing. Whatever we observe or measure have

W. Ziarko and Y. Yao (Eds.): RSCTC 2000, LNAI 2005, pp. 624–631, 2001.

different values at different points of time[10]. The NN mechanism for handling time variant sequences is much like that of finite automatons. It has internal status and feedback structures[11].

In this paper we analyze the mathematical function of the complex integral neural network[7][8]. We introduce the neuron model and the neuron behavior in section 2 and section 3; In section 4, we study the function of the neural network with one hidden layer; In section 5, we study the cases which have many hidden layers; And finally, we conclude the paper.

2. The Integral Neuron Model Based on Complex Domain

Figure 1 shows the structure of a complex integral neuron. It is based on complex domain, so its weights, input and output signals are all taken from complex domain. Besides, the model has two other features. One is that the inputs and outputs of the neuron are all time series rather than some fixed numbers; Another is that the output signals feedback to itself. The detailed process is as follows:

When $t = 0$, the input signals are $f_{1i}(0)$, $f_{2i}(0)$, .., $f_{mi}(0)$. They are multiplied with the weights Z_1, Z_2, .., Z_m respectively, then calculate the sum $\Sigma(f_{mi}(0)*Z_m)$, then we get the output $f_o(0)$. The relationship between $f_o(0)$ and $\Sigma(f_{mi}(0)*Z_m)$ depend on the activate function $F(*)$. At the same time, the output $f_o(0)$ multiplies the feed-back weight Z_{m+1} and feeds back to the neuron as a new input signal at $t = 1$. So, when $t = 1$, the input signals are $_{1i}(1)$, $_{2i}(1)$, .., $f_{mi}(1)$ and $_o(0)$. The summation is $\Sigma(f_{mi}(1)*Z_m)+ f_o(0)*Z_{m+1}$ instead of $\Sigma(f_{mi}(1)*Z_m)$. After calculate the activate function $F()$, we get the output $f_o(1)$, and so on and so forth. Generally, the output at $t = n$ depends on the inputs at $t=n$ ($f_{1i}(n)$, $f_{2i}(n)$, .., $f_{mi}(n)$) and the output at $t=n-1$ ($f_o(n-1)$).

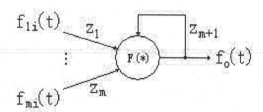

Figure 1. The structure of a complex integral neuron.

Though different activate function can be chosen, in the rest of the paper we always choose it as the identical function.

3. The function of the neuron

For simplifying the neuron model, we first consider the neuron, which only has one input end (See figure 2).

In figure 2, the input series of the neuron is $f_i(t)$. The output series is $f_o(t)$. The input weight and feedback weight are z and $ż$ respectively. Suppose t_{n-1}, t_n are two adjacent time points, we get:

$$f_o(t_n) = f_i(t_n) \cdot z + f_o(t_{n-1}) \cdot z' \tag{1}$$

where z and ' are complex numbers, f(t), f(t) are complex series.   The following theorem shows the function of this model.

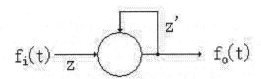

Figure 2. The neuron which only has one input end.

Theorem 1.   Suppose $z' = \exp(-i\omega)$ and z= 1, then the output $f_o(t_n)$ is the value of the Discrete Fourier Transform of the series $f_i(t_0)$, $f_i(t_1)$, .., $f_i(t_n)$ at the frequency ω, where n is the number of the sampling points.

Proof: Let $f_i(t) = f_n$ when t = n. We first proof

$$f_o(n) = \sum_{m=0}^{n} f_{n-m} \cdot (z')^m \tag{2}$$

It is obviously when n = 0 and n = 1.

Suppose it is true when n = k-1, that is

$$f_o(k-1) = \sum_{m=0}^{k-1} f_{k-1-m} \cdot (z')^m$$

Then

$$f_o(k) = f_k + (\sum_{m=0}^{k-1} f_{k-1-m} \cdot (z')^m) \cdot z' \qquad by\ (1)$$

$$= f_{k-0} \cdot (z')^0 + \sum_{m=1}^{k} f_{k-m} \cdot (z')^m \qquad replace\ m+1\ by\ m$$

$$= \sum_{m=0}^{k} f_{k-m} \cdot (z')^m$$

So the formula (2) holds true, that means the output is the part sum of the unilateral Z transform based on the reversed series $\{f_n\}$.   If the time is enough long, the part sum is very close to the value of the Z transform of the input series at the point $z = (z')^{-1}$.   In the case of $z' = \exp(-i\omega)$, the part sum

$$\sum_{m=0}^{n} f_{n-m} \cdot (z')^m$$

equals to:

$$\sum_{m=0}^{n} f_{n-m} \cdot e^{-i\omega m}$$

That is the part sum of the discrete Fourier Transform of the input series. If the time is enough long, the part sum can be very close to the value of the Fourier Transform of the input series at certain frequency point $\omega = \text{Arg}(z)$.
End of proof.

The experimental results[8] show that this kind of neuron can be used as a filter or an amplitude indicator. The following theorem also is interesting.

Theorem 2. Suppose $f(t) \in L^2$, and for all $n \in N$, $f(n) = f(n)$, $F(\omega) = \int_0^\infty f(t) e^{-i\omega t} dt$, if the sampling frequency is large than ω_0, and $z = \exp(-i\omega_0)$, then

$$F(\omega_0) = \lim_{n \to \infty} \frac{f_o(n)}{n\omega_0}$$

Proof: Because $f(t) \in L^2$, the integral
$F(\omega) = \int_0^\infty f(t) e^{-i\omega t} dt$ is convergent.
The DFT form is:

$$\frac{1}{2\pi} \sum_{m=0}^{T_0} f(m) \cdot e^{-i\omega_0 m} = F(\omega_0)$$

By the theorem 1 we get:

$$f_0(n) = \sum_{m=0}^{n} f(m) \cdot e^{-i\omega_0 m}$$

$$= \sum_{m=0}^{T_0} f(m) \cdot e^{-i\omega_0 m} + \dots + \sum_{m=(k-1)T_0+1}^{kT_0} f(m) \cdot e^{-i\omega_0 m} + \sum_{m=kT_0+1}^{n} f(m) \cdot e^{-i\omega_0 m}$$

$$= 2\pi \cdot k \cdot F(\omega_0) + \sum_{m=kT_0+1}^{n} f(m) \cdot e^{-i\omega_0 m}$$

Let $n_0 = n - kT_0$, and $n_0 < T_0$
Because $\omega_0 = 2\pi/T_0$, and $n = kT_0 + n_0$, we get $k \approx n/T_0 = n\omega_0/2\pi$.
When $n \to \infty$, $k = n\omega_0/2\pi$, so

$$\lim_{n \to \infty} \frac{f_o(n)}{n\omega_0} = \lim_{n \to \infty} \frac{2\pi \cdot k \cdot F(\omega_0) + \sum_{m=kT_0+1}^{n} f(m) \cdot e^{-i\omega_0 m}}{n\omega_0}$$

$$= \lim_{n \to \infty} \left(\frac{n\omega_0 \cdot F(\omega_0)}{n\omega_0} + \frac{\sum_{m=kT0+1}^{n} f(m) \cdot e^{-i\omega_0 m}}{n\omega_0} \right)$$

$$= \lim_{n \to \infty} \frac{n\omega_0 \cdot F(\omega_0)}{n\omega_0} \quad \left(\lim_{n \to \infty} \frac{\sum_{m=kT_0+1}^{n} f(m) \cdot e^{-i\omega_0 m}}{n\omega_0} = 0 \right)$$

$$= F(\omega_0)$$

End of proof

Let us consider the general situation of multi-input neuron (See figure 1). The output series should be:

$$f_o(t) = \int_{t_0}^{t} \sum_{k=1}^{m} (f_{ki}(t) z_k) \cdot z'_{m+1} dt \qquad (3)$$

From the formula (3) we can see the values of the weights of the neural inputs have different function compare to that of the self-feedback weight. The former gives each input signal an adjustment of phase and/or amplitude, while the latter determines the impulse response at some certain frequency point of the signals adjusted. Because we do not change the frequency value when adjusting the phase and/or amplitude, we can then amplify/diminish the original signal and prevent/facilitate the offset of two or more signals.

4. Network with one hidden layer

Function Storing

For understanding what information is stored by the weight values, we let input series to be the δ function. The δ function is defined as follows:

$$\delta(t) = \begin{cases} 1 & t = 0 \\ 0 & t \neq 0 \end{cases}$$

If we put it as the input signals of the neuron on figure 2, we can easily see that the output series is $z \cdot \exp(-in\omega)$, where $z' = \exp(-i\omega)$. That means the weights memorize the function $z \cdot \exp(-in\omega)$. Combine a set of neuron as figure 3, the function they memorize are $\Sigma(z_k \cdot \exp(-in\omega_k))$. Because any function which satisfy Dirichlet condition can be represented as a Fourier series, It means that a set of hidden layer neuron can store a large set of functions.

The interested fact is that the same set of neurons can memorize different functions in different channels or at different outputs.

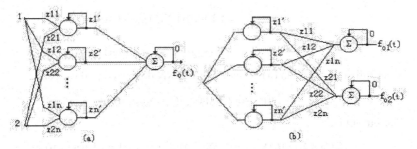

Figure 3. (a)A set of neurons stored different functions in different channels.(b)A set of neurons produce different functions at different outputs.

In figure 3 (a), the set of neurons memorized different functions in different channels. That is: If we put the δ function at Ist input end, the output series is

$\Sigma(z_{1k} \cdot \exp(-in\omega_k))$; If we put the δ function at 2nd input end, the output series is $\Sigma(z_{2k} \cdot \exp(-in\omega_k))$. In figure 3 (b), the set of neurons produce different functions at different outputs. That is: If we input the δ function, the output series $f_{o1}(t)$ and $f_{o2}(t)$ are $\Sigma(z_{1k} \cdot \exp(-in\omega_k))$ and $\Sigma(z_{2k} \cdot \exp(-in\omega_k))$ respectively.

Convolution

The above study shows that the output would be the memorized function $f(t)$ if the input function were δ function. In many cases, the input function is any time series function $g(t)$. What the output will be? The following theorem tells us which calculation the neurons take on functions $f(t)$ and $g(t)$.

Theorem 3. Suppose $f(t)$ is a function, which is memorized by a set of, neurons with the structure of figure 3, and $g(t)$ is a input function. The output function is the convolution of $f(t)$ and $g(t)$: $f_o(s) = \Sigma f(n)g(s-n)$.

Proof: We first consider the neuron that figure 2 described. The function it stored is:

$$g(n) = \Sigma(z \cdot \exp(-in\omega))$$

We will prove

$$f_o(s) = \Sigma f(n)g(s-n) = \Sigma(f(n) \Sigma(z \cdot \exp(-i(s-n)\omega))) \qquad (4)$$

When $s = 0$

$$f_o(0) = f(0)(z) = f(0)g(0) \qquad \text{(formula (4) is true)}$$

and $s = 1$

$$f_o(1) = f(1)(z) + f(0)(z \cdot z') = f(0)g(1) + f(1)g(0) \quad \text{(formula (4) is true)}$$

Suppose $f_o(k-1) = \Sigma f(n)g(k-1-n)$ holds true, then

$$f_o(k) = f(k)(z) + f_o(k-1) \cdot z'$$

$$= f(k)g(0) + (\Sigma f(n)g(k-1-n)) \cdot z'$$

$$= f(k)g(0) + \Sigma f(n)g(k-1-n) \cdot z'$$

$$= \Sigma f(n)g(k-n) \qquad (\text{ for } g(k-1-n) \cdot z' \text{ is } g(k-n))$$

So for all $s \in N$, formula (4) is true. For the NN described in the figure 3,

$$g(n) = \Sigma(z_k \cdot \exp(-in\omega_k)) = \Sigma g_k(n).$$

The output is:

$$f_o(s) = \Sigma\{\Sigma f(n)g_k(s-n)\} = \Sigma\{f(n)(\Sigma g_k(s-n))\} = \Sigma f(n)g(s-n)$$

End of proof

Theorem 3 shows that the integral NN with one hidden layer neurons can memorize a time variant function $f(t)$, which, together with the input series $g(t)$, produce the convolution series of the functions $f(t)$ and $g(t)$. Applying the time shifting property and conjugate property[7], this structure can also calculate the value

of the correlation function series. This is because:
$Cor(f,g)(\tau) = \int f(t)g(t+\tau)dt = \int f(t)g^*(-t-\tau)dt = (f*g)(-\tau)$, and we can get $g^*(t)$ from $g(t)$ by replace $z_1, z_2, ..,z_n$ with $z^*_1, z^*_2, ..,z^*_n$.

5. The behavior of the Integral NN with many hidden layers

In section 3, we indicated that a single neuron could perform Fourier Transform. Here we will illustrate that the many hidden layers NN can perform the Fourier Transform with many variants. We just give the example of two hidden layers as it is drawn in figure 4.

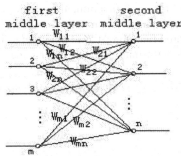

Figure 4. The NN with two hidden layers

Theorem 4. Suppose the feedback weights of first hidden layer neurons and second hidden layer neurons are $\exp(-in\omega_l)$ and $\exp(-in\psi_k)$ respectively, where $l=1,2, ..,m$ and $k=1,2, ..,n$. The output of kth neuron of second layer will be $F(\omega_l,\psi_k)$, if $w_{1k}= 1$, and $w_{jk}= 0$ for all $j\neq 1$, where $F(\omega,\psi)$ are two variants DFT of input series $f(x,y)$: and the inputs series of first hidden layer are $f_l(x,l)$, $l=1,2,..,m$.

$$F(\omega,\psi) = \sum_{t_2=0}^{t_{20}}(\sum_{t_1=0}^{t_{10}} f_l(t_{10} - t_1,l)\cdot e^{-i(\omega t_1 +\psi t_2)})$$

Proof: By the theorem 1, the output series from 1st neuron of the first layer is:

$$y_{ol}(t_{10}) = \sum_{t_1=0}^{t_{10}} f_l(t_{10} - t_1,l)\cdot e^{-i\omega_l t_1}$$

So, the output of kth neuron of second layer is:

$$y_{ok}(t_{10},t_{20}) = \sum_{t_2=0}^{t_{20}}(\sum_{i=1}^{m} w_{ij} \cdot y_{oi}(t_{10}))e^{-i\psi_k t_2}$$

$$= \sum_{t_2=0}^{t_{20}}(\sum_{i=1}^{m} w_{ij} \sum_{t_1=0}^{t_{10}} f_i(t_{10} - t_1,i)e^{-i\omega_i t_1})e^{-i\psi_k t_2}$$

$$= \sum_{t_2=0}^{t_{20}}(\sum_{t_1=0}^{t_{10}} f_l(t_{10} - t_1,l)\cdot e^{-i\omega_l t_1})\cdot e^{-i\psi_k t_2} \qquad only \ \ w_{lk} = 1$$

$$= \sum_{t_2=0}^{t_{20}} \sum_{t_1=0}^{t_{10}} f_l(t_{10} - t_1,l)\cdot e^{-i(\omega_l t_1 +\psi_k t_2)}$$

$$= F(\omega_l,\psi_k)$$

End of proof

6. Conclusion

In this paper, we introduce an Integral Neural Network based on complex domain. It is based on complex domain and the inputs and outputs are all time series. We describe the model of this kind of neuron, analyze the behavior of a single neuron, and indicate that in certain conditions it performs the calculation of Fourier Integration.

In the case of the NN with one hidden layer, the weights memorize a time series function. If the input function is δ function, the memorized function f(t) will be retrieved at the output. If the input function is g(t), what we obtain from output is the convolution series of the stored function f(t) and the input function g(t).

In the case of the NN with many hidden layers, the outputs are multi-variant DFT in certain conditions.

With the above features, this kind of neural network is useful in many application fields.

Reference

1. Iku Nemoto, Makoto Kubono, "Complex Associative Memory", Neural Networks, Vol. 9, No. 2 (1996) 253-261.
2. T. Nitta, "An Extension of the Back-Propagation Algorithm to Complex Numbers", Neural Networks, Vol. 9, No. 2 (1996) 1392-1414
3. Stanislaw Jankowski, andrzej Lozowski, and Jacek M. Zurada, "Complex-Valued Multistate Neural Associative Memory", IEEE Trans. on Neural Networks, Vol.7, No.6, (Nov. 1996) 1491-1496
4. Srinivasa V. Chakravarthy and Joydeep Ghosh, "A Neural Network-based Associative Memory for Storing Complex-valued Patterns", Systems, Man, and Cybernetics, 1994, Intel. Conf., 2213-2218
5. Widrow, B., McCool, J., & Ball, M., "The complex LMS algorithm", Proceedings of the IEEE, 63(4) (1975) 719-720
6. A. Hirose, "Proposal of fully complex-valued Neural Networks", In Proceedings of the International Joint Conference on Neural Networks, Vol. 4 (1992) 152-157
7. Pan Yong, Li lei and Shi Hongbao, "An Integral Neural Network Based on Complex Domain", Advances in Computer Science and Technology, Proceeding of the Fifth International Conference for Young Computer Scientists, Vol.1 (1999) 537-539
8. Pan Yong, Li lei and Shi Hongbao, "An Integral Neural Network Based on Complex Domain and its applications", Proceeding of International Symposium on Domain Theory (1999)
9. Price, D et al, Pairwise Neural Network Classifiers with Probabilistic Outputs, NIPS (1996) 1009-1116
10. Ulbricht C., Dorffner G., Canu S., Guillemyn D., Marijuan G., Olarte J., Rodriguez C., Martin I., Mechanisms for Handling Sequences with Neural Networks, Intelligent Engineering Systems through Artificial Neural Networks, Volume 2, ASME Press, New York, 1992.
11. Ulbricht C., Handling Sequences with a Competitive Recurrent Network, Proceedings of the International Joint Conference on Neural Networks, Baltimore, 1992.

On Trace Assertion Method of Module Interface Specification with Concurrency *

Ryszard Janicki and Yan Liu**

Department of Computing and Software,
McMaster University,
Hamilton, ON, L8S 4K1 Canada
janicki@mcmaster.ca, Yan_Liu-YLIU1@email.mot.com

Abstract. The trace assertion method is a formal state machine based method for specifying module interfaces [1,9]. It can be seen as an alternative to algebraic specification technique. We extend the sequential model presented in [9] by allowing simple concurrency.

1 Introduction

It is a well established fact that state machine (not necessary finite) models and algebraic models are equivalent ([2,4]). This relationship differs for different machines and algebras, but the general idea of relationship may be illustrated as follows:

$$\underbrace{\delta(p, a) = q}_{state\ machine} \quad \Leftrightarrow \quad \underbrace{a(p) = q,}_{algebra}$$

where δ is a transition function of a state machine with a as a function name, and $a(p)$ is a function named a applied to p.

Very often automata models are better suited for specifying and analyzing concrete software systems, while algebraic models are better suited for defining more abstract and general theories. This is exactly the case for algebraic specification versus *trace assertion method* (see [9]). The trace assertion method was first formulated by Bartussek and Parnas in [1], as a possible answer for some problems with algebraic specifications [3,15], like specifying a bounded stack (bounded modules in general). It also can avoid the problem of overspecification in model-oriented specifications, e.g. [9]. Since its introduction the method has undergone many modifications [5,11,9]. In recent years, there has been an increased interest in the

---

* Partially supported by NSERC and CITO Research Grants
** current affiliation: Motorola, 1501 West Shure Dr., Arlington Heights, Illinois 60004, USA

W. Ziarko and Y. Yao (Eds.): RSCTC 2000, LNAI 2005, pp. 632–641, 2001.

trace assertion method [9, 14]. Despite man y important industry applications, solid mathematical foundations of trace assertion method ha ve
only very recently been provided, see [9], our major reference. The model
presented in [9] does not include concurrency. In this paper we add some
concurrency to the trace assertion method.

The trace assertion method is based on the following postulates:
(1) *Information hiding* [12, 13] is a fundamental principle for specification, so we describe only those features of a module that are externally
observable;
(2) *Sequences* are simple and po werful tool for specifying abstract *objects*;
(3) *Explicit equations* are preferable over *implicit equations* like those of
algebraic specifications;
(4) *State machines* are simple and po werful tools for specifying *modules*.

The fundamental difference between algebraic specification and trace
assertion method is that algebraic specification supports *implicit equations*, while trace assertion method uses *explicit equations* only.

The areas of applications for the algebraic specifications are different
than for the trace assertion method. The algebraic specification is better
suited for defining abstract data types in programming languages (as
SML, LAR CH, etc., see [15]). The trace assertion method is better suited
for specifying complex interface modules as for instance comm unication
protocols [5, 14]. A very wide bibliography concerning the Trace Assertion
Method can be found in [9].

2   Introductory Examples

W e shall consider the follo wing simple modules: Queue, Drunk Queue,
Very Drunk Queue, Concurren t Queue and Concurrent Drunk Queue. The
Queue module pro vides four access programs: *insert*(i) - which inserts an
integer i to the rear of the queue, *remove* - which takes no argument and
remo ves the first element of the queue, *front* - which takes no argument
and returns the value of the first element of the queue, *rear* - which takes
no argument and returns the value of the last element of the queue.

Since a trace specification describes only those features of a module
that are externally observable, the question arises what an atomic observation is. Following [9], we assume that an atomic observ ation is a pair
(*access_program*(*arguments*), *value_returned*), written as
access_program(*arguments*) : *value_returned*. No argument and no returned value is represented by *nil*, however we also adopt a convention
of omitting *nil*, in particular as arguments. Hence, the Queue module

has the following atomic observations, called *call-responses*: $insert(i):nil$, $remove(nil):nil$, $front(nil):a$, $rear(nil):b$, or, when nil's are omitted: $insert(i)$, $remove$, $front:a$, $rear:b$, where a is the value of first element of the queue, and b is the value of the last element in the queue.

Intuitively, a state of the queue is determined by the finite sequence of integers, the last element of the sequence represents the rear of the queue, and the first represents the beginning of the queue. Note that every sequence of properly used access programs leads to exactly one state. For instance $insert(4).insert(1).remove.insert(7)$ and $insert(1).insert(7)$ both lead to the state $\langle 1,7 \rangle$. They could be seen as equivalent and we can choose for instance the trace $insert(1).insert(7)$ as a *canonical trace* representing the state $\langle 1,7 \rangle$.

Module Drunk Queue is the same as Queue except that access program *remove* behaves differently, namely: if the length of the queue is one it removes the first element; if the length is greater than one it removes either the first element or the first *two* elements of the queue. Now the trace $insert(4).insert(1).remove.insert(7)$ may lead to two states: $\langle 1,7 \rangle$ or $\langle 7 \rangle$. However, each state is unambiguously described by an appropriate trace built from *insert* calls.

The Very Drunk Queue has two "drunk" access programs *remove*, which works identically as in case of Drunk Queue, and $insert(i)$, which enters an integer i either once or twice to the queue. In this case the trace $insert(1).insert(7)$ leads to $\langle 1,7 \rangle$, $\langle 1,1,7 \rangle$, $\langle 1,7,7 \rangle$, or $\langle 1,1,7,7 \rangle$.. The canonical traces, interpreted as traces that can unambiguously describe states, *cannot* be defined in this case. The model presented in this paper does not work for the cases like the Very Drunk Queue. Such cases have been extensively discussed in [9] (see an example of a Very Drunk Stack) and the theory presented below can easily be extended to cover them.

The Concurrent Queue has the same access programs as Queue, but *simultaneous* calls are allowed, for instance if a queue is not empty, a simultaneous call of $insert(i)$ and $remove$ is allowed, as well as a simultaneous call of $insert(i)$ and $front$, or $remove$ and $rear$. Simultaneous calls might be represented by *steps* like $\{insert(5), remove\}$, and it is more convenient to use *step-traces* to represent the observations. For instance the step-trace $\{insert(1)\}.\{insert(5), front:1\}.\{remove, rear:5\}$ leads to the state $\langle 5 \rangle$.

The Concurrent Drunk Queue has "drunk" *remove* and allows simultaneous calls.

3 The Model

3.1 Type of Concurrency

W e assume that executions (observations) of concurrent behaviours can fully be modeled by *step-sequences* (or, equivalently stratified posets). This means we assume simultaneity is observable and, when restricted to single concurrent history, it is also transitive. We also assume the a possibility of simultaneous execution of a and b implies a possibility of execution in the order a followed by b, and in the order b followed by a (see for instance [6] for discussion of various models of concurrency). We are fully aware of the restrictions imposed by the model we have chosen. Its basic advantage is simplicit y, and yet ability to model a wide spectrum of systems. W e hope the simplicit y of the model will help to adopt the model quickly by the current industrial users of the Trace Assertion Method.

3.2 Alphabet

What formally constitutes an alphabet from whic h the traces are built?

Let f be the name of an access program and let $input(f)$ and $output(f)$ be the sets of possible argumen t and result values. The *signature* $sig(f)$ is the triple:

$$sig(f) = (f, input(f), output(f)).$$

W e assume that neither $input(f)$ nor $output(f)$ are empty by having $nil \in input(f)$ and $nil \in output(f)$ as default. For example:

$sig(insert) = (insert, integer, \{nil\}), sig(remove) = (remove, \{nil\}, \{nil\}),$
$sig(front) = (front, \{nil\}, integer), sig(rear) = (rear, \{nil\}, integer).$

For a finite set E of access program names, the *signature* $sig(E)$ is the set of all signatures of $f \in E$:

$$sig(E) = \{sig(f) \mid f \in E\}.$$

Given E, the *call-response alphabet* Δ_E is the set of all possible triples, written $f(x){:}g$ of access program names, argumen ts, and return values:

$$\Delta_E = \{f(x){:}g \mid f \in E, x \in input(f), y \in output(f)\}.$$

W e adopt the convention of omitting nil in signatures. For example, for the queue modules we have $E = \{insert, remove, front, rear\}$ and:

$\Delta_E = \{insert(i) \mid i \in integer\} \cup \{front{:}i \mid i \in integer\} \cup$
$\{rear{:}i \mid i \in integer\} \cup \{remove\}.$

For a given set E of access program names, w e also define the *call alphabet* Σ_E and the *response alphabet* $\mathcal{O}_E$:

$$\Sigma_E = \{f(x) \mid f \in E, x \in input(f)\},$$
$$\mathcal{O}_E = \{d \mid \exists f \in E. \, d \in output(f)\}.$$

Note that the sequences and step-sequences of call-response event occurrences are what is really observed.

3.3 Trace Assertion Specification

For every set X, let $\mathcal{S}(X) = \{A \mid \emptyset \neq A \subseteq X \wedge A \text{ is finite}\}$. Elements of $\mathcal{S}(X)$ will be called *steps*, while elements of $\mathcal{S}(X)^*$ are called *step-sequences*. For instance, if $X = \{a, b, c\}$, then $\{a, b\}.\{b\}.\{a, b, c\} \in \mathcal{S}(X)^*$ is a step-sequence. Traditionally λ denotes the empty step-sequence. If is a set of call a *step-trace* is $X = \Delta_E$ for some E.

For every set X, let $Rel(X) = \{R \mid R \subseteq X \times X\}$, and for every symmetric $R \in Rel(X)$, let $cliques(R) \subseteq \mathcal{S}(X)$, the set of all cliques of R, be the set defined as follows: for every $x \in X$, $\{x\} \in cliques(R)$ and for every finite $A = \{x_1, ..., x_k\} \subseteq X$, $A \in cliques(C)$ iff $(x_i, x_j) \in R$ for $i \neq j$.

In principle a Trace Assertion Specification is an automaton with call-response events as an alphabet (might be infinite), and some sequences of call-response events (traces) as states (again might be infinite). However, the automaton is *finitely defined*, in the sense that the number of *explicit* equations that define elements of alphabet, states and the transition function is finite. For practical applications is is also important that the number of these equations is small and they are relatively simple. The fact that the expressions are explicit (as oppose to algebraic specification where implicit equations are more natural) is extremely important from the application viewpoint, even though it is not very significant fact as far as the theory is concerned (see [9] for details).

Formally a *Concurrent Full Trace Assertion Specification* is a tuple:

$$CFTA = (sig(E), \mathcal{C}, \delta, \delta_c, \mathcal{K}, enabled, t_0),$$

where: • E is the set of names of *system calls*, $|E| < \infty$,
• $sig(E)$ is the *signature* defined by E,
• $\mathcal{C} \subseteq \mathcal{S}(\Delta_E)^*$ is the set of *canonical step-traces* (state descriptors),
• $\delta : \mathcal{C} \times \Delta_E \to 2^{\mathcal{C}}$ is the *sequential transition function*, and $\delta^* : \mathcal{C} \times \Delta_E^* \to 2^{\mathcal{C}}$ is a standard extension of δ onto Δ_E^* (see [4]). In general an automaton that is a frame for CTA is non-deterministic, so the range of δ is defined as $2^{\mathcal{C}}$, see [9],
• $\delta_c : \mathcal{C} \times \mathcal{S}(\Delta_E) \to 2^{\mathcal{C}}$ is the *concurrent transition function*, and $\delta_c^* : \mathcal{C} \times \mathcal{S}(\Delta_E)^* \to 2^{\mathcal{C}}$ is a standard extension of δ onto $\mathcal{S}(\Delta_E)^*$,
• $\mathcal{K} \subseteq \mathcal{C} \times \Sigma_E$ is a *competence* set, if $(c, \alpha) \notin \mathcal{K}$, then applying the system call α at the state c is an erroneous/exceptional behaviour, like for

instance the *remove* call at empty queue,

• *enabled* : $\mathcal{C} \to 2^{\mathcal{S}(\Delta_E)}$ is the mapping that defines concurrency; it states what *steps* are *enabled* at each state (canonical trace),

• $t_0 \in \mathcal{C}$ is the *initial* (canonical) *state*,

and the following conditions are satisfied:

1. for all $c \in \mathcal{C}$, $\delta_c^*(t_0, c) = \{c\}$,
2. for all $c \in \mathcal{C}$, and all $S_1, S_2 \in \mathcal{S}(\Delta_E)$,
   $S_1 \subseteq S_2 \in enabled(c) \Rightarrow S_1 \in enebled(c)$,
3. for all $c \in \mathcal{C}$, if $\{\alpha, \beta\} \in enabled(c)$, then $\delta^*(c, \alpha.\beta) = \delta^*(c, \beta.\alpha)$,
4. for all $c \in \mathcal{C}$, and all $A \in \mathcal{S}(\Delta_E)$, if $A = \{\alpha_1, ..., \alpha_k\} \in enabled(c)$ then
   $\delta_c(c, A) = \delta^*(c, \alpha_1. \ ... \ .\alpha_k)$, otherwise $\delta_c(c, A) = \emptyset$,
5. for all $c \in \mathcal{C}$ and all $a{:}d \in \Delta_E$, if there exists $A \in enabled(c)$ such that
   $a{:}d \in A$ and $|A| \geq 2$ then $(c, a) \in \mathcal{K}$.
6. for all $c \in \mathcal{C}$ and all $a \in \Sigma_E$ there exists $d \in \mathcal{O}_E$ such that $\delta(c, a{:}d) \neq \emptyset$,
7. for all $c \in \mathcal{C}$ and all $\alpha \in \Delta_E$, $\delta(c, \alpha) \neq \emptyset \iff \{\alpha\} \in enabled(c)$.

The condition (1) guarantees that the states are correctly and uniquely defined by canonical traces. The second condition says that every non-empty subset of an enabled step is also an enabled step at the given state c. This means we *do not* enforce maximal concurrency (see [7]). The condition (3) enforces the rule that simultaneous executions of $\{\alpha, \beta\}$ implies a possibility of execution in the order α followed by β, and in the order β followed by α. The condition (4) defines the concurrent transition δ_c by the sequential transition δ. As a matter of fact, the concurrent transition function δ_c is redundant, since it is fully described by δ and *enabled*, however it makes the theoretical considerations and definition easier and more readable. However in the concrete examples it is usually omitted (see an example in Figure 1). The fifth condition states that concurrent activity is restricted to normal non-erroneous behaviour. Any exceptional activity must be sequential. This follows from the suggestions of practitioners who recommend not to mix concurrency with erroneous behaviour, since the results might become difficult to handle. The condition (6) is based on the observation that we cannot practically forbid to use system calls "illegally" (there is always a possibility that somebody will try to apply *remove* to empty queue), so the specification should be able to handle such cases. The last condition states that δ and *enabled* do not contradict.

The functions δ and δ_c can be decomposed into δ^N, δ^{err}, and δ_c^N, δ_c^{err}, as follows. For all $c \in C$ and all $a{:}d \in \Delta_E$, we have:

$$\delta^N(c, a{:}d) = \begin{cases} \delta(c, a{:}d) & (c, a) \in \mathcal{K} \\ \emptyset & (c, a) \notin \mathcal{K} \end{cases} \text{ and } \delta^{err}(c, a{:}d) = \begin{cases} \emptyset & (c, a) \in \mathcal{K} \\ \delta(c, a{:}d) & (c, a) \notin \mathcal{K} \end{cases}.$$

The conditions (4) and (5) guarantee that
$\delta_c^N(c, \{\alpha_1, ..., \alpha_k\}) = \delta^{N*}(c, \alpha_1. ... \alpha_k)$, and if $A \in enabled(c)$ and
$\delta_c^{err}(c, A) \neq \emptyset$, then $A = \{\alpha\}$ is a singleton, and $\delta_c^{err}(c, \{\alpha\}) = \delta^{err}(c, \alpha)$.

Lemma 1.

1. $\delta = \delta^N \cup \delta^{err}$ and $\delta^N \cap \delta^{err} = \emptyset$,
2. $\delta_c = \delta_c^N \cup \delta_c^{err}$ and $\delta_c^N \cap \delta_c^{err} = \emptyset$.   ∎

The functions δ^N, δ_c^N are called *normal* transition and *normal concurrent* transition functions, while the function δ^{err} is called an *exceptional* transition function. Due to the condition (5) the function δ_c^{err} is of a little use. The concurrent full trace assertion specification $CFTA$ restricted to the function δ^N is called *concurrent trace assertion specification*, denoted CTA, while $CFTA$ restricted to δ^{err} is called and *enhancement* of CTA and denoted ETA. Lemma 1 allows us to write (informally, but true), $CFTA = CTA + ETA$. For concrete examples, CTA (i.e. the functions δ^N, δ_c^N) should be *specified first*, and an enhancement should be added later. Lemma 1 and the condition (7) guarantee that such approach is sound.
The enhancement ETA is called *plain* if $\delta^{err}(c, \alpha) \neq \emptyset$ implies there are c_1 and α_1 such that $\delta^N(c, \alpha_1) \neq \emptyset$ and $\delta^N(c_1, \alpha) \neq \emptyset$. Non-plain enhancement means that there are some special error recovery states and some separate error recovery procedure ([9]). Our example in Figure 1 has a plain enhancement.
We say that $CFTA$ is *deterministic* iff for all $c \in C$ and all $\alpha \in \Delta_E$, $|\delta(c, \alpha)| \leq 1$. Note that this implies $|\delta_c(c, A)| \leq 1$, for every state c and step A. From the examples introduced in Section 2, Queue and Concurrent Queue are deterministic, the remaining are not-deterministic. The concept of determinism defined above corresponds to the concept of determinism used in automata theory (see [9]).
For a given $CFTA$, let $sim : C \to Rel(\{\alpha \mid \{\alpha\} \in enabled(c)\})$ be defined as follows: $(\alpha, \beta) \in sim(c) \iff \{\alpha, \beta\} \in enabled(c)$.
For every c, $sim(c)$ defines *simultaneity* relation at the state c.

Lemma 2. $A \in enabled(c) \iff A \in cliques(sim(c))$.   ∎

From the above lemma and the condition (3) of the $CFTA$ definition, it follows that we may equivalently define $CFTA$ as $CFTA = (sig(E), \mathcal{C}, \delta, \mathcal{K}, sim, t_0)$ with appropriate changes of the constraints (1) - (6). No definition is better than the other. For the theory the definition with δ_c and $enabled$ seems to be better (see [10]), for specifying the concrete examples δ_c is almost never explicitly specified, for some cases using $enabled$ is better, for others sim is better (compare [10]).

The constraints (1) - (6) have to be proven for every concrete example. They are an essential part of a specification, the part which is frequently called an *obligation proof* in software engineering. If the specification is thoroughly thought of, those proofs are usually easy, but they may be labour consuming, if the specification is complex itself. The use of some automatic theorem provers as PVS or IMPS is highly recommended [10].

3.4   Specification Format

To be useful in practice, the trace assertion technique must provide some reliable, readable and easy to use specification format. This issue is completely irrelevant from the theoretical view-point, but very important if the technique is going to be used outside academia. The details of a specification format are to be found in [9, 10]. It uses heavily Tabular Expressions (see [8, 13] for more details), for simple cases it appears to be self-explained (see Figure 1).

The technique described above (slightly changed to fit to one page, usually tabular expressions are also used to describe $enabled$) is illustrated in Figure 1, which presents a Full Trace Assertion Specification for a Concurrent Queue. The symbol "%" in the definition of δ indicates the parts that define δ^{err}, i.e. exceptional behaviour. Figure 1 provides only the first part of a specification. The second one, "the obligation proof" is not provided. It is relatively easy, but not so short so it is omitted. An interested reader is referred to [10].

4   Final Comment

We extend the theory of [9] by allowing simple concurrency. The work can be further extended in several aspects. This is not a general model of concurrency since simultaneity here is transitive. For more complex non-sequential models a possible delay between a call and its response

Syntax of Access Programs

Name	Argument	Value	Action-response Form	Full Action-response Form
$Front$		$integer$	$Front{:}d$	$Front{:}d$
$Rear$		$integer$	$Rear{:}d$	$Rear{:}d$
$Insert$	$integer$		$Insert(a)$	$Insert(a){:}nil$
$Remove$			$Remove$	$Remove{:}nil$

Canonical Step-traces

t is *canonical* $\iff (t = \lambda \lor t = \{Insert(a_1)\}.\ldots.\{Insert(a_k)\})$, where $1 \leq k \leq size$.
$t_0 = \lambda$, i.e. empty step-sequence.

Enabled

if $c = \lambda$ then $enabled(c) = \{ \{Insert(x)\} \mid x$ is an integer$\}$.
if $c = \{Insert(a)\}.t_1.\{Insert(b)\} \land |c| = size$ then
$\qquad enabled(c) = \{ \{Remove\}, \{Front{:}a\}, \{Rear{:}b\}, \{Rear{:}b, Remove\} \}$.
if $c = \{Insert(a)\}$ then
$\quad enabled(c) = \{\{Remove\}, \{Front{:}a\}, \{Rear{:}a\}\} \cup \{\{Insert(x)\} \mid x$ is an integer$\} \cup$
$\quad \{\{Insert(x), Remove\} \mid x$ is an integer$\} \cup \{\{Insert(x), Front{:}a\} \mid x$ is an integer$\}$.
if $c = \{Insert(a)\}.t_1.\{Insert(b)\} \land |c| < size$ then
$\quad enabled(c) =$
$\quad \{ \{Remove\}, \{Front{:}a\}, \{Rear{:}b\}, \{Rear{:}b, Remove\}, \{Front{:}a, Rear{:}b\} \} \cup$
$\quad \{ \{Insert(x)\} \mid x$ is an integer$\} \cup \{ \{Insert(x), Remove\} \mid x$ is an integer$\} \cup$
$\quad \{ \{Insert(x), Front{:}a\} \mid x$ is an integer$\}$.

Trace Assertions

$\delta(t, \{Front{:}d\}) =$

Condition	Trace Patterns	Result
	$t = \{Insert(d)\}.t_1$	$\{t\}$
% $d = nil$	$t = \varepsilon$	$\{\lambda\}$

$\delta(t, \{Rear{:}d\}) =$

Condition	Trace Patterns	Result
	$t = t_1.\{Insert(d)\}$	$\{t\}$
% $d = nil$	$t = \varepsilon$	$\{\lambda\}$

$\delta(t, \{Insert(a)\}) =$

Condition	Result
$length(t) < size$	$\{ t.\{Insert(a)\} \}$
% $length(t) = size$	$\{t\}$

$\delta(t, \{Remove\}) =$

Trace Patterns	Result
$t = \{Insert(b)\}.t_1$	$\{t_1\}$
% $\quad t = \varepsilon$	$\{\lambda\}$

Dictionary

$size$: the size of the queue
$length(t)$: the length of the trace t

Fig. 1. Full Trace Assertion Specification for Concurrent Bounded Queue Module

must be modeled. It would be interesting to see how the model looks like for the general causality model or "true-concurrency" model. Also the concept of refinement is not considered here. Extending this model to multi-object case also seems to be a challenge. One may have noticed that proving that the concurrency law is obeyed for more complex module is very labor consuming but usually rather easy. A tool that can do it in an automatic way would be a great help.

References

1. W. Bartussek, D.L. Parnas, Using Assertions About Traces to Write Abstract Specifications for Software Modules, Proc. 2nd Conf. on European Cooperation in Informatics, *Lecture Notes in Computer Science 65*, Springer 1978, pp.211-236.
2. P. M. Cohn, *Universal Algebra*, D. Reidel 1981.
3. H. Ehrig, B. Mahr, *Fundamentals of Algebraic Specification 1*, Springer-Verlag, 1985.
4. S. Eilenberg, *Automata, Languages and Machines*, vol A, Academic Press, 1974.
5. D.M. Hoffman, The Trace Specification of Communication Protocols, *IEEE Transactions on Computers 34*, 12 (1985), pp.1102-1113.
6. R. Janicki, M. Koutny, Structure of Concurrency, *Theoretical Computer Science* 112 (1993), 5-52.
7. R. Janicki, P. E. Lauer, M. Koutny, R. Devillers, Concurrent and Maximally Concurrent Evolution of Non-Sequential Systems, *Theoretical Computer Science* 43 (1986), 213-238.
8. R. Janicki, D. L. Parnas, J. Zucker, Tabular Representations in Relational Documents, in C. Brink, G. Schmidt (Eds.): *Relational Methods in Computer Science*, Springer-Verlag, 1997.
9. R. Janicki, E.Sekerinski, Foundations of the Trace Assertion Method of Module Interface Specification, *IEEE Transactions on Software Engineering*, to appear, also SERG Report No. 376, McMaster University, Hamilton, Canada, July 1999, available at http://www.crl.mcmaster.ca/SERG/serg.publications.html
10. Y. Liu, The Trace Assertion Method of Module Interface Specification with Concurrency, Master Thesis, McMaster University, Hamilton, Canada, 1999, also SERG Report No. 385, McMaster University, Hamilton, Canada, May 2000, available at http://www.crl.mcmaster.ca/SERG/serg.publications.html
11. J. McLean, A Formal Foundations for the Abstract Specification of Software, *Journal of the ACM*, 31,3 (1984), pp.600-627.
12. D. Parnas, A Technique for Software Module Specification with Examples, *Comm. of ACM*, 15,5 (1972), pp.330-336.
13. D. Parnas, J. Madey, M. Iglewski, Precise Documentation of Well-Structured Programs, *IEEE Transactions on Software Engineering*, 20, 12 (1994), pp.948-976.
14. A. J. van Schouwen, On the road to practical module interface specification, A lecture presented at McMaster Workshop on Tools for Tabular Notations, McMaster University, Hamilton, Ontario, Canada 1996.
15. M. Wirsing, Algebraic Specification, in J. van Leeuwen (ed.): *Handbook of Theoretical Computer Science*, Vol 2., Elsevier Science Publ., 1990, pp.675-788.

A Proposal for Including Behavior in the Process of Object Similarity Assessment with Examples from Artificial Life

Kamran Karimi [1], Julia A. Johnson [2] and Howard J. Hamilton [1]

[1] Department of Computer Science
University of Regina
Regina, Saskatchewan
Canada S4S 0A2
{karimi,hamilton}@cs.uregina.ca

[2] Département de mathématiques et d'informatique
Université Laurentienne
Sudbury, Ontario
Canada P3E 2C6
julia@cs.laurentian.ca

Abstract. The similarity assessment process often involves measuring the similarity of objects X and Y in terms of the similarity of corresponding constituents of X and Y, possibly in a recursive manner. This approach is not useful when the verbatim value of the data is of less interest than what they can potentially "do," or where the objects of interest have incomparable representations. We consider the possibility that objects can have behavior independent of their representation, and so two objects can look similar, but behave differently, or look quite different and behave the same. This is of practical use in fields such as Artificial Life and Automatic Code Generation, where behavior is considered the ultimate determining factor. It is also useful when comparing objects that are represented in different forms and are not directly comparable. We propose to map behavior into data values as a preprocessing step to Rough Set methods. These data values are then treated as normal attributes in the similarity assessment process.

1   Introduction

Data is usually considered simply the raw material to be processed. In this view, one receives the data, possibly from a database, looks at it, maybe modifies it and then returns it to a database if needed. In this view, there is a clear separation between the code that processes the data, and the data that is being processed. However, there are problems when "what the data can do," and not "the way they look," is of real interest.

In this paper we propose allowing objects to have behavior, and show that this opens the door for Rough Set [7] techniques to be applied to fields such as Artificial Life [5]. The method suggested in the paper involves representing behavior as a single data value or a set of data values for input to standard Rough Set methods for classification and decision making. These are then treated as if they are constituent parts of the objects. This preprocessing step allows us to retain compatibility with traditional Rough Set methods.

W. Ziarko and Y. Yao (Eds.): RSCTC 2000, LNAI 2005, pp. 642–646, 2001.

Assessing the similarity of two data sets, also commonly referred to as objects, without necessarily having any thing to do with the Object Orientation principles, is an important and common operation. Classification of objects is one example of the usefulness of measuring similarity. A concept is expressed by a set of objects that incorporate that concept. In the presence of uncertainty, Rough Set bounds this target set H by two sets, a lower approximation $\underline{H}$, and an upper approximation $\overline{H}$ such that we have $\underline{H} \subseteq H \subseteq \overline{H}$. The Rough Set theory has found many practical applications. Similarity measures include graph measures of Semantic Relatedness [3] for disambiguation of natural language expressions, Correlation measures [1] for calculating the relatedness between word pairs, and Information Theoretic techniques [2] for measuring object associations.

When using Rough Sets to assess the similarity of two objects, researchers usually focus on the parts that make up those objects. Let $x = o(x_1, x_2, ..., x_k)$ denote an object x constructed from sub-objects $x_1, x_2, ..., x_k$. The usual approach involves computing a function f to measure the similarity of two objects x and y in terms of the similarity of their components, i.e., similarity$(x, y) = f($similarity$(x_1, y_1), ...,$ similarity$(x_2, y_2), ...,$ similarity$(x_k, y_k))$. This is a static approach. Because each part of an object can have behavior (like a function, which has a source code, and also a run-time behavior), we wish to include this behavior in the process too, essentially along the lines of Object Oriented programming. This is a dynamic approach.

The rest of this paper is organized as follows. In section 2 Artificial Life is briefly introduced and the reader is told why classification methods that use the verbatim values of an object are not of much use there. An example of when behavior is of the ultimate importance is provided. Section 3 presents a guideline to measure behavior and translate it to a single value, or a set of values, which can then be used by ordinary Rough Set techniques, thus retaining compatibility with existing methods and application software. Section 4 concludes the paper.

2   An Artificial Life Problem

Artificial Life is concerned with the study of systems that behave as if they are alive. In most cases the systems are pieces of software, usually called *creatures*, that *live* in an artificial environment. Each creature can be considered a plan that when executed, affects its environment. A simulator can generate new creatures from scratch randomly, or by applying genetic operations of mutation and crossover to existing creatures. Rules of the environment are enforced on the creatures, and pre-defined fitness measures are used as a guide in creating the next generation. Thousands of generations are tried, and the creatures usually evolve to display certain characteristics that help them survive by conforming to the rules of the environment as much as possible. The rules determine the physics of the artificial world, and dictate how "normal" the creatures will behave when compared to the real world. Considering the random elements present in this process, it is no wonder that *spontaneous* emergence of behavior is one of the key characteristics observed in an Artificial Life environment. It is usually very hard to predict how the creatures will evolve. One usual behavior is that *herds* of creatures show up. Members of each herd have a lot of resemblance to each other, and differ substantially from members of other herds. Artificial Life techniques have been used to *breed* programs that perform useful functions [4].

In this paper, *behavior* is defined as the side effects of *interpreting* data. This interpretation is domain dependent, and can for example be the same as the execution of code produced by an automatic code generator. The definition of behavior can be generalized to include static data too. If there is an easily detectable relationship between the representational format of an object and the effects of its interpretation in the environment, then there is no need to interpret the object. If the data are not interpreted, then behavior is defined as their verbatim values.

If the simulated environment is non-trivial then there is no direct correspondence between the source code of a creature and its behavior. The reason is that in a non-trivial system, on one hand there is more than one way to cause the same effects, and on the other hand executing seemingly similar, but not identical pieces of code can have very different results. In general, a behavior measurement procedure might have to be told to look for specific patterns of interest in certain locations of the system. This problem is greatly reduced when moving to object oriented programming, where global data is more contained and manageable, and completely disappears in a functional programming environment, where global state does not exist.

Comparing behavior is of paramount importance in fields like Artificial Life. One concrete problem is the classification of creatures produced automatically. Because there are thousands of creatures at any time in an Artificial Life simulator, and their behavior may change from one generation to the next, it is very difficult to do the classification manually. One example problem is the classification of the creatures into hunters and non-hunters. Consider an imaginary artificial world, where plant food is created randomly by the simulator. The simulator ages the creatures at regular intervals, which makes them weaker. When they have passed a threshold of weakness, they die and are converted into plants. Creatures all start as peaceful vegetarians, but after a while some may begin developing the traits associated with hunting, like attacking others. This results in the attacked creatures becoming weaker, and thus dying sooner. Such behavior could develop simply because it may be rewarding for the creatures that display them: As the number of competing creatures reduces, there is more food to eat. After a while, they may learn that it is a good idea to hang around weak animals. Still another trait would be to attack weaker animals and then wait, which would probably be the most rewarding behavior.

In this example, there is no explicit hunting behavior, because all the creatures do is eat plants. However, their behavior in the last case may very well be considered to closely resemble that of hunters. Behaviors like attacking other creatures (which makes them weak), waiting near old creature (because they will die in a short time), and moving fast (to chase other creatures) are some of the condition attributes that can be used to help in the classification of the creatures into hunters and non-hunters. Another attribute, creature size, is of doubtful value, but can be considered if the expert looking at the simulator thinks there is a correlation between it and hunting. Using a Rough Set paradigm, one can come up with Table 1 for the creatures of this artificial world. Note that there are no variables anywhere in the simulator to tell us if a creature attacks others, or waits near old creatures, or moves fast, and the values should be extracted from the creatures' behavior.

| # | Condition Attributes | | | | Decision Attribute |
	Attacks?	Waits near old creatures?	Moves fast?	Creature size	Hunter?
1	yes	yes	yes	small	yes
2	yes	yes	no	small	yes
3	no	no	yes	big	no
4	no	yes	yes	small	no
5	yes	no	yes	big	yes
6	no	no	no	small	no
7	no	no	no	small	no
8	no	yes	yes	small	yes

Table 1. Condition attributes used to determine if the given animals are hunters

Table 1 uses intuitive notions about how a hunter should behave. For example, it is clear that animals 1 and 2 are smart hunters. They attack others, and wait near old (weak) animals, which increases their chance of finding food in a short time, as old animals die sooner. This does not necessarily mean that they hang around the same creature that they attacked, though. Animal number 5, on the other hand, is a stupid hunter because it does attack others, but being a fast mover, does not wait to use the results of its efforts. Animal 4 can be compared to a vulture. It does not attack others, but does wait near old animals. Animal 8 also acts like a vulture, but is classified as a hunter, which is counter-intuitive. This could be the result of an error on the part of the expert who did the classification.

The above table gives the following indiscernability classes: $\{1\}$, $\{2\}$, $\{3\}$, $\{4, 8\}$, $\{5\}$, $\{6, 7\}$. Following Standard Rough Set techniques gives us $\underline{H} = \{1, 2, 3\}$ and $\bar{H} = \{1, 2, 4, 5, 8\}$. If we change the value of creature size for creature 8 from big to small, then we get the following indiscernability classes: $\{1\}$, $\{2\}$, $\{3\}$, $\{4\}$, $\{5\}$, $\{6, 7\}$, $\{8\}$. Deleting the size attribute gives the original indiscernability classes. This hints that creature size is redundant. This is also intuitive, as in nature the physical size does not determine if an animal hunts others.

3   Mapping Behavior

The usual way of comparing two objects is to directly compare the values of their corresponding parts, and then use some statistical or heuristic function to come up with a measure of similarity. This method is used in many applications. Using a compatible way to measure behavior will enable us to continue to use the same programs and methods. This can be achieved by the introduction of a mapping function which takes interpretable data, and produces a value or a set of values. These can then be used in the similarity assessment procedure. In general the result of interpreting the data may depend on the global state, and the interpretation may change this state. More formally, we define a function f such that:

- $f(\varphi, \alpha) = \{\varphi, \alpha\}$ where α is a data structure that is not interpreted and φ is the global state. The global state does not change and the return value is α itself.
- $f(\varphi, \beta) = \{\varphi', \sigma\}$ where β is interpretable data, and σ is a measure of changes resulting from interpreting β. φ is the starting state and φ' is the resulting state.

This ensures that the same method can be applied to objects with and without inherent behavior. It can also be applied when an object has behavior that is to be ignored for some reason, in which case β is treated like α. Mapping behavior back to the form of data values (α or σ) makes it unnecessary to introduce new terms and techniques, and allows us to retain compatibility with existing methods.

f is domain dependent and should be defined by the experts of the domain. An example for automatic code generation in the functional programming paradigm is that f is simply the result of executing the function β. For a neural network, f provides the input and allows the network to produce its output.

In Table 1, the value of a condition attribute such as "Attacks other creatures" is obtained by a function $f_1(\varphi_1, \beta)$, with φ_1 being the global state of the simulator at the time the function starts execution, and β being the representation of the creature. The result is a possible change in the simulator (leading to the global state φ_2), plus a return value from the set {yes, no}. Similar functions (f_2, f_3, ...) should be used to get the other condition attributes.

4   Conclusion

We have suggested taking a broader look at data in the similarity assessment process. We propose allowing data to have behavior, and using this behavior to measure the similarity of two objects. The main point to consider is that comparing parts of an object based solely on their data values may not reveal the complete picture. When the behavior of an object is more important than its representational format, then the data should be interpreted and the results should be included in the similarity assessment process. The conceptually simple technique of expressing behavior in terms of the results of its execution allows for the easy addition of behavior to existing similarity assessment systems. This makes it possible for standard, well-understood methods to be applied to domains such as Artificial Life, where systematic ways of comparison and classification are lacking.

References

1. Dent, M. and Mercer, R. E., A Comparison of Word Relatedness Measures, *Pacific Association for Computational Linguistics*, pp. 270-275, 1999.
2. Jiang J. J. and Conrath, D. W., From Object Comparison to Sematic Similarity, *Pacific Association for Computational Linguistics*, pp. 256-264, 1999.
3. Johnson. J. A., Semantic Relatedness, *Computers & Mathematics with Applications*, pp. 51-63, 1995.
4. Koza, J. R., Genetic Programming: A Paradigm for Genetically Breeding Populations of Computer Programs to Solve Problems, *Stanford University Computer Science Department Technical Report STAN-CS-90-1314*, p.117, June 1990.
5. Levy, S., Artificial Life: A Quest for a New Creation, Pantheon Books, 1992.
6. Marven, C. and Ewers, G., *A Simple Approach to Digital Signal processing*, John Wiley & Sons, 1996.
7. Pawlak, Z., Rough Sets: Theoretical Aspects of Reasoning About Data, Kluwer Academic Publications, 1991.

How Can Help an Inductive Approach in the Resolution of a Problematic α Problem ?

Philippe Levecq and Nadine Meskens

Catholic University of Mons, Chaussée de Binche 151, 7000 Mons – Belgium
philippe.levecq@fucam.ac.be

Abstract. Some organizations are commissioned to boost every kind of innovation in favoring collaboration between small businesses and scientific circles. Our research aims at developing a decision-aid tool to help intermediate organizations in their search for innovative enterprises and at determining if these enterprises are receptive to collaboration with a university. This helping tool consists of a set of decision rules thanks to which the enterprises are selected. This set of rules was established with the rough set method. The problem we have to face comes under the problematic P.α and the originality of the paper is that we will show the impact of the choice of decision rules on the type I error and thus on the percentage of objects incorrectly classified.

1 Introduction

In the context of the present international competition, the innovative nature of an enterprise is often a determining and not insignificant advantage. Unfortunately this innovative will is most of the time checked by a lack of human, material as well as financial means. Scientific circles – and more particularly universities – generally have these means. They are moreover willing to put them at the enterprise disposal. The function of some intermediate organizations is initiating and developing collaborations between these two partners. These intermediate organizations cannot materially get in touch with each enterprise and it is important for them to be able to contact the enterprises that are the most likely to develop any kind of collaboration. Our research focuses on segmenting the market of firms to contact. From the characteristics of an enterprise, we have to be able to determine if this enterprise will be receptive to collaboration. In order to carry out the segmentation of the enterprises, a decision-aid tool has been developed. This helping tool consists of a set of decision rules thanks to which the best enterprises are selected, namely ones the most able to develop a fruitful collaboration with a university. This set of rules was established with the rough set method. The application is given in the second part. Section 3 constitutes the originality of this paper. Considering our problem as a P. α problematic problem, we show the impact of the choice of decision rules on the type I error and thus on the percentage of objects incorrectly classified.

W. Ziarko and Y. Yao (Eds.): RSCTC 2000, LNAI 2005, pp. 647–651, 2001.

2   Application of the Inductive Approach

In this section, we analyze the problem of the selection of enterprises using the rough set method (explanation about this method can be find in [1,2]). This study was done using ROSE software [3]. In our case study, the goal of the rough set method is to discover relationships between objects from the information system and pre-defined sorting rules. These relations will express in a "if... then..." decision rules form the expertise of the decision maker on the basis of sorting examples. The enterprises of our sample have thus been classified into 2 groups in order to constitute the decision attribute. Classification of new enterprises will be done on the basis of the decision rules generated.

The first step of the analysis consists in the creation of the decision table. We present hereafter the condition attributes taken into account in order to explain the collaboration potential of an enterprise. Our decision table includes 31 condition attributes grouped into 4 different themes :
- Geographical situation : localization (A1);
- Branch of industry : branch of industry (A2), number of enterprises per sector (A3), evolution of the number of enterprises per sector (A4), Number of workers per sector (A5), evolution of the number of workers per sector (A6);
- Size : type of enterprise (A7), total of the balance sheet (A8), evolution of the total of the balance sheet (A9), average staff (A10),evolution of the average staff (A11);
- Financial situation : corporate performance (profit or loss) (A12), Evolution of the corporate performance (profit or loss) (A13), added value (A14), evolution of the added value (A15), cash-flow (A16), evolution of the cash-flow (A17), stockholder's equity (A18), evolution of stockholder's equity (A19), working capital (A20), evolution of the working capital (A21), added value per worker (A22), evolution of the added value per worker (A23), financial costs/VA (A24), evolution of the financial costs/VA (A25), output of long-lasting resources (A26), evolution of the output of long-lasting resources (A27), liquidity in the strict sense of the word (A28), evolution of the liquidity in the strict sense of the word (A29), own capital/total of liabilities (A30), evolution of stockholder's equity /total of liabilities (A31).

From the available database, 200 firms were sorted into two groups according to whether they could be considered, a posteriori, as enterprises worth contacting or not. It should be noticed that the research is based on small (and medium sized) businesses situated in the province of Hainaut in Belgium.

The information table is constituted by a set of 200 enterprises described by 31 condition attributes corresponding to the 31 selected criteria and by one decision attribute representing the sorting group of the firm.

In order to exploit this table, it is necessary to discretizise the raw data. On the basis of the raw data and different discretization factors, we are now able to build the decision table that will be used by ROSE.

The rough set analysis constructs minimal subsets (reducts) of independent criteria ensuring the same quality of sorting as for the whole set of condition attributes. This leads us to the building of 266.088 different reducts. It was evidently impossible to test all of them to find the best one. So we chose to construct one reduct manually. We first inserted into the attributes list constituting the reduct the attribute ensuring alone the best quality of sorting. We then proceeded by successive adding of attributes in the list in order to maximize the quality of sorting of the combination of all attributes in this list until we attained a final quality of sorting equal to one. This leads to the creation of 4 reducts. Several validation tests highlighted that the following one was the best choice : {A2, A3, A15, A16, A22, A24}.

These reduct was used to generate a minimal set of rules that cover all the objects from the reduced decision table. The number of rules generated is equal to 84. You will find hereafter the interpretation of one of these rules : "IF the *added value* has gone through one positive evolution followed by a negative one, with a higher level in 1995 than in 1997, THEN, the enterprise studied can be considered as worth being contacted. One can effectively suppose that an enterprise whose products are going out of fashion or whose profit margin is lower, would want, for example, to develop new products".

The major objective of the study is to use the sorting rules discovered from the decision table to support new sorting decisions. The 84 sorting rules generated have thus to be validate. A cross-validation test was carried out. We realized 4 validation tests on our sample of 200 enterprises. Table 1 contains the average percentages of objects correctly classified or not.

Table 1. Validation tests results

| Effective belonging to : | Estimated belonging to the group of: | |
	Good prospects	Bad prospects
Good prospects	61,5% – [H]	9% – [M_2]
Bad prospects	21% – [M_1]	8,5% – [H]

Results of type H express correct classification. Results of type M express incorrect classification. M_1 represents type I error and M_2 type II error. In that case, the results are very satisfying since 70 % of firms have been correctly classified. In order to reduce the type I error, we have taken into account the particularity of the problem we have to solve which is a problematic P.α kind of problem.

4 Problematic P.α

The problem we have to face comes under problematic P.α [4]. Indeed, our objective is to choose, within a database, a non ordered subset of enterprises likely to develop a fruitful collaboration with a university. As we are only interested in the 'best' firms, we will use only the decision rules leading to the conclusion 'Good Prospects'. From the database which enables us to generate the decision rules and realize a validation test (see table 1), we can observe that 82,5% of firms have been classified as good prospects. However, we can also note that the type I error is about 21%.

It is clear that when the decision maker applies the decision rules to a new database, all the enterprises selected can not be contacted at a same time. The idea is thus to evaluate the impact of the choice of rules on the selection of enterprises and consequently to identify the influence of the choice of rules on the percentage of correctly classified enterprises.

For the initial analysis of our problem by the rough set method, the algorithm used to generated the decision rules was an algorithm inducing the minimal set of rules covering the entire database. Now, we propose to generate the rules from an algorithm inducing a satisfactory set of rules. This category of algorithms gives as a result the set of decision rules which satisfy a given a priori user's requirements. The measure of quality we choose to use is relative to the strength of the rules. We present in table 2 the main results obtained. The meaning and the interpretation of each column of the table are the following :

1. This column indicates the strength of the rules. It specifies the minimum percentage of objects from the database the rule has to cover to be generated;
2. This column points out the total number of rules generated as a function of the minimum percentage of objects to be covered by each rule. For example, if each rule has to cover at least 1% of objects from the database, the total number of decision rules generated is 726. As we could expect, we notice that the total number of rules generated decreases depending on the required strength of the rules;
3. Identically to column (2), this one gives us the number of decision rules leading to the conclusion 'Good Prospect';
4. Similarly to the way it was calculated in the previous section, this column gives us the percentages corresponding to type I error (wrong classification in the set of good prospects). For example, if during the validation tests we use only decision rules covering at least 3% of objects from the database, we can notice that 20,7% of the total number of enterprises have been incorrectly classified into the set of 'Good Prospects'. It is interesting to observe that the stronger the rules are, the less the percentage of error is high;
5. The values included in this column indicate the number of firms classified into the set of 'Good Prospects' using the rules from column (3). As expected, this number decreases continuously. The more the rules have to be strong the more the number of classified firms diminishes. This is quite normal because, using fewer decision rules, a more and more enterprises can not be classified any more, whether in the set of 'Good Prospects' or in the set of 'Bad Prospects'..
6. Finally, this column gives us the percentage of objects incorrectly classified into the set of 'Good Prospects'. For example, if we use rules covering at least 1% of the objects from the database, the percentage of incorrectly classified objects relative to the total number of objects classified in the set 'Good Prospects' (177 – see (5)), is equal to 28. All the elements of this column allow us to make an interesting observation : the stronger the generated decision rules are, the lower the percentage of incorrectly classified firms. This means that using the strongest rules, the decision maker will be able to select a subset of enterprises, smaller certainly, but increasing his chances of selecting the best ones and thus decreasing the risk of a

bad classification. Then, if we use decision rules covering 11 % of enterprises from the database, 22 firms are classified as 'Good Prospects' with no risk of error.

Table 2. Main results of the analysis

(1)	(2)	(3)	(4)	(5)	(6)
1	726	317	24.8	177	28
2	240	165	23.8	171	27
3	110	35	20.7	139	26.5
4	42	19	15.6	86	21
5	27	4	11.3	64	18
6	10	4	10.9	53	14
7	3	1	9	33	12
8	3	1	8	32	9.5
9	1	1	3.3	25	4
10	1	1	3.3	25	4
11	1	1	0	22	0

On the basis of the results obtained, one recommendation for the decision maker will be to first use the strongest decision rules in order to create a first subset of firms and to work on it until it is exhausted and then enlarge the set of rules successively in descending order of strength of the rules.

5 Conclusion

The aim of this paper was the building of a decision model constituting a decision aid tool allowing   intermediate organizations to optimize the collaboration possibilities between enterprises and universities. This model is represented by a set of decision rules allowing the selection of the enterprises most likely to develop a fruitful collaboration with a university. Decision rules were generated using rough set analysis. The rough set method was used first to find the minimal subset of attributes ensuring an optimal quality of sorting and second to generate a set of decision rules.

References

[1] Pawlak Z. : Rough Sets : Theoretical Aspects of Reasoning about Data, Kluwer Academic Publishers, Dordrecht/Boston/London, 1991.
[2] Slowinsky R. (ed.) : Intelligent Decision Support Handbook of Application and Advances of the Rough Sets Theory. Kluwer Academic Publishers, 1992.
[3] B.Predki, Sz.Wilk, : Rough Set Based Data Exploration Using ROSE System , Foundations of Intelligent Systems, Lecture Notes in Artificial Intelligence, vol. 1609, Springer-Verlag, Berlin (1999), 172-180.
[4] B. Roy : Methodologie Multicritère d'Aide à la Décision, Economica, 1985.

A Timed Petri Net Interpreter:
An Alternative Method for Simulation

Trong Wu and Shu-chiung Huang

Department of Computer Science
Southern Illinois University Edwardsville
Edwardsville, Illinois 62026-1656
twu@siue.edu

Abstract. This paper reports on the design and implementation of a timed Petri net interpreter. Currently, several Petri net simulators written in the Pascal and C languages are available. However, our approach is to use an expert system language called CLIPS to write an interpreter to execute Petri nets. The major difference between a rule-based expert system language like CLIPS and languages such as Ada, C, or Pascal is that the rules of CLIPS can be activated concurrently, while the statements of other languages are sequential. In this project, we first design a Petri net language; programs written in a Petri net language can describe Petri net behavior. Then, we will design and write an interpreter in the CLIPS language that can execute Petri net programs. The CLIPS language is a data driven language, and the interpreter can search for enabled transitions for firing. With this approach, we can avoid complicated data structures and their implementations.

1 Introduction

Simulation analysis is an effective approach to solving problems, because it can obtain the desired information without much cost or the inconvenience of manipulating the real world system. The definition of simulation given by S. V. Hoover and R. F. Perry [11] indicates, "simulation is a process of designing a mathematical or logical model of a real system and then conducting computer-based experiments with the model to describe, explain, and predict the behavior of the real system." From this definition, we see that a good model is an essential element for any simulation analysis. Simulators are frequently used to model systems that involve concurrent activities. Some simulation languages provide special tools for modeling system activities. GPSS has its own flowchart presentation, SLAM uses its network graphs, and SIMON uses activity diagrams. One important drawback of these tools is their lack of generality [26, 27].

A Petri net is a general formal modeling tool that can be used to describe and analyze the flow of information in a system. It is a powerful modeling tool particularly for the representation of asynchronous, and concurrent activities [2, 4, 7, 15, 21, 24]. By using Petri net models, one can gain several advantages over other models. The basic principle is easy to understand, the ability to extend basic Petri net

W. Ziarko and Y. Yao (Eds.): RSCTC 2000, LNAI 2005, pp. 652-660, 2001.

models is quite flexible, and they can describe the dynamic behavior of systems. The design of a Petri net interpreter using the CLIPS language [10] is a new approach to simulation. Several Petri net simulators, written in Pascal and C languages, already exist [4, 8, 14, 22] so a natural question to ask is, "why another one?" The answer and the reason for the development of the interpreter, is that none of the existing simulators can exactly model Petri nets. The interpreter developed here comes closer to this goal.

A Petri net is an asychronized, concurrent, and token driven system.   Since a procedural programming language is statement-oriented languages and executed sequentially, any Petri net simulator written in a procedural programming language hardly simulates a system precisely.   Our approach is totally different.   In this research, the CLIPS programming language was chosen to implement the interpreter because CLIPS is one of the production based programming languages consisting of rules and a database against which those rules are compared. The CLIPS language is a concurrent, and data driven language that can describe Petri net behavior. The use of the CLIPS language not only eases the design of the interpreter, but also helps in design of the Petri net language (PNL).

In this paper, we will describe the design of a PNL which can represent Petri net models.   Then, we develop an interpreter that will be used to execute the programs written in the PNL.   The results from the output of the interpreter will supply information on the system performance. This approach will provide an alternative for simulation. The concept of using a Petri net interpreter for simulation is given below:

Fig. 1.1 Petri net interpreter, an alternative method of simulation.

2 Petri Nets

The theory of Petri nets was first developed by Carl Adam Petri in 1962 to model those systems with interacting concurrent components [23].   Due to its natural representation, Petri nets have been adopted by a wide collection of systems for modeling.   For more detailed descriptions see [3, 5, 16, 18, 20, 25]. The original Petri net only allowed a single arc between places and transactions.   A transition can fire if the input places contain tokens [23].   The theory of Petri nets has been further developed by different researchers with various motivations; therefore, a number of variant theories about Petri nets have appeared and continue to grow.   In this paper, we adopt a formal definition of Petri net is given by Peterson and his rules [23].

2.1   Petri Net Graphs

A Petri net graph is a directed symbolic graph that contains places, tokens, transitions, and arcs [23].   A place is denoted by a circle.   A dot in a circle means this place contains a token. A Petri net with tokens is called a marked Petri net. The occurrence of tokens indicates the state of the net. A transition is denoted by a bar which controls the flow of tokens between places. These places and transitions are connected by directed arcs. If an arc leads from a place to a transition, then this place is an input of the transition. If an arc leads to a place from a transition, then this place is an output of the transition. A Petri net executes by firing transitions that remove tokens from its input places and deposits tokens to its output places.

2.2   Time and Timed Petri Nets

The two main techniques for handling time within Petri nets [9, 12, 16, 19] are: Ramchandani's timed Petri nets or Holliday's Generalized timed Petri nets [12] and Merlin's time Petri nets [1]. A timed Petri net represents time by attaching a finite duration to transitions that are called deterministic firing times. A time Petri net uses two real numbers to form an interval, minimum and maximum time, associated with each transition. Time can be either attached with transitions or places or both. This project will implement timed Petri nets. If all inputs of the transition contain sufficient tokens, then this transition will remove the input tokens immediately and deposit the output tokens after the processing time.

3   Design of a Petri Net Language

In this research project, we have developed a Petri net language (PNL). Programs written in the PNL will have a one to one correspondence with Petri net graphs. Once we have defined the Petri net graph for a given system, we can convert the graph into a program. A program written in PNL consists of two files, a structure file and an initial data file. The structure file expresses transition structure within the Petri net, while the initial data file indicates the initial marking in that net. We can evaluate the performance for distinct initial markings in different initial files. The Syntax of PNL consists of two files, the structure file and the initial data file as shown below:

1   Structure file:

\<structure file\>	::= \<transition list\>
\<transition list\>	::= \<transit item\>|\<transit item\> \<transition list\>
\<transit item\>	::= TRANSITION \<transition number\> TIME \<time\> INPUT \<input\> OUTPUT \<output\>
\<input\>	::= \<place list\>
\<output\>	::= \<place list\>
\<place list\>	::= \<place item\>| \<place item\> \<place list\>
\<place item\>	::= \<place number\>
\<place number	::= [POSITIVE INTEGER]
\<transition number\>	::=[POSITIVE INTEGER]
\<time\>	::=[POSITIVE INTEGER]

2. Initial data file:

\<initial data file\>	::= \<place list\>

4   The Clips Language

The C Language Production System (CLIPS) language is an Expert System Language developed by the Lyndon B. Johnson Space Center [10]. The basic elements of CLIPS are:

1. Fact-list: global memory for data
2. Knowledge-base: contains all the production or rules
3. Inference engine: controls overall production execution

CLIPS is a data driven language. Its programs consist of facts and rules. The facts are the data required for execution, and the job of the inference engine is to determine which rules should be executed. Like OPS5 [6], CLIPS is called a production language or a rule-based language. The main difference between a rule-based language and an imperative language such as Ada, C, FORTRAN, PL/1, or Pascal is that programs written in a rule-based language are data-driven programs, so programs cannot be executed without facts. Another difference is that the rules execute in parallel, while the other languages are sequential in nature [10].

4.1   Fact-List

In CLIPS, the **assert** command is used to put data in the fact-list; the **retract** command is used to remove data from the fact-list. A fact contains one or more fields enclosed in a parentheses.

To assert two facts: (assert (This is a test.))
                    (assert (Any more tests?))
then these two facts are placed in the fact-list as below:
                    f-1(This is a test.)
                    f-2(Any more tests?)
To **retract** a fact we need to specify the fact index of the fact. For example, (retract 1) will remove the fact (This is a test.) from the fact-list.

4.2   Knowledge Base

The structure of a rule is similar to an if then statement in a procedural language. The format of a rule is:
    (define rule_name
            (condition_1); pattern 1
            (condition_2); pattern 2
                . . . ; left-hand side (LHS) of the rule
            (condition_n)
    ⇒
            (action_1)
            (action_2)
                . . .; right-hand side (RHS) of the rule
            (action_m)

CLIPS compares the LHS of the set of rules against facts in the fact-list. If all the conditions of a rule match the facts, the rule is allowed to invoke actions in the RHS.

Example:

(defrule animal)
   ?fact-1 ← (cat);?fact-1 is a variable for cat
   ?fact-2 ← (dog) ;?fact-2 is a variable for dog
⇒
   (retract ?(fact-1 ?fact-2)
      (assert (animals are cats and dogs)))

This short program does the following things: if there is a "cat" and a "dog" in the fact-list, then it removes "cat" and "dog" from the fact-list and asserts "animals are cats and dogs" in the fact-list.

5   The Interpreter

The interpreter is written in the CLIPS programming language. The algorithm for the interpreter is given as follows:

1. Read the user's program: the interpreter reads the user's program that is written in Petri net language, including a structure file and an initial data file. Both files are read line by line, and then we assert each line into the fact-list until the end of the file.
2. Ask the user to choose between a time-oriented or place-oriented simulation: This information will determine the way to terminate the program. A time-oriented simulation will execute the program in a given amount of time, and a place-oriented simulation will execute the program by monitoring the number of inserted tokens until a fixed number is reached in a certain place, then the program will terminate.
3. Set and maintain a global clock: First set the clock to zero and assert this into the fact-list. When no rules matched, the global clock increases by one.
4. Add processing time and current time: the interpreter sums each processing time of the transitions and the current time to be the firing time of transitions, and then inserts this information into the fact-list.
5. Check all transitions that are enabled to fire in the fact-list: If there are enabled transitions, all input tokens of transitions are in the fact-list, then go to step 7.
6. Check unenabled transitions: At step four, the interpreter calculates the firing time for all transitions, but some transitions are not enabled at step five, therefore we need to remove that firing time information from the fact-list.
7. Retract input tokens of enabled transitions from the fact-list: When a transition is enabled, it removes input tokens from the fact-list immediately.
8. Add output tokens of firing transitions into the fact-list: If there are times that are the same as the current time, the interpreter inserts output tokens into the fact-list.

9. Repeat step 3 to step 8 until the end of execution: For a time-oriented simulation, the interpreter checks the global clock to terminate. For a place-oriented simulation, the interpreter decreases the number of tokens, given by the user, by one whenever the token is inserted until the number becomes zero.

6   Macintosh Interface for Clip

This interpreter is written in CLIPS 5.0 and run on an Apple Macintosh computer, this section provides a brief discussion of some menu commands needed to run the user's programs. Commands can be entered in the dialogue window.

6.1 The File Menu

The CLIPS file menu that includes the following commands:

New: This command will open a window named "untitled" for editing.

Open: This command allows the user open a text file for editing.

Load: This command allows the user to load a file into the knowledge base.

Save: This command saves the file in the active edit window.

Save as: This command allows the file in the active edit window to be saved under a new name.

Quit: This command exits CLIPS.

The Execution Menu

The CLIPS Execution Menu that includes the following commands:

Reset: This command is needed before each run so that it can initialize the fact-list.

Run:This command is to execute programs.

Option: Under this window, we need to choose "Fact Duplication", because it is possible to have multiple tokens in   a node (place).

Once we are in the dialogue window, we first choose "Fact Duplication" under the execution menu. This allows multiple tokens in the same place, then we load the interpreter into the knowledge base, Before each run, it is necessary to reset the fact-list.

7   Execution and Output

There are two ways to execute a program written in the Petri net language using the interpreter developed in this project, at execution time or at monitored place. The formats for both methods are given below:

By Execution Time:

Please enter the name of structure file:

Execution is by time or by node?

(time/node)(time/node)

Enter the duration of execution:

Enter the name of initial file:

Enter the input transition(s):

Enter the output transition(s):

By Monitored Place:

Please enter the name of structure file:

Execution is by time or by node?

Enter duration of execution:

Enter node to be traced:

Enter the name of initial file:

Enter the input transition(s):

The name of structure file is the structure file of the Petri net. The name of initial file is the data file of initial tokens. If the user chooses to execute by time, the duration of execution is the total amount of time allocated, otherwise the duration of execution by place will run until some amount of tokens have been inserted into that place. Both methods, by time and node, will result in the same output. The output from the interpreter provides some important information about the system performance. The output information includes the following three lists:

1. A list consists of times and firing transitions: this list provides the time when each transition is fired. From this list, one can obtain the percentage of system idle time and the percentage of system busy time.
2. A list of transitions, total number of firings, and average time between fires: this list tells how busy the system is for each transition. From this list, one can determine the bottleneck of the system. Also, one can obtain the busy time and idle time for each transition.
3. A summary list that includes total time, number of input jobs, number of finished jobs, number of unfinished jobs, and average time to complete a job (the throughput).

8.   Conclusion and Future Research

In this research project, we have developed a Petri net language that is based on our observation of a variety systems and their timed Petri net models. The language is simple and small. It is easy to understand and easy to program. The Petri net language we have developed exactly characterizes the dynamic activities of Petri nets. Therefore, any system that can be modeled by a timed Petri net described in this paper can be directly implemented by a program in the Petri net language.

In simulation, the most difficult process is model validation and model verification. A Petri net is a dynamic system that models a system directly; the execution of Petri nets will perform validation and verification at the same time. The use of the Petri net language allows the simulation of a Petri net model without conversion to other models [26] and without the difficulty of using traditional languages. Also, we have successfully developed a Petri net interpreter by using a rule-based expert system language, CLIPS. This eased the difficulty of utilizing sophisticated data structures and their complicated implementation. The main advantage of using the CLIPS language is that CLIPS can be activated in parallel, so we can simulate Petri nets more effectively. Other than the examples presented in this paper, we have tested several more complicated Petri nets. We have found that the interpreter executes a Petri net program that exactly reflects the execution of the Petri net model of the physical system. We believe that the approach chosen to develop an interpreter is an alternative method of simulation.

The statistics we collected include: the firing time, the total number of firings, the average firing time of each transition, the number of input jobs, finished jobs, unfinished jobs, and the total simulation time. For the further research, there are several points to investigate: one is to attach time to places so that both transitions and places can be active concurrently. Also, we will consider stochastic firing times with different statistical distributions for transitions, to make the effect of the simulation more realistic [13, 17] Finally, some additional features like priorities or colors may be added to the system to make the Petri net more fruitful.

Reference

1. B. Berthomieu and M. Diaz, "Modeling and verification of time dependent systems using time Petri nets," *IEEE Transactions on Software Engineering*, V.17, No. 3, pp.259-273, March 1991.

2. R. Bauman and T. A. Turano "Production based language simulation of Petri nets," *Simulation*, pp.191-198, November. 1986.

3. K. M. Chandy and J. Misra, "Distributed simulation: A case study in design and verification of distributed programs," *IEEE Transactions on Software Engineering*, V. SE-5, no.5, pp.440-452, September 1979.

4. G. Ciardo and K. S. Trivedi, "SPNP: The stochastic Petri net package (version 3.1)," *InternationalWorkshop on Modeling Analysis and Simulation of Computer and Telecommunication Systems*, pp.390-391, January 1993.

5. G. Chiola and A. Ferscha, "Distributed simulation of Petri nets," *IEEE Parallel & Distributed Technology*, pp.33-37, August 1993.

6. C. L. Forgy, 1981 OPS5 User's Manual. *Technical Report CMU-CS-81-135* Department of Computer Science, Camegie Mellon University, Pittsburgh, Penn 15213.

7. R. M. Fujimoto, "Parallel discrete event simulation," *Comm. ACM*, V. 33, no.10, pp.30-53, October 1990.

8. W. Garbe, Stochastic Petri Net Simulator, available on InterNet, <antje@tudurz.urz.tu-dresden.de>

9. C. Ghezzi, D. Mandrioli, S. Morasca, and M. Pezze, "A general way to put time in Petri nets," *ACM Workshop on Software Specification and Design*, pp.60-67, 1989.

10. J. C. Giarratano, *CLIPS User's Guide*, Lyndon B. Johnson Space Center, V-4.3, June 1988.

11. S. V. Hoover and R. F. Perry, *Simulation*, Addison Wesley Publishing Company, Inc., 1989.

12. M. A. Holliday and M. K. Vernon, "A generalized timed Petri net model for performance analysis," *IEEE Transactions on Software Engineering*, Vol. SE13, December 1987.

13. C. Lin and D. C. Marinescu, "Stochastic high-level Petri nets and applications," *IEEE Transactions on Computers*, Vol. 37, No.1, pp.815-825, July 1988.

14. C. Lindemann, "DSPN express: A software package for the efficient solution of deterministic and stochastic Petri nets," *International Workshop on Modeling Analysis and Simulation of Computer and Telecommunication Systems*, pp.373-374, January 1993.

15. D. Mandrioli, R. Zicari, C. Ghezzi, and F. Tisato, "Modeling the Ada task system by Petri nets," *Computer Languages*, V.10, No.1, 1985.

16. L. March and P. LePare, "Defining the semantics of languages for programmable controllers with synchronous processes," *Control Eng. Practice*, V. 1, No.1, pp.79-84, 1993.

17. M. A. Marsan, G. Balbo, A. Bobbio, G. Chiola, G. Conte, and A. C. Cumani, "On Petri nets with stochastic timing," *Proc. Int. Workshop Timed Petri Nets*, pp.80-87, July 1985.

18. J. Misra, "Distributed discrete-event simulation," *ACM Computing Surveys*, V. 18, No.1, pp.39-65, March 1986.

19. M. K. Molly, "Discrete time stochastic Petri nets," *IEEE Transactions on Software Engineering*, V. SE-11, pp.417-423, 1985.

20. M. K. Molly, "Performance analysis using stochastic Petri nets," *IEEE Transactions on Computers*, V. C-31, pp.913-917, September 1982.

21. J. Noe, "Nets in Modeling and Simulation," *Lecture Notes in Computer Science*, Berlin: Springer-Verlag, 1980.

22. G. Nutt, Petri Net Simulator, available on InterNet, <nutt@pawnee.cs.colorado.edu>

23. J. L. Peterson, *Petri Net Theory and The Modeling of System*, Prentice-Hall, Inc., Englewood Cliffs, New Jersey, 1981.

24. C. V. Ramamoorthy and G. S. Ho, "Performance evaluation of asynchronous concurrent systems using Petri nets," *IEEE Transactions on Software Engineering*, V. 6, pp.440-449, September 1980.

25. J. Sifakis, *Use of Petri nets for performance evaluation*, Lecture Notes in Computer Science, Berlin: Springer-Verlag, 1980.

26. A. A. Q. Taqi, et al, "Acomparative study between Petri net and SLAM," *Simulation*, pp.339-344, November 1992.

27. A. A. Torn, "Simulation graphs: A general tool for modeling simulation designs," *Simulation* pp.187-194, December 1981.

An Agent-Based Soft Computing Society

Chengqi Zhang, Zili Zhang, and Ong Swee San
School of Computing and Mathematics
Deakin University, Geelong Victoria 3217, Australia
{chengqi, zili, ong}@deakin.edu.au

Abstract. Soft computing (SC) techniques such as fuzzy logic (FL), neural networks (NN), and genetic algorithms (GA) are complementary. Each SC technique has particular computational properties that make them suited for particular problems and not for others. Thus, in solving complex, real-world problems, we need to incorporate some SC techniques into the application systems to increase the systems' "intelligence". In this paper, we first propose an agent-based framework for integrating SC techniques into practical application systems. We then discuss the design and implementation of a platform independent soft computing support environment based on the framework. We call such an environment *agent-based soft computing society*. Such a society can facilitate the design of truly robust, flexible and adaptive hybrid intelligent systems.

1 Introduction

Soft computing is a term that describes a collection of techniques capable of dealing with imprecise, uncertain or vague information. SC is not a single methodology. Rather, it is a consortium of computing methodologies that collectively provide a foundation for the conception, design and deployment of intelligent systems. The principal members of SC are fuzzy logic (FL), neural network (NN), genetic algorithm (GA) etc. SC technologies such as FL, NN, and GA are complementary rather than competitive. While these SC techniques have produced encouraging results in particular tasks, certain complex problems cannot be solved by a single SC technique alone. Each SC technique has particular computational properties that make them suited for particular problems and not for others. For example, in our ongoing project entitled *"Financial Investment Advisor Using Intelligent Agent Technolgies"*, the NN was used as a pattern watcher for stock market; the GA was used to predict interest rate; and the approximate reasoning based on FL was used to evaluate clients financial risk tolerance ability etc. Thus, in solving complex, real-world problems, we need to incorporate some SC techniques into the application systems.

An agent is an encapsulated computer system that is situated in some environment and that is capable of flexible, autonomous action in that environment in order to meet its design objectives[13]. Recently, N. R. Jennings gave a qualitative analysis to provide the intellectual justification of precisely why agent-based

W. Ziarko and Y. Yao (Eds.): RSCTC 2000, LNAI 2005, pp. 661–668, 2001.

systems are well suited to engineering complex softw are systems[14]. Based on the analysis, he argued that: (1) Agent-orien ted approaches can significantly enhance our ability to model, design and build complex, distributed softw are systems; (2) As well as being suitable for designing and building complex systems, the agent-orien ted approach will succeed as a mainstream softw are engineering paradigm.

We notice that there are many agent-based application systems demonstrate that agent-based systems are a useful and pow erful solution technology. How ev er, these developments also show that designing and building agent systems is difficult. At present, there are tw o major technical impediments to the widespread adoption of agent technology[15]: (1) the lack of a systematic methodology enabling designers to clearly specify and structure their applications as multi-agent systems; and (2) the lack of widely av ailable industrial-strength multi-agent system toolkits.

With these observations in mind, we propose a general-purpose agent-based soft computing societ y. Such a societ y can be applied to complex, real-world problems that can be modeled with multi-agent, and at the same time, different SC techniques must be employ ed to solve the problems. With the support of the soft computing agent societ y, the multi-agent system developers need only to build the domain-specific parts and construct the ontologies used in the specific application field–rather than re-inv enting the wheel as often happens at the moment. In this paper, we will discuss the design and implementation of such a soft computing agent society as well as other relevant issues.

This researc h was initially motivated b y our ongoing financial inv estment advisor project. In this project, we adopted a multi-agent system architecture. A multi-agent system approach is natural for financial inv estment advisor because of the multiplicit y of information sources and different expertise that must be brought to bear to produce a good recommendation (such as a stock buy or sell decision). Meanwhile, there are many successful applications of SC technologies in financial sector[3][4][5]. With these observations in mind, we integrated some SC technologies in to our financial inv estment advice multi-agent system. We hav e discussed the approaches to incorporating SC technologies into the financial inv estment planning multi-agent systems[1][2]. The emphasis of this paper is to extend our approaches used in the project and try to provide a universal framework to incorporate different soft computing technologies into multi-agent systems.

The remainder of the paper is structured as follows. Section 2 is the frame-w ork of the agent-based soft computing society. Section 3 is the design and implementation details whic h include technologies used to develop the SC agent society, behaviors of different agen ts in the society, modeling and implementation details of these different kinds of agents, and an example of the societ y. Section 4 is a brief evaluation of the SC agent societ y. Finally Section 5 is the concluding remarks.

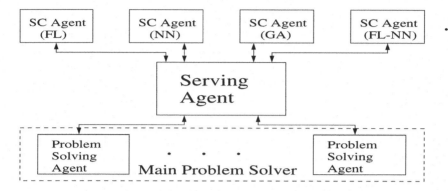

Fig. 1. Framework of Agent-Based SC Society

2   Framework of Agent-Based SC Society

Because each SC technique has particular strengths and weaknesses and that they cannot be applied universally to every problem, this has encouraged the hybridization of these SC techniques. In recent years, an increasing number of researchers have been working in the field of hybrid systems in an attempt to find new ways to integrate two or more technologies to tackle complex real world problems[6][7]. Some of the research work involved in multi-agent systems. Some typical hybrid multi-agent systems include the MIX multi-agent platform[9], the IMAHDA architecture[10], and the PREDICTOR system ([6], Chapter 9) etc.

By analyzing these hybrid multi-agent systems, we find out that the way for integrating SC technologies into multi-agent systems in these systems is to embed the SC technologies in each individual software agent, and did not use any middle agents. Such approaches have the following limitations: (1)It is impossible to embed many SC technologies within a single agent. Otherwise, the agents will be overloaded. In many applications, the agents in multi-agent systems should be kept simple for ease of maintenance, initialization, and customization; (2)It is not flexible to add more SC technologies to or delete some unwanted one from the multi-agent systems. For example, one software agent may be equipped with fuzzy logic, the other with neural network etc. In such a way,one agent can only have one SC capability. If we want the agent to possess two or more SC capabilities, we must modify the implementations.

To overcome the drawbacks of current used approaches, we propose a new approach to constructing intelligent hybrid systems. Figure 1 shows the framework. A complete system under this framework (we call it *agent-based soft computing society*) consists of a set of problem solving agents, soft computing agents, and a serving agent of these two kinds of agents. Here,problem solving agents are agents without SC capability. They are at the front end of a multi-agent system. Soft computing agents are at the back end of a multi-agent system. They provide problem solving agents with soft computing capabilities. The serving agent is a special kind of middle agents. It is similar to the facilitators discussed in [12].

Compared with those hybrid multi-agent systems described above, our framew ork has three crucial characteristics that differentiate our work from others: (1) Each problem solving agent can easily access all the SC techniques available in the system; (2) The presence of the serving agent in our framework allows adaptive agent organization; (3) Overall system robustness is also facilitated through the use of the serving agent. F or example, if a particular SC service provider (SC agent) disappears, a requester agent (problem solving agent) can find another one with same/similar capabilities by interrogating the serving agent.

3   Design and Implementation of the Society

3.1   T ec hnology Platform

The most important design criterion of this serving agent is platform independent. All system components are developed using the Java or other technologies which are platform independent. The linkages among the components or v arious agents are provided by the Knowledge Query and Manipulation Language (K QML)[12], which encapsulates all the necessary message passing and communication capabilities which are needed within our framework.

The implementation of our framework is under the support of Ja vaAgent T emplateLite (JATLite). JATLite is a set of ligh tweight Ja va pac kages being developed at Stanford University that can be used to build multi-agent systems. JATLite pro vides a set of fully functional templates. It is written en tirely in the Java language that supports the construction of softw are agents that communicate using a peer-to-peer protocol. For more information on JATLite, see *http://java.stanford.edu/java_agent.*

3.2   Behaviors of Different Types of Agents

Our SC agent society has three types of agents (see Figure 1): *problem solving* agents, *serving* agents, and *soft computing* agents. The key component of the framework is the serving agent. The behavior of each kind of agent is described below:

- **Problem Solving Agent** It is application-specific, i.e., it has its own knowledge base; It must have some meta-knowledge about when it needs the help of soft computing agents (e.g., pre or post processing some data); It can ask soft computing agents to accomplish some subtasks.
- **Soft Computing Serving Agent** It works as an agent name server (ANS) and matchmaker of the capabilities of SC agents; It keeps track of the names, ontologies, and abilities of all registered soft computing agents in the system; It can reply the query of problem solving agent with appropriate soft computing agent's name and ontology .

- **Soft Computing Agent** Each soft computing agent can provide service for problem solving agents with one or some kind of combined soft computing tec hnologies; It can send bak the processed results to problem solving agents; It must advertise its abilities to the serving agent.

All problem solving agents or SC agents must register and connect to the serving agent.

Each problem solving agent has its own domain-specific knowledge bases w ell as meta-knowledge about when to use soft computing agents. The serving agent records the capabilities, ontologies, and names etc. of all the SC agents in a multi-agent system. The scenario goes as follows:

A tcertain stage of the problem solving process, the problem solving agent sends a KQML message using *recommend-one*performative to the serving agent according to its meta-knowledge. The serving agent then retrieves its SC agent database and replies with an appropriate SC agent's name and ontology which has the capability asked for using *reply* performative. After that, the problem solving agent communicates with the SC ag t directly for a specific problem: The problem solving agent provides the SC agent some parameters according to the *ontology*, and the SC agent sends the final results to the problem solving agent.

Under our framework, the types of problems the problem solving agents can solv e depend on their domain-specific knowledge.

3.3   Modeling and Implementation

The three kinds of agents described in Section 3.2 hav e different models. Figure Figure 2 shows the internal structures of the three kinds of agents.

As we can see in Figure 2, all the agents hav e a common part–KQML Message In terpreter (KMI). That is because we use KQML for inter-agent communication. Also because of this, we call the three kinds of entities in Figure 1 "agents"[11]. The KMI represents the interface betw een KQML router and agents. Once an incoming KQML message is detected, it will be passed to the KMI. The KMI transfers incoming KQML messages into a form that agents can understand. The Implementation of KMI is based on JATLite KQMLLay er T emplates.

Both problem solving agents and soft computing agents have ontology interpreter. They need to decrypt and process the *:content* part of the KQML message when they solve a problem. There is no ontology interpreter in the serving agent because it does not care about the *:content*. When implementing, we maintain an **Ontology Interpreter HashTable**. The table instructs an agent to locate the necessary in terpreter for every on tology in the KQML message. It should be noted that an interpreter can be located at any Internet site other than the place where the agent resides.

The domain kno wledge in problem solving agents usually is not sufficient to solve a problem. They need the help of soft computing agents. The meta-kno wledge in problem solving agents tell them when to ask for helps of soft computing agents.

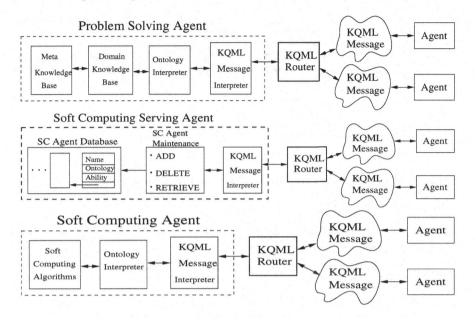

Fig. 2. Agent Models in the Society

The SC agent maintenance module in serving agent has three functions: Add an entry that contains theSC agent's name, ability,and ontology to the database; Delete an entry from the database; and retrieve the database to find out SC agents with specific ability.

For the soft computing algorithms in soft computing agerts, if the agent is under our control, it will be built using KQML as a communication language. If not, we use Java Native method to connect the legacy system to our agent.

3.4   Example

We present an example of how the agent-based soft computing society is used in the determination of a user's investment policy in our ongoing project.

The problem solving agent receives "Determining Investment Policy (aggressive or conservative)" goal in messages coming from other problem solving agents or from a user interface directly. To make such a decision, the problem solving agent needs the information about the user's *risk tolerance* (RT) ability, the falling or rising of *interest rates* (P_1), the state of the *stock market* (P_2), and *unemployment rate* (P_3) etc. The problem solving agent has rules in its domain knowledge base such as

$$If \ RT \ is \ H \ and \ P_1 \ is \ B_1 \ and \ \ldots \ then \ IP \ is \ C$$

where C is a fuzzy subset indicating the aggression or conservation of the investment polcy. H and B_1 are also fuzzy subsets. The problem solving agent also

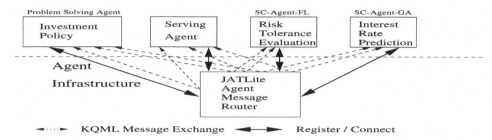

Fig. 3. Practical Architecture for Determining Investment Policy

has meta knowledge such as *using SC agents to evaluate user's RT* and *using SC agents to predict P_1* etc.

Thus, the problem solving agent sends K QML messages using *recommend-one* performative to the serving agent.

The serving agent then retrieves its SC agent database and replies with an appropriate SC agent's name and ontology which has the capability ask ed for using *reply* performative. In our system, there are a risk tolerance ability evaluation agent based on fuzzy logic and an interest rate prediction agent based on genetic algorithms.

The problem solving agent then communicates with SC_agent_FL (for risk tolerance evaluation) and SC_agent_GA (for interest rate prediction) directly . They decrypt and process the parameters or results (the *:content* part of the K QML messages) by using the ontology interpreters.

After the problem solving agent obtains the results of RT and P_1 etc. from corresponding SC agents, it can infer the conclusion about the user's investment policy according to its domain knowledge. The practical architecture we adopted is shown in Figure 3 (under the support of JATLite).

4   Evaluation of the SC Agent Society

In our framework, the implementations are platform independent. The serving agent is general purpose, thus can be used in any applications. T o tailor the SC agent society for other specific applications, we need to develop the domain-specific kno wledgebases as well as the meta knowledge bases of the problem solving agents. In the meantime, we also need to construct the ontologies used in the specific applications. It is easy to wrap the legacy soft computing programs and con vert them to "agents" by using Java Native Method and JATLite Templates. Intelligent hybrid (multi-agent) system developers can use our framework for reference. They can easily follow the ideas and construct their own application systems with flexibility.

5   Concluding Remarks

We presented a flexible framework–the agent-based soft computing society to construct multi-agent application systems that need to incorporate different kinds of soft computing technologies into them. Our framework has three crucial characteristics that differentiate our work from others: (1) Our approach makes every problem solving agent easily access all the SC technologies available in the system; (2)The presence of the serving agent in our framework allows adaptive agent organization, that is, our framework has the ability to add and delete SC agents dynamically as needed; (3) Overall system robustness is also facilitated through the use of the serving agent.

Such an SC agent societ y facilitates the design of robust, flexible and adaptive h ybrid in telligen tsystems–we can build intelligen th ybrid multi-agent systems based on the society, rather than from scratch.

T o facilitate the construction of multi-agent application systems in finance, w e are currently constructing the ontologies used in finance.

References

1. Z. Zhang and C. Zhang, Approaches to Incorporating Soft computing Technologies into Softw are Agents, Proceedings of ICONIP'99, IEEE Press, 1999, 952-957.
2. Z. Zhang and C. Zhang, A Serving agent for Integrating Soft Computing and Soft w are Agents, Proceedings of Australia AI'99, Springer, 1999, 476-477.
3. G. J. Deboeck (Ed.), Trading on the Edge–Neural, Genetic, and Fuzzy Systems for Chaotic Financial Markets, Wiley, 1994.
4. R. J. Bauer, Genetic Algorithms and investment Strategies, Wiley, 1994.
5. R. A. Ribeiro, H. J. Zimmermann, R. R. Yager, and J. Kacprzyk (Ed.), Soft Computing in Financial Engineering, Physica-Verlag, 1999.
6. S. Goonatilake and S. Khebbal (Eds.), Intelligen t Hybrid Systems, Wiley, 1995.
7. L. R. Medsker, Hybrid Intelligen t Systems, Kluw er Academic Publisher, 1995.
8. L. C. Jain and R. K. Jain (Eds.), Hybrid Intelligen t Engineering Systems, World Scien tific, Singapore, 1997.
9. M. Hilario, C. Pellegrini, and F. Alexandre, Modular Integration of Connectionist and Symbolic Processing in Knowledge-based Systems, in: *Int. Symposium on Integrating Knowledge and Neural Heuristics*, Pensacola, Florida, 1994, 123-132.
10. R. Khosla and T. Dillon, Engineering In telligen tHybrid Multi-Agent Systems, Kluw er Academic Publishers, Boston, 1997.
11. M. R. Genesereth and S. P. Ketc hpel, Softw are Agents, Commun. ACM, Vol.37, No.7, 1994, 48-53.
12. T. Finin, Y. Labrou and J. Mayfield, K QMLan Agen t Communication Language, in *J. M. Bradshaw (ed.), Software A gents* AAAI Press/ The MITPress, Menlo Park, CA, 1997, 291-316.
13. M. Wooldridge, Agent-Based Softw are engineering, IEE Proc. Softw are Engineering, Vol. 144, No. 1, 1997, 26-37.
14. N. R. Jennings, On Agent-Based Softw are Engineering, Artificial Intelligence, Vol. 117, 2000, 277-296.
15. N. R. Jennings, K. Sycara, and M. Wooldridge, A Roadmap of Agent Research and Development, Autonomous Agents and Multi-Agent Systems, Vol. 1, 1998, 7-38.

Author Index